Magnetic Resonance in Perspective

Highlights of a Quarter Century

Magnetic Resonance in Perspective

Highlights of a Quarter Century

Edited by
Wallace S. Brey
Department of Chemistry
University of Florida
Gainesville, Florida

Academic Press
San Diego New York Boston London Sydney Tokyo Toronto

This book is printed on acid-free paper. ♾

Academic Press, Inc.
A Division of Harcourt Brace & Company
525 B Street, Suite 1900, San Diego, California 92101-4495

United Kingdom Edition published by
Academic Press Limited
24-28 Oval Road, London NW1 7DX

Library of Congress Cataloging-in-Publication Data

Magnetic resonance in perspective : highlights of a quarter century / edited by Wallace S. Brey.
p. cm.
Includes index.
ISBN 0-12-133145-8 (alk. paper)
1. Nuclear magnetic resonance spectroscopy. I. Brey, Wallace S.
QD96.N8M335 1996
543'.0877--dc20 95-47358
CIP

PRINTED IN THE UNITED STATES OF AMERICA
96 97 98 99 00 01 QW 9 8 7 6 5 4 3 2 1

Contents

Contributors

The affiliations given on the title pages of the articles were those of the authors when the articles were originally published. The addresses listed below are the current addresses for the senior authors of the articles. Numbers in parentheses indicate the pages on which the authors' contributions begin.

E. Raymond Andrew (1), Department of Physics, University of Florida, P.O. Box 118440, Gainesville, Florida 32611-8440.

H. Barkhuijsen (491), Applied Physics Department, Delft University of Technology, P.O. Box 5046, 2600 GA Delft, The Netherlands.

Ad Bax (285, 389, 399, 411, 655), Laboratory of Chemical Physics, National Institute of Diabetes and Digestive and Kidney Diseases, National Institutes of Health, Building 5, Room 126, Bethesda, Maryland 20892.

H. J. C. Berendsen (117), Laboratory of Physical Chemistry, University of Groningen, Zernikelaan, Groningen, The Netherlands.

Geoffrey Bodenhausen (91, 161, 467, 487), NMR Program, National High Magnetic Field Laboratory, Florida State University, 1800 Paul Dirac Drive, Tallahassee, Florida 32306-4005.

Paul T. Callaghan (557), Department of Physics and Biophysics, Massey University, Private Bag 11222, Palmerston North, New Zealand.

Josef Dadok (55), Department of Chemistry, Carnegie Mellon University, 4400 Fifth Avenue, Pittsburgh, Pennsylvania 15213-3890.

Richard R. Ernst (71, 197, 313, 403, 415, 467, 487, 571), Lab Physikalische Chemie, Eidgenössische Technische Hochschule—Zentrum, Universitatstrasse 22, CH-8092 Zürich, Switzerland.

Jens Frahm (509, 529, 563), Max-Planck-Institut für biophysikalische Chemie, Biomedizinische NMR Forschungs Gmbh, Am Postfach 2841, D-37018 Göttingen-Nikolausberg, Germany.

Jack H. Freed (539, 597), Department of Chemistry, Baker Laboratory, Room G75-R, Cornell University, Ithaca, New York 14853-1301.

Ray Freeman (25, 91, 161, 167, 231, 235, 275, 281, 285, 309, 383, 395), Department of Chemistry, University of Cambridge, Lensfield Road, Cambridge CB2 1EP, United Kingdom.

Allen N. Garroway (523), Polymer Diagnostics Section, Chemistry Division, Naval Research Laboratory, Code 6122, Washington, DC 20375-5320.

Axel Haase (509, 529), Physikalisches Institut der Universität Würzburg, Lehrstuhl für biophysik, Ep 5, Am Hubland, D-97074 Würzburg, Germany.

Thomas Herrling (337), Forschungsstelle für ortsauflösende Meßtechnik e. V., Geb. 19.6, Rudower Chaussee 6, 0-1199, Germany.

Howard D. W. Hill (25), Varian Associates, 611 Hansen Way, Palo Alto, California 94303.

David I. Hoult (97, 215), The Institute for Biodiagnostics, National Research Council of Canada, 435 Ellice Avenue, Winnipeg, Manitoba, Canada R3B 1Y6.

Ralph E. Hurd (669), General Electric Medical Systems, 47697 Westinghouse Drive, Fremont, California 94539.

James S. Hyde (321), Biophysics Research Institute, The Medical College of Wisconsin, 8701 Watertown Plank Road, P.O. Box 26509, Milwaukee, Wisconsin 53226.

Thomas L. James (443), University of California at San Francisco Magnetic Resonance Laboratory, Department of Pharmaceutical Chemistry, The University of California at San Francisco, 926 Medical Science, San Francisco, California 94143-0446.

Reinhold Kaiser (9), Department of Physics, University of New Brunswick, Box 4400, Fredericton, New Brunswick, Canada.

W. Karthe (107), Fraunhofer Institute for Applied Optics and Precision Engineering, Friedrich Schiller University, Fürstengraben 1, 07743 Jena, Germany.

Lewis E. Kay (655), Department of Medical Genetics, University of Toronto, Toronto, Ontario, Canada M5S 1A8.

Anil Kumar (71, 415), Department of Physics and Sophisticated Instrument Facility, Indian Institute of Science, Bangalore 560 012, India.

Malcolm H. Levitt (231, 235, 275, 281, 309, 367), Fysikalisk Kemi, Stockholms Universitet, Stockholm S106 91, Sweden.

Gary E. Maciel (389, 411), Department of Chemistry, Colorado State University, Fort Collins, Colorado 80523.

Peter Mansfield (141), Magnetic Resonance Center, Department of Physics, University of Nottingham, University Park, Nottingham NG7 2RD, United Kingdom.

Dominique Marion (655), Institut de Biologia Structurale, CNRS-CEA, 41 Avenue des Martyrs, 38027 Grenoble Cedex 1, France.

Beat U. Meier (487), Laboratorium für Physikalische Chemie, ETH-Zentrum, CH-8092 Zürich, Switzerland.

Gareth A. Morris (91, 167), Department of Chemistry, University of Manchester, Oxford Road, Manchester M13 9PL, United Kingdom.

Alexander Pines (43, 577), Department of Chemistry, University of California, Berkeley, California 94720.

George K. Radda (305), Department of Biochemistry, University of Oxford, South Parks Road, Oxford OX1 3QU, United Kingdom.

Alfred G. Redfield (87), Department of Biochemistry, The Martin Fisher School of Physics, Brandeis University, Waltham, Massachusetts 02254-9110.

Rex E. Richards (97), 13 Woodstock Close, Oxford OX2 8DB, England.

D. J. Ruben (329), Arrhenius Laboratory, Stockholms Universitet, 10691 Stockholm, Sweden.

Jacob Schaefer (61), Department of Chemistry, Washington University, Campus Box 1134, One Brookings Drive, St. Louis, Missouri 63130-4899.

A. J. Shaka (383, 395, 577), Department of Chemistry, University of California—Irvine, Irvine, California 92717-2025.

Ole Winniche Sørensen (607), Novo Nordisk A/S, Novo Allé, DK-2880 Bagsvaerd, Denmark.

E. O. Stejskal (61), Department of Chemistry, North Carolina State University, Box 8204, Raleigh, North Carolina 27695.

John S. Waugh (43, 491), Department of Chemistry, Room 6-235, Massachusetts Institute of Technology, Cambridge, Massachusetts 02139.

Don E. Woessner (263), Department of Radiology, Nell & Rogers Biomedical MR Center, University of Texas Southwestern Medical Center, 5801 Forest Park Road, Dallas, Texas 75235-9085.

Kurt Wüthrich (113, 313, 415), Institüt für Molekularbiologie und Biophysik, Eidgenössische Technische Hochschule-Hönggerberg, CH-8093 Zürich, Switzerland.

Preface

The field of magnetic resonance had its origin half a century ago, with the reports of the first successful experiments in condensed matter to yield resonance responses of magnetic nuclei and of unpaired electrons. Of course, scientists already had quite a good model of the magnetic properties of these particles, based initially on observations such as the splittings associated with the Zeeman and Stark effects in rotational spectra. The terms "fine structure" and "hyperfine structure" were applied to these splittings produced by electrons and nuclei, respectively, in spectral lines. Measurements of magnetic susceptibility and of the deflection of atomic particles in an inhomogeneous magnetic field gave quantitative information about magnetic moments of atomic particles.

Not all materials, however, are amenable to volatilization under high vacuum, and therefore physicists struggled to devise NMR experiments on condensed phases, with the aim of measuring more precisely, and possibly with higher sensitivity, the nuclear gyromagnetic ratio, an important numerical value for the assessment of the validity of theories of nuclear structure.

The best-laid plans often go awry, and so it was with the nuclear quantitative measurements. As is very well known now, the electrons interfere with the interaction between the applied magnetic field and the nuclear magnet. This effect was discovered when two different signals for the copper nucleus appeared in an experiment in which there was a solution of copper sulfate in a probe containing copper wire. The distressing effect of electrons became the chemical shift, and the discovery and interpretation of the indirect spin–spin interaction were soon to follow.

The *Journal of Magnetic Resonance* came into being in 1969, just about halfway in the history of magnetic resonance. At that time, superconducting magnets and Fourier transform methods for NMR were just on the horizon, and many contributions were still being made to the theoretical foundations of magnetic resonance as well as to the collections of empirical results which were to guide the future interpretation of spectral results.

The general idea of the *Journal* was the brainchild of two Academic Press editors, Stanley F. Kudzin and Thomas Lanigan. The privilege of defining the scope of the publication was accorded to the *Journal* editor, who wrote: "It is our intention that this publication carry all those sorts of material which will be of value to the individual doing research in which new aspects of the field are being uncovered or in which magnetic resonance is being applied to the study of the behavior of matter." Thus, the compass has been as wide as possible, serving the interests of all disciplines and all scientists whose work might be related to magnetic resonance, and including a range of subjects from theory through instrumentation and methodology to analysis and correlation of spectral results. The first year saw the publication of six issues with 693 pages, and the institutional subscription rate was $25. In 1995, twenty-four issues have appeared, with a total of 2880 pages and about 65% more material per page.

The history of the *Journal* thus covers the second half of that of magnetic resonance, a period of seemingly exponential growth in techniques and applications. Of course, there have been short intervals when one might have been tempted to share the view of the patent office head of the nineteenth century who averred that everything had been invented. However, new spurts of ideas always came pouring forth in a short while, belying such pessimism. And over the course of this time, the *Journal* has been privileged to have as members of the Editorial Board a number of outstanding leaders in the field. Among these have been Sir Rex Richards, who probably more than any other individual provided the impetus making possible the widespread use of superconducting magnets; Richard Ernst, winner of the Nobel prize; and Ray Freeman, a continuing source of novel techniques and insights in NMR.

As we look back over the period since 1969, the innovations and changes have been truly outstanding. On the one hand, there has been the development of NMR spectrometers utilizing the higher magnetic fields, with adequate homogeneity made possible chiefly through the availability of superconducting magnets. The accompanying enhancements in sensitivity and resolution are developments which some areas of electron spin resonance have also been able to utilize since the speed of electronics has caught up. At the same time, there has been the introduction of pulsed methods, made possible in practice by the availability of fast computers to control experiments and by the development of mathematical methods of processing the response to pulses. These have been combined into a literal multitude of combinations of two-, three-, and four-dimensional experiments. And, of course, the entire field of imaging and localized spectroscopy has come into being, assuming a prominence in terms of integrated numbers of lines of magnetic force and of monetary investment equal to or surpassing that of the older areas.

A very interesting aspect of the development of magnetic resonance is the extent to which cross-fertilization has occurred between various subfields. An outstanding example of this is represented by the last paper in this volume in which the use of gradients in imaging suggested the use of gradients in high-resolution spectroscopy. There have been many other channels of cross-fertilization, some of which have occurred only very recently. Thus, NMR spectroscopists who work in the solid state have adopted two-dimensional techniques from their high-resolution brethren, and electron spin resonators have been using pulsed spin-echo approaches, devised by NMR spectroscopists.

This volume presents *Journal* papers which we—the Editorial Board, the Editor, and several other interested spectroscopists who offered suggestions—believe to represent contributions to the field that are of enduring significance. The choice of content has not been easy, and limitations of space have made it impossible to include all the papers that do merit selection. Authors not represented in the volume should not feel that absence of their papers represents a judgment on the quality of their work.

In choosing the order of papers in the book, we might have attempted to classify them in subject-matter categories. Instead, they are arranged in chronological order, partly because that is a simpler arrangement, but primarily because this order affords the reader a better idea of how the field has developed over time. This arrangement also gives an indication of the way progress has spread from area to area of magnetic resonance. A table has been provided listing the papers that have been included under each of various topical categories, so that a reader may, if desired, follow the development in a single field.

Finally, a few personal words of reminiscence. The editor has himself been associated with magnetic resonance for quite a few years. The first time he learned about it was in 1951 during a lecture by Pierce Selwood in Philadelphia. Selwood's principal interest was the development of the relationship of the magnetic susceptibility of supported metal catalysts to their degree of dispersion on the surface, and his lecture was part of a series sponsored by the Philadelphia Section of the American Chemical Society on various aspects of catalysis. Although the lecture proper did not deal with NMR, subsequent questions by several eager petroleum-company chemists about this "new technique having something to do with protons" led to a demonstration of nuclear precession with the aid of the lecturer's pointer. Selwood had sent a student to Harvard to measure the relaxation times of protons in organic liquids in contact with the catalyst surfaces, and had succeeded in establishing a correlation of the surface relaxivity with the amount of metal surface exposed to the liquid.

A few years later, in 1954, on his way to an American Chemical Society meeting in New York, the editor attended an instrument workshop held in the Commercial Museum in Philadelphia. At one of the half-day sessions, he had the opportunity to watch a demonstration of a working NMR spectrometer, presented by the engineers and scientists in the Varian organization who had been in charge of its development, including, in person, one of the Varian brothers. The particular achievement of interest at that time was the ability to resolve the multiplet structures of the ethanol resonances, a development permitted by the introduction of sample spinning.

The first NMR instrument in the University of Florida Laboratories, one of the earliest of the 60 megahertz spectrometers, arrived in 1958. Life for the spectroscopist was not as easy back then, for there were no shim coils to adjust the magnetic field homogeneity, alterations in field being limited to those produced by stressing the magnet yoke with a screw and ratchet device. The operator was forced to move the sample around in the magnet gap to find the location of optimum homogeneity. Since there were no radiofrequency synthesizers, all spectra were obtained in field-sweep mode. And, of course, there was no field-frequency lock circuit, so that multiplet splittings were evaluated by sweeping the field back and forth through resonance several times with an audio sideband applied, and averaging the distances between peaks on the chart paper. The first electron resonance spectrometer in the Florida labs was assembled from X-band wave guide and homebuilt control circuitry for the klystron tube.

The capabilities of both NMR and EPR instrumentation have come a very long way since the 1950s and indeed quite a long way since the early 1970s when the *Journal* was young. The Editor and the publishers have generated this book with the hope that it will be a source of information of lasting value to magnetic resonators, that it will give readers who have not lived through all this period a feeling for how the complex array of art, science, and technology now available has developed, and that it will bring back some pleasant memories to those who began their careers early enough to have experienced the vast changes that have taken place in the past quarter century.

Wallace S. Brey

Outline of Topics

JOURNAL OF MAGNETIC RESONANCE **2**, 259–266 (1970)

NMR Spectra of Reorienting Nuclear Pairs in Solids: Application to Conformational Changes

E. R. ANDREW* AND J. R. BROOKEMAN

Department of Physics and Astronomy, University of Florida, Gainesville, Florida 32601

Received December 9, 1969; accepted January 6, 1970

The nuclear magnetic resonance spectrum is calculated for a solid containing relatively isolated pairs of nuclei where there is rapid reorientation of the pairs between two fixed directions. Expressions are given for the profile of the spectrum for polycrystalline material as a function of the angle between two orientations. The second moment of the spectrum is derived and also the linewidth between singularities.

Reorientation of nuclear pairs may be encountered with conformational changes of cyclic molecules in solids. The theory is applied to solid cyclobutane to examine the hypothesis that ring inversion is responsible for the narrowing of the proton magnetic resonance spectrum in the low temperature phase. Ring inversion between the two equivalent conformers reorients the methylene proton pairs between two fixed directions. The theory is consistent with the reduced second moment and with the doublet fine structure separation.

INTRODUCTION

The purpose of this paper is to investigate the changes in the nuclear magnetic resonance spectra of a solid containing relatively isolated nuclear pairs when there is rapid reorientation of the pairs between two fixed orientations. Although the argument is general it is prompted by consideration of possible conformational changes in puckered ring molecules in solids. If such molecules contain methylene groups, conformational changes of the kind known to take place in the liquid state will cause the proton pair vectors to reorient between two fixed directions in the crystals, and if this motion is rapid enough the proton magnetic resonance spectrum of the solid will be affected.

CALCULATION OF SPECTRA

Consider a solid containing relatively isolated identical nuclear pairs of spin $\frac{1}{2}$. The directions in the system are indicated stereographically in Fig. 1. Each nuclear pair has two equally probably orientations denoted by R_1 and R_2, whose bisector C is taken as the pole of the stereogram; the angle between the two directions R_1 and R_2 is 2γ. The angles between the direction of the applied magnetic field **H** and R_1, R_2 and C are θ_1, θ_2, and α, respectively; ϕ is the azimuth angle of **H**. It should be noticed that we are not necessarily assuming that the two initial sites of the nuclei in each pair and their two final sites shall be coplanar. In practical applications of the theory it may often happen

* Visiting Professor of Physics, on leave from University of Nottingham, England.

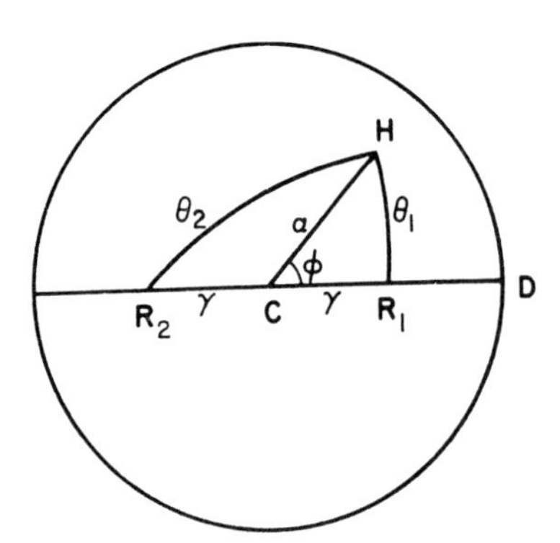

FIG. 1. Stereogram showing relevant directions. R_1, R_2 are the two directions taken by each nuclear pair vector with equal probability; C is their bisector. **H** is the applied magnetic field.

that the four sites are in fact coplanar, but this is not a necessary assumption of the theory.

The spectrum is determined (*1*) by treating the nuclear dipolar interaction as a perturbation on the Zeeman interaction between the nuclei and the applied magnetic field. For one nuclear pair in the system the truncated dipolar Hamiltonian is

$$\mathscr{H}_d = \tfrac{1}{2}\gamma_1\gamma_2\hbar^2 r^{-3}\,(\mathbf{I}_1 . \mathbf{I}_2 - 3I_{1z}I_{2z})\,(3\cos^2\theta - 1), \qquad [1]$$

where γ_1, γ_2 are the magnetogyric ratios of the nuclei in the pair, **r** is the internuclear vector, and θ is the angle between **r** and **H**. If **r** moves rapidly between the two directions R_1 and R_2, spending equal times in each orientation and none in any other, we are required (*2*, *3*) to replace $\mathscr{H}_d$ by its average $\overline{\mathscr{H}}_d$ over the two orientations before calculating the energy eigenvalues of the system and evaluating the spectrum in the vicinity of the resonance frequency. Since the angle θ is the only time-dependent variable in Eq. [1] we find that the spectrum for an array of similarly oriented nuclear pairs consists of two lines of equal intensity (*1*, *2*) at field strengths

$$h = \pm\, h_0\,\overline{(3\cos^2\theta - 1)}, \qquad [2]$$

where

$$h = H - H^*, \qquad [3]$$

and H^* is the resonant magnetic field strength for the frequency of the applied electromagnetic radiation, taken to be fixed. In Eq. [2]

$$h_0 = \tfrac{3}{2}\,\mu r^{-3} \qquad [4]$$

where μ is the magnetic moment of both nuclei in each pair if they are identical. If the two nuclei in each pair are not identical, then

$$h_0 = \mu r^{-3} \qquad [5]$$

where μ is now the magnetic moment of the nonresonant nuclei.

From Eq. [2] we find that the two spectral lines occur at

$$\begin{aligned} h &= \pm\tfrac{1}{2}h_0[(3\cos^2\theta_1 - 1) + (3\cos^2\theta_2 - 1)] \\ &= \pm\, h_0(3\cos^2\alpha\cos^2\gamma + 3\sin^2\alpha\sin^2\gamma\cos^2\phi - 1). \end{aligned} \qquad [6]$$

It is worth noting a few special cases.

(a) If **H** is parallel to the bisector C of the directions R_1 and R_2, then $\alpha = 0$ and

$$h = \pm h_0 (3\cos^2\gamma - 1), \qquad [7]$$

which is the same spectrum as we would get if there were no motion. This is correct since this situation is equivalent to 2-fold reorientation about the bisector C, and 2-fold reorientation does not affect $\mathscr{H}_d$, and is well known not to affect the spectrum directly, in this special case.

(b) In the same way there is no motional narrowing if $\gamma = \pi/2$. In this case Eq. [6] reduces to

$$h = \pm h_0 (3\sin^2\alpha\cos^2\phi - 1) = \pm h_0 (3\cos^2\delta - 1), \qquad [8]$$

where δ is the angle between **H** and the direction D in Fig. 1 with which R_1 is now coincident.

(c) If $\gamma = \pi/4$ Eq. [6] yields

$$h = \pm \tfrac{1}{2} h_0 (1 - 3\sin^2\alpha\sin^2\phi) = \pm \tfrac{1}{2} h_0 (1 - 3\cos^2\varepsilon), \qquad [9]$$

where ε is the angle between **H** and the normal to the plane R_1, R_2 and C in Fig. 1. Thus as ε varies h runs through just half the range of values encountered in absence of the motion.

Most experimental work is carried out with polycrystalline material, and the spectrum therefore consists of a superimposition of pairs of lines given by Eq. [6] for crystallites having values of α and ϕ corresponding to an isotropic distribution of crystallite orientations. Before calculating the form of the spectrum which this superimposition generates we first calculate its second moment M_2 about $h = 0$. For one crystallite we have from Eq. [6]

$$M_2(\alpha, \phi) = h_0^2 (3\cos^2\alpha\cos^2\gamma + 3\sin^2\alpha\sin^2\gamma\cos^2\phi - 1)^2. \qquad [10]$$

For an isotropic distribution of crystallites we therefore have

$$M_2 = \frac{1}{4\pi}\int_0^\pi \sin\alpha\, d\alpha \int_0^{2\pi} M_2(\alpha, \phi)\, d\phi$$

$$= \tfrac{4}{5} h_0^2 (1 - \tfrac{3}{4}\sin^2 2\gamma), \qquad [11]$$

using Eq. [10]. In the absence of reorientational motion the second moment of the isotropically distributed identical pairs is readily found to be

$$M_{2,0} = \tfrac{4}{5} h_0^2. \qquad [12]$$

The reorientational motion therefore reduces the contribution of the pairs to the second moment of the spectrum by a factor

$$M_2/M_{2,0} = (1 - \tfrac{3}{4}\sin^2 2\gamma). \qquad [13]$$

The reduction is greatest when $\gamma = \pi/4$; in this case the reduction ratio is $\tfrac{1}{4}$. It is worth noting that when $\gamma = \pi/4$, the two directions R_1 and R_2 are at right angles. Reorientation between these two directions is therefore indistinguishable magnetically from 4-fold reorientation about a tetrad axis normal to R_1 and R_2, and it is well known *(2)* that this reduces the second moment in such a case by a factor $\tfrac{1}{4}$ for polycrystalline materials. It should also be noted that the reduction factor in Eq. [13] is the same for γ and $(\pi/2) - \gamma$, which is physically obvious.

We now derive the spectral profile for a polycrystalline assembly of reorienting pairs. From each crystallite we obtain a pair of lines given by Eq. [6]. We consider first the positive sign in Eq. [6]; the negative sign generates an exactly similar spectrum mirrored in $h = 0$, obtained by replacing h_0 by $-h_0$ throughout. For convenience we re-express Eq. [6] as

$$h^* = h + h_0 = h_1 \cos^2 \alpha + h_2 \sin^2 \alpha \cos^2 \phi, \qquad [14]$$

where

$$h_1 = 3h_0 \cos^2\gamma \quad \text{and} \quad h_2 = 3h_0 \sin^2 \gamma. \qquad [15]$$

First holding α constant we find the spectrum generated by all azimuth angles ϕ. Describing the spectral shape by the normalized function $g_\alpha(h^*)$ we have

$$g_\alpha(h^*) = \frac{1}{\pi}\left|\frac{d\phi}{dh^*}\right| = (2\pi)^{-1}(h^* - h_1 \cos^2 \alpha)^{-\frac{1}{2}}(-h^* + h_1 \cos^2 \alpha + h_2 \sin^2 \alpha)^{-\frac{1}{2}} \qquad [16]$$

using Eq. [14]. This expression for $g_\alpha(h^*)$ applies for h^* in the range

$$h_1 \cos^2 \alpha \leqslant h^* \leqslant h_1 \cos^2 \alpha + h_2 \sin^2 \alpha, \qquad [17]$$

which, from Eq. [14], is the full range of values which h^* takes on; outside this range $g_\alpha(h^*)$ is zero. We now sum the spectra over all crystallite orientations α, describing the resultant spectrum with the normalized function $g(h^*)$:

$$g(h^*) = \int_{\alpha_1}^{\alpha_2} g_\alpha(h^*) \sin \alpha \, d\alpha. \qquad [18]$$

The range of integration lies within $0 \leqslant \alpha \leqslant \pi/2$, but only for those parts of this range for which $g_\alpha(h^*)$ is non-zero for the particular value of h^*. These ranges are

$$\frac{\pi}{2} \geqslant \alpha \geqslant \cos^{-1}\left(\frac{h^*}{h_1}\right)^{\frac{1}{2}} \quad \text{for} \quad 0 \leqslant h^* \leqslant h_2 \qquad [19]$$

$$\cos^{-1}\left(\frac{h^* - h_1}{h_1 - h_2}\right) \geqslant \alpha \geqslant \cos^{-1}\left(\frac{h^*}{h_1}\right)^{\frac{1}{2}} \quad \text{for} \quad h_2 \leqslant h^* \leqslant h_1. \qquad [20]$$

For convenience we have taken $h_1 > h_2$ ($\gamma < 45°$); however if $h_1 < h_2$ we merely interchange h_1 and h_2. Substituting from Eq. [16] into Eq. [18] and rearranging we find for the range [19]

$$g(h^*) = (2\pi)^{-1} h_2^{-\frac{1}{2}} (h_1 - h^*)^{-\frac{1}{2}} \int_0^{\pi/2} (1 - k^2 \sin^2 \psi)^{-\frac{1}{2}} d\psi \qquad [21]$$

$$= (2\pi)^{-1} h_2^{-\frac{1}{2}} (h_1 - h^*)^{-\frac{1}{2}} K(k), \qquad [22]$$

where

$$k^2 = \frac{h^*(h_1 - h_2)}{h_2(h_1 - h^*)}, \qquad [23]$$

and $K(k)$ is the complete normal elliptic integral of the first kind of argument k. Similarly for the range [20], we obtain

$$g(h^*) = (2\pi)^{-1} h^{*-\frac{1}{2}} (h_1 - h_2)^{-\frac{1}{2}} \int_0^{\pi/2} (1 - k^{-2} \sin^2 \psi)^{-\frac{1}{2}} d\psi \qquad [24]$$

$$= (2\pi)^{-1} h^{*-\frac{1}{2}} (h_1 - h_2)^{-\frac{1}{2}} K(k^{-1}). \qquad [25]$$

From Eqs. [22] and [25] with Eqs. [14] and [15] the spectrum is therefore given by:

for $\quad h_0(3\sin^2\gamma-1) \geqslant h \geqslant -h_0$: $\quad g(h) = (2\pi)^{-1}h_2^{-\frac{1}{2}}(h_1-h_0-h)^{-\frac{1}{2}}K(k) \qquad [26]$

and for

$h_0(3\cos^2\gamma-1) \geqslant h \geqslant h_0(3\sin^2\gamma-1)$: $\quad g(h) = (2\pi)^{-1}(h+h_0)^{-\frac{1}{2}}(h_1-h_2)^{-\frac{1}{2}}K(k^{-1}) \qquad [27]$

together with two similar equations in which h_0 is replaced by $-h_0$ and the sense of the inequalities is reversed. For $\gamma = 0$ these equations correctly reduce to the expressions given by Pake (*1*) for static nuclear pairs. As mentioned above Eqs. [26] and [27] apply for $0 \leqslant \gamma \leqslant \pi/4$; for $\pi/4 \leqslant \gamma \leqslant \pi/2$ we replace γ with $(\pi/2)-\gamma$ in [26] and [27]. The functional form of the spectrum is similar to those encountered with nonaxially symmetrical chemical shift tensors and Knight shift tensors in polycrystalline solids (*4*), and with first order quadrupole splitting with non-zero asymmetry parameter of the electric field gradient tensor in polycrystalline solids (*5*). This relationship becomes apparent when one considers that in the present problem two equal dipolar tensors are being added with their symmetry axes inclined at an angle 2γ. Except when 2γ is $\pi/2$ the resultant tensor is nonaxially symmetric.

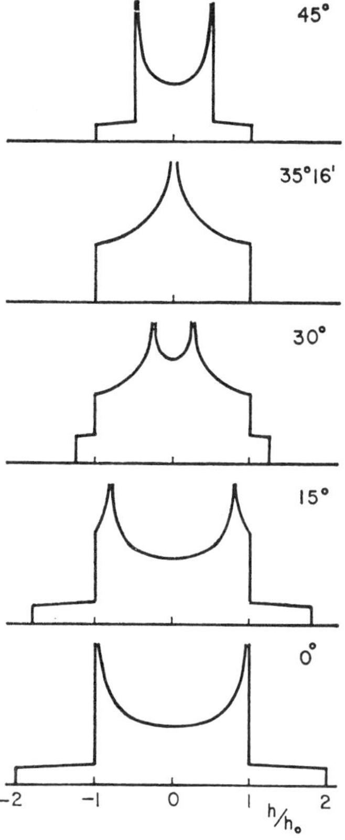

FIG. 2. Calculated spectra for polycrystalline material containing nuclear pairs reorienting between two directions subtending an angle 2γ. Spectra are shown for values of γ of 0°, 15°, 30°, 35° 16′, 45°. Spectra for values of γ between 45° and 90° are identical with those for $90°-\gamma$.

Using tables of complete elliptic functions (*6*) examples of the spectra have been evaluated as shown in Fig. 2. It is seen that singularities occur at

$$h = \pm h_0 (1-3\sin^2\gamma). \quad [28]$$

If the further broadening of the spectra by interactions between neighbouring nuclear pairs is not too great, the spectra will exhibit maxima close to these singularities. Defining a spectral linewidth ΔH as the interval between the two singularities we thus find

$$\frac{\Delta H}{\Delta H_0} = |1-3\sin^2\gamma|, \quad [29]$$

where ΔH_0 is the linewidth in the absence of reorientational motion, and is equal to $2h_0$. The two singularities coincide at $h = 0$ when $\gamma = \sin^{-1} 1/\sqrt{3} = 35°16'$.

DISCUSSION

The analysis just presented applies generally when relatively isolated nuclear pairs reorient between fixed directions. It is likely to be encountered with methylene proton pairs when ring inversion of cyclic molecules occurs in the solid state and by way of example we consider the theory in relation to solid cyclobutane C_4H_8. From electron diffraction (*7*), infra-red (*8*), and liquid-state high-resolution NMR (*9*) work, the molecule is believed to be nonplanar, having D_{2d} symmetry, with a dihedral angle of 33° to 36° (angle between two CCC planes in the molecule). The potential barrier of 1.3 kcal/mole (448 cm^{-1}) (*8*) against ring inversion between the two equivalent conformers is relatively low, as compared for example with about 10.8 kcal/mole (*10, 11*) for cyclohexane in the liquid state. It therefore seems possible that ring inversion may take place in solid cyclobutane even though additional constraints will be imposed by the neighboring molecules in the crystal. Such conformational changes would reorient the proton pair vectors in the four methylene groups of each molecule between two directions inclined at an angle 2γ equal to the dihedral angle of the molecule.

The proton magnetic resonance spectrum of solid cyclobutane has been examined by Hoch and Rushworth (*12*), who found in the low temperature phase that the spectrum narrowed with increasing temperature; the second moment was 20.4 G^2 below 130°K and reached a reduced value of 11.9 G^2 just below the phase transition temperature, 146°K. The authors interpreted the reduction in terms of rapid 4-fold reorientation about the tetrad axes of the molecules which were taken to be effectively planar. However, sincei t is now clear that the molecules are nonplanar (*7, 8, 9*) the spectral narrowing needs reinterpretation. There was, moreover, a difficulty in explaining the doublet structure of the resonance spectrum at 146°K which had a separation of 4.7 G between maxima, whereas the theoretical separation for the planar molecule is $2h_0$, which is 6.8 G.

Let us now examine the hypothesis that the spectral narrowing is caused by ring inversion. The doublet spectral width between singularities is given by Eq. [29], and taking 6.8 G for ΔH_0 and 34° for the dihedral angle which is 2γ, we find

$$\Delta H = \Delta H_0 \, |1-3\sin^2\gamma| = 5.1 \text{ G}, \quad [30]$$

which agrees quite well with the observed value of 4.7 G, especially when one recalls

(*1*, *12*) that the broadening effect of neighbors outside the pairs tends to reduce the peak separation somewhat. The contribution of the pairs to the second moment is reduced by the factor given in Eq. [13] (which is 0.766 for $2\gamma = 34°$) to 7.1 G^2. Using the C–C and C–H bond lengths and HCH bond angle found by Dunitz and Schomaker (*13*) together with the dihedral angle of 34°, the effect of ring inversion on the remaining intramolecular proton interactions has been calculated by established procedures (*14*), and is found to be 0.7 G^2, making a total intramolecular second moment of 7.8 G^2. The observed second moment (*12*) is 11.9 G^2, thus leaving 4.1 G^2 to be accounted for by the reduced intermolecular contribution. The intermolecular contribution for the rigid lattice array was estimated to be 6.8 G^2 (*12*), so that 4.1 G^2 represents a reduction to 60%. This is a higher proportion than is found for rotational motion (*14*), but seems not an unreasonable figure for ring inversion which causes smaller changes in the intermolecular pair vectors. Ring inversion does therefore seem to be consistent with the reduced spectral second moment and with the doublet structure separation. It must be emphasized however that this does not prove that this is in fact the operative spectral narrowing mechanism since it is possible that other mechanisms might also fit the facts, and we now comment on some other related possibilities.

We have assumed that the two conformations are equally probable for each molecule, but without knowledge of the crystal structure we cannot be sure that this is so. One could carry through an analysis similar to that done in this paper, but more complicated, in which a nuclear pair takes one orientation with probability p and another with probability $(1-p)$, where in this particular application p would be a function of temperature.* This would most probably yield a linewidth greater than given by Eq. [28], reducing to that value as p tends to $\frac{1}{2}$. Since Eq. [28] gives a value if anything larger than that observed it suggests that if ring inversion is the operative mechanism of spectral narrowing the two conformations do have almost equal probability at 146°K. Solid cyclobutane has a λ-type specific heat anomaly (*15*) starting at about 120°K and ending abruptly at 146°K. It is possible to interpret this in terms of a progressive cooperative disordering of conformational preference at each molecular site, so that even though p may exceed $\frac{1}{2}$ at low temperature it decreases to $\frac{1}{2}$ in the vicinity of the transition temperature 146°K, where our theory is being applied.

Another form of motion which might be conceived is ring inversion accompanied by reorientation through $\pi/2$ about the D_{2d} axis. This motion would leave the molecule in the same relationship between conformation and environment, but would only reduce the linewidth (*2*) by the factor $\frac{1}{2}(3\cos^2\gamma - 1)$ to 5.9 G, which seems too large to be correct.

* Note added in proof. This analysis has now been carried out and will be the subject of a later paper.

REFERENCES

1. G. E. Pake, *J. Chem. Phys.* **16,** 327 (1948).
2. H. S. Gutowsky and G. E. Pake, *J. Chem. Phys.* **18,** 162 (1950).
3. A. Abragam, "The Principles of Nuclear Magnetism," Oxford Univ. Press (1961).
4. N. Bloembergen and T. J. Rowland, Acta Metallurgica **1,** 731 (1953).
5. M. H. Cohen and F. Reif, *Solid State Phys.* **5,** 321 (1957).

6. E. Jahnke, F. Emde and F. Losch, "Tables of Higher Functions," McGraw-Hill (1960).
7. A. de Meijere, *Acta Chem. Scand.* **20,** 1093 (1965).
8. J. Laane and R. C. Lord, *J. Chem. Phys.* **48,** 1508 (1968).
9. J. B. Lambert and J. D. Roberts, *J.A.C.S.* **87,** 3884 (1965).
10. F. A. Bovey, F. B. Hood, E. W. Anderson, and R. L. Kornegay, *J. Chem. Phys.* **41,** 2041 (1964).
11. F. A. L. Anet, M. Ahmad and L. D. Hall, *Proc. Chem. Soc.* **1964,** 145.
12. M. J. R. Hoch and F. A. Rushworth, *Proc. Phys. Soc.* **83,** 949 (1964).
13. J. D. Dunitz and V. Schomaker, *J. Chem. Phys.* **20,** 1703 (1952).
14. E. R. Andrew and R. G. Eades, *Proc. Roy. Soc.* **216,** 398 (1953).
15. G. W. Rathjens and W. D. Gwinn, *J.A.C.S.* **75,** 5629 (1953).

JOURNAL OF MAGNETIC RESONANCE 3, 28–43 (1970)

Coherent Spectrometry with Noise Signals

R. KAISER

Physics Department, University of New Brunswick, Fredericton, N.B., Canada

Received March 28, 1970; accepted April 3, 1970

Coherent spectrometry with noise signals is described as Fourier transform spectrometry with deterministic signals which are samples from a stochastic process. Input–output relations needed for the processing of measured signal records are derived from the theories of linear physical systems and of sampled functions. The apparatus for obtaining noise-stimulated nuclear-magnetic-resonance spectra is described and results are presented for the spectrum of o-dichlorobenzene. A qualitative comparison is made with Fourier transform spectrometry based on pulsed signals, and the use of pseudorandom binary signals is briefly considered.

INTRODUCTION

A spectrum gives the functional dependence of some output quantity of a physical system upon some independent variable, usually energy or a closely related variable such as frequency or wavenumber. Conservation of energy requires an input to the system, and it is useful to distinguish between coherent and incoherent spectrometry in which use is made or is not made, respectively, of the coherence properties between output and input signals of the system. White noise as produced by an idealized random process has very little coherence in itself and is often thought to be unsuitable for coherent spectrometry. Indeed, optical emission and absorption spectrometry, for which noise-like input and output signals are used (white noise indicates analogy to white light) is of the incoherent type since only input and output intensities are measured. Although the noise signals at input and output may have very little coherence in themselves, the fact that the output is produced by the system from the input forces the two signals to be coherent with each other so that coherent spectrometry with noise signals is perfectly feasible if the cross correlation or the corresponding cross power density functions between input and output signals are measured. A method for coherent nuclear-magnetic-resonance spectrometry with noise signals is described in this work.

It is helpful to consider coherent noise spectrometry from a different viewpoint. In conventional nuclear magnetic resonance spectrometers, a sufficiently weak sinusoidal signal of radiofrequency f and amplitude $X(f)$ is applied to the spin system which is situated in a magnetic field, and its sinusoidal response of amplitude $Y(f)$ is mixed with the input sinusoid in a phase sensitive detector. The output of this detector, after normalization, gives the real part $V(f)$ (absorption mode) or imaginary part $U(f)$ (dispersion mode) of the spin system's frequency response function $R(f) = Y(f)/X(f)$ at frequency f. One method to obtain the entire NMR spectrum is to scan the frequency f of the sinusoid sufficiently slowly (*1*) through the range of interest. For example, a typical high-resolution proton spectrum of a diamagnetic liquid shows resonance peaks of width $\Delta f \simeq 0.3$ Hz so that with $\dot{f} \simeq (\Delta f)^2$ approximately one hour is required to scan the frequency through the 600 Hz wide range of proton resonance frequencies. The measurement time would clearly be shortened and the instrumental stability requirements could be relaxed if several sinusoids of different frequencies could be used simultaneously. If n sinusoids were spaced at equal frequency intervals through the bandwidth B of the spectrum, each would need to scan only a range B/n and the measurement time would be shortened by a factor n. As n is increased, a limit is reached when the scan range B/n becomes equal to the spectral resolution Δf (0.3 Hz in the example given). A scan is then no longer necessary and the measurement time is reduced to the transient decay time $1/\Delta f$. The maximum number of channels that can be usefully employed in this multichannel spectrometer is then $B/\Delta f$.

It is clear that the success of a multichannel operation depends on the n signals not to interfere with each other within the system whose spectrum is to be measured. This condition is satisfied by linear systems. It is well known that nuclear spin systems are not linear since they do show interference between two or more simultaneously applied sinusoids in the form of double resonance, Overhauser, saturation and other effects. However, most systems including nuclear spin systems can be approximated by linear equivalents provided that the several simultaneously applied sinusoids are sufficiently weak, and it will be assumed throughout this work that this condition is satisfied. Basically, in referring to a system's frequency-response function $R(f)$ or spectrum $V(f)$ or $U(f)$ it is implied that the system is approximated by its linear equivalent, and effects caused by the breakdown of this approximation are generally referred to as perturbations of the basic spectrum.

The input and output signals of the multichannel spectrometer without scan were described above as $n = B/\Delta f$ superimposed sinusoids equally spaced at intervals Δf through the frequency range B. This is essentially a description in the frequency domain. The signal as a function of time which is actually applied to the spectrometer input is the Fourier synthesis of the n sinusoids, and the system spectrum is obtained from the spectrometer output signal by Fourier analysis. All n sinusoids which make up the input signal should have the same amplitude for all parts of the spectrum to be stimulated with equal intensity, but in addition the relative phase angles of these sinusoids need to be specified to determine the time dependence of the input signal. Without specification of the phase angles one can only state that the signals must be periodic in time with period $T = 1/\Delta f$ in order to have discrete frequency components at intervals Δf. There are obviously infinitely many ways to choose the n phase angles. For example, they might be selected to produce destructive interference between the n sinusoids over most of the period T and an impulsive signal over the remaining time. The merit of an impulsive signal is that the rf modulator reduces to a simple electronic switch, and the resulting method (with some compromise concerning the frequency independence of the power density) has been described (*2*) as Fourier transform spectrometry. Another possibility would be to make a random selection of the n phase angles. This leads to the use of a sample signal record from a stochastic process and to the coherent noise spectrometry method described here. Again, even for a process with a band-limited white-power-density spectrum, the frequency independence of the excitation is compromised to some extent since it holds only for the expectation value of the periodogram and a single sample record may show considerable variance (*3*). A third type of input signal, resulting from a different, deterministic, choice of the n phase angles, will be mentioned briefly in the Conclusion of this paper after the input–output relations for linear systems have been reviewed in Section II and the experimental procedure has been described in Section III.

THEORY

Consider the time-independent, realizable, stable, linear system (*4*) shown in Fig. 1. Let the input signal be a sinusoid of frequency f with complex amplitude $X(f)$

$$x(t) = \text{Re}\,\{X(f)\cdot\exp(2\pi ift)\} \qquad [1]$$

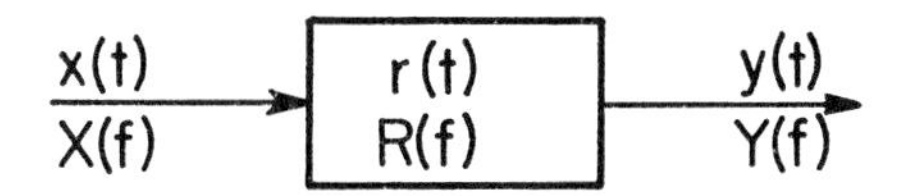

FIG. 1. Signals and response functions for a two-port linear system.

where Re $\{\ldots\}$ denotes the real part of the expression in brackets. The output is then a similar sinusoid of identical frequency f with complex amplitude $Y(f)$

$$y(t) = \text{Re}\,\{Y(f)\cdot\exp(2\pi ift)\} \qquad [2]$$

and the complex frequency-response function is defined by

$$R(f) = V(f) - iU(f) = Y(f)/X(f). \qquad [3]$$

In nuclear magnetic resonance, the real part $V(f)$ of the frequency response function is referred to as the absorption mode spectrum of the spin system and the imaginary part $U(f)$ is the dispersion mode spectrum. Eqs. [1] to [3] apply to the method of measuring the spectrum $R(f)$ with the usual frequency scan.

Since the system is linear by hypothesis, the response to a linear superposition of input signals of the form given by Eq. [1], i.e.

$$x(t) = \int_{-\infty}^{+\infty} X(f)\cdot\exp(2\pi ift)\,d \qquad [4]$$

will be the corresponding superposition of output signals

$$y(t) = \int_{-\infty}^{+\infty} Y(f)\cdot\exp(2\pi ift)\,df \qquad [5]$$

with

$$Y(f) = R(f)\cdot X(f). \qquad [6]$$

Eqs. [4] and [5] describe the input $x(t)$ and output $y(t)$ signals in terms of their Fourier transforms $X(f)$ and $Y(f)$ which are given by

$$X(f) = \int_{-\infty}^{+\infty} x(t)\cdot\exp(-2\pi ift)\,dt \qquad [7]$$

$$Y(f) = \int_{-\infty}^{+\infty} y(t)\cdot\exp(-2\pi ift)\,dt. \qquad [8]$$

Eqs. [6] to [8] form the basis of the method to be described here for measuring the spectrum $R(f)$ by stimulating the system with a stochastic input signal. Eq. [6] describes the action of the system in the frequency domain as a multiplication of the input amplitude spectrum $X(f)$ with the system spectrum $R(f)$ to give the output spectrum $Y(f)$. The corresponding relation in the time domain is obtained by substitution of Eqs. [6] and [7] into [5]:

$$y(t) = \int_{-\infty}^{+\infty} r(t-\tau) \cdot x(\tau)\, d\tau \qquad [9]$$

where

$$r(t) = \int_{-\infty}^{+\infty} R(f) \cdot \exp(2\pi i f t)\, df \qquad [10]$$

so that

$$R(f) = \int_{-\infty}^{+\infty} r(t) \cdot \exp(-2\pi i f t)\, dt. \qquad [11]$$

Eq. [9] shows that the system forms the output signal $y(t)$ by convolving the input signal $x(t)$ in the time domain with the system function $r(t)$ given by Eq. [10] The physical significance of $r(t)$ becomes apparent when excitation of the system with an impulse at time zero is considered:

$$x(t) = \delta(t). \qquad [12]$$

The resulting output then becomes, from Eq. [9],

$$y(t) = r(t) \qquad [13]$$

and $r(t)$ is therefore referred to as the system's impulse response function. Eqs. [11] to [13] form the basis of Fourier transform spectrometry with input pulses.

The signals $x(t)$ and $y(t)$ are real functions of time and it follows then from Eqs. [7], [8], [13], and [11] that

$$X^*(f) = X(-f); \quad Y^*(f) = Y(-f); \quad R^*(f) = R(-f) \qquad [14]$$

where the asterisk indicates the conjugate complex. In particular, the real part of the frequency-response function, i.e. the absorption mode spectrum, is an even function of frequency

$$V(-f) = V(f) \qquad [15]$$

while the imaginary part, i.e. the dispersion mode spectrum, is an odd function of frequency

$$U(-f) = -U(f). \qquad [16]$$

With $R = V - iU$, Eq. [10] can therefore be written in the form

$$r(t) = r_e(t) + r_0(t) = \int_{-\infty}^{+\infty} V(f) \cdot \exp(2\pi i f t)\, df - i \int_{-\infty}^{+\infty} U(f) \cdot \exp(2\pi i f t)\, df \qquad [17]$$

with

$$r_e(t) = \int_{-\infty}^{+\infty} V(f) \cdot \exp(2\pi i f t)\, df = 2 \int_{0}^{+\infty} V(f) \cdot \cos(2\pi f t)\, df = r_e(-t) \qquad [18]$$

and

$$r_0(t) = -i \int_{-\infty}^{+\infty} U(f) \cdot \exp(2\pi i f t)\, df = 2 \int_{0}^{+\infty} U(f) \cdot \sin(2\pi f t)\, df = -r_0(-t). \qquad [19]$$

Eqs.[18] and [19] show that $r_e(t)$, the cosine Fourier transform of the absorption mode spectrum, is an even function of time, while $r_0(t)$, the sine Fourier transform of the dispersion-mode spectrum, is an odd function of time. Eqs. [18] and [19] may be inverted to obtain

$$V(f) = \int_{-\infty}^{+\infty} r_e(t) \cdot \exp(-2\pi i f t)\, dt = 2 \int_{0}^{+\infty} r_e(t) \cdot \cos(2\pi f t)\, dt \quad [20]$$

$$U(f) = i \int_{-\infty}^{+\infty} r_0(t) \cdot \exp(-2\pi i f t)\, dt = 2 \int_{0}^{+\infty} r_0(t) \cdot \sin(2\pi f t)\, dt. \quad [21]$$

The symmetry properties expressed in Eqs. [14] to [21] for the impulse response function $r(t)$ and spectrum $R(f)$ apply also to the signals x, y, and their Fourier transforms X, Y. However, the impulse response function differs from the signals in that it must satisfy the causality condition: The system response, Eq. [13], to an impulse, Eq. [12], must be zero before the occurrence of the impulse, i.e.

$$r(t) = 0 \quad \text{for } t < 0. \quad [22]$$

The even and odd parts of $r(t)$ in Eq. [17] must cancel for negative t. It follows that they are equal to each other for positive t, i.e.

$$\tfrac{1}{2} r(t) = r_e(t) = r_0(t) = 2 \int_{0}^{+\infty} V(f) \cdot \cos(2\pi f t)\, df = 2 \int_{0}^{+\infty} U(f) \cdot \sin(2\pi f t)\, df \quad \text{for } t > 0. \quad [23]$$

Eq. [23] shows that the causality condition establishes a relation between the real and imaginary parts of the system frequency-response function, i.e. between the absorption and dispersion mode spectra $V(f)$ and $U(f)$. Before looking at the details, we need to consider the point $t = 0$. Eqs. [17] to [19] show that

$$r(0) = r_e(0) = 2 \int_{0}^{+\infty} V(f)\, df = 2I; \qquad r_0(0) = 0$$

where I is the total area under the absorption mode spectrum. This area is not zero in a nuclear-magnetic-resonance spectrum so that Eq. [23] does not hold for $t = 0$. To avoid certain difficulties (5) it is convenient to extend Eq. [23] to include the point $t = 0$ by shifting the baseline of the absorption mode spectrum over the frequency range of interest such that $I = 0$. The error committed thereby can easily be compensated later by adding this shift back to $V(f)$. (Poles in $V(f)$ or $U(f)$ should be split off similarly but we need not be concerned therewith since they cannot occur for a stable system.) We shall assume that $V(f)$ has been modified in this manner so that

$$r(0) = r_e(0) = r_0(0) = 0.$$

We then have, for all t,

$$r_e(t) = \text{sgn}(t) \cdot r_0(t) \quad [24]$$

where

$$\text{sgn}(t) = t/|t|, \qquad \text{sgn}(0) = 0$$

has the Fourier transform $1/i\pi f$. The Fourier transform of Eq. [24] becomes with Eqs. [20] and [21]

$$V(f) = \left(\frac{1}{i\pi f}\right) * (-iU) = \frac{-1}{\pi} \int_{-\infty}^{+\infty} \frac{U(\varphi)}{f - \varphi}\, d\varphi = \frac{-2}{\pi} \int_{0}^{+\infty} \frac{U(\varphi)}{f^2 - \varphi^2}\, \varphi\, d\varphi \quad [25]$$

where use has been made of the fact that the Fourier transform of the product of two functions is equal to the convolution (indicated by the asterisk) of the Fourier transforms of the two functions (compare Eqs. [6] and [9]). From $r_0(t) = \text{sgn}(t) \cdot r_e(t)$ we obtain similarly

$$U(f) = \frac{1}{\pi f} * V = \frac{1}{\pi} \int_{-\infty}^{+\infty} \frac{V(\varphi)}{f-\varphi} d\varphi = \frac{2f}{\pi} \int_{0}^{+\infty} \frac{V(\varphi)}{f^2-\varphi^2} d\varphi. \quad [26]$$

The integral transforms of Eqs. [25] and [26] are known as Hilbert transforms, and their application to the real and imaginary parts of a system frequency-response function (or its logarithm) are the familiar Kramers–Krönig relations (*6*). The derivation given here shows that these relations are a consequence of the causality requirement for the system-impulse-response function.

The fact that the real and imaginary parts of $R(f)$ (except for a base line shift) are determined by each other can be put to use in signal processing operations. First, if a measurement yields both the real and imaginary parts of $R(f)$ with not completely correlated noise, the *a priori* knowledge of the Kramers–Krönig relations leads to a signal/noise improvement. In particular, if the noise in the real and imaginary parts is uncorrelated, the measurement of the imaginary part is equivalent to an uncorrelated second measurement of the real part. The simplest way to insure that full advantage is taken of the Kramers–Krönig relations is to compute the Fourier transform as in Eq. [10] of the measured complex $\tilde{R}(f) = \tilde{V}(f) - i\tilde{U}(f)$ after it has been complemented for $f<0$ according to Eq. [14]. If the noise in the real and imaginary parts is not suitably correlated, the resulting $\tilde{r}(t)$ will not be zero for $t<0$. By setting $r(t<0) = 0$ before transforming back by means of Eq. [11] only noise but no system information is suppressed and the resulting $R(f)$ necessarily satisfies the Kramers–Krönig relations Eqs. [25] and [26]. In fact, since usually only the absorption mode spectrum $V(f)$ is required, the right hand part of Eq. [20] may be used for the back transformation wherein $2r_e(t) = r(t)$ for $t \geqslant 0$.

Second, it frequently happens that improper phase adjustment in electronic circuits of the rf receiver results in a measured frequency response function $\tilde{R}(f)$ which is related to the desired $R(f)$ through

$$R(f) = \tilde{R}(f) \cdot \exp(i\theta) \quad [27]$$

The desired absorption mode spectrum $V(f)$ is then given by

$$V(f) = \tilde{V}(f) \cdot \cos\theta + \tilde{U}(f) \cdot \sin\theta \quad [28]$$

so that $\tilde{U}(f)$ is required in addition to $\tilde{V}(f)$. Spectrometers operating with the frequency (or field) sweep method normally give only $\tilde{V}(f)$ or $\tilde{U}(f)$ but not both. To get $\tilde{U}(f)$ from the measured $\tilde{V}(f)$ several procedures are available (5):

(a) The convolution of $\tilde{V}(f)$ with $1/\pi f$ may be computed according to Eq. [26];

(b) $\tilde{r}_e(t)$, obtained from Eq. [18], may be used to form $\tilde{r}_0(t) = \text{sgn}(t) \cdot \tilde{r}_e(t)$ followed by the transformation Eq. [21] to yield $\tilde{U}(f)$;

(c) $\tilde{r}_e(t)$, obtained from Eq. [18], may be used to form $\tilde{r}(t) = \{\text{sgn}(t)+1\} \cdot \tilde{r}_e(t)$ followed by the transformation Eq. [11] to yield $\tilde{R}(f) = \tilde{V}(f) - i\tilde{U}(f)$. (This procedure, of course, gives no signal/noise improvement. The truncation of $\tilde{r}_e(t)$ for $t<0$ results here in a loss of signal as well as noise.) Since this procedure gives $\tilde{R}(f)$, one may prefer to use Eq. [27] in place of [28] depending on the computational facilities.

R. Ernst has shown (5) that methods (b) and (c) offer computational advantages over (a) when the fast Fourier transform algorithm (7) is used on a digital computer.

The phase error θ in Eqs. [27] and [28] was considered to be a constant. However, this is not always true in Fourier transform or noise spectrometry. It will be seen below that it is desirable to pass the stochastic input and output signals through analog filters in order to limit their energy density spectra. These filters give rise to a frequency dependent differential phase shift.

$$\theta(f) = \theta_0 + \theta_1 \cdot f + O(f^2) \tag{29}$$

The linear term causes a time shift of the impulse response function

$$r(t) = \tilde{r}(t + \theta_1/2\pi) \tag{30}$$

while the higher order terms produce dispersion effects in the impulse response function, i.e. they cause a wavepacket or a sudden transient to spread in time. In order to compensate these effects we find it useful to consider the analytical signal

$$a(t) = r(t) - i\rho(t) \tag{31}$$

associated with the impulse response function $r(t)$. It is obtained by supplementing $r(t)$ with its quadrature signal given by the Hilbert transform

$$\rho(t) = \frac{1}{\pi} \int_{-\infty}^{+\infty} \frac{r(\tau)}{t - \tau} \, d\tau \tag{32}$$

The signals $a(t)$, $r(t)$, $\rho(t)$ are in the same generic relation as $\exp(-i\omega t)$, $\cos \omega t$, $\sin \omega t$ respectively. Instead of computing $\rho(t)$ from Eq. [32], which would correspond to method (a) above, we may proceed in a manner analogous to (b) or (c). Method (c) is most convenient since it yields directly the analytical signal $\tilde{a}(t)$ as the inverse Fourier transform, Eq. [10], of $\{1 + \text{sgn}(f)\} \cdot \tilde{R}(f)$. This is simply the measured frequency-response function $\tilde{R}(f)$ with $I \neq 0$, i.e. without shift of the baseline in the absorption mode spectrum, truncated for negative frequencies and doubled for positive frequencies. The constant and linear terms of Eq. [29], substituted into Eq. [27], result in

$$a(t) = \exp(i\theta_0) \cdot \tilde{a}(t + \theta_1/2\pi) \tag{33}$$

and can easily be compensated. The higher order terms cause a precursor to the sudden transient at time $\theta_1/2\pi$ in $\tilde{a}(t)$. We find that an acceptable compensation can be achieved by truncation of this precursor, and we correct for the three terms of the phase function Eq. [29] by shifting the function $\tilde{a}(t)$ an amount $\Delta t = \theta_1/2\pi$ so that its absolute maximum occurs at $t = 0$, truncating for $t < 0$, and rotating by multiplication with $\exp(i\theta_0) = \tilde{a}^*(\theta_1/2\pi)/|\tilde{a}(\theta_1/2\pi)|$ so that $a(0) = |\tilde{a}(\theta_1/2\pi)|$. At the same time, $a(t)$ may be multiplied by a suitable window function to affect smoothing or resolution enhancement by the corresponding convolution with the spectrum. The truncation at $t < 0$ insures that $R(f)$, obtained by back transformation with Eq. [11] from $a(t)$, satisfies the Kramers–Krönig relations. As was explained earlier, this leads to a signal/noise improvement in the case that $\tilde{V}(f)$ and $\tilde{U}(f)$ were measured with uncorrelated noise. We executed these operations on a digital computer where $\tilde{a}(t)$ is represented by a sequence of samples. The sampled structure limits the resolution with which the corrections can be applied without interpolation. It has been our experience that the resolution is generally adequate for the time shift and truncation, but the angle θ_0 changes in rather large steps from one sample to the next so that some interpolation is required. In the present work this was done manually but we see no principal difficulty in implementing a programmed interpolation procedure.

The implementation of these processing methods on a digital computer leads to additional constraints. They arise from the need to represent continuous signal functions and their Fourier and Hilbert transforms by finite-length sequences of discrete samples. The resulting requirements are easily deduced from the sampling theorem (*4*): The operation of sampling a continuous function $g(x)$ at equidistant intervals Δx to form an infinite sequence

$$g_s(x) = |\Delta x| \cdot g(x) \cdot \sum_{l=-\infty}^{+\infty} \delta(x - l \cdot \Delta x) \qquad [34]$$

corresponds to a periodic repetition of its Fourier transform $G(\xi)$

$$G_p(\xi) = G(\xi) * \sum_{k=-\infty}^{+\infty} \delta(\xi - k \cdot P) = \sum_{k=-\infty}^{+\infty} G(\xi - k \cdot P) \qquad [35]$$

with period

$$P = 1/\Delta x. \qquad [36]$$

If this operation is to be reversible, i.e. if the information contained in $g(x)$ is to be preserved through the sampling operation so that $g(x)$ may be reconstructed from the samples, it must be possible to reconstruct $G(\xi)$ from its periodic repetition $G_p(\xi)$. This will evidently be possible if the interference produced in the overlap regions of the periodic repetitions does not destroy information. A sufficient, although not necessary, condition is that there be no overlap. Overlap may be avoided if $G(\xi)$ is constrained to be zero outside some range of width W

$$G(\xi) = 0 \quad \text{for } |\xi - \xi_0| \geqslant W \qquad [37]$$

by choosing the repetition period greater than this width

$$P \geqslant W \qquad [38]$$

which, from Eq. [36], requires a sampling interval

$$\Delta x \leqslant 1/W \qquad [39]$$

Effects caused by intereference in the overlap regions because of violation of relation [39] are the well known aliasing phenomena. For the work described here, these considerations apply to the input signal $x(t)$, to the output signal $y(t)$, and to the system impulse response function $r(t)$ or its analytical version $a(t)$, as well as to their Fourier transforms $X(f)$, $Y(f)$ and $R(f)$. Consider, for example, the input signal. It is formed by modulating a monochromatic rf carrier with a stochastic signal $x(t)$, and if the first order sidebands created by the modulation and used in the subsequent demodulation are the only important ones, the width W of its Fourier transform in Eqs. [37] to [39] is given by twice the bandwidth B of the amplitude-spectrum $X(f)$ of the modulating-signal $x(t)$. Eq. [39] shows that this bandwidth must be restricted so that the condition

$$2B \leqslant 1/\Delta t \qquad [40]$$

is satisfied where Δt is the sampling interval for the signal $x(t)$. On the other hand, the bandwidth $2B$ must cover at least the frequency width of the NMR spectrum if all lines in the spectrum are to be stimulated, so that the NMR spectrum width is a lower limit for the required sampling rate $1/\Delta t$ of the analog to digital converter. In practice, it is desirable to utilize only the modulation sideband on one side of the rf carrier and to place the carrier some distance away from the NMR spectrum so that in fact B should be greater than the width of the NMR spectrum. Allowance should also be made in Eq. [40] for the gradual attentuation of the analog filter used to limit the bandwidth B.

The Fourier transform of the sampled input signal

$$x_s(t) = |\Delta t| \cdot x(t) \cdot \sum_{l=-\infty}^{+\infty} \delta(t-l\cdot\Delta t) \qquad [41]$$

is, from Eq. [35]

$$X_p(f) = \sum_{k=-\infty}^{+\infty} X(f-k\cdot F) \qquad [42]$$

with period

$$F = 1/\Delta t \geqslant 2B. \qquad [43]$$

$X_p(f)$ is a continuous function and $x_s(t)$ has infinitely many samples so that neither function is suitable for digital computer techniques. Clearly, $X_p(f)$ should be sampled at some suitable interval Δf to generate $X_{ps}(f)$, and since its periods do not overlap it should be sufficient to sample only one period so that only a finite number of samples results. Eq. [42] shows that a single period is given by $X(f)$. However, some caution is required because $X_{ps}(f)$ produced by sampling $X_p(f)$ is not necessarily identical with $X_{sp}(f)$ produced by periodically repeating the sample sequence $X_s(f)$. In detail, there results

$$\begin{aligned} X_{ps}(f) &= |\Delta f| \cdot X_p(f) \cdot \sum_{l=-\infty}^{+\infty} \delta(f-l\cdot\Delta f) \\ &= |\Delta f| \cdot \sum_{k=-\infty}^{+\infty} X(f-k\cdot F) \cdot \sum_{l=-\infty}^{+\infty} \delta(f-l\cdot\Delta f) \end{aligned} \qquad [44]$$

$$\begin{aligned} X_{sp}(f) &= \sum_{k=-\infty}^{+\infty} X_s(f-k\cdot F) \\ &= |\Delta f| \cdot \sum_{k=-\infty}^{+\infty} X(f-k\cdot F) \cdot \sum_{l=-\infty}^{+\infty} \delta(f-k\cdot F-l\cdot\Delta f) \end{aligned} \qquad [45]$$

and for these functions to be identical we require

$$F/\Delta f = n \qquad [46]$$

to be an integer. Either function is then given by repetition of the sequence of n samples taken from one period of $X_p(f)$

$$\{X_n\} = |\Delta f| \cdot X_p(f) \cdot \sum_{l=1}^{n} \delta(f-l\cdot\Delta f). \qquad [47]$$

Sampling $X_p(f)$ at intervals Δf has the effect of replicating its inverse Fourier transform $x_s(t)$

$$x_{sp}(t) = \sum_{k=-\infty}^{+\infty} x_s(t-k\cdot T) \qquad [48]$$

with period

$$T = 1/\Delta f \qquad [49]$$

as given by the sampling theorem. The arguments that led to Eqs. [37] to [39] show that in order to conserve information in the Δf sampling operation, $x(t)$ must be constrained to be zero outside of an interval of duration not greater than T so that repetition with period T does not result in overlap. Eqs. [49] and [43] substituted in [46] show that

$$T/\Delta t = n \tag{50}$$

so that $x_{sp}(t) = x_{ps}(t)$ analogous to Eqs. [44] and [45], and both functions may be represented by the sequence of n samples taken from one period of $x_p(t)$

$$\{x_n\} = |\Delta t| \cdot x_p(t) \cdot \sum_{l=1}^{n} \delta(t - l \cdot \Delta t). \tag{51}$$

Eqs. [47] and [51] give the representation of the Fourier pair $x(t)$ and $X(f)$ by two sequences of n discrete samples. The relation between the sequences $\{x_n\}$ and $\{X_n\}$ is the discrete Fourier transform (7). It must be kept in mind that this representation implies periodicity with periods T and F related to the sampling intervals Δt and Δf by Eqs. [43] and [49], and that the ratio n defined by [46] or [50] must be an integer. If $x(t)$ is real, then $n = 2BT$ is the minimum number of values required to characterize the waveform of bandwidth B and duration T. This number is often referred to as the number of degrees of freedom of the signal. Although the sequence values $\{X_n\}$ are then generally complex they satisfy the relation $X_{n-j} = X_j^*$ which follows from Eq. [14], so that there are also n independent values in the frequency-domain characterization. For the system frequency-response-function $\{R_n\}$ an additional relation, the Hilbert transform Eq. [25] or [26], connects the real and imaginary parts so that there are only $n/2$ independent values. Similarly, the analytical impulse response $a(t)$ is represented by n complex samples $\{a_n\}$ half of which are zero (for $t<0$) and the real and imaginary parts of the remaining half, representing $r(t)$ and $\rho(t)$ of Eq. [31], are related by the Hilbert transform Eq. [32] so that only $n/2$ independent values remain. The system is thus characterized by $n/2$ independent values in the time domain or in the frequency domain. In the time domain these may be taken as the $n/2$ samples representing $r(t)$ for $t \geqslant 0$, or in the frequency domain they may be taken as the $n/2$ samples representing the absorption mode spectrum $V(f)$ for $f>0$.

It is interesting to note the redundancy of the method. In order to determine the $n/2$ values characterizing the system, $2n$ measurements are made for the samples $\{x_n\}$ and $\{y_n\}$ of input and output signals, and for maximum efficiency these should all be independent. Now, n degrees of freedom are subsequently lost in forming the ratio $R(f) = Y(f)/X(f)$, and the remaining factor 2 which arises from the Kramers–Krönig relations, i.e. from the causality condition, is used properly for improvement of the signal/noise ratio. In contrast, the pulsed Fourier transform method with widely spaced pulses measures $r(t)$ directly with $n/2$ samples. It should be realized, however, that this is still highly redundant since the usual objective in high resolution NMR spectroscopy is the determination of chemical shifts and coupling constants, and there are much fewer than $n/2$ of these in a typical NMR spectrum.

EXPERIMENTAL

For this initial attempt to demonstrate the feasibility of NMR spectrometry with random signals, a strong spectrum with some closely spaced lines and with a variety of line intensities covering a small frequency bandwidth was desirable. The spectrum of ortho-dichlorobenzene is well known (8) and answers these requirements. A sample of o-dichlorobenzene with a few per cent of tetramethylsilane was prepared by distillation and vacuum degassing in a 5 mm o.d. glass sample tube. Measurements were made on a Varian V4302 spectrometer operating at 56.4 MHz, and at this frequency the bandwidth B of the o-dichlorobenzene spectrum is about 35 Hz.

Figure 2 shows a block diagram of the apparatus. The stochastic input signal was derived from a random-noise generator by filtering and field modulation. The generator produced white noise with a flat power density spectrum in the range 10 Hz to 20 KHz and with approximately Gaussian amplitude probability density up to a crest factor of 3.5. The generator output was passed through a second order filter with corner frequency 100 Hz and damping factor d near 0.7 ($d = 1$ corresponds to critical damping). The filter circuit provides two outputs as shown in Fig. 3: output 2 has the usual low-pass response while output 1 is proportional to the time-derivative of output 2 to compensate the $1/\Omega$ dependence of the modulation index for magnetic field modulation with frequency Ω (*9*). The magnetic field of the NMR spectrometer was stabilized by locking with a feedback control loop a sideband produced by frequency modulation of the rf carrier to the center of the tetramethylsilane reference line in dispersion mode. The sideband frequency of the lock system was adjusted so that the rf carrier was 60 Hz to the low field side of the center of the o-dichlorobenzene

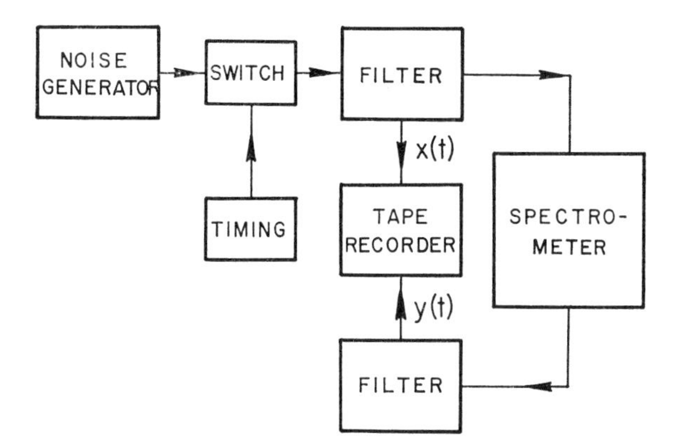

FIG. 2. Block diagram of Noise Spectrometer.

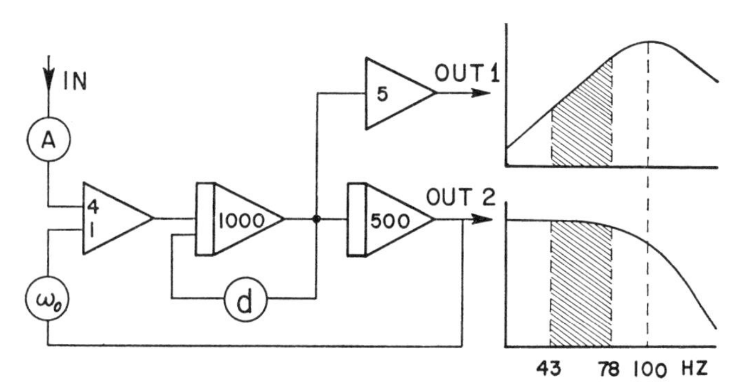

FIG. 3. Filter circuit. A = gain, ω_0 = corner frequency, d = damping coefficient. The graphs indicate the filter response as a function of frequency.

spectrum so that this spectrum fell within the 43 to 78 Hz range of the noise sideband shaded in Fig. 3. The spectrometer receiver contains a rf phase sensitive demodulator whose output was passed through a second filter, identical in construction to that shown in Fig. 3 with nominally the same corner frequency and damping, before being recorded on tape together with the input signal on two tracks of an instrumentation tape recorder. The signals were later reproduced from the tape and processed off line through the analog signal interface of an IBM 360/50 computer. Choosing a different tape speed for playback permitted the time scale of the computer to be matched to that of the experiment.

The one-sided bandwidth of the signals $x(t)$ and $y(t)$ entering the processing procedure is restricted by the two filters to 100 Hz so that Eq. [43] requires a repetition period F in the frequency domain of at least 200 Hz. The choice $F = 375$ Hz was made to avoid folding signals at freqencies just above the filter corner frequency into the signal frequency range. With this choice, frequency components above 275 Hz would be the lowest ones to be folded into the filter passband, and with an asymptotic filter attenuation of 40 db/decade these are attenuated by at least 18 db (19 to 21 db over the spectrum range). The sampling interval in the time domain for the signals $x(t)$ and $y(t)$ is then $\Delta t = F^{-1} = 2.67$ msec. A choice of the number of samples, n, now determines Δf and T through Eqs. [46] and [50]. It is convenient to have a few sample points within a half width of the resonance lines in the absorption mode spectrum, and from this point of view a sampling interval $\Delta f \simeq 0.05$ Hz would seem suitable. This would require $n = 7500$ samples/record, but since the full speed advantage of the fast Fourier transform algorithm (7) is realized only with n equal to some power of 2, $n = 8192$ was chosen so that the sampling interval in the frequency domain became $\Delta f = 0.046$ Hz and the record length in the time domain $T = 21.8$ sec. It was mentioned following Eq. [51] that the use of the discrete Fourier transform implies that the signals are periodic with period T in the time domain. If the continuous noise signals $x(t)$ and $y(t)$ were simply sliced into records of length T, the periodic extension of each record would cause errors in an interval of duration equal to the system memory time around the record boundaries. Consider, for example, a record boundary at $t = 0$. According to the convolution Eq. [9] the system response $y(t)$ immediately following the boundary is determined, in part, by the stimulation signal $x(t)$ just previous to the boundary. The error arises because the input $x(t<0)$ from the previous random noise record, which actually did evoke the response $y(t>0)$, is different from the input $x(t-T)$ implied by the periodic extension of the current noise record $x(t)$. To avoid this error, the input signal $x(t)$ was set to zero by the relay switch shown in Fig. 2 for a duration at least equal to the relaxation time of the spin system which plays the role of the system memory time. In fact, the relay switch was periodically turned on for 14 seconds and off for 14 seconds and a signal record of duration $T = 21.8$ sec was started each time the relay turned on so that 7.8 sec was allowed for the memory time of the spin system.

Several signal traces are presented in Figs. 4 and 5 to illustrate the processing operations. These traces were obtained by recording the corresponding digital sample sequences through a smoothing filter. Traces a) and b) of Fig. 4 show 2.8. sec long segments of the random input and output signals $x(t)$ and $y(t)$, respectively. Graph c) gives the output amplitude spectrum $|Y(f)|$ over the 45 Hz wide frequency range

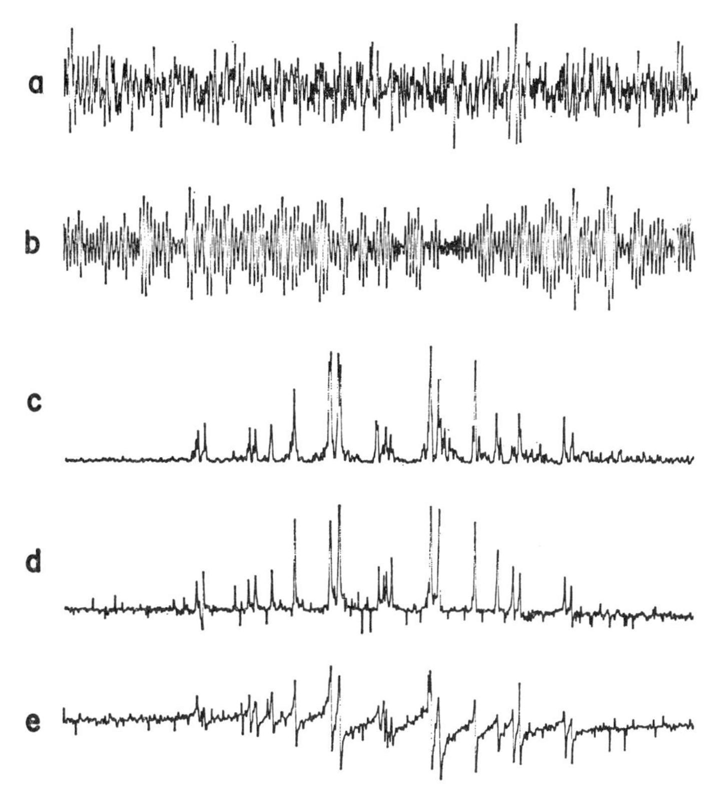

FIG. 4. Processing of a single signal record.
a) 2.8 sec segment of input signal record $x(t)$,
b) 2.8 sec segment of output signal record $y(t)$,
c) modulus of the output spectrum $|Y(f)|$ for the 45 Hz wide frequency band containing the NMR spectrum,
d) real part $V(f)$ (absorption mode) and
e) imaginary part $U(f)$ (dispersion mode) of the noise stimulated NMR spectrum $R(f)$.

containing the NMR spectrum. Eq. [6] yields

$$|Y(f)| = |X(f)| \cdot \{V^2(f) + U^2(f)\}^{\frac{1}{2}}$$

so that $|Y(f)|$ gives the modulus of the NMR spectrum provided that $|X(f)|$ is constant. Freeman (*10*) has recently pointed out that the presentation of an NMR spectrum in form of its modulus is undesirable because this presentation distorts weak resonance lines falling in the wing of a strong line (a weak line then appears essentially in dispersion mode). Furthermore, for the random input signal used here, $|X(f)|$ is a random variable whose variance is of the same order as its expectation value and it can hardly be considered as a constant. Nevertheless, Fig. 4c does show similarity with the conventional absorption spectrum of o-dichlorobenzene of Fig. 5e. Traces d and e of Fig. 4 show the real part $V(f)$ and imaginary part $U(f)$ of the spectrum $R(f)$ computed from Eq. [6] by complex division, term by term, of the digital output spectrum $\{Y_n\}$ by the input spectrum $\{X_n\}$. The signal/noise ratio is obviously low but this is not surprising since the traces represent information in a 100 Hz

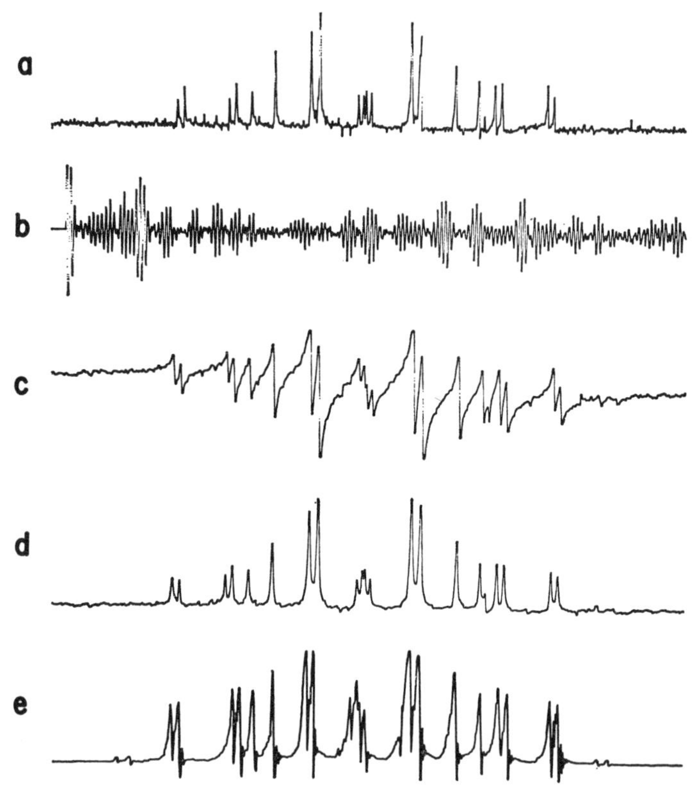

FIG. 5. Processing of the average of three signal records.
a) real part $V(f)$ of averaged spectrum,
b) 2.5 sec segment of impulse response function $r(t)$ of the spin system,
c) imaginary part (dispersion mode) and
d) real part (absorption mode) of smoothed, averaged NMR spectrum,
e) absorption mode spectrum of o-dichlorobenzene taken with conventional field sweep.

bandwidth gathered with a single record of 22 sec duration. The noise in the real and imaginary parts of $R(f)$ is independent, and traces d and e therefore do not form a Hilbert transform pair.

As a first step to improve the signal/noise ratio, two more records were processed in the same manner and the average of the three resulting spectra $R(f)$ was computed. Fig. 5a shows the real part of the averaged spectrum. The improvement is evident in comparison with Fig. 4d and it is gratifying to see the two lines at the symmetry center of the spectrum resolved. A computer simulation of the spectrum of o-dichlorobenzene at 56.4 MHz with the parameters given by Grant *et al.* in Reference (*8*) yields 0.34 Hz for the separation of these two lines. The analytical signal $a(t)$ defined in Eq. [31] was then computed and the effects of the differential phase shift were compensated as described in connection with Eq. [33]. If the x and y signal channels were identical, the differential phase shift would be zero (or at least frequency independent) and compensation would not be required. With our apparatus, most of the differential phase shift originated in a time delay between the x and y samples caused by the multiplexer at the input to the analog to digital converter. Figure 5b shows the corrected system-impulse-response function $r(t)$. The imaginary part $\rho(t)$ of the analytical signal is not presented here because it differs from the real part only by a 90° phase shift of the carrier sinusoid under the same envelope. The line width in the averaged spectrum of Fig. 5a is a little greater than 2 sampling intervals Δf. Convolution with a Lorentzian curve of width $\varphi = 2\Delta f \simeq 0.1$ Hz therefore gives some smoothing without too much loss of resolution, and this was effected by multiplying $a(t)$ with $\exp(-t/\tau)$ where $\tau = (\pi\varphi)^{-1} \simeq 3.5$ sec corresponding to $n/2\pi \simeq 1300$ sampling intervals Δt. The resulting spectrum is shown in traces c and d of Fig. 5. Since $a(t)$ was truncated for $t<0$, the dispersion mode spectrum Fig. 5c and absorption mode spectrum Fig. 5d are necessarily Hilbert transforms of each other. Trace e finally shows the spectrum of the same o-dichlorobenzene sample obtained with the conventional field sweep method on the same spectrometer. In comparing the noise in Figs. 5d and e one should consider that the scan through the 35 Hz wide spectrum in trace e required 11 minutes, and that the three records from which trace d has been derived were gathered in 66 seconds. Furthermore, the useful width of the NMR spectrum generated by the noise processing covers about twice the 35 Hz width of the o-dichlorobenzene spectrum, and one may argue that for equivalent resolution the scan rate should not exceed 0.1 Hz/3 sec so that an equivalent scan time of 35 minutes could be considered. However, any such comparison is not valid in the present case because the stimulating rf field strength had not been optimized for either the conventional scan nor the noise procedure.

CONCLUSION

The noise power density in the rf sideband used to stimulate the spin system in these experiments was probably well below the value required for maximum signal/noise ratio in the NMR spectrum. An attempt to increase the noise power density encounters two difficulties: The rf carrier amplitude must be increased if the modulation index is to be kept small, and this leads to Bloch–Siegert shifts and other double resonance effects since the rf carrier frequency is only 60 Hz distant from the center of the 35 Hz wide spectrum. Of course, the carrier could be placed farther away from the spectrum but the filter bandwidth must then be increased leading to a greater sampling rate and more computational effort. On the other hand, if the modulation index is increased, mixing occurs between the noise modulation and the sinusoidal modulation required for the spectrometer field lock loop, and this makes it difficult to maintain a stable operation.

It appears possible to avoid these difficulties by using balanced modulation with a suitable periodic pseudorandom signal instead of truly random bandlimited noise. Periodic, maximum length, binary signals (*11*) have a crest factor of unity and a uniform energy spectrum, and thus form a suitable class of signals for this application. They are easily generated by means of a shift register with suitable feedback, their binary nature simplifies the design of the modulator, they have already been used for noise decoupling (*12*) and their application to spectrum stimulation is being explored (*13*). The use of periodic, pseudorandom signals makes the relay switch redundant in the block diagram of Fig. 2, because the system will reach a steady state in which the response is periodic with the period of the stimulus. The amplitude spectrum of a maximum length binary sequence is easily calculated (*11*), and the signal can be arranged to have an essentially constant amplitude modulus over a given frequency range. This calculation would only need to be executed once since the input signal would repeat in subsequent records. If the field control loop in the spectrometer is sufficiently stable the computational effort can be further reduced by averaging the output signals $y(t)$ over the several records rather than the spectra $R(f)$ as was necessary with the truly random signals used in this work. We are presently exploring the use of such pseudorandom and other input signals.

We conclude with a brief and qualitative comparison of Fourier transform spectrometry with noise stimulation and with impulsive stimulation. The sensitivity improvement essentially results from multichannel operation in either method. The noise method has the advantage of using signals with low crest factor and avoids the electronic requirements associated with the large dynamic range of impulsive signals. However, the free induction decay measured with the pulse method reflects the motion of the spin system under the action of the unperturbed Hamiltonian starting from the initial state produced by the pulse so that double resonance effects cannot occur (although Overhauser effects may be generated by the pulses), while the noise method measures the motion under continuous stimulation by a Hamiltonian with stochastic time dependence which does give double resonance effects (*14*). This difference vanishes in the linear approximation used here but it must be considered for a detailed evaluation of the sensitivity improvement which may be achieved with either method under stronger excitation.

ACKNOWLEDGMENT

Financial support of this work through operating and equipment grants from the National Research Council of Canada is gratefully acknowledged.

REFERENCES

1. R. R. Ernst and W. A. Anderson, *Rev. Sci. Instrum.* **36,** 1696 (1965).
2. R. R. Ernst and W. A. Anderson, *Rev. Sci. Instrum.* **37,** 93 (1966).
3. W. B. Davenport and W. L. Root, "Random Signals and Noise," McGraw–Hill, New York, 1958;
J. S. Bendat and A. G. Piersol, "Measurement and Analysis of Random Data," John Wiley and Sons, Inc., New York, 1966.
4. E. A. Guillemin, "Theory of Linear Physical Systems," John Wiley and Sons, Inc., New York, 1963;
R. N. Bracewell, "The Fourier Transform and its Applications," McGraw–Hill, New York, 1965.
5. R. R. Ernst, *J. Mag. Resonance* **1,** 7 (1969).
6. A. Abragam, "The Principles of Nuclear Magnetism," Chapter III, Appendix, Oxford Univ. Press, Oxford, 1961.
7. See special issue on fast Fourier transform, *IEEE Trans. on Audio and Electroacoustics*, **AU-15,** no. 2 (1967).
8. D. R. Whitman, *J. Chem. Phys.* **36,** 2085 (1962);
D. M. Grant, R. C. Hirst, and H. S. Gutowsky, *J. Chem. Phys.* **38,** 470 (1963).
9. W. A. Anderson, *Phys. Rev.* **102,** 151 (1956).
10. R. Freeman, private communication.
11. S. W. Golomb, "Shift Register Sequences," Holden–Day, San Francisco, 1967.
12. R. R. Ernst, *J. Chem. Phys.* **45,** 3845 (1966).
13. R. R. Ernst, private communication.
14. R. R. Ernst and H. Primas, *Helv. Phys. Acta* **36,** 583 (1963).

JOURNAL OF MAGNETIC RESONANCE 4, 366–383 (1971)

Phase and Intensity Anomalies in Fourier Transform NMR

RAY FREEMAN AND H. D. W. HILL

Varian Associates Analytical Instrument Division, Palo Alto, California 94303

Received February 1, 1971; accepted February 4, 1971

Carbon-13 Fourier transform NMR spectra are reported which exhibit gross discrepancies in signal amplitude and phase that vary in a cyclic manner as a function of the frequency offset from resonance. These anomalies occur when the interval between pulses is made short in comparison with the spin-spin relaxation times, and they are associated with the establishment of a steady-state response where there are finite macroscopic transverse components of magnetization at the end of the pulse interval. The regular pulse sequence has the property of refocussing the isochromatic magnetization vectors that have been dispersed by field inhomogeneity, giving rise to a spin echo at the time of the next pulse. These echoes and the phase and intensity effects may be masked by short-term instabilities in the field/frequency ratio, or by incoherence introduced by heteronuclear noise decoupling. A simple device is proposed which effectively suppresses these anomalies by introducing a small random delay in the timing of the radiofrequency pulses.

1. INTRODUCTION

High resolution nuclear magnetic resonance spectra are normally examined by sweeping the excitation frequency through the region of resonance absorption, or by sweeping the applied static field at constant frequency. It has recently been demonstrated (*1*) that significant improvements in sensitivity can be achieved by deriving NMR spectra as the Fourier transform of the free induction signal after a strong radiofrequency pulse (*2*). To improve the signal-to-noise ratio, a large number of free induction signals are accumulated coherently and the resulting time-averaged signal transformed, usually on a digital computer. The highest sensitivity is attained when the pulses are repeated at the highest rate compatible with the required resolution, which is determined by T_0^{-1}, where T_0 is the period over which the free induction signal is sampled. Under the influence of a regular sequence of radiofrequency pulses a steady-state response is established where the nuclei are still far from thermal equilibrium when the next pulse in the sequence occurs. Ernst and Anderson (*1*) and Ernst (*3*) have derived the equations that describe this steady state under conditions where field inhomogeneity broadening can be neglected. They show that when the free induction signals are observed under steady-state conditions where there are finite net transverse components of magnetization just before each pulse, the lines of the transformed spectrum exhibit anomalous intensities and a variable degree of dispersion-mode character even if the spectrometer phase is correctly set for absorption. The suggested remedy (*1*) for avoiding these artifacts is to increase the interval between pulses until it exceeds about three times the longest spin-spin relaxation time in the sample. Unfortunately, this is in conflict with the requirements of high sensitivity, particularly in the common practical case where field inhomogeneity causes a much more rapid decay of the free induction signal than spin-spin relaxation. (For example, the normal practice in ^{13}C spectroscopy is to operate with pulse intervals of the order of 1 second, whereas the spin-spin relaxation times may well be an order of magnitude longer.)

Experimental results are reported here for ^{13}C resonances with long relaxation times which show gross variations of intensity and phase as a cyclic function of the frequency offset from resonance. The treatment of Ernst and Anderson is extended to take into account field inhomogeneity broadening, and it is then shown that the steady-state regime has the inherent property of refocussing field inhomogeneity effects so that a spin echo begins to form just before each radiofrequency pulse. These echoes are related to the 90°–90° echoes observed by Hahn (*4*). Finally, a simple method is suggested for eliminating the phase and amplitude anomalies while still benefiting from the steady-state regime.

2. STEADY-STATE BEHAVIOR OF SPINS SUBJECTED TO A REPETITIVE SEQUENCE OF RF PULSES

The response of a nuclear spin system to a regular sequence of identical pulses has been treated by Ernst and Anderson (*1*) based on the Bloch equations. This treatment is extended by relaxing some of the restrictions and by taking account of the influence of finite field inhomogeneity broadening. The aim of these calculations is to study the effect of finite macroscopic transverse components of magnetization at the time the next pulse in the sequence is applied. It will be shown that field inhomogeneity effects alone do not eliminate these transverse components; the steady-state regime causes a refocussing of the dispersed spin isochromats to give a spin echo (*5–7*).

The motion of the spin magnetization is considered in a frame of reference rotating about the direction of the static field Z in synchronism with the radiofrequency ω_1. A regular sequence of rf pulses is applied in the form of a field H_1 directed along the X axis for a duration τ seconds, the counterrotating component of angular frequency $2\omega_1$ in this frame having negligible influence. For the sake of simplicity, H_1 is assumed to be strong with respect to all frequency offsets from resonance, so that the effective radiofrequency field is always equal to H_1 and directed along the X axis. Relaxation during the pulse is neglected.

It is assumed that a steady-state response will eventually be established where the motion of the nuclear magnetization vectors will have the same period T_p as the imposed pulse sequence (rather than some integral multiple of this period). This appears to be borne out in most cases in practice. Each pulse rotates the spin vectors about the X axis through a "flip angle" given by

$$\alpha = \gamma H_1 \tau. \quad [1]$$

The magnetization immediately after the pulse $\mathbf{M}^+$ is thus related to that before, $\mathbf{M}^-$, through the rotation operator:

$$\mathbf{M}^+ = R_x(\alpha)\mathbf{M}^-. \quad [2]$$

Between the pulses the spins precess freely through an angle

$$\Theta = 2n\pi + \theta = \omega_0 T_p, \qquad [3]$$

where ω_0 is the offset from exact resonance with the radiofrequency. During this period spin-spin and spin-lattice relaxation occurs according to the operator

$$S(T_p, T_1, T_2) = \begin{bmatrix} E_2 & 0 & 0 \\ 0 & E_2 & 0 \\ 0 & 0 & E_1 \end{bmatrix}, \qquad [4]$$

where E_1 and E_2 are abbreviations for $\exp(-T_p/T_1)$ and $\exp(-T_p/T_2)$. Thus if $\mathbf{M}^*$ is the magnetization at the end of the free precession period,

$$\mathbf{M}^* = R_z(\Theta)S(T_p, T_1, T_2)\mathbf{M}^+ + (1-E_1)\mathbf{M}_0, \qquad [5]$$

$\mathbf{M}_0$ is the magnetization at thermal equilibrium. For a steady-state solution $\mathbf{M}^* = \mathbf{M}^-$. Thus the three components before and after the pulse are found to be

$$M_x^- = M_0(1-E_1)[E_2 \sin\alpha \sin\theta]/D, \qquad [6]$$

$$M_y^- = M_0(1-E_1)(E_2 \sin\alpha \cos\theta - E_2^2 \sin\alpha)/D, \qquad [7]$$

$$M_z^- = M_0(1-E_1)[1 - E_2 \cos\theta - E_2 \cos\alpha(\cos\theta - E_2)]/D, \qquad [8]$$

$$M_x^+ = M_x^-, \qquad [9]$$

$$M_y^+ = M_0(1-E_1)[(1-E_2\cos\theta)\sin\alpha]/D, \qquad [10]$$

$$M_z^+ = M_0(1-E_1)[E_2(E_2-\cos\theta)+(1-E_2\cos\theta)\cos\alpha]/D, \qquad [11]$$

where

$$D = (1-E_1\cos\alpha)(1-E_2\cos\theta)-(E_1-\cos\alpha)(E_2-\cos\theta)E_2. \qquad [12]$$

Except for a change in the sign (*8*) of M_x^+, Eqs. [9] and [10] are identical with those derived by Ernst and Anderson (*1*).

Although these coordinates appear to be rather complicated functions of the relaxation parameters E_1 and E_2, the flip angle α and the excess precession angle θ, in fact, the locus of the tips of the isochromatic magnetization vectors has a quite simple shape both before and after the pulse. For a given flip angle both loci are circles if $T_1 = T_2$, and ellipses if $T_1 > T_2$, and thus give ellipses when projected onto the XY plane. The YZ plane is of course a plane of symmetry for all these figures. If $T_1 = T_2$, the circular locus for $\mathbf{M}^+$ is inclined with respect to the XZ plane at an angle ψ where

$$\tan\psi = (\cos\alpha - E_1)/\sin\alpha. \qquad [13]$$

Before the pulse the circle is inclined at $\psi+\alpha$. The radii of these circles are given by

$$R = M_0 E_1 \cos\psi/(1-E_1). \qquad [14]$$

The center of the $\mathbf{M}^+$ circle subtends an angle $\eta+\psi$ at the origin with respect to the Y axis, where

$$\tan\eta = E_1 \sin\alpha/(1-E_1^2). \qquad [15]$$

Some of these features are illustrated by the two plots of XY projections of these loci calculated for $T_p = 1$ second, $T_1 = 5$ seconds, $T_2 = 5$ seconds, and flip angles 0.6 and 0.9 radian (see Fig. 1).

If the flip angle is set to the special condition

$$\cos\alpha = E_1, \qquad [16]$$

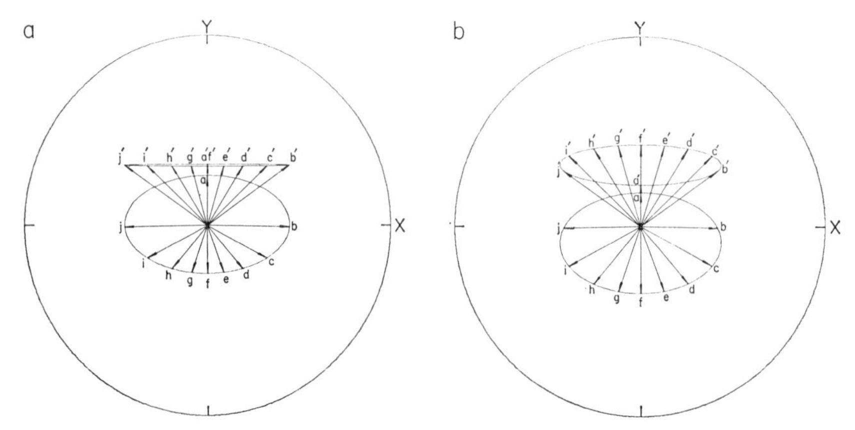

FIG. 1. Solutions of the steady-state equations for the parameters $T_p = 1$ sec, $T_1 = T_2 = 5$ sec, represented as the projections on the XY plane of ten typical magnetization vectors (a through j before the pulse, a' through j' after the pulse), calculated at equal intervals of θ from 0 to 2π. The locus of the tips of these vectors is seen to be an ellipse with a nonuniform distribution of the vectors as a function of θ. Projection (a) has been calculated for a flip angle $\alpha = 0.6$ radian, very close to α_E where M_y^+ is independent of θ. Projection (b) is for a larger flip angle $\alpha = 0.9$ radian. During the free precession period a typical vector c' breaks up into a large number of isochromatic components precessing at different rates due to field inhomogeneity, but these refocus again into vector c at the time of the next pulse.

the Y component of magnetization after the pulse is independent of θ,

$$M_y^+ = M_0 \sin \alpha/(1+E_1). \qquad [17]$$

This special setting will henceforth be referred to as the Ernst angle α_E. When spin-spin relaxation is sufficiently rapid so that $E_2 \ll 1$, the transverse magnetization just before the pulse, M_x^- and M_y^-, can be neglected, and then it is readily seen from Eq. [10] that the maximum signal is obtained when $\alpha = \alpha_E$. The present treatment specifically considers conditions where there *are* appreciable transverse components of magnetization at the end of the interval T_p, in which case α_E is not the optimum flip angle (see Section 2.4).

For a flip angle α_E the locus of the tips of the magnetization vectors after the pulse is a circle or an ellipse in a plane parallel to the XZ plane, and thus appears as a straight line when projected on the XY plane. Figure 1(a) has been calculated for a flip angle very close to α_E to illustrate this effect. For flip angles less than α_E the $\mathbf{M}^-$ and $\mathbf{M}^+$ ellipses are tilted in the same sense, whereas for flip angles greater than α_E they are tilted in opposite senses with respect to the XZ plane.

The magnetization vectors are not uniformly distributed around these ellipses as a function of θ, but are bunched together somewhat around the condition $\theta = \pi$. If the projection of a typical $\mathbf{M}^+$ vector in the XY plane makes an angle ε with the Y direction, then

$$\tan \varepsilon = M_x^+/M_y^+ = E_2 \sin \theta/(1-E_2 \cos \theta). \qquad [18]$$

A typical vector before the pulse makes an angle ε', where

$$\tan \varepsilon' = M_x^-/M_y^- = \sin \theta/(\cos \theta - E_2). \qquad [19]$$

Thus the directions of all vectors before and after the pulse are independent of the flip angle α [compare Fig. 1(a) and 1(b)]. The angle ε represents the dephasing of the detected signal after the pulse and is the source of the phase anomalies observed in the transformed spectrum. The bunching of vectors around the condition $\theta = \pi$ is most pronounced at very small flip angles, and in the extreme case only a few vectors with θ values near zero have any appreciable Y component of magnetization. Since θ is a periodic function of offset across the transformed spectrum with a relatively high frequency $1/T_p$ Hz, severe bunching leads to very marked intensity anomalies.

2.1. *Phase Anomalies*

At this point it is necessary to make an important distinction between the time for which a given free induction signal is observed T_0 and the interval between successive rf pulses T_p. Two new exponential decay factors may then be defined:

$$E_1^0 = \exp(-T_0/T_1), \tag{20}$$

$$E_2^0 = \exp(-T_0/T_2). \tag{21}$$

Suppose the free induction signal may be described in terms of a single frequency component ω_0:

$$f(t) = M_{xy}^+ \cos(\omega_0 t + \phi) \exp(-t/T_2), \tag{22}$$

where M_{xy}^+ is the XY component of magnetization immediately after the pulse, and hence $M_x^+ = M_{xy}^+ \sin \phi$ and $M_y^+ = M_{xy}^+ \cos \phi$. The Fourier cosine transform will give the continuous spectrum,

$$C(\omega) = \frac{-1}{(2T_0)^{1/2}} M_y^+ \left\{ \frac{T_2}{1+(\Delta\omega T_2)^2} [E_2^0 \cos(\Delta\omega T_0) - 1] - \frac{\Delta\omega T_2^2}{1+(\Delta\omega T_2)^2} E_2^0 \sin(\Delta\omega T_0) \right\}$$
$$- \frac{1}{(2T_0)^{1/2}} M_x^+ \left\{ \frac{T_2}{1+(\Delta\omega T_2)^2} E_2^0 \sin(\Delta\omega T_0) + \frac{\Delta\omega T_2^2}{1+(\Delta\omega T_2)^2} [E_2^0 \cos(\Delta\omega T_0) - 1] \right\}, \tag{23}$$

where $\Delta\omega$ has been written for $\omega - \omega_0$. Introducing the abbreviation

$$K = (1-E_1) \sin \alpha / [(1 - E_1 \cos \alpha)(1 - E_2 \cos \theta) - (E_1 - \cos \alpha)(E_2 - \cos \theta) E_2], \tag{24}$$

then $M_x^+ = KM_0 E_2 \sin \theta$ and $M_y^+ = KM_0(1 - E_2 \cos \theta)$, and Eq. [23] may be rewritten

$$C(\omega) = \frac{KM_0}{(2T_0)^{1/2}} \left\{ \frac{T_2}{1+(\Delta\omega T_2)^2} [1 - E_2 \cos \theta - E_2^0 \cos(\Delta\omega T_0) + E_2 E_2^0 \cos(\theta + \Delta\omega T_0)] \right.$$
$$\left. + \frac{\Delta\omega T_2^2}{1+(\Delta\omega T_2)^2} [E_2 \sin \theta + E_2^0 \sin(\Delta\omega T_0) - E_2 E_2^0 \sin(\theta + \Delta\omega T_0)] \right\}. \tag{25}$$

The first term of Eq. [25] represents the absorptive part and the second term the dispersive part. If the acquisition time is set sufficiently long in comparison with T_2, then (since $T_p \geqslant T_0$)

$$E_2, E_2^0 \ll 1, \tag{26}$$

and the dispersive part can be neglected and there will be no phase anomalies in the transformed spectrum. Moreover, K will become independent of θ and there will be no intensity anomalies (*1*).

The present treatment concentrates on cases where condition Eq. [26] is not fulfilled, in the belief that many natural spin-spin relaxation times are significantly longer than the pulse intervals dictated by sensitivity requirements. In practice certain instrumental shortcomings (such as short-term field/frequency instability or incoherence effects arising from heteronuclear noise decoupling) attenuate the net transverse magnetization irreversibly in the interval between pulses or cause phase variations from one transient to the next, tending to mask the anomalies. However, it is to be expected that newer NMR instruments will have higher inherent stability, and then the transverse magnetization will persist long enough to affect phase and intensity in the spectrum.

Even if the acquisition time is short compared with T_2, the dispersive part of Eq. [25] vanishes if the spectrometer is designed to operate with an acquisition time T_0 identical with the pulse interval T_p. The cancellation arises if the Fourier transform is performed digitally, the discrete sampling of the transformed spectrum obscuring the dispersive term. A total of N samples of the free induction signal $f(t)$ are taken at regular time intervals T_0/N secs, resulting in discrete samples of $C(\omega)$ at regular frequency intervals $\omega = 2n\pi/T_0$ rad/sec, where $n = N/2$. Suppose that $T_0 = T_p = T$ and, therefore, $E_2 = E_2^0$, then

$$\theta + \Delta\omega T = \omega_0 T + (\omega - \omega_0)T = \omega T = 2n\pi. \qquad [27]$$

This sampling rate is unique in that the dispersive term of Eq. [25] vanishes (*9*) since

$$\sin\theta = -\sin(\Delta\omega T), \quad \text{and} \quad \sin(\theta + \Delta\omega T) = 0. \qquad [28]$$

Even with this condition $T_p = T_0$ there are still intensity variations across the spectrum due to the dependence of K upon θ. In most practical spectrometers, however, T_p is set slightly longer than T_0, mainly to allow the recovery of the receiver after a strong rf pulse. Even a very small discrepancy between T_0 and T_p causes $\sin\theta$ and $\sin(\Delta\omega T_0)$ to get out of step, introducing significant dispersive components in some parts of the transformed spectrum (*10*).

2.2. *Intensity Anomalies*

In Fourier transform NMR three different phenomena may influence the relative intensities of lines arising from the same number of nuclear spins:

(a) Differences in relaxation time T_1.
(b) Inadequate digitization along the frequency axis.
(c) Intensity anomalies arising in the steady-state regime.

At a sacrifice in sensitivity, the intensity variations due to relaxation effects can be minimized by operation at small flip angles α, or long pulse intervals T_p. They reflect similar intensity variations observable in conventional swept spectra, and are outside the scope of the present investigation. Intensity errors attributable to coarse sampling of the transformed spectrum arise when the free induction signal does not decay to a sufficiently low level at the end of the acquisition time T_0, that is to say, the transformed line width is comparable with or less than the digitization interval $1/T_0$ Hz. These are just the conditions necessary to study (c) the intensity anomalies caused by a finite net transverse component of magnetization just before each pulse in the steady-state mode.

A simple device serves to separate out these two effects in order to study (c) alone. Before transformation, the free induction signal is multiplied by a decaying exponential function with time constant short enough to outweigh its natural decay. This has the effect of broadening the lines in the transformed spectrum to the point where there are several sample points per line width and the digitization errors are then minimal.

With such an artificially damped free induction signal, any spin-echo effects (see Section 2.3) are minimized and the intensity anomalies may be predicted in terms of the Y component of magnetization immediately after the pulse M_y^+. The extent of these anomalies is greatest at small flip angles, where bunching of the isochromatic magnetization vectors around the condition $\theta = \pi$ is very marked, and these vectors contribute only a very small Y component after the pulse. Only the very few vectors near $\theta = 0$ contribute significantly to the detected signal. Hence in a practical case where $T_p = 0.5$ second, NMR signals would only have appreciable intensities if they happened to fall at an offset $\omega_0/2\pi$ that is near an exact multiple of 2 Hz. When the flip angle is set equal to α_E (see Eq. [16]) M_y^+ becomes independent of θ and the intensity anomalies disappear. For much larger flip angles, the $\mathbf{M}^+$ ellipse is tilted in the other sense and it is the vectors around the condition $\theta = \pi$ that have large Y components of magnetization M_y^+, reaching a broad maximum at $\theta = \pi$ and a minimum at $\theta = 0$.

Figure 2 illustrates these effects by plotting M_y^+ as a function of θ for a small flip angle $\alpha_E/6$, the Ernst angle α_E, and a large flip angle $6\alpha_E$. It also illustrates the extent of the dephasing, which from Eq. [18] is known to be independent of flip angle. This calculation has ignored field inhomogeneity effects entirely. Instrumental broadening, if it occurs through spatial inhomogeneity of the applied magnetic field, moderates these phase and intensity anomalies by taking a weighted average over a range of

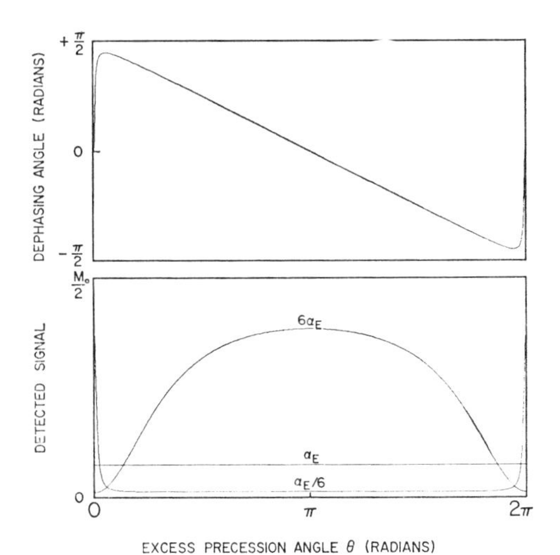

FIG. 2. Steady-state solutions for the Y component of magnetization after the pulse (M_y^+) and the dephasing angle (ε) as a function of the excess precession angle θ, for parameters appropriate to the ^{13}C resonance of carbon disulphide ($T_p = 0.5$ sec, $T_1 = 44$ sec, $T_2 = 38$ sec). The detected signal is independent of θ for a flip angle $\alpha = \alpha_E$, but shows sharp maxima at $\theta = 0$ and 2π for a small flip angle $\alpha = \alpha_E/6$, and a broad maximum at $\theta = \pi$ for a large flip angle $\alpha = 6\alpha_E$. The dephasing is the same function of θ irrespective of flip angle. Field inhomogeneity effects have been neglected.

θ values. Nevertheless, the general features of Fig. 2 have been observed experimentally in the ^{13}C spectrum of carbon disulphide (see Section 4.1).

2.3. *Field Inhomogeneity Refocussing Effects*

The macroscopic distribution of the applied field across the sample can be represented as a large number of isochromatic magnetization vectors with some suitable distribution function $f(\omega_0)$. For the sake of simplicity, this is often assumed to be Lorentzian,

$$f(\omega_0) = T_2^*/\{\pi[1+(\omega_0^0-\omega_0)^2 T_2^{*2}]\}, \qquad [29]$$

where ω_0^0 is the offset corresponding to the center of the line. Equations [6]–[12] represent the steady-state solutions for each individual isochromat. An interesting consequence of these equations is that although field inhomogeneity causes destructive interference as these vectors lose phase, the repetitive pulses have the intrinsic property of refocussing them at the time of the next pulse in the sequence. This is clearly related to the spin echoes observed by Hahn (*4*) after a 90°–τ–90° pulse sequence. A spin echo appears at time 2τ and is negative-going [see Fig. 3(a)]. If a third 90° pulse is imposed at the center of this echo, two further (negative) echoes are excited at times 3τ and 4τ [Fig. 3(b)]. From the results of applying four and then five regularly spaced 90° pulses [Figs. 3(c) and (d)] it is possible to visualize the establishment of a steady state where each 90° pulse is preceded by a negative "half echo" and followed by a positive "free induction decay".

This refocussing is a direct consequence of setting up the steady-state regime with pulse intervals comparable with T_2. In the rotating frame the motion of the isochromats has been shown to be such that the vectors are directed around the circumference of an ellipse immediately after the rf pulse. If a steady state is established, the vectors immediately *before* the pulse must follow a similar locus, defined by

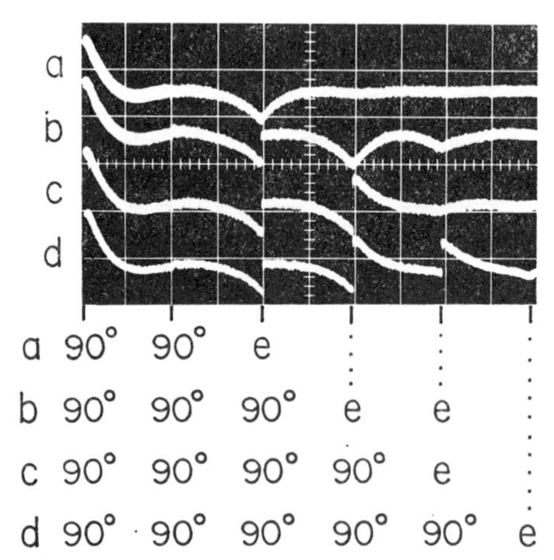

FIG. 3. Proton spin echoes from water observed when trains of equally-spaced 90° pulses are applied to the spin system at thermal equilibrium. (a) Two pulses induce a (negative) Hahn echo. (b) With a sequence of three 90° pulses, the third pulse is applied at the center of the Hahn echo, and two further negative echoes are generated. (c) When four pulses are applied, the behavior of the magnetization just before and just after the fourth pulse closely resembles the steady-state behavior. (d) A sequence of five pulses.

rotation of the first ellipse through $-\alpha$ radians about the X axis (see Fig. 1). The refocussing reaches a maximum at the time of the pulse and generates a negative Y component.

The effect can be calculated starting with Eqs. [9] and [10] and imposing a Lorentzian distribution of isochromats according to Eq. [29]. Suppose an arbitrary flip angle $\alpha = 1$ radian is chosen, together with an arbitrary offset ω_0^0 of the line center such that this particular isochromat precesses five complete revolutions plus an excess angle $\theta = 1$ radian in the period T_p. Figure 4 illustrates numerical calculations of the free induction signal between pulses for several values of the field inhomogeneity parameter (T_2^*/T_p). Note that the decay due to inhomogeneity is always reversed in time to give a negative echo at T_p. Long relaxation times have been assumed (with $T_p = 0.5$ second, $T_1 = 44$ seconds, and $T_2 = 38$ seconds, parameters corresponding to the reported values for ^{13}C in carbon disulphide at 15 MHz (*11*)). The detailed shapes of these signals also depend somewhat on the flip angle α and the excess precession angle θ for the line center, but the general

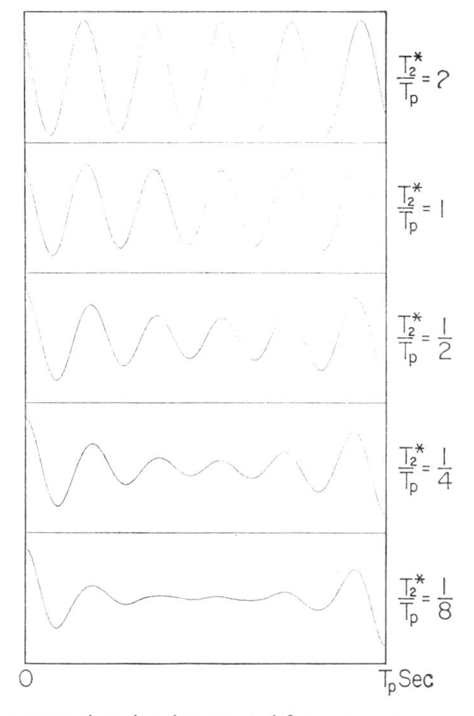

FIG. 4. Steady-state free precession signal computed for parameters appropriate to ^{13}C in carbon disulphide ($T_p = 1$ sec, $T_1 = 44$ sec, $T_2 = 38$ sec) and an arbitrary flip angle $\alpha = 1$ radian. Between the pulses the nuclei precess through five revolutions plus an excess angle of 1 radian. Different degrees of broadening by field inhomogeneity have been introduced by calculating the integral over a range of θ values, weighted by a Lorentz function of half-width $1/T_2^*$. In each case the dispersal of isochromats by the inhomogeneous field is refocussed at time T_p to form a spin echo, showing the predicted phase reversal. As a result, the conditions immediately preceding the pulse are independent of field inhomogeneity.

behavior is always the same. Refocussing increases the total integrated signal and thus improves the sensitivity of Fourier transform NMR in the steady-state mode (*12*). When the steady-state Fourier transform experiment is used for progressive saturation studies to determine spin-lattice relaxation times (*7*), these spin echo effects must be inhibited.

An important consequence of the refocussing effect is that field inhomogeneity does not influence the steady-state conditions, since M_x^- and M_y^- are unaffected by the field. This is a justification for the treatment of field inhomogeneity simply as an integration over a range of θ values, each with steady-state conditions determined by Eqs. [6]–[12].

2.4. Optimum Flip Angle

The highest sensitivity in Fourier transform experiments is obtained only if the flip angle is correctly adjusted for the maximum detected signal in the steady-state regime. It is clear from Eqs. [9] and [10] that M_x^+ and M_y^+ both have the same dependence on the flip angle α, and it is readily shown (*1*) that the optimum flip angle has a complicated dependence on θ:

$$\cos\alpha_{\mathrm{opt}} = \frac{E_1 + E_2(\cos\theta - E_2)/(1 - E_2\cos\theta)}{1 + E_1 E_2(\cos\theta - E_2)/(1 - E_2\cos\theta)}. \quad [30]$$

Since it is important to know whether the maximum is broad or narrow, this is bes illustrated by plotting M_y^+ as a function of α for several values of θ [Fig. 5(a)]

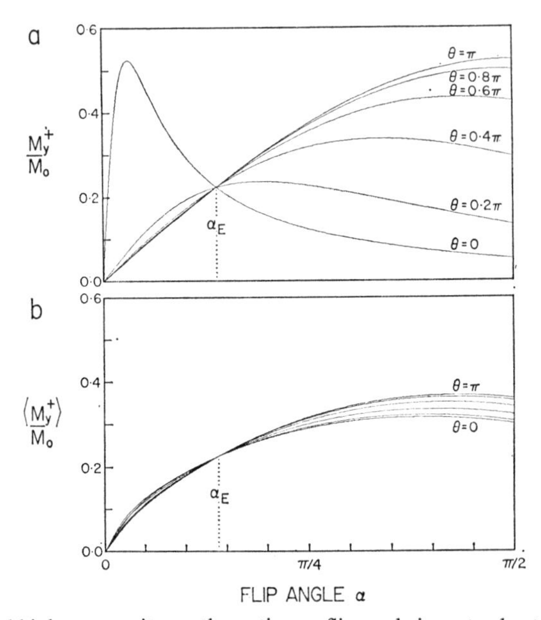

FIG. 5. Effect of field inhomogeneity on the optimum flip angle in a steady-state Fourier transform experiment, assuming $T_p = 1$ sec, $T_1 = T_2 = 10$ sec. (a) The dependence of M_y^+ on flip angle varies widely as a function of θ when field inhomogeneity is neglected. All the curves in this family cross at $\alpha = \alpha_E$. (b) When inhomogeneity is taken into account by integration over a Lorentzian line with $T_2^* = 0.4$ sec, the dependence on flip angle is much more uniform, the optima being broad and all centered in the region of $\alpha = 0.4\pi$.

assuming suitable values for the parameters $T_p = 1$ second, $T_1 = T_2 = 10$ seconds. The optimum flip angle is quite small and quite critical for $\theta = 0$, while larger values of θ demand much larger flip angles and exhibit much broader maxima. Clearly, there is no single setting of α that will be near optimum for all θ values, and although $\alpha = \alpha_E$ gives a signal that is independent of θ, this is well below the optima that may be achieved.

Introduction of field inhomogeneity broadening moderates this problem considerably, since the detected signal can now be considered to be a suitably weighted integral over a range of values of θ. Figure 5(b) illustrates the dependence of M_y^+ on flip angle for the same parameters as Fig. 5(a) but with a Lorentzian distribution of static field with $T_2^* = 0.4$ second. For all values of θ_0 (the precession angle at the center of the line) there is now a broad maximum in the region near $\alpha = 0.4\pi$ (*12*).

3. EXPERIMENTAL

3.1. *Equipment*

Carbon-13 spectra were recorded on a Varian XL–100–12 spectrometer fitted with a VFT–100 Fourier transform accessory. Field/frequency regulation was achieved through a deuterium internal lock, using strong signals in an effort to minimize residual instabilities that might cause destructive interference and hence an accelerated decay rate when several free induction signals were accumulated. Proton noise decoupling was not used, since this has been shown to attenuate spin echo effects and accelerate the Bloch decay (*7*). The spectrometer was operated under the control of a small digital computer which was also used for data accumulation and fast Fourier transformation. The experimental details have been given elsewhere (*13*).

The proton spin echoes shown in Fig. 3 were obtained by Dr. D. R. Ware of this laboratory on a home-built spin-echo spectrometer.

3.2. *Elimination of Phase and Intensity Anomalies*

Variation of phase and intensity across the spectrum as a function of the excess precession angle θ is a direct result of operating with a pulse interval T_p that is shorter than or comparable with the spin-spin relaxation times of the sample. (This condition is assumed to be imposed throughout for reasons of sensitivity.) Pulses are thus applied while there is a finite net transverse component of magnetization M_{xy}^-, even in the presence of severe inhomogeneity of the field, since these effects are refocussed. The two possible remedies appear to be to destroy this transverse magnetization, or to accumulate the steady-state signals in such a way as to average the anomalies to zero. The problem is to introduce some form of incoherence into the nuclear resonance condition, through, for example, an incoherent rf source, heteronuclear noise decoupling, an unstable field/frequency lock, or field gradient pulses (made random in amplitude to prevent refocussing). If such devices operate while the free induction signal is being recorded, the decay is accelerated and the transformed spectrum has broadened lines. This undesirable feature could be avoided by applying very strong, random field gradient pulses in the short interval after acquisition of the transient signal but before the next pulse; unfortunately, this would adversely affect the operation of internal field/frequency regulation schemes. A solution which does not lead to any broadening is to introduce a random delay into the timing of the rf pulses. This timing was under control of the spectrometer computer, and the program was therefore modified to provide three possible timing modes: (a) normal, (b) "random pulse delay", (c) "scrambled steady state". In the normal mode, the pulse interval T_p was held quite stable, the maximum variations being of the order of microseconds. Modification (b) introduced a randomly generated delay of a few milliseconds on the timing of every rf pulse, making use of the last data word from the analog-to-digital converter as a source of noise. As a result, any resonance line with an offset $\omega_0/2\pi$ greater than a few hundred Hz experiences essentially random variations in θ over the entire range 0 to 2π radians. This has the effect of preventing the establishment of a steady state as far as the transverse components of magnetization M_x and M_y are concerned, eliminating the phase and intensity anomalies associated with steady-state operation.

Modification (c) allows a steady-state regime to be established by maintaining a stable pulse interval T_p for a predetermined number of pulses (typically 64). Then a new value for T_p is introduced by adding or subtracting a randomly-generated delay which is then held fixed for another batch of pulses in order to establish a different steady state with new settings of θ for every line in the spectrum. This mode benefits from the improved signal strength obtainable in the steady-state regime, and yet "scrambles" the phase and intensity anomalies by averaging the results over many random values of θ.

These pulse-delay techniques have the advantage of averaging over a range of θ values without broadening the lines. If field inhomogeneity broadening is comparable with the inverse pulse interval $1/T_p$, the phase and intensity anomalies are attenuated by this averaging mechanism, reducing the need for incoherence in the pulse timing.

4. RESULTS

4.1. *The Problem*

The periodic variation of phase and intensity as a function of the excess precession angle θ may best be illustrated with a resonance line of long spin-spin relaxation time. Carbon disulphide has a single narrow ^{13}C resonance line, and is reported to have $T_1 = 44$ seconds, $T_2 = 38$ seconds at 15 MHz (*11*). The intensity and phase of this resonance line were studied as a function of θ by varying the frequency offset control of the spectrometer in 0.2 Hz steps; with the pulse interval set at $T_p = 0.5$ second the complete range of 2π radians corresponded to an offset range of 2 Hz. The experiments were carried out at three settings of the flip angle, one corresponding as nearly as possible to $\alpha = \alpha_E$, one much smaller ($\alpha_E/6$), and one much larger ($6\alpha_E$). A total of 500 free induction signals were averaged and then multiplied by an exponentially decaying function with time constant 0.4 second in order to broaden the transformed line and ensure that any intensity anomalies could *not* be attributed to inadequate digitization along the frequency axis (there were 14 samples per line width). Under these conditions the signal intensity in the transformed spectrum is determined predominantly by the Y component immediately after the pulse, M_y^+, and the simple calculation illustrated in Fig. 2 may be used to predict the signal as a function of θ. The results are shown in Fig. 6. Marked phase and intensity discrepancies are clearly evident and show a periodicity corresponding to the pulse interval $T_p = 0.5$ second. Note that at the small flip angle ($\alpha_E/6$), the signal is very

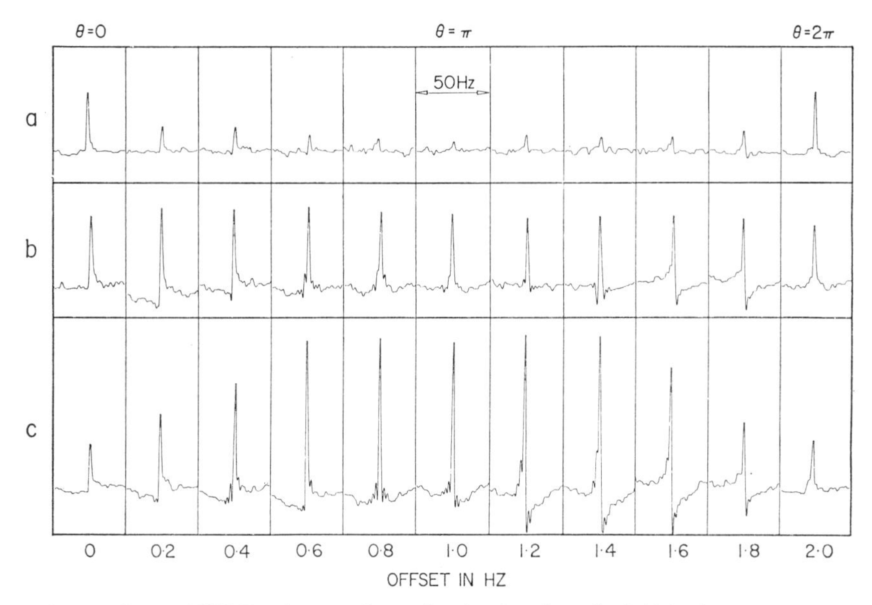

FIG. 6. Experimental ^{13}C Fourier transform signals of carbon disulphide observed in the steady-state regime as a function of offset from resonance (with respect to an arbitrary zero). Phase and intensity anomalies are evident even though the spectrometer gain and phase settings were fixed throughout. (a) With a small flip angle $\alpha = \alpha_E/6$, the signal intensity has sharp maxima at $\theta = 0$ and 2π, but very weak signals between. (b) With the special condition $\alpha = \alpha_E$, the intensity becomes essentially independent of θ. (c) With $\alpha = 6\alpha_E$, the signals are strong near $\theta = \pi$ and weak near $\theta = 0$ or 2π, the extrema being rather broad. The phase anomalies are also cyclic functions of offset, antisymmetrical about $\theta = \pi$, and apparently independent of flip angle. These results are to be compared with the calculations of Fig. 2, except for the broadening effect of field inhomogeneity.

weak at all offsets except those near the conditions $\theta = 0$ or 2π where it exhibits a sharp maximum. As predicted, there is little variation in intensity for $\alpha = \alpha_E$, whereas for $\alpha = 6\alpha_E$ there is a broad maximum centered about $\theta = \pi$ with minima near $\theta = 0$ and 2π.

Throughout these experiments the spectrometer phase controls and the phase information supplied to the Fourier transform program were held constant at settings believed to correspond to the pure absorption mode. The observed phase anomalies appear to be antisymmetrical about the condition $\theta = \pi$ as predicted, and as far as can be judged are the same for all three settings of the flip angle. Since the phase is a very rapidly changing function of θ near $\theta = 0$ or 2π, it is difficult to adjust the offset with sufficient precision to achieve the expected pure absorption mode for $\theta = 0$ or 2π. If some allowance is made for the effects of field inhomogeneity, these results are seen to reflect all the general trends predicted in Fig. 2 (where broadening by the magnet was neglected).

There seems little doubt that such gross intensity variations would lead to considerable confusion in the interpretation of Fourier transform spectra of unknown materials; because of the phase anomalies the problem would be compounded if integration were to be attempted. Fortunately, these shortcomings are readily corrected by the simple techniques described below.

4.2. The Remedy

Pulsing the rf excitation at some randomly chosen phase of the nuclear precession, followed by accumulation of several free induction signals in good time registration eliminates these cyclic variations of phase and intensity across the transformed spectrum. This is borne out by the experimental Fourier transform ^{13}C spectra of (enriched) methyl iodide shown in Fig. 7. The three spectra within each set (I or II) have all been recorded at a fixed value of the frequency offset, and the offset has been changed slightly between set I and set II. The intensity anomalies are evident as deviations from the 1 : 3 : 3 : 1 ratio expected for a methyl resonance, and the phase anomalies as deviations from the pure absorption mode. Spectra recorded with a fixed setting of the pulse interval T_p [I(a) and II(a)] show marked discrepancies in intensity and phase. Random delays were then introduced into the timing of all rf pulses so that the transverse magnetizations M_x and M_y could not establish steady-state conditions. This restores the absorption mode character and approximate 1 : 3 : 3 : 1 intensity ratio [I(b) and II(b)], but at the cost of a reduction in overall intensity. The two spectra were then reexamined in the scrambled steady-state mode where the pulse timing was only changed after each "batch" of 64 pulses, but kept fixed within each batch. Intensity and phase were again restored to normal [I(c)

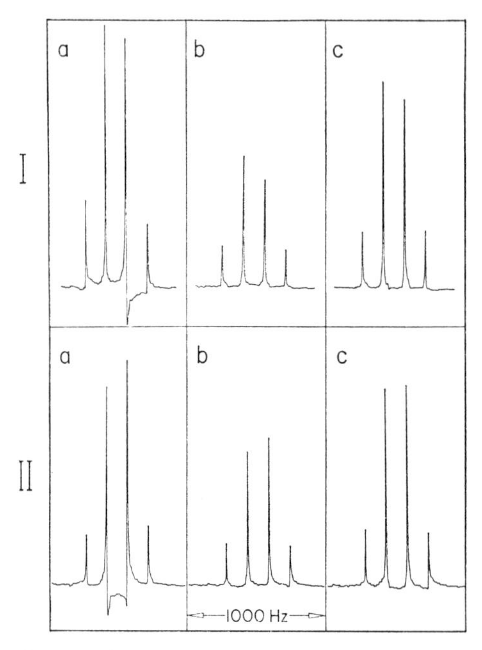

FIG. 7. Carbon-13 Fourier transform spectra of methyl iodide observed in a steady-state regime with $T_p = 0.4$ sec. The three spectra in set I were recorded at a fixed offset which was changed slightly for set II. (a) With regularly timed pulses there are evident phase and intensity anomalies. (b) With a random delay in the timing of each pulse, the anomalies are to a large extent eliminated, but the overall intensity is reduced. (c) With a random delay introduced for each "batch" of 64 pulses, but held constant within each batch, the steady-state regime is established and the anomalies suppressed without significant loss of intensity.

and II(c)] while the overall intensity was improved because the transverse magnetization components M_x^- and M_y^- contribute to the total detected signal once the steady state has been established.

One requirement of the random pulse delay and scrambled steady-state techniques is that the radiofrequency should be set up sufficiently far from the nearest nuclear precession frequency that the product of a typical pulse delay ΔT_p and the smallest offset ω_0 corresponds to a variation of θ of the order of 2π radians or more. This is readily achieved with ΔT_p of the order 10^{-2} second and the minimum offset $\omega_0/2\pi$ of the order of 100 Hz.

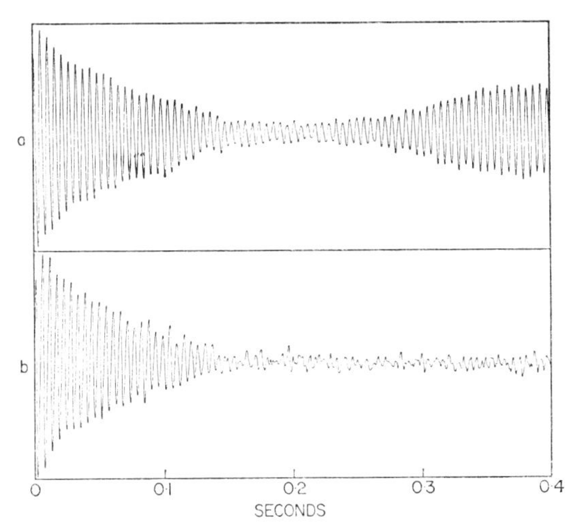

FIG. 8. Steady-state spin echoes observed with the free induction signal of ^{13}C in enriched methyl iodide, coherently decoupled from protons. Each trace represents the average of 256 transients. Trace (a) has been recorded with a stable setting of the pulse interval, and a clear refocussing effect is evident. In trace (b) the time-averaged echo has been suppressed by the introduction of a random delay into the timing of each pulse.

Experimental evidence for the refocussing of field inhomogeneity effects in the steady state is provided by the free induction signals observed for the ^{13}C resonance of methyl iodide coherently decoupled from protons. Figure 8(a) shows the regrowth of the signal towards the end of the pulse interval, achieving optimum refocussing at the time of the next pulse. A total of 256 free induction signals was accumulated. Figure 8(b) shows the inhibition of echo formation when a random delay was introduced into the timing of each pulse. Prevention of spin echo effects is important in the use of the Fourier transform method to measure spin-lattice relaxation times by an analog of the progressive saturation method (*7*).

5. DISCUSSION

The intensity and phase variations across a Fourier transform spectrum predicted by Ernst and Anderson (*1*) have been shown to be quite severe in selected examples of ^{13}C spectra. These effects disappear if the sample has sufficiently short transverse relaxation times, and they are of course moderated by magnet-field inhomogeneity. These anomalies may also be suppressed almost entirely by incoherence effects in the spectrometer which accelerate the decay of the free induction signal or cause destructive interference when many signals are averaged. Instrumental effects of this kind may explain why the intensity and phase anomalies have not proved to be so pronounced or so widespread as to excite special comment, since it has become common practice to employ proton noise decoupling in a routine fashion for ^{13}C spectroscopy (*14*), and Fourier transform spectrometers built for high sensitivity often do not require the highest degree of field/frequency regulation.

The same instrumental shortcomings can inhibit the refocussing effect which gives rise to spin-echo formation at the end of the pulse interval (*7*), another phenomenon which escaped notice until quite recently (*5*, *6*), for the reported steady-state echoes from methyl iodide (Fig. 8) were observed to be fairly rapidly attenuated at longer pulse intervals. The refocussing process is of course most dramatic in a magnetic field of poor homogeneity (see, e.g., the lowest trace in Fig. 4), a condition where the phase and intensity anomalies tend to be masked. The spin echoes illustrated for methyl iodide were obtained in a relatively inhomogeneous field. Refocussing is fundamental to the dynamics of the steady-state response. The macroscopic sample will contain many separate volume elements related by the fact that although they precess through different angles Θ in the pulse interval, they have the same *excess* precession angle θ (Eq. [3]). These isochromats are dispersed by field inhomogeneity but refocussed into a single vector at the end of the pulse interval. This temporary dispersal causes destructive interference and loss of signal intensity near the center of the pulse interval. However, the conditions immediately prior to the pulse are independent of field inhomogeneity and it is thus permissible to analyse the steady state according to the treatment of Ernst and Anderson (*1*) and take into account the magnet broadening after the fact.

The same is not true of the incoherence introduced by the random pulse delay technique, and it must be emphasized that the spin dynamics have not been analysed for this case; it seems doubtful if a steady-state response can be defined for a pulse sequence applied at random phases of the nuclear precession if $T_2 \gg T_p$. On the other hand, except for the relatively short time required to establish the steady state, the scrambled steady-state mode is correctly described by the equations of Section 2.

Another complication in the analysis of the steady-state regime arises in Fourier transform spectroscopy of nuclei such as protons and fluorine that exhibit homonuclear spin-spin coupling. The formation of spin echoes would be expected to be perturbed by homonuclear coupling, since the pulses affect both groups of coupled nuclei, inducing a modulation of the amplitude of the echoes (*15*) which in the general case of a multispin system can be extremely complex (*16*). Nevertheless, intensity anomalies have been observed in the proton Fourier transform spectra of the aldehyde groups of 3-bromothiophene-2-aldehyde and furan-2-aldehyde that are believed to have the same origin as the anomalies reported here (*17*). In these experiments, only a single free induction transient was recorded, but the initial conditions before the pulse were determined by a previous steady-state regime set up to maintain the internal proton field/frequency lock. The pulse interval was then 8 seconds, and θ varied from 0 to 2π radians within a frequency range of only 0.125 Hz.

In the steady-state mode, the transverse components of magnetization just before the pulse have the effect of augmenting the detected signal, and since they are not destroyed by field inhomogeneity, the sensitivity of the Fourier transform experiment can be significantly higher than that predicted by analyses (*18*) that assume a complete decay of the free induction signal between pulses (*12*). This reduces the relative advantage to be gained by more intricate refocussing schemes such as the "driven equilibrium Fourier transform" method [DEFT] (*19*) or "spin echo Fourier transform" [SEFT] (*20*). The refocussing that occurs in the steady-state response to a regular sequence of pulses enhances the sensitivity by an amount comparable with the DEFT and SEFT techniques (*12*).

Free precession signals are commonly multiplied by a function that decays with time in order to deemphasize the tails of the decay where the signal-to-noise ratio is poor, optimum sensitivity being achieved with a function that corresponds to the use of a matched filter in the frequency domain (*3*). It has, therefore, become customary to use an exponentially decaying function with the same time constant as that of the free induction decay. However, because of the spin echo effect in the steady-state regime, this function would no longer appear to be optimum; instead, each point on the free induction signal should be weighted with the signal-to-noise ratio at that point, a function which initially decays but then increases again with time. Since there would be a large discontinuity at the end of the transient, the need for a suitable apodization function would then be more critical.

In a multiline Fourier transform spectrum the occurrence of a marked phase anomaly is often indicative of a line with long T_2, or a line not appreciably affected by heteronuclear noise decoupling, such as the ^{13}C resonance of a quaternary carbon atom. (In a single trace it may be difficult to identify an intensity anomaly.) In principle, it should be possible to devise a more quantitative estimate of the spin-spin relaxation time based on the intensity effects, but in practice the analysis would appear to be too cumbersome for a general method.

It would be a pity to be forced to degrade spectrometer resolution to ensure that these phase and intensity anomalies were averaged out. Fortunately, the scrambled steady-state technique suppresses these effects without line broadening, and there seems to be no reason why it should not be incorporated in a routine fashion in the program of a computer-controlled Fourier transform spectrometer (*21*). Otherwise, if there is a continued improvement of instrument stability over the years, such undesirable effects would be expected to become a much more common occurrence.

ACKNOWLEDGMENTS

The authors would like to thank Dr D. R. Ware for obtaining the proton spin echoes shown in Fig. 3, and Dr W. C. Jankowski for bringing to their attention a persistent phase problem in the ^{13}C Fourier transform spectrum of dipyridyl, the spur to starting this study.

REFERENCES

1. R. R. Ernst and W. A. Anderson, *Rev. Sci. Instr.* **37,** 93 (1966).
2. I. J. Lowe and R. E. Norberg, *Phys. Rev.* **107,** 46 (1957).
3. R. R. Ernst, "Sensitivity Enhancement in Magnetic Resonance" in "Advances in Magnetic Resonance," Vol. 2, Academic Press, New York, 1966.
4. E. L. Hahn, *Phys. Rev.* **80,** 580 (1950).

5. This effect was mentioned in a presentation by J. S. Waugh at the 11th Experimental NMR Conference held at the Mellon Institute, Pittsburgh, Pennsylvania, on April 22nd, 1970 (unpublished).
6. P. Waldstein and W. E. Wallace, *Rev. Sci. Instr.* (in press). These authors were kind enough to communicate their results prior to publication.
7. R. Freeman and H. D. W. Hill, *J. Chem. Phys.* **54,** April 1st (1971).
8. This sign change reflects a different choice of the sense of precession but has no significant effect on any of the observable quantities.
9. However, even if $T_0 = T_p = T$, if the free precession signal is weighted before transformation with some function $f(t)$ so that the decay is no longer described by exp $(-t/T_2)$, this changes the integral that gave rise to Eqs. [23] and [25], and there can be no cancellation of the dispersive term.
10. This contrasts with the conclusion of Ernst and Anderson [Ref. (*1*)] since they assume $T_0 = T_p$.
11. R. R. Shoup and D. L. VanderHart, *J. Amer. Chem. Soc.* (in press).
12. P. Waldstein and W. E. Wallace have reached the same conclusion independently; see Ref. (*6*).
13. W. Bremser, H. D. W. Hill, and R. Freeman, *die Messtechnik*, **79,** 14 (1971).
14. R. R. Ernst, *J. Chem. Phys.* **45,** 3845 (1966).
15. E. L. Hahn and D. E. Maxwell, *Phys. Rev.* **88,** 1070 (1952).
16. A. Abragam, "The Principles of Nuclear Magnetism", Clarendon Press, Oxford, 1961; R. Freeman and H. D. W. Hill, *J. Chem. Phys.* **54,** 301 (1971).
17. R. Freeman and R. C. Jones, *J. Chem. Phys.* **52,** 465 (1970).
18. J. S. Waugh, *J. Mol. Spectrosc.* **35,** 298 (1970).
19. E. D. Becker, J. A. Ferretti, and T. C. Farrar, *J. Amer. Chem. Soc.* **91,** 7784 (1969); W. E. Wallace, *J. Chem. Phys.* (in press).
20. A. Allerhand and D. W. Cochran, *J. Amer. Chem. Soc.* **92,** 4482 (1970).
21. Field/frequency regulation schemes that employ a *homonuclear* (pulsed) internal lock would require resynchronization of the pulse timing for the reference line, but the timing would remain essentially random for the other lines. See R. Freeman and H. D. W. Hill, *J. Chem. Phys.* **51,** 3140 (1969).

JOURNAL OF MAGNETIC RESONANCE **8**, 354–365 (1972)

Quantitative Aspects of Coherent Averaging. Simple Treatment of Resonance Offset Processes in Multiple-Pulse NMR*

ALEXANDER PINES† AND JOHN S. WAUGH

Department of Chemistry and Research Laboratory of Electronics, Massachusetts Institute of Technology, Cambridge, Massachusetts 02139

Received January 28, 1972

A simple zero-order treatment of averaging by resonance offset fields in multiple-pulse NMR is presented. If the resonance offset is smaller than the inverse rf cycle time but larger than the local fields, then the average dipolar Hamiltonian has the same form as the truncated dipolar Hamiltonian in the rotating frame, and is scaled by an easily calculated factor. From this fact, several general conclusions are drawn and the possibility of line narrowing, spin locking and time reversal is discussed for several pulse sequences. For the phase-alternated two-pulse sequence it is demonstrated explicitly by a calculation of truncated second moments that behavior at and far from resonance should be radically different, indicating that caution should be exercised in the analysis of such experiments.

INTRODUCTION

The problem of calculating the response of a coupled spin system to an arbitrary radiofrequency excitation is, in general, an extremely complex one. It is well known, however, that if the excitation fulfils certain conditions of periodicity and duration, an enormous simplification ensues, taking the form of what we have called "coherent averaging" (*1*). This theory has formed a powerful tool for the description and design of a variety of effects in pulsed NMR, including spin locking (*2–4*), line narrowing and high resolution NMR in solids (*5–7*), spin echoes (*8*), steady-state behavior (*9*), and scaling of inhomogeneous shifts (*10*).

It was noticed in several of these experiments that when pulse trains were applied with the rf carrier frequency substantially displaced from the Larmor frequency of the spin system, i.e., with a "resonance offset," a new phenomenon manifested itself in the form of additional averaging, for example enhanced line narrowing. This phenomenon has recently been explained and, in fact, has been shown to possess a number of useful properties (*11*).

Basically, what happens, in qualitative terms, is the following: in the presence of rf excitation the coupling between the spins becomes modulated with a period equal to that of the excitation itself, in an appropriate reference frame (*1*, *12*). In the limit of strong modulation we perform a time average of this time-dependent coupling and obtain an average coupling which effectively governs the response of the spin system.

* This work was supported in part by the National Institutes of Health and in part by the National Science Foundation.

† Present address: Department of Chemistry, University of California, Berkeley, CA 94720.

The process of "truncation" of the dipolar interaction in the presence of a large static external field (*13–16*), for example, is a particular case of this approach (*11*). Now, consider what happens if there is a large resonance offset. In the rotating frame, this appears as a static magnetic field and, as in the case of truncation just mentioned, this field will provide an additional modulation of the coupling, which may be averaged in the appropriate limit (*1*). Thus, the behavior of pulsed NMR experiments near and far from resonance should be distinctly different, as has indeed been observed (*11*).

In this communication, we wish to present a simple treatment of this phenomenon in a form more amenable to an intuitive understanding and quantitative application to a variety of NMR experiments than that of our previous work. This presentation is warranted by the fact that these multiple-pulse techniques are becoming increasingly useful in studies of chemical problems, and resonance offset effects form an integral component in their usefulness and in their understanding.

The approach here is simple in the sense that only zero-order (*1*) averaging effects are considered, and insofar as is possible, the notation and tools used should be quite familiar. Higher-order effects are more conveniently treated in terms of irreducible tensor operators, since we are interested in the transformations of such operators under the rotations induced by the rf excitation. This will be deferred to a later detailed exposition of these experiments.

In section II a brief review of pertinent theory is presented and a general description of averaging due to resonance offset fields is derived. In section III this is applied to some simple multiple-pulse sequence and some of its uses and limitations are discussed. Details on the exact steps involved in the calculations of average Hamiltonians in various representations are not given, since this has been presented several times. The procedure is simply mentioned and the results written down directly. A concise review may be found in a forthcoming treatise on "magic-angle" experiments (*7*).

GENERAL THEORY

Hamiltonian and Frame of Reference

We begin by writing down a Hamiltonian for the system. We select as an example of a spin coupling the dipolar interaction, since this is the most important and widely encountered for solids of interest to chemists. The results are easily generalized, as we shall see. We write the Hamiltonian in the laboratory frame in frequency units:

$$\mathscr{H}_L(t) = \mathscr{H}_0 + \mathscr{H}_1(t) + \mathscr{H}_d. \tag{1}$$

$\mathscr{H}_0$ is the Zeeman interaction with applied static field along the z axis

$$\mathscr{H}_0 = -\omega_0 I_z. \tag{2}$$

$\mathscr{H}_1(t)$ the applied rf excitation

$$\mathscr{H}_1(t) = -2\omega_1(t) I_x \cos[\omega t + \phi(t)], \tag{3}$$

where $\omega_1(t)$ and $\phi(t)$ describe the amplitude and phase modulation of the rf excitation applied at frequency ω, and

$$\mathscr{H}_d = -\sum_{i<j} \frac{\gamma^2 \hbar}{r_{ij}^3} \left[\frac{3(\boldsymbol{I}_i \cdot \boldsymbol{r}_{ij})(\boldsymbol{I}_j \cdot \boldsymbol{r}_{ij})}{r_{ij}^2} - \boldsymbol{I}_i \cdot \boldsymbol{I}_j \right] \tag{4}$$

is the dipolar interaction.

We now perform the customary transformation to a coordinate system rotating at frequency ω about the z axis. An average is then performed over a time period of $2\pi/\omega$ and this removes the time-dependent terms in $\mathscr{H}_1(t)$ and $\mathscr{H}_d$ in this frame. All this is well known; it corresponds to discarding the counter-rotating component of the rf field and the nonsecular terms of the dipolar interaction (or truncation) and is just a special case of coherent averaging. It is clearly legitimate whenever $\mathscr{H}_0$ is much larger than $\mathscr{H}_1$ and $\mathscr{H}_d$. Our Hamiltonian in the rotating frame thus assumes the form

$$\mathscr{H}_R^0(t) = \mathscr{H}_\Delta + \mathscr{H}_1(t) + \mathscr{H}_D^0, \qquad [5]$$

where R indicates the rotating frame.

$\mathscr{H}_\Delta$ is the resonance offset:

$$\mathscr{H}_\Delta = -\Delta\omega I_z; \qquad \Delta\omega = (\omega_0 - \omega) \qquad [6]$$

$$\mathscr{H}_1(t) = -\boldsymbol{\omega}_1(t)\cdot\boldsymbol{I}, \qquad [7]$$

$$\mathscr{H}_d^0 = \sum_{i<j} b_{ij}(3I_{iz}I_{jz} - \boldsymbol{I}_i\cdot\boldsymbol{I}_j); \qquad b_{ij} = \frac{-\gamma^2\hbar}{r_{ij}^3} P_2(\cos\theta_{ij}). \qquad [8]$$

The form of $\boldsymbol{\omega}_1(t)$ depends on the particular experiment at hand. Note that for brevity we sometimes use the same notation for terms of $\mathscr{H}$ in different frames, to conform with accepted practice—the frame will always be specified if this is done. The ^{0} superscript indicates a truncation or zero-order average of the particular term.

If the spin system is initially described in the rotating frame by the density matrix $\rho_R(0) = \rho_L(0)$, then at time t it has evolved to a state described by

$$\rho_R(t) = U_R^0(t,0)\,\rho_R(0)\,U_R^{0\dagger}(t,0), \qquad [9]$$

here $U_R^0(t,0)$ is the effective time development operator in the rotating frame, given by

$$U_R^0(t,0) = T\exp\left[-i\int_0^t \mathscr{H}_R^0(t')\,dt'\right] \qquad [10]$$

and T is a time-ordering operator. It is an expansion of U in which we shall be interested.

Coherent Averaging Effects

We now prepare to take account of the modulation that the first two terms in Eq. [5] produce in the third. To this end we assume:

(i) $\mathscr{H}_1(t)$ is cyclic and periodic with period t_c, and t is restricted to integer values of t_c, $t = Nt_c$

(ii) $t_c \ll t_\Delta, T_2$; $t_\Delta = 2\pi/\Delta\omega$, [11]

(i) has been discussed in detail (*1*) and (ii) will allow us to perform a factorization of [10] in two steps (*11*). Employing the conditions above we obtain to a good approximation factoring out $\mathscr{H}_1(t)$ in a straightforward application of the theory:

$$U_R^0(t,0) = \bar{U}_{TR}^0(t,0) = \exp(-it\bar{\mathscr{H}}_{TR}^0), \qquad [12]$$

where

$$\bar{\mathscr{H}}^0_{TR} = \frac{1}{t_c}\int_0^{t_c} U_1^\dagger(t,0)\,[\mathscr{H}^0_d + \mathscr{H}_\Delta]\,U_1(t,0)\,dt$$
$$= \bar{\mathscr{H}}^0_d + \bar{\mathscr{H}}_\Delta = \bar{\mathscr{H}}^0_d - \overline{\Delta\omega} I_{\bar{\mu}}; \qquad U_1(t,0) = T\exp\left[-i\int_0^t \mathscr{H}_1(t')\,dt'\right]. \quad [13]$$

Note that restriction [11(i)] applies only to calculations involving time evolution as in Eq. [12] and not, of course, to calculations of expansion terms as in Eq. [13]. $\bar{\mathscr{H}}^0_{TR}$ is the average Hamiltonian; T stands for "toggling" (*12*) since $\bar{\mathscr{H}}^{0'}_{TR}$ is precisely the average Hamiltonian in an interaction (toggling) frame defined by $\mathscr{H}_1(t)$; (due to restriction 11(i), $\bar{U}^0_{TR}$ is thus the effective evolution operator in the rotating frame). $\bar{\mu}$ refers symbolically to the average direction $\bar{\boldsymbol{\mu}}$ in the rotating frame along which the spins are quantized (*12*). This is easily visualized by taking a unit spin vector $\boldsymbol{\mu}$ along the z axis and applying $\mathscr{H}_1(t)$ in the rotating frame. Then we define:

$$\bar{\boldsymbol{\mu}} = \frac{\int_0^{t_c} U_1^\dagger(t,0)\,\boldsymbol{\mu}\,U_1(t,0)\,dt}{\left|\int_0^{t_c} U_1^\dagger(t,0)\,\boldsymbol{\mu}\,U_1(t,0)\,dt\right|}. \quad [14]$$

For example if $\boldsymbol{\omega}_1(t)$ consists of 90° phase-alternated δ-pulses along the x axis then $\bar{\boldsymbol{\mu}} = (1/\sqrt{2})(\boldsymbol{j}+\boldsymbol{k})$. This is illustrated in Fig. 1. $\overline{\Delta\omega}$ is analogously the average resonance offset over this cycle:

$$\overline{\Delta\omega} = \frac{\Delta\omega}{t_c}\left|\int_0^{t_c} \boldsymbol{\mu}(t)\,dt\right|. \quad [15]$$

We now add to [11] (i)–(ii) the further restriction:

$$\text{(iii)}\ t_{\bar{\Delta}} \ll T_2; \qquad t_{\bar{\Delta}} = 2\pi\overline{\Delta\omega}^{-1}. \quad [11]$$

This allows us, exactly as above, to factor the operator in [12] and then take an average over one cycle of the interaction with the static field $-\overline{\Delta\omega} I_{\bar{\mu}}$. What we are doing is in fact a truncation of the average dipolar interaction $\bar{\mathscr{H}}^0_d$ due to the presence of a large static field along the $\bar{\mu}$ axis. To a good approximation then:

$$\bar{U}^0_{TR}(t,0) = \bar{U}_\Delta(t,0)\,\bar{U}^{00}_{DTR}(t,0), \quad [16]$$

where

$$\bar{U}_\Delta(t,0) = \exp(-it\bar{\mathscr{H}}_\Delta) \quad [17]$$

$$\bar{U}^{00}_{DTR}(t,0) = \exp(-it\bar{\mathscr{H}}^{00}_{DTR}) \quad [18]$$

$$\bar{\mathscr{H}}^{00}_{DTR} = \bar{\mathscr{H}}^{00}_d = \frac{1}{t_{\bar{\Delta}}}\int_0^{t_{\bar{\Delta}}} \bar{U}^\dagger_\Delta(t,0)\,\bar{\mathscr{H}}^0_d\,\bar{U}_\Delta(t,0)\,dt. \quad [19]$$

Here the additional D subscript stands for "doubly rotating"—$\bar{\mathscr{H}}^{00}_{DTR}$ is precisely the average Hamiltonian in a frame which now rotates about the $\bar{\mu}$ axis at frequency $\overline{\Delta\omega}$ (*12*). Superscript ^{00} indicates the double averaging or truncation.

Now using Eqs. [13] and [19] and changing the order of integration over t_c and $t_{\bar{\Delta}}$ we have

$$\bar{\mathscr{H}}^{00}_{DTR} = \frac{1}{t_c t_{\bar{\Delta}}} \int_0^{t_c} dt'' \int_0^{t_{\bar{\Delta}}} dt' \, \bar{U}_{\Delta}^{\dagger}(t',0)\, U_1^{\dagger}(t'',0)\, \mathscr{H}_d^0\, U_1(t'',0)\, \bar{U}_{\Delta}(t',0), \qquad [20]$$

but using [13] and [17] it is easily found that

$$\frac{1}{t_{\bar{\Delta}}} \int_0^{t_{\bar{\Delta}}} dt' \, \bar{U}_{\Delta}^{\dagger}(t',0)\, U_1^{\dagger}(t'',0)\, \mathscr{H}_d^0\, U_1(t'',0)\, \bar{U}_{\Delta}(t',0) = \mathscr{H}^0_{d\bar{\mu}} P_2[\bar{\boldsymbol{\mu}} \cdot \boldsymbol{\mu}(t'')] \qquad [21]$$

where

$$\mathscr{H}^0_{d\bar{\mu}} = \sum_{j<i} b_{ij}(3I_{i\bar{\mu}} I_{j\bar{\mu}} - \boldsymbol{I}_i \cdot \boldsymbol{I}_j). \qquad [22]$$

$\mathscr{H}^0_{d\bar{\mu}}$ is of course just the "secular" part of $\mathscr{H}_d^0$, the part that commutes with $\bar{\mathscr{H}}_{\Delta}$. With this notation, for example, $\mathscr{H}_d^0 \equiv \mathscr{H}^0_{dz}$. Thus, we obtain finally for the average Hamiltonian (excluding $\bar{\mathscr{H}}_{\Delta}$) putting [21] into [20]:

$$\bar{\mathscr{H}}^{00}_{DTR} = \bar{\mathscr{H}}^{00}_{d} = \mathscr{H}^0_{d\bar{\mu}} \bar{P}_2[\bar{\boldsymbol{\mu}} \cdot \boldsymbol{\mu}(t)]. \qquad [23]$$

As usual the bar denotes an average over the rf cycle:

$$\bar{P}_2(\bar{\boldsymbol{\mu}} \cdot \boldsymbol{\mu}(t)) = \frac{1}{t_c} \int_0^{t_c} P_2[\bar{\boldsymbol{\mu}} \cdot \boldsymbol{\mu}(t)]\, dt, \qquad [24]$$

where $\bar{\boldsymbol{\mu}}$ is defined in [14] and $\boldsymbol{\mu}(t)$ is the integrand. We prefer to leave [23] in the rotating frame since normal detection methods (*17*) correspond to measurements in this frame and not in the more suitable tilted frame along $\bar{\boldsymbol{\mu}}$ (*14*).

Discussion

Equation [23] expresses a simple yet remarkable result. It says that no matter how complicated the cycle of rf excitation, if it is applied with a resonance offset fulfilling conditions 11(i)–(iii) then the average dipolar Hamiltonian is just the truncated Hamiltonian itself along an average axis in the rotating frame. This is depicted pictorially in Fig. 1. Obviously in a frame tilted (*14*) with its z axis along $\bar{\boldsymbol{\mu}}$, $\bar{\mathscr{H}}_d^{00}$ is exactly proportional to [8]. In order to calculate this average Hamiltonian it is thus not necessary to calculate $\bar{\mathscr{H}}^0_{TR}$ in the intermediate step—we need only: (i) find $\bar{\boldsymbol{\mu}}$ and write down $\mathscr{H}^0_{d\bar{\mu}}$ immediately and (ii) calculate $\bar{P}_2$ as in [24], so no manipulations are necessary on the spin variables. This fact, although rather obvious with a little reflection, is obscured in the original work (*11*) and with complicated cycles the manipulation of spin operators becomes unwieldy. The form of [23] allows a simple understanding of these experiments and permits us to draw more general conclusions (as we shall see for example in the discussion of spin locking). Of course in the event that we wish to enquire about the behavior at resonance ($\Delta\omega = 0$) this treatment is not valid and $\bar{\mathscr{H}}^0_{TR}$ must be calculated separately.

These results are easily generalized. If our interaction is not dipolar but is given, say, by a Hamiltonian $\mathscr{H}_n^0$ whose effective spin part transforms like the zero-order component of an n'th rank irreducible tensor in the rotating frame, then under the same conditions:

$$\bar{\mathscr{H}}_n^{00} = \mathscr{H}_{n\bar{\mu}}^0 \bar{P}_n[\bar{\boldsymbol{\mu}} \cdot \boldsymbol{\mu}(t)]. \qquad [25]$$

For example the isotropic chemical shift treated by Ellett and Waugh (*10*) has a spin part which transforms like a first-rank irreducible tensor and indeed their average Hamiltonian is a special case of our results. There, however, $\bar{\mathscr{H}}_{TR}^0$ commutes with $\bar{\mathscr{H}}_{\Delta}^0$ (since $\boldsymbol{I}$ transforms like $\boldsymbol{\mu}$) and thus the result is independent of resonance offset.

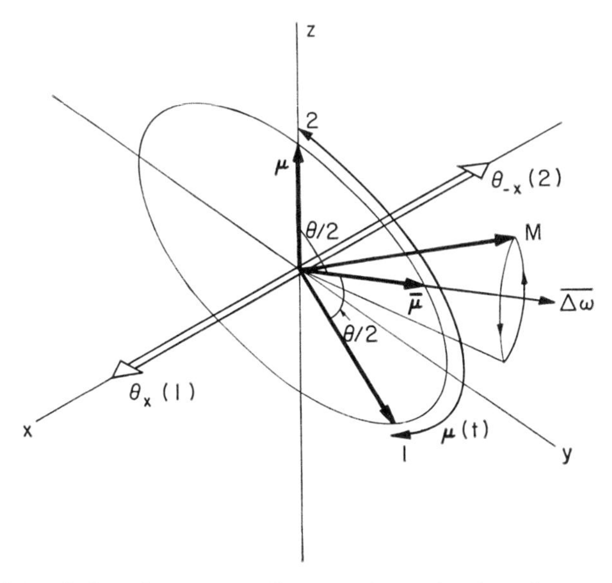

FIG. 1. Pictorial description of resonance offset averaging under the pulse sequence of Fig. 2(a). The pulses are imagined to be δ pulses ($t_w = 0$) and are depicted by arrows along x and $-x$ which nutate the unit magnetization vector $\boldsymbol{\mu}$ alternately by angles θ° and $-\theta^\circ$. Thus $\boldsymbol{\mu}(t)$ switches between positions 1 and 2 and its average direction is given by the unit vector $\bar{\boldsymbol{\mu}}$. The average offset field $\overline{\Delta\omega} = \cos(\theta/2)\Delta\omega$ points along $\bar{\boldsymbol{\mu}}$ and thus any magnetization M will precess on the average about this axis with frequency $\overline{\Delta\omega}$. This gives rise to the scaling of chemical and inhomogeneous shifts. The factor $\bar{P}_2[\bar{\boldsymbol{\mu}} \cdot \boldsymbol{\mu}(t)]$ in Eqs. [23], [24] is simply evaluated for this sequence as $P_2(\cos\theta/2)$, so the average dipolar interaction is scaled by $\bar{P}_2[\bar{\boldsymbol{\mu}} \cdot \boldsymbol{\mu}(t)] = P_2(\cos\theta/2)$.

One nice feature of [23] is that it allows a quantitative comparison with experiment as, for example, in the work of Lee and Goldburg (*15*). In the case of irradiation at resonance, $\bar{\mathscr{H}}_d^0$ usually has some form different from $\mathscr{H}_d^0$ and lineshape comparisons are difficult to reconcile with theory. Thus, we have to date mostly satisfied ourselves with qualitative or semiquantitative conclusions, except in special cases. In the present case, however, arguments can be made more precise and quantitative.

Note that this treatment is not applicable to resonance offset experiments such as those of Lee and Goldburg (*15*) and Barnaal and Lowe (*16*), since there, requirement 11(ii) is violated ($t_c \sim t_\Delta$). In these cases the first transformation must be made not by $U_1(t,0)$ as in our case, but by the full effective field (*14*):

$$U_e(t,0) = \exp[-it(\mathscr{H}_1 + \mathscr{H}_\Delta)].$$

In the present case the fact that H_1 is not exactly perpendicular to H_0 would be accounted for by correction terms in the expansion of $\bar{U}_R^0$, which are not considered here.

REPRESENTATIVE PULSE SEQUENCES

It now remains for us to specify the form of $\omega_1(t)$ in [7] and to write down some representative results. We select only simple examples to illustrate the general approach to application of the theory. Other examples will undoubtedly be investigated for fun by the interested reader.

Two-Pulse Cycle

Figure 2(a) shows the form of $\omega_1(t)$ for a phase-alternated two-pulse sequence. By inspection (see Fig. 1) we find immediately that $\bar{\boldsymbol{\mu}}$ is in the $y-z$ plane and makes an angle of $\theta/2$ with the z axis. Defining the duty cycle δ by

$$\delta = 2t_w/t_c \tag{26}$$

and using Eq. [23] we find trivially for a pure dipolar interaction

$$\bar{\mathscr{H}}_d^{00} = \mathscr{H}_{d\bar{\mu}}^{0}\{[3p_\delta(\theta) + 1]/4\}$$
$$p_\delta(\theta) = (1-\delta)\cos\theta + \delta(\sin\theta/\theta). \tag{27}$$

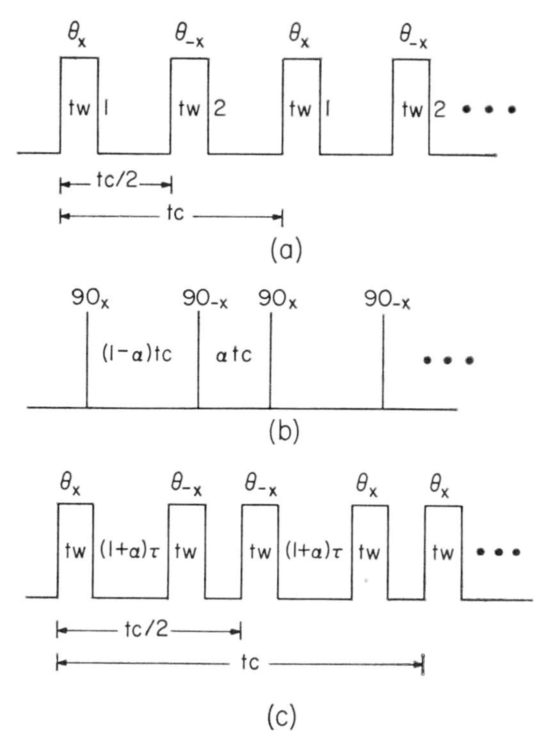

FIG. 2. Pulse sequences discussed in the text. Although not carried out here, these sequences may be symmetrized (*31*) by a redefinition of the cycle to eliminate some correction terms to the average Hamiltonian.

If there is an inhomogeneous or chemical shift term in the rotating frame Hamiltonian:

$$\mathscr{H}_c^0 = -\sum_i \sigma_{izz} I_{iz} \qquad [28]$$

where $\mathscr{H}_c^0$ is again a truncated form of the full chemical shift $\mathscr{H}_c$, then $\bar{\mathscr{H}}_{DTR}^{00}$ will contain another term given by:

$$\bar{\mathscr{H}}_c^{00} = \mathscr{H}_{c\bar{\mu}}^0 \, p_\delta(\theta/2) \qquad [29]$$

with p_δ defined in [27]; this is in agreement with previous results (*10*).

Line Narrowing

Figure 3 shows a calculated plot of $(3p_\delta(\theta) + 1)/4$ as a function of θ for several values of the duty factor δ. For $\delta < 0.75$ we see that $\bar{\mathscr{H}}_d^{00}$ can be made to vanish by an appropriate selection of θ and thus leads to a simple technique for line narrowing. For $\delta = 0$,

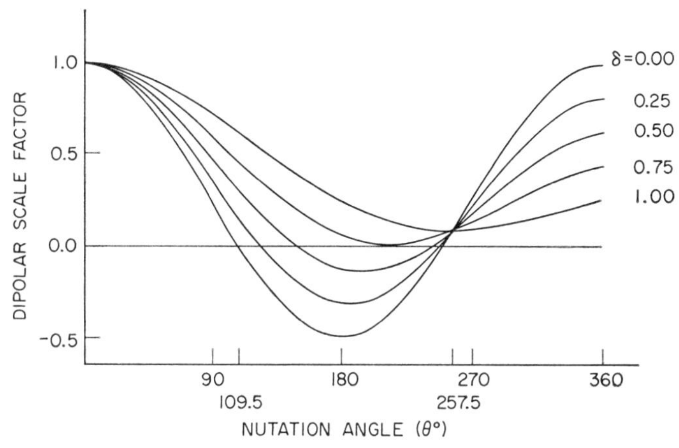

FIG. 3. Plot of the dipolar scale factor $\bar{P}_2[\bar{\boldsymbol{\mu}}\cdot\boldsymbol{\mu}(t)] = 1/4[3p_\delta(\theta) + 1]$ vs. θ for several values of the duty factor δ in the phase alternated pulse sequence [Fig. 2(a)]. The sequence produces line narrowing and is applicable for high resolution NMR in solids when $\bar{P}_2 = 0$, e.g., at 109.5° for $\delta = 0$ (the phase-alternated tetrahedral experiment). For $\bar{P}_2 < 0$ the Hamiltonian becomes "negative" yielding the conditions for time reversal.

i.e., δ pulses, we need $\theta = \theta_t$ where θ_t is the tetrahedral angle (109°28′); this is just the previously described PAT sequence (*11*). However, the present analysis shows that even for $\delta > 0$ the coupling can be made to vanish with $\theta > \theta_t$, thus eliminating the term in Eq. [21] of reference (*10*).

The fact that $\theta = \theta_t$ explains the limited line narrowing in Fig. 2 of the above work. With careful adjustment of $\theta > \theta_t$ decay times exceeding 1 msec for the ^{19}F nuclei in CaF have actually been attained for this experiment (*18*). Interestingly, it appears from Fig. 2 that employing $\delta \sim 0.75$ should be superior to the PAT experiment since the region of line narrowing is then markedly less sensitive to the exact value of θ and should thus be less sensitive to any inhomogeneities in the rf field. However, in this range $\overline{\Delta\omega}$ is also reduced so there is no large gain in cycle time. An experimental check of the full curves in Fig. 2 will be interesting and should provide an additional verification of this simple theory. Note that for any solution of $\tan\theta = \theta$, e.g., $\theta = 257.5°$, the line narrowing is predicted to be independent of the duty factor δ.

Magic Echoes

Magic echoes, which appear after homogeneous free induction decays in solids, were first reported by Rhim, Pines and Waugh (*8*). To remind the reader, magic echoes are produced in the following way: following the decay of magnetization due to $\mathscr{H}_d^0$, a strong rf perturbation is applied under which the effective Hamiltonian is given by $k\mathscr{H}_d^0$ with $k < 0$. This induces a negative time development which recalls the previous history of the spin system and produces an echo.

Looking at Fig. 2, we see that for small δ, we can make $3p_\delta(\theta) + 1 < 0$, i.e., $\bar{\mathscr{H}}_d^{00}$ "negative," thus yielding the necessary conditions for time reversal (*8*). The observation of magic echoes should provide verification for this aspect of the theory. These could be produced simply by applying a train of phase alternated pulses "sandwiched" between a pair of phase shifted ξ and $-\xi$ pulses as in the analogous on-resonance experiments (*8*).

Note that this is another manifestation of the simple form taken by the average Hamiltonian $\bar{\mathscr{H}}_{DTR}^{00}$. In a general pulse experiment, say an on-resonance phase-alternated sequence, the effective Hamiltonian has some other form except for special cases like 90° pulses (*3*) (*vide infra*) and time reversal becomes much more restricted.

Spin Locking

The form of $\bar{\mathscr{H}}_{DTR}^{00}$ allows another general conclusion. Since

$$[I_{\bar{\mu}}, \bar{\mathscr{H}}_{DTR}^{00}] = 0, \qquad [30]$$

we have all the prerequisites for spin locking of the $\bar{\mu}$ component of magnetization. We point out that [30] is the correct criterion to employ for spin locking; the presence of an average or mean H_1 field is neither sufficient nor a necessary condition as is still sometimes erroneously assumed.

We see, then, that the observation of $T_\parallel$ and $T_\perp$ (*20*) should be a general phenomenon in any resonance offset multiple-pulse experiment. If we start with the magnetization along an arbitrary axis in the rotating frame, we may separate it into components perpendicular to and parallel to $\bar{\boldsymbol{\mu}}$. The perpendicular component will then precess about $\bar{\boldsymbol{\mu}}$ and decay by spin–spin processes with a time constant $T_\perp = \bar{P}_2^{-1} T_2$ where $\bar{P}_2$ is given by [24] and T_2 is the normal transverse relaxation time. For $\bar{P}_2 \to 0$ the decay will of course be dominated by correction terms. The parallel component will be spin locked and will change only by spin–lattice relaxation in the rotating frame (*12*, *14*, *21–24*). The effects of spin–lattice relaxation are very interesting in these experiments and will be treated in detail elsewhere.

If the effective field along $\bar{\boldsymbol{\mu}}$, i.e., in $\bar{\mathscr{H}}_\Delta$, is inhomogeneous (for example from an inhomogeneous H_0 or from inhomogeneous shifts as in a polycrystalline solid) then $T_\perp$ may be dominated by this inhomogeneity if $\bar{P}_2$ is small. This effect can be demonstrated quite dramatically by production of inhomogeneous rotary spin echoes (*25*) when $\bar{\boldsymbol{\mu}}$ is at the magic angle (i.e., $\bar{P}_2 = 0$) as in the experiments of Rhim and Kessemeier (*26*) and Pines, Rhim and Waugh (*27*). In the latter experiment, the nature of the two components of magnetization is shown quite clearly using the off-resonance four-pulse technique (*5*).

The simple behavior we have outlined above is a peculiar characteristic of the resonance offset experiment. If we enquire into the behavior of the spin system under the pulse sequence of Fig. 2(a) at resonance ($\Delta\omega = 0$) we find from [13]

$$\bar{\mathscr{H}}_d^0 = -\tfrac{1}{2}\mathscr{H}_{dx}^0 + \tfrac{3}{2}p_\delta(\theta)\sum_{i<j} b_{ij}[\cos\theta(I_{iz}I_{jz} - I_{iy}I_{jy}) - \sin\theta(I_{iz}I_{jy} + I_{iy}I_{jz})]. \quad [31]$$

In this case an analysis of the spin system response is difficult and comparison with the normal unperturbed behavior can be made only on the basis of moments of the decays. For example the second moments defined by (*28*)

$$\langle\omega^2\rangle_\mu = \frac{Tr(\bar{\mathscr{H}}_d^0, I_\mu)^2}{Tr I_\mu^2} \quad [32]$$

are given for [31] by

$$\langle\omega^2\rangle_x = p_\delta^2(\theta)\langle\omega^2\rangle_0 \quad [33]$$

$$\langle\omega^2\rangle_y = \tfrac{1}{4}[p_\delta^2(\theta) + 2\cos\theta\, p_\delta(\theta) + 1]\langle\omega^2\rangle_0 \quad [34]$$

$$\langle\omega^2\rangle_z = \tfrac{1}{4}[p_\delta^2(\theta) - 2\cos\theta\, p_\delta(\theta) + 1]\langle\omega^2\rangle_0, \quad [35]$$

where $\langle\omega^2\rangle_0$ is the normal high field truncated second moment

$$\langle\omega^2\rangle_0 = -\frac{Tr(\mathscr{H}_d^0, I_x)^2}{Tr I_x^2}. \quad [36]$$

For $\theta = 90°$, $\delta \sim 0$ we have the experiment of Waugh and Huber (*29*) which produces a prolonged decay only along the x axis as expected from [33]. As we have seen though, this picture changes drastically as we go off resonance.

We take this opportunity to note that [33] also gives the second moment of a spin echo decay (with constant δ) produced by a 90_x pulse followed by a θ_y pulse since this is one cycle of the appropriate pulse sequence. This important problem has been dealt with separately (*9*).

Ideal 90° Pulses

Since the simple properties of 90° δ-pulses have a special appeal and in fact most proponents of multiple-pulse NMR have been primarily obsessed with such pulses, we enquire here into the possibility of using them for experiments similar to those mentioned above. A general two-pulse sequence composed of such pulses is shown in Fig. 2(b) with α variable from 0 through 1. If this sequence is applied with the appropriate resonance offset then, from [23]

$$\bar{\mathscr{H}}_d^{90} = \mathscr{H}_{d\bar{\mu}}^0\left[1 - \frac{3}{2}\frac{\alpha(1-\alpha)}{\alpha^2 + (1-\alpha)^2}\right], \quad [37]$$

where $\bar{\boldsymbol{\mu}}$ is in the $y - z$ plane and

$$\boldsymbol{\mu}\cdot\bar{\boldsymbol{\mu}} = \alpha/[\alpha^2 + (1-\alpha)^2] \quad [38]$$

and therefore $\bar{\mathscr{H}}_d^{90}$ cannot be made to vanish for any real α. Thus using only 90° δ-pulses we conclude that we must resort to a cycle containing more than two pulses to achieve line narrowing. It is interesting that the maximal line narrowing of 0.25 occurs at $\alpha = 1/2$ which gives just the pulse sequence of Waugh and Huber (*29*).

Four-Pulse Cycle

There are many possible four-pulse cycles. The most well-known to pulsed NMR spectroscopists is the four-pulse four-phase WAHUHA cycle (*5*). Here we treat a four-pulse cycle employing only *two* phases shown in Fig. 2(c). Defining δ as in [26] we find again using [23]:

$$\bar{\mathscr{H}}_d^{00} = \mathscr{H}_d^0\left[\tfrac{1}{2}(1-\alpha)(1-\delta) + \tfrac{1}{4}(1+\alpha)(1-\delta)(3\cos^2\theta - 1) + \frac{\delta}{4}\left(1 + \frac{3\sin\theta\cos\theta}{\theta}\right)\right] \quad [39]$$

which reduces for $\theta = \pi/2$ to

$$\bar{\mathscr{H}}_d^{00} = \mathscr{H}_d^0[1 - 3\alpha(1-\delta)]/4. \quad [40]$$

For ideal δ pulses, $\delta = 0$, we see that we can achieve an effective line narrowing, $\bar{\mathscr{H}}_d^{00} = 0$, for $\alpha = 1/3$; this yields precisely the same timing as that of the WAHUHA cycle and has been verified experimentally, yielding decay times of 2 msec on the ^{19}F spins of CaF_2. Just as in this latter cycle the effects of finite pulse width can be compensated for. If we wish to retain 90° pulses then this is easily done by varying α (as long as $\delta < 2/3$) using [40]. The observed change in effective decay times going off resonance is quite marked for this pulse sequence. Magic echoes may be produced just as in the two-pulse cycle by making $\bar{\mathscr{H}}_d^{00}$ "negative."

SUMMARY

We have attempted to present a clear picture of the additional averaging effects produced by resonance offset fields in multiple-pulse NMR, and have illustrated this with some simple examples. In particular, the theory shows that behavior at, and far from resonance may be distinctly different, and that calculations made disregarding the resonance offset field lead to erroneous results (*30*). In addition, several uses of this phenomenon including line narrowing, spin locking, and magic echoes have emerged. We conclude by pointing out two other possible applications of this phenomenon in the future.

(i) Design of more efficient pulse sequences for line narrowing and magic echoes. The discussion in this paper has centered on zero-order effects, but there are equally profound higher-order effects which can be accounted for by the theory. These effects may in some cases outweigh the conclusions drawn from symmetry considerations alone in designing multiple-pulse experiments (*7, 8, 27, 31, 32*).

(ii) This type of experiment provides a simple means of producing effective static fields in the rotating frame with arbitrary directions and magnitudes, and with modified dipolar interactions, while being able to observe the magnetization between pulses. These facts give it an appealing potential for application to double-resonance experiments (*33, 34*) which have stirred up some interest among chemists since their recent adaptations to high resolution NMR of dilute spins in solids (*35–37*).

ACKNOWLEDGMENT

We are grateful to J. D. Ellett, Jr. for many helpful conversations and for his aid with some aspects of the theory and experiments.

REFERENCES

1. U. Haeberlen and J. S. Waugh, *Phys. Rev.* **175,** 453 (1968).
2. E. D. Ostroff and J. S. Waugh, *Phys. Rev. Lett.* **16,** 1097 (1966).
3. J. S. Waugh, C. H. Wang, L. M. Huber, and R. L. Vold, *J. Chem. Phys.* **48,** 662 (1968).
4. P. Mansfield and D. Ware, *Phys. Lett.* **22,** 133 (1966).
5. J. S. Waugh, L. M. Huber, and U. Haeberlen, *Phys. Rev. Lett.* **20,** 180 (1968).
6. M. Mehring, A. Pines, W.-K. Rhim, and J. S. Waugh, *J. Chem. Phys.* **54,** 3239 (1971).
7. M. Mehring and J. S. Waugh, *Phys. Rev.* (in press).
8. (a) W.-K. Rhim, A. Pines, and J. S. Waugh, *Phys. Rev. Lett.* **25,** 218 (1970); (b) W.-K. Rhim, A. Pines, and J. S. Waugh, *Phys. Rev.* **3,** 684 (1971).
9. A. Pines and J. S. Waugh (to be published).
10. J. D. Ellett, Jr. and J. S. Waugh, *J. Chem. Phys.* **51,** 2851 (1969).
11. U. Haeberlen, J. D. Ellett Jr., and J. S. Waugh, *J. Chem. Phys.* **55,** 53 (1971).
12. U. Haeberlen and J. S. Waugh, *Phys. Rev.* **185,** 420 (1969).
13. J. H. Van Vleck, *Phys. Rev.* **74,** 1168 (1948).
14. A. G. Redfield, *Phys. Rev.* **98,** 1787 (1955).
15. (a) W. I. Goldburg and M. Lee, *Phys. Rev. Lett.* **11,** 255 (1963); (b) M. Lee and W. I. Goldburg, *Phys. Rev.* **140,** A1261 (1965).
16. D. Barnaal and I. J. Lowe, *Phys. Rev. Lett.* **11,** 258 (1963).
17. J. D. Ellett Jr., M. G. Gibby, U. Haeberlen, L. M. Huber, M. Mehring, A. Pines, and J. S. Waugh, "Advances in Magnetic Resonance," Vol. 5, Academic Press, New York, 1971.
18. A. Pines, J. D. Ellett Jr., and J. S. Waugh (unpublished).
19. (a) I. Solomon, *C.R.H. Acad. Sci.* **248,** 92 (1950); (b) S. R. Hartman and E. L. Hahn, *Bull. Amer. Phys. Soc.* **5,** 498 (1960).
20. $T_{\parallel}$ and $T_{\perp}$ refer to the two components of magnetization respectively parallel and perpendicular to the effective field in the rotating frame.
21. I. Solomon and J. Ezratty, *Phys. Rev.* **127,** 78 (1962).
22. B. N. Provotorov, *Z. Eks. Teor. Fiz.* **41,** 1582 (1961).
23. M. Goldman, "Spin Temperature and Nuclear Magnetic Resonance in Solids," Oxford University Press, London, 1970.
24. H. Betsuyaku, *J. Phys. Soc. Japan* **30,** 641 (1971).
25. I. Solomon, *Phys. Rev. Lett.* **2,** 301 (1959).
26. W.-K. Rhim and H. Kessemeier, *Phys. Rev.* **3,** 3655 (1971).
27. A. Pines, W.-K. Rhim, and J. S. Waugh, *J. Magn. Resonance* (in press).
28. A. Abragam, "The Principles of Nuclear Magnetism," Clarendon, London, 1961.
29. J. S. Waugh and L. M. Huber, *J. Chem. Phys.* **47,** 1862 (1967).
30. This is similar to the trouble encountered when there are microscopic inhomogeneities stronger than the spin–spin interactions. The effects of the inhomogeneities on the spins cannot be accounted for independently, but modulate the spin–spin interactions, making additional truncation or averaging of these interaction terms a necessity: D. Hone, V. Jaccarino, T. Ngwe, and P. Pincus, *Phys. Rev.* **186,** 291 (1969).
31. P. Mansfield, *J. Phys. C* **4,** 1444 (1971).
32. This fact probably explains the observation by us that there is no increase in line narrowing efficiency going from the four-pulse cycle of Ref. 5 to the symmetrical four-pulse cycle of Ref. 8b.
33. S. R. Hartmann and E. L. Hahn, *Phys. Rev.* **128,** 2042 (1962).
34. F. M. Lurie and C. P. Slichter, *Phys. Rev.* **133,** A1108 (1964).
35. H. E. Bleich and A. G. Redfield, *J. Chem. Phys.* **55,** 5405 (1971).
36. P. Mansfield and P. K. Grannell, *J. Phys. C* 1197 (1971).
37. (a) A. Pines, M. G. Gibby, and J. S. Waugh, *Bull. Amer. Phys. Soc.* **16,** 1403 (1971); (b) A. Pines, M. G. Gibby, and J. S. Waugh, *J. Chem. Phys.* **56,** 1776 (1972).

JOURNAL OF MAGNETIC RESONANCE **13**, 243–248 (1974)

Correlation NMR Spectroscopy*

In 1965 Ernst and Anderson (*1*) showed that, under suitable conditions of excitation, a nuclear spin system can be treated as a linear system, thereby showing a way to obtain an order of magnitude in enhancement of sensitivity in high-resolution pulsed NMR experiments. However, the problems associated with the high peak radio frequency power required for adequate excitation bandwidth and the special filtering required to define the receiver bandwidth led to the exploration of excitations other than pulses. One of those was the stochastic excitation proposed by Ernst (*2*) and Kaiser (*3*). Some of the properties desired in an ideal excitation are low peak power, readily variable bandwidth, and uniform power spectral density in the specified bandwidth. We thus require a function which has an autocorrelation function of the form $\sin(x)/x$. A fast linear sweep meets all of these requirements almost ideally.[1] We have previously reported the use of a fast linear sweep for signal to noise enhancement (*4*) where the spectrum was obtained by cross-correlation of a continuous wave spectrum recorded in fast passage with a single reference line recorded under the same conditions. As we learned later, a similar technique was used by Peterson (*5*) for filtering NMR spectra. Here we wish to discuss the use of a calculated reference line, clarify the physical interpretation of the experiment, and present one experimental example of the use of the technique.

If we express the time function which describes a circularly polarized radio frequency field which has a frequency varying at a constant rate (a), we obtain Eq. [1]:

$$u_i(t) = A\exp(jat^2/2). \qquad [1]$$

The Fourier transform of $u_i(t)$ is given by Eq. [2],

$$\int_{-\infty}^{\infty} u_i(t)\exp(-jwt)dt = A(2\pi/a)^{1/2}\exp[j(\pi/4 - w^2/2a)] \qquad [2]$$

and implies a constant energy spectral density equal to $A^2 2\pi/a$. If we limit the sweep so that w runs from w_p to w_q, then the spectrum of $u_i(t)$ is given by Eq. [3]:

$$\text{F.T.}\{u_i(t)\} = A(\pi/2a)^{1/2}\exp[j(\pi/4 - w^2/2a)]\cdot[\text{erf}(z_q) - \text{erf}(z_p)], \qquad [3]$$

where erf(z) is the error function with a complex argument and $z_n = (2a)^{-1}(w_n - w)\exp(-j\pi/4)$. Investigation of Eq. [3] will show that the power spectrum is approximately constant for w between w_p and w_q and approximately zero outside that interval. The transition regions near w_p and w_q are of the order of $1/|t_p - t_q|$ in width. For the purposes

* Presented in part at the 5th International Conference on Magnetic Resonance in Biological Systems, New York, NY, December 1972, and at the 14th Experimental NMR Conference, Boulder, CO, April 1973.

[1] Another way to obtain such a function is to synthesize it from the prescribed power spectral density as was suggested by B. L. Tomlinson and H. D. W. Hill, *J. Chem. Phys.* **59**, 1775 (1973).

of analysis, it will be easier to follow the experiment for a single sweep from $w = -\infty$ to $w = +\infty$ and then investigate the influence of a finite sweep width.

The fast linear sweep excitation can be used in two ways. First, it can replace the pulse of a standard Fourier transform pulse spectrometer or the stochastic excitation of a stochastic resonance Fourier transform spectrometer. In order to accumulate repeated spectra, we need a phase lock between the swept frequency and the constant reference frequency supplied to the synchronous detector. The spectrum $[H(w)]$ is recovered from the Fourier transform of the response $[U_0(w)]$ through the relationship given by Eq. [4],

$$H(w) = U_0(w)/U_i(w), \qquad [4]$$

where $U_i(w)$ can be readily calculated from Eq. [2]. Second, it can replace the slow sweep in a conventional continuous wave spectrometer which in addition incorporates a cross-correlator. This configuration has many new and interesting properties which follow

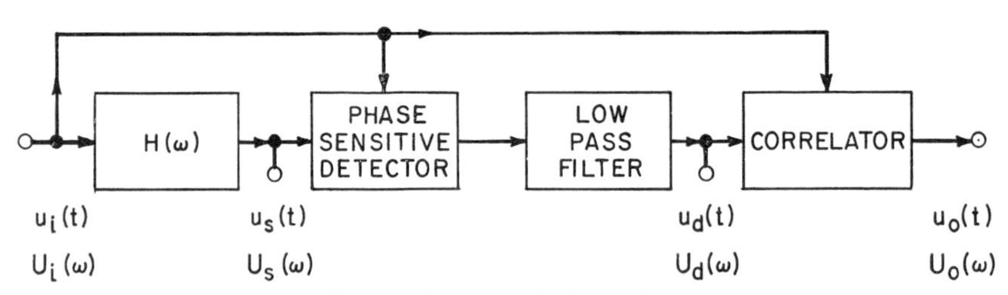

FIG. 1. A block diagram for a practical correlation spectrometer.

from the analysis of signal flow in the spectrometer represented by Fig. 1. The time functions at the indicated points in the spectrometer are related through the expressions given in Table 1. The corresponding relationships in the spectral representation are also given there, but in a condensed notation in which * represents the complex conjugate and ⋆ represents correlation. From the expressions for $u_0(t)$, we see that such

TABLE 1

SIGNALS PRESENT IN A PRACTICAL CORRELATION SPECTROMETER IN TIME AND FREQUENCY DOMAINS

k	$u_k(t)$	$U_k(w)$
i	$\exp(jat^2/2)$	$(2\pi/a)^{1/2}\exp(j\pi/4 - jw^2/2a)$
s	$\int_{-\infty}^{\infty} u_i(x)h(t-x)dx$	$U_i(w)\cdot H(w)$
d	$u_i(t)\cdot u_s^*(t)$	$U_i(w) \star U_s(w)$
o	$\int_{-\infty}^{\infty} u_i(x)u_d^*(x+t)dx =$	$U_i(-w)\cdot U_d^*(-w) =$
	$(2\pi/a)^{1/2}\exp(j\pi/4)H(at) =$ $c\cdot H(w)$	$(2\pi/a)^{3/2}\exp(j\pi/4)h(-w/a) =$ $(2\pi c/a)h(-t)$

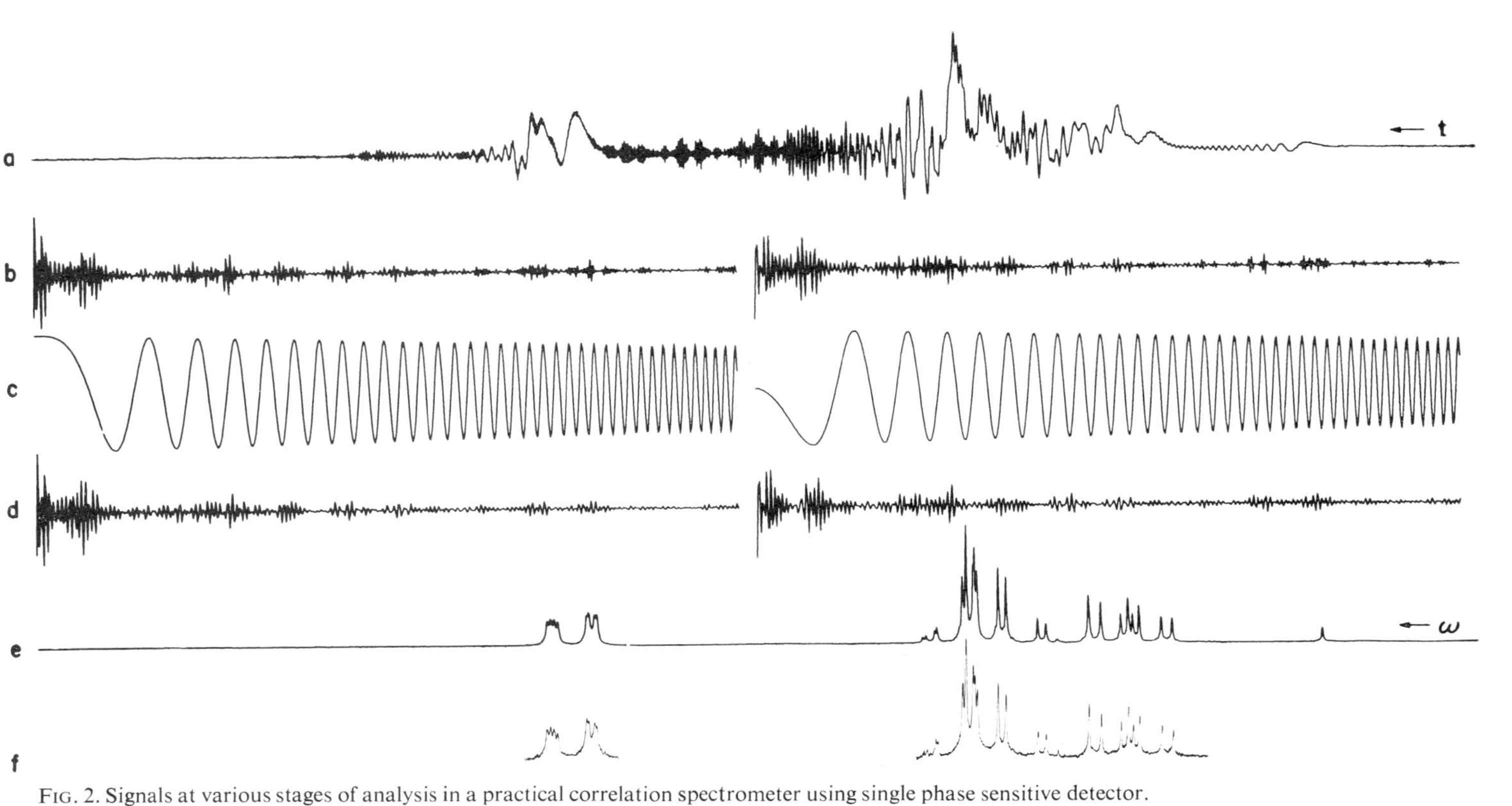

FIG. 2. Signals at various stages of analysis in a practical correlation spectrometer using single phase sensitive detector.

(a) Thirty accumulated scans of the NMR spectrum of *o*-chloronitrobenzene, 50 mg/ml in $CDCl_3$, TMS. Sweep width = 290 Hz, sweep time = 12 sec, time constant = 10 msec $[\mathrm{Re}\, u_d(t)]$.

(b) Fourier coefficients of (a). The first 500 coefficients are plotted. The left half are the cosine terms, the right half are the sine terms. $[(1/2)U_d(w)]$.

(c) Terms derived from Eq. [2] displayed as in (b). $[\exp(-j(w^2/2a))]$.

(d) The product of (b) and the complex conjugate of (c) displayed as in (b). $[h^*(w/a)]$.

(e) The inverse transformation of (d), a real function plotted at 4096 points. $[\mathrm{Re}\, H(w)]$.

(f) A slow passage spectrum of the same sample as (a) with sweep time ~5000 sec.

a device can yield $H(w)$ directly. That function is related to Green's function $h(t)$ for the system by a Fourier transformation.

In a practical spectrometer, one might decide to accumulate $u_d(t)$ as a sampled function and average it over repeated scans of $u_i(t)$ in order to improve the signal to noise ratio. In such a case, $u_d(t)$ must be coherent from sweep to sweep and this is guaranteed by the synchronous detector. Once one has the function $u_d(t)$ in sampled numerical

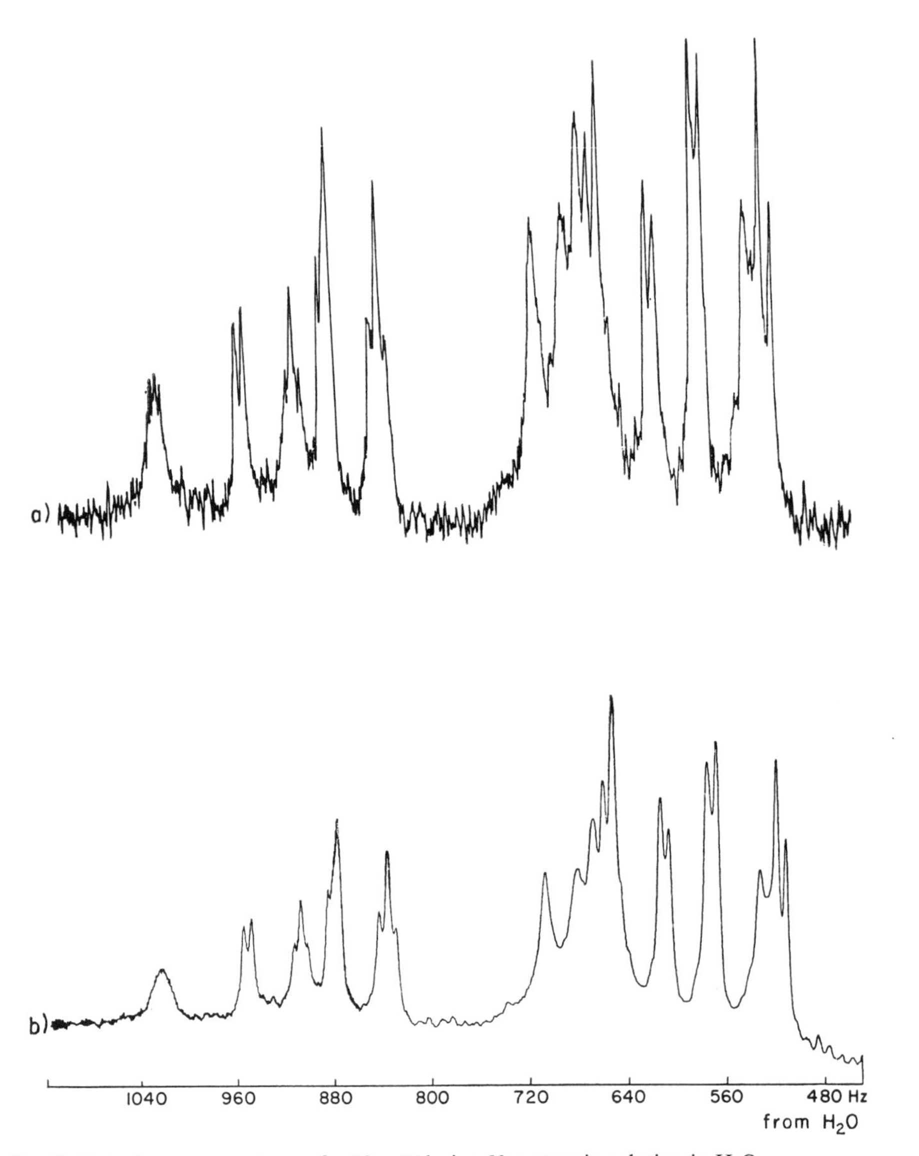

FIG. 3. Part of proton spectrum of a 10 m*M* lysine–Vasopressin solution in H_2O.
(a) Single scan CW, 900 Hz in 250 sec.
(b) 83 scans at 3 sec in correlation mode, sensitivity enhanced by applying calculated reference line of 0.75 Hz line width.

form, the correlation is readily accomplished using Fourier transformations, as indicated in Table 1, and $U_i(-w)$ as given by Eq. [2]. We obtain $U_0(w)$ as the intermediate product. It has the same form and meaning as $h(t)$ has in a pulsed NMR experiment, and therefore at this point, we may use any of the resolution enhancement functions or filter functions which are routinely used in pulsed Fourier transform spectroscopy. The simplest treatment could be an exponential filtering. Then we would recognize the complex conjugate of the Fourier transform of a reference line at zero frequency with a linewidth of $2w_2$ in the partial product of $U_i(-w)$ and the filter function of the form $\exp(-w_2 w/a)$. The response of each line in the spectrum is transformed by the synchronous detector to a Jacobsohn–Wangsness decay (*6*), so each line suffers the same distortion by the low-pass filter. If the filter is used only to eliminate signals which violate the Nyquist criterion, the distortion is negligible in most instances. An incorrect setting of the phase at the synchronous detector by ϕ_d may be compensated by multiplication of the intermeditae product by $\exp(-j\phi_d)$. The transition to a discrete Fourier transformation poses no problems if the Nyquist criterion is imposed on $u_a(t)$ by a suitable cutoff frequency on the low-pass filter, and the signal from the last line decays to a small value before the sweep ends. In the derivation of the equations in Table 1, we have assumed a complex signal, implying a spectrometer which contains two phase detectors. It is easy to prove that a single detector will supply the same information with a loss in signal to noise ratio of 3 db.

Proton spectra were recorded at 250 MHz on the time sharing spectrometer at the NIH Facility for Biomedical Research in Pittsburgh, Pennsylvania (*7*). The spectrometer was operated in the linear frequency sweep mode with an internal homonuclear lock on TMS. The sweep was under program control using a Xerox Data Systems (XDS) Sigma 5 computer with a 15-bit digital to analog converter. The data were collected at the output of a single Princeton Applied Research Model 121 synchronous detector using the 12 db per octave low-pass filter in that device, and an XDS Model MD 51 analog to digital converter. A total of 4096 sample points were collected on each sweep and accumulated in memory. The data were analyzed[2] using a fast Fourier transform program by Singleton (*8*). The resulting spectrum was displayed on a Cal Comp Model 564 drum plotter. The slow passage spectrum was recorded by the continuous wave technique for comparison purposes. The results are shown at various stages of the analysis in Fig. 2.

We feel that this new technique, which has been used now for over one year in our laboratory, offers several advantages over pulse techniques. The most important of them is the readily created rectangular power spectrum with practically no power outside the sweep width. This makes the accumulation of spectra in the presence of strong solvent lines without overdriving the dynamic range of the spectrometer rather easy (Fig. 3). If the sweep time chosen is of the order of the relaxation time T_1, then the overall optimum sensitivity of the Correlation NMR Spectroscopy is similar to the sensitivity of Pulsed FT Spectroscopy. The same is true for the Fast Sweep FT Spectroscopy based on Eq. 4. We hope to explore these and other questions in a full paper.

ACKNOWLEDGMENTS

We would like to thank the Division of Research Resources of the National Institutes of Health for

[2] A Fortran subroutine which transforms $U_a(w)$ to $U_0(w)$ is available from RFS on request.

continued support through Grant No. RR00292, and one of us (RFS) would like to thank the Sarah Mellon Scaife Foundation for a generous unrestricted grant.

REFERENCES

1. R. R. Ernst and W. A. Anderson, *Rev. Sci. Instr.* **37,** 93 (1966).
2. R. R. Ernst, *J. Magn. Resonance* **3,** 10 (1970).
3. R. Kaiser, *J. Magn. Resonance* **3,** 28 (1970).
4. J. Dadok and R. F. Sprecher, Paper presented at the 13th Experimental NMR Conference, Asilomar, CA, April 1972.
5. G. A. Petersson, Thesis, California Institute of Technology, 1970.
6. B. A. Jacobsohn and R. K. Wangsness, *Phys. Rev.* **73,** 942 (1948).
7. J. Dadok, R. F. Sprecher, A. A. Bothner-By, and T. Link, Paper presented at the 11th Experimental NMR Conference. Pittsburgh, PA, April 1970.
8. R. C. Singleton, *IEEE Trans. Audio and Electroacoust.* **AU-17,** 93 (1969).

Josef Dadok
Richard F. Sprecher

Carnegie-Mellon University
Pittsburgh, Pennsylvania 15213

Received November 9, 1973

JOURNAL OF MAGNETIC RESONANCE 14, 160–169 (1974)

Comparisons of Quadrature and Single-Phase Fourier Transform NMR

E. O. STEJSKAL AND JACOB SCHAEFER

Monsanto Company, Corporate Research Department, 800 N. Lindbergh Boulevard, St. Louis, Missouri 63166

Received November 14, 1973

Reflections of lines in time-averaged Fourier transform NMR spectra obtained by imperfect quadrature detection can be removed either by 90° rf phase shifting of the excitation pulse together with simple data routing, or by phase and amplitude manipulations of the two imperfect free induction decays after completion of data accumulation. The sensitivity of the resulting reflection-free spectra is comparable to that of spectra obtained by single-phase detection using a rf crystal filter to remove aliased noise. However, since the quadrature experiment involves data sampling rates half as fast as the comparable single-phase experiment, distortions associated with the finite recovery time of the spectrometer are less severe. Even for fast-recovery spectrometers, asymmetric spectral distortions in Fourier transform experiments involving moderately broad lines can occur. In quadrature detection schemes, these distortions are less severe since they are not folded through the entire spectral width as occurs in single-phase experiments.

INTRODUCTION

A quadrature NMR experiment involves the simultaneous detection of the absorption and dispersion components of the magnetization. This procedure results in several clear advantages (*1*) over more conventional single-phase experiments. For example, in Fourier transform experiments, the excitation pulse can be placed in the center of the anticipated spectrum for the more efficient use of the available rf power. Ideal quadrature detection then eliminates the aliasing of NMR lines which would occur in a single-phase experiment in this situation. Aliasing of noise is also eliminated leading to a $\sqrt{2}$ improvement in sensitivity over an otherwise comparable conventional single-phase experiment. Since the excitation pulse can be placed within the spectrum (as opposed to either the extreme left or right), a wide variety of long pulse experiments (pulse width $\gtrsim 1$/spectral width) can be simply performed for either suppression of solvent lines without cross-relaxation effects (*2*) or for the investigation of the structure of heterogeneous line shapes (*3*).

Despite its use to advantage in some laboratories (*1, 4*), quadrature detection for high-resolution Fourier transform NMR has not received widespread acceptance. This situation can probably be attributed to two factors. The first is the fear that unless the absorption and dispersion channels of a quadrature detector are identically matched in amplitude, and exactly out of phase by 90°, the reflections (*1*) of lines which would result would hopelessly complicate the spectrum. The second factor is the feeling that the $\sqrt{2}$ gain in sensitivity can be more simply achieved in a single-phase experiment by the use of an appropriate rf crystal filter to eliminate that noise which would normally fold over and be detected in a conventional single-phase experiment (*5*).

In this paper we show that the reflections in quadrature experiments (using unmodified, commercially available components) are at the 1% level. Furthermore, we show that even these reflections can be removed either by straightforward phase shifting of the exciting rf pulses together with appropriate routing of the outputs of the two channels of the quadrature detector (*6*), or by simple manipulations of the two imperfect free induction decays after completion of data acquisition. We also show that, since the sensitivity achieved by quadrature detection can be matched by single-phase detection (with rf crystal filter) only by doubling the sampling rate used to acquire the free induction decay, for a sufficiently wide spectral width (related to the recovery time of the spectrometer) the single-phase detection will produce spectral distortions while the quadrature detection will not. Finally, we show that even for a perfect spectrometer, with infinitely fast recovery time, sizeable distortions in transformed spectra can still result when dealing with moderately broad Lorentzian lines ($T_2 \lesssim 50$ times the data sampling time). These distortions result from the incorrect characterization of the dispersion of the line and are generally more serious in single-phase experiments due to the spectral folding required to remove reflections.

EXPERIMENTAL DETAILS

Natural abundance ^{13}C Fourier transform NMR experiments were performed at 22.6 MHz using a modified Bruker HFX-90 spectrometer. The modifications included addition of field-frequency stabilization by an external ^{19}F lock (to permit use of microcells) and replacement of the intermediate frequency section of the receiver (to permit use of quadrature detection). The quadrature phase detector consisted of a Merrimac PCM-3 miniature phase comparator (Merrimac Research and Development, Inc., West Caldwell, NJ) operating at 2.05 MHz, specially adjusted by the manufacturer to within 1° of quadrature at the center of the passband and no worse than ±3° over the entire passband of 200 kHz. Each of the two outputs of the phase comparator were independently amplified by separate Princeton Applied Research PAR 113 preamplifiers, and digitized by a Nicolet SD 81/2 dual digitizer, the latter a part of a Nicolet 1085 data system. Audio filtering (with cutoff frequency matched to sampling time) of the inputs to the dual digitizer was provided by the independent channels of a Multimetric tunable four-pole Butterworth filter.

Complex Fourier transforms were performed by modifications of computer software provided by Nicolet Instruments. Two kinds of data routing schemes were used in these experiments. Both schemes involved a simple two-pulse cycle. In the first scheme,[1] after the first pulse, the absorption and dispersion components of the magnetization were the outputs of the A and B channels of the quadrature detector and were acquired in the first and second quadrants of the Nicolet 1085, respectively. The contents of these quadrants were then transferred to the third and fourth quadrants by automatic software command. After the second pulse (whose rf phase was shifted by 90°), the A and B outputs correspond to what had originally been the dispersion and the negative of the

[1] This routing program was written by Dr. James W. Cooper, Nicolet Instruments Corp., Madison, Wisconsin.

absorption component, respectively, and these were again acquired in the first and second quadrants, but were subtracted and added to the contents of the fourth and third quadrants, respectively. (See the top row of Fig. 1.) Thus, the first and second quadrants of the 1085 act as a buffer with the actual accumulation of data in the third and fourth quadrants. This simple two-pulse cycle is sufficient to make the two channels of the quadrature detector essentially equivalent, except for a 90° phase shift.[2]

In the second routing scheme, no buffer is required. After the first pulse, the absorption and dispersion components of the magnetization (A and B outputs of the quadrature detector) were acquired in the first and second halves of a Nicolet 1085 and

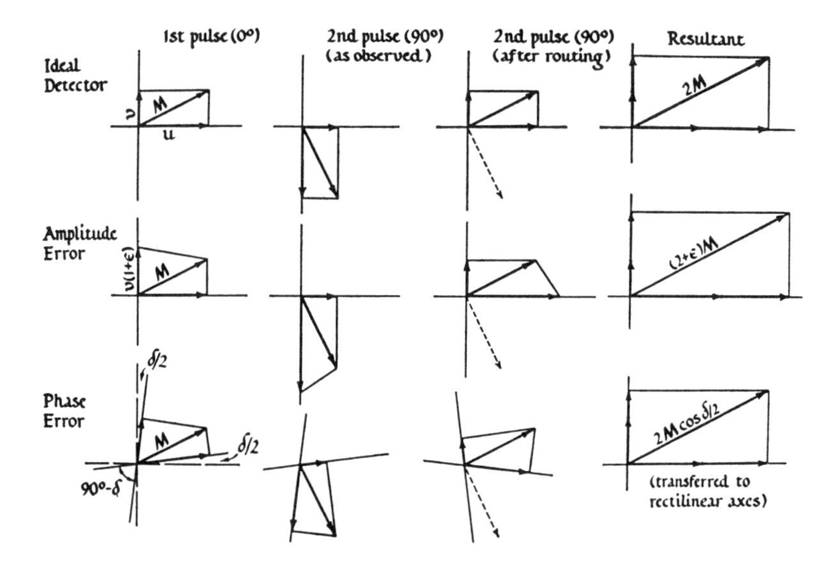

FIG. 1. Graphical representation of cancellation of quadrature errors by rf phase shifting and data routing.

immediately exchanged, with a sign change of what had been the absorption so as to prepare it for the second part of the sequence. After the second pulse (whose rf phase is shifted by 90°), the A and B outputs of the quadrature detector correspond to what had originally been the dispersion and the negative of the absorption, respectively, and these were added to the contents of the first and second halves of memory. (Again, the relationships between the signals following the two pulses are shown schematically in the simpler but equivalent routing scheme of the top row of Fig. 1.) The accumulated contents of the first and second halves were again exchanged, with a sign change of what had been in the second half of memory, thereby completing the cycle. The switching of halves of data back and forth allows the computer to accept data directly from the detector without a buffer (which the scheme of the top row of Fig. 1 does not permit) and ensures coherent accumulation of the signal as well as cancellation of quadrature errors. The cancellation of quadrature errors is illustrated in Fig. 1. The cancellation is complete unless *both* the quadrature detector and the rf phase shifter are imperfect, or

[2] A suggestion made to the authors by Prof. R. E. Richards and Dr. David Hoult, Oxford, England.

unless *both* phase and amplitude errors are present, in which case residual second-order errors remain. The switching of data requires 48 μsec per data location pair, or, for two 8K free induction decays, about a 400-msec delay. The first routing scheme, using the buffer and our present software, reduces this delay time by about 20%. For most applications, this kind of a delay time, amounting to about one-fourth of the time spent in data acquisition to produce a 5-kHz spectrum, is not a severe limitation.

Single-phase experiments were performed by using only one channel of the quadrature detector. Single-phase experiments with a rf crystal filter were performed by filtering the input to one channel of the quadrature detector using a 2.05-MHz crystal filter with a bandwidth of 7 kHz and characteristics typical of 4-pole Butterworth filters (McCoy Electronics Co., Mt. Holly Springs, Pa.). A 5-mm ^{13}C insert was used in these experiments (with a 10-μsec 90° pulse) accepting 100-μliter microcells (Wilmad Glass Co., Inc., Buena, N. J.).

RESULTS AND DISCUSSION

The natural abundance ^{13}C NMR spectra of 100 μliters of neat oleic acid obtained using single-phase detection with a rf crystal filter, using quadrature detection, and using conventional single-phase detection are shown in Fig. 2. The signal to noise ratio

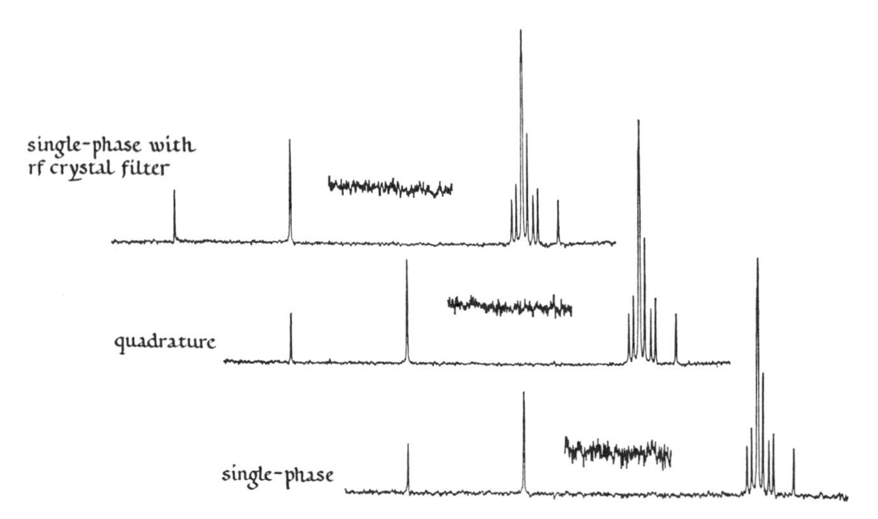

FIG. 2. Natural abundance ^{13}C NMR 5-kHz spectra of 100 μliters of neat oleic acid obtained by Fourier transform techniques using different detection schemes. Each spectrum is the result of the accumulation of 128 scans, obtained with 90° rf pulses and a 3-sec pulse repetition time.

of the spectra obtained by the first two detection schemes are comparable, and about 1.5 times greater than that obtained by conventional single-phase detection. Except for the relative sensitivity, all three spectra are of comparable quality and appearance. Since there is no aliased noise in a quadrature experiment, the use of a rf crystal filter *with* quadrature detection will not produce further improvements in sensitivity.

The quadrature experiment resulting in the spectrum of Fig. 2 utilized data routing and a two-pulse cycle to eliminate any slight differences between the two channels of the quadrature receiver. Actually, even without the two-pulse cycle, a quadrature detector accurate to 1° in phase and 1% in amplitude (reasonable specifications for commercially available components) does not produce visible reflections in spectra of the quality of Fig. 2.

Visible reflections do occur when phase errors are on the order of 15°, or when relative amplitude errors between the two outputs of the quadrature detector are on the order of 30%. These situations are illustrated in Fig. 3. The reflections of the stronger lines are now comparable in size to some of the weaker lines. The phases of the reflected lines depend upon their origin as either phase or amplitude errors, and upon the phasing of the original spectrum. These reflections can be removed by performing the original quadrature experiment using the two-pulse cycle described above. They can also be

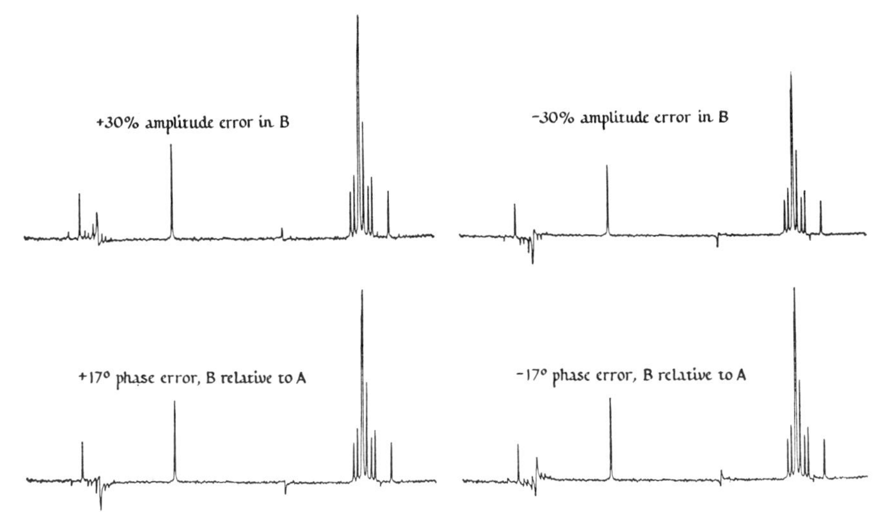

FIG. 3. Complex Fourier transforms of imperfect quadrature data of oleic acid.

removed from the imperfect data by simple amplitude and phase corrections of the free induction decays before Fourier transformation.

To see the origin of this correction process, consider the complex Fourier transforms of Fig. 4. The top spectrum contains the 10-kHz real and imaginary components resulting from a complex Fourier transform of only the absorption (A) output of a quadrature detector. The free induction decay associated with the dispersion has been zeroed. The spectrum is reflected about its center, with the reflected lines 180° out of phase with the true lines. When only the dispersion (B) output of the quadrature detector is transformed, a similar spectrum results, except that now the reflected lines have the same phase as the true lines. The sum of these two spectra is a 10-kHz spectrum, containing real and imaginary components, but with no reflections, and with a $\sqrt{2}$-improved sensitivity. Considering the cancellation of reflections, it is clear that small reflected lines in spectra obtained by complex Fourier transforms of quadrature data can occur if the absorption and dispersion free induction decays are not identical in amplitude, or exactly out of phase by 90°. If, therefore, reflections do occur in a spectrum, it is possible to remove them by returning to the original free induction decays and either generating *one* new component by shifting one of the old components into the other (by the constant phase correction software command normally used with the transformed data), or by changing the size of one component relative to the other, or by a combination of both of these procedures. Thus, in effect, a spectrum with reflections equal in magnitude but opposite in phase to the reflections in the original spectrum can be generated. The sum of the two will be a spectrum free of all reflections.

This postcorrection removal of reflected lines is essentially equivalent to removal by

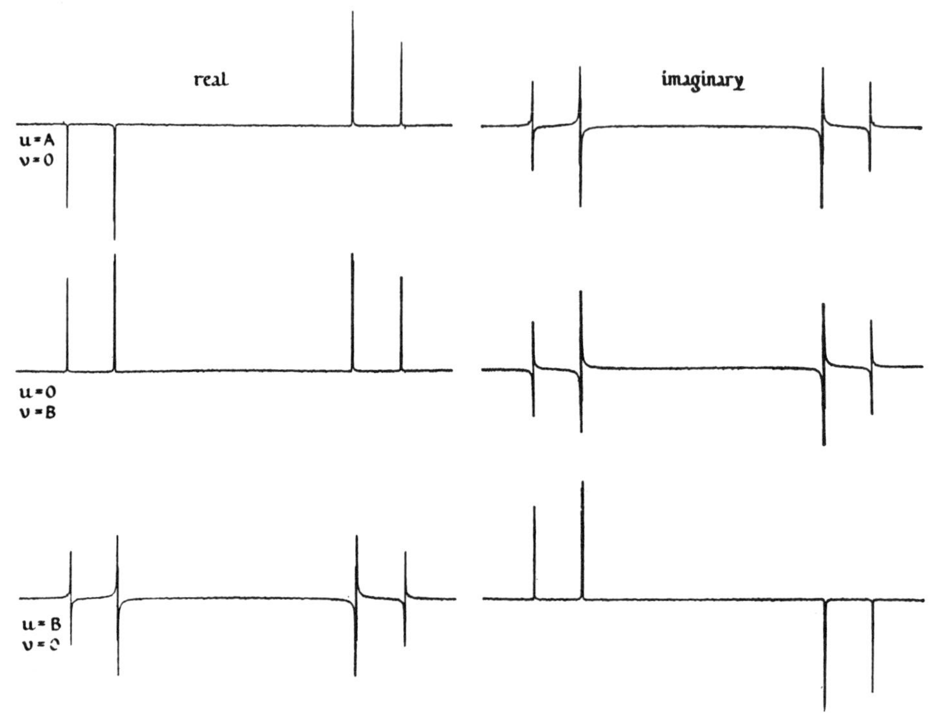

FIG. 4. Complex Fourier transforms of quadrature data (with one component intentionally set to zero) of propylene glycol, $HOCHCH_3CH_2OH$.

rf phase shifting and the two-pulse cycle. In both cases, the resulting spectra are free of reflections and, to the extent that the corrections were sizeable, reduced in sensitivity from what would be obtained from a perfect quadrature detector. (The two-pulse cycle will also partially remove systematic noise within the spectrometer, while the post-correction scheme will not. Furthermore, the two-pulse cycle will remove worst possible quadrature errors, e.g., one channel with zero output, while the postcorrection scheme will not.) One can envision an iterative procedure for postcorrection removal of reflected lines, under computer software control, which is analogous to the familiar computer-controlled constant and linear phase corrections of transformed data. Of course, if the imperfections of a particular quadrature detector are stable, once the imperfections have been determined, phase-shifting and amplitude corrections of the free induction decays will be the same from one spectrum to the next.

Actually, at least four schemes for the removal of reflected lines are possible: (1) rf phase shifting and the two-pulse cycle with its necessary delays while accumulated data are being rearranged; (2) computer-controlled postcorrections, which involve no delays during data accumulation, but do involve some time spent in correcting the final spectrum; (3) rf phase shifting and the two-pulse cycle with a synchronized switching circuit external to the data system to do the routing, which avoids the delay to rearrange data; and (4) rf phase shifting but without the two-pulse cycle. In this latter scheme, a 90° phase shift and a change in data routing would be introduced only after

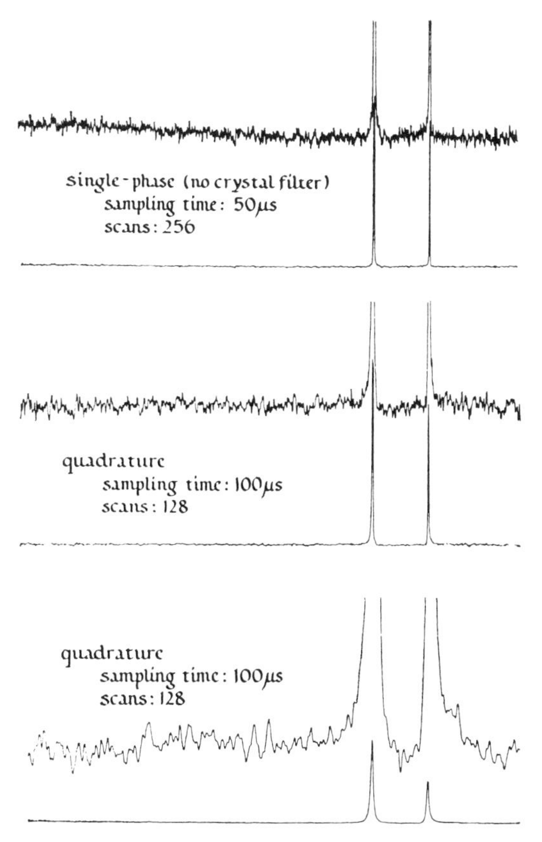

FIG. 5. Natural abundance ^{13}C NMR 10-kHz spectra of 100 μliters of neat propylene glycol obtained by Fourier transform techniques using different detection schemes and zero linear phase corrections.

half the desired number of scans had been completed. This routing avoids delays during data accumulation and requires no postcorrection manipulations. A spectrum free of reflections is obtained if the imperfections in the quadrature detector are constant throughout the experiment.

A complex Fourier transform algorithm is generally used to treat single-phase data. In this situation, what is done is to put the same data in both the absorption and dispersion channels. (Slight variations of this procedure can also be used.) The resulting complex Fourier transform will produce a spectrum containing real and imaginary components (thereby permitting phase corrections), and free of reflections, but only if the spectrum and its noise are folded back on themselves. This is illustrated by the sum of the middle and bottom spectra of Fig. 4, and is a familiar result. A key feature of the folding procedure is that it necessarily reduces the spectral width by one-half, relative to a quadrature experiment, even though the data acquisition sampling rates are identical (*1*): Consequently, in order for a single-phase experiment to produce the same spectral width as a quadrature experiment, a twofold-faster sampling rate must be used. This means that one will reach a sampling rate which is faster than the recovery time of the spectrometer sooner in doing a single-phase experiment than in a quadrature experiment. Thus, distortions in spectra from imperfections in data due to limitations in the spectrometer are necessarily more common in single-phase experiments. For our high-resolution spectrometer, for example, the limit in distortion-free data acquisition sampling time is about 100 μsec. With a careful choice of filters and delay times, we can obtain a distortion-free 10-kHz spectrum using quadrature detection and a 100-μsec sampling time, but not using single-phase detection and a 50 μsec sampling time. This situation is illustrated by the distorted sloping baseline of the upper spectrum of Fig. 5, compared to the flat baselines of the bottom two spectra, the latter obtained by quadrature detection.

Even for a perfect spectrometer, with an infinitely fast recovery time, coupled to a perfect data acquisition system, sizeable distortions can result in the Fourier transformed spectra of moderately broad Lorentzian lines. These distortions occur because of the foldover of the extensive wings of the dispersion of a Lorentzian line. If the spectrometer is operated in such a way that delays occur in the beginning of data acquisition relative to the rf excitation pulse, the linear phase correction which is necessary to ensure that all lines in the absorption spectrum will have the same phase regardless of frequency will, in general, also transfer some of the distortion of the dispersion of the spectrum into the absorption component. The net result is a distorted baseline, as illustrated in Fig. 6.

This spectrum was generated from a complex Fourier transform of artificial quadrature data. Artificial, noise-free data were produced by placing different baseline offsets in the usual absorption and dispersion quadrature locations in the computer memory and then performing a linear phase change. This produces sine and cosine functions in the two memory halves, respectively, whose frequency depends upon the size of the phase change. (The production of these modulations is, in fact, similar to the production of the baseline distortions themselves, the latter essentially due to a few incorrect data points resulting from the incompletely characterized dispersions of Lorentzian lines.) An exponential multiplication of the sine and cosine functions with an appropriate time constant, followed by Fourier transformation, then led to the Lorentzian line of the desired width. The phase of the line was chosen by various combinations of constant and linear phase corrections, just as would be done in phasing any line in a spectrum following Fourier transformation. Except for three very specific values of the linear phase correction, obvious distortions in the baseline occur. The most pronounced distortions occur at one end of the spectrum when the center of the line is at the other end. This is a consequence of the way spectral information outside the displayed spectral width is folded back in quadrature spectra, i.e., a line too far to the right reappears on the left.

The baseline distortions are more serious in single-phase experiments because the

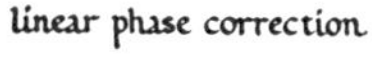

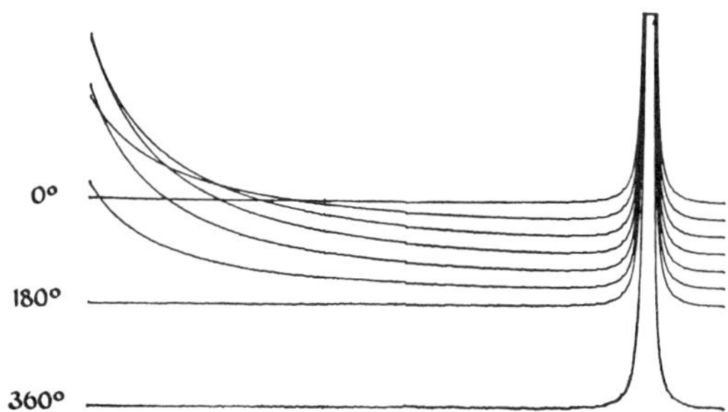

FIG. 6. Complex Fourier transforms of simulated quadrature data of a single Lorentzian line under different phase corrections. Between zero and 180°, the increment in the linear phase correction is 30°.

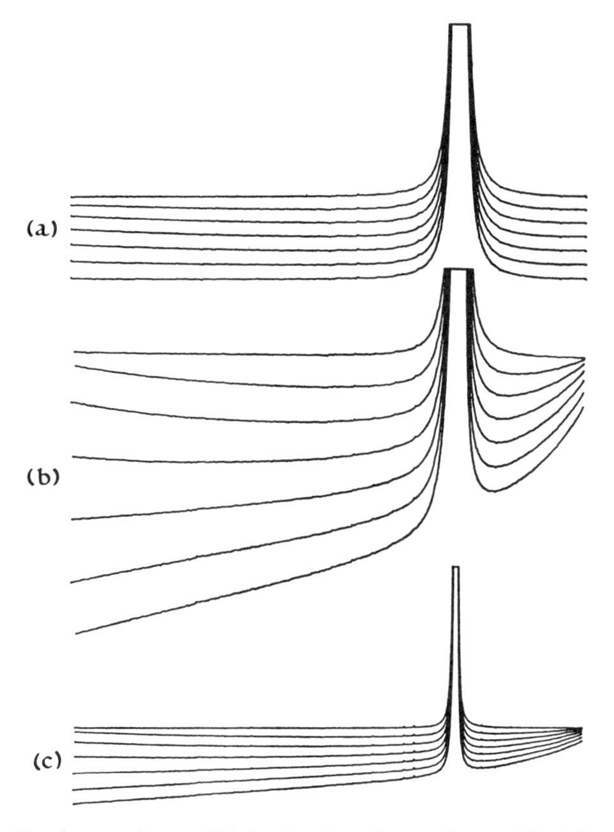

FIG. 7. Complex Fourier transforms of (a) simulated quadrature data and (b, c) simulated single-phase data of a single Lorentzian line with a linear phase correction varying from 0° to 180°, by 30° increments. The set of spectra shown in (a) is taken from the right-hand half of Fig. 6.

folding of the spectrum folds the distortions back upon the spectrum one more time. This point is illustrated in Fig. 7. The top set of spectra are the right-hand halves of quadrature spectra of Fig. 6, on an expanded frequency scale. The middle and bottom sets of spectra show Lorentzian lines of comparable and smaller widths, respectively, but obtained from Fourier transformation of single-phase data. The baseline distortions are pronounced in the latter two sets, while in the former they consist only of innocuous offsets of the baseline. Unlike the situation for quadrature detection, distortions in single-phase detected spectra occur for all nonzero values of the linear phase correction. The baseline distortions associated with the single-phase experiment contain excursions on the order of 1% of the height of the baseline to peak maximum, when T_2 of the line is comparable to 50 times the data sampling time. This kind of excursion is about two times the noise level of the bottom spectrum of Fig. 5. Baseline distortions such as these can, therefore, be real problems, especially in experiments involving single-phase detection of weak, broad Lorentzian lines in the presence of strong lines. They are not much of a problem in experiments involving either quadrature or single-phase detection of narrow lines, or distributions of narrow lines, of comparable intensities. Naturally, these distortions may, in fact, be small compared to those resulting from imperfections in the spectrometer itself, many of which can lead to distorted baselines.

REFERENCES

1. J. D. ELLETT, M. G. GIBBY, U. HAEBERLEN, L. M. HUBER, M. MEHRING, A. PINES, AND J. S. WAUGH, *Advan. Magn. Resonance* **5,** 117 (1971).
2. A. G. REDFIELD AND R. K. GUPTA, *J. Chem. Phys.* **54,** 1418 (1971).
3. J. SCHAEFER, *Macromolecules* **5,** 427 (1972).
4. A. G. REDFIELD AND R. K. GUPTA, *Advan. Magn. Resonance* **5,** 82 (1971).
5. A. ALLERHAND, R. F. CHILDERS, AND E. OLDFIELD, *J. Magn. Resonance* **11,** 272 (1973).
6. E. O. STEJSKAL AND J. SCHAEFER, *J. Magn. Resonance* **13,** 249 (1974).

JOURNAL OF MAGNETIC RESONANCE **18**, 69–83 (1975)

NMR Fourier Zeugmatography

ANIL KUMAR, DIETER WELTI, AND RICHARD R. ERNST

Laboratorium für Physikalische Chemie, Eidgenössische Technische Hochschule, 8006 Zürich, Switzerland

Received August 2, 1974

A new technique of forming two- or three-dimensional images of a macroscopic sample by means of NMR is described. It is based on the application of a sequence of pulsed magnetic field gradients during a series of free induction decays. The image formation can be achieved by a straightforward two- or three-dimensional Fourier transformation. The method has the advantage of high sensitivity combined with experimental and computational simplicity.

I. INTRODUCTION

P. C. Lauterbur (*1*) has recently described an ingenious technique to determine one-, two- or three-dimensional images of the distribution of magnetic moments in a macroscopic sample. He calls such an image a zeugmatogram. The potential use of this method includes the measurement of the spatial distribution of a given nuclear species in living tissue and the determination of its relaxation times with the possibility of localizing cancerous parts in a living organism (*2*, *3*). The most important nuclei to be detected are the protons of water in biological materials.

Lauterbur's method (*1*) is based on the application of linear magnetic field gradients in different directions in a series of experiments. Each of the resulting spectra of a single resonance line represents a distribution function of the nuclei as a function of the local magnetic field strength. It can be considered as a projection of the three-dimensional nuclear spin density onto the axis along which the linear field gradient has been applied. From a sufficient number of such projections onto different axes, it is possible to partially reconstruct the two- or three-dimensional spin density function by means of well-known image reconstruction techniques (*4–7*). The inherent spatial resolution is determined by the number of independent projections obtained and by the natural width of the resonance line under consideration. The various linear gradients along different directions can be generated either by means of a set of suitable gradient coils or, more easily, by means of a single gradient coil and step-by-step rotation of the sample.

A modification of the Lauterbur technique (*1*) can be obtained by the application of Fourier spectroscopy techniques (*8*). For each gradient setting, a free induction decay signal (FID) is recorded. Except for improved sensitivity and for the elimination of one computational step (many image reconstruction techniques require as an intermediate step the Fourier transformation of the spectrum (*4*, *7*)), this method does not show advantages over the Lauterbur procedure and it will not be mentioned further.

In this paper, an alternative technique is described which is remarkable for its experimental and computational simplicity and by its inherent high sensitivity. It is based on the application of a sequence of pulsed orthogonal linear field gradients to the sample during the FID. The spatial spin density function can then be reconstructed by a straightforward two- or three-dimensional Fourier transformation. One of the important features of this method is the homogenous error distribution over the entire frequency range such that low and high frequency components can be reconstructed with equal accuracy. The method can easily be implemented on a small on-line computer.

II. MATHEMATICAL ANALYSIS OF THE TECHNIQUE

Although the experimental examples presented in this paper and probably many of the future applications of this technique will be confined to the two-dimensional imaging of a sample, the theory will be developed for the more general three-dimensional case.

The principle of the technique is explained by means of Fig. 1. At time $t = 0$, an FID is generated by means of a short 90° pulse. In the course of this decay, three orthogonal

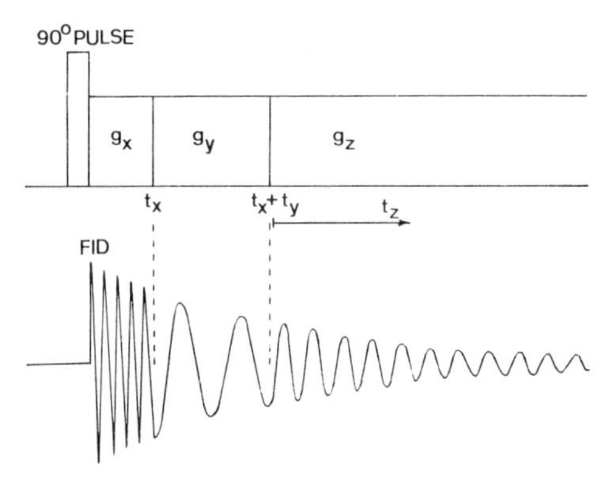

FIG. 1. Diagram depicting the principle of the zeugmatographic method. The FID is recorded during the third time-interval as a function of t_z. For a three-dimensional zeugmatogram, usually N^2 such FID signals will be needed for a complete set of values for x- and y-gradients. N is the number of samples in an FID.

linear magnetic field gradients, g_x, g_y, and g_z, are applied in succession. The z-component of the local magnetic field is then given by

$$H_z(\mathbf{r}) \begin{cases} = H_0 + g_x x, & \text{for} \quad 0 < t < t_x \\ = H_0 + g_y y, & \text{for} \quad t_x < t < t_x + t_y \\ = H_0 + g_z z, & \text{for} \quad t_x + t_y < t. \end{cases} \qquad [1]$$

The FID is sampled in the third time-interval as a function of $t_z = t - (t_x + t_y)$. It is at the same time a function of the preceding time-intervals t_x and t_y. It will be denoted by $s(\mathbf{t}) = s(t_x,t_y,t_z)$. The experiment is repeated for a full set of equally spaced t_x and t_y values. It will be shown in this section that the three-dimensional Fourier transform of $s(\mathbf{t})$ is a measure for the spatial spin density function $c(\mathbf{r}) = c(x,y,z)$ and provides a three-dimensional image of the sample.

The observed signal $s(\mathbf{t})$ is a composite of the contributions from the various volume elements of the sample and can be written as

$$s(\mathbf{t}) = \iiint c(\mathbf{r})\, s(\mathbf{r},\mathbf{t})\, dv \quad [2]$$

where $s(\mathbf{r},\mathbf{t})\, dv$ is the contribution from the volume element $dv = dx\, dy\, dz$ at position $\mathbf{r}$. For a single resonance, the function $s(\mathbf{r},\mathbf{t})$ can easily be found by solving Bloch equations. After phase-sensitive detection with the frequency ω_1, the signal is given by

$$s(\mathbf{r},\mathbf{t}) = M_0 \cos\{(\Delta + \eta_x x)t_x + (\Delta + \eta_y y)t_y + (\Delta + \eta_z z)t_z\} \cdot \exp\{-(t_x + t_y + t_z)/T_2\}. \quad [3]$$

The resonance offset in the absence of a field gradient is given by $\Delta = -\gamma H_0 - \omega_1$. The field gradients $\eta_k = -\gamma g_k$ are measured in frequency units. The setting of the phase-sensitive detector has been assumed arbitrarily to produce a cosine signal. It can easily be shown that the function $|\bar{c}(\mathbf{r})|$, which is plotted in a Fourier zeugmatogram, is independent of this arbitrary phase setting.

The three-dimensional Fourier transform of $s(\mathbf{t})$ is denoted by $S(\boldsymbol{\omega}) = S(\omega_x,\omega_y,\omega_z)$ and is given by

$$S(\boldsymbol{\omega}) = \iiint s(\mathbf{t}) \exp(-i\boldsymbol{\omega}\mathbf{t})\, dt_x\, dt_y\, dt_z. \quad [4]$$

It is again a composite of the contributions from the various volume elements,

$$S(\boldsymbol{\omega}) = \iiint c(\mathbf{r})\, S(\mathbf{r}, \boldsymbol{\omega})\, dv. \quad [5]$$

Here, $S(\mathbf{r},\boldsymbol{\omega})$ is the Fourier transform of $s(\mathbf{r},\mathbf{t})$ and is calculated to be

$$\begin{aligned} S(\mathbf{r},\boldsymbol{\omega}) = \tfrac{1}{2}\{&G(\Delta + \eta_x x - \omega_x)\, G(\Delta + \eta_y y - \omega_y)\, G(\Delta + \eta_z z - \omega_z) \\ &+ G(-\Delta - \eta_x x - \omega_x)\, G(-\Delta - \eta_y y - \omega_y)\, G(-\Delta - \eta_z z - \omega_z)\} \end{aligned} \quad [6]$$

with the complex line shape function

$$G(\omega) = A(\omega) + iD(\omega) = \frac{M_0/T_2}{(1/T_2)^2 + \omega^2} + i\frac{M_0\omega}{(1/T_2)^2 + \omega^2}. \quad [7]$$

The second term in Eq. [6] which describes the contribution of the resonance near $-\Delta$ can be neglected whenever the linewidth is small compared to Δ.

Equation [6] shows that the following identity holds,

$$S(\mathbf{r}, \boldsymbol{\omega}) = S(\mathbf{0}, \boldsymbol{\omega} - \eta\mathbf{r}) \quad [8]$$

where η is a diagonal matrix with the elements η_x, η_y, and η_z. One obtains for $S(\boldsymbol{\omega})$

$$S(\boldsymbol{\omega}) = \iiint c(\mathbf{r})\, S(\mathbf{0}, \boldsymbol{\omega} - \eta\mathbf{r})\, dv. \quad [9]$$

The frequency variable $\boldsymbol{\omega}$ will now be replaced by a spatial variable $\mathbf{r}'$ with

$$\boldsymbol{\omega} = \Delta\mathbf{I} + \eta\mathbf{r}'. \quad [10]$$

$\mathbf{I}$ is the vector (1, 1, 1). Then, one obtains

$$S(\boldsymbol{\omega}) = S(\Delta\mathbf{I} + \eta\mathbf{r}') = \bar{c}(\mathbf{r}') = \iiint c(\mathbf{r})\, S(\mathbf{0}, \Delta\mathbf{I} + \eta(\mathbf{r}' - \mathbf{r}))\, dv. \quad [11]$$

This integral is clearly a three-dimensional convolution integral. It represents a "filtered" spin density function $\bar{c}(\mathbf{r}')$ obtained from the original spin density function $c(\mathbf{r})$ by a convolution with the lineshape function $S(\mathbf{0}, \Delta\mathbf{I} + \eta\mathbf{r})$. By means of Eq. [6] and neglecting the contribution of the resonance near $-\Delta$, one obtains finally

$$\bar{c}(\mathbf{r}') = \tfrac{1}{2}\iiint c(\mathbf{r})\, G(\eta_x(x - x'))\, G(\eta_y(y - y'))\, G(\eta_z(z - z'))\, dv. \quad [12]$$

The filtered spin density function $\bar{c}(\mathbf{r}')$ is a complex function. Its real and its imaginary parts both contain products of absorption- and dispersion-like parts and can have positive and negative function values. It is, therefore, advisable to compute and plot the absolute value $|\bar{c}(\mathbf{r}')|$ rather than plotting $\mathrm{Re}\{\bar{c}(\mathbf{r}')\}$ or $\mathrm{Im}\{\bar{c}(\mathbf{r}')\}$. For a sufficiently narrow resonance line or for sufficiently strong gradients η_x, η_y, and η_z, $|\bar{c}(\mathbf{r}')|$ is a good measure for $c(\mathbf{r}')$ itself. In Section IV, a modified technique is described that permits complete separation of absorption and dispersion mode signals.

In principle, it is also possible to utilize a quadrature phase detector that produces at its output $s(\mathbf{t})$ as well as the quadrature component $s'(\mathbf{t})$ which is given by equations similar to Eqs. [2] and [3] where the cosine function is replaced by a sine function. A linear combination of the two signals permits complete elimination of the contributions of the resonance near $-\Delta$. But the absorption and dispersion parts are not separated and the final result is equivalent to Eq. [12] except for an improvement of the sensitivity by a factor $\sqrt{2}$.

It is a major feature of the described technique that it does not involve one-dimensional projections of the three-dimensional spin density and that the Fourier transform of the spin density is directly measurable (except for the filtering caused by the natural lineshape of the NMR signal). Many of the image reconstruction techniques which can be used for the Lauterbur procedure (*1*) utilize the fact that the Fourier transform of a one-dimensional projection of the spin density represents a one-dimensional cross section of the three-dimensional Fourier transform of the spin density function (*7*). All the cross sections that can be obtained in this way pass through the point $\boldsymbol{\omega} = \mathbf{0}$. The density of the obtained samples, therefore, is maximum for $\boldsymbol{\omega} = \mathbf{0}$ and decreases for increasing $|\omega|$. To obtain equally spaced samples representing the Fourier transform, it is at first necessary to go through an interpolation procedure. This is a prerequisite for the execution of the inverse Fourier transformation that produces the desired image. This ultimately implies that the low frequency components are obtained with higher precision than the high frequency components of the zeugmatogram. Therefore, the coarse features are better represented than the details. In some cases, this may be no disadvantage, and it may, in particular cases, even be desirable. This feature is inherent and is independent of the reconstruction procedure used. It also occurs in direct algebraic reconstruction techniques (*6*) that do not involve a Fourier transformation.

In the described Fourier technique, on the other hand, an equal sample spacing of the Fourier transform is automatically obtained. The error distribution of a Fourier zeugmatogram is therefore homogenous over the spatial and over the covered frequency range, in contrast to the Lauterbur procedure. Coarse features and details are obtained with the same accuracy.

III. EXPERIMENTAL PROCEDURE AND RESULTS

The predominant problem in practical applications of Fourier zeugmatography is the economy of data storage. Three-dimensional zeugmatograms can be obtained in exceptional cases only because of the enormous amount of data required. Therefore, only the two-dimensional case will be discussed.

The experiments to be described in this section have been performed with standard equipment available in our laboratory. It has not been optimized for this particular purpose and could be improved in many respects. The set-up is indicated in Fig. 2. It consists of a Varian high resolution 15-in. electromagnet with an 11-gradient shim system, a Bruker SXP4-100 high power pulse spectrometer with a Bruker single coil probe assembly for high power pulse experiments, and a Varian 620/L-100 computer system equipped with 12 k memory, a fast 12-bit analog-to-digital converter and a number of execute lines for the connection to the spectrometer.

The linear magnetic field gradients have been generated by means of the x- and z-gradient shim coils of the Varian shim system. The currents necessary to produce the linear field gradients have been generated by a set of external stable power supplies and

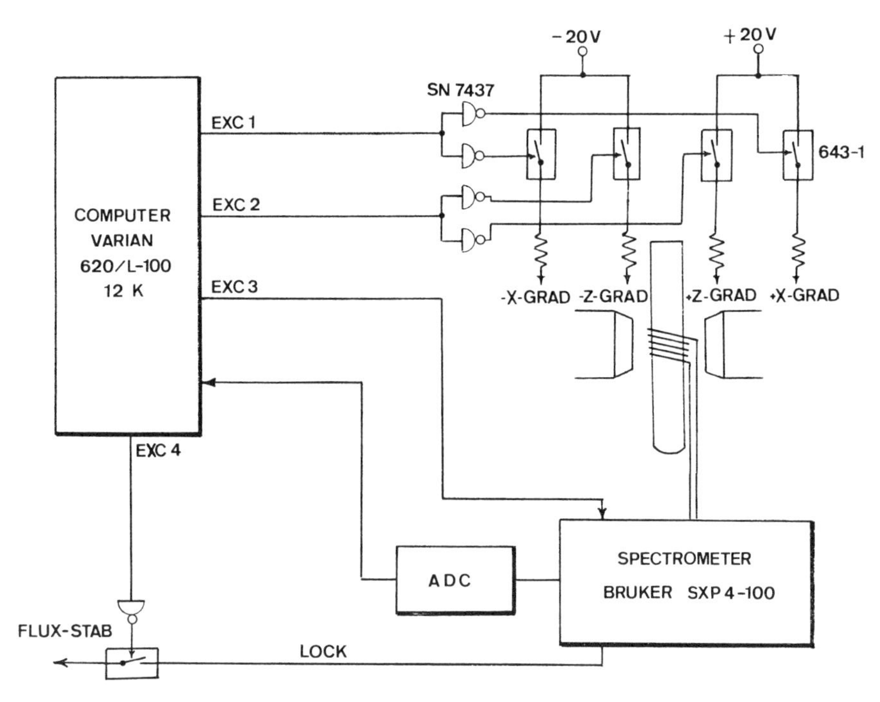

FIG. 2. Block diagram of the experimental set-up.

are computer-controlled by means of a series of very fast solid state DIP relays (Teledyne Relays 643-1) with a response time of less than 10 μsec. Currents of approximately 70 mA are necessary to generate field gradients of 1,000 Hz/cm.

The magnetic field has been stabilized by a Bruker B-SN 15 external pulsed proton lock with a long term stability of 1 Hz. To prevent a disturbance of the lock by the applied field gradients, it was necessary to interrupt the control loop during the application of the field gradients for approximately 100 msec. This did not affect the field stability.

The maximum number of samples representing the zeugmatogram is limited by the available memory size of the computer. In general, a quadratic image with $N \times N$ samples is desired. It is then necessary to record $2N$ FID's and to digitize each FID into $2N$ samples. To permit the use of a fast Fourier transform routine, N is usually selected to be a power of two.

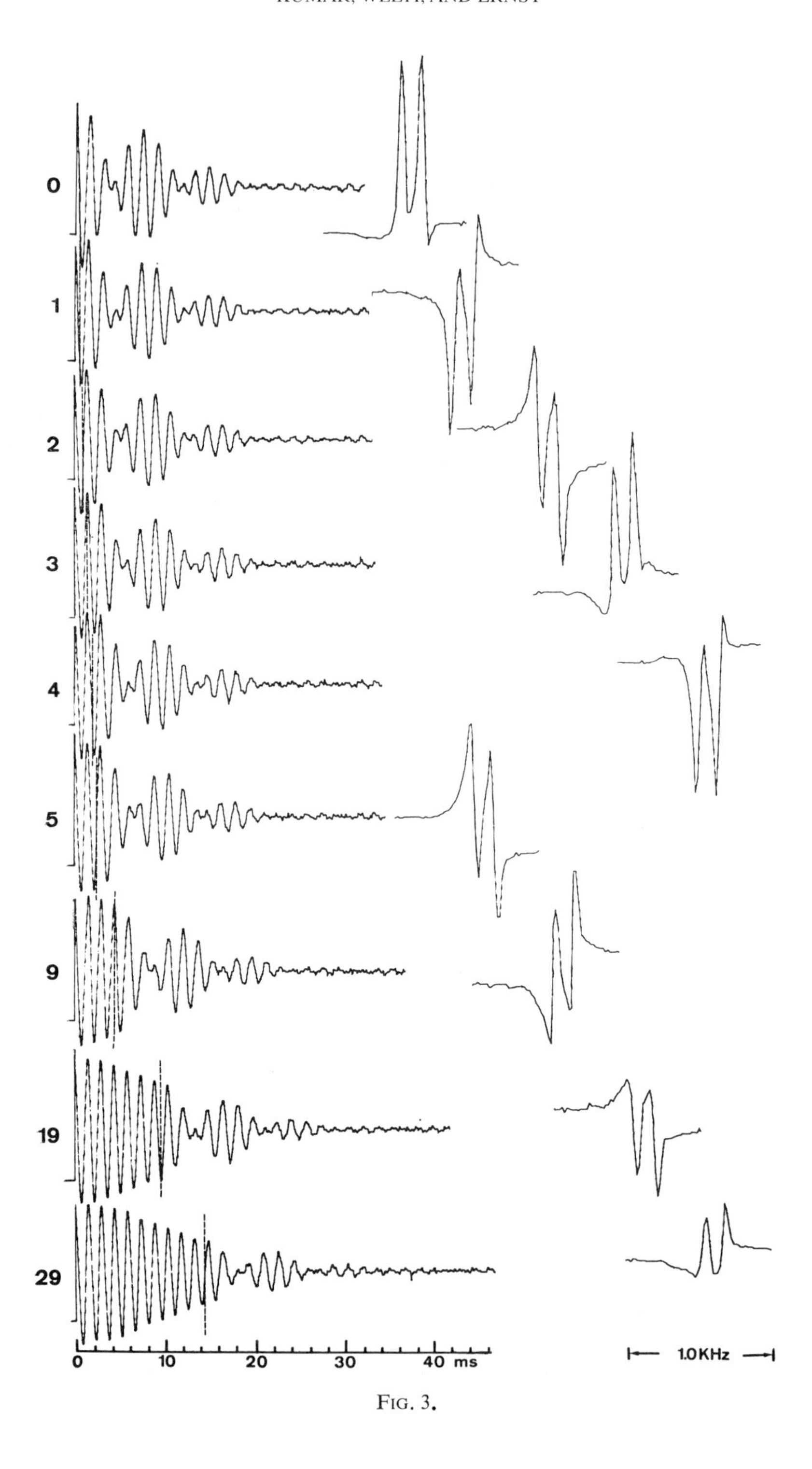

FIG. 3.

A well-known procedure to obtain a finer representation of a Fourier transform is the addition of a set of zeros to the array to be transformed (9). A simple method that requires only $N(N+2)$ memory locations but produces an $N \times N$ zeugmatogram is the following one. N FID's consisting of N samples each are recorded. To perform the first Fourier transformation, the N samples representing the FID k, $\{s_{k0}, s_{k1}, \ldots, s_{kN-1}\}$, are transferred to a separate memory block and are augmented by N zero values, $\{s_{k0}, s_{k1}, \ldots, s_{kN-1}, 0, 0, \ldots, 0\}$. The Fourier transform consists then of N complex values $\{\mathbb{S}_{k0}, \mathbb{S}_{k1}, \ldots, \mathbb{S}_{kN-1}\}$. The real parts, $\{R_{k0}, R_{k1}, \ldots, R_{kN-1}\}$, are retained only and are stored back in place of the original FID. After transformation of all FID's, the matrix $\{R_{kj}\}$ is transposed, $\{R_{kj}\} \rightarrow \{R_{jk}\}$, and each row, augmented by N zero values, is Fourier-transformed a second time. The absolute values of the N^2 complex Fourier coefficients are then utilized for the plot of a two-dimensional zeugmatogram. It can easily be shown that the neglect of the imaginary part after the first Fourier transformation does not cause any loss of information nor does it deteriorate the sensitivity.

The limited number of samples available to represent each FID calls for a careful selection of the center frequency, of the strength of the applied field gradients and of the sampling rate, such that the spatial resolution is sufficient without violating the sampling theorem and avoiding frequency foldover which can seriously distort a zeugmatogram. It must be remembered that dispersion-like parts as well must be represented. Dispersion mode signals have a much higher tendency to cause problems with frequency foldover than absorption mode signals because of the much broader wings of the former.

The number of samples N has been selected to be 64. This results in a total of 4096 sample values. The time required for one complete experiment including the data transformation is 8 min and the plotting of the 64×64 zeugmatogram on the teletype requires another 7 min.

The NMR samples that were used to demonstrate the principle of Fourier zeugmatography consisted of two parallel glass capillary tubes filled with H_2O. For Figs. 3 and 4, the two capillaries, with an inner diameter of 1.0 mm and a separation of the centers by 2.2 mm, were surrounded with D_2O. The sample was positioned in the magnet gap such that the capillary tubes were parallel to the y-axis and the line joining the centers of the two tubes was parallel to the z-axis. Figure 3 shows a series of typical FID's and their first Fourier transform. During the first time-interval of length t_x, a linear gradient of 500 Hz/cm was applied along the x-axis. The two capillary tubes are then in the same local field and the FID remains unmodulated as is demonstrated by Fig. 3. During the second time-interval, a gradient of 700 Hz/cm was applied along the z-axis. It causes the two tubes to be in different local fields and it is responsible for the modulation of the FID as well as for the doublet structure of the first Fourier transform with respect to t_z, shown on the right-hand side of Fig. 3. The phase and amplitude of the signal after the first Fourier transformation map out the FID at the end of the first time-interval. A

FIG. 3. Nine typical FID's selected out of a complete set of 64 signals obtained for pulsed linear field gradients along the x- and z-axis. The corresponding Fourier transforms are shown on the right-hand side. The numbers on the ordinate represent the time intervals in terms of sampling cycles during which the x-gradient was on. The broken vertical lines in the FID's indicate the point in time when the x-gradient was stopped and the z-gradient was switched on. At the same time, the recording of the FID was started. The sample consisted of two parallel capillary tubes arranged such that their centers were lying on the z-axis. The sampling interval was 0.5 msec giving a total spectral width of 1 kHz.

second Fourier transformation with respect to t_x then yields the final two-dimensional Fourier zeugmatogram shown in Fig. 4. For this map, the total intensity range has been divided into eight equal intervals and a teletype character assigned to each interval. The intensity intervals are indicated in increasing order by the symbols (blank), •, *, A, B, C, D, and E, respectively. This assignment is used for all zeugmatograms given in this paper.

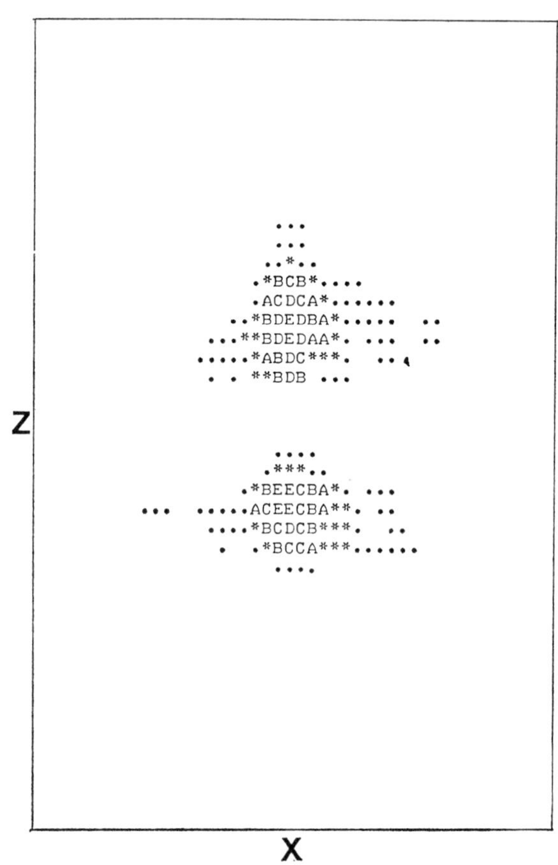

FIG. 4. The Fourier zeugmatogram obtained from the data partially shown in Fig. 3. The absolute value $|\bar{c}(r')|$ of the spin density function is plotted as a function of x and z. Only the central section of the 64 × 64 zeugmatogram is shown.

Figures 5 and 6 show a series of FID signals along with their first Fourier transforms and the Fourier zeugmatogram computed therefrom for the same experimental parameters and for the same sample as used for Figs. 3 and 4, except that the two gradients have been interchanged in time. The first gradient is now along the z-axis and the second along the x-axis. In this case, phase and amplitude of the first Fourier transform clearly show the beats caused by the different local fields of the two capillaries during the first time interval. Figures 4 and 6 represent images of the same sample effectively rotated by 90°. The two zeugmatograms do not exactly match due to experimental imperfections which will be discussed below.

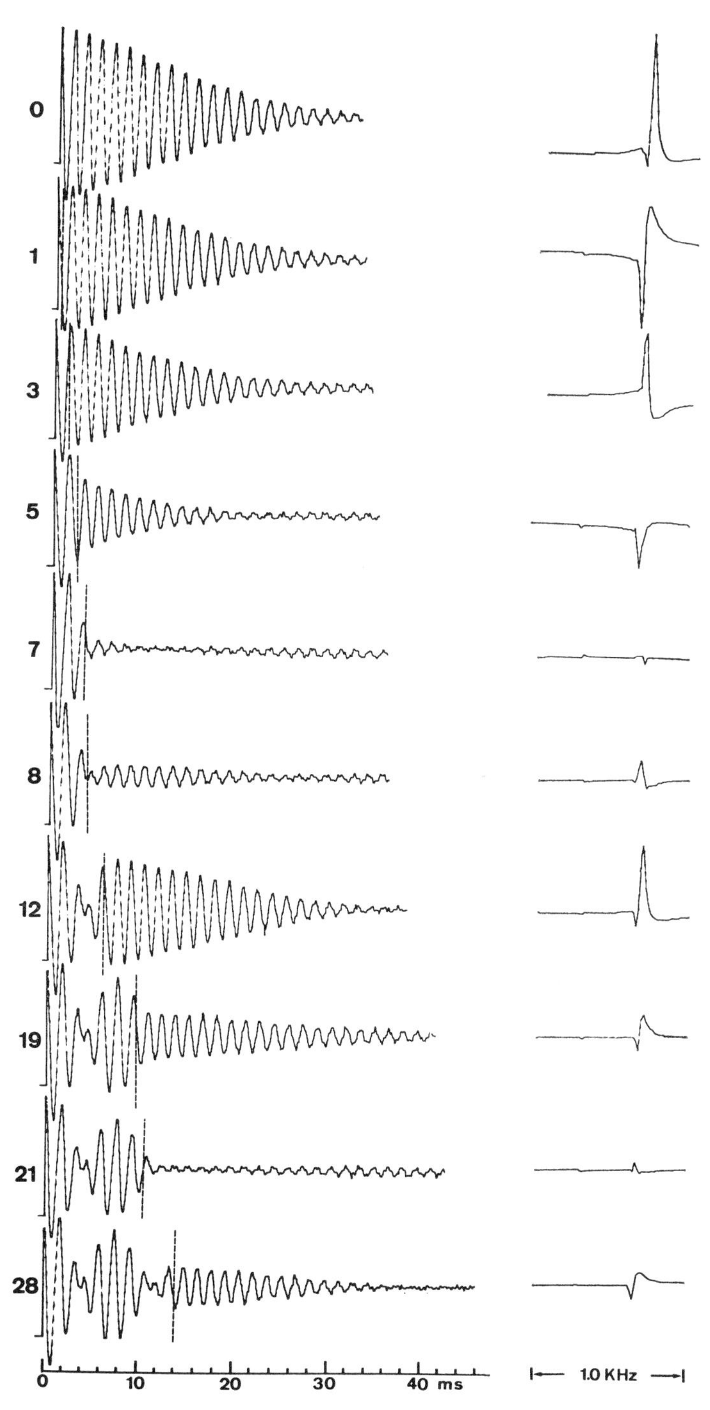

FIG. 5. Ten typical FID signals and their Fourier transforms obtained for the same conditions as in Fig. 3, but with the x- and z-gradients interchanged in time. The FID is recorded in this case during the time the x-gradient is on.

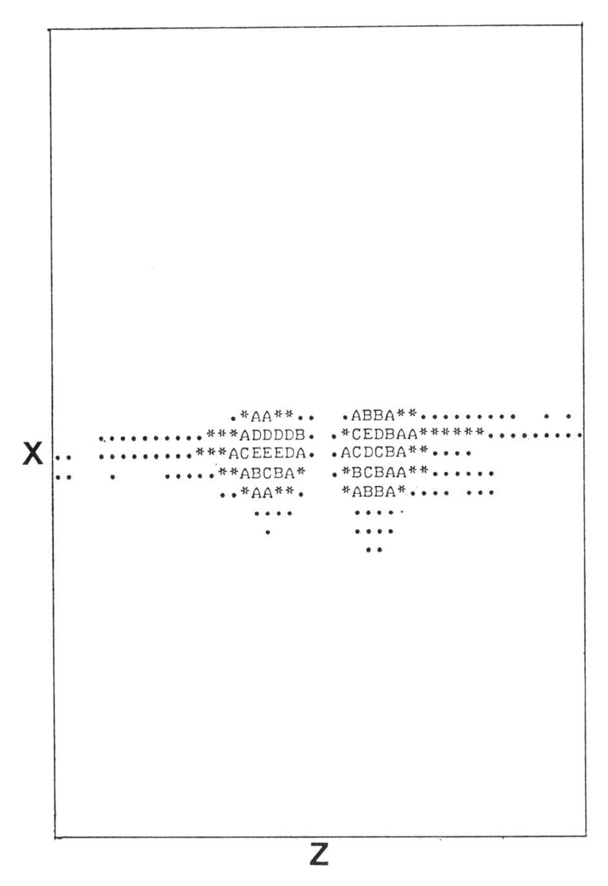

FIG. 6. The Fourier zeugmatogram calculated from the data partially shown in Fig. 5. The x- and z-axis are interchanged as compared to Fig. 4. Only the central section of the 64×64 zeugmatogram is shown.

Figure 7 gives the zeugmatogram of another sample consisting of two capillary tubes, with an inner diameter of 1.3 mm and a separation of the centers of 2.6 mm, placed in the magnet gap such that the line joining the centers was making an angle of about 30° to the z-axis. This sample had no D_2O outside the capillaries and the change in susceptibility required a retuning of the basic magnetic field homogeneity. This figure demonstrates the two-dimensional resolution of Fourier zeugmatography.

The zeugmatograms shown in this paper were obtained mainly to demonstrate the principle of the technique and therefore a very simple sample geometry was used. With future applications in view, it may be worthwhile to point out some of the problems that were encountered. These problems are most likely responsible for some of the spread and for the limited resolution of the zeugmatograms shown. We believe that it is possible to improve the images by paying attention to the following points.

(a) *Linearity and homogeneity of the gradients.* It is obvious that in any zeugmatographic technique the linearity and the homogeneity of the applied field gradients is of crucial importance. Very often, the shim coils provided in commercial spectrometer

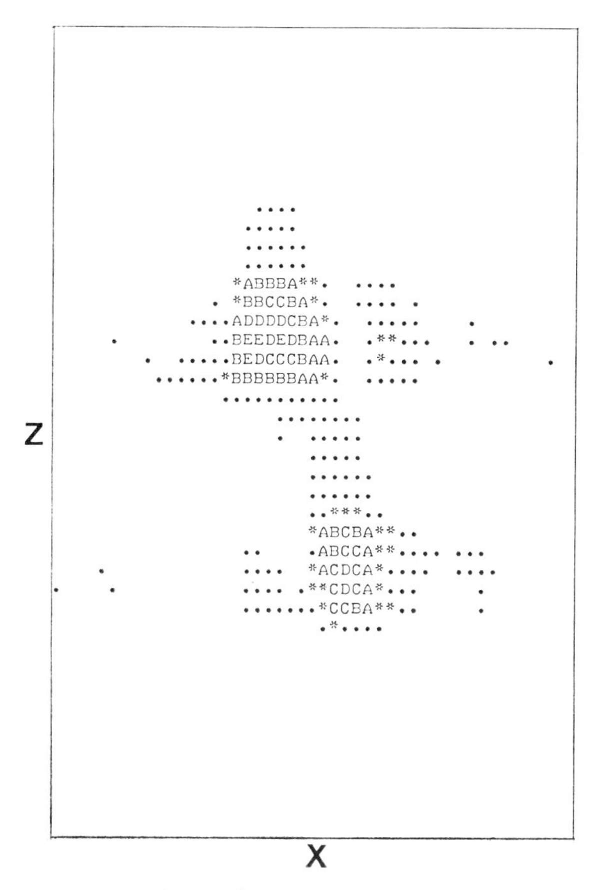

FIG. 7. Fourier zeugmatogram of a sample consisting of two parallel capillaries with their centers lying on a line making an angle of about 30° to the z-axis. The first x-gradient was 600 Hz/cm and the second z-gradient was 700 Hz/cm. The absolute value $|\bar{c}(r')|$ of the spin density function is plotted as a function of x and z. Only central section of the 64 × 64 zeugmatogram is shown.

systems produce gradients that are not of sufficient linearity over the entire sample volume. It may, therefore, be advisable to add special gradient coils with improved linearity (*10*).

(*b*) *Rise and fall time of the gradients.* The described pulsed version of zeugmatography relies on the instantaneous application and removal of field gradients. Eddy currents in pole caps and metal shields can cause response-time problems that will result in serious distortions of the zeugmatogram. Special arrangements of gradient coils which minimize the rise and fall time of the gradients have been described by Tanner (*10*).

(*c*) *Susceptibility problems.* In many samples, the local magnetic field is already inhomogenous due to variations of the susceptibility and due to the particular shape of the sample. It is then necessary to apply sufficiently strong gradients to overcome these "natural" field gradients.

The pictorial representation could certainly be improved for visual effects by more sophisticated means, for example, by means of a computer-controlled display.

IV. EXTENSIONS OF THE TECHNIQUE

Many possible extensions of the described technique exist. They have not yet been pursued in great detail. This section gives a brief description of some of the possibilities.

(a) *Recording of a Pure Absorption Mode Zeugmatogram*

It may be desirable to completely separate absorption- and dispersion-like parts as the absorption mode signal has an inherently higher resolution than the absolute value signal. The pure absorption mode may often produce a more accurate zeugmatogram. This separation can be achieved by the following modification of the basic technique. For each set of values t_x and t_y, a series of four different experiments is performed and the four resulting FID's are averaged:

$$s_{av}(\mathbf{t}) = \tfrac{1}{4}\{s^{+++}(\mathbf{t}) + s^{++-}(\mathbf{t}) + s^{+-+}(\mathbf{t}) + s^{-++}(\mathbf{t})\}. \qquad [13]$$

The four experiments differ by the signs of the applied field gradients and by the position of the reference frequency ω_1 above or below resonance (negative or positive frequency offset Δ). The notation of Eq. [13] is explained in Table 1. For this technique, it is not only necessary to switch gradients between the three phases of the FID but also to

TABLE 1

GRADIENTS AND RESONANCE OFFSETS IN THE FOUR REQUIRED EXPERIMENTS

	Applied gradients η			Resonance offset Δ		
Time interval	t_x	t_y	t_z	t_x	t_y	t_z
$s^{+++}(\mathbf{t})$	η_x	η_y	η_z	$\|\Delta\|$	$\|\Delta\|$	$\|\Delta\|$
$s^{++-}(\mathbf{t})$	η_x	η_y	$-\eta_z$	$\|\Delta\|$	$\|\Delta\|$	$-\|\Delta\|$
$s^{+-+}(\mathbf{t})$	η_x	$-\eta_y$	η_z	$\|\Delta\|$	$-\|\Delta\|$	$\|\Delta\|$
$s^{-++}(\mathbf{t})$	$-\eta_x$	η_y	η_z	$-\|\Delta\|$	$\|\Delta\|$	$\|\Delta\|$

change the sign of the resonance offset Δ. This can be achieved by an appropriate change of the dc magnetic field (during the off-time of the field-frequency lock) or, better, by a sudden change of the reference frequency of the phase-sensitive detector, ω_1. Care must be taken to retain phase coherence during this frequency switching.

By means of Eq. [3] and utilizing the trigonometric addition theorems, one obtains for $s_{av}(\mathbf{r},\mathbf{t})$

$$s_{av}(\mathbf{r},\mathbf{t}) = M_0 \cos((\Delta + \eta_x x)\,t_x)\cos((\Delta + \eta_y y)\,t_y)\cos((\Delta + \eta_z z)\,t_z) \cdot \exp\{-(t_x + t_y + t_z)/T_2\}. \qquad (14)$$

In this expression, the three variables t_x, t_y, and t_z are separated. The three-dimensional Fourier transformation is now executed step-by-step, and, after each transformation, the imaginary part is eliminated. This is equivalent to a three-dimensional cosine transformation and gives the result

$$S_{av}(\omega) = \iiint c(\mathbf{r})\, S_{av}(\mathbf{r},\omega)\, dv \qquad (15)$$

with

$$S_{av}(\mathbf{r},\omega) = \tfrac{1}{8}\{A(\Delta + \eta_x x - \omega_x) + A(-\Delta - \eta_x x - \omega_x)\}$$
$$\cdot\{A(\Delta + \eta_y y - \omega_y) + A(-\Delta - \eta_y y - \omega_y)\}$$
$$\cdot\{A(\Delta + \eta_z z - \omega_z) + A(-\Delta - \eta_z z - \omega_z)\}. \quad [16]$$

This is a three-dimensional absorption mode signal and is the desired result. The contribution of the negative resonances can again be neglected, in general, and the function is converted into a function of $\mathbf{r}'$, $\bar{c}_{av}(\mathbf{r}') = S_{av}(\omega)$, in analogy to Eqs. [11] and [12], and plotted as a function of x, y and z.

In two dimensions, a series of only two experiments is necessary to determine $s_{av}(\mathbf{t})$:

$$s_{av}(\mathbf{t}) = \tfrac{1}{2}\{s^{++}(\mathbf{t}) + s^{+-}(\mathbf{t})\}. \quad [17]$$

It is obvious that this modified technique is more complicated. It has not yet been used, and it is not clear in which cases it is worth the effort.

(b) Recording of a Two-Dimensional Cross Section

It has been mentioned that a full three-dimensional zeugmatogram requires an amount of data that goes beyond the capacity of most small computers. The two-dimensional mapping described in Section III, on the other hand, does not provide distinct cross sections but rather a projection of the three-dimensional spin density onto a two-dimensional plane. In many circumstances, a true cross section would be more desirable.

A cross section can be obtained by the following technique. A quadratic or possibly higher order gradient is applied along the third direction, e.g., the y-axis. Then, only the volume elements near $y = 0$ will appreciably contribute to the signal amplitude. The remaining contributions will be smeared over a much larger spectral region and may be disregarded. For an additional smearing, it is also possible to average several FID's for various field gradients along the y-axis. For the selection of another cross section perpendicular to the y-axis, the sample has to be moved along the y-axis.

A limited confinement to a cross-sectional area is also possible by using very short receiver and transmitter coils.

(c) Improved Resolution with Limited Core Memory

An improved representation of a zeugmatogram may be obtained in the following way when the memory space is limited. Each FID is sampled to obtain M sample values which are Fourier-transformed to produce a spectrum containing M samples. The sampling rate may be selected such that the interesting part of the spectrum covers only a small portion of the total spectrum. N significant, not necessarily equally spaced, samples of the spectrum are selected and stored for the subsequent second Fourier transformation. The number of relevant samples, N, can be considerably smaller than M. All other sample values can be discarded to save memory space. A similar procedure for the second Fourier transform is unfortunately not possible.

(d) Measurement of Relaxation Times

The measurement of the two- or three-dimensional distribution of spin–lattice relaxation times can be achieved by straightforward extensions of the inversion-recovery technique (*11*) or of the saturation recovery method (*12*). The spin system is prepared at $t = -T$ by means of a 180° pulse or by means of a saturating burst of rf pulses. The zeugmatogram obtained with a 90° pulse at $t = 0$ is then a measure for the spatial dependence of the recovery of the z-magnetization during the time T and permits determination of $T_1(\mathbf{r})$. An adaptation of the progressive saturation technique (*13*) is also feasible.

(*e*) *Related Techniques*

A somewhat related technique has been described by Mansfield and Grannell (*14*). It is called NMR diffraction and its aim is the determination of the periodic structure of a solid by applying a single linear field gradient and recording the FID signal under high resolution conditions. This technique could also be generalized by applying a sequence of pulsed field gradients in the same manner as described in the present paper.

Two further techniques which also involve time-dependent magnetic field gradients have recently been described by Hinshaw (*15*). Of particular interest seems to be his "sensitive point method" as it permits picking out the signals originating from a distinct point within a three-dimensional object.

It should also be mentioned that the technique of Fourier zeugmatography is remotely related to pulse-pair Fourier spectroscopy as proposed by Jeener (*16*). In this technique, which has a completely different aim, two 90° rf pulses are applied at $t = 0$ and at $t = t_1$. The free induction decay signal after the second pulse is then a function of the two time parameters t_1 and $t_2 = t - t_1$, $s(t_1, t_2)$. It resembles the signal obtained in two-dimensional Fourier zeugmatography. Its two-dimensional Fourier transform produces a two-dimensional spectrum that contains information of the kind usually obtained in double resonance experiments. It does not give information about the spatial distribution of nuclear spins. But the same experimental set-up and particularly the same computer programs can be used for both techniques.

More sophisticated extensions are conceivable, like the measurement of the spatial distribution of flow by measuring the echo height in a spin–echo experiment in an inhomogenous magnetic field. Various double resonance techniques can also be combined with zeugmatography, for example, to single out the contributions of one particular resonance line in a more complex spin system. Fourier zeugmatography has the potential to adopt many of the well-known pulse techniques presently in use in high resolution NMR of liquids and of solids.

ACKNOWLEDGMENTS

This work has partially been supported by the Swiss National Foundation of Science. The receipt of a preliminary manuscript by J. Jeener and G. Alewaeters is gratefully acknowledged. A discussion with P. Lauterbur has considerably helped to clarify several points.

REFERENCES

1. P. C. Lauterbur, *Nature* (*London*) **242,** 190 (1973).
2. R. Damadian, *Science* **171,** 1151 (1971).
3. I. D. Weisman, L. H. Bennett, L. R. Maxwell, M. W. Woods, and D. Burk, *Science* **178,** 1288 (1972).
4. D. J. De Rosier and A. Klug, *Nature* (*London*) **217,** 130 (1968).
5. R. Gordon and G. T. Herman, *Comm. ACM* **14,** 759 (1971).
6. G. T. Herman, A. Lent, and S. W. Rowland, *J. Theor. Biol.* **42,** 1 (1973).

7. A. Klug and R. A. Crowther, *Nature* (*London*) **238,** 435 (1972).
8. R. R. Ernst and W. A. Anderson, *Rev. Sci. Instr.* **37,** 93 (1966).
9. E. Bartholdi and R. R. Ernst, *J. Magn. Resonance* **11,** 9 (1973).
10. J. E. Tanner, *Rev. Sci. Instr.* **36,** 1086 (1965).
11. R. L. Vold, J. S. Waugh, M. P. Klein, and D. E. Phelps, *J. Chem. Phys.* **48,** 3831 (1968).
12. J. L. Markley, W. J. Horsley, and M. P. Klein, *J. Chem. Phys.* **55,** 3604 (1971); W. A. Anderson, R. Freeman, and H. Hill, *Pure Appl. Chem.* **32,** 27 (1972).
13. R. Freeman and H. D. W. Hill, *J. Chem. Phys.* **54,** 3367 (1971).
14. P. Mansfield and P. K. Grannell, *J. Phys. C Solid State Phys.* **6,** L422 (1973).
15. W. S. Hinshaw, *Phys. Lett.* **48A,** 87 (1974).
16. J. Jeener and G. Alewaeters, private communication (1973).

JOURNAL OF MAGNETIC RESONANCE **19,** 114–117 (1975)

Dynamic Range in Fourier Transform Proton Magnetic Resonance*

The problem of detecting weak proton magnetic resonances in the presence of nearby strong peaks is especially difficult when Fourier transform NMR is attempted because signals from all classes of spins, weak and strong, occur simultaneously. R. K. Gupta and A. G. Redfield (*1*, *2*) partially solved this problem by employing a long, weak observation pulse, as first used by S. Alexander (*3*), which acts as a 45 to 90° pulse for resonances of interest while not flipping a solvent resonance at some specific frequency. We have used the technique to observe resonances less than 200 Hz from H_2O resonances at 5-m*M* concentration, the main problem being H_2O spinning-sidebands (*2*).

Attempts to apply this method to very broad resonances of large proteins proved difficult because of a baseline curvature produced by the several moles/liter of protein-proton spins which are off-resonance from the spectral window selected by the filtering system. This curvature is not removable by filtering, and removing it by artificial computer baseline flattening (*1*) is not useful for broad resonances. Therefore, we sought a way to flip over protons resonating in a band of frequencies while leaving those in another band relatively unaffected. Two elegant methods (*4*, *5*) have been described which already do this, but the method described below was simpler for us.

One new pulse sequence was guided by expediency, a need to have the pulse sequence confined to a small time (for T_1 measurements), and the principle that the null in the effectiveness of a pulse sequence would nearly coincide with the null in its Fourier transform. This is true in the limit of low rf amplitude, but leads in practice to only small errors compared to a precise (effective field) treatment which takes account of the non-linearity of the Bloch equations. Thus a long 90° pulse, of amplitude H_1, frequency f_0 Hz, and length $\tau = \pi/2\gamma H_1$ (where γ is the gyromagnetic ratio), has a Fourier transform which is proportional to $\sin[2\pi(f-f_0)\tau]/(f-f_0)$. This transform has a null at frequencies $\pm 1/\tau$ away from the carrier frequency, whereas the null in the effectiveness of the pulse for flipping nuclei occurs about $0.97/\tau$ away (*2*). Two very narrow pulses spaced $\frac{1}{2}\tau$ apart will have a Fourier transform $\cos[\pi(f-f_0)\tau]$ with a first null at the same frequency as the long pulse of length τ.

If these two waveforms are added in the proper relative strength and phase, it is possible to produce a broad null by arranging to have the slopes of the Fourier transforms of the two components equal and opposite at the null. As an approximation to this sequence, guided by the fact that it is much easier to produce 180° phase shifts of pulses than to vary their amplitude, we evolved the symmetric sequence shown in Fig. 1(a), in which the third and seventh tenth of the pulse is shifted by 180° relative to the rest of it. We call this a "2–1–4 pulse" after the lengths of its first three intervals. Its Fourier transform is compared with that of the simple long pulse in Fig. 1(b). This pulse can be viewed as a simple long pulse of amplitude H_1, minus two short pulses of amplitude $2H_1$, the latter being a convenient approximation to two very short strong pulses.

* Partially supported by Grants GU3852 and GP37156 from the National Science Foundation and GM20168 from the U.S. Public Health Service.

The amplitude H_1 must be 5/3 that of the simple long pulse, for the same flipping angle at the center frequency.

In practice the solvent water peak is placed a distance in frequency 0.97 to $0.99\tau^{-1}$ away from the transmitter carrier frequency, and the rf amplitude is approximately that which results in a roughly 44° pulse for spins resonant at the carrier frequency. A phase and amplitude correction must be applied to the complex Fourier transform to compensate for the use of this pulse. We find it adequate to use the same phase correction (*1*) used to correct for distortion produced by the simple (*3*) long pulse of length τ, which is stored in our computer. In practice the 2–1–4 pulse, or the simple long

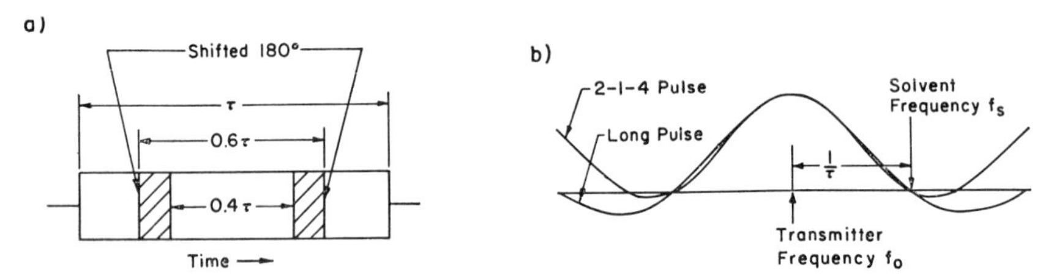

FIG. 1. (a) The 2–1–4 pulse. (b) The Fourier transform of the simple long pulse of length τ compared with that of the 2–1–4 pulse. The two transforms have been normalized to have the same amplitude at center frequency.

pulse, lead to a roughly 100-fold decrease in the water signal at 90 MHz. Efficient audio filtering (*1*) gives a further 10 to 100 fold decrease in the solvent signal going into the A/D converter, so that this signal is only a few times thermal noise for typical bandwidths. At the output of the filter there is a large initial transient due to the out-of-band signal from the solvent and the majority of protein protons. This latter protein transient can be minimized more efficiently with the 2–1–4 pulse until it is only 10 to 100 times noise. The time intervals of the parts of the pulse are fine-tuned, together with the carrier frequency, to produce the smallest transient signal and/or flattest baseline. The adjustments for the 2–1–4 pulse are much less critical than for the long pulse.

A number of spectra obtained by these techniques are shown in Fig. 2, obtained with a Bruker WH-90 spectrometer modified along the lines described previously (*1*). The usefulness of the 2–1–4 pulse in daily research is not fully evident from Fig. 2; the difference was more dramatic before some overloading in the signal amplifying system was eliminated. Thus this type of pulse apparently decreases the linearity requirements of the spectrometer amplifier. The pulse length in Fig. 2(a) was chosen to give the flattest baseline, and is also that which gives the smallest solvent free induction decay. Changing the pulse length by only 0.2% would produce a 45° tilt in the baseline, and at least as much baseline curvature as that in Fig. 2(a). The spectra in Fig. 2(b) and (c) are comparable in signal-to-noise ratio to protein spectra published previously (Fig. 1(b) of Ref. (*2*)) and may also appear comparable in baseline flatness. However, the previously published spectra were subjected to a baseline-flattening computer routine, whereas the spectra of Fig. 2 have not been arbitrarily flattened or tilted except insofar as this is possible by trimming pulse lengths. All spectra were computer corrected as described previously (Ref. (*1*), Eqs. [5] and [6]) for phase and amplitude distortions produced by the filtering and long pulse, but the computer baseline flattening routine outlined in Ref. (*1*) was not used.

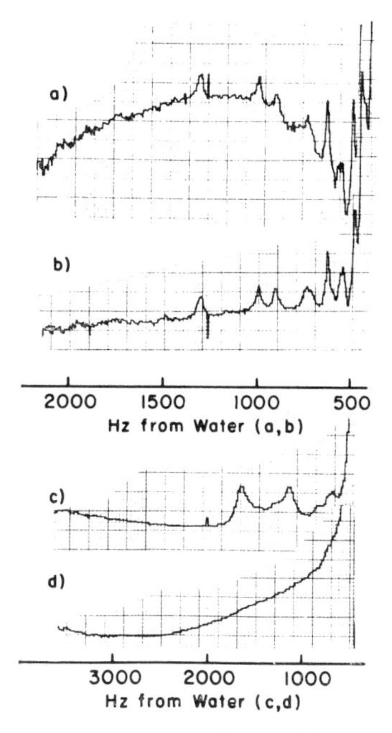

FIG. 2. Spectra of hemoglobin and cytochrome-c dissolved in 80–90% H_2O, obtained in 5 minutes. (a) Contact shifted downfield resonances of 2 m*M* ferricytochrome-c, pH7, using a simple 777 μsec long pulse. 1500 transients, transmitter frequency 1250 Hz from water, 1024 input points (total of real and imaginary), sampling rate 2560 complex points/sec. (b) The same sample and conditions, except that the 2–1–4 pulse (lengths 156, 78, 312 μsec) was used. The pulse amplitude was increased 4 dB over that used in the first spectrum, to maintain a roughly 45° pulse at center band. Most of the resolved peaks are resonances of single spins. (c) Downfield resonances of 1.5 millimolar (tetramer) deoxyhemoglobin, stripped of diphosphoglycerate. 2–1–4 pulse (lengths 99, 58, 208 μsec), 2800 transients, transmitter frequency 2000 Hz from water, 512 input points, sampling rate 5120 points/sec. Each of the two prominent peaks is believed to be the resonance of a methyl group of which there are two per tetramer. This spectrum was obtained in a nonspinning 10 mm NMR tube. (d) Spectrum of oxyhemoglobin. The sample preparation and volume were exactly the same as in (c), except that the sample was not deoxygenated; the spectrometer pulse length and amplitude, and detector phase settings were not touched between (c) and (d).

The hemoglobin spectra (Figs. 2(c) and (d)) demonstrate the limitations as well as the strengths of the technique. The spectrum of Fig. 2(c) was obtained after trimming the pulse parameters to give the flattest downfield baseline. Fig. 2(d) was then obtained without any readjustment of the spectrometer, using a similar sample which had not been deoxygenated and which should have had no resonances beyond 1000 Hz. The baseline slope probably arose from a sizeable shift in the aromatic part of the spectrum of oxy- vs. deoxyhemoglobin. It could have been eliminated by a slight trimming of the pulse length, and was relatively insensitive to the detector phase setting.

A further problem arises when T_1 or other double irradiation experiments are performed, due to the solvent signal produced by the preobservation pulse. This gives a signal which can block the receiver electronics. We have obtained satisfactory results by the saturation-recovery method using a long (0.1 sec), weak pulse for preirradiation, a spoil pulse (*6*), and slow gating-off (in 5 msec) of the preirradiation pulse rather than the usual fast gating. Slow gating-off tends to return the solvent magnetization adiabatically to the dc field direction, thereby reducing its transverse magnetization and the resulting signal. Baseline wiggles which seem to arise from solvent can be reduced by degrading the field homogeneity.

There seems to be a misconception that the long pulse method is only applicable to spectral regions which are narrow and/or far from water. On the contrary, a single run at a transmitter frequency f_0 typically covers a range from about $(f_0 - f_s)/3$ to $5(f_0 - f_s)/3$ away from the solvent frequency f_s, in other words from 50 to 250 Hz (or 100 to 500 Hz) away from solvent for high resolution work, or either 400 to 2000 or 800 to 4000 Hz away from water for the runs of Fig. 2. Thermal noise is increased at the edges of the spectra, as can be seen at the left side of the spectra of Fig. 2. Therefore, 20–30% of the theoretical maximum spectral width, as determined by the sampling rate, is relatively useless and is automatically discarded by our instrument. The large increase at the right side of the spectra of Fig. 2 is the edge of the amide and aromatic region. Observation of lines close to the solvent resonance is limited solely by confusion with spinning sidebands of the latter; Refs. (*1*) and (*2*) incorrectly stated that solvent spinning sidebands are eliminated by the long pulse. In our spectrometer, with concentrations of a few millimolar of small molecules in H_2O, using 5-mm NMR tubes, lines more than 100 Hz from water are easy to observe. Lines as close as 50 Hz are observable with difficulty. The present communication is not intended to show the capabilities of these methods for small molecules, and readers interested in such problems should recognize that, because of the relatively narrower lines for small molecules, the quality of spectra is better than might be thought from a casual look at Fig. 2. The long pulse methods could probably be profitably combined with water saturation methods (*7*) for observing resonances very close to solvent.

REFERENCES

1. A. G. Redfield and R. K. Gupta, *Adv. Magnetic Resonance* **5,** 81 (1971).
2. A. G. Redfield and R. K. Gupta, *J. Chem. Phys.* **54,** 1418 (1971).
3. S. Alexander, *Rev. Sci. Instr.* **32,** 1066 (1961).
4. J. Dadok and R. F. Sprecher, *J. Magn. Resonance* **13,** 243 (1974).
5. B. L. Tomlinson and H. D. W. Hill, *J. Chem. Phys.* **59,** 1775 (1973).
6. R. L. Vold, J. S. Waugh, M. P. Klein, and D. O. E. Phelps, Jr., *J. Chem. Phys.* **48,** 3831 (1968).
7. See, for example, S. L. O. Patt and B. D. Sykes, *J. Chem. Phys.* **56,** 3182 (1971), and E. S. Mooberry and T. R. Krugh, *J. Magn. Resonance* **17,** 128 (1975).

A. G. Redfield[1]
Sara D. Kunz
E. K. Ralph[2]

Department of Physics
Brandeis University
Waltham, Massachusetts 02154

Received December 18, 1974

[1] Also at the Department of Biochemistry and the Rosenstiel Basic Medical Sciences Research Center, Brandeis University.

[2] On sabbatical leave from the Chemistry Department, Memorial University of Newfoundland, St. Johns, Newfoundland, Canada.

JOURNAL OF MAGNETIC RESONANCE **23**, 171–175 (1976)

A Simple Pulse Sequence for Selective Excitation in Fourier Transform NMR

Pulse-excited Fourier transform NMR has such clear advantages in sensitivity and for the study of time-dependent phenomena that only recently has it become necessary to reconsider the question of *selective* excitation of individual resonances, so readily achieved in the earlier continuous-wave spectrometers. Tomlinson and Hill (*1*) have described an elegant pulse technique which "tailors" the pattern of excitation frequencies to any desired shape, and have applied it to the problems of solvent peak suppression, homonuclear Overhauser effect measurements (*2*), and selective spin–spin relaxation studies (*3*). However, the technique makes considerable demands on instrumentation and computer capacity.

Not all applications call for this degree of sophistication; often a much simpler irradiation pattern would suffice—for example, a single narrow band of frequencies. This can be achieved in a particularly simple way by means of a regular sequence of short, identical radiofrequency pulses. The present treatment is confined to the transient response; the steady-state response to such a stimulus, involving relaxation, has been analyzed elsewhere (*4*). Consider the nuclear magnetization vectors in a frame of reference rotating at the frequency f_0 of the radiofrequency source. For simplicity, relaxation may be neglected, and the radiofrequency pulses are assumed to be "hard" in the sense that $\gamma H_1/2\pi > |\Delta f|$ for all significant offsets Δf from f_0. Each pulse turns the magnetization vector through a small flip angle α radians, while in the interval τ seconds between the pulses, each vector precesses through an angle $\theta = 2\pi\tau\Delta f$ radians. The key to the method proposed is a pulse repetition rate set to the condition $\theta = 2n\pi$ radians, that is, $\Delta f_n = n/\tau$, where n is an integer. At these particular offsets Δf_n, the effect of the pulses is a cumulative tipping motion toward the XY plane of the rotating frame, whereas at all other offsets free precession robs the pulses of their cumulative effect, and the overall flip angle is small. Formally the excitation may be regarded as consisting of narrow components at the radiofrequency f_0 and at a harmonically related series of sidebands separated by n/τ Hz from f_0. The experiments described here use only the first sideband, where the magnetization vector precesses through one complete cycle between pulses.

Simple computer simulations based on the Bloch equations confirm that for a very narrow band of frequencies near the $\theta = 2n\pi$ condition, the magnetization vector moves in a succession of zig-zag steps, in which the steps caused by pulses outweigh those caused by precession, so that the resultant trajectory approximates an arc close to the vertical YZ plane, extending down into the equatorial XY plane, as with a single conventional $\pi/2$ pulse. Such a vector induces a maximum NMR signal in the receiver coil and this signal is close to the pure absorption mode condition. For all other offsets, precession between pulses carries magnetization vectors away from the YZ plane so that they execute small cyclic excursions in the vicinity of the positive Z axis, generating only weak signals. Some typical trajectories projected into the XY plane are shown in

Fig. 1. Only the excess precession angle is shown, the amount by which θ exceeds an integral number of revolutions. The selectivity of the method can be appreciated in Fig. 2, where the absorption mode signal is plotted against offset Δf, and it is seen to improve with the number m of pulses in the sequence for a fixed total flip angle $m\alpha$. As a practical example a train of 50 pulses, lasting 100 msec, significantly excites a band of frequencies spanning only 10 Hz at an offset $\Delta f = 500$ Hz.

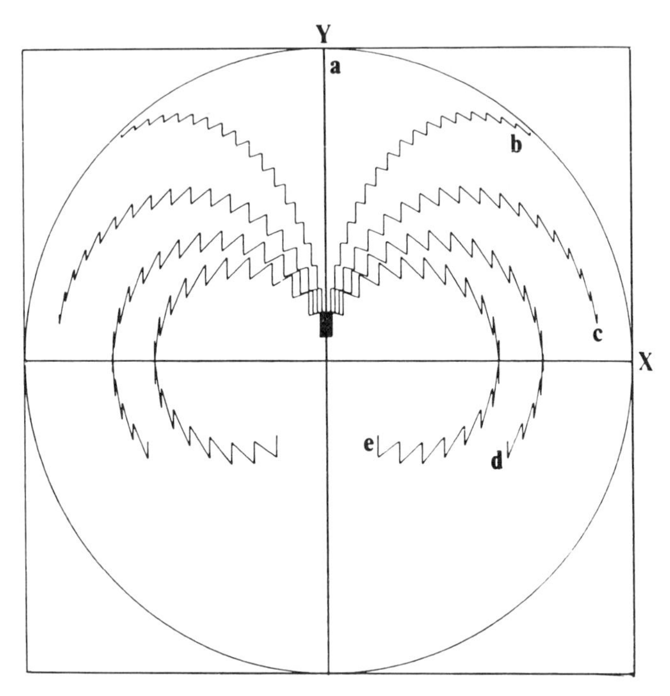

FIG. 1. Trajectories of the tips of magnetization vectors with different frequencies, projected onto the XY plane, during a sequence of 20 pulses of flip angle $\pi/40$ radians, spaced 2 msec apart. Only the excess precession (over 2π radians) is shown, for reasons of clarity. The offsets Δf of the magnetizations from the radiofrequency f_0 are, to the right of the Y axis, (a) 500, (b) 505, (c) 510, (d) 515, and (e) 520 Hz; and similarly to the left of the Y axis, 495, 490, 485, and 480 Hz, respectively.

A number of possible applications of this method have been explored, including solvent peak suppression, selective relaxation-time measurements (*5*, *6*), and hole burning (*7*, *8*); it is hoped to publish details elsewhere. For the purposes of this communication the following illustration must suffice: the investigation of proton-coupled carbon-13 spectra by separation into subspectra from individual carbon sites, a suggestion recently mentioned by Ernst (*9*). Carbon-13 spectra are only occasionally investigated under proton-coupled conditions, partly because of poor sensitivity and partly because the complex overlapping proton spin multiplet structure often makes interpretation extremely difficult, despite the fact that these coupling constants surely carry valuable structural information. This important problem has been attacked recently by the two-dimensional Fourier transform method (*9*, *10*).

A relatively simple model system is provided by a high-field section of the carbon-13 spectrum of 1-dimethylamino-2-methylpropene. Experiments were carried out on a conventional Fourier transform spectrometer (Varian CFT-20) with minor software and hardware modifications, using a moderately low radiofrequency level $\gamma H_1/2\pi = 1.8$ kHz. Figure 3a shows the high-field region recorded under conditions of proton noise decoupling, showing the three methyl peaks at frequencies P, Q, and R. The remaining

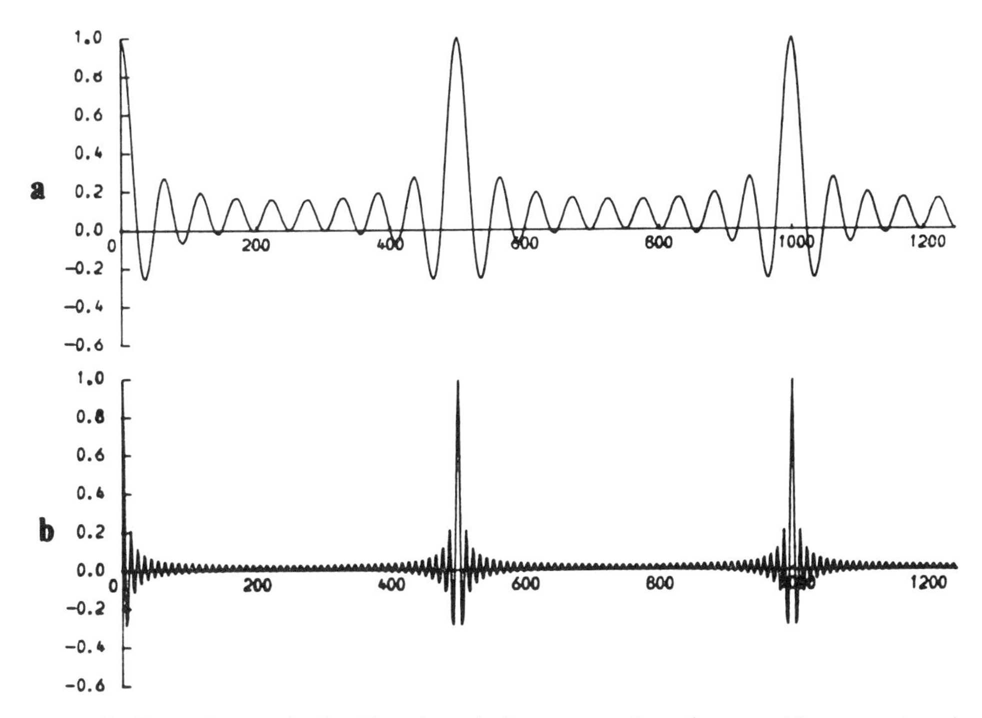

FIG. 2. Detected magnetization M_y at the end of a sequence of m pulses spaced 2 msec apart, each of flip angle $\pi/2m$ radians, as a function of offset Δf Hz for (a) $m = 10$ and (b) $m = 50$ pulses. The plot was obtained by a simulation based on the Bloch equations, neglecting relaxation, and as expected is qualitatively similar to the real part of the Fourier transform of the pulse sequence. The vertical scale is normalized to the condition $M_y = 1$ after an isolated "hard" $\pi/2$ radians pulse.

traces, (b)–(e), were obtained by means of the gated decoupling technique (*11*) in order to retain the nuclear Overhauser enhancement without losing the proton multiplet structure. During the noise irradiation of the protons, a sequence of 70 pulses of 2 microseconds' duration was applied so as to excite the carbon-13 methyl resonance at P ($\Delta f = 514$ Hz) in a selective manner to give a cumulative flip angle $m\alpha = \pi/2$. At the end of this pulse train the decoupler was gated off and the magnetization then evolved with time according to the *spin-coupled* Hamiltonian, permitting the acquisition of a free induction decay due to the multiplet structure of resonance *P* only. Fourier transformation gave the proton-coupled subspectrum illustrated in Fig. 3b; similar experiments with the excitation centered at Q and at R produced the subspectra 3c and 3d. The conventional proton-coupled spectrum after a single $\pi/2$ pulse (Fig. 3e) is the superposition of these subspectra with additional lines from the acetone-d_6 lock material and the tetramethylsilane reference.

Clearly the method is applicable to larger spin systems with a more complex system of overlapping multiplets. The sensitivity achieved for these subspectra equals that for the conventional full proton-coupled spectrum. Where long waiting times are required between acquisitions, these intervals could be used to excite and observe other resonances selectively without significantly affecting the relaxation of the resonance first observed. The modest instrumental and data processing requirements of the method,

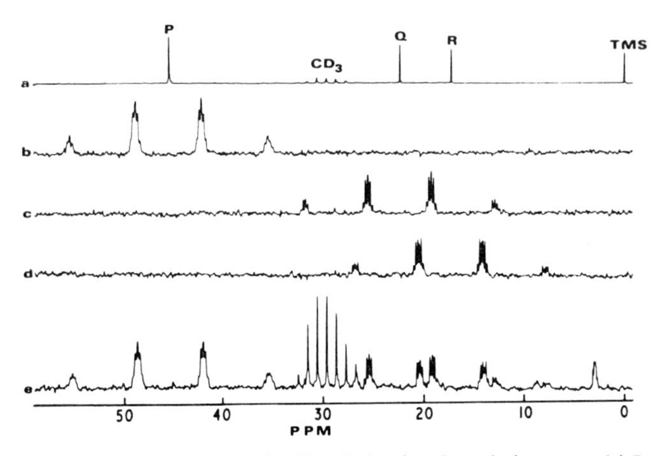

FIG. 3. The high-field regions of spectra of 1-dimethylamino-2-methylpropene. (a) Proton-decoupled spectrum; (b) to (d) multiplet subspectra corresponding to the three methyl resonances P, Q, and R, respectively; and (e) full proton-coupled spectrum, showing overlapping and interference from acetone-d_6 and tetramethylsilane.

coupled with the good sensitivity and resolution obtainable, make it an effective technique for the study of coupled carbon-13 spectra.

The proposed selective excitation technique is expected to be applicable to a wider range of problems, and exploratory experiments on some of these have been mentioned above. Other anticipated applications include excitation within the magnet line width to obtain information about natural linewidth, transient (*2*, *12*), and steady-state homonuclear Overhauser effect studies without the need for a second radiofrequency source, and three-dimensional mapping of the spin distribution in a sample in a strong imposed field gradient (*13–15*).

A number of improvements to the simple basic technique are possible. The use of a more complex pulse sequence, derived from the superposition of several regular pulse trains with different periods τ, allows the simultaneous excitation of more than one specific resonance. Alternatively the shape of the excitation function (Fig. 2), and hence the degree of selectivity obtained, may be modified by width, phase, or amplitude modulation of the radiofrequency pulses used.

ACKNOWLEDGMENT

The authors gratefully acknowledge several helpful discussions with Dr. M. J. T. Robinson, who also kindly prepared the sample of 1-dimethylamino-2-methylpropene.

REFERENCES

1. B. L. Tomlinson and H. D. W. Hill, *J. Chem. Phys.* **59,** 1775 (1973).
2. R. Freeman, H. D. W. Hill, B. L. Tomlinson, and L. D. Hall, *J. Chem. Phys.* **61,** 4466 (1974).
3. H. D. W. Hill, private communication.
4. R. Freeman and H. D. W. Hill, *J. Magn. Resonance* **4,** 366 (1971).
5. R. Freeman and H. D. W. Hill, "Dynamic Nuclear Magnetic Resonance" (L. M. Jackman and F. A. Cotton, Eds.), Chap. 5, Academic Press, New York, 1975.
6. R. Freeman and S. Wittekoek, *J. Magn. Resonance* **1,** 238 (1969).
7. N. Bloembergen, E. M. Purcell, and R. V. Pound, *Phys. Rev.* **73,** 679 (1948).
8. R. Freeman and B. Gestblom, *J. Chem. Phys.* **48,** 5008 (1968).
9. L. Muller, A. Kumar, and R. R. Ernst, *J. Chem. Phys.* **63,** 5490 (1975).
10. J. Jeener and G. Alewaeters, private communication.
11. R. Freeman and H. D. W. Hill, *J. Magn. Resonance* **5,** 278 (1971).
12. I. Solomon, *Phys. Rev.* **99,** 559 (1955).
13. P. C. Lauterbur, *Nature (London)* **242,** 190 (1973); *Pure Appl. Chem.* **40,** 149 (1974).
14. A. N. Garroway, P. K. Grannell, and P. Mansfield, *J. Phys. Soc. C* **7,** L457 (1974); P. K. Grannell and P. Mansfield, *Phys. Med. Biol.* **20,** 477 (1975).
15. A. Kumar, D. Welti, and R. R. Ernst, *Naturwiss.* **62,** 34 (1975); *J. Magn. Resonance* **18,** 69 (1975).

Geoffrey Bodenhausen
Ray Freeman
Gareth A. Morris

Physical Chemistry Laboratory
Oxford University
Oxford, England

Received April 6, 1976

JOURNAL OF MAGNETIC RESONANCE **24**, 71–85 (1976)

The Signal-to-Noise Ratio of the Nuclear Magnetic Resonance Experiment

D. I. HOULT AND R. E. RICHARDS

Department of Biochemistry, University of Oxford, South Parks Road, Oxford OX1 3QU, England

Received March 5, 1976

A fresh approach to the calculation of signal-to-noise ratio, using the Principle of Reciprocity, is formulated. The method is shown, for a solenoidal receiving coil, to give the same results as the traditional method of calculation, but its advantage lies in its ability to predict the ratio for other coil configurations. Particular attention is paid to the poor performance of a saddle-shaped (or Helmholtz) coil. Some of the practical problems involved are also discussed, including the error of matching the probe to the input impedance of the preamplifier.

INTRODUCTION

An important reason for the development of high-resolution NMR spectrometers employing superconducting, rather than iron, magnets has been the belief that the signal-to-noise ratio (S:N) available is proportional to the Larmor frequency to the three-halves power (*1*, *2*), and thus that an increase of, for example, three times in frequency from 90 to 270 MHz would bring as a return an improvement of a factor of 5.2 in S:N. It is always difficult accurately to compare the performances of different instruments as a variety of factors (to be discussed further) come into play; nevertheless, it is generally felt by those with experience of superconducting systems that the improvement obtained is disappointing—for example, about 2 for the frequencies quoted. As it is acknowledged that the electronics needed at very high and ultrahigh frequencies are "difficult," the accusing finger has tended to point in the engineers' direction. However, recent developments in the design of low-noise amplifiers (*3*) and the general improvement in frequency-changing techniques, due mainly to the increased use of hot carrier diodes, have caused us to consider afresh a fundamental dictum of NMR., the $\frac{3}{2}$ power law, and to examine anew the derivation of the formula, *which was derived prior to the advent of superconducting systems.*

PRIMARY AND SECONDARY CONSIDERATIONS

The usual formula for the signal-to-noise ratio available after a 90° pulse is given by (*1*, *2*)

$$\Psi_{\mathrm{rms}} = K\eta M_0 (\mu_0 Q \omega_0 V_c / 4FkT_c \Delta f)^{1/2}, \qquad [1]$$

where K is a numerical factor (~1) dependent on the receiving coil geometry; η is the "filling factor," i.e., a measure of the fraction of the coil volume occupied by the sample; M_0 is the nuclear magnetization which is proportional to the field strength B_0; μ_0 is the permeability of free space; Q is the quality factor of the coil; ω_0 is the Larmor angular frequency; V_c is the volume of the coil; F is the noise figure of the preamplifier; k is Boltzmann's constant; T_c is the probe (as opposed to sample) temperature; and Δf is the bandwidth (in Hertz) of the receiver.

We may consider that the primary factors involved in any analysis of S:N are those contained within the equation. Secondary factors, such as whether or not quadrature detection is used, the availability of Fourier transform techniques, sweep rate or pulsing rate, the use of decoupling or Overhauser effect, though of great importance, are not of such a fundamental nature and will therefore not be considered further.

For all its usefulness, Eq. [1] is not a "fundamental" equation. It contains four unknowns, K, η, Q and F, only two of which (Q and F) are easily measurable. The definition of filling factor $\eta = V_s/2V_c$ (V_s is the sample volume) may well be satisfactory for a solenoidal coil, but its validity for other coil configurations must be questioned. Further, the equation contains little information as to the dependency of S:N on various physical parameters; for example, if we quadruple the coil volume while keeping η constant, do we obtain only a doubling of S:N, or does the change in the coil dimensions alter K and Q also? Table 1 shows how complex the use of Eq. [1] may be. The interac-

TABLE 1

INTERACTION OF TERMS IN THE TRADITIONAL EQUATION FOR SIGNAL-TO-NOISE RATIO AS GIVEN BY EQ. [1]

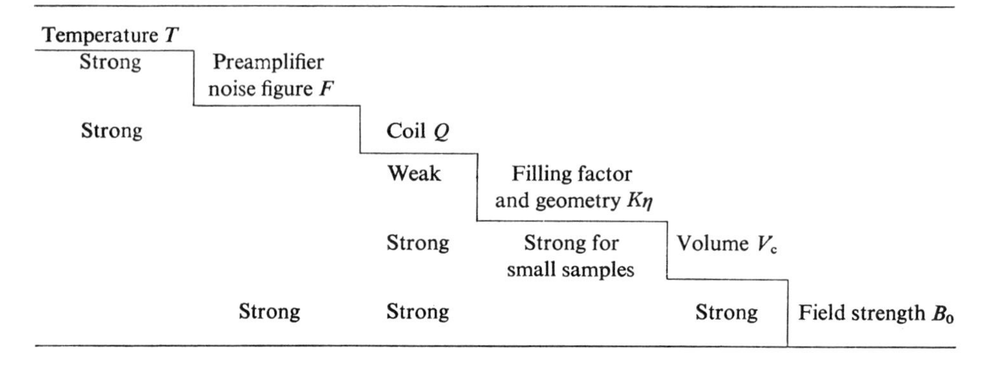

Temperature T					
Strong	Preamplifier noise figure F				
Strong		Coil Q			
		Weak	Filling factor and geometry $K\eta$		
		Strong	Strong for small samples	Volume V_c	
	Strong	Strong		Strong	Field strength B_0

tions between the various factors are manifold and mostly strong. This is partly a consequence of the method of calculation. The reader is referred to Ref. (*1*, *2*) for further details, but it is perhaps worth quoting Abragam who states that "the above calculation gives only an order of magnitude."

THE PRINCIPLE OF RECIPROCITY

Clearly it would be of use to formulate a different method of calculation which gives a direct insight into the various factors involved and which removes some of the interactions inherent in Eq. [1]. This may be done to a reasonable extent by invoking the principle of reciprocity. Consider the induction field $\mathbf{B}_1$ produced by a coil C carrying unit current (See Fig. 1). Obviously, the field at point A is much stronger than at point B. Intuitively, one would expect therefore that if a magnetic dipole $\mathbf{m}$ were placed at point A and set rotating about the z axis, the alternating signal it induced in the coil would be much greater than that induced by the same dipole placed at point B. This is indeed the case, and it may easily be shown that the induced emf is given by

$$\xi = -(\partial/\partial t)\{\mathbf{B}_1 \cdot \mathbf{m}\}, \quad [2]$$

where $\mathbf{B}_1$ is the field produced by the unit current at $\mathbf{m}$. It follows that for a sample of volume V_s, which has been recently subjected to a 90° pulse, we need only know the

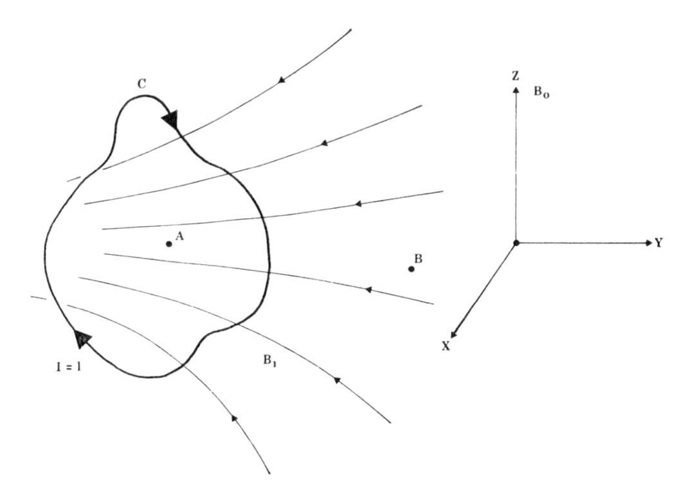

FIG. 1. The induction field $\mathbf{B}_1$ produced by a coil C carrying unit current.

value of $\mathbf{B}_1$ at all points in the sample to be able to calculate the emf induced in the coil. Thus, if $\mathbf{M}_0$ lies in the xy plane,

$$\xi = -\int_{\text{sample}} (\partial/\partial t)\{\mathbf{B}_1 \cdot \mathbf{M}_0\}\, dV_s. \quad [3]$$

The calculation of $\mathbf{B}_1$ is feasible for most shapes of coil; of course, if $\mathbf{B}_1$ may be considered to be reasonably homogeneous over the sample volume, the calculation is considerably simplified as the integration of Eq. [3] becomes trivial, giving

$$\xi = K\omega_0 (B_1)_{xy} M_0 V_s \cos \omega_0 t, \quad [4]$$

where K is an "inhomogeneity factor" which may if necessary be calculated, $(B_1)_{xy}$ is the component of $\mathbf{B}_1$ perpendicular to the main field B_0, and phase has been neglected. The magnetization M_0 is given by

$$M_0 = N\gamma^2 \hbar^2 I(I+1) B_0 / 3kT_s, \quad [5]$$

where N is the number of spins at resonance per unit volume, γ is the magnetogyric ratio, and T_s is the sample temperature. As $\omega_0 = -\gamma B_0$ it follows from Eqs. [4] and [5] that the EMF induced in the coil is proportional to the square of the Larmor frequency.

THE NOISE

Having laid the basis of the calculation for the emf induced by the nuclear magnetization in the receiving coils, we now consider the noise. In a correctly designed system, this should originate solely from the resistance of the coil. As the dimensions of the coil are inevitably much less than the wavelength of the radiation involved, the radiation resistance is negligible, and it should therefore be possible to predict the thermal noise present purely on the basis of the equation

$$V = (4kT_c \Delta f R)^{1/2}. \qquad [6]$$

Here, T_c is the temperature of the coil and R its resistance. Unfortunately, the calculation of R may not be performed accurately, and it is solely this factor which leads to uncertainty in the theoretical prediction of signal-to-noise ratio. At the frequencies of most interest in NMR spectroscopy (say >5 MHz) the "skin effect" associated with the magnetic field generated by a current ensures that that current flows only in regions of the conductor where there is also a magnetic flux. Thus, for a long, straight cylindrical conductor, the current flows in a skin on the surface. This situation is easily amenable to calculation and yields the result that

$$R = (l/p)\,(\mu\mu_0\,\omega_0\,\rho(T_c)/2)^{1/2}, \qquad [7]$$

where l is the length of the conductor; p is its circumference, μ is the permeability of the wire; and $\rho(T_c)$ is the resistivity of the conductor, which is of course a function of temperature. However, the situation is considerably more complicated if the conductor is not cylindrical, and worse still, if there are many conductors in close proximity, as is, in effect, the case with a coil, the magnetic field created by the current of one conductor influences the distribution of current in another. This "proximity effect" (*4*), which is also manifest when conductors such as the silvering on a dewar, or even the sample itself, are close to the coil, normally tends to reduce the surface area over which current is flowing, and thus the resistance is increased from that calculated from Eq. [7] by a factor ζ of about 3. Attempts have been made to calculate ζ, but only for a single-layer solenoid may any confidence be placed in the results (*5*).

THE SIGNAL-TO-NOISE RATIO

By combining Eqs. [4] to [7] we may arrive at an equation for the signal-to-noise ratio,

$$\Psi_{\mathrm{rms}} = \frac{K(B_1)_{xy}\,V_s\,N\,\gamma\,\hbar^2\,I(I+1)}{7.12\,kT_s}\cdot\left(\frac{p}{FkT_c l\zeta\Delta f}\right)^{1/2}\cdot\frac{\omega_0^{7/4}}{[\mu\mu_0\,\rho(T_c)]^{1/4}}. \qquad [8]$$

Note that the proximity factor ζ and the noise figure F of the preamplifier have been included. At first sight, this equation appears unmanageable, but this is in fact not the case. First, the unknowns η and Q of Eq. [1] are absent; they have been replaced by a single function ζ, which, from experience, is reasonably well known, and second, the number of factors in Eq. [8] which are variable in a given experimental situation is small. These factors are:

(a) $K(B_1)_{xy}$ the effective field over the sample volume produced by unit current flowing in the receiving coil.

(b) p the perimeter of the conductor.

(c) l the length of the conductor.

The above factors are dependent only on coil geometry.

(d)	T_c	the temperature of the coil.
(e)	$\rho(T_c)$	the material from which the coil is made.
(f)	F	the quality of the preamplifier.

It is of interest to note that the frequency dependence is to the power of 7/4. This is not a new conclusion; it has also been postulated by Soutif and Gabillard (*6*) and it must be stressed that, *for a solenoid*, in no way is Eq. [8] in contradiction with Eq. [1]. Figure 2 shows an experimental plot of Q versus frequency for a set of solenoids all wound in the same manner with the same overall dimensions. This plot clearly recon-

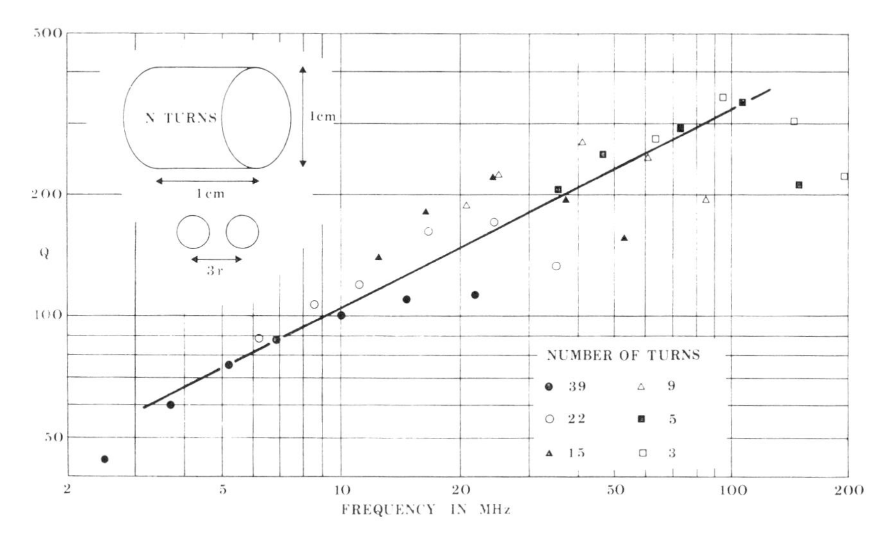

FIG. 2. The Q of a set of solenoids. The dimensions of each solenoid are the same, but the number of turns varies with frequency. The winding geometry is shown in the inset, and gives the optimum Q (*4*). For many turns, and well below the self-resonant frequency, $Q \propto f^{1/2}$.

ciles the two equations and the derivation of Eq. [1] from Eq. [8] is shown in Appendix 1. The major advantage of the calculation from first principles which results in Eq. [8] is that it is applicable to any coil geometry, for it is found that ζ changes but little with a change of configuration provided the separation of the windings is small in comparison with the overall dimensions of the coil. Of particular interest are saddle-shaped (or Helmholtz) coils used mainly with superconducting instruments and solenoidal coils used predominantly with conventional machines. Let us therefore compare the performance of these two configurations.

SADDLE-SHAPED AND SOLENOIDAL COILS

The factors of interest are (a) to (c) above, and so, assuming the same volume of sample is used in each case, the essential part of Eq. [8] is

$$\Psi \propto V_s (B_1)_{xy}/R^{1/2} \qquad \text{or} \qquad \Psi \propto V_s (B_1)_{xy} (p/l)^{1/2}. \qquad [9]$$

Our first task therefore is to calculate $(B_1)_{xy}$ for the two coils. While this is trivial for a many-turn solenoid, it is not for the saddle-shaped coils, and the essence of the calculation is indicated in Appendix 2. Let a be the radius of the coils, and $2g$ the lengths, as shown in Fig. 3.

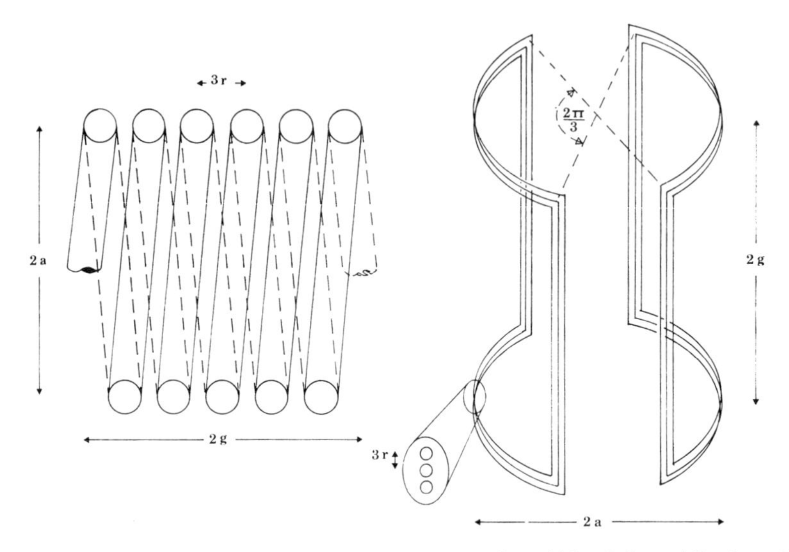

FIG. 3. The two winding geometries considered. The angular width of the saddle-shaped coils is 120°, as this value gives the best homogeneity, and the width of the windings is approximately $g/5$.

Saddle-shaped

$$(B_1)_{xy} = \frac{n\sqrt{3}\,\mu_0}{\pi}\left\{\frac{ag}{[a^2+g^2]^{3/2}} + \frac{g}{a[a^2+g^2]^{1/2}}\right\},$$

$$n \geqslant 1.$$

Solenoid

$$(B_1)_{xy} = \frac{\mu_0 n}{2}\frac{1}{[a^2+g^2]^{1/2}},$$

$$n \gg 1.$$

Typically, $a \simeq g$, and hence

$$(B_1)_{xy} = 0.585\, n\mu_0/a. \qquad (B_1)_{xy} = 0.354\, n\mu_0/a.$$

We must now calculate p and l. Let us assume that wire is used in both cases. We then have that

$$l \simeq 8na\{(g/a) + (\pi/3)\}. \qquad l \simeq 2\pi an.$$

When $a \simeq g$,

$$l \simeq 16.4\, na. \qquad l \simeq 6.3\, na.$$

The length of wire per unit turn is thus greater for the saddle-shaped coil by a factor of about 2.6, a significant fact which will be considered later. The radius r of the wire used is dependent upon the manner in which the coils are wound, but assuming that a planar structure is used, the n turns must fit in a length $2g$ for the solenoid and in a width $g/5$ approximately for the saddle-shaped coils. To optimize the performance of each coil with respect to proximity effect, the distance between the centers of each turn should be roughly $3r$ (*4*), and so, assuming this value, we have that for $a \simeq g$,

$$3r(n-1) \simeq a/5, \qquad\qquad 3r(n-1) = 2a,$$

$$\therefore\ p \simeq 2\pi a/15(n-1). \qquad\qquad \therefore\ p = 4\pi a/3(n-1).$$

Thus, for $n \gg 1$,

$$p \simeq 0.42a/n, \qquad\qquad p \simeq 4.2\,a/n,$$

$$\therefore\ \psi \propto 0.094\,\mu_0\,V_s/a. \qquad\qquad \therefore\ \psi \propto 0.29\,\mu_0\,V_s/a. \qquad [10]$$

Thus the performance of the solenoid would appear to be approximately three times better than that of the saddle-shaped coils. It is difficult to explain this result on the basis of the traditional formula as represented by Eq. [1]. This is not surprising, as there is a critical assumption in the derivation (shown in Appendix 1) involving the energy stored by current flowing in the coil. Briefly, this assumption is that half the energy is stored, via the field $\mathbf{B}_1$, within the confines of the coil and that the field is homogeneous within those confines. While this is broadly true for a solenoid, due to the continuity and closed form of the winding, it is certainly not true for the saddle-shaped coils, where the open structure dictates that much of the magnetic energy is stored in flux which lies close to the wires and which does not pass through the sample. A further deficiency of Eq. [1] concerns the length of a 90° pulse when the probe is of the single-coil variety. One would be led to believe that the length was dependent only on $Q^{-1/2}$. Typically a saddle-shaped coil has a lower Q than a solenoid, and so the 90° pulse length should be a little longer (up to, say, 60%) but Eq. [10] show the inadequacy of this statement. This may be seen from the following argument.

90° PULSE LENGTHS

When the probe is matched to the transmitter (a point to be considered later), the power supplied W is dissipated entirely in the resistance R of the coil. The current flowing through the coil is thus given by

$$I = (W/R)^{1/2}.$$

Now the $\mathbf{B}_1$ field employed in the calculations to date has been derived for unit current. It follows that when the transmitter is on, the irradiating field $(B_1^*)_{xy}$ is given by

$$(B_1^*)_{xy} = I(B_1)_{xy} = (B_1)_{xy}(W/R)^{1/2}.$$

Thus, from Eq. [9],

$$(B_1^*)_{xy} \propto \mathrm{W}^{1/2}\,\psi/V_s. \qquad [11]$$

The 90° pulse length is thus a direct measure of the S:N obtainable from a single coil system and if the preceding calculations are correct, one would expect the length to be about three times longer for the saddle-shaped coils. To check the results obtained, experiments were performed at two frequencies: 20 MHz, where the condition $n \gg 1$ holds, and 129 MHz, where it does not. Table 2 shows the data collected, which are in good agreement with the theory. Finally, it is of interest to note that the results of Eq. [10] are independent of the number of turns on the coil. This is equivalent to saying that

TABLE 2

EXPERIMENTAL RESULTS[a] SHOWING THE SUPERIOR PERFORMANCE OF A SOLENOIDAL RECEIVING COIL AS COMPARED TO A SADDLE-SHAPED COIL WHEN USING THE SAME SAMPLE: A SPHERE 7.5 MM IN DIAMETER

Frequency	Coil type	Number of turns	Q	Height (mm)	Radius (mm)	90° Pulse[b] (μsec)	Signal[c] (volts)
129 MHz, ^{31}P	Saddle-shaped	2	210	7	4	27	1.0
	Solenoid	4	300	6	4	9	2.6
			Ratio 1.4			Ratio 3	Ratio 2.6
20 MHz, ^{1}H	Saddle-shaped	6	80	10	5	46	0.78
	Solenoid	18	208	10	5	18	2.20
			Ratio 2.5			Ratio 2.6	Ratio 2.8

[a] All values obtained are accurate to better than 10%.

[b] No comparison is intended between the performances at the two frequencies as the two transmitters are of different powers.

[c] The signals are measured relative to a constant noise background.

the Q of a coil is mainly dependent on the overall dimensions of that coil rather than the number of turns within those dimensions. That this is broadly true may be seen from Fig. 2.

DEPENDENCIES

From the various equations derived, it is possible to determine how the signal-to-noise ratio varies as parameters are altered. With regard to coil geometry, the latter is usually determined by the homogeneity of the main magnetic field B_0. Thus, for example, a superconducting magnet with a field of 11 tesla might have a homogeneity of one part in 10^9 over 1 ml, whereas with a 4 tesla magnet the equivalent volume might be 25 ml. For simplicity, we shall assume that coil length and diameter are approximately equal, though it should be stressed that this ratio is not necessarily optimal for the best utilization of the main field homogeneity nor, in the case of a saddle-shaped coil, does it produce the best B_1 homogeneity (*7*). Thus, from Eq. [10] we can see that if *all* linear dimensions are scaled in the same manner, the S:N varies as a^2 or as $V^{2/3}$. Assuming a frequency dependence of $\omega_0{}^{7/4}$, it is thus easy to show that the low-field magnet gives 25% *better* signal to noise than its high-field counterpart. It has of course been assumed that 25 ml of sample is available. In most biochemical applications of NMR, such extravagence would be unthinkable and so, for higher sensitivity, a higher field is still called for.

The major variable in Eq. [8] not yet considered is temperature. While the sample temperature T_s may well be fixed by the chemistry of the system under observation, there is no such limitation on the temperature T_c of the probe, and if the latter is cooled, significant improvements may be obtained. From Eq. [10], *for a constant sample volume*, the signal-to-noise ratio varies only as a^{-1}, and so the insertion of a dewar vessel between the coil and the sample may well only decrease ψ by 30%. On the other hand, at liquid nitrogen temperature (77 K), T_c is a quarter of room temperature and thus, if the preamplifier has an excellent noise figure, the noise is reduced by a factor of 2 on cooling. However it must not be overlooked that the conductivity of the conductor is temperature dependent, and for copper, $\rho(T_c)$ is one-tenth of its room-temperature value at 77 K. The net gain in signal-to-noise ratio obtainable by cooling the probe is therefore about 2.5 (Note, however, that the 90° pulse length should decrease by only about 25%.) Of course, if the sample may also be cooled, further improvement is possible.

LIMITATIONS OF THE FORMULA

So far, little indication has been given of the limitations of Eq. [8]. These are predominantly twofold. First, the calculation breaks down below about 5 MHz, where, for an average size of sample, the radius of the wire used becomes comparable with the skin depth. Second, the calculation breaks down when the distributed capacitance between the turns on the coil is sufficiently large at the frequency of interest to change the phase of the emf induced in one part of the coil relative to another. In its extreme manifestation, this effect causes self-resonance, a condition whereby the coil resonates without an external tuning capacitor. This should be avoided, and the onset of self-resonance may be seen in each of the curves of Fig. 2, where, with increasing frequency, the Q of each coil begins to drop away from the $f^{1/2}$ line. To avoid this effect, the number of turns on the coil must be decreased with increasing frequency. There is a limit of course to this procedure; for the saddle-shaped coil, the limit is a single turn. In this case, the formula still holds, provided attention is paid to the fact that the leads from the coil and the links between the two sections contribute appreciable resistance. A further modification may be to connect the two halves in parallel rather than series. This situation may also be accounted for and allows satisfactory use up to about 300 MHz. Also, it is usual for the conductor to be foil rather than wire with such a configuration.

With a solenoid, the calculation breaks down when the condition $n \gg 1$ is not obeyed and this may be seen in Fig. 2 to occur for frequencies in excess of 150 MHz. However, a single-turn solenoid fabricated from foil is a special case for which provision can be made, and whereas a single-turn saddle-shaped coil is useful up to about 200 MHz, a single-turn solenoid is useful to 600 MHz. The reason for this behavior lies in the fact, mentioned earlier, that a saddle-shaped coil requires 2.6 times the length of conductor per unit turn of a solenoid, and thus has greater inductance and self-capacitance per unit turn. This is ironic when it is remembered that it is only at the high frequencies made available by superconducting systems that the saddle-shaped configuration is required. It is the authors' opinion that the disappointing signal-to-noise ratio experienced with superconducting systems is a direct consequence of the use of saddle-shaped coils, and the construction of a spectrometer to work at a frequency of 470 MHz being undertaken in this laboratory presents a considerable challenge, as the only satisfactory coil configuration so far found is solenoidal.

A further limitation is the amount of space available for the construction of the coil. If there is appreciable coupling, magnetic or electrostatic, between the coil and, say, a shield or a strongly conducting sample (for example, a saline solution), then in general, the S:N ratio will be degraded. The mechanism by which the degradation takes place is dependent upon geometry and frequency; the field generated by unit current may be lessened by induction, proximity effect may change the coil resistance, and there may be resistive losses in the coupled element. The simplest way of monitoring these effects is to measure the Q of the coil in free air and then to observe the change when the coil is in place in the probe. It should preferably be less than 10%, and a good rule of thumb for obtaining this value is that no conductor should be closer to the coil than the largest dimension of the latter. The effects of coupling are particularly noticeable at low temperatures. It is quite easy to obtain Q's of over 600 at 77 K in free liquid nitrogen, but another matter in the confines of the probe. Of particular importance is coupling to conductors at room temperature. The component of resistance introduced into Eq. [6] by this coupling carries with it a temperature four times greater than that of liquid nitrogen and its noise contribution is thus disproportionately large. It might be added that the construction of a low-temperature probe with a room-temperature sample is no easy matter, and that to date, we do not have a reliable system.

THE PREAMPLIFIER AND TRANSMITTER

The only factor not so far considered is the noise figure of the preamplifier. For the frequency range 50 to 500 MHz the best semiconductor now available is probably a gallium arsenide field effect transistor. The authors have described elsewhere the design and construction of a preamplifier with a noise figure of 0.3 db at 129 MHz (*3*), and it

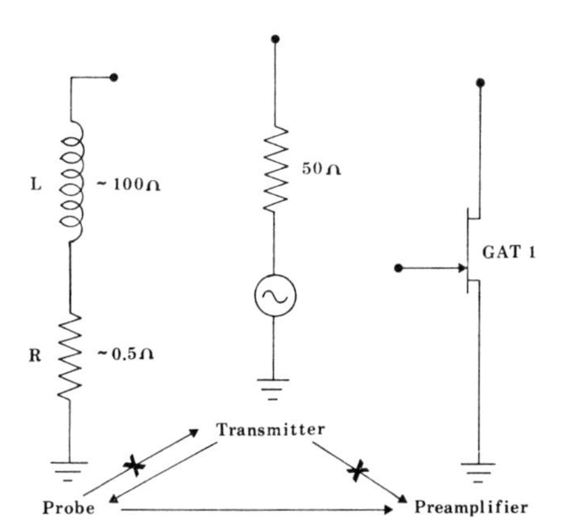

FIG. 4. The three elements probe, transmitter, and transistor must be interfaced in such a way that the signal-to-noise ratio is not degraded and the transistor is not damaged.

remains therefore to consider in what manner the probe coil and the preamplifier can be interfaced in order to obtain the best noise performance. If a single-coil probe is used, there is also the problem of interfacing to the transmitter, while protecting the preamplifier from the pulses and conserving the noise performance. As this is the most difficult situation likely to be encountered, it is to this that we turn our attention. The problem is illustrated in Fig. 4. Considering first the interface between probe coil and transmitter, it is obvious that at the frequency of interest, the impedance of the coil $Z \sim 0.5 + j\ 100$ ohms $[j = (-1)^{1/2}]$ must be transformed in such a manner as to power match the transmitter. As is well known, an essentially lossless transformation may be effected with the aid of the circuit of Fig. 5, provided high-Q capacitors are used. However, a word of warning is required here. It is desirable to keep the leads from the coil

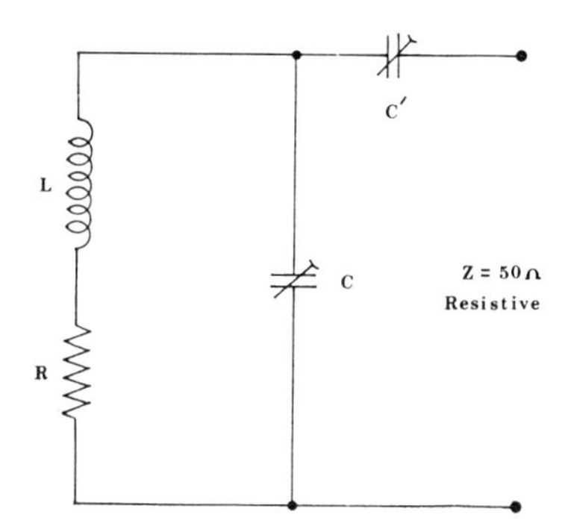

FIG. 5. Impedance transformation using a tuned circuit. The value of C is slightly less than that required to tune to resonance and the matching capacitance C' is given by $C' \simeq (C/50Q\ \omega_0)^{1/2}$.

to the capacitors as short as possible to minimize the resistance R. In this situation, the capacitor C is physically close to the coil and unfortunately, many high-Q variable capacitors are ferromagnetic. The main field homogeneity is thus disturbed.

Turning now to the interface between the coil and the F.E.T. one must ask under what conditions the transistor gives its best performance. Robinson (*8*) has considered this problem and has shown that the optimum noise figure is obtained when the input capacitance is almost tuned out and the signal source has a source impedance which is resistive and given by

$$R_{SO} \simeq 1.6 f_T / f g_m, \qquad [12]$$

f_T is the figure of merit for the F.E.T. and is given by

$$f_T = g_m / 2\pi C_j, \qquad [13]$$

where g_m is the transconductance of the device and C_j is the junction capacitance. Alternatively,

$$R_{SO} \simeq 1.6/2\pi f C_j. \qquad [14]$$

With a junction capacitance of 2 pF and a frequency of $f = 129$ MHz, $R_{SO} \simeq 980\ \Omega$. A practical value obtained with the amplifier of Ref. (*3*) gave $R_{SO} = 800\ \Omega$. On the other hand, the input impedance of the F.E.T. with its junction capacitance tuned out is resistive, and given by

$$R_{in} \simeq g_m/(2\pi f C_j)^2. \qquad [15]$$

At 129 MHz, and taking the values $C_j = 2$ pF, $g_m = 15$ mA/V, we have $R_{in} \simeq 5.7$ kΩ. Obviously, the source and the F.E.T. are grossly mismatched powerwise when they are noise matched, and it follows that one should never tune a probe by looking for the maximum signal from the receiver. The signal may well be a maximum; the signal-to-noise ratio most certainly will not be.

The probe has been matched to 50 Ω resistance in order to power match to the transmitter. It follows, to make a noise match to the F.E.T., that 50 Ω must be transformed to R_{SO} within the preamplifier. This, and the tuning out of the gate capacitance C_j, is easily accomplished using the transformation properties of tuned circuits and the reader is referred to Ref. (*3*) for further details. In general, the greater the ratio f_T/f, the better the noise performance. However, in the pursuit of excellence one must beware, particularly at low frequencies, of making the optimum source impedance higher than is practical. Above a value of about 2 kΩ, losses in the transforming device must also be taken into account. Hence to obtain very low noise figures, it may be necessary to cool the preamplifier.

Finally, the preamplifier must be protected from the potentially destructive transmitter pulses, and if a class A transmitter is used, there must be a gate of some sort which prevents the injection of noise from transmitter to receiver. Nor must the protection or the gate degrade the noise performance of the system. As one progresses to higher frequencies, crossed diodes (*9*) become increasingly ineffective due to their junction capacitance—typically 4 pF. Not only does such a large value allow noise to pass from the transmitter; it also ensures that any diodes used to protect the F.E.T. form, above say 50 MHz, a major part of the tuning capacitance. This may, depending on the diode, be disastrous, for after passing heavy current, many diodes exhibit a change of capacitance which lasts many milliseconds. This change can ruin the noise performance of a tuned amplifier and even cause it to oscillate. A far more elegant way to protect the preamplifier is to use PIN diodes (*10*) but unfortunately, PIN diodes have a low resistance to radio frequencies when they are passing a heavy direct current (say, 2 Ω at 40 mA). A PIN diode in the line from the probe to the preamplifier therefore introduces shot noise and can degrade the noise figure of the receiver by as much as 4 db. Fortunately, it is possible to construct a PIN diode circuit which not only protects the preamplifier (60 db isolation) from the transmitter, but which also has all the diodes "off" when the spectrometer is receiving signal. Details may be found in Ref. (*11*).

CONCLUSION

The authors have attempted to provide a direct physical picture of the factors governing the signal-to-noise ratio in an NMR experiment. It has been shown that the signal received from a sample by a set of coils is directly proportional to the magnetic field that would be created at the sample if unit current were passed through the coils, while the noise present in the coils has been shown to be purely a function of the coil resistance. This simple argument allows a direct comparison of the efficiency of different coil configurations to be made. For example, while unit currents flowing through saddle-shaped and solenoidal coils create similar B_1 fields, and the coils thus receive similar signals from the sample, the resistance of a saddle-shaped coil is considerably larger than that of a solenoid and so the signal-to-noise ratio is much less. The correct manner of interfacing between the probe, the transmitter and the preamplifier has been discussed, and attention has been drawn to the importance of noise matching the probe to the amplifying device used, and the distinction between noise and power matching. Finally, the problems of protecting the receiver from the transmitter pulses and noise have been considered, and the use of PIN diodes advocated as a solution.

APPENDIX 1: THE EQUIVALENCE OF THE PRESENT AND THE TRADITIONAL FORMULATIONS

From Eqs. [4] and [6], the signal-to-noise ratio may be written as

$$\Psi_{\mathrm{rms}} = K\omega_0 (B_1)_{xy} M_0 V_s/(8kT_c R\Delta f)^{1/2}. \qquad [16]$$

To convert to the traditional formula of Eq. [1], we must find a relationship between the energy stored in the B_1 field, which is a measure of the coil inductance, and the value of $(B_1)_{xy}$ at the sample. If over the sample volume, the B_1 field is predominantly homogeneous and in the xy plane, we may say that $(B_1)_{xy} \simeq B_1$ for the sample. The energy stored in the sample volume is given by

$$E = \frac{1}{2\mu_0} \int_{\mathrm{sample}} B_1^2\, dV \simeq (B_1)_{xy}^2 (V_s/2\mu_0). \qquad [17]$$

The inductance of the coil is given by

$$L = (1/\mu_0) \int_{\mathrm{all\ space}} B_1^2\, dV. \qquad [18]$$

If, following Hill and Richards (*2*), we define the filling factor as

$$\eta = \int_{\mathrm{sample}} B_1^2\, dV \Big/ \int_{\mathrm{all\ space}} B_1^2\, dV, \qquad [19]$$

then from Eqs. [17] to [19],

$$(B_1)_{xy} \simeq (\mu_0 \eta L/V_s)^{1/2}$$

or $\qquad [20]$

$$K(B_1)_{xy} = \bar{K}(\mu_0 \eta L/V_s)^{1/2},$$

where mean and root mean square inhomogeneity factors have been introduced, $K \simeq \bar{K} \simeq 1$.

Substituting in Eq. [16] and adding the noise figure of the preamplifier we thus obtain

$$\Psi_{\mathrm{rms}} = KM_0 \left[\left(\frac{\omega_0 L}{R}\right) \frac{\eta \mu_0 V_s}{8kT_c F\Delta f}\right]^{1/2}. \qquad [21]$$

For the special case of a solenoid, it may be shown that

$$\int_{\mathrm{coil\ volume}\ V_c} B_1^2\, dV \simeq \tfrac{1}{2} \int_{\mathrm{all\ space}} B_1^2\, dV.$$

Thus if the field within the solenoid is homogeneous

$$\eta \simeq V_s/2V_c. \qquad [22]$$

Substitution in Eq. [21] gives Eq. [1],

$$\Psi_{\mathrm{rms}} = K\eta M_0 (\mu_0 Q\omega_0 V_c/4FkT_c \Delta f)^{1/2},$$

which is valid only for a solenoid.

APPENDIX 2: THE FIELD AT THE CENTER OF A SADDLE-SHAPED COIL

The vector magnetic potential **A** at point P due to an element of arc **ds** is given by

$$d\mathbf{A} = (\mu\mu_0 I/4\pi)\,(d\mathbf{s}/v), \qquad [23]$$

where $v = |\mathbf{p} - \mathbf{a}|$ is the distance of P from **ds** (see Fig. 6). For the special case of P at

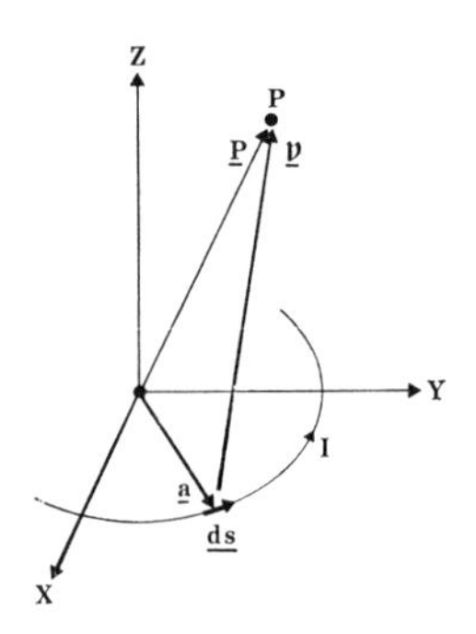

FIG. 6. The coordinate system.

the center of the coil of Fig. 3, we have that

$$v = (a^2 + g^2)^{1/2}$$

and further, the contributions to $(B_1)_{xy}$ from all four arcs add. Thus we have that

$$(\mathbf{B}_1)_{\text{arcs}} = \text{curl}\left\{\frac{\mu\mu_0 I}{\pi}\int_{-\pi/3}^{+\pi/3} \frac{(-a\sin\phi)\,\mathbf{i} + (a\cos\phi)\,\mathbf{j}}{(a^2+g^2)^{1/2}}\,d\phi\right\}$$

where **i** and **j** are unit vectors in the x and y directions.

$$\therefore \quad (\mathbf{B}_1)_{\text{arcs}} = \text{curl}\left\{\frac{\sqrt{3}\,\mu\mu_0 Ia}{\pi\,(a^2+g^2)^{1/2}}\mathbf{j}\right\};$$

$$\therefore \quad (\mathbf{B}_1)_{\text{arcs}} = -\frac{\sqrt{3}\,\mu\mu_0 I}{\pi}\frac{ag}{[a^2+g^2]^{3/2}}\mathbf{i}. \qquad [24]$$

The potential due to one of the four verticals of the coil is given by

$$\mathbf{A} = \int_{-g}^{+g} \frac{\mu\mu_0 I}{4\pi\,(a^2+z^2)^{1/2}}\,d\mathbf{z}$$

and the field produced by them, which is parallel to that produced by the arcs, is given by

$$(\mathbf{B}_1)_{\text{verticals}} = \frac{\sqrt{3}\,\mu\mu_0 I}{2\pi}\frac{\partial}{\partial a}\left\{\int_{-g}^{+g} \frac{dz}{(a^2+z^2)^{1/2}}\right\}\mathbf{i}$$

$$= \frac{\sqrt{3}\,\mu\mu_0 I}{\pi}\frac{\partial}{\partial a}\left\{\sinh^{-1}\left(\frac{g}{a}\right)\right\}\mathbf{i}; \qquad [25]$$

$$\therefore \quad (\mathbf{B}_1)_{\text{verticals}} = -\frac{\sqrt{3}\,\mu\mu_0 I}{\pi}\frac{g}{a(a^2+g^2)^{1/2}}\mathbf{i}.$$

Hence, from Eqs. [24] and [25], the field at the center of a saddle-shaped coil which is passing unit current is given by

$$(B_1)_{xy} = \frac{\sqrt{3}\,\mu\mu_0}{\pi}\left\{\frac{ag}{(a^2+g^2)^{3/2}} + \frac{g}{a(a^2+g^2)^{1/2}}\right\}. \qquad [26]$$

By using the type of analysis, briefly indicated above, for a point P which is off-center, the total magnetic field can be analyzed in a series of spherical harmonics, from which it may be shown that the optimum homogeneity is obtained when the angular width of the coil is 120°, as shown in Fig. 3, and the length is twice the diameter.

REFERENCES

1. A. ABRAGAM, "The Principles of Nuclear Magnetism," pp. 82–83, Clarendon Press, Oxford, 1961.
2. H. D. W. HILL AND R. E. RICHARDS, *J. Phys. E, Ser. 2* **1,** 977 (1968).
3. D. I. HOULT AND R. E. RICHARDS, *Electron. Lett.* **11,** 596 (1975).
4. F. E. TERMAN, "Radio Engineer's Handbook," 1st ed., pp. 77–85, McGraw–Hill, New York, 1943.
5. A summary of Butterworth's extensive work on this subject is given by B. B. AUSTIN, *Wireless Eng. Exp. Wireless* **11,** 12 (1934).
6. M. SOUTIF AND R. GABILLARD, "La Résonance paramagnétique nucléaire" (P. Grivet, Ed.), 1st ed., pp. 149–161, Centre National de la Recherche Scientifique, Paris, 1955.
7. D. I. HOULT, D. Phil. Thesis, Oxford, 1973.
8. F. N. H. ROBINSON, "Noise and Fluctuations in Electronic Devices and Circuits," Chaps. 11, 12, 13, Clarendon Press, Oxford, 1974.
9. I. J. LOWE AND C. E. TARR, *J. Phys. E, Ser. 2* **1,** 320 (1968).
10. K. E. KISMAN AND R. L. ARMSTRONG, *Rev. Sci. Instrum.* **45,** 1159 (1974).
11. D. I. HOULT AND R. E. RICHARDS, *J. Magn. Resonance* **22,** 561 (1976).

JOURNAL OF MAGNETIC RESONANCE **24**, 201–204 (1976)

Digital Filtering with a Sinusoidal Window Function: An Alternative Technique for Resolution Enhancement in FT NMR

ANTONIO DE MARCO AND KURT WÜTHRICH

Institut für Molekularbiologie und Biophysik, Eidgenössische Technische Hochschule, 8093 Zürich–Hönggerberg, Switzerland

Received February 13, 1976

As an alternative to convolution difference techniques for resolution enhancement in the ^{1}H NMR spectra of proteins, digital filtering of the FID with a sinusoidal window function is suggested. As an illustration, this "sine bell routine" was applied to the high-field region of the ^{1}H NMR spectrum at 360 MHz of the basic pancreatic trypsin inhibitor.

Even with the highest currently available magnetic fields, the spectral resolution in the ^{1}H NMR spectra of proteins and other macromolecules is limited by the mutual overlap of the resonance lines of individual groups of protons (Fig. 1A). In FT NMR,

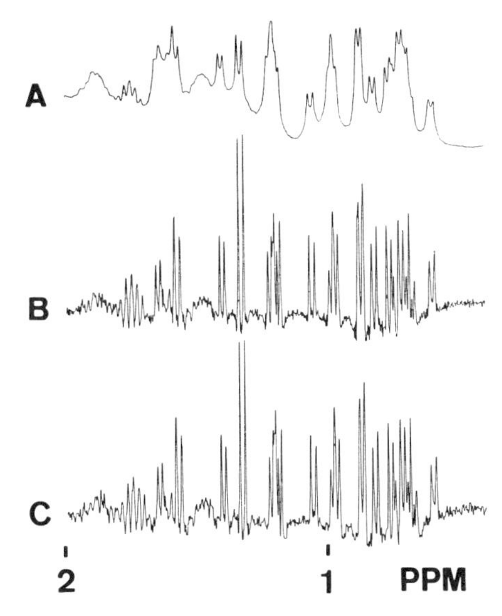

FIG. 1. Aliphatic region of the FT ^{1}H NMR spectrum at 360 MHz of the basic pancreatic trypsin inhibitor, 0.01 *M* solution in D_2O, 225 scans, 4000 Hz spectral width, 2 sec acquisition time, 16*K* of memory in the time domain. (A) Normal spectrum. The natural linewidth $\Delta\omega_{1/2}$ is approximately 20 rad sec^{-1}. (B) Same spectrum as (A) after digital filtering of the FID with the sine bell routine. (C) Same spectrum as (A) after application of the convolution difference routine with $\tau_1 = \infty$, $\tau_2 = 0.31$ sec, $K = 1$.

an artificial resolution enhancement can be obtained by suitable manipulation of the FID (*1*); several resolution-enhancement routines were proposed in the past (*2–4*). For work with proteins, convolution difference techniques (Fig. 1C) have so far generally been preferred (*5–9*). The present paper suggests digital filtering of the FID with a sinusoidal window function as an alternative technique, which yields a comparable resolution enhancement with comparable or lesser side effects of lineshape distortion and reduced signal:noise ratio (Fig. 1B). This "sine bell" routine seems attractive because of the simplicity of its practical applications.

In the sine bell routine, the free induction decay is multiplied by a sinusoidal function with zero phase and a period of twice the acquisition time AT. For this operation, we used the "Hanning Window" from the Lab-1180 General Signal Averaging Package, Nicolet Instrument Corporation. The lineshape after this procedure is given by

$$L(\mathrm{SIN}) \propto \int_0^{AT} \exp(-t/T_2^*)\cos(\Delta\omega t)\sin(\pi t/AT)\,dt, \qquad [1]$$

where T_2^* is the characteristic time for the decay of the transverse magnetization and $\Delta\omega$ the difference between the frequency considered and the resonance frequency. It is readily apparent that analogous to the convolution difference method, the broad components are drastically reduced compared to the sharp ones. Moreover, there should be essentially no truncation effects since the free induction decay is forced to be zero at time AT. Neglecting terms in $\exp(-AT/T_2^*)$, lineshape function [1] becomes

$$L(\mathrm{SIN}) \propto \left(\frac{\pi T_2^{*2}}{AT}\right)\frac{1+(\pi T_2^*/AT)^2-\Delta\omega^2 T_2^{*2}}{[1+(\pi T_2^*/AT)^2-\Delta\omega^2 T_2^{*2}]^2+4\Delta\omega^2 T_2^{*2}}. \qquad [2]$$

In the following, lineshape function [2] is compared with the Lorentzian line, with the lineshape obtained in convolution difference spectra. This comparison will mainly focus on the linewidth and the signal intensity I_0 at the resonance frequency, and the line distortion will also be considered. For practical reasons we take the width $\Delta\omega_{1/2}^0$ of the distorted lines obtained from the sine bell routine and the convolution difference routine as one-half of the half-width $\Delta\omega^0$ at the intensity zero (Fig. 2); the distortion D is defined as $|d/I_0|$ (Fig. 2). The resolution enhancement RE with respect to the Lorentzian is defined as $1/\Delta\omega_{1/2}^0 \cdot T_2^*$. Comparing the sine bell lineshape with the Lorentzian, we have from Eq. [2],

$$RE(\mathrm{SIN}) = \frac{2}{[1+(\pi T_2^*/AT)^2]^{1/2}}, \qquad [3]$$

$$\frac{I_0(\mathrm{SIN})}{I_0(\mathrm{LOR})} = \frac{\pi T^*/AT}{1+(\pi T_2^*/AT)^2}, \qquad [4]$$

$$D(\mathrm{SIN}) = \frac{1}{4}\frac{1+(\pi T_2^*/AT)^2}{1+[1+(\pi T_2^*/AT)^2]^{1/2}}. \qquad [5]$$

As shown in Fig. 3, RE(SIN) reaches a plateau close to 2 for relatively small values of AT/T_2^*, where the distortion is already close to its minimum value of $\frac{1}{8}$, but the relative intensity is still in a favorable range.

In the convolution difference method (*2*), the three parameters T_A, T_B, and K, where

$1/T_A = (1/T_2^*) + (1/\tau_1)$ and $1/T_B = (1/T_2^*) + (1/\tau_2)$, can be adjusted to obtain a suitable compromise of optimal resolution enhancement with acceptable signal intensity and line distortion. To obtain a meaningful comparison with the sine bell routine, we selected T_A and K so that the RE(CD) was at a maximum i.e., $T_A = T_2^*$ and $K = 1$, and then

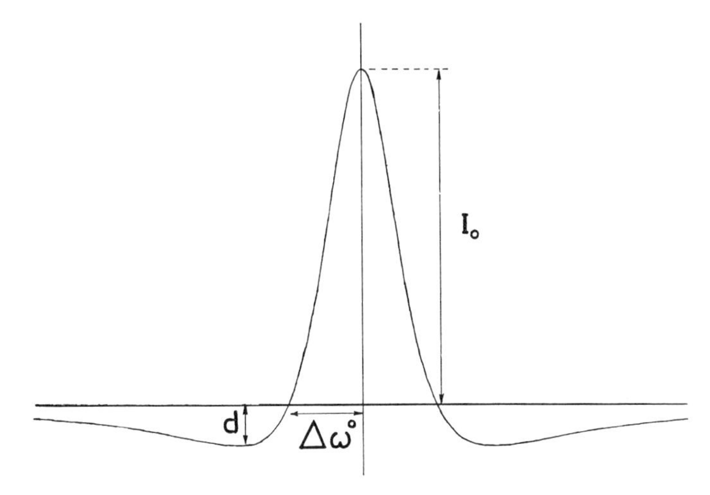

FIG. 2. Definition of the signal intensity I, the linewidth $\Delta\omega^0$, and the distortion $D = |I_0/d|$ for the resonance lines obtained with the resolution-enhancement routines.

adjusted τ_2 so that $I_0(\text{CD}) = I_0(\text{SIN})$. Assuming that $T_2^* = 0.05$ sec, which is quite typical for ^{1}H NMR spectra of proteins, we found that for an acquisition time $AT = 1$ sec, $RE(\text{SIN})/RE(\text{CD}) = 1.08$; with $AT = 2$ sec, $RE(\text{SIN})/RE(\text{CD}) = 1.04$. In both examples, the line distortions for the two routines were approximately the same and equal to ≈ 0.12. As an illustration, the two routines were applied to the aliphatic region of the basic pancreatic trypsin inhibitor, which is a "miniprotein" with molecular weight 6500 (*8*) (Fig. 1).

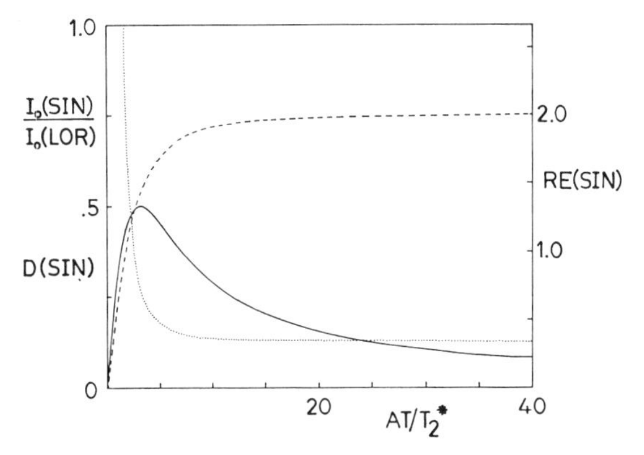

FIG. 3. Plots versus AT/T_2^* of the lineshape parameters resulting from the application of the sine bell routine. ——, intensity at centerband I_0(SIN) relative to the intensity in the Lorentzian lineshape I_0(LOR); -----, resolution-enhancement factor RE(SIN) relative to the Lorentzian lineshape; ....., distortion D(SIN) relative to the intensity at centerband.

The foregoing considerations show that for work with proteins the sine bell routine [1], which does not require selection of any arbitrary parameters besides the experimental acquisition time AT, yields a somewhat improved resolution enhancement as compared to the convolution difference routine applied with optimal choice of the parameters T_A, T_B, and K. For the sake of completeness it may be added that the sine bell routine can also be used in a form which includes an arbitrary parameter $K_S \geqslant 0$ corresponding essentially to the parameter $K \leqslant 1$ in the convolution difference method. The resulting lineshape is given by

$$L(\mathrm{SIN}) \propto \int_0^{AT} \exp(-t/T_2^*) \cos(\Delta\omega t)[K_S + \sin(\pi t/AT)]\,dt. \quad [6]$$

When this appears desirable, e.g., when working with very dilute solutions or with rare nuclei, K_S can be adjusted to obtain a suitable compromise of acceptable signal:noise ratio and resolution enhancement. On the other hand, since the characteristics of the sine bell window tend to suppress truncation effects, no undue line distortions result when the FID is deconvoluted by multiplication with $e^{t/\tau}$ prior to the application of the sine bell routine. RE greater than 2 can thus in practice be obtained by successive applications of the two routines.

The availability of convoluted spectra (Figs. 1B and 1C) in addition to the normal spectrum (Fig. 1A) is of considerable practical interest for studies of macromolecular systems. Since the broad components of the spectrum are eliminated in the altered data (Eq. [1]), the sharp lines are strongly emphasized. As a consequence the multiplet patterns of the narrow components are more readily recognized; spin decoupling thus becomes a more reliable technique for identification of the constituent spin systems, e.g., of individual amino acid residues in proteins (*5–8*). The convoluted spectra are also more readily amenable for a systematic determination of chemical shifts and spin–spin coupling constants. In certain cases it may further be of interest to determine the numbers of nuclei giving rise to broad and narrow resonances, respectively. In the normal spectrum real broad resonances cannot readily be distinguished from broad lines arising as a consequence of mutual overlap of many sharp lines. Comparison with the altered spectra, however, can in general provide this information.

ACKNOWLEDGMENT

Financial support by the Schweizerischer Nationalfonds (Project 3.1510.73) and by the Italian N.R.C. (stipend to A.DeN.) is gratefully acknowledged.

REFERENCES

1. R. R. Ernst, *Advan. Magn. Resonance* **2,** 1 (1966).
2. I. D. Campbell, C. M. Dobson, R. J. P. Williams, and A. V. Xavier, *J. Magn. Resonance* **11,** 172 (1973).
3. W. B. Moniz, C. F. Poranski, Jr., and S. A. Sojka, *J. Magn. Resonance* **13,** 110 (1974).
4. N. N. Semendyaev, *Stud. Biophys.* **47,** 151 (1974).
5. I. D. Campbell, S. Lindskog, and A. I. White, *J. Mol. Biol.* **90,** 469 (1974).
6. I. D. Campbell, C. M. Dobson, G. Jeminét, and R. J. P. Williams, *FEBS Lett.* **49,** 115 (1974).
7. L. R. Brown, A. De Marco, G. Wagner, and K. Wüthrich, *Eur. J. Biochem.* **62,** 103 (1976).
8. G. Wagner, A. DeMarco, and K. Wüthrich, *J. Magn. Resonance* **20,** 565 (1975).
9. E. Oldfield, R. S. Norton, and A. Allerhand, *J. Biol. Chem.* **250,** 6381 (1975).

JOURNAL OF MAGNETIC RESONANCE **24**, 343–366 (1976)

Proton Magnetic Relaxation and Spin Diffusion in Proteins

A. KALK AND H. J. C. BERENDSEN

Laboratory of Physical Chemistry, The University of Groningen, Zernikelaan, Groningen, The Netherlands

Received May 10, 1976

A theory has been worked out for the influence of cross relaxation between protons on the relaxation behavior of the protons in a protein. The longitudinal relaxation is strongly influenced in large proteins and at high frequencies, the effect being particularly evident above 10,000 molecular weight at frequencies exceeding 200 MHz. The longitudinal relaxation rates of individual protons tend to become equal as a result of spin diffusion and are dominated by the relaxation rate of rotating methyl groups that act as relaxation sinks. The magnetization decay curves in a pulsed T_1 experiment are often nonexponential. The saturation behavior becomes complicated and the nuclear Overhauser effect becomes less specific. The transverse relaxation of methyl protons is also considered. The theory has been tested on papain and ribonuclease at 100 MHz. These results and those of others are in good agreement with the theory. Determination of motional correlation times and local mobilities based on T_1/T_2 ratios and frequency-dependent relaxation data are often unreliable.

INTRODUCTION

The availability of Fourier transform techniques on high-resolution NMR spectrometers has opened the possibility of studying the longitudinal relaxation behavior of protons in a protein. In deuterated solvents in the absence of paramagnetic species, this relaxation is governed by the mutual dipole–dipole interactions of the protein protons, and should in principle provide valuable information on interproton distances and intramolecular motions.

In large molecules at high frequencies the longitudinal relaxation of an individual proton is no longer the result of the combined spin–lattice contributions from the fluctuating dipole couplings to its neighbors, but is largely influenced by *cross-relaxation* effects (*1–3*). Cross relaxation causes a mutual exchange of spin magnetization, often at a rate larger than the spin–lattice relaxation of the protons. In solids this effect leads to "spin diffusion."

In proteins, the longitudinal relaxation of protons (*4–7*) has been characterized by single relaxation times T_1. In several cases the interpretation of such measurements has been based on the assumption that T_1 is determined by the sum of the relaxation contributions from dipole–dipole interactions with neighboring protons, for which the standard equations as a function of correlation time (*8*) are valid.

We give here a simple theoretical treatment of the longitudinal relaxation in a protein, including the effects of cross relaxation (*9*). The transverse relaxation, the saturation behavior, and the nuclear Overhauser effect (NOE) are considered as well. In addition we report measurements on proton relaxation at 100 MHz of the enzymes papain and ribonuclease. The usual 180°–t–90° pulse sequence was employed for T_1 measurements, while saturation studies were carried out in the continuous wave mode. A satisfactory explanation for both our own results and those of others can be given. It appears that in large proteins at high frequencies, rotating methyl groups form the major sinks for spin–lattice relaxation.

THEORY

Relaxation Theory for Proteins

The relaxation of a system of coupled spins is governed by the equations of motion of the quantum-mechanical density matrix describing the statistical behavior of all spins. The complexity of a system of coupled protons, as in a protein, is such that no useful results can be expected from this approach. One therefore has to rely on approximations that are based on generalizations of the behavior of a spin pair, the latter having been treated in full by Solomon (*1*). Such generalizations to multiple spin systems (*2*) assume that effects due to the correlation of the motion of different dipolar pairs (*cross correlation*) can be ignored. Although in principle not justified, it has been shown by numerous calculations (*10*) that the neglect of cross correlation is a reasonable assumption as long as the overall motion is isotropic. In fact, the use of pairwise addition is a general practice in the study of NOE (*2*).

Consider a protein molecule as a system of proton spins $\mathbf{I}_i$ which are mutually coupled by dipole–dipole interactions. These dipole interactions fluctuate because of the isotropic rotational diffusion of the molecule as a whole; we shall assume that, with the exception of methyl rotation, no intramolecular motion occurs; i.e., the protein can be considered as a rigid molecule. The rotational diffusion is characterized by a correlation time τ_c which is short compared to the inverse of the dipolar interaction (measured in units of angular frequency), so that the usual relaxation theory applies, but which is not short compared to the inverse Larmor frequency.

The longitudinal relaxation of the magnetization I_{zi} of the ith spin in the absence of an rf driving field can be written in generalized form as

$$\frac{dI_{zi}}{dt} = -R_{1i}(I_{zi} - I_{0i}) - \sum_j R_{tij}(I_{zi} - I_{zj}). \quad [1]$$

Here, R_{1i} is the local spin–lattice relaxation of the ith spin (equal to the inverse of the relaxation time T_{1i}):

$$R_{1i} = \tfrac{3}{10}\gamma^4 \hbar^2 \sum_j \frac{1}{r_{ij}^6}\left[\frac{\tau_c}{1+(\omega\tau_c)^2} + \frac{4\tau_c}{1+(2\omega\tau_c)^2}\right], \quad [2]$$

where r_{ij} is the distance between the ith and jth nuclei, ω is the Larmor frequency, and γ and $\hbar$ have their usual meanings. The quantity I_0 is the equilibrium value of I_z, and

$$R_{tij} = \tfrac{1}{10}\frac{\gamma^4 \hbar^2}{r_{ij}^6}\left[\tau_c - \frac{6\tau_c}{1+(2\omega\tau_c)^2}\right]. \quad [3]$$

This term, which governs the cross relaxation between spins i and j, follows from the relaxation theory for spin pairs of Solomon (*1*). R_t is negative, and for a spin pair equal to $-R_1/3$ for $\omega\tau_c \ll 1$ it passes through zero at $\omega\tau_c = 1.118$ and becomes much larger than R_1 for $\omega\tau_c \gg 1$.

For the transverse relaxation rate R_{2i}, Solomon (*1*) derived expressions similar to those for the longitudinal relaxation. Although in general nonexponential behavior can be expected, in the two important limiting cases of like or unlike spins, the behavior of I_{xi} can be described by a single relaxation rate (in the absence of a driving rf field),

$$dI_{xi}/dt = -R_{2i} I_{xi}, \qquad [4]$$

with

$$R_{2i} = (1/20)\gamma^4 \hbar^2 \sum_j (1/r_{ij}^6) f(\tau_c). \qquad [5]$$

Here $f(\tau_c)$ has different forms depending on whether the spins i and j are *alike* in the sense that their shift difference is small compared to their linewidths, or *unlike*, meaning that the shift difference is large compared to both the linewidth and the scalar spin coupling J_{ij}:

$$\text{alike: } f^a(\tau_c) = 9\tau_c + 15\tau_c/(1 + \omega^2\tau_c^2) + 6\tau_c/(1 + 4\omega^2\tau_c^2), \qquad [6]$$

$$\text{unlike: } f^u(\tau_c) = 5\tau_c + 9\tau_c/(1 + \omega^2\tau_c^2) + 6\tau_c/(1 + 4\omega^2\tau_c^2). \qquad [7]$$

For $\omega\tau_c \ll 1$, the ratio R_2 (unlike)/R_2 (like) is $\frac{2}{3}$, whereas for $\omega\tau_c \gg 1$ this ratio is $\frac{5}{9}$. For cases which fulfill neither the condition for like nor the condition for unlike spins, no explicit theory is available, but intermediate values for R_2 can be expected.

Using Eqs. [1]–[7], it is possible to calculate the relaxation rates in a protein if the distances between the protons are known and an assumption is made about τ_c. In the coefficients R_{1i}, R_{2i}, and R_{tij}, only interactions with nearest neighbors are important since the contributions to these quantities are proportional to r^{-6}. The transverse relaxation (Eq. [4]) therefore only depends on these nearest-neighbor interactions. This is not true for the longitudinal relaxation. If $\omega\tau_c \gg 1$, the R_{tij}'s become much larger than R_{1i}. Hence the rate of transfer of spin energy between protons becomes much larger than the rate of energy exchange with the lattice. This has the effect that the longitudinal relaxation rates of all protons tend to the same value R_1 given by the average

$$R_1 = N^{-1} \sum_i R_{1i}, \qquad [8]$$

where N is the total number of protons. As we shall see, proteins represent a borderline case where this spin diffusion is neither negligible nor extremely fast, being more pronounced for larger proteins and higher frequencies.

A Simple Model for the Longitudinal Relaxation in Proteins

A rough picture of longitudinal relaxation can be obtained by making a few simplifications. The first is that only CH protons are considered: Relaxation measurements on proteins are carried out usually in D_2O, in which protons of COOH, NH_2, and OH and most of the NH protons are exchanged for deuterons. Second, we divide the protons into two groups of dipolar pairs: *Group A* contains dipolar pairs at the distance of 1.78 Å, which is the distance between the protons of a CH_2 or CH_3 group. *Group B* consists of pairs with a looser coupling, with interproton distances of 2.5 Å or larger. The distance between protons on an aromatic (six-membered) ring is 2.48 Å. The distance between two protons in an aliphatic chain in the staggered conformation and Van der Waals' distance of closest approach for hydrogen atoms are approximately 2.5 Å.

As a result of the different arrangements of neighboring protons, the R_{1i}'s of the protons differ. A proton at 1.78 Å gives an R_1 which is seven to eight times larger than the R_1 caused by coupling at 2.5 Å. Protons involved in dipolar pairs of Group A also can have further contributions to their R_{1i} by looser coupling. R_{tij} values also are larger for proton pairs of Group A than for those of Group B. The effect of cross relaxation within a CH_2 group (or CH_3 group) is hardly observed, because the proteins within such a group mostly have the same Larmor frequency and the relative differences in their R_{1i} due to couplings outside the groups are small. Thus the rate of the observable spin diffusion is determined by the diffusion caused by dipolar interaction of protons at distances of 2.5 Å or greater. From inspection of the papain molecular model (*11*) it became clear that for almost every proton another proton could be found at about 2.5 Å. Therefore we shall take the R_{tij} calculated for $r = 2.5$ Å as a measure for the rate of cross relaxation.

To evaluate the influence of this limited spin diffusion on the relaxation behavior of coupled spins having different spin–lattice relaxation rates, we examined a simple model. In this model two proton spins **I** and **S** are considered with different R_{1i} values, given as R_{1I} and R_{1S}. The rate of cross relaxation is given by R_t. The equations of motion for I_z and S_z are given by Eq. [1] and, using the equality of equilibrium magnetizations, $I_0 = S_0$, can be written as

$$dI_z/dt = -(R_{1I} + R_t)(I_z - I_0) + R_t(S_z - S_0), \qquad [9a]$$

$$dS_z/dt = -(R_{1S} + R_t)(S_z - S_0) + R_t(I_z - I_0). \qquad [9b]$$

The general solutions of these equations yield expressions for I_z and S_z which are sums of two exponentials

$$(I_0 - I_z)/I_0 = A\exp(-\lambda_1 t) + B\exp(-\lambda_2 t), \qquad [10a]$$

$$(S_0 - S_z)/S_0 = C\exp(-\lambda_1 t) + D\exp(-\lambda_2 t), \qquad [10b]$$

in which

$$\lambda_{1,2} = \tfrac{1}{2}(R_{1S} + R_{1I} + 2R_t) \pm \tfrac{1}{2}[(R_{1S} - R_{1I})^2 + 4R_t^2]^{1/2}. \qquad [10c]$$

The factors A, B, C, and D depend on R_{1I}, R_{1S}, R_t, and the initial conditions (at $t = 0$). In a 180°–t–90° experiment, I_z and S_z are inverted by the 180° pulse (to $-I_0$ and $-S_0$, resp.), after which the recovery of I_z and S_z to their equilibrium values follows during a time t. We solved Eqs. [9] for some special cases. Taking $R_{1I} = 5R_{1S}$, R_t was varied relative to the R_{1i}'s. Logarithmic plots of $(I_0 - I_z)/I_0$ and $(S_0 - S_z)/S_0$ as a function of time $(t \cdot R_{1I})$ are given in Fig. 1. It is shown that the cross-relaxation mechanism becomes active when I_z and S_z begin to differ, giving concave curves for spins **I** (with the largest R_{1i}) and slightly convex curves for spins **S**. If the effect of cross relaxation is still limited, the tangents to the relaxation curves at $t = 0$ are $-R_{1I}$ and $-R_{1S}$ for spins **I** and **S**, respectively. In this way information about the different proton environments can be obtained. If cross relaxation is rapid, such information is lost. In that case [curve (d) of Fig. 1] the R_{1i} values are equal and are given by Eq. [8]. We also see from Fig. 1 that the difference in relaxation behavior of spins **I** and **S** tends to disappear if R_t becomes larger than R_{1I}. The convex curves for spins **S** are found to be hardly distinguishable from straight lines (also for other R_{1I}/R_{1S} ratios), although they extrapolate to incorrect amplitudes at $t = 0$.

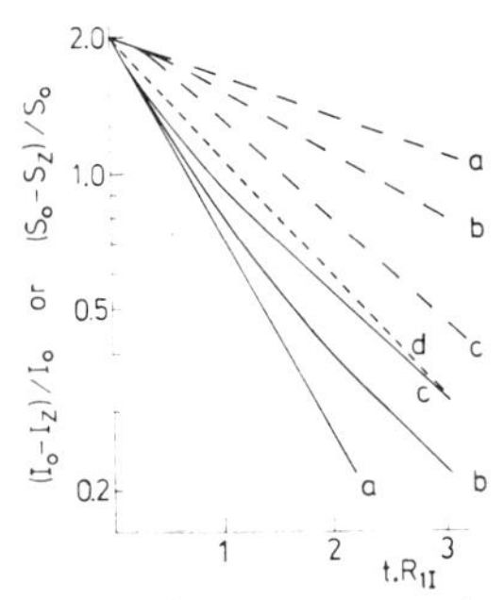

FIG. 1. Relaxation of two coupled spins **I** (full-drawn curves) and **S** (broken curves) after inversion of the magnetization by a 180° pulse at $t = 0$, for different values of the cross-relaxation rate R_t. The intrinsic relaxation rate R_{1I} of spin **I** is five times that of spin **S** (R_{1S}). (a) $R_t = 0$, (b) $R_t = 0.2R_{1I}$, (c) $R_t = R_{1I}$, (d) $R_t = \infty$.

Methyl Proton Relaxation

When cross relaxation is appreciable, those groups within the protein that provide the most effective relaxation give a relatively strong contribution to the overall spin–lattice relaxation. The methyl groups, which are known to rotate rapidly in solid alkanes (*12, 13*) and also in amino acid crystals (*14*), can provide effective spin–lattice relaxation sinks. Extrapolation to 31°C yields a methyl rotational correlation time (τ_r) of 10^{-10} sec in solid valine and leucine and of 2×10^{-11} sec in solid decane, whereas τ_r for methyl rotation of leucine in polymixin in solution (*5*) was also found in this range (3×10^{-11} sec). In solid alanine the methyl rotation is somewhat slower (10^{-9} sec), while in methionine the rotation is probably faster. In proteins we expect the rate of methyl rotations to be intermediate between the values in a solid and in a small peptide such as polymixin in solution, i.e., $\tau_r \lesssim 10^{-10}$ sec.

The relaxation of a rapid rotating methyl group attached to a macromolecule that itself is subject to a slower rotational diffusion has been treated with various approximations. Woessner (*15*) considered the relaxation of one proton to be the independent sum of the relaxation contributions of the two protons, with which it is coupled, thus neglecting the cross correlation between the motions of the proton pairs. If the macromolecular rotational motion is isotropic with correlation time τ_c and the methyl rotational motion is characterized by a correlation time τ_r (for a three-position random jump mechanism the average jump rate from one position to either of the adjacent positions equals $1/3\tau_r$), R_1 is given for each dipole–dipole interaction by

$$R_1 = 0.3(\gamma^4 \hbar^2/r^6)[\tfrac{1}{4}f_1(\tau_c) + \tfrac{3}{4}f_1(\tau_{c3})], \qquad [11a]$$

where $f_1(\tau)$ is given by

$$f_1(\tau) = \tau/(1 + \omega^2\tau^2) + 4\tau/(1 + 4\omega^2\tau^2) \qquad [11b]$$

and

$$\tau_{c3}^{-1} = \tau_c^{-1} + \tau_r^{-1}. \qquad [11c]$$

The factor $0.3\gamma^4\hbar^2/r^6$ in Eq. [11a] is equal to 5.372×10^9 sec^2 for a proton distance of 1.78 Å. In an isolated methyl group each proton is coupled to two other protons yielding a value of R_1 twice that given by Eq. [11a]. The decay of magnetization, both longitudinal and transverse, is a simple exponential when cross correlation is neglected.

It has been shown both theoretically (*16*, *17*) and experimentally (*18–20*), however, that cross-correlation effects cannot always be neglected for anisotropically reorienting methyl groups. The cross correlation gives rise to deviations from simple exponential behavior. A general theory for anisotropically reorienting methyl groups has been given by Werbelow and Marshall (*17*). We have applied their theory for the special case of a methyl rotation with correlation time τ_r superimposed on an isotropic rotational

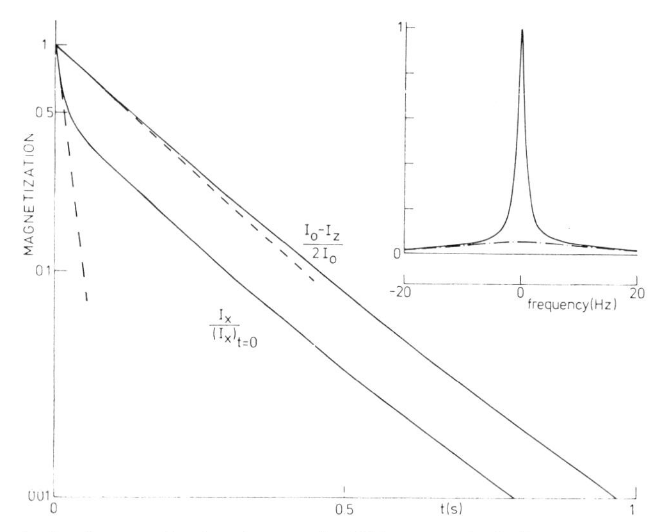

FIG. 2. The calculated magnetization decays of a CH_3 group, including cross correlation, for the longitudinal relaxation after a 180° pulse $(I_0 - I_z)/2I_0$ and for the transverse relaxation. The correlation time for overall rotation (τ_c) is 10^{-8} sec and for methyl rotation (τ_r) is 10^{-10} sec. The tangents to these curves are obtained by neglecting cross correlation. The insert shows the lineshape of the methyl resonance, of which the broad component is given by the dash–dot curve.

diffusion with correlation time τ_c. In this theory the rotational diffusion constants for the symmetric top on which the CH_3 rotation is superimposed are $D_{\parallel} = D_{\perp} = (6\tau_c)^{-1}$ and the diffusion constant of the internal CH_3 rotation $D_{\text{int}} = (4\tau_r)^{-1}$ (*21*). In Fig. 2 the decay curves are plotted for the longitudinal and the transverse magnetization and compared with the results of Woessner's theory in which cross correlation is neglected, for a typical protein case (papain: $\tau_c = 10^{-8}$ sec and $\tau_r = 10^{-10}$ sec). The longitudinal relaxation shows a nearly exponential behavior with a slightly lower R_1 than expected from Woessner's theory. We found this result to be typical for the range of relaxation times usually encountered in proteins.

It is of interest to note that the transverse relaxation deviates much more strongly from a single exponential and from the theoretical value based on neglect of cross correlation than the longitudinal relaxation does. In Fig. 2 the lineshape is plotted for a methyl resonance with the same set of parameters as for the transverse relaxation given in the same figure. The line consists of three contributions: one narrow line (half-width, 1.84 Hz; 44.7% of the integral), one broad line (half-width, 27.43 Hz; 50.7% of the integral), and one small contribution (0.86 Hz; 4.8%). In the limiting case of long τ_c and short τ_r, which is almost fulfilled for proteins, the origin of the broad component is quite clear. A system of three equivalent $I = \frac{1}{2}$ spins gives rise to a spin quadruplet ($I = \frac{3}{2}$) and two spin doublets ($I = \frac{1}{2}$). Dipole coupling between the spins gives in first order a perturbation of the spin levels proportional to $3m^2 - I(I+1)$. Hence the secular perturbation, which is the dominant term for the transverse relaxation in the case of long τ_c, only affects the transitions between the $m = \frac{1}{2}$ and $m = \frac{3}{2}$ and between the $m = -\frac{1}{2}$ and $m = -\frac{3}{2}$ states of the quadruplet. These transitions correspond to 50% of the total intensity of the absorption line. Thus the methyl resonance will consist of two components, each of about half-intensity. The broad component is determined by the overall motion of the protein with

$$R_2 = (9/20)\,\gamma^4 \hbar^2 \tau_c / r^6, \qquad [12]$$

as follows from Werbelow and Marshall's theory for $\omega\tau_c \gg 1$ and $\tau_r \ll \tau_c$. This corresponds to *twice* the width given by Woessner's theory (*15*). The narrow component is still a superposition of two Lorentzians, the exact shape depending on ω, τ_c, and τ_r. It can be shown, however, that for proteins with $\tau_c > 10^{-8}$ sec at 100 MHz the narrow linewidth is approximated, roughly within a factor of 2, by

$$R_2' \simeq 2\gamma^4 \hbar^2 \tau_r / r^6. \qquad [13]$$

For $\tau_r = 10^{-10}$ sec, $R_2' \simeq 3$ sec^{-1}, corresponding to a linewidth of less than 2 Hz. This means that the contributions of next neighbors to the linewidth of the narrow component of the methyl resonances are dominant in all practical cases. For example, a proton at 2.5 Å, such as the proton on the neighboring carbon atom, gives a contribution of 10 sec^{-1} if $\tau_c = 10^{-8}$ sec.

We note that the initial rate of decay of the transverse methyl relaxation is equal to the value derived from Woessner's theory (*15*), because terms relating to cross correlation cancel for $t = 0$. This also implies that broadening effects on methyl resonances in fast-exchanging ligands are described correctly by the "classical" theory.

In the presence of spin diffusion between methyl groups, a nonexponential longitudinal relaxation can also result from the fact that cross relaxation can only take place between methyl spin states of the same symmetry (*22*). Apart from the fact that in rotating samples the effect is greatly reduced (*23*), symmetry restrictions are only valid between methyl groups mutually and not between methyl protons and other protons. Coupling to neighboring nonmethyl protons has been shown to reduce the deviations from exponential behavior of methyl longitudinal relaxation (*18*, *19*).

We may thus conclude that the longitudinal relaxation of methyl groups in proteins is nearly exponential and deviates only slightly (10–20%) from R_1 values calculated on the basis of Woessner's theory (*15*). On this basis we have calculated the T_1's ($T_1 = R_1^{-1}$) for methyl protons for some typical values of the correlation time τ_r of the methyl rotation (Fig. 3). For comparison T_1 is plotted also for a methylene group and for a proton pair at Van der Waals' distance. For typical methyl correlation times around 10^{-10} sec the methyl relaxation becomes dominant in larger proteins. The importance of cross relaxation is shown by the dash–dot curve representing T_t, the inverse of the cross-relaxation rate R_t between a proton pair at Van der Waals' distance. At 100 MHz and for proteins up to 20,000 MW the cross relaxation is of marginal to moderate importance. For larger proteins at 100 MHz, and at 300 MHz also for smaller proteins, the cross relaxation is expected to be of considerable influence on the longitudinal relaxation. The methyl groups in a 20,000 MW protein at 300 MHz provide an order of magnitude more effective sink than the methylene protons. In the limit of strong cross relaxation (long τ_c and high frequency) the R_1 of all protons simply becomes the average R_1 of the methyl protons multiplied by the ratio of the number of methyl

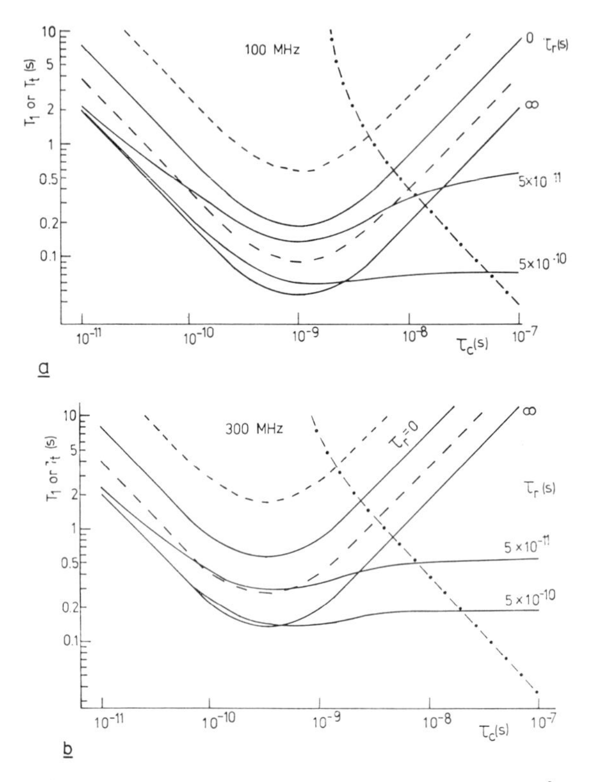

FIG. 3. T_1 (R^{-1}) as a function of τ_c for the interaction of a proton pair at 2.5 Å (- - - -), the protons in a CH_2 group (— —), and the protons in a CH_3 group (———), according to Woessner (*15*) for different values of τ_r. The cross-relaxation time T_t (R_t^{-1}) is given for the interaction of a dipolar pair at 2.5 Å in its positive range (·—·—·). (a) 100 MHz, (b) 300 MHz.

protons to the total number of protons. Assuming a τ_r of 10^{-10} sec, we calculate for a variety of proteins (with 24–31% of protons in methyl groups) a limiting T_1 of 0.8 to 1.1 sec.

Temperature and Frequency Dependence

The *temperature dependence* of the relaxation times also differs from what is expected in the absence of cross relaxation. The transverse relaxation time increases at higher temperatures with an activation energy determined by the viscosity of the medium, as expected. The longitudinal relaxation time of rigid groups decreases with temperature in the absence of cross relaxation, while the T_1 of methyl groups is determined by both the τ_c and τ_r, according to Eqs. [11]. This leads to a behavior that is almost temperature independent, or to a slight increase with temperature of the methyl T_1 in the region of interest (Fig. 4), while in the limit of large τ_c, T_1 increases with temperature, with the activation energy (12–20 kJ/mole) of methyl rotation (*14*).

In the presence of sufficiently strong cross relaxation, the temperature dependence of the experimental T_1 of those rigid groups for which $T_t < T_{1i}$ and which are coupled to methyl protons is largely determined by T_t, while in the limiting case of strong cross relaxation, T_1 is determined by the spin–lattice relaxation of the methyl sinks proper.

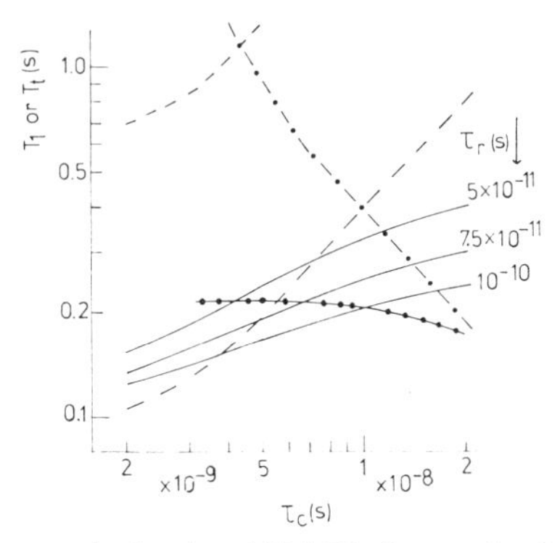

FIG. 4. Plots of T_1 (R_1^{-1}) vs τ_c, calculated at 100 MHz for a pair of proton spins at 2.5 Å (----), for a CH_2 proton pair (— —), and for CH_3 protons for some fixed τ_r values (———). The curve -•-•-•- represents T_1 vs τ_c for a CH_3 group, if the τ_c change is caused by temperature change, assuming $\tau_r = 10^{-10}$ sec and $\tau_c = 10^{-8}$ sec at 20°C in D_2O, together with an activation energy of 12.7 kJ/mole for the CH_3 reorientation about the threefold axis. T_t (R_t^{-1}) vs τ_c for proton dipolar coupling at 2.5 Å is given as ·–·–·–.

In both cases the measured T_1 of the rigid groups is expected to increase with temperature, contrary to what is expected in the absence of cross relaxation. Hence if such a behavior is found experimentally, one is not justified in concluding that the group itself is subject to fast internal motion.

The *frequency dependence* of T_1 cannot be used to derive a value of τ_c in the presence of spin diffusion. Because of cross relaxation the increase with frequency of T_1 for rigid groups becomes much less than expected. Hence the ratio of T_1's for two frequencies is closer to unity than expected on the basis of simple theory and the values for τ_c derived on that basis will be smaller than the actual τ_c. For methyl groups the intrinsic T_1 is only slightly frequency dependent; the cross relaxation in this case causes an increase in methyl T_1 with frequency. In the limit of rapid spin diffusion all protons tend to have the same frequency-independent T_1, as long as $\omega\tau_r < 1$.

Saturation Behavior and Overhauser Effect

Since cross relaxation has a profound influence on the relaxation behavior of protons in rigid proteins, it is expected that the saturation behavior in steady-state cw experiments is influenced as well. When single $R_1 = T_1^{-1}$ and $R_2 = T_2^{-1}$ can be defined, the resonance intensity of the absorption mode is (*24*)

$$v = \gamma H_1 T_2/[Z^{-1} + (\omega_0 - \omega)^2 T_2^2], \qquad [14a]$$

where H_1 is the amplitude of the rotating irradiating field at frequency ω and the saturation factor is

$$Z = [1 + (\gamma H_1)^2 T_1 T_2]^{-1}. \qquad [14b]$$

The transverse relaxation is described with sufficient accuracy by a single R_2 (Eqs. [4]–[7]), except in the case of methyl groups, where two components can be distinguished, each with its own R_2 (Eqs. [12, 13]). The longitudinal relaxation requires further attention.

In the Appendix it is shown that, if one spin is irradiated, its saturation behavior is equivalent to that given by Eq. [14], with an effective R_1' that includes contributions from

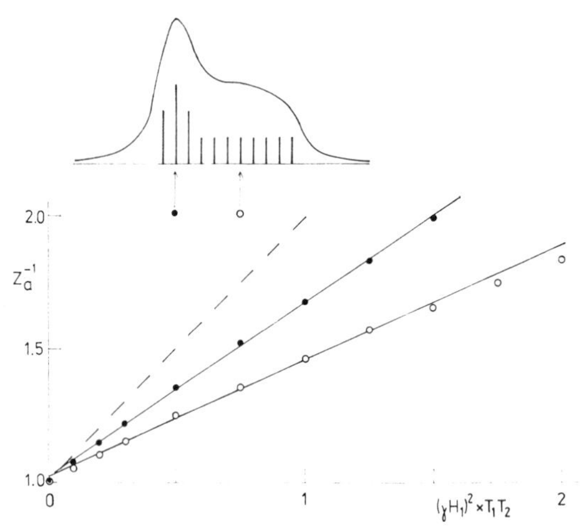

FIG. 5. Plots of Z_a^{-1} vs $(\gamma H_1)^2 T_1 T_2$, calculated for the positions in the constructed spectrum, given above the plots. The stick spectrum is shown, together with the envelope, obtained by summation of the Lorentzian-broadened lines with equal T_1 and T_2. The spacing between the "sticks" is half of the half-width of the individual Lorentzian lines. Z^{-1} for these Lorentzians is given by the dashed curve. The straight lines are drawn approximately through the calculated points.

the relaxation rates of other protons and depends on the rates of cross relaxation (Appendix, Eq. [A7]). The R_1' determined from saturation measurements in the presence of noticeable cross relaxation is always larger than the R_1 determined from a 180°–t–90° pulse experiment, provided that nonselective pulses are used, and also larger than the R_1 in the absence of cross relaxation.

The reliability of relaxation parameters derived from saturation experiments is strongly influenced by the overlap of resonances. In Fig. 5 the inverse of the apparent saturation factor Z_a, as determined from H_1 divided by the signal amplitude, is plotted versus $(\gamma H_1)^2$ for an overlapping set of resonances with equal T_1 and T_2, as given in the insert of Fig. 5. The dotted line shows the saturation behavior of the top of a single line in the spectrum. It is clear that the apparent $T_1 T_2$, given by the slopes of the curves, can be as much as a factor 2 to 3 lower than the real values, while the deviation from ideal saturation behavior (a straight line) is hardly observable. Also, when resonances with different $T_1 T_2$ values overlap, an almost linear relation between Z_a^{-1} and $(\gamma H_1)^2$ is found.

The values for T_1T_2 can be calculated for several simple models. In the absence of cross relaxation, T_1T_2 goes through a shallow minimum near $\omega\tau_c = 1$ and becomes constant for long τ_c. When cross relaxation occurs, however, $T_1'T_2$ decreases with increasing τ_c. This decrease is relatively stronger and is also evident at lower τ_c for isolated protons coupled to CH_2 or CH_3 groups; it may be as large as a factor of 10.

For CH_3 groups T_1T_2 is quite sensitive to the rate of CH_3 rotation, especially in the range of τ_r around 10^{-10} sec. In the limit of fast rotation T_1T_2 is 16 times larger than in the absence of rotation. In the latter case T_1T_2 is about 10^{-3} sec^2 at 100 MHz (its frequency dependence is proportional to ω^2). Measurement of methyl saturation can be used to provide a rough estimate of τ_r, even in the presence of resonance overlap and cross relaxation. The latter does not appreciably influence the effective T_1 of methyl protons.

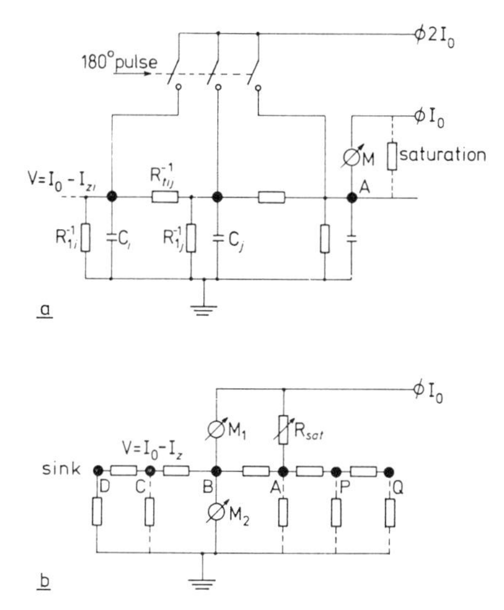

FIG. 6. Equivalent circuit for the z magnetization of a system of coupled spins, in which each spin is represented by a capacitor of unit value. The voltage on each capacitor represents the deviation of the z magnetization from its equilibrium value. Intrinsic T_1's are represented by leak resistors, and cross-relaxation times by resistors connecting different capacitors. In (a) the meter M measures the amplitude of spin A in a cw experiment; increased saturation is simulated by a decrease in meter resistance. In (b) an Overhauser experiment, measuring B and irradiating A, is simulated. M_1 measures the NOE, M_2 the differential NOE.

The general effect of cross relaxation on the NOE is to diminish the specificity of the latter. In the extreme case in which the cross relaxation is dominant and a complete saturation of the irradiated spectral component is achieved, a homogeneous negative NOE will be observed throughout the spectrum. In practice, however, some specificity may still be observed, but the relation between NOE and proton distances is neither straightforward nor simple. In general it can be stated that in the presence of relaxation sinks, protons in a coupled chain between irradiated proton and a sink show a specificity: protons close to the irradiated proton have a larger NOE and protons near the sink show little NOE.

A system of coupled spins can be represented by an electrical analogy, which provides insight into its relaxation and saturation behavior. In Fig. 6a the equivalent circuit for a normal cw measurement, as well as a 180°–t–90° pulse experiment, is given. Each proton is represented by a capacitor of value 1. The resistor across the ith capacitor has the value R_{1i}^{-1} and the resistor between the ith and jth capacitors has the value R_{tij}^{-1}. The circuit represents Eq. [1], when the voltage on the ith capacitor is $I_0 - I_{zi}$. Thus the meter in Fig. 6a measures I_{zA}, the total signal intensity of the resonance of proton A, in a low-power cw experiment or after Fourier transformation of a 90° pulse. When a 180° pulse is given, the effect is to invert all I_z, which is achieved by charging all capacitors to voltage $2I_0$ at time $t = 0$. Saturation of the cw measurement at proton A is equivalent to inserting a resistor across the meter M. An Overhauser experiment is exemplified in Fig. 6b, where the capacitors have been omitted. The irradiation on proton A is represented by R_{sat}; the signal I_{zB} is measured by M_1 on proton B. Meter M_2 measures the differential NOE, i.e., a difference spectrum between irradiated and unirradiated conditions (at A). In the example of Fig. 6b, the protons B, C, and D will have different NOE, but P and Q have nearly the same NOE, becuase they are both saturated. Although the NOE can be calculated for any specific model of proton couplings, such calculations will be rather tedious.

EXPERIMENTAL

Papain was isolated (*25*) from dried papaya latex (Chas. Zimmerman & Co., England) and fractionated (*26*) to give mercuripapain that could be activated to 100% activity. Excess of solutes was removed by extensive dialysis against doubly distilled water. H_2O was replaced by D_2O (Merck, Darmstadt, >99.75% D) using ultrafiltration apparatus (Amicon, Lexington, Mass.). Final solutions of mercuripapain (0.8 mM) were in 0.3 M perdeuteroacetate buffer pH 3.9 (meter reading) in D_2O and contained a slight amount (~0.2 mM) of $HgCl_2$.

Bovine pancreatic ribonuclease A (RNase) was obtained from Miles–Seravac (grade 1, gel-filtered on sephadex G-25 in 0.1 M acetic acid and lyophilized) and from Worthington (phosphate-free, RAF grade). Final solutions of RNase (2 mM) in 0.2 M perdeuteroacetate buffer/D_2O, pH = 5.5 (meter reading) were made using the methods of Benz *et al.* (*4*), followed by removal of some residual impurities by ultrafiltration.

Proton NMR was carried out on a Varian XL100-15 spectrometer, provided with a pulse unit in combination with a Varian 620/f computer. Twelve-millimeter sample tubes were used, with internal deuterium lock. Chemical shifts were determined relative to the methyl resonances of sodium 2,2-dimethyl-2-silapentane-5-sulfonate (DSS) in D_2O (downfield shifts given by positive numbers).

The spin–lattice relaxation of the protein protons at high resolution was studied by the usual $(180°–t–90°–T)_n$ pulse sequence, followed by Fourier transformation. The delay times T were four times the estimated T_1 or longer. The "relaxation spectra" were sometimes obtained in the differential mode, utilizing a $(90°–T–180°–t–90°–T)_n$ pulse sequence and subtraction of the second free induction decay from the first in each sequence. The ultimate signal amplitudes after Fourier transformation then are equal to $A_\infty - A_t$, in which A_t is the signal amplitude from a $(180°–t–90°–T)_n$ sequence and A_∞ the signal amplitude from that sequence for $t \to \infty$ or without the 180° pulses.

Saturation studies were carried out in the continuous wave mode under slow passage conditions. The histidine C_2 resonances were accumulated on a Varian C-1024 computer. The rf field strength H_1 was determined according to Leigh (*27*). Nuclear Overhauser experiments were carried out on mercuripapain by difference spectroscopy. The spectrum of the aromatic region obtained with simultaneous strong irradiation (with maximum decoupling power) at a position outside the absorption region (more than 700 Hz upfield) was first accumulated for 25 scans on the C-1024 computer and then 25 spectra obtained under strong irradiation at 2.2 ppm were subtracted.

RESULTS AND INTERPRETATION

Pulse Experiments

The relaxation of the proton spins in mercuripapain at 31°C to their equilibrium distribution after a 180° pulse is illustrated in Fig. 7, given for various times t. The

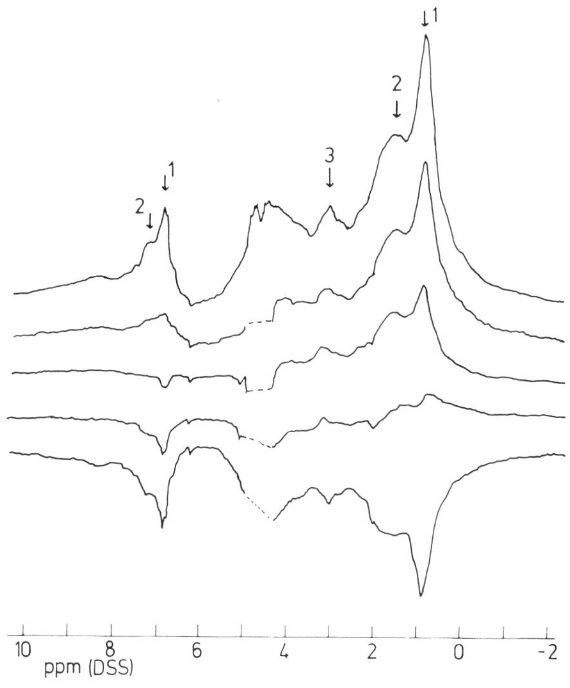

FIG. 7. Relaxation spectra of mercuripapain for times between 180° and 90° pulses of 0.05, 0.2, 0.3, 0.5 and 2.5 sec (bottom to top). The number of accumulations was 100 in each case.

relaxation for RNase at 31°C and pH 5.5 is shown in Fig. 8 in the differential mode. Plots were made of $\log(A_\infty - A_t)$ vs t at the positions indicated by the arrows in Figs. 7 and 8 for mercuripapain and RNase, respectively. The same was done for the resonance maxima of the distinguishable histidine protons in RNase.[1] These plots are given in Figs. 9 and 10. Contributions to the spectrum (*30*) at the positions indicated by the arrows in Figs. 7 and 8 are as follows. In the aliphatic region: (1) at 0.90 ppm (Aliph 1): CH_3 protons of Leu, Val, Ile, and (partially) Thr; (2) at about 1.5 ppm (Aliph 2): predominantly CH_2 protons, but also the CH_3 protons of Ala; (3) at 3 ppm (Aliph 3):

[1] The assignment of C_2–H resonances used here is that given originally by Meadows *et al.* (*28*). Recent experiments (*29*) indicate that the assignments to the C_2-12 and C_2-119 protons must be mutually exchanged.

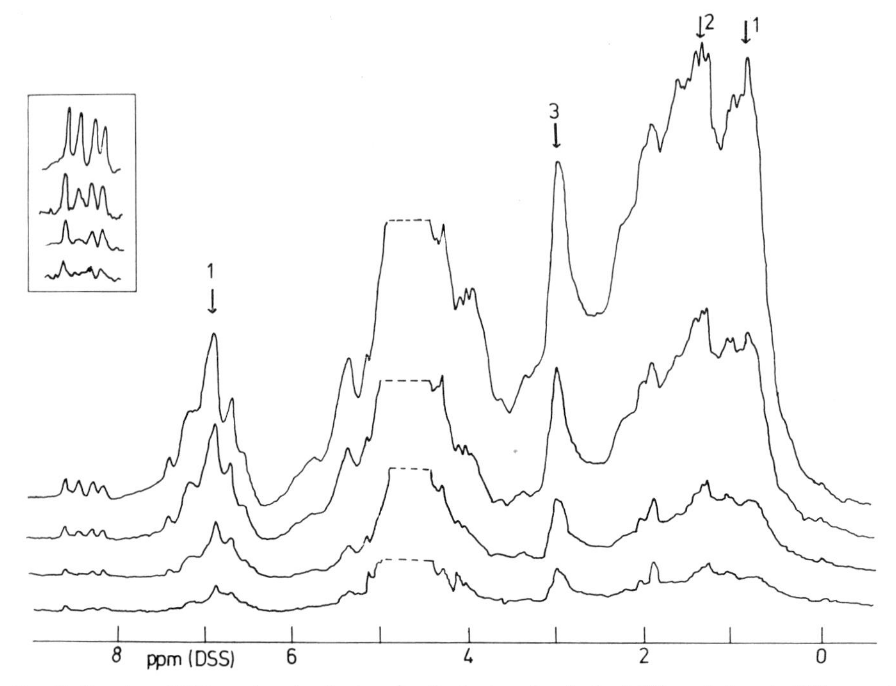

FIG. 8. Spectra of RNase, showing the longitudinal relaxation in the "differential mode." The number of full sequences was 126. The values of t (from top to bottom) are 0.025, 0.2, 0.5, and 0.8 sec. The five resonances at the low-field side (see also insert) are assigned to C_2-105, C_2-12, C_2-119, C_2-48, and C_4-105 histidine protons, from left to right, respectively.

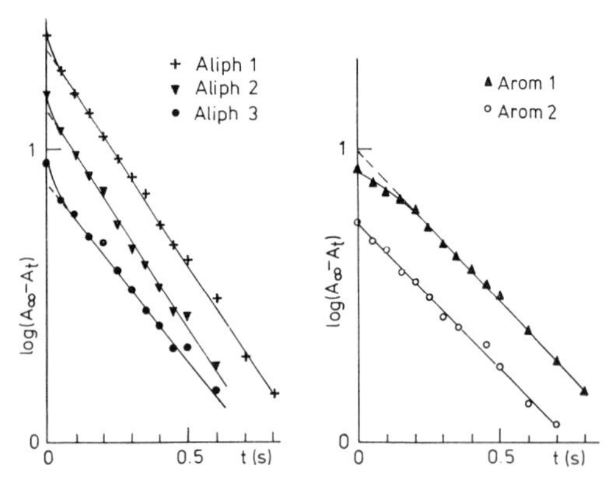

FIG. 9. Plots of $\log(A_\infty - A_t)$ versus time for mercuripapain at 31°C. The peaks are identified in Fig. 7.

exclusively CH_2 protons. In the aromatic region: at 6.8 ppm for RNase (Arom 1) and at 6.8 and 7.2 ppm for papain (Aroms 1 and 2, respectively). Aliphs 1, 2, and 3 contain the Group A type of dipolar interactions; the aromatic protons belong to Group B.

For papain no obvious difference in the (bulk) relaxation behavior was found upon activation of mercuripapain with cysteine (10 mM) and EDTA (1 mM). Neither was

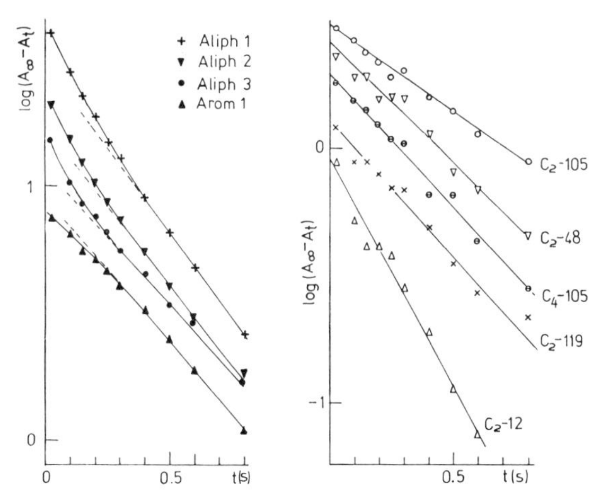

FIG. 10. Experiment similar to that in Fig. 9 for RNase at 31°C. Some curves have been displaced vertically for clarity.

an effect observed at the indicated positions when oxygen was removed from the papain solution.

As can be seen from Figs. 9 and 10 some of the magnetization decays are not single exponentials, as predicted from the simple two-spin model in the Theory section. The curves for protons with a large R_{1i} (Group A) are concave, whereas some curves of protons with smaller R_{1i} (aromatic protons) are convex. Nonexponential decays can also arise from a mere summation of exponential decays due to different R_{1i}'s or to cross correlation (CH_3 groups). Such summations only give concave curves. So the occurrence of convex curves gives a direct indication for the presence of spin diffusion. Because of the rather poor signal-to-noise ratio of the data for the distinguishable histidine protons, it is not possible to determine whether the decays are single exponentials or not. The expected deviations from exponentiality in the latter case are not large anyway. The initial differences between the relaxation rates of the bulk protons tend to equalize at longer values of t, although not completely. This again agrees with *limited* spin diffusion. The final relaxation rates of the Aliph 3 protons are almost equal to those of the aromatic protons. This is reasonable since the Aliph 3 protons consist to a large extent of β-CH_2 protons of aromatic, cysteine, and asparagine residues (*30*); these protons will play a major role in the relaxation of the loosely coupled nearby aromatic and α-CH protons.

Although some of the decays are not single exponentials, an estimate of the average T_1's was obtained by making a least-squares fit to the relaxation data according to

$$\ln(A_\infty - A_t) = \ln 2A_\infty - t/T_1. \qquad [15]$$

The T_1's determined in this way are compiled in Table 1 for the aliphatic and aromatic protons, and in Table 2 for the distinguishable histidine protons in RNase. For papain, T_1 values determined from the time t_0 for zero passage ($A_t = 0$), using $T_1 = t_0/\ln 2$, are also given. The discrepancy between the T_1 values from t_0 and those from the least-squares fit is in agreement with the shapes of the decay curves.

TABLE 1

T_1 VALUES (sec) FOR PAPAIN AND RNASE AT 100 MHz

Protein	Temperature (°C)	Aliph 1	Aliph 2	Aliph 3	Arom 1	Arom 2
Papain						
LSF[a]	31	0.31	0.31	0.36	0.46	0.43
ZP[b]	31	0.24	0.23	0.29	0.51	0.44
LSF	46	0.25	0.24	0.27	0.45	0.40
RNase						
LSF	31	0.27	0.30	0.36	0.42	

[a] Lease-squares fit. [b] Zero passage.

The overall correlation time τ_c for the enzyme molecules can be estimated from the initial relaxation rates and the T_1 values with the aid of Fig. 3a. A value between 6 and 10 nsec is plausible. For RNase such a value agrees with the τ_c found from fluorescence polarization (*31*) and electric birefringence studies (*32*) of 4 to 8 nsec. An estimate of τ_c can be made using the Stokes–Einstein relation for a rigid sphere:

$$\tau_c = 4\pi\eta r^3/3kT. \qquad [16]$$

X-Ray crystallography shows dimensions for RNase of 38 × 28 × 22 Å (*33*) and for papain, 48 × 36 × 36 Å (*34*). From Woessner's theoretical treatment (*35*) we conclude that for axial ratios less than 2 a reasonable accuracy is obtained if in Eq. [16] r is replaced by the geometric mean radius of the three axes ($r^3 = r_1 r_2 r_3$). Thus a τ_c of

TABLE 2

T_1 AND T_2 VALUES (sec) FOR HISTIDINE PROTONS IN RNASE[a]

	C_2-105	C_2-12	C_2-119	C_2-48	C_4-105
T_1 (100 MHz), this work[b]	0.63 (0.03)	0.24 (0.01)	0.43 (0.02)	0.49 (0.04)	0.41 (0.02)
T_1 (100 MHz), Ref. (*4*)[c]	0.77 (0.03)	0.28 (0.02)	0.40 (0.02)	0.51 (0.03)	0.50 (0.02)
T_1, as above, degassed	0.98 (0.02)	0.35 (0.03)	0.49 (0.02)	0.56 (0.01)	0.58 (0.02)
T_1 (220 MHz), Ref. (*7*)[d]	1.09 (0.02)	0.92 (0.04)	1.07 (0.03)	1.14 (0.07)	
T_2 (100 MHz), this work[b]	0.07 (0.01)	0.07 (0.01)	0.07 (0.01)	0.06 (0.01)	0.07 (0.01)

[a] Values in parentheses give standard errors from a least-squares fit for T_1 and estimated errors for T_2.
[b] pH 5.5, 31°C.
[c] pH 5.5, 33°C.
[d] pH 5.7, 18°C.

nsec for RNase and a τ_c of 8 nsec for papain are obtained, using $\eta = 1.0 \times 10^{-3}$ N sec m^{-2} for D_2O at 31°C.

As is seen from Table 2, there is an appreciable difference in the T_1's of individual histidine protons. About 20% of the relaxation results from paramagnetic interaction with dissolved oxygen, which is quite effective because all histidine rings can interact with the solvent. The remaining differences in T_1 values were explained by Benz *et al.* (*4*) by different dipolar interactions with protons in the surroundings of the histidine and by local mobility. These authors were led to the conclusion that the rotational correlation times for the histidines are less than 1 nsec, on the basis of the temperature dependence of T_1. In fact, all their observations are consistent with a rigid protein model with cross relaxation.

The values of T_2 (70 msec) exhibit only a slight spread compared with the spread in T_1. A T_2 of 70 msec is consistent with an interaction with two unlike protons at 2.5 Å and a τ_c of 10 nsec, while the contribution of protons farther away brings the estimated τ_c closer to the range of 6–10 nsec, as previously estimated. A contribution to the transverse relaxation from exchange between the protonated and uncharged state at a limited rate is very unlikely because such exchange broadening would be expected to depend on pH; we found T_2 to be the same at pH 5.5 (in the titration region) and at pH 3.7. Thus it is unlikely that the histidines will exhibit local mobility at a rate higher than the overall τ_c. In the absence of cross relaxation, T_1 at 100 MHz is now expected to be 0.8–1.2 sec for all histidine protons. The T_1 found by Benz *et al.* (*4*) in the degassed sample for the C_2-105 protons indeed lies within that range, but the influence of cross relaxation brings the T_1 values of the other His protons down to a value closer to the average of the whole molecule. At 220 MHz this effect is much stronger and all protons tend to have the same T_1. For comparison, without cross relaxation T_1 of the histidine protons is expected in the range 3.9–5.8 sec at 220 MHz. The values observed by Wasylishen and Cohen (*7*) for RNase at 18°C are close to the expected T_1 values derived from the average relaxation rate at 220 MHz. For $\tau_c = 8$ nsec and τ_r is 5×10^{-11} sec, this T_1 can be calculated from Eq. [11] to be 0.92 sec. In this case the CH_3 groups, which constitute 24.4% of all protons, contribute 52% to the average relaxation rate, the CH_2 protons (47.7% of all protons) contribute 31%, and the remainder is due to the average loose coupling, assumed to be equivalent to two protons at a distance of 2.5 Å. At lower τ_c, T_1 decreases because the contribution of the CH_2 protons to the relaxation increases; at lower τ_r, T_1 increases because the relaxation contribution of the CH_3 protons decreases. Thus a value of 0.8 sec for T_1 is found for $\tau_c = 6$ nsec and $\tau_r = 5 \times 10^{-11}$ sec. From the Stokes–Einstein relation we find $\tau_c = 4$ nsec at 18°C. It seems, both from T_1 and T_2, and also from the importance of the cross relaxation at 31°C and 100 MHz, that the real τ_c is larger than the Stokes–Einstein value by a factor of 1.5 to 2.5. This is probably due to the marked deviation from spherical symmetry of RNase.

The fact that the longitudinal relaxation of the C_2-105 proton is less affected by spin diffusion than that of the other C_2 protons is due to the position of His 105 slightly outside the molecule. Near His 105, many atoms have ill-defined positions in the crystal structure of ribonuclease-*S* (*36*). This might reflect local mobility that can reduce the rate of spin diffusion within the His 105 proton environment. Moreover, this proton environment is rather isolated from the rest of the molecule by the sparsely protonated backbone. In contrast, the C_2 protons of His 12 and His 119 closely approach protons which in turn are close to the hydrophobic core of the protein, providing a more efficient spin diffusion path. A similar argument applies to C_2-48, while the C_4-105 is relaxed directly through its own β protons.

The *temperature dependence* of the longitudinal relaxation in mercuripapain was tested by measurements at 31 and 46°C. Plots of $\log(A_\infty - A_t)$ versus time are given in Fig. 11. The T_1 values, as determined from a least-squares fit, are listed in Table 1. The average relaxation rates for the aliphatic protons are larger at the higher tempera-

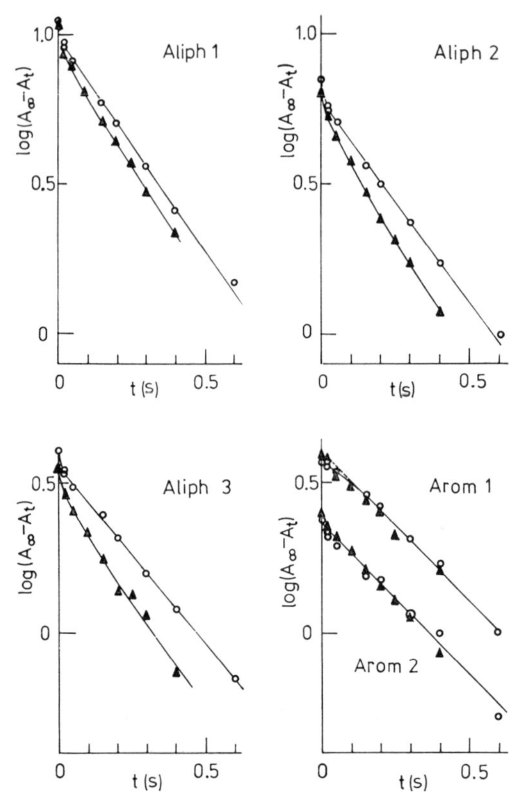

FIG. 11. Plots of $\log(A_\infty - A_t)$ versus time for mercuripapain at 31°C (open symbols) and at 46°C (filled symbols). A is given in arbitrary units.

ture, whereas those of the aromatic protons remain almost the same. By increasing the temperature to 46°C the τ_c is expected to decrease, according to the Stokes–Einstein relation, by about 35%. The intrinsic T_1 values for rigid groups will also decrease by 35% (cf. Fig. 4), whereas the transfer relaxation times T_t will increase. Thus T_t/T_{1i} will increase; i.e., R_t/R_{1i} will decrease. A comparison with Fig. 1 shows that the measured relaxation behavior is in accordance with the theory. At the higher temperatures the spin diffusion becomes smaller and the relaxation times tend more to the intrinsic T_1 values. For aliphatic protons these are lower, but for aromatic protons the decrease of the intrinsic T_1 is counteracted by the decreasing "magnetization leakage" by spin diffusion.

We notice that the change in T_1 for the methyl protons (Aliph 1) is small relative to that for the CH_2 protons. T_1 (CH_2) has a tendency to drop below T_1 (CH_3). This is clearly shown also in the theoretical curves of Fig. 4, where the T_1 (CH_2) line crosses the T_1 (CH_3) line at a τ_c of 5.5 nsec, where $T_1 = 0.2$ sec. This is very close to the observed value, indicating that the assumptions made in Fig. 4 with respect to the rates of methyl rotation and cross relaxation are reasonable. At 46°C the τ_c then must be about 6 nsec, yielding a τ_c of 8 nsec at 31°C, in agreement with the Stokes–Einstein relation. For the approximately spherical papain molecule the Stokes–Einstein relation is apparently more reliable than for RNase.

Saturation Studies

The steady-state saturation behaviors of papain and RNase were analyzed at the same positions as in the case of the T_1 experiments. Plots of Z_a^{-1} against $(\gamma H_1)^2$ are

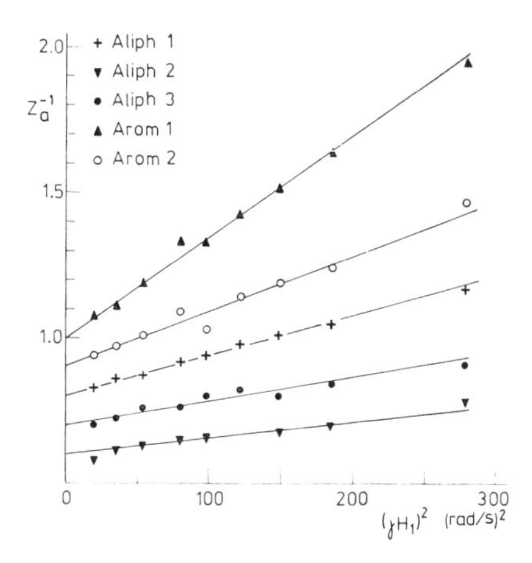

FIG. 12. Determination of the apparent $T_1'T_2$ from saturation studies, according to Eq. [17], for papain at 31°C. The scale of Z_a^{-1} refers to the data of the upper curve. The other curves have been displaced for clarity.

given in Fig. 12 for papain and in Fig. 13 for RNase. Apparent $T_1'T_2$ values were determined by a least-squares fit according to

$$Z_a^{-1} = C_0 \gamma H_1 / A = 1 + (\gamma H_1)^2 T_1' T_2, \tag{17}$$

where A is the signal amplitude and C_0 is a constant, which is determined from extrapolation to $H_1 = 0$ of a straight line through the data points.

The apparent $T_1'T_2$ values are given in Table 3. The histidine C_2 protons of RNase showed only a slight overlap at the highest power settings; for these protons the apparent $T_1'T_2$ values will be close to the real values.

The apparent $T_1'T_2$ values for Aliphs 2 and 3 in RNase are consistent with the value expected for methylene protons, which are not strongly influenced by cross-relaxation. The values for the histidine C_2 and the aromatic protons are lowered by cross relaxation. For the aromatic protons this reduction is stronger than for histidine protons, because the overlap in the spectrum may cause a reduction by a factor of 2 to 3. For His 105, which is less strongly influenced by cross relaxation, the $T_1'T_2$ value approaches that

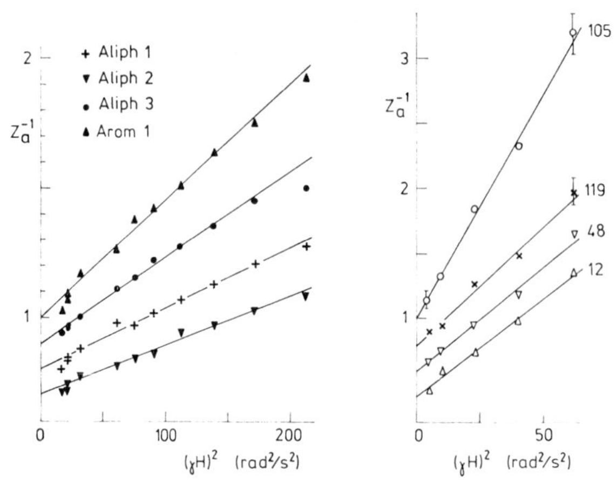

FIG. 13. Saturation studies, similar to those in Fig. 12, for RNase at 31°C. The figure on the right refers to the histidine C_2 protons.

TABLE 3

APPARENT $T_1'T_2$ VALUES (sec²) FOR MERCURIPAPAIN AND RNASE[a]

	Aliph 1	Aliph 2	Aliph 3	Arom 1	Arom 2
Mercuripapain	1.4×10^{-3}	0.6×10^{-3}	0.8×10^{-3}	3.4×10^{-3}	1.8×10^{-3}
RNase	2.6×10^{-3}	2.0×10^{-3}	3.2×10^{-3}	4.7×10^{-3}	

RNase His C_2–H	Saturation studies	T_1T_2[b]
C_2-105	$(3.6 \pm 0.2) \times 10^{-2}$	$(4.4 \pm 0.8) \times 10^{-2}$
C_2-12	$(1.5 \pm 0.1) \times 10^{-2}$	$(1.8 \pm 0.3) \times 10^{-2}$
C_2-119	$(1.9 \pm 0.2) \times 10^{-2}$	$(2.8 \pm 0.6) \times 10^{-2}$
C_2-48	$(1.8 \pm 0.2) \times 10^{-2}$	$(2.9 \pm 0.8) \times 10^{-2}$

[a] Standard errors are 10% or less. Temperature, 31°C.
[b] T_1 from pulse experiments, T_2 from linewidths.

expected for a proton coupled to two protons at 2.5 Å without cross relaxation (about 5×10^{-2} sec²).

The apparent $T_1'T_2$ for the methyl protons (Aliph 1) indicates a rotation of the methyl groups, even when a factor of 2 to 3 is allowed for resonance overlap. The value of τ_r must be in the neighborhood of 10^{-10} sec. The lower value for the methyl $T_1'T_2$ in papain is due to the longer τ_c of papain. The methylene protons in papain have a lower value of $T_1'T_2$ than in RNase. This is a direct indication of the occurrence of a stronger spin diffusion in papain. It is reemphasized that quantitative conclusions cannot be accurately obtained from saturation studies.

In Fig. 14 an Overhauser experiment in the differential mode is given for the aromatic region in papain at 100 MHz. The difference between the normal spectrum [curve (a)] and the spectrum obtained with strong radiation at 2.2 ppm is given as curve (b). As expected, a rather aspecific NOE is observed. Its magnitude of about 50% (the base

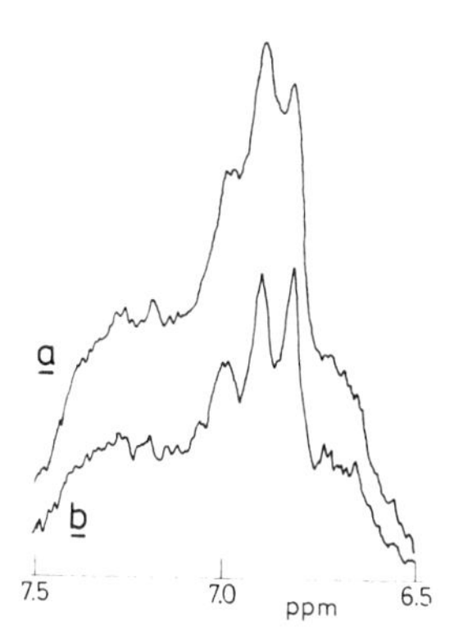

FIG. 14. Curve (a) is a part of the aromatic proton region of mercuripapain, whereas (b) is the difference between that normal spectrum and the spectrum obtained with strong irradiation at 2.2 ppm. The base line in each case coincides with the level of the ppm scale.

lines for both spectra are the same) implies that for the aromatic protons $\Sigma R_t \gtrsim R_1$, which means according to Eqs. [2 and 3] that $\tau_c \gtrsim 4$ nsec. Some specificity is observed in the region of 6.8–7.0 ppm, but we have not attempted to interpret such specificities in terms of the molecular structure.

DISCUSSION

The experimental results are all consistent with the theory given, assuming a rigid protein and rapid rotation of methyl groups. In no instance has there been any need to invoke local fast motion to explain the magnitude and frequency or temperature dependence of relaxation times, or the ratio of transverse and longitudinal relaxation rates. It is highly probable that also in other cases conclusions on local mobilities using relaxation data are incorrect if cross relaxation is not taken into account. Another example is the two-subunit enzyme triosephosphate isomerase (*5*) with a molecular weight of 50,500. Reported T_1 values for CH_3 protons and aromatic protons at 270 MHz are equal (0.75 sec) within experimental accuracy. The T_1 values are only twice those at 90 MHz, where they are mutually different. With the known amino-acid composition (*37*), and with the assumptions for the couplings as made before, we calculated the average T_1 value to be 0.79 sec, for a correlation time of methyl rotation of 10^{-10} sec and for an overall correlation time τ_c of 2.3×10^{-8} sec, which follows from the Stokes–Einstein relation, assuming a spherical molecule.

The assumption that rotating methyl groups provide the only effective relaxation sinks for large proteins at high frequencies seems justified from the correct results obtained with this model. Methyl protons comprise roughly 25% of all protons in a protein.

There is evidence, both from X-ray diffraction and from carbon-13 NMR (*38*), that in some cases side chains have ill-defined positions and show a higher mobility. In particular hydrophilic side chains in contact with the solvent, such as lysine, show such behavior. For the effects of cross relaxation, such side chains are not very important because they do not provide effective sinks themselves and spin diffusion along a mobile chain is not effective.

It is also well known from studies on basic pancreatic trypsin inhibitor (*39*) that some aromatic side chains (tyrosine and phenylalanine) rotate, or at least make 180° flips about the C_β–C_γ bond. Lifetimes between such flips were found in the range 10^{-1} to 8×10^{-4} sec. In fluorotyrosine alkaline phosphatase (*40*) it was found that only motions about the C_α–C_β bond could be present with correlation times larger than 10^{-6} sec. Such motions are all very much slower than the rotational correlation time of the molecule as a whole and do not influence the longitudinal relaxation rate of protons.

If the protein contains paramagnetic centers, such as spin labels or paramagnetic metal ions, these provide additional and quite effective relaxation sinks. Paramagnetic effects have been described by Andree (*41*) and will not be further considered here.

Important implications of cross relaxation arise if relaxation measurements are carried out on exchanging ligands binding to a macromolecule (*41*). While the ligand is bound to the macromolecule, it exchanges magnetization with the macromolecule at a rate, given by the cross relaxation, which is very high in large proteins. Thus the T_1 of the ligand is strongly influenced and is not simply a result of the weighted average of the relaxation rates of a bound and a free species.

The interpretation of T_1 and NOE data under conditions where cross relaxation relatively strong is no longer straightforward. Both T_1 and NOE tend to become less specific and therefore less useful. However, the *initial* decay of magnetization is much more dependent on the local environment than the overall decay. This fact can be used, in combination with frequency-selective 180° pulses and selective saturation, to provide experimental data on proton environment that can be helpful for spectral assignments. The straightforward use of T_1 data to detect local mobilities is certainly not justified for frequencies above 100 MHz and molecular weights above 20,000.

APPENDIX

To evaluate the effect of cross relaxation on saturation,we consider the relaxation for I_z given in Eq. [1], which can be rewritten as

$$dI_{zi}/dt = -(R_{1i} + \sum_{j \neq i} R_{tij})(I_{zi} - I_{0i}) + \sum_{j \neq i} R_{tij}(I_{zj} - I_{0j}). \qquad [A1]$$

If we define

$$R_{tii} \equiv -(R_{1i} + \sum_{j \neq 1} R_{tij}), \qquad [A2]$$

Eq. [A1] reduces to

$$dI_{zi}/dt = \sum_j R_{tij}(I_{zj} - I_{0j}). \qquad [A3]$$

In the case of a driving field applied to spin k only, the Bloch equation for I_{zk} becomes

$$dI_{zk}/dt = -\gamma H_1 v + \sum_i R_{tki}(I_{zi} - I_{0i}), \qquad [A4]$$

where v is the resonance intensity of the absorption mode. Equation [A3] remains valid for $i \neq k$. In a steady-state experiment, $dI_z/dt = 0$ for all i and Eq. [A3] can be solved for $I_{zi} - I_{0i}$, $i \neq k$:

$$\sum_{j \neq k} R_{tij}(I_{zj} - I_{0j}) = -R_{tik}(I_{zk} - I_{0k}), \qquad i \neq k,$$

or

$$I_{zi} - I_{0i} = -\sum_{j \neq k} Q_{tij}^{-1} R_{tjk}(I_{zk} - I_{0k}), \qquad i \neq k. \qquad [A5]$$

Here Q_t^{-1} is the inverse of the matrix Q_t, which is equal to the matrix R_t in which the kth row and column have been removed. If Eq. [A5] is inserted into Eq. [A4] we obtain

$$dI_{zk}/dt = -\gamma H_1 v + [R_{tkk} - \sum_{i \neq k} R_{tki} \sum_{j \neq k} Q_{tij}^{-1} R_{tjk}](I_{zk} - I_{0k}). \qquad \text{[A6]}$$

This has exactly the same form as the normal Bloch equation for a single-spin I_{zk}, with an effective longitudinal relaxation rate R'_{1k} given by

$$R'_{1k} = R_{1k} + \sum_{i \neq k} R_{tki} + \sum_{i \neq k} R_{tki} \sum_{j \neq k} Q_{tij}^{-1} R_{tjk}. \qquad \text{[A7]}$$

Since the Bloch equations for the transverse components are not influenced, the saturation behavior is equivalent to that given by Eq. [14].

The effective R'_1 for saturation behavior reduces in the case of a two-proton system (**I** and **S**) to

$$R'_{1I} = R_{1I} + [(R_t \cdot R_{1S})/(R_t + R_{1S})], \qquad \text{[A8]}$$

but with more than two spins a matrix inversion is required.

REFERENCES

1. I. Solomon, *Phys. Rev.* **99**, 559 (1955).
2. J. S. Noggle and R. E. Schirmer, "The Nuclear Overhauser Effect. Chemical Applications," Academic Press, New York, 1971.
3. I. D. Campbell and R. Freeman, *J. Magn. Resonance* **11**, 143 (1973).
4. F. W. Benz, G. C. K. Roberts, J. Feeney, and R. R. Ison, *Biochim. Biophys. Acta.* **278**, 233 (1972).
5. J. H. Coates, K. A. McLaughlan, I. D. Campbell, and C. E. McColl, *Biochim. Biophys. Acta* **310**, 1 (1973).
6. Y. Arata, R. Khalifah, and O. Jardetzky, *Ann. N.Y. Acad. Sci.* **222**, 230 (1973).
7. R. E. Wasylishen and J. S. Cohen, *Nature (London)* **249**, 847 (1974).
8. A. Abragam, "The Principles of Nuclear Magnetism," Clarendon Press, Oxford, 1961.
9. A. Kalk, Thesis, University of Groningen, 1975.
10. P. S. Hubbard, *Phys. Rev.* **109**, 1153 (1958) and **111**, 1746 (1958); G. W. Kattawar and M. Eisner, *Phys. Rev.* **126**, 1054 (1962); P. S. Hubbard, *Phys. Rev.* **128**, 650 (1962); P. M. Richards, *Phys. Rev.* **132**, 27 (1963); L. K. Runnels, *Phys. Rev. A* **134**, 28 (1964).
11. J. Drenth, J. N. Jansonius, R. Koekoek, and B. G. Wolthers, *Advan. Protein Chem.* **25**, 79 (1971).
12. J. W. Anderson and W. P. Schlichter, *J. Phys. Chem.* **69**, 3099 (1965).
13. K. van Putte, *J. Magn. Resonance* **2**, 216 (1970).
14. E. R. Andrew, W. S. Hinshaw, and M. G. Hutchins, *J. Magn. Resonance* **15**, 196 (1974).
15. D. E. Woessner, *J. Chem. Phys.* **36**, 1 (1962).
16. R. L. Hilt and P. S. Hubbard, *Phys. Rev. A* **134**, 392 (1964); P. S. Hubbard, *J. Chem. Phys.* **51**, 1647 (1969).
17. L. G. Werbelow and A. G. Marshall, *J. Magn. Resonance* **11**, 299 (1973).
18. M. F. Baud and P. S. Hubbard, *Phys. Rev.* **170**, 384 (1968).
19. K. van Putte, *J. Magn. Resonance* **5**, 367 (1971).
20. J. F. Rodrigues de Miranda and C. W. Hilbers, *J. Magn. Resonance* **19**, 11 (1975).
21. F. Noack, *in* "NMR Basic Principles and Progress" (P. Diehl, E. Fluck, and R. Kosfeld, Eds.), Vol. 3, Springer–Verlag, Berlin, 1971.
22. S. Emid and R. A. Wind, *Chem. Phys. Lett.* **27**, 312 (1974).
23. R. A. Wind and S. Emid, *Phys. Rev. Lett.* **33**, 1422 (1974).
24. A. Carrington and A. D. McLauchlan, "Introduction to Magnetic Resonance," Harper and Row, New York, 1967.
25. J. R. Kimmel and E. L. Smith, *Biochem. Prep.* **6**, 61 (1958).
26. L. A. Æ. Sluyterman and J. Wijdenes, *Biochim. Biophys. Acta* **200**, 593 (1970).

27. J. S. LEIGH, *Rev. Sci. Instrum.* **39,** 1594 (1968).
28. D. H. MEADOWS, O. JARDETZKY, R. M. EPAND, H. H. RUTERJANS, AND H. A. SCHERAGA, *Proc. Nat. Acad. Sci. U.S.A.* **60,** 766 (1968).
29. J. H. BRADBURY AND J. SENG TEH, *J. Chem. Soc. Chem. Commun.* 936 (1975); J. L. MARKLEY, *Biochemistry* **14,** 3547 (1975); J. H. MARKLEY, *Accounts Chem. Res.* **8,** 70 (1975); D. J. PATEL, L. L. CANNEL, AND F. A. BOVEY, *Biopolymers* **14,** 987 (1975).
30. C. C. McDONALD AND W. D. PHILLIPS, *Biochemistry* **12,** 3170 (1973).
31. D. M. YOUNG AND J. T. POTTS, *J. Biol. Chem.* **238,** 1995 (1963).
32. S. KRAUSE AND C. T. O'KONSKI, *Biopolymers* **1,** 503 (1963).
33. G. KARTHA, J. BELLO, AND D. HARKER, *Nature* **213,** 862 (1967).
34. J. DRENTH, J. N. JANSONIUS, AND B. G. WOLTHERS, *J. Mol. Biol.* **24,** 449 (1967).
35. D. E. WOESSNER, *J. Chem. Phys.* **37,** 647 (1962).
36. H. W. WYCKOFF, D. TSERNOGLOU, A. W. HENSON, J. R. KNOX, B. LEE, AND F. W. RICHARDS, *J. Biol. Chem.* **245,** 305 (1970).
37. J. C. MILLER AND S. G. WALEY, *Biochem. J.* **122,** 211 (1971).
38. V. GLUSHKO, P. J. LAWSON, AND F. R. N. GURD, *J. Biol. Chem.* **247,** 3176 (1972).
39. K. WÜTHRICH AND G. WAGNER, *FEBS Lett.* **50,** 265 (1975).
40. W. A. HULL AND B. D. SYKES, *Biochemistry* **13,** 3431 (1974).
41. P. J. ANDREE, Thesis, University of Groningen, 1975.

JOURNAL OF MAGNETIC RESONANCE 27, 101–119

Planar Spin Imaging by NMR

P. MANSFIELD AND A. A. MAUDSLEY

Department of Physics, University of Nottingham, University Park, Nottingham NG7 2RD, U.K.

Received July 30, 1977

A new method of spin-density imaging by NMR is described which allows simultaneous observation and differentiation of signals arising from spins distributed throughout a thin layer or plane within the specimen. The method, which is based on selective rf irradiation of the sample in switched magnetic field gradients, can produce visual pictures considerably faster than previously described line-scan imaging methods. Some simple examples of two-dimensional images obtained by the method are presented.

1. INTRODUCTION

During the last few years a number of methods have been described for producing nuclear magnetic resonance images related to the spin density distribution in solids and liquids (*1–4*). Of particular interest is the application of these imaging techniques in the study of biological material on a microscopic and a macroscopic scale.

Microscopic imaging at the cellular level offers a new method of studying localized relaxation times and diffusion processes within a single cell *in vitro*. Studies *in vivo* of water transport in plants is another interesting possibility.

At the macroscopic level, proton spin imaging *in vivo* could have clinical applications as a lower hazard alternative to X rays for medical imaging in man (*5, 6*). Known differences in the spin–lattice relaxation time between normal and malignant tissue (*7, 8*) might possibly be exploited for the early detection of cancer (*6*).

In all cases the important factors that comprise a general quality factor for the image are spatial resolution, signal/noise ratio, and the picture-formation time. Clearly, for a given imaging system, there is always some trade-off between signal/noise and picture-formation time. In an ideal system, all information from the spin distribution would arrive at a rate determined essentially by the experimenter and in this situation, the trade-off mentioned above would be straightforward. However, in practice it may not be possible to control the data input rate because of factors inherent in the imaging method itself. In addition, the information input may not be in a usable form, or it may take a considerable time to unravel the input or get it into a usable form. All the methods of imaging currently proposed attempt to reach the ideal, but they invariably fall short in one sense or another.

In this paper, which is an amplification and extension of our recent letter (*9*), we describe a new variant of the selective irradiation method of Garroway *et al.* (*2*), which allows simultaneous observation and differentiation of signals arising from spins distributed throughout a plane, or a set of planes within a three-dimensional object. By

ISSN 0022-2364

this means, we show that the speed of image formation can be increased by an order of magnitude or so over that of the single line-scanning method of imaging recently demonstrated by Mansfield *et al.* (*10*). This speed is vitally important at both the microscopic and macroscopic extremes of specimen size if we are to see NMR imaging usefully applied to living systems. Experimental results demonstrating the imaging method together with some examples of planar images are also presented.

2. GENERAL DESCRIPTION OF METHOD

2.1. *Selective Excitation in a Three-dimensional Object*

Let the specimen be placed in a static magnetic field B_0 which defines the y axis of quantization of the nuclei. We now describe three successive stages of selective irradiation and signal observation, (A), (B), and (C). This procedure closely follows that

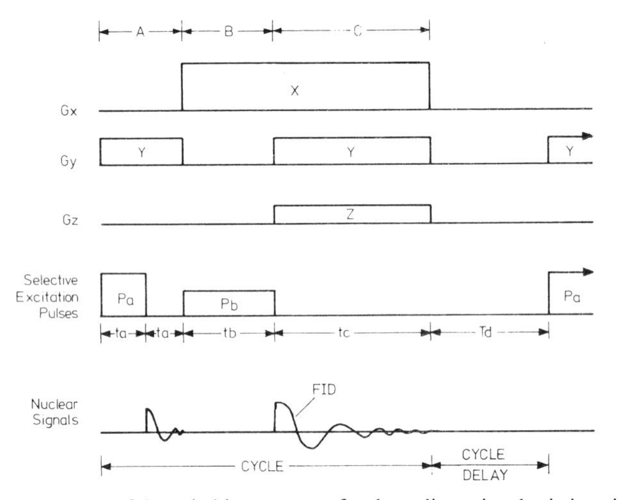

FIG. 1. Diagram of the switching sequence for three-dimensional spin imaging.

of Mansfield *et al.* (*10*). The order in which the various selections are performed is important in the practical realization of these imaging schemes. We describe a sequence which is less susceptible to pulse misalignments and relaxation effects. The particular order is indicated in Fig. 1, but other pulse permutations are also possible.

(A) In period A (Fig. 1), the gradient G_y is switched on and a selective irradiation pulse P_a is applied for time t_a in order to saturate regions of the specimen lying between a set of slices each of thickness Δy regularly spaced at $y = y_0 + mb$ (m an integer) from the origin with grid separation b (Fig. 2). These slices comprise undisturbed layers (or planes) of magnetization in equilibrium with the static magnetic field B_0. After a further time t_a, the disturbed spins decay, producing an FID signal. If this signal is sampled and Fourier transformed, it yields the projection profile of the saturated regions of the sample and could be used for alignment purposes.

(B) We now concentrate on these slices of undisturbed spins (Fig. 2). In phase B, G_x alone is switched on and a second multiple-slit irradiation pattern P_b selectively excites the spins in a series of layers normal to the x axis for a time t_b. For a cylindrical sample as sketched in Fig. 3, these are a series of discs of thickness Δx spaced a apart. The

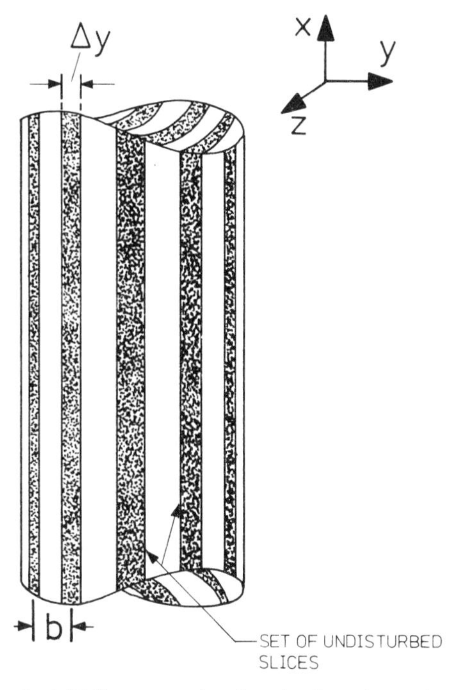

FIG. 2. Diagram showing the initially saturated regions in phase A together with the undisturbed slices in an extended cylindrical sample.

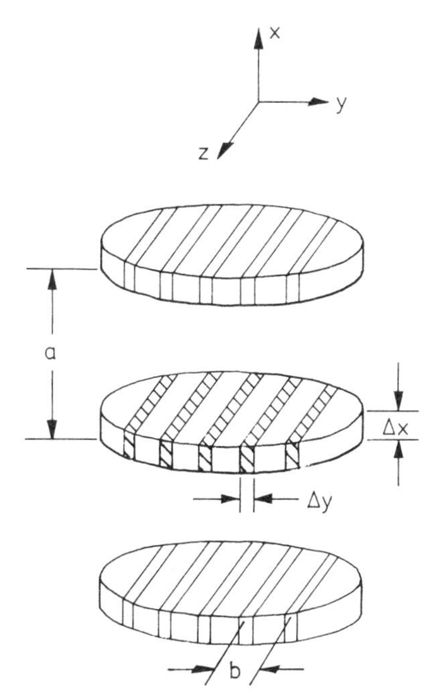

FIG. 3. Diagram showing the discs of magnetization selected in phase B. The active spin regions producing the FID signal are indicated by cross hatching.

shaded strips correspond to the active spin regions. The general undisturbed spin distribution is determined by the preparatory tailored pulse sequence, the details of which are discussed below.

The excitation pulses referred to here and previously could nutate the initially undisturbed spins through any angle θ, but here we shall think of θ as being 90°.

(C) At the end of the (B) excitation pulse, all three gradients are switched on so that the precessing spins experience the combined effects of G_x, G_y, and G_z. (For imaging in a single plane, only G_y and G_z are required in the read period C.) The free induction decay (FID) from all volume elements $mx\Delta y\Delta z$ spaced at $y = y_0 + mb$ in the planes $x = x_0 + la$ is observed and Fourier transformed to give a spin-density distribution within the solid $\rho(x, y, z)$.

In the simplest cyclic arrangement (Fig. 1), the spin system is left to recover in the cycle delay time T_d before repeating the sequence. In Section 4, however, we describe modifications to this procedure which enhance (or alternatively minimize) the effect on the FID of the various relaxation times.

2.2. *Principle of Operation*

The application of a magnetic field gradient to a three-dimensional or even a two-dimensional continuous distribution of spins will not ordinarily allow all elements of the distribution to be uniquely assigned magnetically. However, if they could be uniquely assigned then the "absorption" lineshape would in a single trace reveal the entire spin density distribution.

By a process of selective irradiation, a discrete lattice structure can be superimposed on the otherwise continuous spin distribution. That is to say, we can arrange that we observe only those spins lying on a well-defined lattice structure, the dimensions of which are controlled by selective excitation, etc.

For a given orthorhombic lattice, it is easy to convince oneself that all points can be uniquely assigned a definite frequency by application of *one* appropriately chosen gradient only. This process thus gives spot (point) checks, as it were, of the entire three-dimensional spin-density distribution. Spins between lattice points do not contribute to the observed signals. Enlargement of the points to include surrounding spins imposes constraints, as we shall see later, which may make it easier to look at a single plane, or a selected few planes, rather than the entire three-dimensional object. This may also be desirable from a data-handling and -storage point of view.

2.3 *Theory*

The initial selection and preparation of the spin system in period (A) involves the gradient G_y and rf pulses which nutate some of the spins through 90°. Precisely which spins are affected will depend on the magnitude of the field gradient and the spectral distribution of the perturbing tailored rf pulse.

For simplicity† we may represent the combined effects of such a tailored pulse and field gradient by a spatially selective operator $(1 - \hat{S}_y)$. If the spin-density distribution is $\rho(xyz)$ then $\hat{S}_y\rho(xyz)$ represents the undisturbed spin distribution while $(1 - \hat{S}_y)\,\rho(xyz)$ is the distribution of spins which receives a 90° nutation pulse. In a similar manner, we introduce a second spatially selective operator $\hat{S}_x$ which selects all the spins nutated

† Detailed calculations indicate that the effect of selective pulses on the spin system is somewhat more complicated than indicated here.

through 90° by the combined action of the second tailored excitation pulse and the new field gradient G_x.

In the following analysis, we assume ideal 90° nutations and no spin–lattice relaxation effects. In this case the operations referred to above are commutative.

Neglecting multiplicative constants, let the initial equilibrium density matrix of the spin system (spin I) in the high-temperature approximation and distributed in volume v be

$$\sigma(0) = \int \delta_0 dv, \tag{1}$$

where $\delta_0 = \rho(xyz)I_y$. The z component of the transverse response signal in the rotating reference frame at time t is given by

$$S_z(t) = \mathrm{Tr}\{\sigma(t)I_z\}, \tag{2}$$

where Tr is the trace or diagonal sum.

After the first 90° pulse, the density operator describing the spin system at time $t_a + t$ becomes

$$\delta_1(t_a + t) = \hat{S}_y\rho(xyz)I_y + (1 - \hat{S}_y)\rho(xyz)e^{i\gamma yG_yI_yt}I_ze^{-i\gamma yG_yI_yt}, \tag{3}$$

where γ is the magnetogyric ratio. The second term in Eq. [3] will give a transverse signal which quickly decays in t_a. Following the second 90° rf pulse, the density operator at time $2t_a + t_b + t$ becomes

$$\begin{aligned}\delta_2(2t_a + t_b + t) = \hat{S}_x \exp(i\gamma tI_y[xG_x + yG_y + zG_z])\tilde{\delta}_1(2t_a)\\ \times \exp(-i\gamma tI_y[xG_x + yG_y + zG_z]),\end{aligned} \tag{4}$$

where the tilde on $\tilde{\delta}_1$ means that I_y is replaced by I_z and I_z by $-I_y$ in δ_1.

It is straightforward to show, by expanding Eq. [4] and substituting into Eq. [2], that the only significant nonvanishing signal following t_b at time t is

$$S(xyzt) = \int \hat{S}_y\hat{S}_x\rho(xyz)\cos\gamma t[xG_x + yG_y + zG_z]\, dv. \tag{5}$$

We have assumed, throughout, a noninteracting spin system which evolves during selective irradiation as though the rf pulse were a pure 90° rotation operator. No spin echo is expected following the second tailored 90° pulse since the field gradient change from $\mathbf{j}G_y$ to $\mathbf{G}$ will in general suppress it, except for those spins lying in a line along y corresponding to the magnetic centers of the gradients G_x and G_z, i.e., when x and z are both zero in Eq. [5]. These spins are likely to give a very small echo signal, which we entirely ignore in this analysis.

(*a*) *Fourier transform nesting.* In a generalization of these experiments we have in mind that the selection processes embodied in the operators $\hat{S}_x$ and $\hat{S}_y$ correspond not to single layers of material, but to multiple layers. We specialize to equally spaced layers of thickness Δx, strips of width Δy, and points of spacing Δz (set by sampling) with spatial periodicities a, b, c. In the limit where the undisturbed spin ranges Δx, Δy approach zero, and for discrete sampling of the distribution along z, we have

$$\hat{S}_x\hat{S}_y\hat{S}_z\rho(xyz) \to \rho(la_1mb_1nc) = \rho_{lmn} \qquad (l, m, n \text{ integers}), \tag{6}$$

where $\hat{S}_z$ is the spatial sampling operator. The effective density, therefore, becomes a discrete distribution ρ_{lmn} corresponding to the lattice points $x = al$, $y = bm$, and $z = cn$.

In this limit, the FID signal, Eq. [5], becomes

$$S = \sum \rho_{lmn} \cos t[l\Delta\omega_x + m\Delta\omega_y + n\Delta\omega_z]\Delta v_{lmn}, \quad [7]$$

where Δv_{lmn} is the volume of spins at a lattice point contributing to the signal, and is assumed to be constant for all points. The angular frequency increments are given by

$$\Delta\omega_x = \gamma a G_x, \qquad \text{etc.} \quad [8]$$

We see from Eq. [7] that if the gradients and lattice constants are chosen so that

$$N\Delta\omega_z \leqslant \Delta\omega_y \leqslant \Delta\omega_x/\mathrm{M} \quad [9]$$

where M and N are the largest values of m and n, respectively, all points in the distribution ρ_{lmn} are uniquely defined in the frequency domain. (Although we are talking here about superimposing a regular lattice by selective irradiation and sampling, the above point is true for a natural orthorhombic lattice.) A single Fourier transformation of S_{lmn} will thus yield in one calculation the complete three-dimensional distribution function ρ_{lmn}.

2.4. *Resolution*

The requirement that all points in the object be simultaneously resolved is more stringent along the z axis. If there are N points spaced c apart, and each point has an extent Δz, then the condition for linearity of the z gradient is (*11*)

$$\Delta z/Nc = \Delta G/G_z, \quad [10]$$

where ΔG is the deviation of G_z from uniformity. In addition, the natural linewidth of the resonance ΔW_{nat} in the static field (which includes static field inhomogeneity, relaxation effects, diffusion in the gradient, and bulk broadening effects of the sample), must satisfy the relationship

$$\Delta W_{\text{nat}} \leqslant \Delta z G_z. \quad [11]$$

Of course, this stringency arises only if one wishes to put more information into a given bandwidth in the frequency domain. If the frequency per point along z is the same as in the line-scan experiment (*10*) then the requirements of both the z gradient coil and the natural linewidth, and hence the static field, are exactly the same.

On the other hand, if data are compressed into a narrow frequency band, they create higher demands on the uniformity of both the gradient coils and the static magnet. The advantage is a narrower bandwidth per picture point, giving an increased signal/noise ratio.

2.5. *Presentation and Readout of Data*

The presentation and readout of data are perhaps best illustrated with a simple example of a three-dimensional image of a cylinder of mobile spins (water). We assume that the spin system has been selectively irradiated and prepared in phases (A) and (B) as described in Section 2.1 and we are about to observe the FID in phase (C).

Now the FID following the (B) phase selection can be read in G_x alone. In this case, all spins within a layer are undifferentiated, but of course the layers *are* differentiated so the Fourier transform would simply be (for three layers, as in Fig. 3) three equal-amplitude spikes (Fig. 4a). However, if readout were done with G_x and G_y on, and in such a way that Eq. [9] is satisfied, namely, that $\Delta\omega_x \geqslant M\Delta\omega_y$, then we obtain the spectrum in Fig. 4b. The lineshape in this case consists of three equal spectra, each comprising the *discrete* projection profile for the cylinder. Finally, if the readout is performed in all three gradients G_x, G_y, and G_z satisfying Eq. [9], we expect a discrete lineshape of constant amplitude, as indicated in Fig. 4c. Each element is ideally

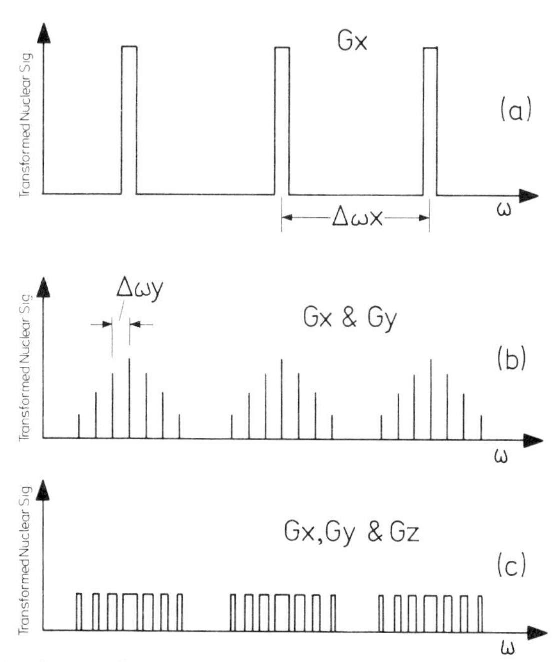

FIG. 4. Expected Fourier transformed signals for a cylindrical sample of homogeneously distributed spins prepared as in Fig. 3 by the selective irradiation procedure. (a) Signal with read gradient equal to G_x only, showing three equal lines corresponding to the signals from the three discs of magnetization as in Fig. 3. (b) Each of the three lines in (a) is split by the addition of G_y. The discrete profile corresponds to the projection of the spin-density distribution of the disc along the y axis. (c) Expected signal on application of all three gradients, as in an imaging experiment. The addition of G_z broadens each line of the discrete spectrum in (b) above. Ideally, the profile of each broadened line is rectangular and of constant height, giving a direct visual picture of successive cross-sectional views through the discs.

rectangular, with the width varying across the spectrum as indicated, and represents a cross-sectional profile of each strip within the layer.

2.6. *Tailored rf Pulses*

Unlike line-scan imaging (*10*) the tailored rf pulse sequences in these experiments have a multiple discrete spectral distribution as illustrated in Fig. 5 for the special cases of (a) rectangular spectral profiles, and (b) uniform amplitude. That is to say, instead of scanning line by line, the entire object is irradiated at the same time. (In certain cases, it is possible to compensate for rf inhomogeneity in the transmitter coil system by changing the spectral profile from a constant amplitude to something which varies with frequency in the desired manner.)

For a multiple discrete rf spectral distribution of L components with constant amplitude $H_{1\omega}$, individual constant widths $m\Delta\omega_p$ (m an integer) and with the center frequency of each component separated by $\Delta\omega l = n\Delta\omega_p$, the time domain pulse must be shaped according to the expression

$$H_1(t_n) = H_{1\omega} m\Delta\omega_p \operatorname{sin c} (\Delta\omega_p m t_n) \sum_0^L e^{i(l\Delta\omega_p t_n)}. \qquad [12]$$

In this expression, the angular frequency per point is given by

$$\Delta\omega_p = 2\pi/N\tau, \qquad [13]$$

in which N is the total number of points in the time domain and τ is their spacing.

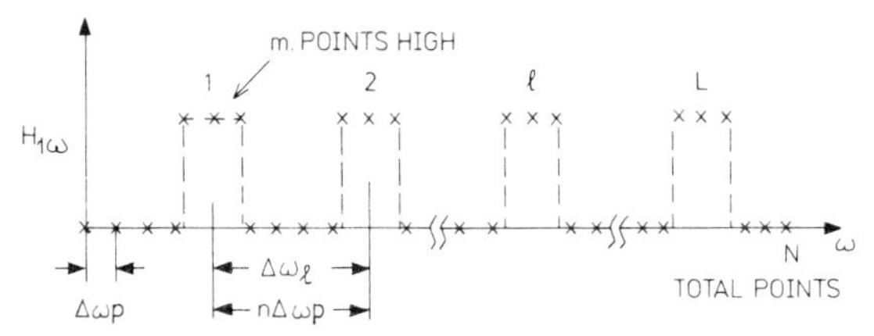

FIG. 5. Diagram of the discrete multiple rectangular spectral distribution for the tailored rf pulse applied in t_y.

Of course, if an additional frequency offset is added so that image fields are nonsecular as far as the resonant spins are concerned (i.e., outside the range of frequencies produced by the field gradient), then we take the real part of the expression and simply compute the cosine transform.

Equation [12] gives the initial rf pulse amplitude

$$H_1(0) = 2\pi L m H_{1\omega}/N\tau, \qquad [14]$$

where $H_1(0)/Lm$ is the field amplitude used to calculate the 90° pulse condition for each irradiated point.

(*a*) *Pulse sequence.* The A irradiation pattern could look like Fig. 5 with m large and $m - n$ small, corresponding to narrow gaps of zero $H_{1\omega}$ intensity, which in turn correspond to the unirradiated and hence undisturbed magnetization in Fig. 2. However, it is technically more convenient because of rf power requirements to produce a complementary radiation pattern corresponding to small m, but this does just the opposite of what is required. That is to say, it disturbs the narrow strips that we wish to leave undisturbed.

One procedure, therefore, is to pulse with rf having a spectral distribution pattern which is complementary to that required. This is followed immediately with a short 90° rf pulse whose rf carrier phase is shifted by 180°. The net effect is thus to tip the magnetization in the narrow strips back up along the y axis, while the undisturbed magnetization in the wider strips is tipped backward by 90° into the x–z plane of the rotating reference frame.

The pulse shaping referred to is achieved in our case by computing on-line the Fourier transform of the desired rf spectral distribution, which is then used to modulate directly the rf pulse envelope (*10*).

It is also possible to approximate the complementary rf spectral distribution by a train of equally spaced short rf pulses. The discrete frequency distribution of such a pulse train has a width inversely proportional to pulse duration, a periodicity inversely proportional to the pulse repetition period, and a discrete linewidth inversely proportional to the pulse train length. The spectral distribution may be reasonably approximated as constant over a restricted frequency range. However, generating a pulse spectrum wider than required, in order to satisfy the constant-amplitude approximation, represents a considerable waste of rf power in the unwanted sidebands and is best avoided, particularly when irradiating live specimens.

2.7. *Effects of Finite Selection Width*

The condition for magnetic uniqueness, expression [9], was derived for point regions of spins within a general three-dimensional object and it ignored the effect of finite volume of the sample at the lattice sites.

If the elemental volume at all lattice sites is $dv = \Delta x \Delta y \Delta z$, then additional constrains on the size of this volume and the magnitude of the gradients arise if all points are to be simultaneously resolved. In this case the inequality, expression [9], is modified and becomes

$$\delta\omega_x + \delta\omega_y + N\Delta\omega_z \leqslant \Delta\omega_y \leqslant (1/M)[\Delta\omega_x - \delta\omega_x - \delta\omega_y], \tag{15}$$

where $\delta\omega_x = \Delta x G_x$ and $\delta\omega_y = \Delta y G_y$.

The advantage of expanding the elemental volume is of course that more of the sample contributes to the observed signal. However, as we shall see, the price paid for this increase in sensitivity is the measurement of an average spin density over the volume, or in other words a decrease in spatial resolution.

We now enquire in detail into the effect of the additional contributions to the FID signal when the discrete lines of the rf spectral distribution, during selection, have a finite width. For simplicity we consider the case of single-plane imaging. That is to say, the FID is read in the two gradients G_y and G_z only. However, the arguments developed will be applicable to more general multiplanar imaging.

Suppose now that we do not have delta functions, but a set M of broad spikes or even rectangular spectral distributions (Fig. 5) which in one gradient G_y produce the discrete density projection profile $f_m(\omega_{mq})$ along the y axis. The mth spike or rectangle can be regarded as being made up of a closely spaced set q of delta functions, each one of which would broaden in combined gradients G_y and G_z to yield the z-axis density function $\rho_z^{mq}(\omega_{mq})$. Thus over the subset q the observed profile will be a broadened, smeared function $\Gamma_m(\omega_{mq})$ which is the weighted sum of individual functions, i.e., the convolution-like function

$$\Gamma_m(\omega_{mq}) = \sum_{q'} \rho_z^{q'm}(\omega_{q'm}) f_m(\omega_{mq'} - \omega_{mq}). \tag{16}$$

We note that, unlike the ordinary convolution function, the broadening function $f_m(\omega_{mq})$ varies in general with m across the projection.

For a closely spaced subset, Eq. [16] reduces to

$$\Gamma_m(\omega) \to \int \rho_z^m(\omega') f_m(\omega' - \omega)\, d\omega', \tag{17}$$

where the discrete variables ω_{mq} and $\omega_{mq'}$ are replaced by the continuous variables ω and ω'.

For a set of well-resolved spikes (or ideally, delta functions) where the extent of the additional broadening produced by G_x does not overlap, we get from Eq. [17]

$$\Gamma_m(\omega) = \rho_z^m(\omega). \quad [18]$$

This is the result already obtained and illustrated (Fig. 6). But Eq. [16] also allows the evaluation of $\rho_z^{mq}(\omega_{mq})$ when $f(\omega)$ is not a delta function, provided the mth cross-sectional profile does not change significantly with q, i.e., if we replace $\rho_z^{mq}(\omega_{mq})$ by $\rho_z^m(\omega_{mq})$, which corresponds to a high degree of short-range spatial correlation. In this case Eq. [17] is a localized convolution integral. The function $\rho_z^m(\omega_{mq})$ can be obtained

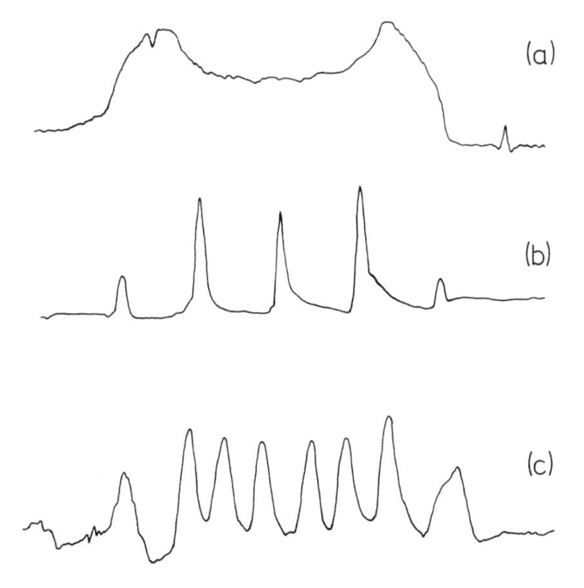

FIG. 6. Experimental results obtained from a mineral oil annulus. (a) Projection of the annulus in a static field gradient $G_y = 0.95$ G cm^{-1}. The signal was averaged 39 times. (b) response profile when the annulus is irradiated with a discrete spectral distribution in G_y only (see text). (c) Response profile of annulus to discrete spectral irradiation as in (b) above, but with additional read gradient $G_z = 0.266$ G cm^{-1} switched on. The signals in (b) and (c) were averaged 128 times.

directly from the Fourier transform of $\Gamma_m(\omega_{mq})$ and the localized projection profile $f_m(\omega_{mq})$ without the additional broadening of G_x.

The procedure for single-plane images, for example, is thus first to measure the discrete projection profile in G_y alone. Next the broadened profile is recorded in both gradients G_y and G_z. Each discrete section of both profiles is inversely Fourier transformed to the time domain and the broadened signal is divided by the corresponding unbroadened signal computed at zero frequency offset. The quotient is then Fourier transformed back to the frequency domain, and the resulting signal represents the true density profile along the z axis.

3. EXPERIMENTAL RESULTS

The results described here were obtained at 15.0 MHz with a computer-controlled spectrometer which has been substantially described elsewhere (*10*). In the present

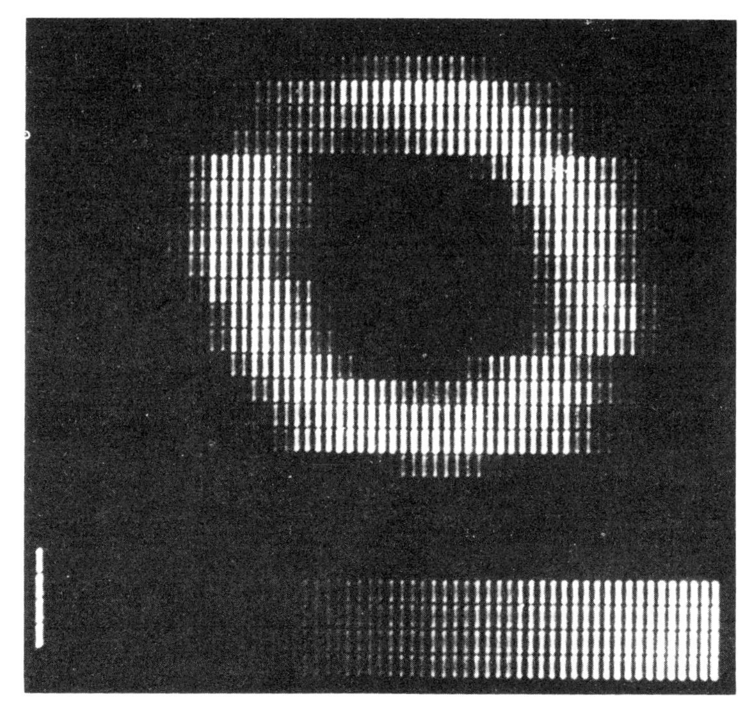

FIG. 7. A cross-sectional proton spin-density image of a mineral oil annulus produced by planar spin imaging using the data of Fig. 6c. An intensity scale corresponding to a 16-level linear density wedge is also included across the bottom of the picture.

experiments, however, an orthogonal transmitter and receiver coil system was used (*6, 12*). The planar imaging method has been tried out for protons in a liquid mineral oil sample in the form of a cylindrical annulus. The measured outer and inner diameters of the annulus were 13.7 and 8.1 mm, respectively. For simplicity, specialization to a

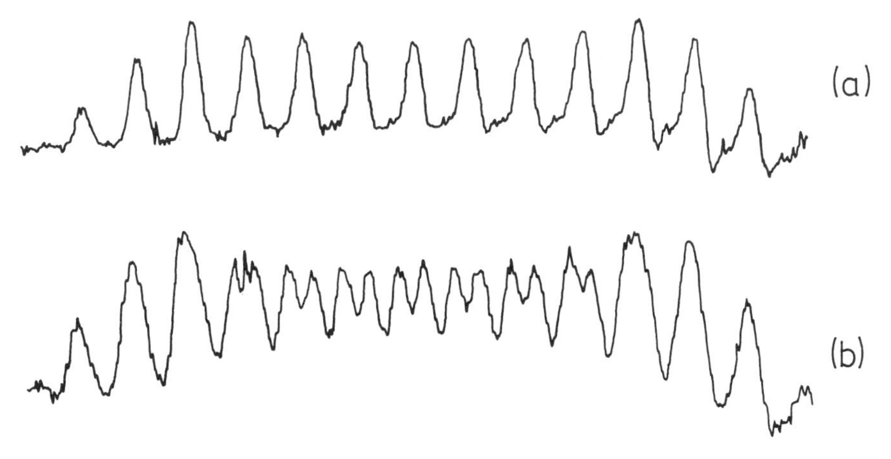

FIG. 8. (a) Discrete projection spectrum of a mineral oil annulus comprising 13 lines. This was obtained by irradiating and reading in G_y only. (b) Response profile as in (a) above, but broadened in the read phase by additional gradient G_z. (See text for details.)

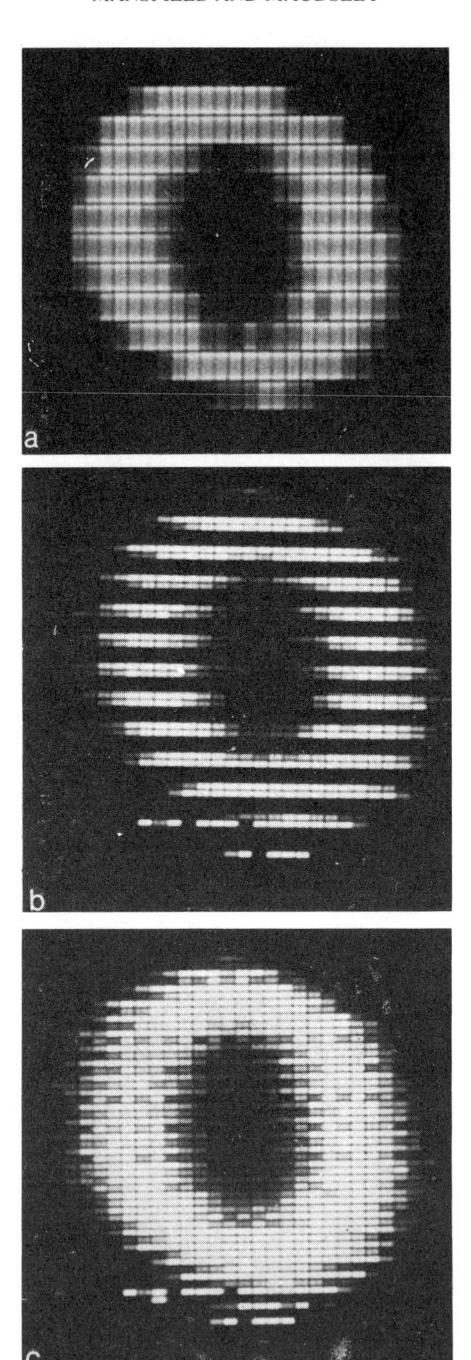

FIG. 9. Planar spin images from protons in a mineral oil annulus showing buildup of the image with a fourfold interlace. (a) Coarse planar image obtained with data similar to those of Fig. 8b. Spot wobbling has been used to fill in the gaps between individual spots. (b) Second stage of the fourfold interlace pictures. The spot size is reduced to avoid overlap. (c) Complete fourfold interlace image. See text for details.

layer of material by selective irradiation is not employed since the sample has cylindrical symmetry along the x axis. The cycle thus starts in phase B (Fig. 1) with G_z switched off and G_y switched on throughout.

The one-dimensional projection of the annulus measured in G_y is shown in Fig. 6a. The sample was selectively irradiated with a tailored pulse corresponding to five equally spaced, equal-intensity rectangular peaks in the rf spectral distribution arranged to just span the sample. That is to say, $L = 5$, $m = 1$, and $n = 20$ (see Fig. 5). The frequency per point, $\Delta\omega_p$, is 855 rad sec^{-1}. The response to this selective irradiation in the gradient G_y is shown in Fig. 6b. As expected, it is the discrete projection profile. If more lines were included, we would end up tracing out the continuous projection profile (Fig. 6a). The width of each spike in Fig. 6b is 195 Hz. About half the broadening of these spikes is due to static magnetic field inhomogeneity, the remainder being ascribed to nonlinearity of the field gradient coils.

The response to the selective irradiation in G_y read in G_y and G_z is shown in Fig. 6c. Each spike of Fig. 6b is broadened by the action of the additional field gradient G_z to yield directly the planar spin distribution in the form of successive cross-sectional views through the oil annulus. The distribution has a fairly coarse grid along the y direction. A finer grid was obtained by an interlacing procedure in which the radiation pattern was shifted up in frequency by one-fourth, one-half, and three-fourths of the frequency spacing between adjacent peaks in the rf spectral distribution. This gives a fourfold finer grid resolution across the specimen and has thus allowed us to produce a visual image. This is shown in Fig. 7 and consists of a 20×40 array made up from four spectra, as in Fig. 6c. Each broadened spectrum is the result of 128 averages. The transformed data are used to modulate the spot intensity of an oscilloscope in a television raster display (*13*). A linear grey scale of 16 levels of spot brightness covering the range black to white is used. The picture resolution in Fig. 7 was limited by a maximum of 256 Fourier-transformed points in the frequency domain. Increasing this number to 512 (1024 in the time domain) has improved both the resolution and the speed with which a planar image can be produced. Fig. 8a shows another discrete one-dimensional projection of the annulus in G_y. In this case the discrete excitation spectrum had 16 equal-intensity peaks, that is to say $L = 16$, $m = 1$, and $n = 8$ (see Fig. 5). The response to this excitation spectrum applied in G_y and read in combined gradients G_y and G_z is shown in Fig. 8b and is an average of 16 spectra. The gradient values, $G_y = 1.86$ G cm^{-1} and $G_z = 0.1$ G cm^{-1}, were arranged so that the excitation spectrum spanned the annulus with some overlap.

The planar image corresponding to Fig. 8b is shown in Fig. 9a and has a resolution of 16 elements along the vertical picture axis. The horizontal picture axis has a total of 32 elements. With a cycle delay of 0.3 sec, this picture took 7.8 sec to produce, which includes 3.0 sec for the 1024-point time domain Fourier transform. Figure 9b shows the second stage of a fourfold interlace, and Fig. 9c the complete fourfold interlace planar image. The whole picture took 37.2 sec to produce and is a 64×32 array. This is 16 times faster than producing an equivalent picture by single line scanning (*10*). The broadening of the spikes in Fig. 8a is due to static field inhomogeneity and nonlinearity of the y axis gradient and we suspect that it is the principal cause of degradation of the spectrum in Fig. 8b. Preliminary tests of the localized deconvolution procedure outlined in Section 2.7 show that the method will remove broadening due to the static field inhomogeneity and gradient coils as well as broadening due to finite width of the rf irradiation pattern. However, full picture enhancement on-line has not so far been tried out because of computer core limitations.

4. IMAGING CYCLE VARIANTS

4.1. T_1 Discrimination

The basic planar imaging method described in Section 2 and Fig. 1 is T_1 selective through the cycle delay T_d and can thus discriminate against regions of the specimen having a spin–lattice relaxation time longer than T_d. However, from the point of view of data accumulation, this delay represents wasted time, and hence a low efficiency in information input, since the FID is zero in this period.

4.2. T_2 and T_1 Discrimination

The transverse signal can be made to persist following the initial FID by recalling the signal in a series of spin echoes. This may be done with T_d zero or short by either (i) periodically reversing all gradients in an extended C phase, or (ii) applying a train of 180° rf pulses in a Carr–Purcell sequence, again in an extended C phase with static gradients. Method (i) is unlikely to give additional echoes from the initially saturated spins in (A) and will work perfectly for single-plane imaging when $G_x = 0$ in the read phase (C). Method (ii) will work satisfactorily for single-plane imaging, but for multiplanar imaging the 180° pulses will affect the initially saturated spins in (A). However, this is unlikely to produce an extra echo for the same reason as that in case (i) above, namely, that the additional gradients G_y and G_z will tend to quench the formation of any echo. The timing of the pulse sequence also does not favor the formation of an observable signal.

The peak echo amplitude decays with time constant T_2. Thus if $T_2 \gg t_b$, many field gradient reversals can be made and many echoes can be produced. These echoes can be suitably co-added to improve the signal/noise ratio over that of the single FID signal.

The echo-averaging process described here allows the transverse decay signal, in effect, to persist for a time T_2 or so. If $T_1 \simeq T_2$ there is no wait period required between the end of signal averaging and the repeat of the A phase in the following cycle. In biological materials, however, T_2 can be less than T_1 and in this case time could be wasted waiting for the spin system to repolarize. Nevertheless this method could find useful application since for $T_2 < T_1$, it will produce images which discriminate against regions of the specimen with short T_2. For $T_2 \simeq T_1$ the method discriminates against regions of short T_1 and thus acts in a manner complementary to that of the basic method described in A above.

4.3. Relaxation–Time-independent Methods

Relaxation time discrimination can be of considerable value in enhancing picture contrast in medical images (*6*). However, for images representing true spin-density variations, relaxation time discrimination can be a problem. We now introduce modifications to the cycle of Fig. 1 which substantially remove relaxation time dependences and improve the data-sampling efficiency.

The modified cycle is shown in Fig. 10 and should run continuously. The A and B phases are the same as indicated in Fig. 1. The read phase C is similar to that described previously, and although one FID and one-half spin echo are shown, this phase could be extended from $2t_c$ to $2nt_c$ (where n is an integer) to include more echoes.

The new feature of the cycle is the store phase D. As indicated, the refocused signal at a spin-echo peak is switched back to the equilibrium position by the same selective excitation pulse P_b but with a 180° rf carrier phase shift denoted P_b^{180}. Any magnetization loss in the read and store periods can be recovered in the delay period. Of

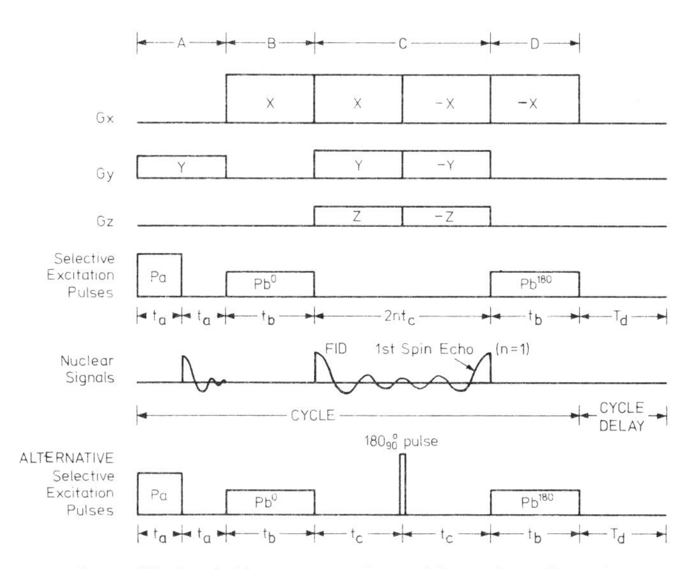

FIG. 10. Diagram of a modified switching sequence for multilayer three-dimensional spin imaging. The main object of this modification is to enable rapid cycling of the sequence in times less than T_1 and T_2. An alternative arrangement of the selective excitation pulses is also shown. In this case, the field gradient reversals are not required.

course, the recovery of magnetization may not exactly balance any losses, so that the initial FID signal amplitude may, after a number of cycles, approach a new equilibrium magnetization, different from the static equilibrium value, whose amplitude is optimized by varying the cycle delay T_d. The important point is that the read signal is substantially independent of both T_1 and T_2 for $2nt_c \ll T_2, T_1$.

In a given cycle, the fraction of time spent reading the signal can be increased by including more echoes. The maximum time $2nt_c$ is, of course, limited to T_2 but it will probably be better to keep n of the order of 1 or 2 because of the relaxational losses in the signal amplitude.

A number of alternative and equivalent excitation pulse arrangements are also shown which use 180° rf pulses to refocus the spin echo. (The gradient reversals are *not* required in this case.) Notice that the 180° carrier phase shift of the second selective pulse P_b is still required. The disadvantage of this arrangement is that more rf power is required, but this could be outweighed in a practical system by the greater simplicity achieved.

Using rf pulses to refocus (without gradient reversals) for the read phase is similar to the driven equilibrium Fourier transform (DEFT) method of signal enhancement described by Becker *et al.* (*14*). A number of modifications to this method proposed by Waugh (*15*) are also incorporated in our Fig. 10. These modifications are to the carrier phasing of the 180° refocusing pulses, with respect to the P_b selective excitation pulse. Thus strictly speaking, the 180°_{90} rf pulses referred to in Fig. 10 should really be selective excitation pulses as well, with the same sharp spectral profile as P°_b and P^{180}_b. The object of introducing these apparent complications in the rf carrier phasing is to compensate the cycle automatically for imperfections due to rf inhomogeneity. If parts of the specimen do not receive an exact 180° rf pulse, then repeated applications of the cycle will in general lead to a deterioration and perhaps complete loss of the signal. For proper compensation, it is better to have $n = 2$, thus producing two spin echoes between P^0_b and P^{180}_b. In the rf pulse version, refocusing can be achieved by using either 180°_{90} rf pulses for all n or 180°_0 (n odd) and 180°_{180} (n even) pulses. Alternate phase reversal in this alternative 180° pulse arrangement produces alternate sign reversals of the spin echoes. Co-addition of these for signal averaging should therefore take account of the sign changes.

(*a*) *A simpler cycle.* Under the right conditions, repetitive selective excitation pulses P_b, as in the simple imaging method of Fig. 1, can be made to produce a nonzero FID in a quasi-equilibrium state, even when the cycle delay T_d is a lot less than the spin–lattice relaxation time T_1. This effect has been described by Carr (*16*) and others (*4, 17, 18*) and is often referred to as steady-state free precession (SSFP). We have observed experimentally that SSFP works for selective irradiation pulses and in switched magnetic field gradients.

As in the previous modified cycle above, an advantage of combining our new imaging method with SSFP is that the image produced is substantially independent of both T_1 and T_2.

5. PICTURE QUALITY AND PERFORMANCE

A useful measure of the performance of a particular imaging scheme is the imaging time T_i (*10*). This is defined as the total time required to produce a picture to a given resolution and signal/noise ratio. The imaging time clearly depends on the imaging method, and indeed, even for the same method, can vary depending on precisely how the data obtained are processed.

For a given three-dimensional spin system, there is a limited amount of information which has to be read out, processed, and displayed. Quite irrespective of the imaging method used to access this information, there are a number of fundamental constraints which allow an ideal imaging time to be calculated.

5.1. *Imaging Time*

Let us divide the imaging volume field, assumed to be a cube, into m^3 volume elements. For each volume element there is a corresponding picture element in the image field and we assume that only the spins in a fraction β of each volume element contribute to the observed signal for each picture point. The signal/noise ratio per picture point, R_p, is given by (*10*)

$$R_p = Af(Q\omega_0^3\beta V/B)^{1/2} = R_{p_0}/\beta^{1/2} \quad [19]$$

where V is the total volume of the sample assumed to fill the resonant receiver coil completely with a distribution of mobile spins, and $f = 1/m^3$ is the filling factor for one picture volume element. The quantities Q, ω_0, and B are, respectively, the receiver coil quality factor, the Larmor angular frequency of the spins, and the bandwidth per picture point. The constant A embodies the spin susceptibility, the receiver temperature and noise figure, and a coil geometry factor. We now assume that data from all m^3 volume elements can be collected simultaneously, but in such a way as to preserve the spatial localization of each element. That is to say, we consider a planar or multiplanar imaging system such as that described in Fig. 10, for example.

The imaging time T_i is given quite generally by

$$T_i = N_A T_c + D_{lmn}, \qquad [20]$$

where N_A is the number of averages of the imaging cycle necessary to achieve the desired signal/noise ratio R, and D_{lmn} is the time required to Fourier transform the data into $l \times m \times n$ picture points. The cycle period T_c (see Fig. 10) is given by

$$T_c = 2t_a + 2t_b + 2nt_c + T_d, \qquad [21]$$

in which $2nt_c$ is the data collection time, t_a and t_b are selective excitation pulse lengths, and T_d is a delay which incorporates the time to compute the Fourier transforms of the selective irradiation excitation spectra. The bandwidth per point B of Eq. [19] can be expressed in terms of a single data-collection interval as

$$B = 2\pi/t_c. \qquad [22]$$

The signal/noise ratio enhancement produced by averaging the signal $2n\mathrm{N}_A$ times is given by

$$R = R_p(2nN_A)^{1/2}. \qquad [23]$$

From Eqs. [19] and [20] it is clear theoretically that the shortest imaging time $T_{i\,\min}$ is achieved when the signal from all the spins in the specimen is sampled all of the time. i.e., when $\beta = 1$ and $T_c = 2nt_c$. Practical imaging systems, which inevitably fall short of this ideal, can, however, be simply compared with it by introducing an imaging efficiency η_i defined by

$$\eta_i = T_{i\,\min}/T_i, \qquad [24]$$

which, from Eqs. [19], [20], and [21], reduces to

$$\eta_i = \alpha\beta, \qquad [25]$$

where $\alpha = 2nt_c/T_c$. The Fourier transformation time, which depends on the type of computing facility available, has been ignored in this calculation.

In the selective irradiation scheme for planar imaging, some space must be allowed between adjacent line elements. This means that β cannot be unity unless interlacing of the picture is used. This is discussed in Section 5.2. The nature of the selective pulses also means that α cannot equal unity. Realistic values of α and β for single-plane imaging as described in this paper would be $\alpha = \frac{1}{2}$ and $\beta = \frac{1}{4}$, yielding an efficiency $\eta_i = 12\%$.

5.2. *Comparison with Line Scanning*

It is of interest to compare the imaging times for single-plane and line-scan imaging under optimum conditions for equal signal/noise ratio and resolution. If we consider pictures comprising equally spaced grids of m lines in both cases, then the necessary gaps between adjacent lines in planar imaging mean in general that β^{l} can be greater than β^{p}, where the superscripts l and p used here and later refer, respectively, to line scan and planar imaging.

An alternative way of leaving unfilled gaps in the planar image is the idea of interlacing, already discussed and used to produce the NMR images presented in Section 3. In this case it is straightforward to arrange that $\beta^{l} = \beta^{p}$, corresponding to the final fine grid required for a given picture resolution. However, the final planar image is then made up of an interlace of i (*integer*) coarser grid planar images produced by selective irradiation with m/i lines. The necessary gap between adjacent rows in each coarse planar image is thus created by the missing $(i - 1)$ lines. In this case and from Eqs. [19], [20], and [23] we obtain for the single-plane imaging of m^2 points in an i-fold interlace

$$T_i^{p} = (i/2n)\left(\frac{R}{R_{p_0}}\right)^2 \frac{T_c^{p}}{\beta^{p}} + iD^{p}_{m^2/i}. \qquad [26]$$

Line scanning the same data in m rows with m points per row yields an imaging time (using the same receiver and bandwidth per point)

$$T_i^{l} = (m/2n)\left(\frac{R}{R_{p_0}}\right)^2 \frac{T_c^{l}}{\beta^{l}} + mD^{l}_{m}. \qquad [27]$$

If we take $T_c^{p} = T_c^{l}$, $\beta^{p} = \beta^{l}$ and discard the Fourier transformation times then the imaging time ratio is

$$T_i^{l}/T_i^{p} = m/i. \qquad [28]$$

Of course, account of the Fourier transformation times may be made by using the exact expressions [26] and [27]. Using fast computers, Fourier transformation of data into an array of 128 × 128 data points can be performed in about 3.5 sec. When the first term in Eq. [26] is comparable to $iD^{p}_{m^2/i}$, then if $iD^{p}_{m^2/i} \simeq md^{l}_{m}$, the imaging time ratio becomes for large m

$$T_i^{l}/T_i^{p} \simeq m/2i, \qquad [29]$$

which still represents a substantial improvement in speed for planar imaging over line scanning. Taking $m = 128$ and a fourfold interlace, Eq. [29] shows that in a typical situation planar imaging is 16 times faster than line scanning. Line scanning under these conditions yields an imaging efficiency η_i of only 0.24%.

6. DISCUSSION AND CONCLUSIONS

We have shown both experimentally and theoretically that, by a process of selective irradiation in switched magnetic field gradients, it is possible to receive and differentiate NMR signals arising simultaneously from spins distributed throughout a thin slice or plane within a specimen. These signals can be used to produce a visual image related to the spin-density distribution throughout the slice. Initial applications of planar imaging are likely to be limited to forming pictures one plane at a time. However, our analysis shows that it is possible to produce simultaneous multiplanar images.

For the most efficient generalized imaging system one would ideally wish to observe all the spins distributed within the specimen all of the time. In the planar image of Fig. 9a only one-fourth of the total number of spins in the slice contributed to the signal. With improved gradient coils and by using localized deconvolution, it should be possible to increase this fraction, to between one-half and one-third. Ways of increasing the data acquisition time are also proposed. Using a version of the DEFT technique (*14, 15*) in conjunction with planar imaging, one could increase the fraction of total time available for data acquisition to between one-fourth and one-half. A similar improvement might be obtained by using planar imaging in combination with SSFP (*17, 18*). Both methods should produce planar images which are substantially independent of relaxation time effects.

The fact that signals arise from selected regions throughout a plane or slice means that planar spin imaging is in principle faster than the single line-scanning method for the same signal/noise ratio and resolution. As an example, we can produce a one-shot planar image comprising a 16×32 data array in 30 msec plus the Fourier transformation time, with a signal/noise ratio of the image of about 4:1. Under optimum line-scanning conditions the same image would take 16 times longer to produce.

ACKNOWLEDGMENT

We are grateful to the Science Research Council for an equipment grant.

REFERENCES

1. P. C. Lauterbur, *Pure Appl. Chem.* **40,** 149 (1974).
2. A. N. Garroway, P. K. Grannell, and P. Mansfield, *J. Phys. C* **7,** L457 (1974).
3. A. Kumar, D. Welti, and R. R. Ernst, *J. Magn. Resonance* **18,** 69 (1975).
4. W. S. Hinshaw, *J. Appl. Phys.* **47,** 3709 (1976).
5. P. Mansfield and A. A. Maudsley, *Phys. Med. Biol.* **21,** 846 (1976).
6. P. Mansfield and A. A. Maudsley, *Brit. J. Radiol.* **50,** 188 (1977).
7. R. Damadian, *Science* **171,** 1151 (1971).
8. J. G. Diegel and M. M. Pintar, *J. Nat. Cancer Inst.* **55,** 725 (1975).
9. P. Mansfield and A. A. Maudsley, *J. Phys. C* **9,** L409 (1976).
10. P. Mansfield, A. A. Maudsley, and T. Baines, *J. Phys. E* **9,** 271 (1976).
11. P. Mansfield and P. K. Grannell, *Phys. Rev.* **12,** 3618 (1975).
12. D. M. Ginsberg and M. J. Melchner, *Rev. Sci. Instrum.* **41,** 122 (1970).
13. T. Baines and P. Mansfield, *J. Phys. E* **9,** 809 (1975).
14. E. D. Becker, J. A. Ferretti, and J. C. Farrar, *J. Amer. Chem. Soc.* **91,** 7784 (1969).
15. J. S. Waugh, *J. Mol. Spectrosc.* **35,** 298 (1970).
16. H. Y. Carr, *Phys. Rev.* **112,** 1693 (1958).
17. R. R. Ernst and W. A. Anderson, *Rev. Sci. Instrum.* **37,** 93 (1966).
18. R. Freeman and H. D. W. Hill, *J. Magn. Resonance* **4,** 366 (1971).

JOURNAL OF MAGNETIC RESONANCE **28,** 471–476 (1977)

Correlation of Proton and Carbon-13 NMR Spectra by Heteronuclear Two-Dimensional Spectroscopy

A great deal of information is contained in the proton and carbon-13 NMR spectra of a given compound, but sometimes it proves necessary to look further to discover which proton and carbon shifts are related by virtue of proximity within the molecule. This presupposes an interaction between them, usually the spin–spin coupling, and the most common technique of chemical shift correlation makes use of a series of coherent off-resonance decoupling experiments (*1, 2*). This communication utilizes an alternative approach based on double Fourier transformation (*3, 4*) that allows information about the proton spectrum to be transmitted to the carbon-13 spins. It is an application of a general experiment recently proposed and realized for the indirect detection of carbon-13 resonance by Ernst and Maudsley (*5, 6*).

It is convenient to consider the proton magnetization vectors in a reference frame rotating in synchronism with the proton transmitter frequency, ν_0. Proton spins with an equilibrium Z magnetization are rotated by a nonselective 90° pulse and aligned along the Y axis of this frame, where they precess freely for a variable time interval t_1. Following the arguments of Ernst (*5*), consider first the simple CH spin system where there are just two transverse proton magnetization vectors $M(\alpha)$ and $M(\beta)$ precessing at frequencies $f(\alpha) = \nu_H - \nu_0 + \frac{1}{2}J(\mathrm{CH})$ and $f(\beta) = \nu_H - \nu_0 - \frac{1}{2}J(\mathrm{CH})$. At time t_1 a second 90° pulse rotates these proton vectors about the X axis, creating longitudinal magnetizations $M_z(\alpha)$ and $M_z(\beta)$. These are equal to the Y components of the precessing vectors just before the second pulse, and are therefore cosine functions of the precession angles built up during t_1,

$$\varphi(\alpha) = 2\pi f(\alpha)\, t_1 \qquad \text{and} \qquad \varphi(\beta) = 2\pi f(\beta)\, t_1. \qquad [1]$$

These Z magnetizations correspond to nonequilibrium population differences across the proton transitions (Table 1) and are modulated as a function of t_1. The populations may be thought of as being coherently "stirred" at the frequencies $f(\alpha)$ and $f(\beta)$. This affects the population differences across the carbon transitions, given by

$$\Delta P(A) = 1 + \tfrac{1}{2}(\gamma_H/\gamma_C)\,[\cos\varphi(\alpha) - \cos\varphi(\beta)], \qquad [2]$$

$$\Delta P(B) = 1 - \tfrac{1}{2}(\gamma_H/\gamma_C)\,[\cos\varphi(\alpha) - \cos\varphi(\beta)]. \qquad [3]$$

This time dependence of the populations introduces an amplitude modulation of the carbon-13 signal excited by a 90° pulse. [No significant flip angle dependence can be detected in these experiments (*7*).] Note that the senses of the two modulations are opposite. The maxima of the cosine functions correspond to the condition that one of the proton vectors has completed an integral number of complete rotations during t_1, so that the net effect of the two 90° pulses represents a population inversion similar to that induced by a *selective* 180° pulse applied to one of the proton transitions.

ISSN 0022–2364

TABLE 1

THE ENERGY LEVELS APPROPRIATE TO A CH SPIN SYSTEM SHOWING THE SPIN POPULATIONS AFTER TWO 90° PROTON PULSES COMPARED WITH THOSE AT BOLTZMANN EQUILIBRIUM

	Spin states		Spin populations[a]	
Level	Carbon	Proton	At equilibrium	At time t_1
1	α	A	$-\Delta - \delta$	$-\Delta \cos \phi(\alpha) - \delta$
2	β	A	$-\Delta + \delta$	$-\Delta \cos \phi(\beta) + \delta$
3	α	B	$+\Delta - \delta$	$+\Delta \cos \phi(\alpha) - \delta$
4	β	B	$+\Delta + \delta$	$+\Delta \cos \phi(\beta) + \delta$

[a] The equilibrium population difference is 2Δ across proton transitions and 2δ across carbon-13 where $\Delta/\delta = \gamma_H/\gamma_C \simeq 4$.

The carbon-13 signal thus contains two modulated components and a constant component, and these are separated by Fourier transformation. Two successive transformations are carried out, the first as a function of the usual acquisition parameter t_2, the second as a function of the period of free precession of the protons, t_1. The result is a two-dimensional spectrum with the frequency axis F_2 representing the conventional proton-coupled carbon-13 spectrum, and the orthogonal axis F_1 the proton spectrum of the carbon-13-substituted molecule. For the two-spin system CH there are four modulated responses; in any given frequency dimension the pairs of responses have opposite intensities and interfere destructively if J(CH) is comparable with the linewidth.

As mentioned by Ernst and Maudsley (*5, 6*), this is the heteronuclear analog of the double-pulse homonuclear experiment proposed by Jeener (*3*).

The Two-Dimensional Spectrum of Ethanol

The application of this technique may be illustrated by the shift correlation spectrum of ethanol, measured as a slightly acidified sample containing deuterochloroform as internal lock material. The Varian CFT-20 spectrometer used a computer program for double Fourier transformation described elsewhere (*8*). Proton pulses of 40-μsec duration were obtained by pulsing the proton decoupler in the coherent mode. Note that the frequency of the decoupler determines the frequency zero of the F_1 dimension. For the sake of simplicity Fig. 1 displays the spectrum in the absolute-value mode (*8*), the relative signs of the intensities having been verified in separate phase-sensitive experiments.

A characteristic feature of Fig. 1 is the strong conventional proton-coupled carbon-13 spectrum that runs along the line $F_1 = 0$ and represents the unmodulated carbon populations. At the F_2 coordinate of each of these responses, a horizontal trace runs out in the F_1 dimension carrying the corresponding *proton* responses. These traces are equivalent to the carbon-13 satellites of a conventional proton spectrum except that the two halves have oppoposite intensities (not apparent in the absolute-value display of Fig. 1). For example, in the left foreground of Fig. 1 there is a trace originating from the methylene protons showing a doublet [J(CH) = 141.3 Hz] of quartets [J(HH) = 7.1 Hz]. In the background of the diagram there is a doublet [J(CH) = 125.5 Hz] of triplets

[J(HH) = 7.1 Hz] from the protons of the methyl group. These responses provide the chemical shifts of the CH_2 and CH_3 protons even though the spectrometer detects only carbon-13 signals, and the proton and carbon shifts are correlated through the direct couplings.

These traces are repeated at the F_2 frequencies corresponding to the carbon-13 lines, but instead of the usual 1 : 2 : 1 intensity ratio for the triplet and 1 : 3 : 3 : 1 intensity ratio for a quartet, these have become 1 : 0 : 1 and 1 : 1 : 1 : 1, respectively, as predicted and

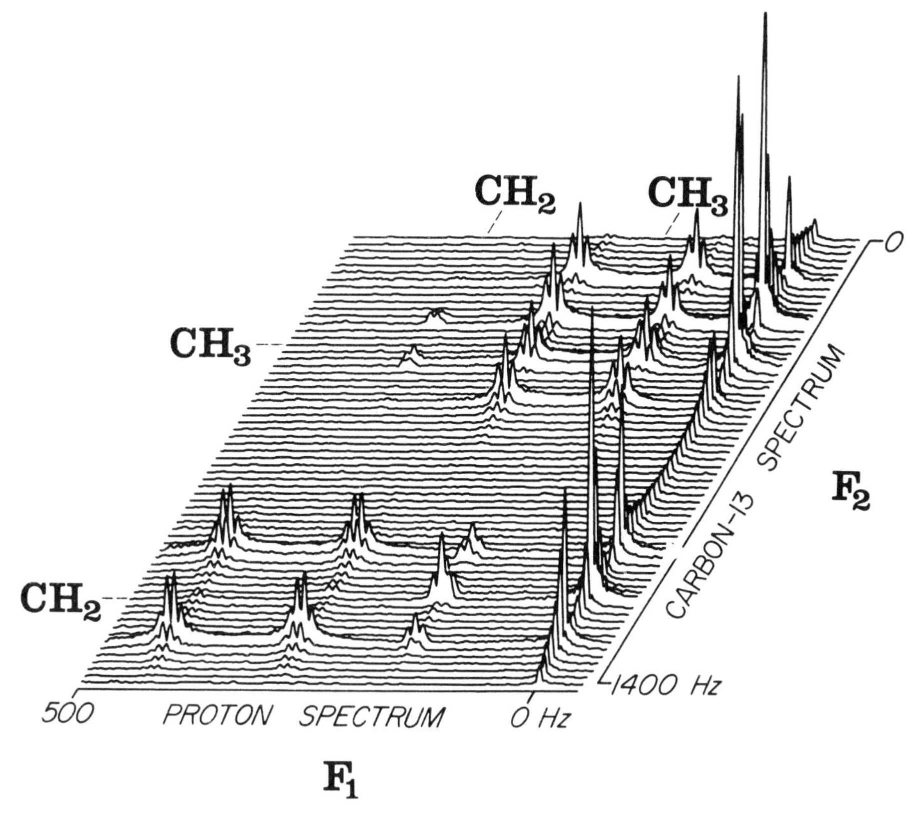

FIG. 1. Two-dimensional spectrum of carbon-13 in natural abundance in ethanol, obtained by a pulse technique which permits the indirect detection of the proton resonance frequencies, and the correlation of the proton and carbon chemical shifts.

demonstrated by Ernst (*5*). A simple extension of the population "stirring" calculations of Table 1 establishes that the expected intensity ratios for a triplet should be −1 : 0 : + 1, the zero arising because the population effects cancel for the intermediate symmetric levels while the antisymmetric levels remain unaffected. Similarly it is easy to show that the quartet intensities are −1 : −1 : +1 : +1. (exactly equivalent results apply for the population *changes* induced by a selective 180° pulse on one of the proton lines.)

Population transfer also occurs through long-range coupling, as illustrated by the three triplets in the right foreground and the four quartets in the left background of Fig. 1. The former are actually doublets [J(CCH) = 4.2 Hz] of triplets [J(HH) = 7.1 Hz] in the F_1 dimension but the long-range coupling is not resolved, although it is large enough to prevent mutual cancellation of these opposed signals. Similarly, in the F_2 dimension there is an unresolved 4.2-Hz splitting (a 1 : 1 : 1 : 1 quartet) and a well-resolved 1 : 2 : 1 triplet due to coupling to the "passive" protons of the methylene group.

The weaker responses in the left background of Fig. 1 belong to the methylene protons of molecules with carbon-13 in the methyl group. Here the long-range coupling [$J(CCH) = 2.5$ Hz] is small enough to cause appreciable cancellation in both frequency dimensions. This leaves an apparent 1 : 3 : 3 : 1 quartet (in F_2) of 1 : 3 : 3 : 1 quartets (in F_2) due to coupling to the "passive" methyl protons.

Simplification of the Spectra

For many of the applications to larger molecules it is important to simplify these two-dimensional correlation spectra, and this can be achieved by greatly reducing the splitting due to protons in both frequency dimensions. Since the algebraic sum of the intensities of the multiplet components of the modulated responses is zero, noise

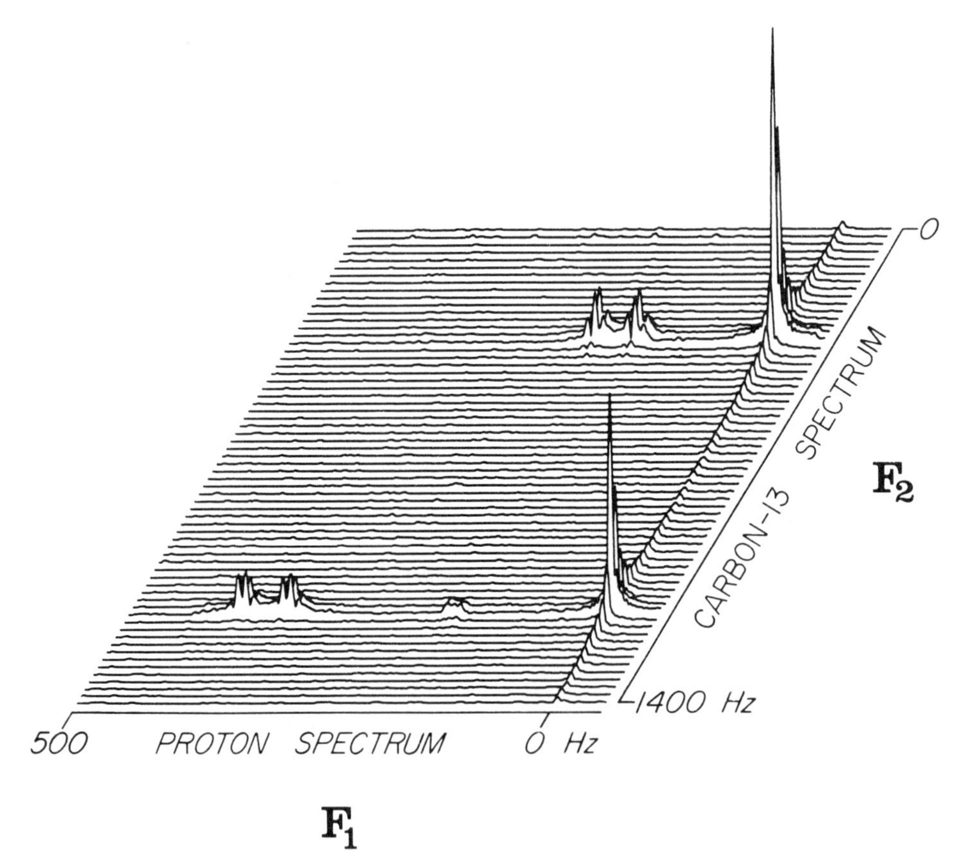

FIG. 2. A two-dimensional spectrum of ethanol similar to Fig. 1, except that the proton–carbon splittings have been coherently decoupled in the F_2 dimension, and scaled down in the F_1 dimension by a factor of 3 by a partial refocusing technique. Note the almost complete elimination of the responses due to the long-range coupling.

decoupling during acquisition of the carbon-13 signal is not feasible. However, since the proton transmitter is normally set off-resonance in this experiment, coherent decoupling can be used, leaving a small residual splitting in the F_2 dimension, sufficient to prevent mutual cancellation of the opposed signals, but not necessarily large enough to be resolved in the two-dimensional display (Fig. 2). An equivalent simplification may be achieved in the F_1 dimension by the introduction of a 180° refocusing pulse on the carbon spins, slightly delayed with respect to $\frac{1}{2}t_1$, so that the refocusing of the α and β proton vectors is not quite complete. In practice this pulse was applied at $2t_1/3$, reducing the splittings by a factor of 3. This partial refocusing acts on the long-range couplings in a similar way, and the reduced splittings become comparable with the linewidths, causing destructive interference. In this way shift correlations based on long-range coupling are eliminated, leaving only correlations based on the large directly bonded couplings, a considerable simplification of the spectrum. Figure 2 shows the ethanol spectrum modified by these two devices, the only strong responses representing populations transferred through $J(CH_3) = 125.5$ Hz and $J(CH_2) = 141.3$ Hz, with a very weak response through the long-range coupling $J(^{13}C \cdot CH_3)$. The center frequencies of these two groups of lines indicate the chemical shifts of the protons with respect to the decoupler frequency ν_0. The relative shift is 195 Hz ($\approx$2.5 ppm at 80 MHz).

Conclusions

The preliminary results with this technique show considerable promise for increasing the information content of carbon-13 spectroscopy by introducing correlated proton shift information. The two-dimensional spectrum may show the detailed coupling scheme as pointed out by Ernst (*5*), as in Fig. 1, or it may be considerably simplified as in Fig. 2. It is primarily the modulated spin populations that are of interest, and since steady-state saturation of the carbon-13 spins affects only the unmodulated signals, the method should be applicable to carbon-13 nuclei with long spin–lattice relaxation times. Sensitivity is thus determined by the spin–lattice relaxation of protons, which is often more favorable than that of carbon-13. Furthermore the population pumping mechanism operates through population differences appropriate to protons (Δ) rather than for carbon (δ), where $\Delta/\delta \simeq 4$, so there is a sensitivity advantage similar to that obtained through the nuclear Overhauser effect.

ACKNOWLEDGMENTS

This research was supported by an equipment grant from the Science Research Council, and a Scholarship from the Salters' Company (G.B.). The idea of detecting nuclear resonance indirectly by observation of the coherent signal transferred to another nucleus was originally suggested and experimentally verified by R. R. Ernst and A. A. Maudsley (*5, 6*). One of the authors (R.F.) is pleased to acknowledge an illuminating discussion with Professor Ernst.

REFERENCES

1. H. J. Reich, M. Jautelat, M. T. Messe, F. J. Weigert, and J. D. Roberts, *J. Amer. Chem. Soc.* **91,** 7445 (1969).
2. L. F. Johnson, Tenth Experimental NMR Conference, Pittsburgh, Pa. 1969.
3. J. Jeener, Ampere International Summer School II, Basko Polje, 1971; Second European Experimental NMR Conference, Enschedé, Holland, 1975.

4. W. P. Aue, E. Bartholdi, and R. R. Ernst, *J. Chem. Phys.* **64,** 2229 (1976); R. R. Ernst, W. P. Aue, P. Bachmann, J. Karhan, A. Kumar, and L. Müller, Nineteenth Congress Ampere, Heidelberg, 1976.
5. R. R. Ernst, Eighteenth Experimental NMR Conference, Asilomar, Calif., 1977; Sixth International Symposium on Magnetic Resonance, Banff, Canada, 1977.
6. A. A. Maudsley and R. R. Ernst, *Chem. Phys. Lett.*, in press.
7. S. Schäublin, A. Höhener, and R. R. Ernst, *J. Magn. Resonance* **13,** 196 (1974).
8. G. Bodenhausen, R. Freeman, R. Niedermeyer, and D. L. Turner, *J. Magn. Resonance* **26,** 133 (1977).

Geoffrey Bodenhausen
Ray Freeman

Physical Chemistry Laboratory,
Oxford University,
Oxford, England.

Received July 19, 1977

JOURNAL OF MAGNETIC RESONANCE, **29**, 433–462 (1978)

Selective Excitation in Fourier Transform Nuclear Magnetic Resonance

GARETH A. MORRIS AND RAY FREEMAN

Physical Chemistry Laboratory, University of Oxford, South Parks Road, Oxford, England

Received July 12, 1977

The applications of frequency-selective excitation methods in Fourier transform NMR are discussed, and a simple technique is described for selective excitation of a narrow frequency region of a high-resolution NMR spectrum in a Fourier transform spectrometer. A regular sequence of identical radiofrequency pulses of small flip angle exerts a strong cumulative effect on magnetizations close to resonance with the transmitter frequency or one of a set of equally spaced sidebands separated by the pulse repetition rate. All other magnetizations precess through an incomplete number of full rotations between pulses, and are caught by successive pulses at an ever changing phase of their precession, which destroys the cumulative effect. The motion of the various nuclear magnetization vectors may be described pictorially according to the Bloch equations, neglecting relaxation during the pulse sequence. A general theory is presented for selective or "tailored" excitation by an arbitrary modulation of the radiofrequency transmitter signal. It confirms earlier conclusions that the frequency-domain excitation spectrum corresponds to the Fourier transform of the transmitter modulation pattern, provided that the NMR response remains linear. The excitation spectra calculated for the selective pulse sequence by these two alternative approaches show good agreement within their respective limitations. A number of practical applications of selective excitation are explored, including solvent peak suppression, the detection of partial spectra from individual chemical sites, selective studies of relaxation and slow chemical exchange, and hole-burning or localized saturation.

1. INTRODUCTION

1.1. Applications of Selective Excitation

Most modern NMR spectrometers for high-resolution work operate in the Fourier transform mode in which the excitation is produced by strong nonselective radiofrequency pulses. In this fashion an important improvement in sensitivity is achieved, transient phenomena such as relaxation may easily be followed, and a good approximation to the slow-passage linewidth is obtained. The majority of the applications of the Fourier transform method have no need for linear sweep or fine adjustment of the transmitter frequency, and such facilities are not normally provided in dedicated Fourier spectrometers. There are, however, a number of experiments in which fine adjustment and selectivity in the frequency domain would be a decided advantage.

The most widely encountered problem is that created by intense solvent resonances (particularly water) which make excessive demands on receiver linearity and on the

0022–2364/78/0293–0433$02.00/0

dynamic range of the analog-to-digital converter. Several of the methods suggested for solvent peak suppression involve selective irradiation, either to saturate the solvent resonance (*1–11*) or to avoid exciting it (*12–15*). One of the principal advantages claimed for the new technique of correlation spectroscopy (*16–18*) is this ability to examine a spectrum without allowing the transmitter frequency to sweep through the strong solvent line.

A severe limitation on the applicability of NMR to large molecules is the complexity that results from the crowding together of many overlapping spin multiplets. This problem has provided the impetus for the development of very high field superconducting solenoids for the study of biologically significant systems. The growing interest in proton-coupled carbon-13 spectra (*20–26*) provides an illustration of this problem, since even in relatively simple molecules there can be severe overlap of spin multiplets. Frequently this can be the most important factor limiting the size of molecule which may be studied. When heteronuclear spin coupling is responsible for the overlap problem, selective excitation methods can provide a solution.

Wideband proton irradiation collapses the multiplet structure of carbon-13 spectra, giving a significant sensitivity improvement at the expense of the loss of all proton–carbon coupling information. This means that carbon-13 chemical shifts are available after relatively short periods of signal averaging. Knowledge of the chemical shifts makes it possible to unravel complex proton-coupled spectra by a simple combination of selective excitation and gated decoupling (*27–30*). The carbon-13 multiplets are collapsed by proton irradiation, leaving one resonance for each chemically distinct site. One such resonance is excited selectively, and the decoupler is then switched off to allow free precession under proton-coupled conditions. Fourier transformation of the resultant free-induction decay generates a multiplet "subspectrum" corresponding to the chosen carbon site. A series of such subspectra may be assembled, one for each chemical shift frequency; their sum is just the normal full proton-coupled spectrum. The effect of the sequence of experiments is to decompose the spectrum into its components, revealing in full all the information previously obscured by overlap.

This technique is clearly applicable to multiplet structure that is partly dipolar in origin, as in spectra of partially oriented molecules in a liquid-crystal phase. Excitation of partial spectra in this fashion can also be important when very high resolution is required over a relatively narrow spectral region, for it permits very fine digitization to be used in the frequency domain without introducing aliasing problems (*30*).

Additional information about proton-coupled carbon-13 multiplets may be obtained by selective pulsed double-resonance techniques, such as selective population transfer (*31–39*). These methods allow the relative signs of proton–carbon coupling constants to be determined using "soft" selective pulses in the proton spectrum. The analogous experiments are well known in proton magnetic resonance (*40*). Selective irradiation is also important in the study of chemical exchange processes by following the transfer of saturated magnetization, first proposed by Forsén and Hoffman (*41–44*).

The study of relaxation mechanisms is another field where the pulse sequence used to perturb the spin system may be required to be frequency selective (*45–59*). Cross-relaxation effects in the spin–lattice relaxation of protons may be studied by comparing the recovery after a selective population inversion pulse with the recovery after a completely nonselective pulse (*60–65*). The analysis is particularly simple if only the initial rates are considered. More detailed insight into cross-relaxation is obtained if the relaxation after multiple selective population inversions is studied. Information on molecular structure and dynamics is available from such experiments.

The determination of spin–spin relaxation times in the presence of homonuclear spin–spin coupling by spin echo methods is greatly hampered by J modulation of the echoes (*66–69*). This modulation normally disappears if the chemically shifted groups are examined individually by means of a pulse sequence in which excitation and refocusing pulses are frequency selective (*69*).

Spin locking, or forced transitory precession (*70–77*), is an attractive alternative method for studying spin–spin relaxation in liquids, avoiding the modulation effects encountered with spin echoes. Since the spin-locking radiofrequency field must be applied for times comparable with the spin–spin relaxation time T_2, and must be strong compared with the range of chemical shifts to be studied, considerable problems with radiofrequency heating of the sample arise. This difficulty is circumvented if a frequency-selective experiment is performed, exciting the chemically shifted spin multiplets one at a time, with a radiofrequency field strong compared with the multiplet splitting but weak compared with the chemical shift difference of the nearest group (*46*, *77*).

Finally there are some important experiments that require the radiofrequency excitation to be restricted to certain geometrical regions of the sample. One such technique is that of spin mapping (*78–84*) or zeugmatography (*85–89*), the determination of the density of nuclei (usually protons in water) inside an object as a function of the spatial coordinates, by examining the NMR signals in the presence of strong applied static field gradients. Several such methods (*80*, *84*) use selective irradiation schemes to saturate or excite selected small regions of the sample. Variation of the irradiation frequency or the magnetic field gradients allows the coordinates of this chosen region to be scanned, thus mapping out the spin density within the sample. The speed with which the selective pulse sequence can be computed is a critical factor in the attainable sensitivity of one embodiment of this experiment (*84*) and for this purpose a simplified sequence would be preferable to the full synthesized excitation technique of Tomlinson and Hill (*14*).

A related technique in the field of high-resolution NMR would be to restrict the effective sample volume by selective excitation in both natural field gradients and superimposed strong applied gradients. The nuclear spin response would be monitored with the imposed gradients removed. If the imposed gradients are chosen so as to match the dominant natural gradients, the sample region excited would correspond to a highly homogeneous field, and the signal from this region would transform to a spectrum where the linewidths were significantly narrower than the normal instrumental width. An equivalent physical reduction in the *actual* size of the sample is not usually possible since the shape and location of the region of high homogeneity are not known, and because of bulk susceptibility changes at the sample boundaries. These experiments are related to localized saturation, or "hole burning" (*90–92*), which has been used for precise measurement of frequency separations in high-resolution double resonance (*91*) and for approximate measurements of natural linewidths (*92*).

1.2. Methods of Selective Excitation

One of the first techniques used to excite a chosen resonance with a pulse while avoiding excitation of a nearby resonance was proposed for spin echo work by

Alexander (*12*). The principle is to use a relatively weak pulse of long duration, calculated to give a 90° flip to the chosen resonance line, but a 360° flip to the neighboring line, which experiences an effective field in the rotating frame $H_{eff} = 4H_1$. Redfield (*13*) has extended this idea to solvent peak suppression in Fourier transform NMR, using a computer to calculate the corrections for the phase and intensity distortions inherent in this method.

For spectra with well-resolved groups of resonance lines, a simpler weak-pulse method may be employed. The radiofrequency intensity is set so low that only the chosen group of lines is affected significantly, while all other groups are so far from resonance ($\Delta H \gg H_1$) that their excitation is negligible (*45–59*). In the limit of extremely weak pulses, it is possible to pick out one component line of a spin multiplet for excitation (*46*). These techniques were developed on conventional "continuous-wave" spectrometers with facilities for fine adjustment of the irradiation frequency; modern Fourier transform spectrometers do not normally have this capability.

For Fourier transform spectrometers Tomlinson and Hill (*14*) have proposed a very general method in which the frequency spectrum of the excitation may be "tailored" to any desired pattern. The operator simply defines the desired excitation spectrum in the frequency domain, and the computer calculates the Fourier transform of this pattern and uses it to modulate the amplitude or the pulse width of a sequence of radiofrequency pulses. This "synthesized excitation" technique has been applied successfully to solvent peak suppression (*14*), transient and steady-state nuclear Overhauser effects (*53*), selective spin locking (*77*), and spin mapping(*84*). It remains the most versatile general method of selective excitation proposed to date, but it does have some drawbacks. The very generality of the frequency-domain excitation pattern requires that considerable computer storage be set aside to retain this information, typically halving the data space available for NMR signals. Furthermore, the provision of a pulse modulation scheme that has the required linearity is by no means trivial. Amplitude modulation requires that the radiofrequency amplifier and probe response be linear, whereas conventional transmitters use "Class C" amplifiers and nonlinear probe networks to achieve good isolation between transmitter and receiver. Linear pulse width modulation is difficult to implement since the probe "*Q*" factor limits the minimum pulse width that can be employed (*93*). The price paid for versatility is a considerable increase in instrumental complexity.

However, a critical examination of applications of selective excitation suggests that complex excitation patterns are seldom needed, for the majority of experiments could be carried out with one (or perhaps two) effectively monochromatic irradiation fields. In conjunction with conventional nonselective pulses and radiofrequency phase inversion, it should be possible to perform the majority of selective excitation experiments without recourse to synthesized excitation. This "effectively monochromatic" irradiation, finely adjustable in frequency, can be achieved in an extremely simple manner with the minimum modification to a conventional Fourier transform spectrometer (*28*). This selective pulse sequence consists of a regular train of identical, short, strong radiofrequency pulses, each with flip angle $\alpha \ll \pi/2$ radians, spaced τ sec apart. Only those resonances that are offset from the transmitter by $\Delta\nu = n/\tau$ Hz, where n is an integer, are excited to a significant extent. The nature of this selectivity is examined in more detail in the next section, but a number of useful results may be derived relatively simply.

Fourier analysis of the pulse sequence shows that if the pulse widths Δt are much less than the pulse interval τ, then the selective pulse sequence is approximately equivalent to the superposition of a number of continuous radiofrequency signals with frequencies spaced $1/\tau$ Hz apart, which can be regarded as "sidebands" symmetrically disposed with respect to the transmitter frequency at ν_0, $\nu_0 \pm 1/\tau$, $\nu_0 \pm 2/\tau$, etc., each of amplitude $H_1 \Delta t/(\Delta t + \tau)$, where H_1 is the pulse amplitude. For a magnetization close to one of these sideband conditions, the pulse sequence acts just like a weak selective pulse. A suitable choice of transmitter frequency ν_0 and repetition rate $1/\tau$ will ensure that only one sideband (normally the first sideband at $\nu_0 \pm 1/\tau$) falls within the spectrum of interest. For nuclear magnetization components that are not in the immediate vicinity of one of the sidebands, the influence of the next-nearest sideband must also be taken into account and this simple picture breaks down. This situation has been treated for slow-passage NMR by Bloch and Siegert (*94*) and Ramsey (*95*).

A more pictorial description of the effects of the selective pulse sequence may be obtained by following the motion of a macroscopic magnetization vector **M** at a frequency offset $\Delta\nu$ Hz from the transmitter. Each pulse nutates **M** through α radians about the X axis of the rotating frame, while each pulse interval allows **M** to precess about the Z axis through $2\pi\Delta\nu\tau$ radians. If relaxation can be neglected during the relatively short total duration of the sequence, the motion is composed only of alternate rotations about the X and Z axes. It is readily seen that if **M** precesses through an integral number of complete revolutions during the interval τ, successive pulses will find **M** in the YZ plane and their overall effect will be a cumulative rotation about the X axis in a number of small steps. This is the same effect as that of a weak field on resonance, the conditions $2\pi\Delta\nu\tau = 2\pi n$ corresponding to the sideband condition $\Delta\nu = n/\tau$ described above. This condition achieves the maximum excitation.

A magnetization at a more general offset $\Delta\nu = (2n\pi + \theta)/2\pi\tau$, will precess through an excess angle θ in each pulse interval, being carried progressively further away from the YZ plane, which robs the pulses of their full cumulative effect. The fate of such components is readily illustrated by computer simulations based on the Bloch equations, neglecting relaxation. Figure 1 shows a family of such trajectories calculated for a sequence of 20 pulses with flip angle $\alpha = \pi/40$ radians and pulse interval 2 msec. The "first sideband" condition is thus $\Delta\nu = 500$ Hz, and the frame of reference is chosen to rotate at $\nu_0 + 500$ Hz, in synchronism with the first sideband. The advantage of this particular rotating frame is that the family of magnetization vectors then precesses through only small "excess" angles θ between pulses, simplifying the trajectories. The vector for $\Delta\nu = 500$ Hz then remains throughout in the YZ plane, so that after twenty pulses it has rotated from the Z axis to the Y axis, corresponding to a perfect selective 90° pulse. Vectors only a few hertz away from this condition are curled away from the YZ plane, while vectors greater than 25 Hz distant execute small approximately circular trajectories near the Z axis, never having a significant component along the X or the Y axis of the rotating frame.

These calculations can be extended to illustrate how the X component (dispersion) and Y component (absorption) of a magnetization **M** at the end of the sequence depend on the offset from the transmitter frequency ν_0 (Fig. 2). The scale has been normalized with respect to the equilibrium magnetization M_0, so that a 90° pulse at the exact sideband condition gives a transverse signal of unit intensity. Such diagrams may be

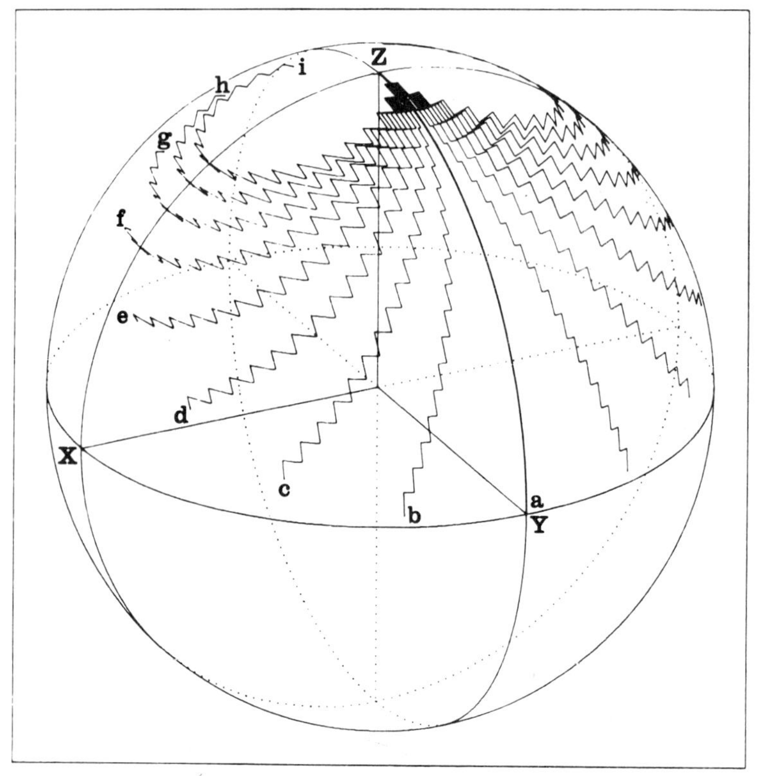

FIG. 1. Trajectories of magnetization vectors computed according to the Bloch equations and viewed in a frame of reference rotating at the "first sideband" frequency, ν_0 + 500 Hz. The excitation was a train of twenty pulses of flip angle $\pi/40$ radians each, spaced 2 msec apart. Trajectory (a) represents magnetization exactly 500 Hz above the transmitter frequency ν_0 and it experiences a cumulative nutation of $\pi/2$ radians with no net precession between pulses. Trajectories (b) through (i) represent increasing offsets in steps of 2.5 Hz up to a total of 520 Hz above ν_0, the unlabeled trajectories representing the corresponding offsets below 500 Hz.

termed "excitation spectra." The weak oscillations in the flanks of the main excitation peak reflect the cyclic excursions noted in Fig. 1. It will be seen later that these excitation spectra are very similar to those obtained by Fourier transformation of the pulse sequence.

1.3. Nomenclature

In the search for a suitable name for the selective pulse sequence, a striking literary parallel with the type of magnetization trajectory discussed in Section 1.2 was discovered. The second section of Dante Alighieri's "La Divina Commedia," the "Purgatorio," describes a journey through Purgatory by Dante and Virgil. Dante's Purgatory consists of a series of ten circular ledges, one above the other, running around a mountain. Souls in Purgatory progress toward Heaven by circumnavigating each ledge before moving up to the next. This motion is the exact reverse of that undergone by the tip of a magnetization vector close to the first sideband condition of a ten-pulse selective pulse sequence, when viewed in a frame of reference rotating at the

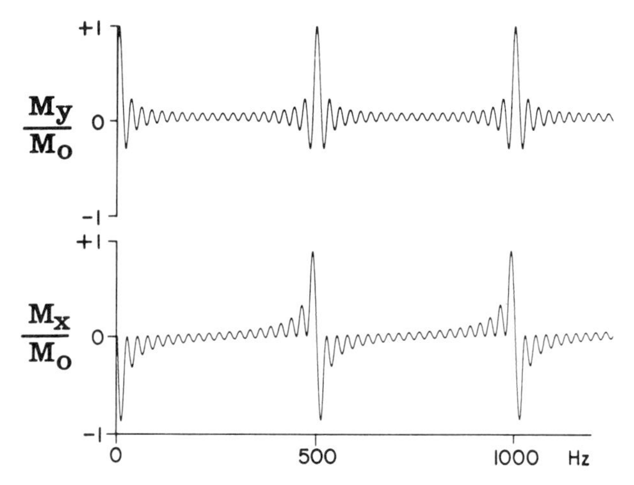

FIG. 2. Frequency-domain excitation spectra corresponding to the pulse sequence used in Fig. 1, showing the Y and X components of magnetization immediately after the last pulse of a selective pulse sequence as a function of the offset from the transmitter frequency. The computer simulation was based on the Bloch equations, neglecting relaxation. Note that the number of weak maxima between the main excitation peaks of M_y is given by m-2, where m is the number of pulses in the sequence.

transmitter frequency. The selective pulse sequence has thus acquired the name DANTE (Delays Alternating with Nutations for Tailored Excitation). A similar relationship has led to the naming of the spin-locking experiments to be described in Sections 2.3 and 3.5 after Dante's Inferno, which has the form of a cone whose apex lies at the center of the earth. The trajectory of a magnetization during an INFERNO experiment (Irradiation of Narrow Frequency Envelopes by Repeated Nutation and Orbiting) lies on the surface of just such a cone (see Fig. 8).

2. THEORY OF SELECTIVE EXCITATION

2.1. General Treatment for an Arbitrary Excitation

The problem of calculating the frequency dependence of the effects of a completely general radiofrequency perturbation $H_1(t)$ may be attacked in one of two basic ways. The first is to use the Bloch equations in the time domain to give an explicit solution for each individual frequency. This becomes progressively more difficult as $H_1(t)$ becomes more complicated, but for excitation methods such as the DANTE pulse sequence it can provide useful insights. It is difficult to derive analytical results in the time domain, and hence computer methods are required to produce quantitative excitation spectra.

The second approach is to solve the problem in the frequency domain. Provided the response of the spin system remains approximately linear, the excitation spectrum of a perturbation $H_1(t)$ will be proportional to its Fourier transform. This result, much used in linear system theory, greatly facilitates the analytical treatment of selective excitation, and provides a ready conceptual link between the form of a pulse sequence and its excitation spectrum.

The concepts of Fourier transformation have been widely used in magnetic resonance (see, for example, Ref. (*2*, *13*, *47*, *97–101*)), but do not appear to have been used in an analytical treatment of transient selective excitation. Various analyses of the steady-state response to repetitive pulse sequences have been performed by other methods (*102–105*). Despite the fact that pulse Fourier transform spectrometers routinely operate in a nonlinear region of the spin response, considerable insight into frequency-selective effects may be achieved by examining the linear response.

The most convenient starting point for a consideration of the selectivity that may be achieved in the frequency domain is the set of phenomenological equations of Bloch (*106*) which describe magnetic resonance effects in terms of magnetization vectors, or isochromats, $\mathbf{M}(\Delta\omega)$. The response to a short "hard" radiofrequency pulse of flip angle α at time $t = 0$ may be described in terms of the magnetization components immediately before the pulse, M_x^-, M_y^-, and M_z^-. The transient solution of the Bloch equations is then

$$M_z(t) = M_z^- \cos\alpha - M_y^- \sin\alpha + M_0\,[1 - \exp(-t/T_1)], \qquad [1]$$

$$M_y(t) = (M_y^- \cos\alpha \cos\Delta\omega t + M_z^- \sin\alpha \cos\Delta\omega t - M_x^- \sin\Delta\omega t)\exp(-t/T_2), \qquad [2]$$

$$M_x(t) = (M_x^- \cos\Delta\omega t + M_y^- \cos\alpha \sin\Delta\omega t + M_z^- \sin\alpha \sin\Delta\omega t)\exp(-t/T_2). \qquad [3]$$

If it can be assumed that $\mathbf{M}$ remains close to the Z axis of the rotating reference frame, so that $M_z \simeq M_0$, and that the flip angle α is small, so that $\sin\alpha \simeq \alpha$ and $\cos\alpha \simeq 1$, then Eqs. [1]–[3] reduce to

$$M_z(t) = M_0, \qquad [4]$$

$$M_y(t) = (M_y^- \cos\Delta\omega t + M_0\,\alpha \cos\Delta\omega t - M_x^- \sin\Delta\omega t)\exp(-t/T_2), \qquad [5]$$

$$M_x(t) = (M_x^- \cos\Delta\omega t + M_y^- \sin\Delta\omega t + M_0\,\alpha \sin\Delta\omega t)\exp(-t/T_2). \qquad [6]$$

This impulse response may be written in terms of the complex transverse magnetization, $M_{xy}(t)$:

$$M_{xy}(t) = M_x(t) + iM_y(t) = (M_{xy}^- + iM_0\,\alpha)\exp(-t/T_2)\exp(-i\Delta\omega t). \qquad [7]$$

An isochromat initially at thermal equilibrium subjected to a regular sequence of m short hard pulses of flip angle α_n at times t_n will generate transverse magnetization at the end of the sequence ($t > t_m$) given by the expression

$$\begin{aligned} M_{xy}(t) = \{&\alpha_m + \alpha_{m-1}\mathbf{exp}[-(t_m - t_{m-1})/T_2]\mathbf{exp}[-i\Delta\omega\,(t_m - t_{m-1})] \\ &+ \cdots + \alpha_1\mathbf{exp}[-(t_m - t_1)/T_2]\mathbf{exp}[-i\Delta\omega(t_m - t_1)]\}iM_0\mathbf{exp}[-(t - t_m)/T_2] \\ &\mathbf{exp}[-i\Delta\omega(t - t_m)], \end{aligned} \qquad [8]$$

where the operator **exp** is defined by

$$\mathbf{exp}(x) = e^x \qquad \text{for } x < 0,$$

$$\mathbf{exp}(x) = 0 \qquad \text{for } x > 0.$$

Expression [8] simplifies to

$$M_{xy}(t) = iM_0 \sum_{n=1}^{m} \alpha_n\,\mathbf{exp}\,[-(t - t_n)/T_2]\,\mathbf{exp}\,[-i\Delta\omega(t - t_n)]. \qquad [9]$$

It is possible to write this in terms of the convolution operation, defined as

$$f(x) \otimes g(x) = \int_{-\infty}^{\infty} f(x')\, g(x - x')\, dx',$$

$$M_{xy}(t) = iM_0 \left\{ \sum_{n=1}^{m} \alpha_n \delta(t - t_n) \right\} \otimes \mathbf{exp}\,(-t/T_2)\, \mathbf{exp}\,(-i\Delta\omega t). \qquad [10]$$

This is an example of a general class of results in linear systems theory, linking the response $R(t)$ to the excitation $E(t)$ by means of the impulse response $h(t)$

$$R(t) = E(t) \otimes h(t). \qquad [11]$$

Equation [10] may be converted into an expression of the form of Eq. [11] by considering a complex radiofrequency excitation $H_1(t) \cos \omega_0 t$, where ω_0 is the angular frequency of the rotating frame of reference. In this frame the perturbing field is $H_1(t)$; it is considered to be composed of an infinite number of impulses dt sec long, so that the impulse corresponding to time t' has a flip angle $\alpha(t') = \gamma H_1(t')dt$. Conversion of the summation in Eq. [10] into an integral gives

$$M_{xy}(t) = iM_0 \int_{-\infty}^{t} \gamma H_1(t')\, \delta(t - t') \otimes \mathbf{exp}\,(-t/T_2)\, \mathbf{exp}\,(-i\Delta\omega t)\, dt'$$

$$= iM_0\, \gamma H_1(t) \otimes \mathbf{exp}\,(-t/T_2)\, \mathbf{exp}\,(-i\Delta\omega t). \qquad [12]$$

In this way the transverse magnetization $M_{xy}(t)$ resulting from an excitation $H_1(t)$ has been expressed in the general form of Eq. [11].

Equation [12] is a simple expression for the transverse magnetization due to an isochromat **M** of frequency $\Delta\omega$ in the rotating frame, as a function of time. It may be expressed as a function of frequency $M_{xy}(\Delta\omega)$ at a particular time t_1 by Fourier transformation of the right-hand side of Eq. [12], using the identity

$$f(0) = \frac{1}{2\pi} \int_{-\infty}^{+\infty} \mathrm{FT}^{-}[f(t)]\, d\omega, \qquad [13]$$

where FT^{-} indicates the Fourier transform operation, and FT^{+} the inverse Fourier transformation (*107*, *108*). This gives for the transverse magnetization at time t_1

$$M_{xy}(\Delta\omega) = \frac{i\gamma}{2\pi} M_0(\Delta\omega) \int_{-\infty}^{+\infty} \mathrm{FT}^{-}[H_1(t - t_1) \otimes \mathbf{exp}\,(-t/T_2)\, \mathbf{exp}\,(-i\Delta\omega t)]\, d\omega. \qquad [14]$$

This permits the convolution theorem (*107*, *108*) to be used to separate the two functions on the right-hand side of Eq. [12]

$$M_{xy}(\Delta\omega) = \frac{i\gamma}{2\pi} M_0(\Delta\omega) \int_{-\infty}^{+\infty} \mathrm{FT}^{-}[H_1(t - t_1)]\, \mathrm{FT}^{-}[\mathbf{exp}\,(-t/T_2)\, \mathbf{exp}\,(-i\Delta\omega t)]\, d\omega. \qquad [15]$$

This may be expanded in terms of standard transforms to give

$$M_{xy}(\Delta\omega) = \frac{i\gamma}{2\pi} M_0(\Delta\omega) \int_{-\infty}^{+\infty} \mathrm{FT}^{-}[H_1(t - t_1)]\, [A(\omega) \otimes \delta(\omega + \Delta\omega)]\, d\omega, \qquad [16]$$

where $\Lambda(\omega)$ has been written for the complex Lorentzian

$$\Lambda(\omega) = (T_2 - i\omega T_2^2)/(1 + \omega^2 T_2^2). \qquad [17]$$

Equation [16] may be rearranged by making use of the identity, easily proved by writing the convolution explicitly and changing the order of integration.

$$\int_{-\infty}^{+\infty} a(\omega)[b(\omega) \otimes c(\omega)]\, d\omega = \int_{-\infty}^{+\infty} [a(\omega) \otimes b(-\omega)]\, c(\omega)\, d\omega, \qquad [18]$$

with the result

$$M_{xy}(\Delta\omega) = \frac{i\gamma}{2\pi} M_0(\Delta\omega) \int_{-\infty}^{+\infty} \{\mathrm{FT}^-[H_1(t - t_1)] \otimes \Lambda(-\omega)\}\, \delta(\omega + \Delta\omega)\, d\omega. \qquad [19]$$

This may be simplified by means of the integral representation of the δ function:

$$M_{xy}(\Delta\omega) = (i\gamma/2\pi) M_0(\Delta\omega) \{\mathrm{FT}^-[H_1(t - t_1)] \otimes \Lambda(\omega)\}_{\omega=\Delta\omega}. \qquad [20]$$

If an excitation spectrum $E(\Delta\omega)$ is defined such that

$$E(\Delta\omega) = (i\gamma/2\pi)\, \mathrm{FT}^-\, [H_1(t - t_1)] \otimes \Lambda(\omega), \qquad [21]$$

then Eqs. [20] and [21] may be simply written:

$$M_{xy}(\Delta\omega) = M_0(\Delta\omega)\, E(\Delta\omega). \qquad [22]$$

If relaxation can be neglected during the period of application of the excitation $H_1(t)$, this leads to a particularly simple result. The transverse magnetization generated from an isochromat at a frequency $\Delta\omega$ is directly proportional to the Fourier component of the excitation $H_1(t)$ at that frequency:

$$M_{xy}(\Delta\omega) = i\gamma M_0(\Delta\omega)\, \mathrm{FT}^-[H_1(t - t_1)]. \qquad [23]$$

Thus the excitation of a spin system as a function of frequency is simply proportional to the Fourier transform of the radiofrequency excitation, provided that the excitation time is short compared to the relaxation times T_1 and T_2, and that the net perturbation at any frequency is small.

Equation [23] may be inverted to produce the pulse sequence necessary for a desired excitation spectrum $E(\Delta\omega)$:

$$H_1(t - t_1) = (1/i\gamma)\, \mathrm{FT}^+[E(\Delta\omega)]. \qquad [24]$$

This is the basis of the "synthesized excitation" experiment of Tomlinson and Hill (*14*); the effect of introducing T_2 relaxation is simply to restore the Lorentzian convolution term to Eq. [20], thus limiting the "sharpness" of any excitation to the natural linewidth $(1/\pi T_2)$. It is of interest to note that as a result, the limiting linewidth obtainable in a selective excitation experiment is governed by T_2 rather than by T_2^*, the decay constant due to static field inhomogeneity. This does not, however, lead of its own accord to any improvement in effective spectral resolution.

The analysis presented above deals only with the Bloch case of noninteracting classical magnetizations. While adequate for most purposes (e.g., weakly coupled spectra, proton-decoupled carbon-13 spectra), a variety of interesting experiments may be envisaged involving strongly coupled systems, for which a more rigorous quantum mechanical approach would be necessary. Decoupling by a DANTE sequence and hole burning in strongly coupled systems fall into the latter category.

2.2. The DANTE Pulse Sequence

For a spin system close to equilibrium, Eq. [22] describes the transverse magnetization generated by an excitation $E(t)$, as a function of frequency. However, in many pulsed NMR experiments it is desirable to generate the maximum possible NMR signal, in which case the spin response is grossly nonlinear. In such cases the analysis presented in the preceding section is no longer valid, although it can often yield valuable qualitative results.

The DANTE pulse sequence is a case in point. Consider first the predictions of the linear approximation applied to a regular sequence of m pulses of radiofrequency field H_1 of duration Δt sec spaced at intervals of τ sec. This excitation can be represented in terms of some standard functions: an infinite train of δ functions $(\Delta t + \tau)$ sec apart [the "shah" function of Bracewell (*107*)] multiplied by a window function of width $m(\Delta t + \tau)$ sec (Bracewell's Π function) and convoluted with a window of width Δt sec (another Π function). The transforms of these standard functions are well known (another shah function and two sinc functions, respectively) and the convolution theorem can thus be used to deduce the transform of the pulse sequence.

In practice Δt is usually much less than τ, so that the last multiplication by the sinc function has very little effect in the frequency region of interest. This leads to a very simple form for the Fourier transform of the DANTE sequence: a series of δ functions $1/\tau$ Hz apart (the "sideband" frequencies mentioned earlier) convoluted with a complex sinc function, the real part of which is $1/m\tau$ Hz wide between zero-crossing points. Figure 3 illustrates how the convolution theorem leads to this result. Note that the transform of the overall window function has both a real and an imaginary part, since,

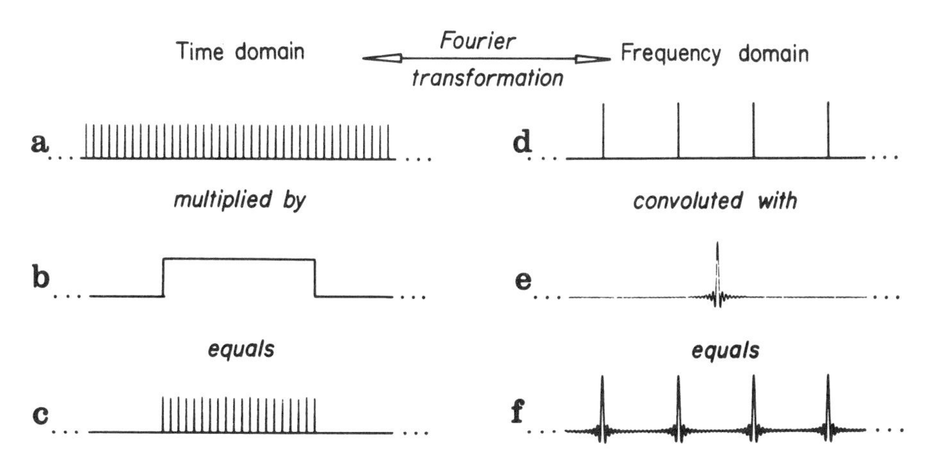

FIG. 3. A graphical illustration of the use of the convolution theorem to derive the transform of a DANTE sequence with $\Delta t \ll \tau$. Since the DANTE sequence is the product of (a) and (b), Fourier transformation of (c) produces (f), the convolution of (d) with (e). Only the real parts of the functions are shown.

as discussed in the previous section, zero time is taken to mean the end of the pulse sequence; only the real parts of the functions are shown in the diagram.

Consider a practical example consisting of ten pulses of flip angle 0.01 radians. If $\tau = 2$ msec, $\gamma H_1/2\pi = 1.59$ kHz and $\Delta t = 1$ μsec; then since $\Delta t \ll \tau$ it is possible to neglect the multiplication by the second sinc function. The spacing of the δ functions is 500 Hz, and these are convoluted by the first sinc function, which has a width of 50 Hz between the first zero crossings. If relaxation can be neglected, the excitation spectrum is given by

$$E(\Delta\omega) = i\gamma H_1[\text{shah}(1/\tau)\otimes\text{sinc}(1/m\tau)]. \qquad [25]$$

For the above parameters the excitation maxima appear at offsets of 0, ± 500 Hz, ± 1000 Hz, etc., and have a magnitude $M_y/M_0 = 0.100$.

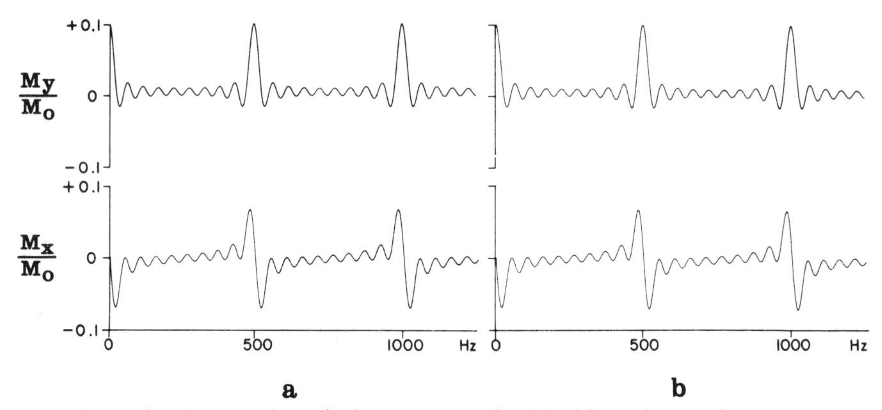

FIG. 4. Comparison between the excitation spectra predicted by (a) Fourier transformation of the pulse sequence according to the linear approximation, and (b) computer simulation based on the Bloch equations, for a ten-pulse DANTE sequence with overall flip angle $m\alpha = 0.1$ radians.

This may be compared with the results of the pictorial approach described in Section 1.2, which predicts a maximum Y magnetization of M_0 sin (0.1) = 0.998 M_0. It can be seen that there is good agreement between the predictions of the two approaches, since for this case the deviation from $M_z \simeq M_0$ is small and the linear approximation is justified. This is illustrated in Fig. 4, which shows excitation spectra obtained by computer simulation according to the Bloch equations, compared with the calculation based on Eq. [25]. Figure 5 shows the improvement in frequency selectivity obtained by increasing m to 50 for the same pulse interval as in Fig. 4 with a proportionately smaller flip angle.

The effects of nonlinearity are made clear in Fig. 6, which shows computer-simulated spectra for $m\alpha = \pi/2$ radians, equivalent to a selective 90° pulse on resonance, and normalized to $M_y/M_0 = 1$. The excitation spectrum predicted by the linear approximation differs mainly in the excitation maxima, which have $M_y/M_0 = \pi/2$ rather than the observed $M_y/M_0 = 1$. Figure 6 shows (a) the excitation spectrum predicted for $m = 20$ pulses according to the linear approximation, (b) the trigonometric sine of (a), and (c) the computer-simulated excitation spectrum. For the DANTE sequence under these conditions it can be seen that the actual excitation spectrum may usefully be approximated by the sine of the predicted linear response:

$$M_{xy}(\Delta\omega) = M_0 \sin[E(\Delta\omega)]. \qquad [26]$$

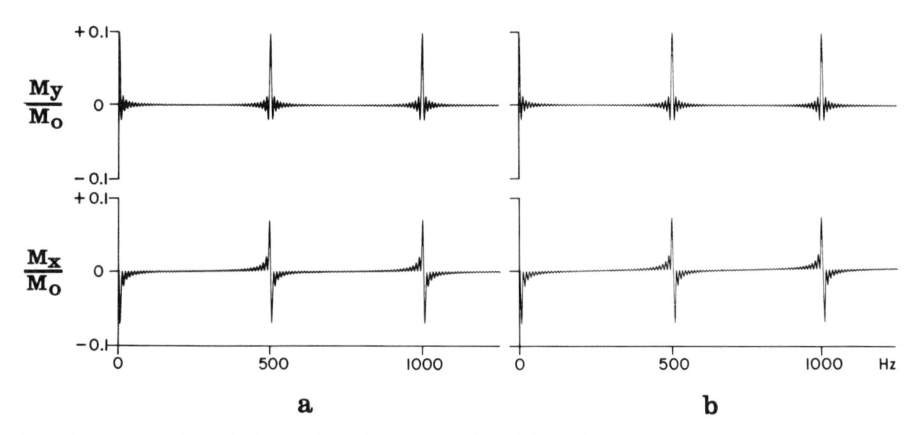

FIG. 5. A comparison similar to that of Fig. 4, but for a fifty-pulse sequence at the same repetition rate, showing the improvement in selectivity in the frequency domain. Note that the duration of the pulse sequence has increased by a factor of 5.

This approximation is particularly effective for the DANTE sequence because the degree of excitation is only large for a few evenly spaced frequencies.

The linear approximation permits two useful practical conclusions to be drawn about the DANTE sequence. First, as comparison between Figs. 4 and 5 shows, the ratio of the excitation at an exact sideband frequency to that approximately midway between sidebands increases with m, the number of pulses in the sequence. Second, the frequency selectivity increases in direct proportion to $m\tau$, the overall duration of the sequence. This can be measured as the width between zeros of the real sinc function, given by $1/(m\tau)$ Hz, or as the width at half-height, which is approximately $0.64/(m\tau)$ Hz. These values are slightly changed when the effects of nonlinearity are taken into account. For a

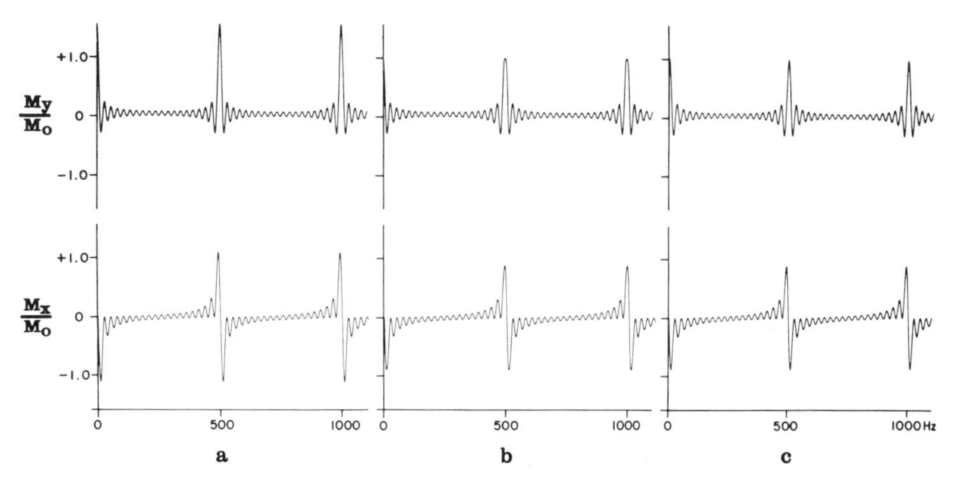

FIG. 6. An illustration of the effect of nonlinearity on excitation spectra for a twenty-pulse DANTE sequence of total flip angle $m\alpha = \pi/2$ radians. (a) The Fourier transform of the pulse sequence according to the linear approximation. (b) The trigonometric sine of (a). (c) The computer simulation based on the Bloch equations. Note that although the general features of (b) and (c) are similar, the main excitation peaks are broader in (b).

twenty-pulse DANTE sequence calculated to give a 90° pulse at exact resonance the width between zeros becomes about $0.92/(m\tau)$ Hz, and the width at half-height, $0.57/(m\tau)$ Hz.

However, since X magnetization appears in spectra as unwanted dispersion mode signals, a more useful guide to the selectivity of the pulse sequence is the absolute-value excitation spectrum $M_{xy}(\omega)$ (See Fig. 7a). For small total flip angles the excitation maxima have a width between zeros of approximately $2/(m\tau)$ Hz, and a half-height width of $1.21/(m\tau)$ Hz. Figure 7a compares the total transverse magnetization M_{xy} and its components M_x and M_y as a function of frequency for the case of a twenty-pulse DANTE sequence.

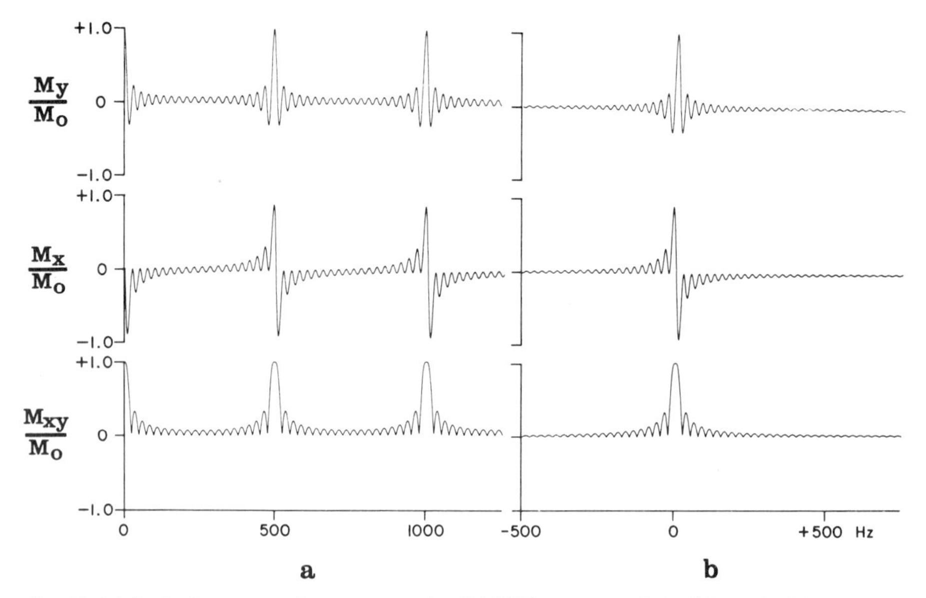

FIG. 7. (a) Excitation spectra for a twenty-pulse DANTE sequence, derived from the Bloch equations. For many purposes the selectivity is best judged from the absolute-value signal M_{xy}. (b) Excitation spectra corresponding to a single weak radiofrequency pulse of length 40 msec, equal to the total duration of the DANTE sequence used in (a), and of the same total flip angle $\pi/2$ radians.

It can be seen from Fig. 6 that away from the excitation maxima the linear approximation gives predictions close to the computer-simulated excitation spectra based on the macroscopic magnetization vector diagram. This is to be expected, since vectors far from the excitation maxima remain close to the Z axis throughout the DANTE sequence, and the linear approximation is justified.

Close to one of the sideband frequencies there is a better approximation than Eq. [26] available. Fourier analysis (rather than Fourier transformation) of the DANTE sequence shows a series of components spaced $1/\tau$ Hz apart in the frequency domain. For magnetization near one of these components, the effect of the DANTE sequence is similar to that of a low-level radiofrequency pulse of intensity $H_1\Delta t/(\Delta t + \tau)$. Figure 7 compares the computer-simulated excitation spectrum for a low-level radiofrequency pulse with that computed for a DANTE sequence with the same overall duration.

Thus for most practical purposes approximations are available which adequately describe the NMR response to a DANTE sequence. For some other purposes, for example, hole burning or solvent peak saturation, a steady-state analysis of the response to a regular sequence of pulses would be applicable (*103*, *105*). For certain applications the pictorial simulations based on the Bloch equations offer the most revealing description of the NMR behavior. A good example is the selective spin-locking experiment using a long DANTE sequence, for here the relaxation effects may be displayed very clearly.

2.3. The INFERNO experiment

Spin locking, or forced transitory precession (*70–76*), is one of the most effective methods of measuring transverse relaxation times in liquids, being considerably more tolerant of instrumental imperfections than spin-echo methods (*69*). The basic

TABLE 1

SEQUENCE OF EVENTS FOR THE INFERNO EXPERIMENT

(1)	Wait for equilibrium
(2)	Pulse of flip angle α about X axis
(3)	Delay τ
(4)	Repeat (2) and (3), m times
(5)	Pulse of flip angle β about Y axis
(6)	Delay τ
(7)	Repeat (5) and (6), n times
(8)	Acquire and transform free-induction decay

experiment consists in applying a 90° pulse to a magnetization, and then "locking" it in the transverse plane by applying a powerful radiofrequency field on resonance, phase shifted with respect to the 90° pulse, so as to align both magnetization and radiofrequency fields along the same axis in the rotating frame. The principal disadvantage of the experiment is that in order to spin lock all the lines in a spectrum it is necessary to employ a field strong compared to the width of the spectrum, leading to problems with transmitter design, sample overheating, or off-resonance effects.

These problems do not arise if resonances are studied individually, since then the only power requirement is that the radiofrequency field be strong compared to the linewidth (or multiplet width). This selective irradiation may be provided by a DANTE sequence, as in the INFERNO experiment (see Section 1.3). This uses two consecutive DANTE pulse sequences, both having the same pulse interval τ. The first, consisting of m pulses of flip angle α, where $m\alpha = \pi/2$, selectively rotates one magnetization down to lie along the Y axis in the rotating frame. The second sequence consists of n pulses of flip angle β, all shifted in radiofrequency phase by 90° with respect to the previous pulses, and provides a selective spin-locking field. In a typical experiment, m might be 50 and n several thousand. The sequence of events in an INFERNO experiment is summarized in Table 1.

As mentioned earlier the most fruitful theoretical approach to use here is to consider the longer DANTE locking sequence as being equivalent to a number of continuous low-level signals at the sideband frequences $\nu_0 + n/\tau$ Hz. Close to resonance, the effects

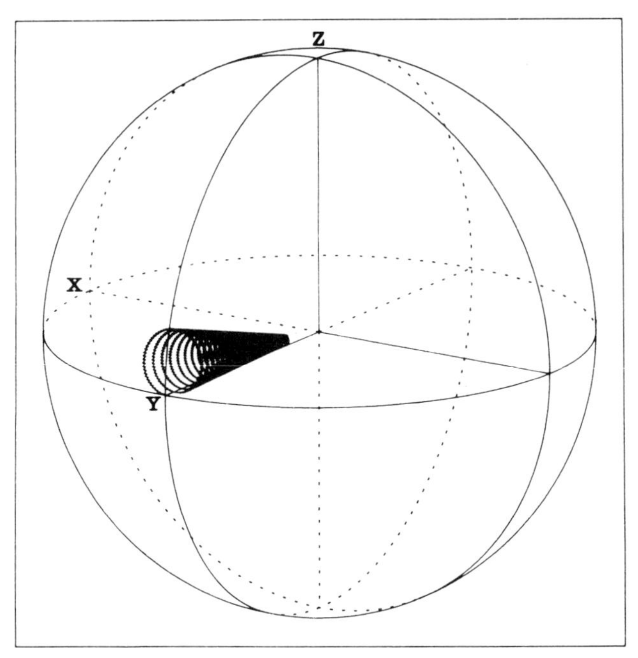

FIG. 8. The computer-simulated trajectory of a magnetization vector in a spin-locking experiment carried out by the INFERNO pulse sequence, viewed in a frame rotating at the "first sideband" frequency, ν_0 + 500 Hz. After the initial selective $\pi/2$ pulse, the spin-locking sequence used τ = 2 msec and $\beta = \pi/30$ radians. The calculation is based on the Bloch equations with $T_1 = T_2$ = 2 sec, an offset of 499 Hz from ν_0, and a spin-locking time of 4 sec.

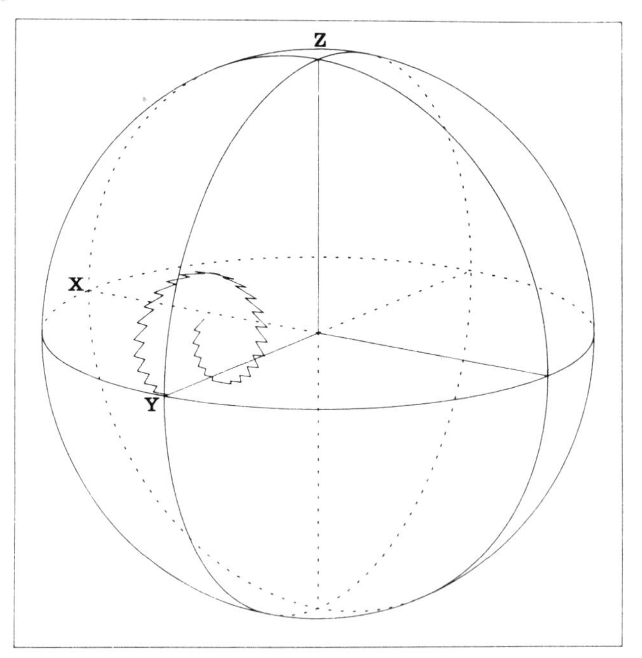

FIG. 9. A computer-simulated trajectory similar to Fig. 8, but with parameters changed in order to emphasize the zig-zig motion of the spin-locked magnetization, reflecting alternating periods of nutation and free precession. T_1 = 10 sec; T_2 = 0.5 sec; $\beta = \pi/12$ radians; τ = 10 msec, and the offset is 99 Hz. The trajectory has been terminated after thirty pulses for the sake of clarity.

of one of these signals will predominate, and the experiment becomes equivalent to using a spin-locking field of intensity equal to the full transmitter power multiplied by the DANTE sequence duty cycle $\Delta t/(\Delta t + \tau)$. Thus by a suitable choice of β and τ any apparent spin-locking field strength may be obtained without changing the experimental hardware.

Figure 8 illustrates the trajectory of a magnetization close to resonance during an INFERNO experiment in a frame of reference rotating at the first sideband frequency, just as in Fig. 1. The effect of using a DANTE sequence rather than a low-level continuous field is to superimpose rapid small oscillations on the normal smoothly decaying spiral trajectory about the tilted effective field in the rotating frame. Figure 9 shows a similar experiment simulated using fewer pulses of larger flip angle in order to emphasize this zig-zag motion. The essence of the spin-locking experiment here is seen to lie in competition between rotation about the Y axis (nutation about the radiofrequency field) and about the Z axis (free precession due to the offset from resonance). It is this competition which prevents the dephasing of isochromats by forcing them to remain close to the Y axis in the rotating frame, counteracting the effects of static field inhomogeneity.

3. APPLICATIONS OF THE SELECTIVE PULSE SEQUENCE

3.1. Instrumental

All spectra were obtained on a Varian CFT-20 spectrometer operating at 20 MHz for carbon-13 and near 80 MHz for protons. The spectrometer had been slightly modified to improve the timing stability of the pulse generator but these changes are probably not essential for the applications described below. The selective pulse sequence requires small flip angles of the order of a few degrees each, and since the CFT-20 is designed to give a minimum pulse width of 1 μsec, it was found advantageous for most experiments to insert a resistive attenuator between the 20-MHz transmitter and the probe. This reduced the radiofrequency level such that a 90° pulse increased from 20 to 150 μsec. A more elegant solution (*77*) might be to generate pulses of small flip angle as a combination of two larger pulses 180° out of phase, so that the net nutation angle represents the difference in pulse widths. This would have the advantage of reducing problems of slow rise time and "phase glitch" effects (*93*).

A very simple software routine times a regular sequence of short pulses at a repetition rate determined by the offset of the chosen line from the transmitter frequency. In a case where the higher-order "sidebands" might accidentally irradiate other lines, it is sufficient to reset the transmitter frequency and readjust the repetition rate accordingly. The procedure is analogous to that used with field- or frequency-modulation sidebands in continuous-wave spectrometers. The selectivity of the irradiation in the frequency domain is determined by the overall duration of the pulse sequence. For example, a selectivity of the order of 1 Hz is achieved with a pulse sequence of duration 1 sec, the number of pulses in the sequence being determined by the repetition rate. In all other respects the spectrometer operates in the conventional manner, the free-induction signal following the new pulse sequence being acquired and transformed in the usual way.

Experiments that use both selective and nonselective pulses could be carried out by switching in the attenuator electronically when required, or by dispensing with it altogether. Irradiation of a broad band of frequencies essentially uniformly except for one chosen narrow band is achieved by means of a nonselective pulse and a selective sequence set 180° out of phase and adjusted to have the same overall flip angle.

Occasionally it is necessary to irradiate *two* arbitrarily chosen narrow regions within a spectrum. This requires the superposition of two independent selective pulse sequences in the time domain. Each sequence is timed to *end* at the same instant, but the sequences begin at different times and have different repetition rates. The control program examines the two superimposed sequences for near coincidences, substituting a single pulse of twice the width.

With these simple combinations many kinds of selective excitation experiment can be explored. Excitation patterns in the frequency domain that are more complicated than these are better implemented by the complete "synthesized excitation" method described by Tomlinson and Hill (*14*).

3.2. Proton-Coupled Carbon-13 Spectra from Individual Sites

The analysis of proton-coupled carbon-13 spectra can present considerable difficulty when spin multiplets from different carbon sites overlap. Since the coupling to directly bonded protons is often quite complicated, this situation is fairly common, and the task of unscrambling the fine structure may require careful manipulation of differential solvent shifts or temperature-dependence studies, as in a recent analysis of cyclopentadiene (*25*). By combining selective excitation and gated decoupling, it is possible to circumvent this problem, and display the multiplet structure from each carbon site as a separate "subspectrum," provided only that the fully decoupled carbon-13 resonances are sufficiently well resolved (*27–30*); a separation of the order of 1 Hz is adequate with normal linewidths.

The first stage of the experiment is to prepare the spin system with a nuclear Overhauser enhancement by noise irradiation of the protons. In the presence of the decoupler each carbon site will give rise to a single carbon-13 resonance line, and one of these is chosen for selective irradiation by a DANTE sequence, the others remaining as Z magnetization. The decoupling field is then removed, allowing the carbon-13 nuclei to precess according to the proton-coupled spin Hamiltonian, and the resulting free-induction signal is recorded. Fourier transformation yields a "subspectrum" containing the multiplet structure associated with only a single carbon site. Repetition of the process produces an array of different subspectra which together comprise the conventional proton-coupled spectrum. A penalty must be paid in sensitivity or instrument time since each subspectrum requires about the same accumulation time as the full coupled spectrum. However, it is seldom necessary to record the entire array of subspectra, and it is often possible to excite suitable subspectra in pairs which have no overlapping lines. An important bonus associated with this method is the significantly improved digital resolution in the frequency domain, a feature that is particularly useful for the study of fine structure (*30*). It arises because a subspectrum requires a much narrower spectral width than the full coupled spectrum, so that the data points may be more finely disposed without folding unwanted lines into the spectrum.

Figure 10 shows an example of the application of this method to the spectrum of menthone. The top trace shows the decoupled carbon-13 spectrum with lines (A) through (I) from the nine different carbon sites. The corresponding subspectra are shown in traces (a) through (i); in each case the selectivity sufficed to remove all interference from other spin multiplets, the solvent, and the lock material. The lowest trace shows the conventional proton-coupled carbon-13 spectrum, and illustrates the

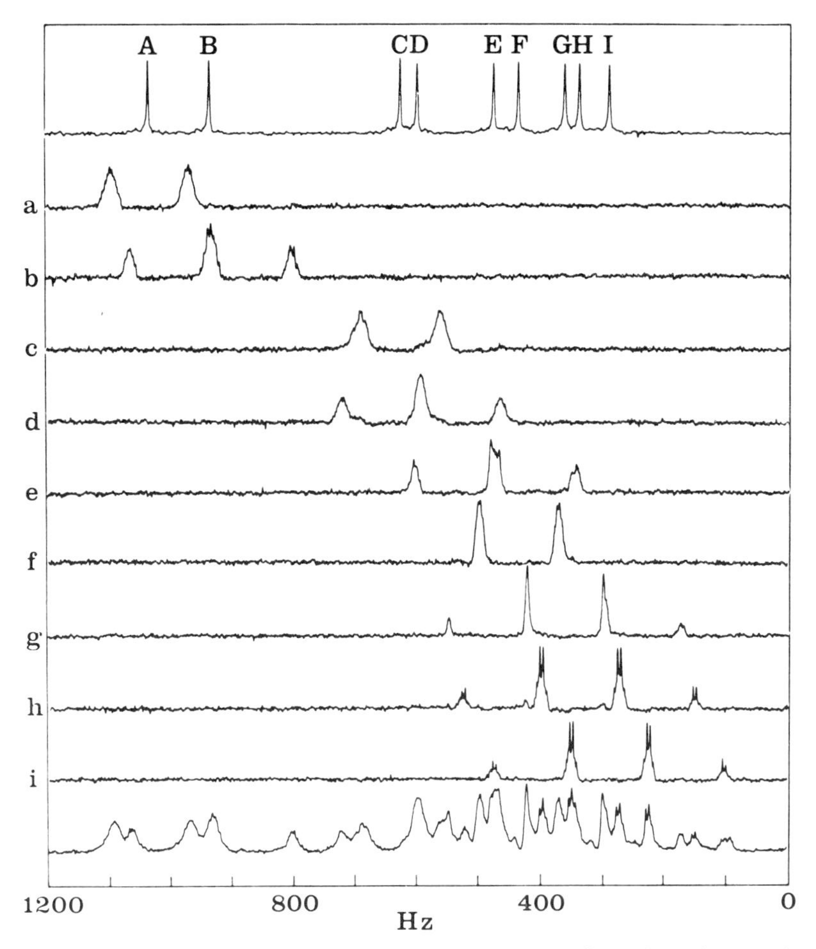

FIG. 10. The conventional proton-decoupled carbon-13 spectrum of menthone (top trace) showing resonances from nine chemically distinct sites. Traces (a) through (i) show proton-coupled subspectra from sites (A) through (I), obtained by selective excitation by a DANTE sequence and gated decoupling. The lowest trace shows the conventional proton-coupled spectrum illustrating the complicated overlap of spin multiplets.

complex overlap problem. Subspectra (h) and (i) have been used to provide evidence about the rotameric equilibrium in menthone (*29*). It will be noticed that some of the spin multiplets show a significant asymmetry about the chemical shift frequency, a consequence of strong coupling in the proton spectrum (*109*).

3.3 Suppression of Unwanted Signals

The investigation of biochemical systems in aqueous solution has been severely hampered by the limitations imposed on the dynamic range when the signal is converted into digital form in a Fourier transform spectrometer, and even with heavy water as solvent, the residual HDO peak can be troublesome. Several ingenious experiments have been suggested to suppress unwanted strong solvent peaks. The selective pulse sequence may be employed in a variety of ways to alleviate such problems. A long regular sequence of pulses may be used to saturate a particular resonance selectively, the rest of the spectrum being examined by a subsequent nonselective pulse. Formally this experiment would be described by the steady-state solution for the response to a regular sequence of pulses (*103*, *105*) provided that the sequence is sufficiently long, rather than the transient solutions presented in Section 2.2. The result is very similar to the application of a continuous weak radiofrequency field. Figure 11 illustrates the application to the proton spectrum of the *cis*-3,5 dimethylpiperidinium ion, showing saturation of the water line by a selective pulse sequence several seconds long. If the duration of the "saturation" sequence is comparable with or shorter than the relevant relaxation times, it is still possible to achieve almost complete elimination of the irradiated line. Although the net nuclear magnetization from the sample may be

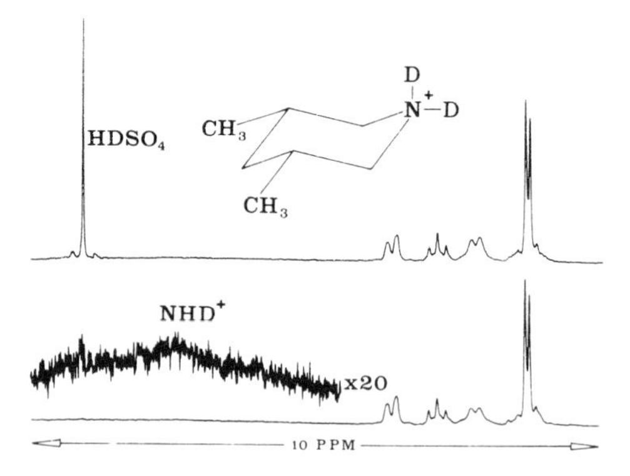

FIG. 11. Proton spectra of the *cis*-3,5 dimethylpiperidinium ion in D_2SO_4 solution, with and without saturation of the solvent proton line by a DANTE sequence of several seconds' duration prior to the usual $\pi/2$ excitation pulse. A weak broad line from NHD^+ is detectable in the trace at increased gain, normally masked by the tail of the strong solvent line.

negligible, local magnetization components are merely dispersed by transient nutations in an inhomogeneous effective field, mainly because the radiofrequency field is spatially inhomogeneous (*5*, *46*, *47*, *110*).

Selective excitation (rather than saturation) may also be used to remove unwanted lines. For example, a selective 90° pulse followed by a nonselective 90° pulse leaves the chosen line with negligible transverse magnetization, but the rest of the spectrum with full transverse components. A rather better variation inverts the radiofrequency phase between the selective and nonselective pulses, which may have arbitrary but equal flip angles. Selective 180° pulses are also useful. Patt and Sykes (*7*) have described a method of eliminating solvent peaks, particularly HDO in biochemical solutions, by inverting all resonances with a nonselective 180° pulse, and then exciting the spectrum at the precise instant that the HDO peak passes through the null condition. The method relies on the faster spin–lattice relaxation of the large solute molecules, but suffers intensity distortions due to different relaxation rates. Substitution of a *selective* 180° pulse would largely circumvent this intensity problem, and improve the sensitivity of the method. This experiment, using a low-level pulse rather than a DANTE sequence, has recently been described by Gupta (*11*).

3.4. *"Hole Burning"*

If a resonance line is inhomogeneously broadened, it is possible in principle to saturate a narrow region of the line but leave the remainder essentially unaffected. The effect of the strong irradiation is essentially confined to a small restricted volume somewhere within the sample, other regions being off-resonance. Bloembergen *et al.* (*90*) describe this phenomenon as burning a hole in the line. The experiment has remained something of a curiosity, but has been employed in double resonance (*91*) and in studies of linewidths in rubber (*92*). The technique requires an extremely stable and finely tunable radiofrequency oscillator, or a suitable sideband of the master oscillator.

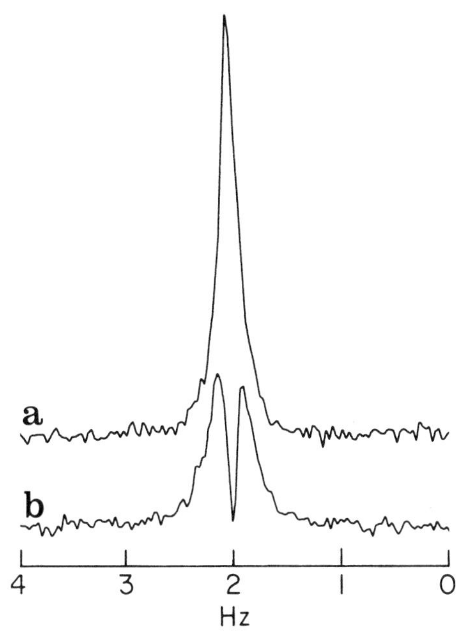

FIG. 12. One of the inner lines of the proton-coupled carbon-13 quartet of methyl iodide, (a) before and (b) after application of a hole-burning DANTE sequence lasting approximately 7 sec. The linewidth is 0.3 Hz, a typical value for the CFT-20, and the width of the hole approximately 0.1 Hz.

In a Fourier transform spectrometer an attractive alternative is a long selective pulse sequence with low duty cycle.

Figure 12 shows a hole burnt in one component of the proton-coupled carbon-13 quartet of methyl iodide by this method. A DANTE sequence of 350 pulses of flip angle $\pi/300$ radians, lasting about 7 sec, was used to burn the hole, followed by an observation pulse of flip angle $\pi/6$ radians. This resulted in a hole approximately 0.1 Hz wide, or about one-third of the limiting instrument linewidth. Because the pulse sequence used lasted rather less than T_1, the hole represents, not the steady-state saturation reported previously (*90–92*), but a highly selective transient excitation of restricted regions of the sample.

Since a hole may never be narrower than the natural linewidth, the method may be used to give a rough indication of differences in natural linewidth for assignment purposes, as well as to provide a simple test for inhomogeneous broadening.

3.5. Selective Spin Locking

The INFERNO experiment described in Section 2.2 may be used to measure spin–lattice relaxation times in the rotating frame $T_{1\varrho}$, for individual lines in a spectrum. In liquids, these relaxation times carry similar information to T_2, and hence have considerable interest. The experiment consists in measuring the magnetization remaining after increasing periods of spin locking, normally produced by incrementing the number n of locking pulses. Two approaches recommend themselves: acquire and transform a series of free-induction decays after increasing periods of spin locking, after the manner of a conventional inversion–recovery experiment; or sample the magnetization during the interpulse delays, giving a signal decaying with time constant $T_{1\varrho}$. The latter "transparent" method is more difficult, but is a very efficient experiment as it records an entire $T_{1\varrho}$ decay at once. Figure 13 shows the results of spin-locking experiments on the proton-decoupled carbon-13 resonances of pyridine, using the

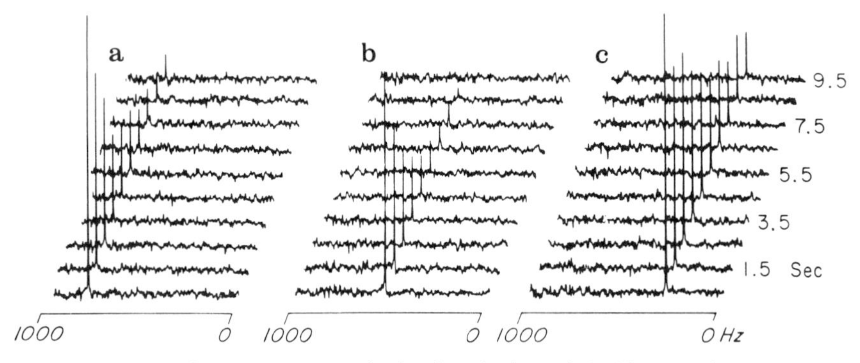

FIG. 13. The decay of transverse magnetization in selective spin-locking experiments on the three proton-decoupled carbon-13 resonances of pyridine, (a) C2, (b) C4, and (c) C3. The same spectral window was used in each case; in a conventional spectrum the three lines appear at offsets of 789, 531, and 301 Hz, respectively. The INFERNO pulse sequence was used with an effective spin-locking field intensity of about 20 Hz.

former of the two methods in order to emphasize the selective nature of the experiment. Because the spin-locking method is less affected by instrumental shortcomings such as radiofrequency field inhomogeneity and pulse irreproducibility, these decays are rather longer than those obtained by conventional spin-echo refocusing experiments on this instrument, although at the present level of experimental technique the results are still sensitive to instrumental shortcomings.

3.6. The Study of Chemical Exchange by Saturation Transfer

Forsén and Hoffman (*41–43*) described a powerful double-resonance experiment for monitoring chemical exchange processes which take place on a time scale comparable with relaxation. In this technique the longitudinal magnetization of a resonance is measured as a function of time after the perturbation of another resonance with which it is exchanging. This perturbation may be the imposition of a saturating field, or a 180° pulse, or various other possibilities.

The experiment has been performed using low-level pulses (*111*) and saturating fields (*112–116*) in Fourier transform NMR. The largest effects are seen when the longitudinal magnetizations of the exchanging sites are monitored after one resonance is selectively inverted. This selective inversion may easily be achieved using a DANTE sequence with cumulative flip angle 180° on resonance. The expressions for magnetization at sites A and B as a function of the delay τ_x between inversion and sampling are

$$M_z^A/M_0^A = 1 - (2k_A/\beta)\{\exp[-\tfrac{1}{2}(\alpha - \beta)\tau_x] - exp[-\tfrac{1}{2}(\alpha + \beta)\tau_x]\}, \qquad [27]$$

$$M_z^B/M_0^B = 1 - (1/\beta)\{(\beta + R_B - R_A)\exp[-\tfrac{1}{2}(\alpha + \beta)\tau_x] + (\beta + R_A - R_B)\exp[-\tfrac{1}{2}(\alpha - \beta)\tau_x]\}, \qquad [28]$$

where

$$\alpha = R_A + R_B, \qquad \beta = [(R_A - R_B)^2 + 4k_A k_B]^{1/2},$$

and

$$R_A = k_A + (1/T_{1A}), \qquad R_B = k_B + (1/T_{1B}).$$

TABLE 2

SEQUENCE OF EVENTS FOR OBTAINING A DIFFERENCE SPECTRUM AFTER SELECTIVE INVERSION

(1)	Wait for equilibrium
(2)	Nonselective 90° sampling pulse
(3)	Acquire free induction decay and add to store
(4)	Wait for equilibrium
(5)	DANTE pulse sequence with 180° flip angle on resonance
(6)	Exchange delay τ_x
(7)	Nonselective 90° sampling pulse
(8)	Acquire free-induction decay and subtract from store
(9)	Improve signal-to-noise by repeating (1) through (8)
(10)	Fourier transform averaged signal to obtain difference spectrum

Thus by monitoring the recovery curves of sites A and B it is possible in favourable cases to extract both the exchange rate constants, k_A and k_B, and the longitudinal relaxation times, T_{1A} and T_{1B}, of the two sites.

In order to obtain cleaner spectra and more reproducible results, a form of difference spectroscopy may be used (*111, 117*). Time averaging is carried out by alternately co-adding the transients following an isolated 90° pulse and subtracting those obtained after selective inversion. This difference signal transforms to a spectrum which is the difference between the equilibrium spectrum and the perturbed spectrum. The overall sequence of events is summarized in Table 2. The method thus yields spectra showing the deviation from equilibrium longitudinal magnetization caused by the selective inversion, allowing the transfer of magnetization to be followed as a function of time, without interference from any overlapping lines.

Figure 14 shows a series of such difference spectroscopy experiments with increasing exchange delays τ_x, performed on the 3,5 carbon resonances of the 1,2,6-trimethyl-piperidinium ion in slightly alkaline solution. This undergoes slow interconversion between the equatorial *N*-methyl and axial *N*-methyl stereoisomers, so that the inversion

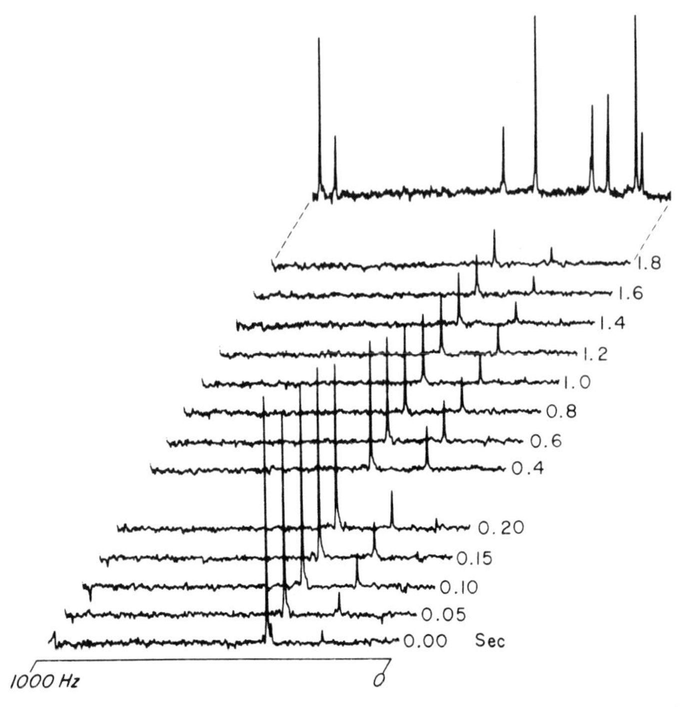

FIG. 14. Study of chemical exchange rates by a selective population inversion experiment, using a DANTE pulse sequence of seventeen pulses of flip angle $\pi/17$ radians, lasting 45 msec. The top trace shows the full proton-decoupled carbon-13 spectrum of the 1,2,6-trimethylpiperidinium ion, containing lines from both the interconverting equatorial and axial *N*-methyl stereoisomers. The difference spectra below were obtained by subtracting the inverted and conventional free-induction decays, leading to spectra showing the decrease in Z magnetization caused by the selective inversion. The time axis represents the delay τ_x after selective inversion of the low-field (equatorial *N*-methyl stereoisomer) line, showing the transfer of negative Z magnetization to the weaker high-field line of the axial stereoisomer.

of the 3,5 carbon resonance of the equatorial species (the left-hand line of Fig. 14) causes transfer of negative Z magnetization to the 3,5 resonances of the axial stereoisomer. The difference spectra thus show a biexponential decay for the equatorial resonance recovering from inversion, while the axial resonance first loses magnetization to the equatorial resonance and then returns slowly to equilibrium. The time scale of Fig. 14 is split in order to show the growth of magnetization exchange, followed by the slower, relaxation-limited return to equilibrium. The conventional spectrum in the diagram illustrates an important advantage of the difference spectroscopy technique—the 3,5 resonance of the axial stereoisomer overlaps with the axial *N*-methyl line, making it difficult to measure accurately the intensity changes of the former. An analysis of the results yields the values $T_{1A} = T_{1B} = 1.1 \pm 0.2$ sec, $k_A = 3.4 \pm 0.5$ sec^{-1}, and $k_B = 1.6 \pm 0.5$ sec^{-1}.

4. DISCUSSION

In the preceding pages some general principles for selective excitation in Fourier transform NMR have been established, and applied to one particular family of experiments based on the DANTE plus sequence. A range of techniques involving transient and steady-state selective irradiation has been shown to be possible on simple conventional instrumentation.

The use of the linear approximation and Fourier transformation methods greatly facilitates the design of selective experiments, by providing an easy link between a desired excitation spectrum and the necessary pulse sequence. The biggest advantage of the DANTE sequence is of course its simplicity; on the CFT-20 the added pulse programming can amount to less than two dozen locations. In the vast majority of cases the characteristic multiple sinc function excitation spectra are quite adequate; however, in more critical situations the use of Fourier transform concepts can allow the selectivity of the DANTE sequence to be enhanced without recourse to the full Tomlinson/Hill synthesized excitation experiment (*14*).

When it is desirable to excite a particularly narrow spectral region, the lobes of the sinc functions in the excitation spectrum can cause problems by giving limited excitation to neighboring resonances. The ideal rectangular "window" function spectrum can be more closely approached by modulating the DANTE sequence with a sinc function envelope (or possibly just a crude approximation to it). Thus by giving each pulse in a DANTE sequence an amplitude (or more simply a width) determined by a preset recipe such that the flip angles of successive pulses follow an approximate $\sin(t)/t$ dependence, a more rectangular window excitation spectrum may be produced.

In wide and crowded spectra the presence of more than one maximum in the excitation spectrum (i.e., the presence of many sidebands) may be a problem. The number of these sidebands may be reduced by phase modulating the pulses in a DANTE sequence. For example, alternating the phases of successive pulses to give alternate positive and negative flip angles leads to an excitation spectrum with maxima not at $\nu_0 + n/\tau$ as in a conventional DANTE experiment, but at $\nu_0 + 1/(2\tau) + n/\tau$. This means in effect that to excite the same frequency a phase-alternated DANTE sequence would use twice the pulse repetition rate of a normal sequence, giving twice the sideband spacing.

The linearity of the Fourier transform operation (the transform of the sum of two functions is the sum of their individual transforms) means that more complex excitation spectra may be produced by adding together different pulse sequences. Thus the construction of a pulse sequence by adding two DANTE sequences with pulse spacing τ_1 and τ_2 leads to an excitation spectrum containing maxima at $\nu_0 + n/\tau_1$ and $\nu_0 + n/\tau_2$. This allows the simultaneous excitation of two independent frequencies, as illustrated in Fig. 15; phase coherence between the two excited resonances is achieved by finishing both DANTE sequences at the same moment.

The combination of excitation by DANTE sequence and gated decoupling, a possibility first envisaged by Müller *et al.* (*118*), is an example of a useful class of experiments having no counterparts in continuous-wave or slow-passage spectroscopy. The time-domain character of Fourier transform spectroscopy makes it possible to change experimental conditions between excitation and observation, allowing correlations to be made between different types of spectrum of the same sample. Thus in the example given in Fig. 10, by exciting a decoupled carbon-13 resonance and measuring the response under proton-coupled conditions it is possible to obtain the multiplet structure associated with one decoupled line. The experiment could equally,

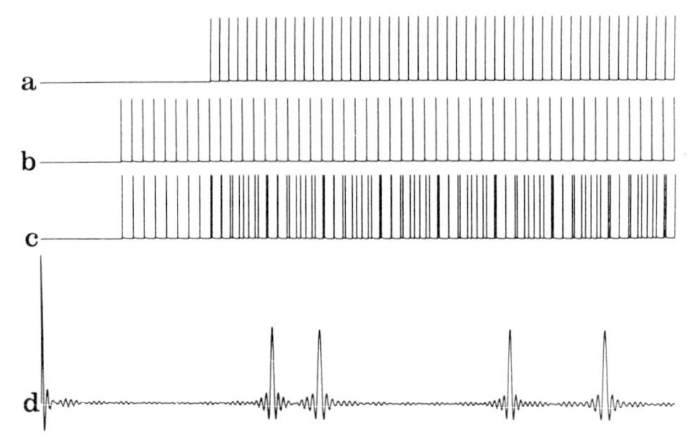

FIG. 15. The superposition (c) of two DANTE pulse sequences (a) and (b) with different repetition rates. Trace (d) shows the frequency-domain excitation spectrum calculated by Fourier transformation of the time-domain pattern (c). This scheme permits selective irradiation of two arbitrary frequencies.

although less usefully, have been performed by exciting one line in the coupled spectrum and acquiring the free-induction decay with proton decoupling. In this way the chemical shift corresponding to any line in the coupled spectrum may be measured.

Naturally the technique is not restricted to carbon-13–proton double resonance; similar experiments could be envisaged for any pair of coupled nuclei, or for homonuclear coupling between widely separated groups of resonances. The analysis of spectra containing two types of heteronuclear coupling, for example, partially fluorinated compounds, could be greatly facilitated by observing coupled carbon-13 subspectra after selective excitation in the presence of proton or fluorine decoupling. In very complex carbon-13 spectra, off-resonance decoupling is frequently used to obtain information about multiplicity. Selective excitation could remove overlap in the off-resonance decoupled spectrum by changing the decoupling conditions from noisy on-resonance to coherent off-resonance.

A more prosaic advantage of being able to examine small regions of coupled spectra at a time is that the available digitization may be more effectively used without interference from folded lines. This has been exploited in some high-resolution measurements on the proton-coupled carbon-13 spectrum of pyridine (*30*).

Several more-complex methods have recently been proposed for decomposing coupled carbon-13 spectra into multiplets from different carbon sites (*118–122*), using the technique of double Fourier transform NMR (*123–125*). For small systems the selective excitation method has the advantages of simplicity and better signal-to-noise ratio, although the latter no longer applies in more complex spectra, since the double Fourier transform techniques acquire information about all carbon sites in a molecule in a single experiment. Some two-dimensional spectroscopy techniques (those based on "*J*" or "spin-echo" spectroscopy) have the added advantage of providing linewidths which in principle are limited not by magnetic field inhomogeneity but by T_2 relaxation. The "proton flip" two-dimensional *J*-spectroscopy experiment can also provide extra information about strong coupling (*122*).

The study of chemical exchange by monitoring the transfer of saturation was first performed in continuous-wave mode by Forsén and Hoffman (*41–43*), but lends itself readily to Fourier transform experiments (*111–116*). Of the many techniques possible, one of the most powerful is to invert a magnetization at site A, and monitor the recovery both of this resonance and that of site B with which it is exchanging. Together with difference spectroscopy, as described in Section 3.6, the use of the DANTE sequence for selective inversion here allows a simple and clean determination of exchange rate constants. The full range of rate constants between those accessible by line shape analysis and those comparable with relaxation rates may be covered. Three site exchange problems (*43*) may be dealt with using single and dual DANTE sequences for inversion.

The measurement of T_2 (or the similar parameter $T_{1\varrho}$) in high-resolution NMR by Fourier transform methods has proved a difficult goal to achieve on a routine basis. One possible response to the problem is to use instrumentally less demanding techniques on individual resonances, rather than try to measure all resonances in a spectrum simultaneously. The INFERNO experiment is simple and effective, and in its "transparent" version, is capable of obtaining results more quickly than nonselective methods when applied to small molecules. Dual or multiple DANTE sequences would enhance the speed of the method by allowing more than one resonance to be spin locked at once.

The hole-burning experiment is a quick and useful test for inhomogeneous broadening. A number of other possible uses exist; for instance it should be feasible to identify regressively connected transitions in coupled systems by burning a hole in one line and seeing which other lines are affected. In more general terms, the selective perturbation of populations in strongly coupled systems can lead to a variety of useful and interesting effects.

All the experiments described in this paper may of course be performed using low-level continuous radiofrequency pulses, rather than DANTE sequences, on spectrometers with weak pulse facilities. The DANTE sequence does, however, have the advantage that total flip angle and selectivity may be controlled independently, without changing the pulse power. The wide range of selective experiments possible has yet to be fully exploited; the DANTE family of techniques represents one approach to performing such experiments on conventional Fourier transform spectrometers. A simple selective excitation program using the DANTE sequence, SELEX MK6, has been written for unmodified CFT-20 spectrometers, and is available from the authors on request.

ACKNOWLEDGMENTS

This work was made possible by an equipment grant and a research studentship from the Science Research Council. The authors gratefully acknowledge several helpful discussions with Dr. H. D. W. Hill and with Dr. M. J. T. Robinson, who also provided the sample of 1,2,6-trimethylpiperidinium chloride.

REFERENCES

1. J. Schaefer, *J. Magn. Resonance* **6,** 670 (1972).
2. J. P. Jesson, P. Meakin, and G. Kneissel, *J. Amer. Chem. Soc.* **95,** 618 (1973).
3. I. D. Campbell, C. M. Dobson, G. Jeminet, and R. J. P. Williams, *FEBS Lett.* **49,** 115 (1974).
4. H. E. Bleich and J. A. Glasel, *J. Magn. Resonance* **18,** 401 (1975).
5. D. I. Hoult, *J. Magn. Resonance* **21,** 337 (1976).
6. N. R. Krishna, *J. Magn. Resonance* **22,** 555 (1976).

7. S. L. Patt and B. D. Sykes, *J. Chem. Phys.* **56,** 3182 (1972).
8. F. W. Benz, J. Feeney, and G. C. K. Roberts, *J. Magn. Resonance* **8,** 114 (1972).
9. E. S. Mooberry and T. R. Krugh, *J. Magn. Resonance* **17,** 128 (1975).
10. T. R. Krugh and W. C. Schaefer, *J. Magn. Resonance* **19,** 99 (1975).
11. R. K. Gupta, *J. Magn. Resonance* **24,** 461 (1976).
12. S. Alexander, *Rev. Sci. Instrum.* **32,** 1066 (1961).
13. A. G. Redfield and R. K. Gupta, *J. Chem. Phys.* **54,** 1418 (1971).
14. B. L. Tomlinson and H. D. W. Hill, *J. Chem. Phys.* **59,** 1775 (1973).
15. A. G. Redfield, S. D. Kunz, and E. K. Ralph, *J. Magn. Resonance* **19,** 114 (1975).
16. J. Dadok and R. F. Sprecher, *J. Magn. Resonance* **13,** 243 (1974).
17. R. K. Gupta, J. A. Ferretti, and E. D. Becker, *J. Magn. Resonance* **13,** 275 (1974).
18. Y. Arata and H. Ozawa, *J. Magn. Resonance* **21,** 67 (1976).
19. J. A. Ferretti and R. R. Ernst, *J. Chem. Phys.* **65,** 4283 (1976).
20. F. W. Wehrli and T. Wirthlin, "Interpretation of Carbon-13 NMR Spectra," Heyden, London, 1976, and references therein.
21. M. Hansen and H. J. Jakobsen, *J. Magn. Resonance* **20,** 520 (1975).
22. L. Ernst, D. N. Lincoln, and V. Wray, *J. Magn. Resonance* **21,** 115 (1976).
23. V. A. Chertkov and N. M. Sergeyev, *J. Magn. Resonance* **21,** 159 (1976).
24. L. Ernst, E. Lustig, and V. Wray, *J. Magn. Resonance* **22,** 459 (1976).
25. V. A. Chertkov, Y. K. Grishin, and N. M. Sergeyev, *J. Magn. Resonance* **24,** 275 (1976).
26. L. Ernst, V. Wray, V. A. Chertkov, and N. M. Sergeyev, *J. Magn. Resonance* **25,** 123 (1977).
27. R. Freeman, Sixth Conference on Molecular Spectroscopy, Durham, England, March 1976; published as Chap. 2 of "Molecular Spectroscopy" (A. R. West, Ed.), Heyden, London, 1977.
28. G. Bodenhausen, R. Freeman, and G. A. Morris, *J. Magn. Resonance* **23,** 171 (1976).
29. R. Freeman, G. A. Morris, and M. J. T. Robinson, *J. Chem. Soc. Chem. Commun.*, 754 (1976).
30. R. Freeman, G. A. Morris, and D. L. Turner, *J. Magn. Resonance* **26,** 373 (1977).
31. K. G. R. Pachler and P. L. Wessels, *J. Magn. Resonance* **12,** 337 (1973).
32. S. Sørensen, R. S. Hansen, and H. J. Jakobsen, *J. Magn. Resonance* **14,** 243 (1974).
33. H. J. Jakobsen, S. A. Linde, and S. Sørensen, *J. Magn. Resonance* **15,** 385 (1974).
34. A. A. Chalmers, K. G. R. Pachler, and P. L. Wessels, *J. Magn. Resonance* **15,** 415 (1974).
35. A. A. Chalmers, K. G. R. Pachler, and P. L. Wessels, *Org. Magn. Resonance* **6,** 445 (1974).
36. S. A. Linde and H. J. Jakobsen, *J. Amer. Chem. Soc.* **98,** 1041 (1976).
37. N.J. Koole, D. Knol, and M. J. A. deBie, *J. Magn. Resonance* **21,** 499 (1976).
38. N. J. Koole and M. J. A. deBie, *J. Magn. Resonance* **23,** 9 (1976).
39. H. J. Jakobsen and H. Bildsøe, *J. Magn. Resonance* **26,** 183 (1977).
40. R. A. Hoffman and S. Forsén, "Progress in Nuclear Magnetic Resonance Spectroscopy" (J. W. Emsley, J. Feeney, and L. H. Sutcliffe, Eds.), Vol. 1, p. 15, Pergamon, Oxford, 1966.
41. S. Forsén and R. A. Hoffman, *J. Chem. Phys.* **39,** 2892 (1963).
42. S. Forsén and R. A. Hoffman, *Acta Chem. Scand.* **17,** 1787 (1963).
43. S. Forsén and R. A. Hoffman, *J. Chem. Phys.* **40,** 1189 (1964).
44. R. A. Hoffman and S. Forsén, *J. Chem. Phys.* **45,** 2049 (1966).
45. R. Freeman and S. Wittekoek, *in* "Proceedings of the 15th Colloque AMPERE Grenoble (1968)," p. 205, North-Holland, Amsterdam, 1968.
46. R. Freeman and S. Wittekoek, *J. Magn. Resonance* **1,** 238 (1969).
47. E. J. Wells and K. H. Abramson, *J. Magn. Resonance* **1,** 378 (1969).
48. R. Freeman, S. Wittekoek, and R. R. Ernst, *J. Chem. Phys.* **52,** 1529 (1970).
49. A. R. Muir and D. W. Turner, *J. Chem. Soc. Chem. Commun.*, 286 (1971).
50. M. F. Augusteijn, W. M. M. J. Bovée, S. Emid, A. F. Mehlkopf, and J. Smidt, *J. Magn. Resonance* **7,** 301 (1972).
51. C. W. M. Grant, L. D. Hall, and C. M. Preston, *J. Amer. Chem. Soc.* **95,** 7742 (1973).
52. A. Briguet, J.-L. Culty, J.-C. Duplan, and J. Delmau, *J. Phys. E.* **7,** 791 (1974).
53. R. Freeman, H. D. W. Hill, B. L. Tomlinson, and L. D. Hall, *J. Chem. Phys.* **61,** 4466 (1974).
54. C. L. Mayne, D. W. Alderman, and D. M. Grant, *J. Chem. Phys.* **63,** 2514 (1975).
55. L. D. Hall and H. D. W. Hill, *J. Amer. Chem. Soc.* **98,** 1269 (1976).
56. C. L. Mayne, D. M. Grant, and D. W. Alderman, *J. Chem. Phys.* **65,** 1684 (1976).

57. D. W. ALDERMAN, J. J. LED, J. PEDERSEN, AND N. F. ANDERSEN, *J. Magn. Resonance* **21,** 77 (1976).
58. C. LAPRAY, A. BRIGUET, J.-C. DUPLAN, AND J. DELMAU, *J. Magn. Resonance* **23,** 129 (1976).
59. W. M. M. J. BOVÉE, *J. Magn. Resonance* **24,** 327 (1976).
60. I. SOLOMON, *Phys. Rev.* **99,** 559 (1955).
61. I. SOLOMON AND N. BLOEMBERGEN, *J. Chem. Phys.* **25,** 261 (1956).
62. I. D. CAMPBELL AND R. FREEMAN, *J. Chem. Phys.* **58,** 2666 (1973).
63. I. D. CAMPBELL AND R. FREEMAN, *J. Magn. Resonance* **11,** 143 (1973).
64. D. CANET, *J. Magn. Resonance* **23,** 361 (1976).
65. R. K. HARRIS AND R. H. NEWMAN, *J. Magn. Resonance* **24,** 449 (1976).
66. E. L. HAHN AND D. E. MAXWELL, *Phys. Rev.* **84,** 1246 (1951).
67. E. L. HAHN AND D. E. MAXWELL, *Phys. Rev.* **88,** 1070 (1952).
68. J. G. POWLES AND A. HARTLAND, *Proc. Phys. Soc.* **77,** 273 (1961).
69. R. FREEMAN AND H. D. W. HILL, *J. Chem. Phys.* **54,** 301 (1971); R. FREEMAN AND H. D. W. HILL, "Dynamic Nuclear Magnetic Resonance Spectroscopy" (L. M. Jackman and F. A. Cotton, Eds.), Chap. 5, Academic Press, New York, 1975.
70. I. SOLOMON, *Compt. Rend.* **248,** 92 (1959).
71. I. SOLOMON, *Compt. Rend.* **249,** 1631 (1959).
72. J. H. STRANGE AND R. E. MORGAN, *J. Phys. C* **3,** 1997 (1970).
73. R. FREEMAN AND H. D. W. HILL, *J. Chem. Phys.* **55,** 1985 (1971).
74. R. FREEMAN, H. D. W. HILL, AND J. DADOK, *J. Chem. Phys.* **58,** 3107 (1973).
75. T. K. LEIPERT, J. H. NOGGLE, W. J. FREEMAN, AND D. L. DALRYMPLE, *J. Magn. Resonance* **19,** 208 (1975).
76. T. K. LEIPERT, W. J. FREEMAN, AND J. H. NOGGLE, *J. Chem. Phys.* **63,** 4177 (1975).
77. H. D. W. HILL, private communication.
78. P. MANSFIELD AND P. K. GRANNELL, *J. Phys. C.* **6,** L422 (1973).
79. W. S. HINSHAW, *Phys. Lett. A* **48,** 87 (1974).
80. A. N. GARROWAY, P. K. GRANNELL, AND P. MANSFIELD, *J. Phys. C* **7,** L457 (1974).
81. W. S. HINSHAW, *J. Appl. Phys.* **47,** 3709 (1975).
82. P. MANSFIELD AND P. K. GRANNELL, *Phys. Rev. B* **12,** 3618 (1975).
83. P. K. GRANNELL AND P. MANSFIELD, *Phys. Med. Biol.* **20,** 477 (1975).
84. P. MANSFIELD, A. A. MAUDSLEY, AND T. BAINES, *J. Phys. E* **9,** 271 (1976).
85. P. C. LAUTERBUR, *Nature* **242,** 190 (1973).
86. P. C. LAUTERBUR, *Pure Appl. Chem.* **40,** 149 (1974).
87. P. C. LAUTERBUR, D. M. KRAMER, W. V. HOUSE, JR., AND C.-N. CHEN, *J. Amer. Chem. Soc.* **97,** 6866 (1975).
88. A. KUMAR, D. WELTI, AND R. R. ERNST, *J. Magn. Resonance* **18,** 69 (1975).
89. R. R. ERNST, *Chimia* **29,** 179 (1975).
90. N. BLOEMBERGEN, E. M. PURCELL, AND R. V. POUND, *Phys. Rev.* **73,** 679 (1948).
91. R. FREEMAN AND B. GESTBLOM, *J. Chem. Phys.* **48,** 5008 (1968).
92. J. SCHAEFER, "Topics in Carbon-13 NMR Spectroscopy" (G. C. Levy, Ed.), Vol. 1, Chichester, Wiley–Interscience, New York, 1974.
93. M. MEHRING AND J. S. WAUGH, *Rev. Sci. Instrum.* **43,** 649 (1972).
94. F. BLOCH AND A. SIEGERT, *Phys. Rev.* **57,** 522 (1940).
95. N. F. RAMSEY, *Phys. Rev.* **100,** 1191 (1955).
96. D. ALIGHIERI, "La Divina Commedia" (G. Petrocchi, Ed.), Einaudi, Turin, 1975.
97. R. KUBO AND K. TOMITA, *J. Phys. Soc. Japan* **9,** 888 (1954).
98. C. P. SLICHTER, "Principles of Magnetic Resonance," Chap. 2, Harper & Row, New York, 1963.
99. R. R. ERNST AND W. A. ANDERSON, *Rev. Sci. Instrum.* **37,** 93 (1966).
100. E. D. OSTROFF AND J. S. WAUGH, *Phys. Rev. Lett.* **16,** 1097 (1966).
101. A. PINES AND J. D. ELLETT, JR., *J. Amer. Chem. Soc.* **95,** 4437 (1973).
102. H. Y. CARR, *Phys. Rev.* **112,** 58 (1958).
103. R. FREEMAN AND H. D. W. HILL, *J. Magn. Resonance* **4,** 366 (1971); A. SCHWENK, *J. Magn. Resonance* **5,** 376 (1971).
104. D. E. JONES AND H. STERNLICHT, *J. Magn. Resonance* **6,** 167 (1972).

105. J. Kronenbitter and A. Schwenk, *J. Magn. Resonance* **25,** 147 (1977).
106. F. Bloch, *Phys. Rev.* **70,** 460 (1946).
107. R. Bracewell, "The Fourier Transform and Its Applications," McGraw–Hill, New York, 1965.
108. D. S. Champeney, "Fourier Transforms and Their Physical Applications," Academic Press, New York, 1973.
109. M. Hansen and H. J. Jakobsen, *J. Magn. Resonance* **10,** 74 (1973).
110. H. C. Torrey, *Phys. Rev.* **76,** 1059 (1949).
111. F. W. Dahlquist, K. J. Longmuir, and R. B. duVernet, *J. Magn. Resonance* **17,** 406 (1975).
112. A. G. Redfield and R. K. Gupta, *Cold Spring Harbor Symp. Quant. Biol.* **36,** 405 (1963).
113. B. E. Mann, *J. Magn. Resonance* **21,** 17 (1976).
114. P. Ahlberg, *Chem. Scr.* **9,** 47 (1976).
115. B. E. Mann, *J. Magn. Resonance* **25,** 91 (1977).
116. B. E. Mann, *J. Chem. Soc. Perkin II* 84 (1977).
117. R. Freeman and H. D. W. Hill, *J. Chem. Phys.* **54,** 3367 (1971).
118. L. Müller, A. Kumar, and R. R. Ernst, *J. Chem. Phys.* **63,** 5490 (1975).
119. G. Bodenhausen, R. Freeman, and D. L. Turner, *J. Chem. Phys.* **65,** 839 (1976).
120. G. Bodenhausen, R. Freeman, R. Niedermeyer, and D. L. Turner, *J. Magn. Resonance* **24,** 291 (1976).
121. L. Müller, A. Kumar, and R. R. Ernst, *J. Magn. Resonance* **25,** 383 (1977).
122. G. Bodenhausen, R. Freeman, G. A. Morris, and D. L. Turner, *J. Magn. Resonance* **28,** 17 (1977).
123. J. Jeener, Ampere International Summer School II, Basko Polje, Yugoslavia, 1971; Second European International NMR Conference, Enschedé, Holland, 1975.
124. W. P. Aue, E. Bartholdi, and R. R. Ernst, *J. Chem. Phys.* **64,** 2229 (1976).
125. G. Bodenhausen, R. Freeman, R. Niedermeyer, and D. L. Turner, *J. Magn. Resonance* **26,** 133 (1977).

JOURNAL OF MAGNETIC RESONANCE **29**, 523–533 (1978)

Sensitivity of Two-Dimensional NMR Spectroscopy

W. P. AUE, P. BACHMANN, A. WOKAUN, AND R. R. ERNST

Laboratorium für Physikalische Chemie, Eidgenössische Technische Hochschule, 8092 Zürich, Switzerland

Received June 27, 1977

The sensitivity of two-dimensional NMR Fourier spectroscopy is analyzed and compared with the sensitivity of one-dimensional Fourier spectroscopy. It is confirmed by experiment that as little as a factor 2 may be lost by going from 1D to 2D spectroscopy.

1. INTRODUCTION

Two-dimensional spectroscopy appears to be a powerful technique to increase the information content of NMR spectra (*1*) as well as to simplify complicated spectra by means of a two-dimensional spread (*2–5*). The computation of a 2D spectrum represented by $M \times N$ data points requires the recording of M free-induction decays each with N sampling points. Data matrices of up to 1024×1024 are necessary for high-resolution applications. The data taking may become a time-consuming process, particularly when signal averaging is necessary for each of the M FID's to obtain sufficient sensitivity. It is therefore mandatory to consider the achievable sensitivity in 2D spectroscopy to judge its practical applicability.

The general relations for the signal-to-noise ratio in two dimensions are given in Section 2. The two distinct cases, 2D resolved and 2D correlated spectroscopy, are analyzed with regard to sensitivity in the Sections 3 and 4, respectively. Section 5, finally, is devoted to a comparison with experimental results.

2. SIGNAL-TO-NOISE RATIO IN 2D SPECTROSCOPY

The signal-to-noise ratio S/N of a 2D spectrum $\{S_{kl}\}$ can be defined in the same manner as in 1D spectroscopy:

$$\frac{S}{N} = \frac{\text{signal peak value}}{\text{noise rms value}} = \frac{\max(S_{kl})}{\sigma_N}. \qquad [1]$$

The discrete 2D data set $\{S_{kl}\}$ shall represent a single reference signal peak with $0 \leq k < M$ and $0 \leq l < N$. It is assumed that the random noise samples N_{kl} are stationary, uncorrelated, and have zero mean,

$$\varepsilon[N_{kl}N_{pq}] = \sigma_N^2 \delta_{k,p}\delta_{l,q} \qquad \text{and} \qquad \varepsilon[N_{kl}] = 0. \qquad [2]$$

The assumption of uncorrelated noise samples in a computed 2D spectrum is justified whenever the original noise samples of the recorded FID's are independent and have equal variances.

0022-2364/78/0293-0523$02.00/0

It is well known that the S/N ratio can be improved by a suitable linear filtering process (*6*). In two dimensions, filtering is represented by the 2D convolution sum

$$S'_{kl} = \sum_{r=0}^{M-1} \sum_{s=0}^{N-1} S_{rs} H_{k-r,\, l-s} \qquad [3]$$

with the 2D weighting function $\{H_{kl}\}$. The optimum weighting function $\{H_{kl}^{\text{matched}}\}$ maximizing sensitivity shall now be determined. This is most conveniently done by the reduction of Eq. [3] to a 1D convolution sum. With the redefinition

$$S_{rs} = S_q \qquad \text{with } q = rN + s$$

and

$$S'_{kl} = S'_p \qquad \text{with } p = kN + l, \qquad [4]$$

one obtains

$$S'_p = \sum_{q=0}^{MN-1} S_q H_{p-q}. \qquad [5]$$

This expression has indeed the form of a one-dimensional convolution sum. The 1D vector $\{S_p\}$ is formed by the sequence of the M rows of the $M \times N$ data matrix.

On the basis of Eq. [5], one can now apply the well-known principles of 1D matched filtering to the signal $\{S_p\}$. In the case of uncorrelated noise samples, the weighting function $\{H_p^{\text{matched}}\}$ of the matched filter is given by (*6*)

$$H_p^{\text{matched}} = S_{-p}, \qquad [6]$$

and the maximum S/N ratio achieved therewith is

$$\left(\frac{S}{N}\right)_{\max} = \frac{\left[\sum_{p=0}^{MN-1} S_p^2\right]^{1/2}}{\sigma_N} = \frac{\left[\sum_{k=0}^{M-1} \sum_{l=0}^{N-1} S_{kl}^2\right]^{1/2}}{\sigma_N}. \qquad [7]$$

It is thus proportional to the total signal energy of the 2D reference peak and the weighting function of the 2D matched filter is equal to the inverted lineshape

$$H_{kl}^{\text{matched}} = S_{-k,\, -l}. \qquad [8]$$

3. 2D RESOLVED NMR SPECTRA

Several techniques have been proposed to simplify complex NMR spectra by utilizing a 2D spread of a 1D proton or carbon spectrum. Chemical shifts (*2*) or scalar coupling constants (*3–5*) as well as dipolar couplings in solids (*7*) can be used as spreading parameters. The characteristic of these 2D spectra is that each line in the original 1D spectrum is represented by one single peak in the 2D plot.

A typical experiment for 2D resolution will be assumed. Transverse magnetization is generated at time $t = 0$ by means of a 90° rf pulse. During the evolution period of length t_1, the magnetization component $M^{(ab)}$ of the transition (ab) precesses under the influence of the average Hamiltonian $\bar{\mathscr{H}}^{(1)}$ with the frequency $\omega_{ab}^{(1)}$. At time $t = t_1$, the

Hamiltonian is changed into $\bar{\mathscr{H}}^{(2)}$ and the magnetization component $M^{(ab)}$ considered continues to precess with the modified frequency $\omega_{ab}^{(2)}$ during the detection period

$$M^{(ab)}(t_1, t_2) = M^{(ab)}(0)\cos(\omega_{ab}^{(1)} t_1 + \omega_{ab}^{(2)} t_2) \times \exp\{-t_1/T_{2\,ab}^{(1)} - t_2/T_{2\,ab}^{(2)}\}, \quad [9]$$

where $T_{2ab}^{(1)}$ and $T_{2ab}^{(2)}$ are usually equal to either T_2 or T_2^*; M FID's are recorded for the values $t_1 = 0, \tau_1/M, \ldots, (M-1)\tau_1/M$ and each FID is represented by N sample values taken at $t_2 = 0, \tau_2/N, \ldots, (N-1)\tau_2/N$. These values form the M by N data matrix $\{M_{rs}^{(ab)}\}$.

The real part of the complex 2D Fourier transform of $\{M_{rs}^{(ab)}\}$ is plotted

$$\begin{aligned} S_{kl}^{(ab)} &= \mathrm{Re}\left\{\sum_{r=0}^{M-1}\sum_{s=0}^{N-1} M_{rs}^{(ab)} \exp\left\{-i2\pi kr\frac{1}{M} - i2\pi ls\frac{1}{N}\right\}\right\} \\ &= \sum_{r=0}^{M-1}\sum_{s=0}^{N-1} M_{rs}^{(ab)} \cos\left\{2\pi kr\frac{1}{M} + 2\pi ls\frac{1}{N}\right\}. \end{aligned} \quad [10]$$

This experiment can be compared with M identical 1D experiments recorded under the same conditions, coadded and Fourier transformed to produce the spectrum

$$\begin{aligned} S_{l}^{(ab)} &= \mathrm{Re}\left\{M\sum_{s=0}^{N-1} M_{0s}^{(ab)} \exp\left\{-i2\pi ls\frac{1}{N}\right\}\right\} \\ &= M\sum_{s=0}^{N-1} M_{0s}^{(ab)} \cos\left\{2\pi ls\frac{1}{N}\right\} \end{aligned} \quad [11]$$

First, the variances σ_{2D}^2 and σ_{1D}^2 of the random noise in the resulting 2D and 1D spectra shall be computed, assuming uncorrelated noise samples $\{n_{rs}\}$. The Fourier-transformed noise in the 2D spectrum is then given by

$$N_{kl} = \sum_{r=0}^{M-1}\sum_{s=0}^{N-1} n_{rs}\cos\left\{2\pi\left(\frac{kr}{M} + \frac{ls}{N}\right)\right\} \quad [12]$$

with the variance

$$\sigma_{2D}^2 = \sigma_n^2 \sum_{r=0}^{M-1}\sum_{s=0}^{N-1} \cos^2\left\{2\pi\left(\frac{kr}{M} + \frac{ls}{N}\right)\right\} = \tfrac{1}{2}\sigma_n^2 M\cdot N \quad [13]$$

and for the 1D spectrum by

$$N_l = \sum_{r=0}^{M-1}\sum_{s=0}^{N-1} n_{rs}\cos\left\{2\pi\frac{ls}{N}\right\} \quad [14]$$

with the variance

$$\sigma_{1D}^2 = \sigma_n^2 \sum_{r=0}^{M-1}\sum_{s=0}^{N-1} \cos^2\left\{2\pi\frac{ls}{N}\right\} = \tfrac{1}{2}\sigma_n^2 M\cdot N. \quad [15]$$

The two spectra have identical variances.

Two cases will now be treated separately, (a) without filtering, and (b) with matched filtering.

(*a*) *Without filtering.* Here, the peak heights of the two spectra can be compared directly as the random noise has the same variance in the two cases. For simplicity and to obtain closed formulas for the peak heights, the summations in the discrete Fourier transforms, Eqs. [10] and [11], will be replaced by integrals from 0 to τ_1 and from 0 to τ_2. Then, the following peak heights are obtained for the 2D case:

$$S^{(ab)}(\omega_{ab}^{(1)}, \omega_{ab}^{(2)}) = M^{(ab)}(0)\tfrac{1}{2}T_{2ab}^{(1)} \cdot T_{2ab}^{(2)} \times [1 - \exp(-\tau_1/T_{2ab}^{(1)})][1 - \exp(-\tau_2/T_{2ab}^{(2)})] \quad [16]$$

and for 1D spectroscopy

$$S^{(ab)}(\omega_{ab}^{(2)}) = M^{(ab)}(0)\tfrac{1}{2}\tau_1 T_{2ab}^{(2)}[1 - \exp(-\tau_2/T_{2ab}^{(2)})]. \quad [17]$$

The sensitivity ratio is given by

$$\frac{(S/N)_{2D}}{(S/N)_{1D}} = \frac{T_{2ab}^{(1)}}{\tau_1}\left[1 - \exp\left(\frac{-\tau_1}{T_{2ab}^{(1)}}\right)\right]. \quad [18]$$

(*b*) *With matched filtering.* Here, the signal energies have to be compared. They can be computed directly in time space. Again using integrals instead of summations, one obtains

$$E_{2D} = \int_0^{\tau_1} dt_1 \int_0^{\tau_2} dt_2 [M^{(ab)}(t_1, t_2)]^2 = \tfrac{1}{8}T_{2ab}^{(1)} \cdot T_{2ab}^{(2)} \times [1 - \exp(-2\tau_1/T_{2ab}^{(1)})][1 - \exp(-2\tau_2/T_{2ab}^{(2)})] \cdot [M^{(ab)}(0)]^2 \quad [19]$$

and

$$E_{1D} = \tfrac{1}{4}\tau_1 T_{2ab}^{(2)}[1 - \exp(-2\tau_2/T_{2ab}^{(2)})] \cdot [M^{(ab)}(0)]^2. \quad [20]$$

For the sensitivity ratio with matched filtering, one then gets

$$\frac{(S/N)_{2D}}{(S/N)_{1D}} = \left(\frac{E_{2D}}{E_{1D}}\right)^{1/2} = \left\{\frac{1}{2}\frac{T_{2ab}^{(1)}}{\tau_1}\left[1 - \exp\left(\frac{-2\tau_1}{T_{2ab}^{(1)}}\right)\right]\right\}^{1/2}. \quad [21]$$

These formulas are discussed and compared with experimental results in the concluding section.

4. 2D AUTOCORRELATED NMR SPECTRA

Two-dimensional correlated NMR spectra are obtained by a two-pulse experiment of the type $P_{\pi/2} - t_1 - P_\alpha - t_2$(1). In general, the two Hamiltonians $\bar{\mathscr{H}}^{(1)}$ and $\bar{\mathscr{H}}^{(2)}$, acting during evolution and detection periods, respectively, can be different from each other, for example, for heteronuclear 2D correlated NMR spectra (*8*). In this section, the discussion will be restricted to 2D autocorrelated spectra, where $\bar{\mathscr{H}}^{(1)} = \bar{\mathscr{H}}^{(2)}$. Such experiments can be utilized to obtain information on the connectivity of transitions in the energy level scheme. They may be considered as alternatives to double resonance experiments.

The application of the second rf pulse P_α at the end of the evolution period causes a partition of the transverse magnetization of a particular transition (mn) among various cross-peaks in a 2D autocorrelated spectrum. The partition factor $P_{(kl)(mn)}$, which determines the intensity of the cross-peak between the transitions (kl) and (mn) in a *phase-sensitive spectrum*, can be computed from Eq. [33] of Ref. (*1*),

$$p_{(kl)(mn)} = \frac{F_{y\,kl}}{F_{y\,mn}} \{R_{lm}R^*_{kn} \pm R_{km}R^*_{ln}\}, \qquad [22]$$

where F_y is the y component of the total angular momentum operator and R is the rotation operator

$$R = \exp\{-i\alpha F_x\} \qquad [23]$$

representing the second rf pulse with flip angle α.

The sign ambiguity in Eq. [22] depends on the selection of the reference phases. The upper sign refers to the plotting of the spectrum component $S^{ss}(\omega_1, \omega_2)$, whereas the lower sign refers to the spectrum component $S^{cc}(\omega_1, \omega_2)$ (*1*).

For a system of K weakly coupled nonequivalent nuclei with spin $\frac{1}{2}$, one obtains a particularly simple expression for $p_{(kl)(mn)}$. Assuming a 90° pulse separating the evolution and detection periods, one finds

$$p = 2^{-K+1} \qquad [24]$$

uniformly for all peaks in the 2D autocorrelated spectrum of weakly coupled spin systems.

For partitioning of the transverse magnetization only the autopeaks and cross-peaks among transitions belonging to the *same* multiplet are relevant. The cross-peaks among transitions belonging to different multiplets are pairwise 180° out of phase and do not contribute to the total signal integral. This gives then a reduction of the signal amplitudes by a factor 2^{K-1}, i.e., by the number of lines within one multiplet.

Two additional numerical factors distinguish the sensitivity of 2D correlated spectroscopy from the sensitivity of 2D resolved spectroscopy. (a) The 90° mixing pulse selects only one phase component of the transverse magnetization and reduces therefore the signal intensity by a factor 2. (b) The rms noise amplitudes for 1D and 2D correlated spectra are related by

$$\sigma_{2D} = \frac{1}{\sqrt{2}} \sigma_{1D}. \qquad [25]$$

The unequal noise is caused by the use of a 2D cosine transform to obtain the 2D correlated spectrum.

In analogy to 2D resolved spectroscopy, one obtains then the following sensitivity ratios:

Without filter:

$$\frac{(S/N)_{2D}}{(S/N)_{1D}} = p \cdot \frac{T^{(1)}_{2ab}}{\sqrt{2}\,\tau_1} \left[1 - \exp\left(\frac{-\tau_1}{T^{(1)}_{2ab}}\right)\right]. \qquad [26]$$

With matched filter:

$$\frac{(S/N)_{2D}}{(S/N)_{1D}} = p \cdot \frac{1}{\sqrt{2}} \left\{\frac{1}{2} \frac{T^{(1)}_{2ab}}{\tau_1} \left[1 - \exp\left(\frac{-2\tau_1}{T^{(1)}_{2ab}}\right)\right]\right\}^{1/2}. \qquad [27]$$

For strongly coupled systems and for $\alpha \neq \pi/2$, the sensitivity will vary from peak to peak in a 2D autocorrelated spectrum, but the above formulas can be used as a guide even for strongly coupled systems.

5. DISCUSSION AND EXPERIMENTAL CONFIRMATION

Sensitivity ratios based on Eqs. [18], [21], [26], and [27] are given in Table 1. The length of the evolution period, τ_1, is an important parameter, both for sensitivity and resolution. It determines directly the resolution in the ω_1 direction. The minimum full linewidth in ω_1 is given by $\Delta_1 = 0.603/\tau_1$. Long τ_1's will enhance resolution. On the other hand, a long τ_1 will stipulate to record signals far out in the tail of the FID. This will reduce the relative sensitivity, as is demonstrated by Table 1. The general incompatibility of maximizing sensitivity and resolution is apparent from these data.

TABLE 1

RELATIVE SENSITIVITY OF 2D FTS NORMALIZED BY THE SENSITIVITY OF A 1D FOURIER EXPERIMENT PERFORMED IN THE SAME TOTAL TIME FOR VARIOUS LENGTHS τ_1 OF THE EVOLUTION PERIOD

$\tau_1/T_{2ab}^{(1)}$	2D resolved FTS without filter	2D resolved FTS with matched filter	2D autocorrelated FTS without filter[a]	2D autocorrelated FTS with matched filter[a]
0	1	1	$0.707 \cdot 2^{-K+1}$	$0.707 \cdot 2^{-K+1}$
0.5	0.787	0.795	$0.556 \cdot 2^{-K+1}$	$0.562 \cdot 2^{-K+1}$
1	0.632	0.658	$0.447 \cdot 2^{-K+1}$	$0.465 \cdot 2^{-K+1}$
2	0.432	0.495	$0.306 \cdot 2^{-K+1}$	$0.350 \cdot 2^{-K+1}$
3	0.317	0.408	$0.224 \cdot 2^{-K+1}$	$0.288 \cdot 2^{-K+1}$
4	0.245	0.353	$0.174 \cdot 2^{-K+1}$	$0.250 \cdot 2^{-K+1}$

[a] For a weakly coupled system of K nonequivalent spins $\frac{1}{2}$ and assuming a mixing pulse with $\alpha = \pi/2$.

Figure 1 gives a comparison of 1D and 2D autocorrelated spectra of a coupled two-spin-$\frac{1}{2}$ system recorded under identical conditions in the same total time. No filtering has been used in either case. The 2D spectrum is shown only in part to permit a better judgment of sensitivity. A complete 2D spectrum with a sixfold increase in sensitivity is shown in Fig. 2. The experimentally determined sensitivity ratio for Fig. 1 is

$$((S/N)_{2D}/(S/N)_{1D})_{\text{exp}} = 0.143.$$

The variance of the random noise has been determined numerically by calculating the mean square deviation in the absence of a signal. The peak intensity in the 1D spectrum has been determined from one of the stronger inner peaks $S_{(12)}$. In the 2D spectrum, the absorptive cross-peak $S_{(24)(12)}$ has been taken as the intensity reference.

The corresponding partition factor $p_{(24)(12)}$ can be computed from Eq. [24] with the result $p_{(24)(12)} = 0.5$. With the experimental ratio $\tau_1/T_2^* = 2.36$, one obtains from Eq. [26] the theoretical sensitivity ratio

$$((S/N)_{2D}/(S/N)_{1D})_{\text{theor}} = 0.136.$$

A similar comparison for a 2D resolved spectrum of a three-spin-$\frac{1}{2}$ system is given in Fig. 3. To circumvent the mixing of absorptive and dispersive components, which

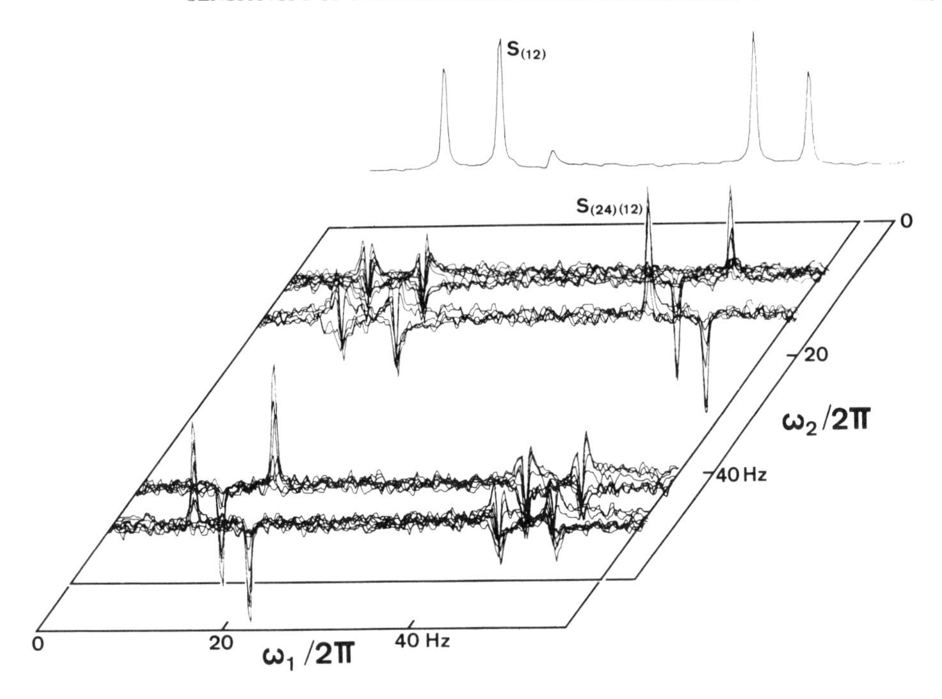

FIG. 1. Two-dimensional correlated spectrum of a solution of 2,3-dibromothiophene in hexafluorobenzene; 256 experiments with 256 data points each have been utilized. A partial spectrum is shown to improve the visualization. The corresponding 1D spectrum shown has been computed on the basis of the same number of experiments, also represented by 256 data points each.

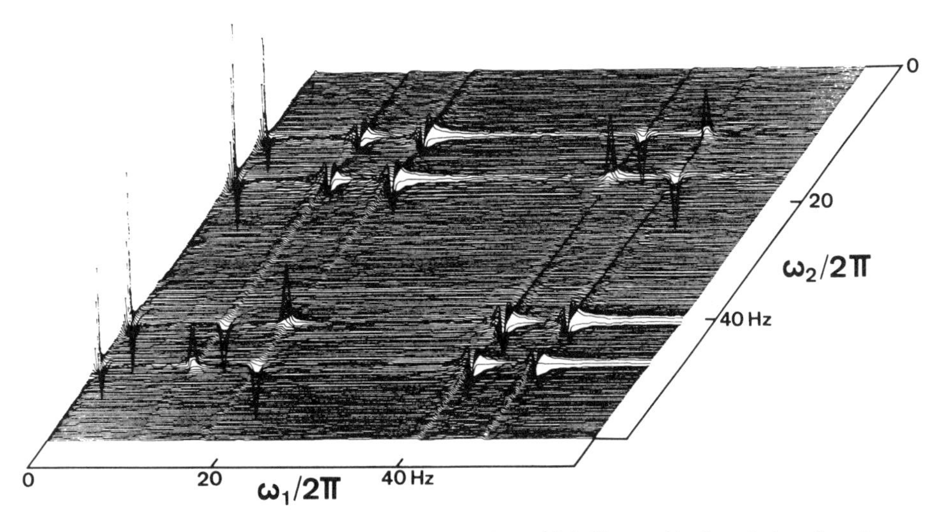

FIG. 2. Complete 2D correlated spectrum of a solution of 2,3-dibromothiophene in hexafluorobenzene with concentration six times that used to obtain Fig. 1.

occurs in phase-sensitive 2D resolved spectra, the 2D spectrum of Fig. 3 has been recorded as an absolute-value spectrum

$$|S_{kl}| = \left| \sum_{r=0}^{M-1} \sum_{s=0}^{N-1} M_{rs} \exp\left\{ -i2\pi \frac{kr}{M} - i2\pi \frac{ls}{N} \right\} \right| . \qquad [28]$$

The corresponding 1D spectrum of Fig. 3 has also been recorded in absolute-value mode to permit a direct comparison. In certain cases, the mixing of absorptive and

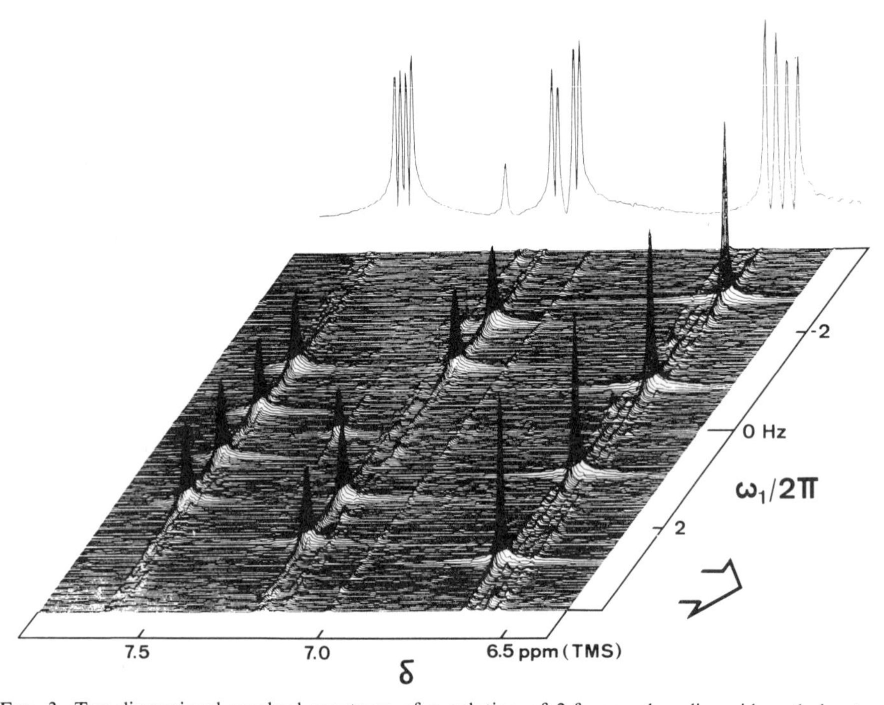

FIG. 3. Two-dimensional resolved spectrum of a solution of 2-furancarboxylic acid methyl ester in hexafluorobenzene and deuterated chloroform; 128 experiments were performed, represented by 512 points each. The signal of the methyl group is not shown. The peak at 7.29 ppm originates from normal chloroform. The director arrow in the lower right corner points in the direction of a conventional 1D spectrum. The corresponding 1D spectrum is the Fourier transform of 128 coadded free-induction decays using 90° pulses.

dispersive components can also be avoided by means of a suitable modification of the experiments (*9*).

Utilizing absolute-value spectra, one has to keep in mind that the character and variance of the random noise are altered by computing absolute values. In the Appendix it is shown that the ratio of the variances of the random noise for absolute-value and phase-sensitive spectra is given by

$$\sigma^2_{\text{absolute-value}} / \sigma^2_{\text{phase-sensitive}} = 0.4292$$

assuming normally distributed random noise without correlation between the two quadrature-phase components.

Comparing the linewidths in the ω_1 direction in Fig. 3, it becomes apparent that the transverse relaxation times for the three protons in 2-furancarboxylic acid methyl ester are not equal. This also affects the sensitivity ratios for the three protons. The relaxation times have been measured with a conventional spin echo experiment. A comparison of theoretical and experimental sensitivity ratios is given in Table 2. Again, the agreement between theory and experiment is satisfactory.

TABLE 2

SENSITIVITY RATIOS FOR THE THREE RING PROTONS IN 2-FURANCARBOXYLIC ACID METHYL ESTER IN A 2D RESOLVED EXPERIMENT IN COMPARISON WITH A CORRESPONDING 1D EXPERIMENT USING THE SAME NUMBER OF EXPERIMENTS

Signal position (ppm)	T_2 (sec)	τ_1/T_2	$(S/N)_{2D}/(S/N)_{1D}$	
			Experimental	Theoretical
6.58	11.3	1.54	0.534	0.510
7.17	4.5	3.87	0.264	0.253
7.61	4.5	3.87	0.274	0.253

The above comparison of 1D and 2D sensitivity has been made for the case without matched filtering. Table 1 shows that using a matched filter in both cases increases the relative sensitivity of 2D spectroscopy by up to 50% and more, depending on the $\tau_1/T_{2ab}^{(1)}$.

In the comparison given, the same number of experiments has been performed for both 1D and 2D spectroscopy. In all cases 90° pulses were employed with long waiting periods between the various experiments to permit complete recovery of the magnetization. The same total performance time has, therefore, been used for the compared spectra.

In practical applications of 1D and 2D spectroscopy, however, it is usually desirable to minimize the performance time by using a shorter waiting time. It has been shown by Waugh (*10*) for 1D experiments that by using 90° pulses and optimizing the repetition rate of experiments, it is possible to gain 90.2% of the maximum achievable sensitivity. This requires, then, a pulse separation $T_{1D} = 1.27 \cdot T_1$ assuming that $T_2^* \ll T_1$ and neglecting possible echo effects.

For 2D experiments, the situation is complicated by the requirement that in all experiments exactly the same initial conditions are used, despite the fact that the perturbation of the system varies from experiment to experiment. To create identical initial conditions, it is advisable to destroy completely the remaining magnetization after each experiment, for example, by the application of a (magnetic field gradient pulse–90° pulse–magnetic field gradient pulse) sequence. The optimum waiting time is again $\tau_W = 1.27 \cdot T_1$, but not including the experiment time in this case. The average experiment separation then becomes

$$T_{2D} = \tau_1/2 + \tau_2 + 1.27 \cdot T_1,$$

neglecting the short time necessary to destroy the magnetization. For $T_2^* \ll T_1$, one obtains $T_{2D} \approx T_{1D}$, but for $T_2^* \approx T_1$, T_{2D} may be somewhat longer than T_{1D}

depending on the resolution to be achieved. Then, the relative sensitivity per unit time of 2D spectroscopy will be further reduced.

It may be astonishing that the sensitivity of a 2D spectrum is not by orders of magnitude smaller than that of a 1D spectrum taken in the same total time although a large set of experiments for different t_1 values has to be performed. But after all, each of these experiments adds to the total information content in a manner similar to the way in which an additional 1D scan will improve the sensitivity of a multiscan average. According to the matched filtering theory, it is the total signal energy, irrespective of whether it originates from M equal or M different experiments, which determines the maximum S/N ratio.

The slight sensitivity loss of 2D resolved spectroscopy originates from the unfavorable sampling of the signal during the detection period after a partial signal decay during the preceeding evolution period. In 2D autocorrelated spectroscopy, it is in addition the factor $p \cdot 1/\sqrt{2}$ which will reduce sensitivity, but, in many circumstances, it will be outweighted by the gain of connectivity information which would otherwise have to be obtained by a multitude of time-consuming double-resonance experiments.

In a strict analogy to 1D spectroscopy, it is also possible in 2D spectroscopy to improve the sensitivity further by a factor $2^{1/2}$ by utilizing dual-phase detection. The quadruple-phase detection, which has recently been proposed for the separation of all four quadrants in a 2D autocorrelated spectrum (*11*), on the other hand, would not further improve sensitivity over that obtained with dual-phase detection.

ACKNOWLEDGMENT

This research has been supported by the Swiss National Science Foundation.

APPENDIX: VARIANCE OF THE RANDOM NOISE IN AN ABSOLUTE-VALUE SPECTRUM

The variance of the noise in an absolute-value spectrum shall be computed for Gaussian noise with the probability density function

$$f(N_i) = [1/\sigma(2\pi)^{1/2}] \exp\{-N_i^2/2\sigma^2\}, \qquad [A1]$$

with $\langle N_i \rangle = 0$. It will be assumed that the noise amplitudes of the absorption and dispersion mode spectra, N_1 and N_2, respectively, are uncorrelated. This is true when the original noise in the time domain, $n(t)$, was white.

The noise amplitude N in the absolute-value spectrum is given by

$$N = (N_1^2 + N_2^2)^{1/2}. \qquad [A2]$$

For the average noise amplitude $\langle N \rangle$, one then obtains

$$\langle N \rangle = \int_{-\infty}^{\infty} dN_1 \int_{-\infty}^{\infty} dN_2 (N_1^2 + N_2^2)^{1/2} f(N_1) f(N_2) \qquad [A3]$$

$$= \sigma(\pi/2)^{1/2}$$

and the average square $\langle N^2 \rangle$ is equal to

$$\langle N^2 \rangle = \langle N_1^2 \rangle + \langle N_2^2 \rangle = 2\sigma^2. \qquad [A4]$$

This gives, finally, for the variance of the absolute value

$$\sigma^2(N) = \langle N^2 \rangle - \langle N \rangle^2 = (2 - \pi/2)\sigma^2 = 0.4292\sigma^2. \qquad [A5]$$

REFERENCES

1. W. P. Aue, E. Bartholdi, and R. R. Ernst, *J. Chem. Phys.* **64,** 2229 (1976).
2. L. Müller, A. Kumar, and R. R. Ernst, *J. Chem. Phys.* **63,** 5490 (1975).
3. W. P. Aue, J. Karhan, and R. R. Ernst, *J. Chem. Phys.* **64,** 4226 (1976).
4. L. Müller, A. Kumar, and R. R. Ernst, *J. Magn. Resonance* **25,** 383 (1977).
5. G. Bodenhausen, R. Freeman, and D. L. Turner, *J. Chem. Phys.* **65,** 839 (1976); G. Bodenhausen, R. Freeman, R. Niedermeyer, and D. L. Turner, *J. Magn. Resonance* **24,** 291 (1976).
6. R. R. Ernst, *Advan. Magn. Resonance* **2,** 1 (1966).
7. R. K. Hester, J. L. Ackerman, B. L. Neff, and J. S. Waugh, *Phys. Rev. Lett.* **36,** 1081 (1976).
8. A. A. Maudsley and R. R. Ernst, *Chem. Phys. Lett.* **50,** 368 (1977).
9. P. Bachmann, W. P. Aue, L. Müller, and R. R. Ernst, *J. Magn. Resonance* **28,** 29 (1977).
10. J. S. Waugh, *J. Mol. Spectrosc.* **35,** 298 (1970).
11. R. R. Ernst, W. P. Aue, P. Bachmann, J. Karhan, A. Kumar, and L. Müller, *in* "Proceedings of the Third Ampere Summer School, Pula, Yugoslavia," 1977.

JOURNAL OF MAGNETIC RESONANCE **33**, 107–111 (1979)

The Measurement of Inhomogeneous Distributions of Paramagnetic Centers by Means of EPR

W. KARTHE AND E. WEHRSDORFER

Department of Physics, Friedrich Schiller University, 69 Jena, Sektion Physik Max-Wien-Platz 1, GDR

Received March 15, 1978

The measurement of inhomogeneous distributions of paramagnetic centers by means of the application of zeugmatography in EPR is discussed. An experiment was performed to test this method.

INTRODUCTION

The investigation of transport phenomena in solids (particularly those of diffusion), metabolic phenomena, or pathological changes in living tissues or the assessment of the quality of a doped crystal necessitates the study of the distribution of special centers in the sample under consideration. The measurements are required not to destroy the investigated subject; i.e., they should be nondestructive. This requirement is met by using X-ray and ultrasonic methods and by the application of radioactive isotopes. Also in this connection, the application of nuclear magnetic resonance is currently widely discussed (*1–5*).

Inhomogeneous distributions of paramagnetic centers have so far been measured by means of electron paramagnetic resonance, and a defined volume of the sample has been removed lamellarly so that the sample has not been available for further measurements. These shortcomings may be overcome by applying NMR methods (the so-called "zeugmatography") to similar methods of EPR. A preliminary experiment to show possible ways of achieving this is described.

METHOD

The power absorbed by a volume $d\tau$ of the sample is

$$p(\mathbf{r})d\tau \propto \chi''(\mathbf{r})H_1^2(\mathbf{r})\,d\tau, \qquad [1]$$

where $p(\mathbf{r})$ is the absorbed power density, $\chi''(\mathbf{r})$ is the imaginary part of the susceptibility, and $H_1(\mathbf{r})$ is the amplitude strength of the radio-frequency field. Since χ'' is proportional to the static susceptibility and thus also to the concentration of the paramagnetic centers $n(\mathbf{r})$, the number of N centers contributing to the intensity of the spectrum is

$$N = \int_{\text{sample}} n(\mathbf{r})\,d\tau. \qquad [2]$$

0022-2364/79/010107-05$02.00/0

When an EPR spectrometer with a conventional reflection cavity is employed, the resonant absorption of the sample is measured by the relative change of the quality factor Q_0:

$$\frac{\Delta Q}{Q_0} = \frac{\int_{\text{sample}} \chi''(\mathbf{r}) H_1^2(\mathbf{r})\, d\tau}{\int_{\text{cavity}} H_1^2(\mathbf{r})\, d\tau} Q_o. \quad [3]$$

Equation [3] can be transformed into

$$\Delta Q/Q_0 \propto \bar{\eta}\chi_0 Q, \quad [4]$$

where $\bar{\eta}$ is a generalized filling factor.

$$\bar{\eta} = \frac{\int_{\text{sample}} c(\mathbf{r}) H_1^2(\mathbf{r})\, d\tau}{\int_{\text{cavity}} H_1^2(\mathbf{r})\, d\tau} \quad [5]$$

and $c(\mathbf{r})$ is a normalized distribution function of the centers.

$$\int_{\text{sample}} c(\mathbf{r})\, d\tau = V_{\text{sample}}, \quad [6]$$

where V_{sample} is the sample volume.

In the case of a homogeneous distribution, Eq. [4] is reduced to

$$\Delta Q/Q_0 \propto \bar{\eta}\chi_0 Q_0, \quad [7]$$

where η is the filling factor defined in the usual way:

$$\eta = \frac{\int_{\text{sample}} H_1^2(\mathbf{r})\, d\tau}{\int_{\text{cavity}} H_1^2(\mathbf{r})\, d\tau}. \quad [8]$$

For an arbitrary distribution function $c(\mathbf{r})$ with a constant field H_1 about the sample, Eq. [4] reduces to

$$\frac{\Delta Q}{Q_0} \propto \chi_0 Q_0 \frac{V_{\text{sample}}}{V_{\text{eff}}}, \quad [9]$$

where V_{eff} is the effective resonator volume.

Only in the two limiting cases determined by Eqs. [7] and [9] can the number of unpaired spins in the sample be determined by using the well-known formula (*6*). This fact requires the choice of special experimental conditions. From the radiofrequency field all those regions in a sample for which the resonance condition is fulfilled simultaneously absorb energy:

$$\omega = \gamma_{\text{eff}} H_0, \quad [10]$$

where ω is the angular frequency of the radiofrequency field, H_0 is the static magnetic field strength, and γ_{eff} is an effective factor which is different for each resonance line of the substance. For example, by choosing a magnetic field with a constant gradient across the sample, one may record the distribution of the centers as a function of H. The nonstationary methods tested in NMR are not suitable for EPR because the relaxation times are often too short. Therefore we discuss only the stationary projection methods (*1*) and spin mapping (*2*). The resonance field strength H_0 is given by

$$H_0^2 = [(H + H_z(\mathbf{r}))^2 + H_y^2(\mathbf{r}) + H_x^2(\mathbf{r})], \qquad [11]$$

where H is the homogeneous variable field in the z-direction and $\mathbf{r}$ is the space vector.

The distribution of the centers in the sample is $n(\mathbf{r})$. Thus a distribution function $S(H)$ of the EPR absorption can be measured as

$$S(H)\,dH = \frac{1}{N}\int_{H}^{H+dH} n(\mathbf{r})\,d\tau$$

$$= \frac{1}{N}\int_{\mathbf{r}} [n(\mathbf{r})/|\mathrm{grad}_{\mathbf{r}'} H|]\,d\tau, \qquad [12]$$

where $d\tau$ is the volume element and N is the total number of spins in the sample; $\mathbf{r}'$ represents the area given by

$$H = -H_z(\mathbf{r}) + [H_0^2 - H_y^2(\mathbf{r}) - H_x^2(\mathbf{r})]^{1/2}. \qquad [13]$$

If the absorption spectrum exhibits the spectral distribution $g(H)$, then a spectrum $I(H)$ can be recorded in a spectrometer of common construction

$$I(H) = \int S(H')g(H - H')\,dH'. \qquad [14]$$

The spectral distribution $g(H)$ can be measured in a homogeneous magnetic field, but the measurement of $\mathbf{H}(\mathbf{r})$ and the retransformation from $S(H)$ to $n(\mathbf{r})$ is more difficult.

This procedure may prove useful if suitable gradients can be generated. For the projection–reconstruction method and for spin mapping as well, a constant gradient is the most favorable:

$$H = H_0 - az; \qquad [15]$$

and thus

$$S(H) = \frac{1}{a}\frac{1}{N}\bar{n}(z), \qquad [16]$$

where $\bar{n}(z)$ is the number of spins in the layer between z and $z + dz$. In the simple case of $g(H) = \delta(H)$ we obtain

$$I(H) = \frac{1}{a}\frac{1}{N}\bar{n}(z)$$

$$= \frac{1}{a}\frac{1}{N}\bar{n}\left(\frac{H_0 - H}{a}\right). \qquad [17]$$

Using the projection–reconstruction method one may record the distribution of centers layer by layer in many directions, and a one-, two-, or even three-dimensional image of the distribution can be obtained by means of the well-known image-reproduction technique. However, the storage and computing time requirements are considerable.

Spin mapping, however, is less expensive, but when one is attempting to gain the same sensitivity as can be gained from employing the projection–reconstruction method, an accumulation of the spectra is indispensable. An estimation of the resolution can be performed for the constant gradient method. Let two volume elements with signals just barely detectable separately have distances Δz. The center of either element produces a symmetrical signal of width ΔH, and therefore from Eq. [15] it follows that

$$\Delta z = \frac{1}{a}\frac{\Delta H}{K} \qquad [18]$$

with $K = 3^{1/2}$ for a Lorentzian shape and $K = (2 \ln 2)^{1/2}$ for a Gaussian one, respectively.

If $\Delta H = 1$ G and $a = 10^3$ G cm^{-1}, then $\Delta z = 10$ μm (Gaussian) or 6 μm (Lorentzian). This resolution is sufficient for many of the investigations mentioned above, but it is insufficient for evaporated, adsorbed, and implanted films or other inner interfaces.

The possibilities of improving the resolution are limited. With the projection–reconstruction method the state-of-the-art sensitivity of the spectrometer of about 10^{10} spins/G requires at least 10^{15} spins/cm^3 to be present if the linear resolution is 1 μm, the linewidth $\Delta H = 1$ G, and the sample dimensions are $5 \times 5 \times 5$ mm^3. Since, in general, broader spectra exist and a signal-to-noise ratio exceeding unity is required, a resolution of less than 1 μm can be obtained only for samples with higher concentrations. In such cases there occur additional and often undesirable interactions such as dipole–dipole or exchange interactions.

EXPERIMENTAL

To test the applicability of the above method to EPR, a simple experiment was performed. A spectrometer of conventional design (of ERS–XQ type designed by the Zentrum für wissenschaftlichen Gerätebau, Academy of Sciences, GDR) with an H_{102} rectangular X-band cavity was used. The field gradient was generated by two coils whose axes were parallel to H_{z0}. The current through the coils was pulsed because of the thermic effects inside the coils, and the EPR signal was recorded only

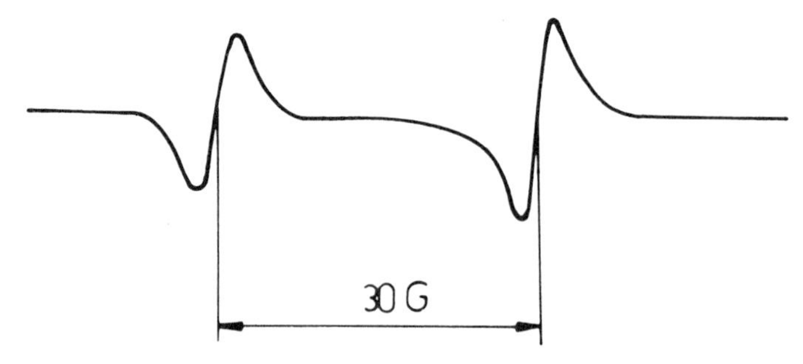

FIG. 1. EPR spectrum of two granules of DPPH at a distance of 200 μm.

during the pulses. Both coils were adjusted in such a way as to give rise to the maximum gradient at the sample and to neglect the other components.

As samples under investigation, two granules of DPPH at a distance of 200 μm in the z-direction were used. By sweeping H, the signals of each granule were recorded separately (Fig. 1), and from Eq. [16], $\bar{n}(z)$ resulted. Experimentally a field gradient of 1.5×10^3 G cm^{-1} could be obtained, and the highest resolution was $\Delta z = 12$ μm. From the line broadening the average diameter of the granules was estimated to be less than 10 μm, and from the intensity ratio of the lines it followed that the spin number of one of the granules exceeded that of the other by about 30%. Further experiments on samples with distributions of paramagnetic centers resulting from diffusion are in progress.

CONCLUSION

Zeugmatography can be employed successfully in EPR if it is possible to gain a resolution of about 1 μm and to generate constant field gradients about the sample. From the experimental point of view the use of a helix instead of a cavity may be more effective. Because of the smaller volume of the helix, higher gradients in each dimension can be obtained.

REFERENCES

1. P. C. Lauterbur, *Nature* **242,** 190 (1973); P. C. Lauterbur, C. S. Dulcey, C. M. Lai, M. A. Failer, W. V. House, D. M. Kramer, C. N. Chen, and R. Dias *in* "Proceedings 18th Ampere Congress, Nottingham, 1974," p. 27; P. C. Lauterbur, D. M. Kramer, W. V. House, and C. N. Chen, *J. Am. Chem. Soc.* **97,** 6866 (1975).
2. W. S. Hinshaw, *Phys. Lett. A* **48,** 87 (1974); W. S. Hinshaw, *in* "Proceedings, 18th Ampere Congress, Nottingham, 1974," p. 433; W. S. Hinshaw, *J. Appl. Phys.* **47,** 3709 (1976).
3. A. N. Garroway, P. K. Grannell, and P. Mansfield *J. Phys. C* **7,** L 457 (1974); P. Mansfield, P. K. Grannell, and A. A. Maudsley, *in* "Proceedings, 18th Ampere Congress, Nottingham, 1974," p. 431; P. Mansfield, A. A. Maudsley, and T. Baines, *J. Phys. E* **9,** 271 (1976).
4. A. Kumar, D. Weltlin, and R. R. Ernst, *J. Magn. Reson.* **18,** 69 (1975).
5. E. R. Andrew, 4th Ampere International Summer School, Pula Yugoslavia, 1976.
6. C. P. Poole, "Electron Spin Resonance" Wiley-Interscience, New York, 1967.

JOURNAL OF MAGNETIC RESONANCE **33**, 183–197 (1979)

Rotating Frame Zeugmatography

D. I. HOULT

Electrical and Electronic Engineering Section, Biomedical Engineering and Instrumentation Branch, Division of Research Services, National Institutes of Health, U.S. Department of Health, Education and Welfare, Public Health Service, Bethesda, Maryland 20014

Received May 15, 1978

A new method of obtaining NMR images is described which retains the inherent sensitivity of the two-dimensional Fourier transform while obviating the need for any changes of field gradient. The conditions applied in the laboratory frame, viz., a homogeneous field plus a field gradient, are also applied in the rotating frame so that the flip angle due to a pulse of length τ is dependent upon position x. The amplitude of the received signal is therefore dependent on τ and x, and these conjugate variables form a Fourier pair allowing resolution in x by variation of τ. Optimization of sensitivity is discussed as well as resolution requirements, and preliminary results are presented demonstrating the viability of the technique.

INTRODUCTION

The essence of the NMR imaging experiment as proposed by Lauterbur (*1*) is that the sample be subjected to a strong, static, and homogeneous magnetic field B_{00} upon which is superimposed a linear field gradient $B_{01}w$, where w is one of the three spatial coordinates x, y or z, and B_{01} is the constant of proportionality. The elementary plane of sample defined by the boundaries $w-\delta w/2$ and $w+\delta w/2$ may then be assigned a unique Larmor frequency ω_w which, of course, is a linear function of position w. The amplitude of the NMR signal received from the plane is a measure of the number of resonant nuclei within that plane, and so the NMR spectrum represents a graph of nuclear concentration versus distance. Now the technique briefly outlined above is one dimensional only, the independent variable being frequency or, in other words, position w, and to extend the method to two or three dimensions, other independent variables must be utilized. As a resonance phenomenon is under consideration, the two conjugate variables frequency and time are clearly indicated, and each, of course, may be partitioned into sections and each section treated as a framework for a pseudoindependent variable. As an example, consider the arrangement of Fig. 1, where the magnetic field is contoured in such a way that each element in the xy plane ($\delta x\,\delta y$) has a distinct Larmor frequency. Unfortunately, such an arrangement is practically impossible, but it does serve to show that one variable can do the work of two. Indeed, Mansfield's method of planar imaging bears some resemblance to this model (*2*). An excellent example of how independent time periods may be used as the necessary independent variables, in

0022-2364/79/010183-15$02.00/0

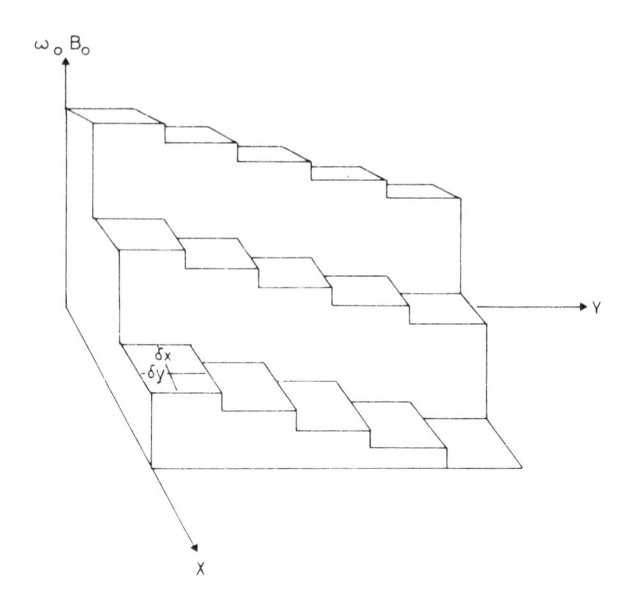

FIG. 1. A hypothetical magnetic field distribution which allows a single variable, in this case Larmor frequency, to be used to define a two-dimensional frame.

conjunction with Fourier transform techniques is contained in Ref. (*3*). This genre also includes Hinshaw's "sensitive point" method (*4*), in which a scanning technique is employed, and the "reconstruction method" (*1*), with which results are collected while a variety of gradients are successively applied. Each method requires that some parameter be changed in time and each has its advantages. The FONAR technique of Damadian *et al.*, although it does not employ the zeugmatographic principle (*5*), would also appear to fall in this category (*6*), as does the method described here.

Now sensitivity, speed, and safety are of great importance if zeugmatography is to be applied to humans, and thus the author has a preference for reconstruction or Fourier transform methods, as they would appear (at least theoretically) to give superior performance in terms of signal-to-noise ratio. However, rapid implementation of both these methods, using Carr–Purcell echoes if necessary, would seem to require rapid, repeated changes of field gradient which induce eddy currents in adjacent conductors *including the sample.* As these are essentially direct currents, Budinger (*7*) has questioned their safety, particularly when present at repetition rates of 10 per second or greater, and thus, some reservations must be made about the use of switched field gradients, quite apart from technical problems caused by eddy-current damping. Now, in general, what applies in the laboratory frame also applies in the rotating frame, and thus it ought to be possible to perform zeugmatography in that frame by subjecting the sample to a strong, homogeneous irradiating field B_{10} upon which is superimposed a linear field gradient $B_{11}\,v$, where v is one of the three laboratory frame coordinates x, y, or z. If desired, one may even carry this principle further to the second rotating frame by applying field modulation at frequency $\omega_1 = \gamma B_{10}$ (*8*), and we now show that this principle enables one to perform two- or three-dimensional zeugmatography without recourse to rapidly switched gradients and while retaining good sensitivity.

ROTATING FRAME ZEUGMATOGRAPHY

Let us assume that the conventional static gradient is applied in the z direction and that at the origin of the laboratory frame, the Larmor frequency is $\omega_{00} = -\gamma B_{00}$. Thus at plane z, the Larmor frequency is $\omega_0 = \omega_{00} - \gamma B_{01} z$. After phase-sensitive detection at frequency ω_{00} (or, if one prefers, in the frame rotating at frequency ω_{00}), the signal originating from the plane at height z has a frequency $\delta\omega_0 = \gamma B_{01} z$, and the signal may be accumulated and Fourier transformed in the usual way to obtain a measure of the nuclear magnetization present in the plane at z. To produce resolution in the second dimension, let us arrange the transmitter coils in a cross-coil probe so that there is a B_1 field gradient in the x direction. Then the frequency of nutation about B_1 in the rotating frame is given by $\omega_1 = -\gamma B_{10} - \gamma B_{11} x$ if $B_1 \gg B_{01} z$, i.e., B_0 effective. To produce resolution in the third dimension, a choice of techniques is open to us. Following Kumar *et al.* (*3*), we can introduce another B_1 gradient, this time in the y direction, and apply the two B_1 fields successively for various times t_x and t_y, or, as mentioned above we can extend the principle by applying field modulation at frequency $-\gamma B_{10} \simeq \omega_1$ with a gradient in the y direction. In addition, we can vary the phases of the various irradiations. The number of permutations and combinations available is clearly rather large so let us consider two possibilities only.

(a) Field Modulation

In the frame rotating at the Larmor frequency let B_1 be applied along the x' axis as usual, commencing at time zero. (Henceforth, primes donate the rotating frame.) The magnetization M_0 then nutates in the $y'z'$ plane with frequency $\omega_1 \simeq -\gamma B_{10}$. In the frame nutating with the magnetization, the effective field in the x' direction is $B_{11}x$, being dependent on position. Field modulation at frequency $\omega_1 = \gamma B_{10}$ gives an alternating field in the z' direction which may be decomposed in the usual way into two counterrotating components, one of which, B_2, moves with the magnetization. This is shown in Fig. 2a. Provided $B_1 \gg B_2$ and $B_2 \gg |B_{11} x|$, the Bloch equations apply normally, and depending on the phase of B_2, we can obtain a spin-locked, or a spin-dispersive, condition in the second rotating frame (*8*). Let us apply B_2 for an integral number of cycles in the dispersive mode; i.e., B_2 is at 90° to the magnetisation. This is shown in Fig. 2b. The magnetization now precesses about B_2 with a frequency $\omega = -\gamma B_2$ in the plane perpendicular to B_2. After an integral number of cycles, as B_2 commences in the y' direction, that plane is the $x'y'$ plane. So, after a time t_y, the magnetization has been flipped an angle $\gamma B_2 t_y$ about y'. This is shown in Fig. 2d.

(b) A Second B_1 Gradient

The same effect can be achieved at the expense of greater probe complexity by applying a B_1 field in the y' direction in the rotating frame, and this is probably

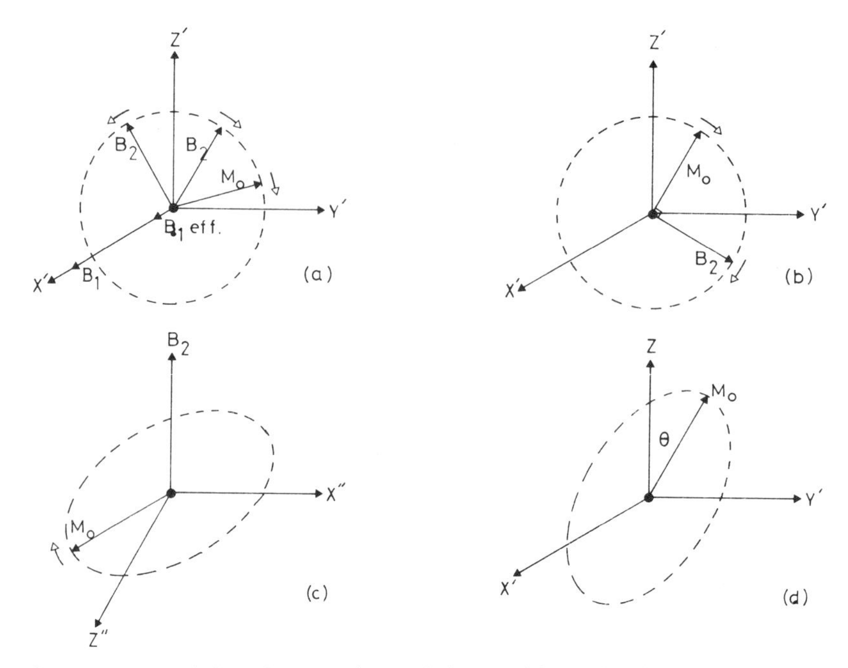

FIG. 2. In a manner entirely analogous to that used when considering the effects of a transverse B_1 field, field modulation in the presence of B_1 irradiation can be considered in a second rotating frame. The modulating field can be considered as two counterrotating components (a) and the phase may be set so that B_2 is orthogonal to the magnetization (b). The magnetization then precesses about B_2 (c) and is in the $x'z'$ plane whenever an integral number of cycles of modulation have been applied (d).

preferable as ω_1 is typically 50 kHz or less, and at lower frequencies the danger of inducing damaging currents in the body increases. In any event, whichever method is used, the applied field includes a gradient in the y direction so that the flip angle is dependent on the y position. We now continue by applying a B_1 field in the y' direction with the aid of the x gradient coils for a time t_x. At the end of the two preparatory periods, the flip angle is given by

$$\Theta = -(\gamma B_{10}(t_x + t_y) + \gamma B_{11}(xt_x + yt_y)), \qquad [1]$$

where we have assumed that the field profile in the y direction is the same as that in the x direction. The magnetization is still in the $x'z'$ plane. Application of a 90° pulse with B_1 in the x' direction now flips the plane of the magnetization into the $x'y'$ plane and the phase of the magnetization relative to the x' axis is $+\Theta$ +90°. The magnetization is allowed freely to evolve in the continuing presence of the main field gradient for time t_z, as shown in Fig. 3. The free-induction decay is therefore given, in the rotating frame, by

$$\xi \propto M_0(x, y, z) \exp[j\{\pi/2 + \gamma B_{10}(t_x - t_y) + \gamma B_{11}(xt_x + yt_y) - \gamma B_{01} z t_z\} - t_z/T_2]. \qquad [2]$$

Three sets of conjugate variables (xt_x), (yt_y), and (zt_z) are present in Eq. [2], and we have thus laid the foundation for the use of a three-dimensional Fourier transform to

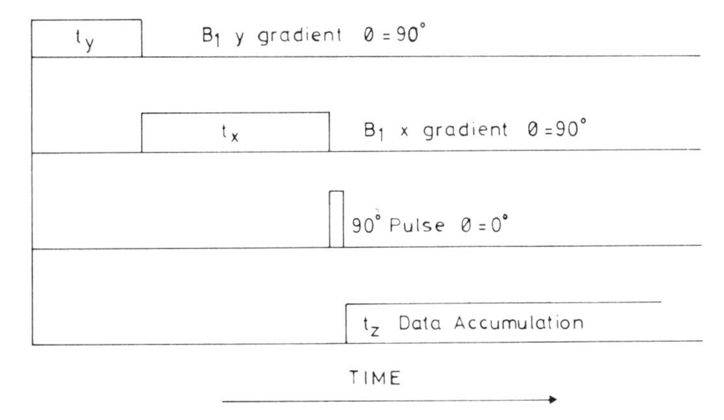

FIG. 3. A timing diagram showing the pulse sequence required to obtain three-dimensional information. The terms t_x, t_y, and t_z are the three independent variables and normally, $t_x, t_y \ll T_2$.

obtain the spin distribution function $M_0\ (x, y, z)$. The quantity T_2 is, of course, the spin–spin relaxation time, and $j = \sqrt{-1}$.

A PRACTICAL EXAMPLE

To demonstrate the feasibility of the method let us restrict the experiment to two dimensions, as excessive computer storage is not then required. We apply the main field gradient in the z direction and the irradiating field gradient in the x direction. The experimental arrangement is shown in Fig. 10. Following a transmitter pulse of length t_x applied in the rotating $+y'$ direction, the magnetization has been flipped an angle $-\gamma(B_{10}+B_{11}x)t_x$ toward the x' axis. The simplest experiment, which, it will be shown, involves a loss of $2^{1/2}$ in sensitivity, is then simply to collect the FID. This is given by

$$\xi \propto -M_0(x, z) \sin \gamma(B_{10}+B_{11}x)t_x \exp[-1/T_2 - j\gamma B_{01}z]t_z. \qquad [3]$$

The amplitude of the FID is dependent on the conjugate variables (xt_x), while the frequency is dependent on the variables (zt_z). We must now choose how to vary t_x. Two constraints apply. First, B_1 must be very much greater than B_0 effective if the two variables x and z are to be truly independent. Second, from the Nyquist sampling theorem, we must collect at least two data points per cycle if there is to be no aliasing. Thus, if $t_x = \rho\tau$, where ρ is an integer, the flip angle $\gamma(B_{10}+B_{11}x)\tau$ should at no point in the sample exceed 180°. Typically, therefore, the flip angle might vary across the sample from 50 to 170°, depending, of course, on the transmitter power available. This is rather wasteful of computer storage, and, if desired, aliasing can be turned to advantage by arranging the transmitter coil so that the flip angle across the sample varies from 63 to 117° for a pulse of length $\tau/3$. Thus, for $t_x = \tau$, the basic flip angle extends from approximately 190 to 350°, which after aliasing becomes 170 to 10° (with a sign inversion), so utilizing fully the computer capabilities.

Whichever method is chosen, we collect FIDs for a set of t_x values, the number depending on the resolution required (see below). The first FID is Fourier

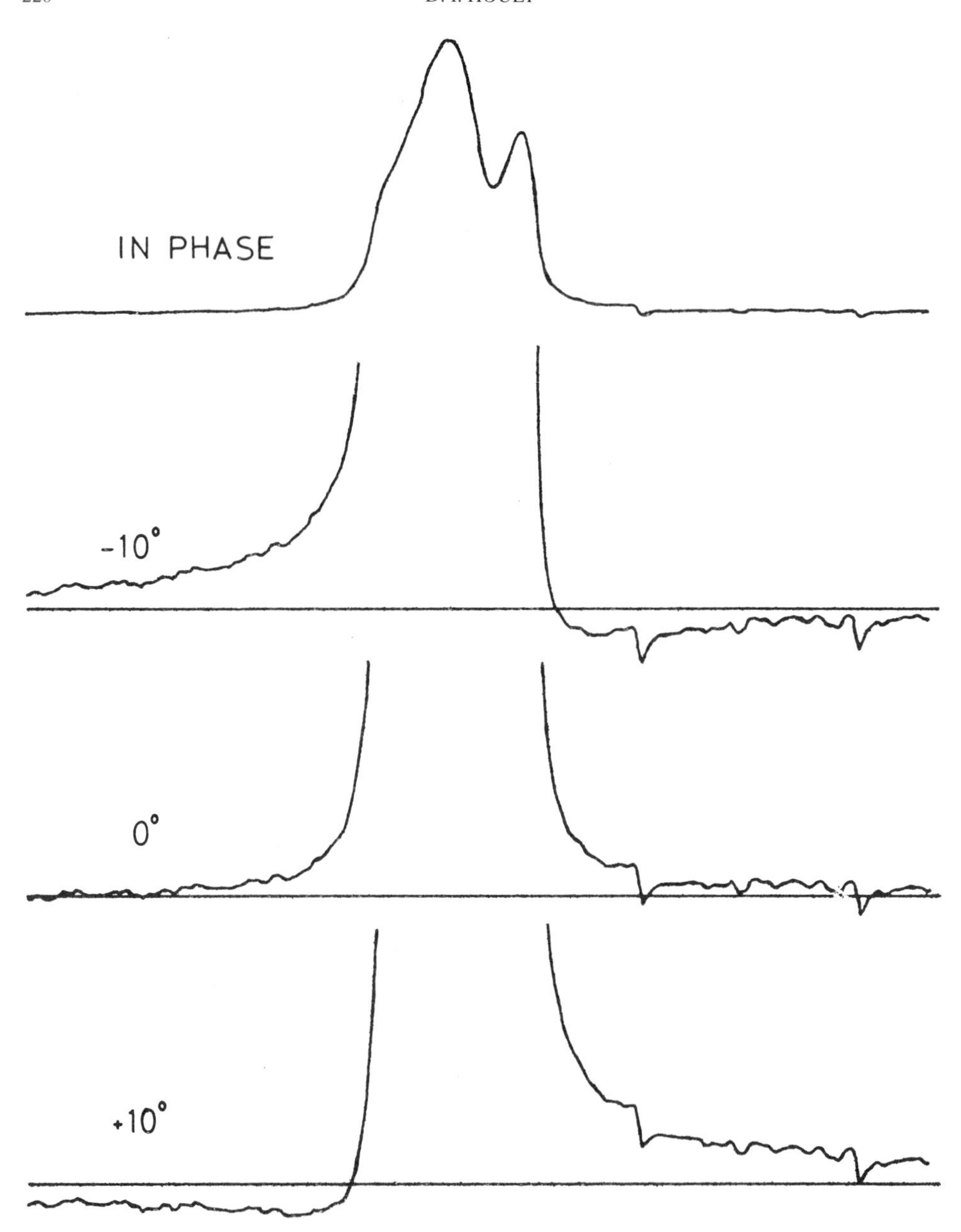

FIG. 4. The spectrum obtained by Fourier transformation of the first FID in the series used to generate the image of Fig. 6. Phasing this spectrum is perfectly straightforward, and the phase correction used is employed for all subsequent spectra.

transformed and the absorption mode spectrum saved. Phasing this spectrum is easy as it does not dip below the baseline (see Fig. 4), and a major advantage of the technique is that no phasing problems arise with the two-dimensional transform. Subsequent FID are similarly transformed *with the same phase correction* and saved,

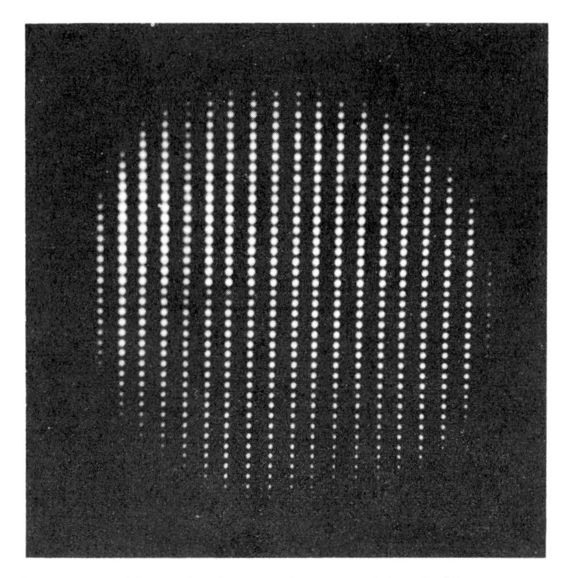

FIG. 5. A cross section, 5 mm thick, of a 6-mm tube containing 85% phosphoric acid. The data were accumulated in about 3 min. The resolution is about 0.4 mm.

so building up a matrix of absorption spectra where the row number ρ represents t_x and the column number the z position. The zeroth row is all zeros, representing a pulse length t_x of zero. After a baseline correction to remove possible phase glitch effects (see below), the sine Fourier transform of each *column* is taken (again, the phase is known; see Eq. [3]). The new matrix so created is the required image which

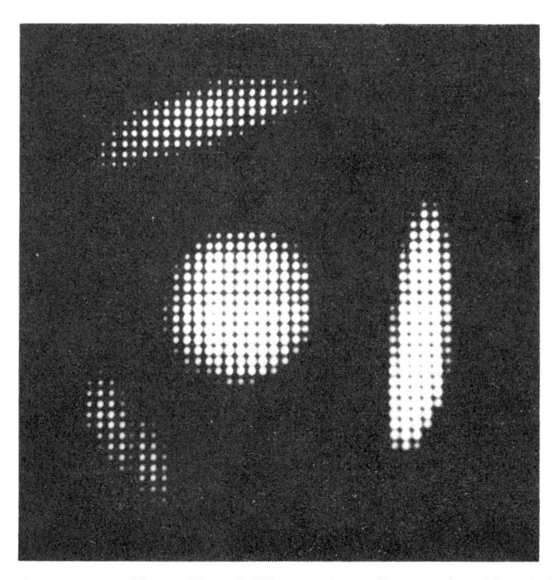

FIG. 6. A "phantom" was created by milling 0.75-mm-deep flats on the side of a Teflon cylinder which slid into a 6-mm tube. In addition a hole was drilled through the center of the phantom, and its length was restricted to 5 mm. The resolution is about 0.2 mm, and the data were accumulated in about 3 min.

may be displayed by scanning the matrix in synchrony with an oscilloscope raster and modulating the display intensity with the contents of the matrix. Two images so obtained are shown in Figs. 5 and 6.

SENSITIVITY

Consider a small element of sample $\delta x\ \delta z$ which is subjected to 90° pulses every $5T_1$ seconds, T_1 being the spin–lattice relaxation time. Let the signal-to-noise ratio of the absorption peak obtained by transformation of a single FID be Ψ. Then after ρ accumulations, the ratio is $(\rho^{1/2}\Psi)$. In the imaging experiment described above, the amplitude of the FID oscillates with increasing ρ. (As the flip angle can be anything, it is essential to wait $5T_1$ between pulses to allow complete recovery of the magnetization, and as this experimental arrangement is not optimal, it can reduce sensitivity by up to 40%. For the moment, however, we must bear this loss.) When we take the sine Fourier transform of the on-resonance matrix column we obtain a peak whose height, because of the sinusoidal modulation of Eq. [3], is $\rho/2$ times that obtained from a single 90° pulse. The noise, during the Fourier transform process, is multiplied by a sine wave and then summed, and the root mean square resultant is therefore $(\rho/2)^{1/2}$ times the noise in a single spectrum. It follows that the final signal-to-noise ratio is $\Psi(\rho/2)^{1/2}$. As such an argument applies to every element in the sample, we conclude that the two-dimensional transform, apart from factors of $2^{1/2}$ and up to 40% loss, gathers information simultaneously from all points in the sample and therefore exhibits excellent sensitivity characteristics.

The factor of $2^{1/2}$ arises because after the pulse of length t_x, we ignore the component of magnetization in the z' direction; Eq. [3] contains only the projection of the magnetization on the x' axis. (This situation may be compared with the difference between single phase and quadrature detection.) If, however, after time t_x, we apply a 90° pulse with B_1 along the x' axis, we flip the z component into the y' direction. This restores the full magnetization to the $x'y'$ plane, and Eq. [3] thus becomes

$$\xi_1 \propto M_0(x, z) \exp[+j\gamma(B_{10}+B_{11}x)t_x + j\pi/2 - j\gamma B_{01}zt_z - t_z/T_2]. \quad [4]$$

The pulse of length t_x now effectively modulates the *phase* of the FID rather than the *amplitude*, and unfortunately the price paid for restoring the sensitivity is that the phase problem, characteristic of two-dimensional transforms, returns. The two-dimensional transform of Eq. [4] mixes the real and imaginary parts of the two transforms together, in the process deforming the image. Thus, from Eq. [4],

$$G_1 = \int_{t_x=0}^{\rho\tau} \int_{t_z=0}^{\infty} \xi_1(t_x, t_y) \exp j(-\omega_x t_x - \omega_z t_z)\, dt_x\, dt_z$$

$$= \left[\frac{\exp(j\,\Delta\omega_x\rho\tau) - 1}{j\,\Delta\omega_x}\right] \left[\frac{T_2}{1 - j\,\Delta\omega_z T_2}\right] M_0(x, z), \quad [5]$$

where $\Delta\omega_x = -\omega_x + \gamma(B_{10}+B_{11}x)$; $\Delta\omega_z = -\omega_z - \gamma B_{01}z$. The real part of G_1 contains, in addition to absorption products, dispersion products also. Two-dimensional transformation of Eq. [3], on the other hand, yields

$$\left[\frac{\sin \Delta\omega_x \rho\tau}{2\,\Delta\omega_x}\right]\left[\frac{T_2}{1-j\,\Delta\omega_z T_2}\right] M_0(x, z), \qquad [6]$$

enabling us to select the absorption mode. (The noise associated with Eq. [6] is $2^{1/2}$ less than that associated with Eq. [5].) To separate the two sections of Eq. [5] we repeat the experiment, but at time t_x we change the phase of the 90° pulse by 180°. The z component of magnetization is now flipped into the $-y'$ direction, giving, in place of Eq. [4],

$$\xi_2 \propto M_0(x, z) \exp[-j\gamma(B_{10}+B_{11}x)t_x - j\pi/2 - j\gamma B_{01} z t_z - t_z/T_2]. \qquad [7]$$

Taking the x conjugate transform (note: *plus* $\omega_x t$),

$$G_2 = \int_{t_x=0}^{\rho\tau}\int_{t_z=0}^{\infty} \xi_2(t_x,t_y) \exp j(+\omega_x t_x - \omega_z t_z)\, dt_x\, dt_z,$$

we obtain

$$G_2 = +\left[\frac{\exp(-j\,\Delta\omega_x \rho\tau)-1}{j\,\Delta\omega_x}\right]\left[\frac{T_2}{1-j\,\Delta\omega\, T_2}\right] M_0(x, z), \qquad [8]$$

and from Eq. [5] and [8],

$$\frac{G_1 - G_2}{2} = \left[\frac{\sin \Delta\omega_x \rho\tau}{\Delta\omega_x}\right]\left[\frac{T_2}{1-j\,\Delta\omega_2\, T_2}\right] M_0(x, z). \qquad [9]$$

Comparison with Eq. [6], allowing for the difference in the noise, shows the $2^{1/2}$ improvement. There is, however, a more subtle improvement. Throughout the course of the experiment, spin–lattice relaxation now takes place, at all points in the sample, only from the $x'y'$ plane. Thus the repetition rate can be increased to the optimal for a 90° pulse (neglecting the first FID) which is once every 1.25 T_1 (*9*), resulting in, at most, a 10% loss in sensitivity. Variation of the repetition rate from experiment to experiment allows, as usual, an estimate of T_1 for each part of the image to be made.

RESOLUTION

Optical analogies are of considerable use in the discussion of resolution. As an example, suppose we wish to resolve two small spheres of sample 1 mm apart in a total sample extending 10 cm in the z direction. The best sensitivity is obtained when each FID is filtered with an exponential of time constant T_2 and integrated during transformation to a limit of $t_z = \infty$; in practice, the limit is $t_z \geq 2.5\ T_2$. The width of the absorption peak in the z direction is then $2/\pi T_2$ Hz, and according to Rayleigh's criterion (*10a*), the two spheres must be separated by at least this frequency if they are to be resolved. Thus the z spectral width must be at least $200/\pi\ T_2$ Hz to cover the entire sample. By Nyquist's theorem, and assuming quadrature detection, we must accumulate one complex data point every $\pi T_2/200$ seconds and so in 2.5 T_2 seconds, 320 data points (160 real, 160 imaginary) have had to be stored in the computer. Most Fourier transform routines employ the Cooley–Tukey algorithm (*11*), and unfortunately this algorithm reduces the available resolution slightly. After transformation, the absorption spectrum contains 160 points covering $200/\pi T_2$ Hz. Obviously, one point every $0.398/T_2$ Hz is insufficient when lines of separation $2/\pi T_2$ Hz($0.637/T_2$ Hz) are present—the lines are not resolved. If the algorithm is to be used, the information present in the dispersion mode spectrum must be transferred to the absorption mode either by zero filling or by accumulating for a minimum of 3.2 T_2. The latter results in 200 points in the absorption spectrum. An example of the improvement in resolution obtained by zero filling is shown in Fig. 7. It is commonly believed that for a resolution of one part in 100, only 100 data points are needed. A little thought shows that in fact one would expect 200 to be needed. Just as, according to Nyquist, at least two points per cycle are needed to define an oscillation, so at least two points per peak are required to define a set of barely resolved spectral lines. To summarize, then, for a resolution of one part in N in the z direction, the spectral width should be at least $2N/\pi T_2$ Hz (with the gradient B_{01} appropriately adjusted to fill that width), and accumulation should last at least 2.5 T_2 with zero filling or 3.2 T_2 if filling is not employed.

Turning now to resolution in the x direction, slightly different conditions pertain. In general, the longest pulse used will normally be much less than T_2, and so little or no decay is associated with the x data. Upon Fourier transformation, the data are therefore truncated, or in other words, diffraction effects are observed (see Fig. 8). At the expense of resolution and sensitivity, the diffraction pattern can be "smoothed out" by applying a filter prior to transformation, just as in the z case. Reference (*12*) contains details. Assuming that filtering is not applied, it has already been shown how, using a nominal 270° pulse as the basis for the experiment, the full x spectral width can be utilized. It remains therefore to determine the number of pulses ρ required to resolve the two small spheres. Because of the oscillations in a sinc function, Rayleigh's criterion is modified slightly, and the usual criterion is that two sinc functions $\sin \alpha/\alpha$ must be separated by at least $\alpha = 270°$ to be resolved. (This corresponds to Abbe's criterion in the theory of optics (*10b*).) Thus, by the ρth pulse of length $\rho\tau$, at least a 270° phase difference must exist between the signals from the two spheres. Thus $\gamma B_{11}\ \delta x \rho\tau \geq 3\pi/2$, where $\delta x = 1$ mm. Now we know that for the full sample width of 10 cm, the spread of the flip angle across the sample is $< 180°$; i.e., $\gamma B_{11}\, x\, \tau < \pi$, where $x = 10$ cms. Thus $\rho > 1.5\ x/\delta x$ or, in this case, 150 pulses. The 150 pulses refer, of course, only to data accumulation, and in view of what has been said about the number of data points required for good resolution in the *display*, some zero-filling upon transformation is essential, particularly as a symmetrical display gives an impression of equal resolution in the two directions. Fewer points are needed in the x accumulation for a given resolution because the data are not decaying. To summarize, if a resolution of one part in N is required in the x direction, at least 1.5 N pulses must be applied. The number of points in the display should be at least 2 N in the x direction, this number being obtained by zero filling. Alternatively, if time spent on the experiment is not important, extra pulses may be used, and a filter applied.

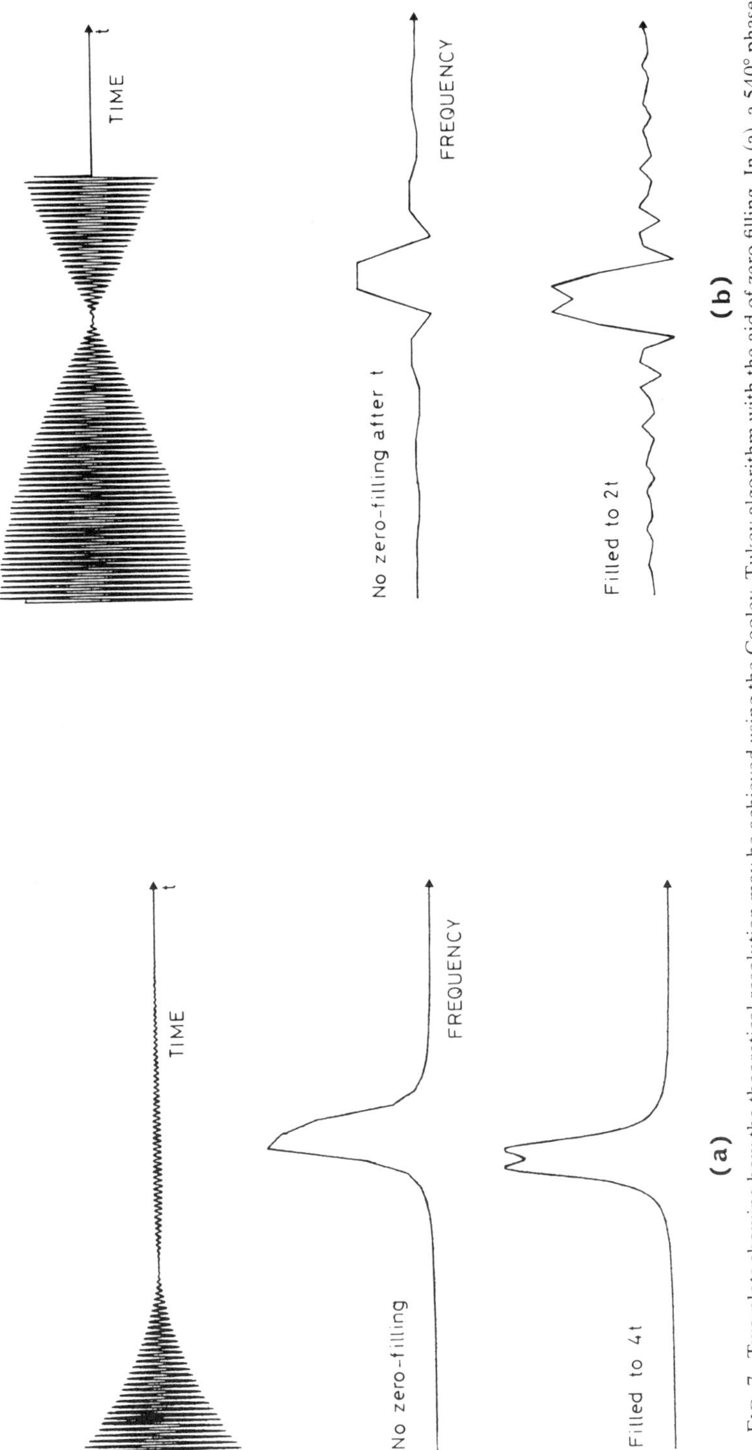

FIG. 7. Two plots showing how the theoretical resolution may be achieved using the Cooley–Tukey algorithm with the aid of zero filling. In (a), a 540° phase difference has developed between the two spectral lines by the end of the accumulation. Fourier transformation using the algorithm does not resolve the lines, but increasing the length of the decay with zeros does. The second spectrum was obtained after the length had been quadrupled. In (b), in the absence of a decay, only a 270° phase difference need develop between two lines for them to be resolved. A small amount of zero filling does not resolve the lines, and to obtain the second, resolved spectrum, the transient had to be increased in length by a factor of 2.67.

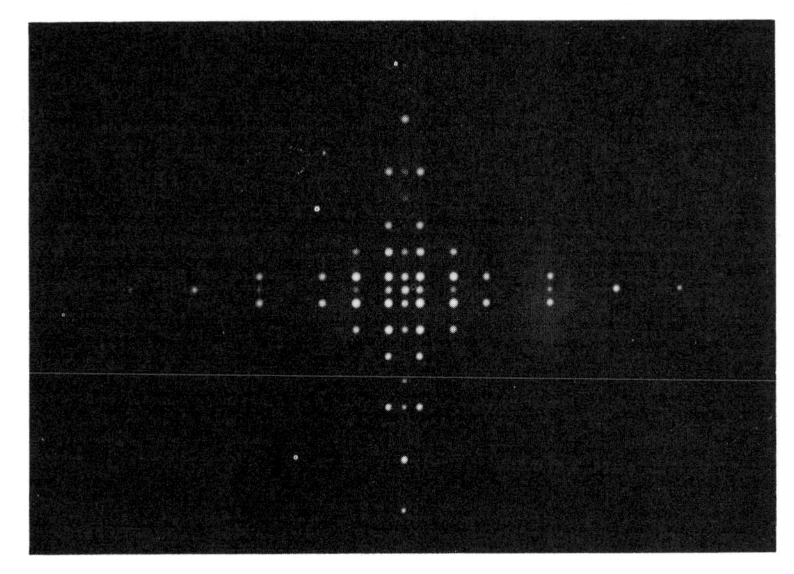

FIG. 8. When the transient is restricted in both time domains and negligible decay occurs, two-dimensional transformation generates a characteristic diffraction pattern, the like of which may be found in any standard optics text (e.g., Ref. (*10*), p. 396). Filtering with, for example, exponentials in the time domains, attenuates the patterns but does not remove the characteristic star shape (*12*).

THE THIRD DIMENSION AND OTHER TOPICS

It would be advantageous to be able to select a plane in a three-dimensional sample and then perform the imaging experiments described above. This can, of course, be accomplished with the aid of switched field gradients (*13*), but an alternative method, which avoids the dangers of switching, is to saturate all but the plane of interest by shaping the B_1 field correctly, a concept due to Damadian *et al.*

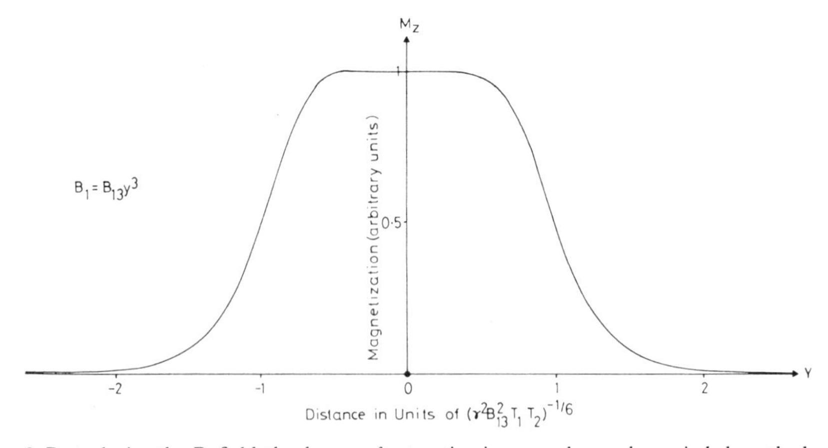

FIG. 9. By tayloring the B_1 field, the degree of saturation in a sample may be varied along the length of the sample. The figure is a plot of magnetization versus length after saturation with a field having a cubic gradient.

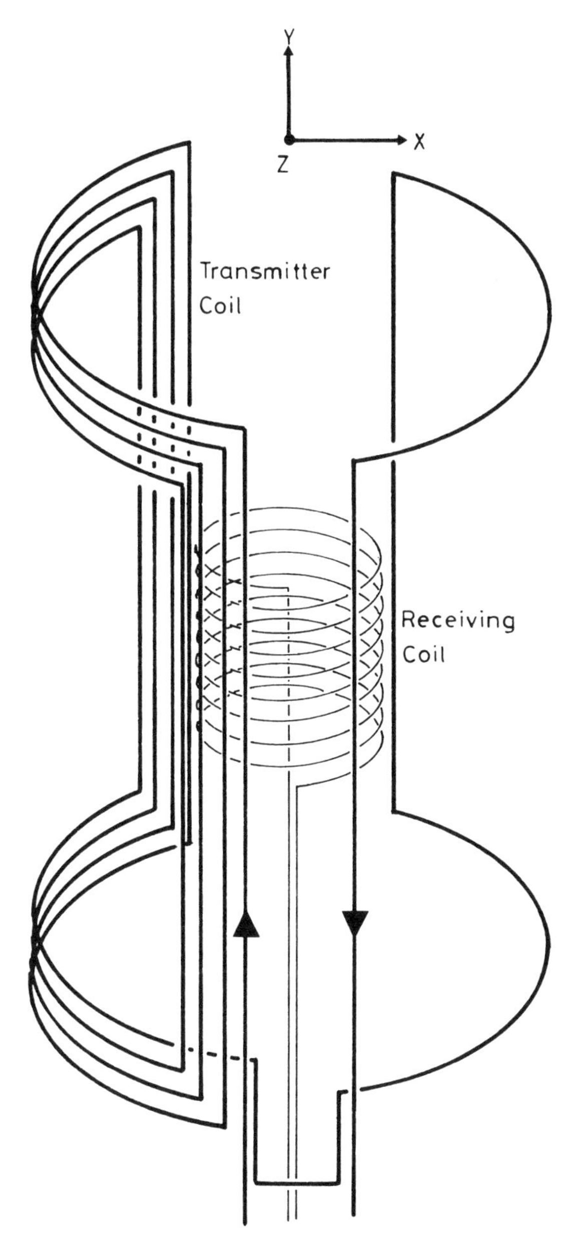

FIG. 10. The probe configuration employed in the experiment. The coils are electrically orthogonal.

(*6*). Suppose, for example, we apply a saturating field which varies as $B_{11}y$. After saturation, the remaining z' component of magnetization is given, from the Bloch equations (*14*), by

$$M'_z = \frac{M_0}{1+\gamma^2 B_{11}^2 y^2 T_1 T_2}, \qquad [10]$$

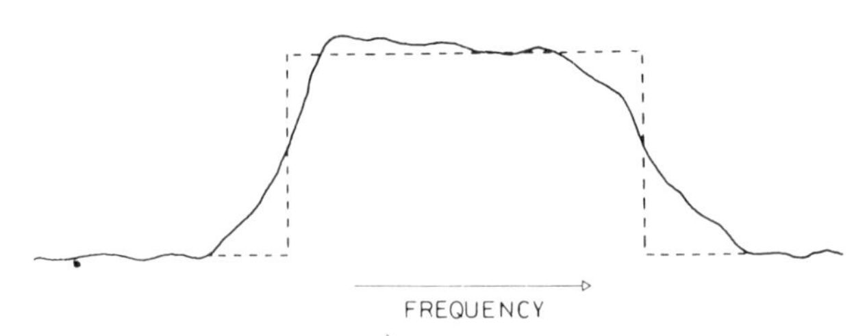

FIG. 11. A line through the image of Fig. 5 when the sample length is not very much less than the transmitter coil length. The unwanted B_1 inhomogeneity distorts the image.

which is Lorentzian in y. A variation of field as $B_{13}\,y^3$ gives a sixth-order response, and many other responses are possible, including the possibility of leaving a column or sphere of sample unsaturated. Figure 9 is a theoretical plot of M'_z when the B_1 saturating field, which should be broadband, varies in a cubic manner.

A point mentioned briefly earlier was the effect of phase glitch (*15*) upon the results. During the pulse of length t_x, the phase glitch at the start of the pulse causes a small amount of magnetization to become spin locked about B_1. This magnetization therefore persists through time t_x and is manifest as a small FID at the end of t_x. After the first Fourier transformation, this spurious signal introduces a baseline error into the data which it is preferable to correct prior to the second transformation. If the sensitivity enhancement technique employing $\pm 90°$ pulses is used, this artifact is largely eliminated. Another artifact may be generated if the transmitter coils are not long in comparison with the sample as shown in Fig. 10. If this is the case, signals from the top of the plane of interest have a different phase from those at the bottom, resulting in partial cancellation of the signals, the cancellation being more pronounced in those regions where B_1 is small. This is shown in Fig. 11.

Finally, it would appear that the technique described is very amenable to acceleration, for one may envisage experiments employing Carr–Purcell echoes where all the data are collected during a single FID. The technical problems involved, particularly in the realm of receiver recovery time, are immense, but in principle, a pulse train of the form $[(n\tau)_{90}-90_0-t_z-180°_{180}-\bar{t}_z-90_0]_{n=0}^{\rho}$ provides all the necessary information. The echo, accumulated during time $\bar{t}_z$, contains, in a time-reversed form, the x conjugate of Eq. [4], i.e., Eq. [5].

CONCLUSION

The report given above is to some extent preliminary, as a considerable amount of specialized hardware and software is required to implement fully the techniques described. Nevertheless, the results obtained show clearly that the method of rotating-frame zeugmatography is viable. The images were obtained with a Nicolet TT14 spectrometer using 85% phosphoric acid ($^{31}P \simeq 24$ MHz) and a modified Nicolet 20-mm probe. The transmitter was an ENI 350L broadband amplifier. The NTCFT program for use with the Nicolet NIC80 computer was modified to collect and store on disk automatically up to 100 transformed spectra, incrementing the pulse length by τ each time. A program was then written to transpose the matrix of results, baseline-correct and Fourier-transform the new rows of data, and display them as a stacked plot, a "whitewashed" plot (small peaks at the back obscured by large peaks at the front), or an intensity-modulated image. Unfortunately, the spectrometer precluded the use of protons as the nucleus of interest.

ACKNOWLEDGMENTS

The author is very grateful to Dr. D. Torchia for the use of his spectrometer and, in particular, for providing a probe which could be modified. He is also indebted to Dr. J. Cohen for the use of computing facilities and to Mr. R. Tschudin for discovering the source of the B_1 phase artifact.

REFERENCES

1. P. C. Lauterbur, *Nature* **242,** 190 (1973).
2. P. Mansfield and A. A. Maudsley, *J. Magn. Reson.* **27,** 101 (1977).
3. A. Kumar, I. Welti, and R. R. Ernst, *J. Magn. Reson.* **18,** 69 (1975).
4. W. S. Hinshaw and P. A. Bottomley, *Nature* **270,** 722 (1977).
5. R. Damadian, U.S. Patent No. 3,789,832, March 17, 1972.
6. R. Damadian, M. Goldsmith, and L. Minkoff, *Physiol. Chem. and Phys.* **9,** 97 (1977).
7. T. F. Budinger, Lawrence Berkeley Laboratory Report No. 5694, University of California, 1978.
8. D. I. Hoult, *J. Magn. Reson.* **21,** 337 (1976).
9. D. I. Hoult, D.Phil. Thesis, Oxford University, 1973.
10. M. Born and E. Wolf, "Principles of Optics," 4th ed., (a) p. 333, (b) p. 419, Pergamon Press, Oxford, 1970.
11. J. W. Cooley and J. W. Tukey, *Math. Comp.* **19,** 296 (1965).
12. G. Bodenhausen, R. Freeman, R. Niedermeyer, and D. L. Turner, *J. Magn. Reson.* **26,** 133 (1977).
13. P. Mansfield, A. A. Maudsley, and T. Baines, *J. Phys. E* **9,** 271 (1976).
14. A. Abragam, "The Principles of Nuclear Magnetism," p. 46, The University Press, Oxford, 1967.
15. J. D. Ellett, Jr., M. G. Gibby, U. Haeberlen, L. M. Huber, H. Mehring, A. Pines, and J. S. Waugh, *in* "Advances in Magnetic Resonance" (J. S. Waugh, Ed.), Vol. 5, p. 142, Academic Press, New York, 1971.

JOURNAL OF MAGNETIC RESONANCE **33**, 473–476 (1979)

NMR Population Inversion Using a Composite Pulse

Several NMR pulse experiments require the accurate adjustment of pulse flip angles to some special condition, often $\gamma H_1 \tau = \pi/2$ or π. Although experimental criteria can be devised to test for the correct adjustment of the pulse length τ, and although there is no difficulty in achieving the requisite fine control of this timing, in practice the spatial inhomogeneity of the radiofrequency field makes it impossible to fulfill the chosen condition over the whole sample. There is a related problem when the excitation frequency is set sufficiently far from resonance that the effective field in the rotating frame is appreciably stronger than H_1 and tilted with respect to the X axis. Chemical shift ranges in present-day magnetic fields are such that it is usually impossible to avoid these resonance-offset effects.

This communication proposes a composite pulse "sandwich" which has the effect of compensating variations in pulse length and errors introduced by the tilting of the effective field, provided that these deviations are not too large. For simplicity the discussion will be restricted to the case of a composite 180° pulse, although the principle is more general. An accurately set 180° pulse that is insensitive to resonance offset and radiofrequency inhomogeneity has obvious applications to the inversion-recovery method of measuring spin–lattice relaxation times.

In a reference frame rotating about the Z axis in synchronism with the transmitter frequency, H_1 is a static field directed along the X axis, and transverse magnetization aligned along the Y axis corresponds to an absorption mode signal. Instead of the conventional inversion pulse 180°(X), a composite group of three pulses is employed, 90°(X), 180°(Y), 90°(X), with the shortest practical delay between pulses, to minimize free precession. Clearly for exact pulse lengths and no off-resonance effects, the overall effect is that of 180°(X), and it is easy to visualize that for a small proportional mis-setting of all three pulses, the rotation about the Y axis compensates the errors in the first and last pulses. This is illustrated by the computer-simulated trajectories of Fig. 1 which represent a family of curves covering the range 80% to 90% of the nominal pulse flip angles. All these trajectories terminate close to the South pole and have very small errors in their Z-components of magnetization. When the missetting of the pulse lengths is such that the excursion during the 180°(Y) pulse is large, then the error in the length of this pulse has an appreciable effect, and the compensation becomes less effective.

Tilting of the effective radiofrequency field due to a resonance offset ΔH is also compensated by this group of three pulses, as illustrated in Fig. 2, which shows a family of trajectories for magnetization vectors with $\Delta H/H_1$ in the range 0.4 to 0.6. This type of display emphasizes the refocussing effect that is achieved, reminiscent of the phenomenon of spin echoes. All the vectors are left much nearer the South pole than they would have been after similarly tilted conventional 180°(X) pulses (the dotted trajectories). For the purposes of an inversion-recovery experiment the

0022-2364/79/020473–04$02.00/0

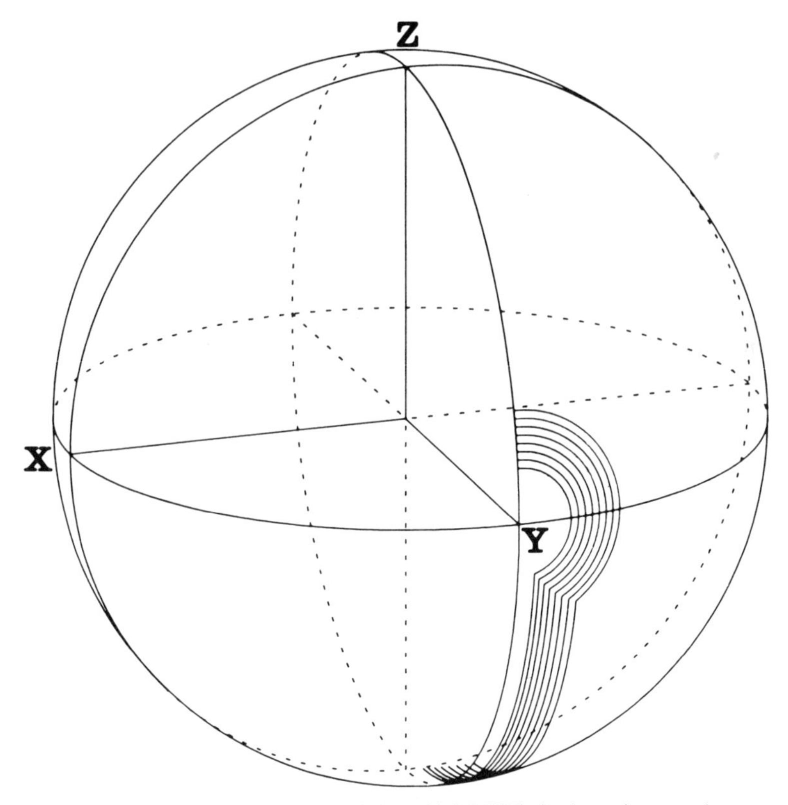

FIG. 1. The use of the composite pulse, 90°(X) 180°(Y) 90°(X), for inverting a nuclear magnetization vector, showing the effect of missetting the pulse lengths. A family of trajectories has been calculated for pulse lengths between 80% and 90% of the nominal values. Note that the ends of the trajectories are much closer to the South pole than corresponding trajectories after a single imperfect 180(X) pulse.

important parameter is the ratio of the Z component of magnetization after the composite 180° pulse, compared with that after an ideal 180° pulse ($-M_0$). It can be shown that this parameter follows a curve that is symmetrical for positive and negative offsets ΔH, although the corresponding trajectories are not symmetrically related.[1] Detailed evaluation of the combined effects of pulse length errors and resonance offsets is deferred to a later publication.

SPIN–LATTICE RELAXATION TIMES BY THE NULL METHOD

A good test of the proposed composite 180° pulse can be made by using it for population inversion in an experiment which evaluates the spin–lattice relaxation time from the null point in the recovery curve. The null may be located by linear interpolation from two determinations of signal intensity S_t which straddle the null condition, and this method has the practical advantage of speed, since no measure-

[1] This could be used to double the effective frequency range of the compensation by setting the transmitter frequency in the center of the spectrum and by using a quadrature detection scheme to distinguish positive and negative frequencies.

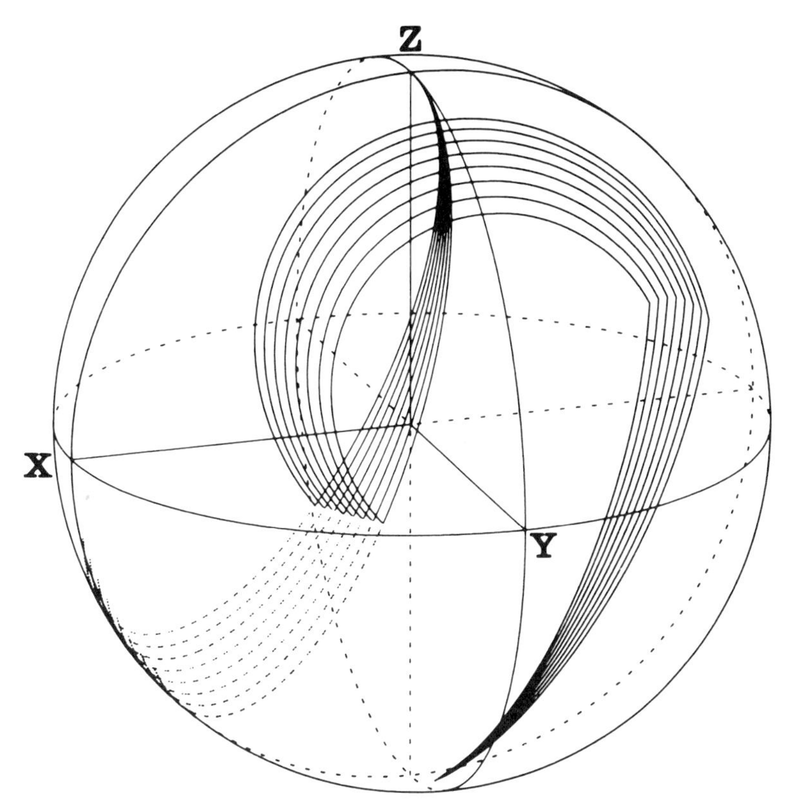

FIG. 2. The effect of the 90°(X) 180°(Y) 90°(X) composite pulse on a family of magnetization vectors at various offsets ΔH from resonance, covering the range where $\Delta H/H_1$ is between 0.4 and 0.6. The pulse lengths have been correctly adjusted for $\Delta H = 0$. Rotations occur about tilted effective fields in the XZ, YZ, and XZ planes, and the net effect is an imperfect refocussing of the magnetization vectors into a region quite near to the South pole. Conventional single inversion pulses for the same resonance offsets cause the vectors to continue along the dotted trajectories which terminate in the lower left rear octant, far from the South pole.

ment of the equilibrium signal S_∞ is required. However the null method is naturally very sensitive to the effects of incomplete inversion and thus provides a searching test of the degree of compensation achieved.

Spin–lattice relaxation measurements were carried out on a Varian CFT-20 spectrometer modified to permit 90° phase shifts of the radiofrequency pulses under computer control (*1*). The test resonance was the high-field carbon-13 line of butan-1,3-diol, for which conventional inversion–recovery measurements with least-squares fitting to an exponential curve had established the "true" relaxation time as 0.78 sec. This method is time-consuming, but is known to be insensitive to pulse errors (*2*), particularly when the phase of the monitoring 90° pulse is alternated (*3*). The composite 180° pulse was then used to invert the spin populations, and the null point bracketed by two determinations and located by linear interpolation. The resulting "apparent" spin–lattice relaxation times were studied as a function of the offset parameter $\Delta H/H_1$. In Table 1 the results are compared with those obtained with the same null-point method after conventional 180° pulses, and it is clear that a

TABLE 1
SPIN-LATTICE RELAXATION TIMES DETERMINED BY THE NULL METHOD

ΔH	$\Delta H/H_1$	Conventional 180° pulse	Composite 180° pulse
0 kHz	0.00	0.63 sec (−19%)[a]	0.76 sec (−3%)
1 kHz	0.09	0.59 sec (−24%)	0.75 sec (−4%)
2 kHz	0.18	0.49 sec (−37%)	0.72 sec (−8%)
3 kHz	0.27	0.42 sec (−46%)	0.72 sec (−8%)
4 kHz	0.36	0.27 sec (−66%)	0.69 sec (−12%)

[a] Systematic error with respect to the value $T_1 = 0.78$ sec measured by careful inversion-recovery expriments.

significant improvement is achieved for this range of offsets. The pulse lengths were carefully adjusted to the 90° and 180° conditions as measured for a line very close to the transmitter frequency. However the composite 180° pulse cannot compensate adequately for the very wide distribution of H_1 field strengths inherent in the single-coil system of this spectrometer, so a sample of restricted length was used (7 mm). This precaution should be unnecessary in a crossed-coil spectrometer.

Work is in progress on a more detailed analysis of composite pulses and on wider applications of this principle. An attractive method for the rapid determination of spin–lattice relaxation times in spectra with many lines has been introduced by Canet (*4*), using inversion-recovery with the waiting time between experiments deliberately set comparable with T_1. Pulse imperfections may then have cumulative effects, and a recent experiment and theoretical analysis (*5*) suggests that they introduce quite large systematic errors on the relaxation times obtained by this method. Self-compensating 180° and 90° composite pulses should help alleviate these problems.

ACKNOWLEDGMENTS

This work was made possible by an equipment grant from the Science Research Council, and was supported by a Research Studentship (M.H.L.).

REFERENCES

1. G. Bodenhausen, R. Freeman, G. A. Morris, R. Neidermeyer, and D. L. Turner, *J. Magn. Reson.* **25,** 559 (1977).
2. R. Freeman, H. D. W. Hill, and R. Kaptein, *J. Magn. Reson.* **7,** 82 (1972).
3. D. E. Demco, P. van Hecke, and J. S. Waugh, *J. Magn. Reson.* **16,** 467 (1974).
4. D. Canet, G. C. Levy, and I. R. Peat, *J. Magn. Reson.* **18,** 199 (1975).
5. H. Hanssum, W. Maurer, and H. Rüterjans, *J. Magn. Reson.* **31,** 231 (1978).

Physical Chemistry Laboratory
Oxford University
South Parks Road
Oxford, OX1 3QZ
England

Malcolm H. Levitt
Ray Freeman

Received November 21, 1978

JOURNAL OF MAGNETIC RESONANCE **38**, 453–479 (1980)

Radiofrequency Pulse Sequences Which Compensate Their Own Imperfections*

RAY FREEMAN, STEWART P. KEMPSELL, AND MALCOLM H. LEVITT

Physical Chemistry Laboratory, Oxford, England

Received September 13, 1979

Radiofrequency pulse sequences are described which have the same overall effect as a single 90 or 180° pulse but which compensate the undesirable effects of resonance offset and spatial inhomogeneity of the radiofrequency field H_1. These "composite" pulses are built up from a small number of conventional pulses which rotate the nuclear magnetization vectors about different axes in the rotating frame, while in the intervals between pulses a limited amount of free precession may be allowed to occur. Insight into the way in which pulse imperfections are compensated is obtained by computer simulation of trajectories of families of nuclear spin "isochromats" representing a distribution of H_1 intensity or resonance offset. Composite 90° pulses are suggested as a method of reducing systematic errors in spin–lattice relaxation times derived from progressive saturation or saturation–recovery experiments, and as the preparation pulse of a spin-locking experiment. A test of the effectiveness of the composite 180° pulse sequence has been made by using it for population inversion in a spin–lattice relaxation measurement where T_1 is derived from the null point in the recovery curve, a technique known to be very sensitive to pulse imperfections.

INTRODUCTION

Pulse excitation has now become an essential part of high-resolution NMR, both for routine operation and for more sophisticated experiments to study relaxation, chemical exchange, multiple-quantum transitions, and two-dimensional Fourier transform spectra. Pulse imperfections therefore merit a closer study, and this work suggests methods for compensating such imperfections by replacing each radiofrequency pulse by a composite pulse—a cluster of closely spaced pulses which achieves the same nominal flip angle, but which has a higher tolerance of the most common imperfections, spatial inhomogeneity of the radiofrequency field and resonance offset effects.

The intensity of the radiofrequency field H_1 varies in different geometrical regions of the sample, normally having the highest intensity near the center of the transmitter coils and falling off in regions remote from the center. This spatial inhomogeneity of H_1 introduces serious errors in the flip angle $\gamma H_1 t_p$ experienced by the nuclei in different volume elements. Attempts to circumvent the problem, by using small bulb samples rather than long cylindrical samples, suffer from poor sensitivity and often

* First presented at the 20th Experimental NMR Conference, Asilomar, Calif., February 1979.

0022-2364/80/060453-27$02.00/0

have difficulties with distortion of the static field H_0 by discontinuities in magnetic susceptibility.

The second type of pulse imperfection arises because H_1 has a finite intensity compared with typical offsets from resonance ΔH. Hence the magnetization vectors rotate about a resultant field H_{eff} which is tilted in the rotating frame. Technological improvements which increase H_1 tend to be matched by developments which increase the applied static field H_0 and hence the chemical shift range. Although quadrature phase detection effectively halves the maximum value of ΔH by allowing the transmitter frequency to be set in the center of the spectrum, resonance offset effects are still a serious problem.

It is proposed to compensate these two kinds of pulse imperfection by replacing the conventional 90 or 180° pulse by a corresponding composite pulse. This is defined here as a sequence of two or more radiofrequency pulses separated by intervals so short that no significant relaxation occurs, and any free precession between pulses is usually of the order of 1 radian or less. In certain sequences this precessional motion is essential to the process of compensation; in others free precession is simply minimized and neglected. It is important to note the assumption, made throughout these calculations, that the initial condition of the nuclear magnetization vector is longitudinal, often the equilibrium magnetization M_0 aligned along $+Z$. This excludes the case of 180° pulses used for refocusing isochromats in the XY plane, as in the Carr–Purcell spin-echo experiment (*1*).

The motion of the nuclear magnetization subjected to a composite pulse may be described by the Bloch equations neglecting relaxation. Figure 1 shows the rotating

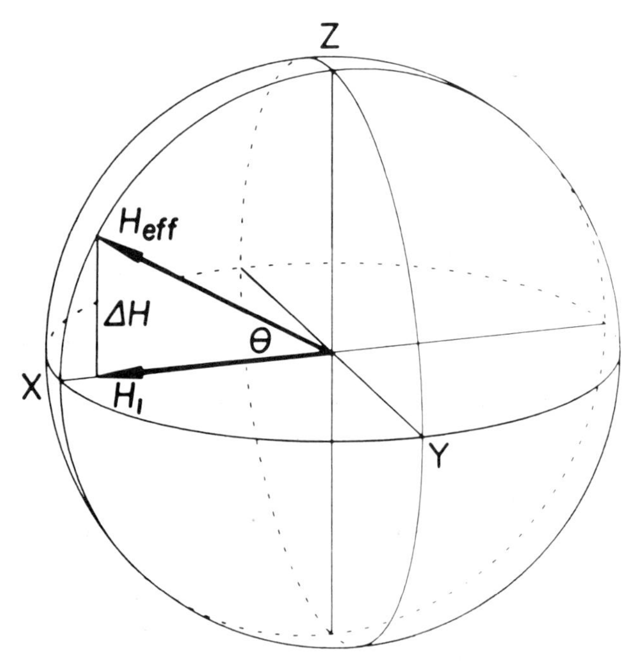

FIG. 1. The effective field operating on nuclear spin magnetization vectors, viewed in a reference frame rotating in synchronism with the transmitter frequency.

frame employed, with H_1 a static vector aligned along the X axis, and the effective field H_{eff} tilted with respect to H_1 through θ in the XZ plane.

$$H_{\text{eff}} = (H_1^2 + \Delta H^2)^{1/2}, \quad [1]$$

$$\tan\theta = \Delta H / H_1. \quad [2]$$

The mechanism by which pulse imperfections are compensated may be visualized by following a family of spin "isochromats" representing either a distribution of values of H_1 or a distribution of values of the resonance offset ΔH. These isochromats are represented by vectors which alternately rotate about H_{eff} during the pulses and precess about the offset field ΔH in the interval between pulses. Although the overall effect of the composite pulse sequence is the solution of a simple problem in spherical geometry, the analytical solution is often complex and uninformative. Therefore the treatment which follows makes extensive use of computer simulation, which readily handles sequences of several pulses and which allows greater insight into the principles involved.

For NMR lines sufficiently close to resonance that $H_1 \gg \Delta H$, a radiofrequency pulse of duration t_p and intensity H_1 nutates the nuclear magnetization through $\gamma H_1 t_p$ radians about the X axis of the rotating frame. In practical cases there is a distribution of values of H_1 in different regions of the sample. In order to be able to describe the various pulses, it is convenient to introduce a nominal value of the radiofrequency field, H_1°, that is representative of the actual intensities near the central region of the transmitter coils. (These regions tend to have the highest intensity H_1 and also the strongest coupling of the NMR signal back to the receiver coil.) A convenient definition of H_1° makes use of a common method of adjusting pulsewidth—searching for the condition $t_{2\pi}$ where the total NMR signal (a weighted average over all sample regions) goes through a null after one complete revolution,

$$\gamma H_1^\circ t_{2\pi} = 2\pi \text{ radians}. \quad [3]$$

(Typical magnetization vectors, of course, nutate through somewhat different angles because of the inhomogeneity of H_1.)

This definition makes it possible to refer to the nominal value of pulse flip angles; for example, a "180° pulse" has $\gamma H_1^\circ t_p = \pi$ radians. In what follows, the quotation marks will be dispensed with, and the terms 90° pulse and 180° pulse will be understood to refer to the nominal value H_1°. Since in practice it is difficult to probe the effects of spatial inhomogeneity of the radiofrequency field directly, some of the experiments to be described use a missetting of t_p to achieve the equivalent of a missetting of H_1.

When the resonance offset ΔH is too large to be neglected in comparison with H_1, then the nuclear magnetization vector nutates through an angle $\gamma H_{\text{eff}} t_p$ radians about an axis which is tilted through θ radians with respect to the XY plane of the rotating frame (Fig. 1). Equation [1] shows that this flip angle will always be greater than that in the on-resonance case. However, it is still convenient to retain the terms "90° pulse" and "180° pulse" on the understanding that they refer to the ideal pulse condition, and that offset and inhomogeneity are simply deviations from the ideal, and indeed are largely compensated by the proposed composite sequences.

In practice both kinds of imperfection operate together and must be treated simultaneously in certain applications. Strictly speaking, this would require a knowledge of the way in which static field inhomogeneities and radiofrequency field inhomogeneities correlate in space. The nature of this correlation is specific to each spectrometer, but fortunately its effect is very small when ΔH is large compared with the instrumental linewidth, and hence it is neglected in all that follows.

In examining the various proposals for composite pulse sequences it is important to clarify whether the compensation acts on radiofrequency inhomogeneity or resonance offset effects, or on both together. Not all the applications require the correction of both types of imperfection.

COMPOSITE 90° PULSES

Compensation for Spatial Inhomogeneity

A conventional 90° pulse, nutating the nuclear magnetization vector about the X axis of the rotating frame, would leave many "isochromats" with finite Z components after the pulse because of the inhomogeneity of the radiofrequency field. Since the pulse length is normally calibrated by observing the aggregate NMR signal, made up principally from sample regions inside the transmitter coil, vectors representing more remote sample regions will show positive Z components of magnetization at the end of the pulse. This can be a source of error in the measurement of spin–lattice relaxation times by the saturation–recovery (*2*) or progressive saturation (*3*) methods. It will be shown that a composite 90° pulse can compensate a significant fraction of these errors. For the purposes of this section, off-resonance effects are neglected.

The proposed sequence is 90°(X) 90°(Y), nutation through 90° about the X axis followed by nutation through 90° about the Y axis, with the interpulse interval kept as short as possible so that negligible free precession takes place. This notation refers to the nominal values of the pulse length, and any variation in flip angle due to spatial inhomogeneity is of course taken to affect all pulses in exactly the same way. Consider a small volume element of the sample which experiences only 80° pulses. There is a residual Z component after the first pulse of $+M_0 \cos 80°$, but a large proportion of this is removed by the 80° rotation about the Y axis. In effect the error in the Z component has been converted into a corresponding phase error with respect to the Y axis, an admixture of dispersion mode.

Errors arising from radiofrequency inhomogeneity are mainly of concern in spin–lattice relaxation experiments carried out by saturation–recovery or progressive saturation methods. The problem is to ensure that *for each volume element of the sample* M_z after the pulse is as close to zero as possible, even though H_1 is different in different sample regions; it is not sufficient to set the overall Z magnetization to zero. An experimental test of the compensation achieved by the composite 90° pulse was therefore made by monitoring the residual Z magnetization immediately after the composite sequence. This determination was made with the sequence

$$P_1\text{–}t\text{–}P_2\text{–Acq.} \qquad (t > T_2^*),$$

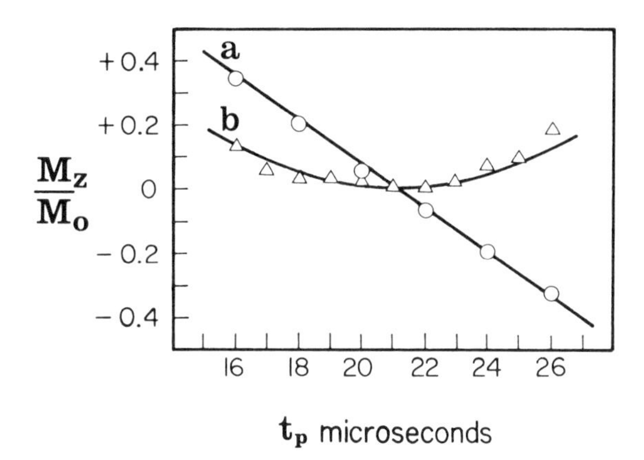

FIG. 2. The residual Z magnetization remaining after (a) a conventional 90°(X) pulse, (b) a composite 90°(X), 90°(Y) pulse. The quantity M_Z is expressed as a fraction of the total equilibrium magnetization M_0. In order to simulate the effect of spatial inhomogeneity in H_1 the pulsewidths t_p have been varied above and below the nominal value (~21 μsec). Note that M_Z after the composite sequence is less sensitive to variation of t_p. The full lines represent calculations based on magnetization trajectories.

where P_2 was in all cases a single "read" pulse with $\gamma H_1^\circ t_p = \pi/2$ radians and P_1 was either a conventional single pulse of flip angle α or a composite $\alpha(X)$, $\alpha(Y)$ sequence where α was varied by varying t_p in order to simulate the effect of spatial inhomogeneity in H_1. The observed signal intensity was plotted as a function of t and extrapolated back to $t = 0$. In order to minimize real radiofrequency inhomogeneity effects, a small spherical bulb sample was used.

When P_1 was a conventional single pulse, the residual Z magnetization showed a strong dependence on pulsewidth t_p, passing through zero near $t_p = 21$ μsec, the condition for $\gamma H_1^\circ t_p = \pi/2$ radians. When P_1 was a composite pulse the Z magnetization went through a shallow minimum as t_p was varied. The experimental points fit very well to the theoretical curves based on simulated magnetization vector trajectories (Fig. 2). It may be inferred from these results that errors in measurements of relaxation times caused by spatial inhomogeneity of the radiofrequency field H_1 should be considerably less serious when a composite sequence is used for saturation.

The compensating effect of the composite 90° pulse can be readily illustrated by computer simulation of trajectories in the rotating frame. In practice the spatial distribution of H_1 intensity is such that volume elements near the center of the transmitter coils have the highest intensity and H_1 falls off for more remote regions of the sample. It is therefore important to compensate for *low* values of H_1, whereas *high* values can normally be neglected. With this kind of distribution a slightly better compensation can be achieved by increasing the flip angle of the second pulse, for example, 90°(X), 110°(Y). Figure 3 illustrates the trajectories of a family of "isochromats" for the 90°(X), 110°(Y) sequence. Note that when the H_1 field intensity is in the range 0.8 to 1.0 H_1°, the trajectories terminate very close to the XY plane.

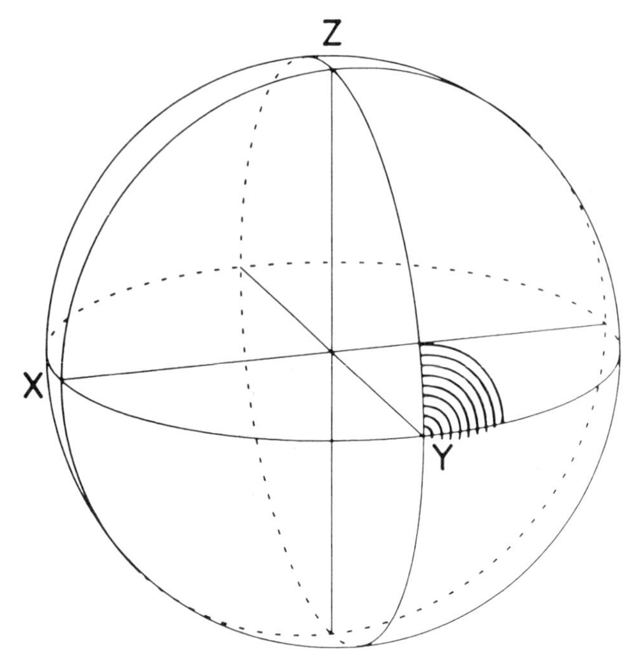

FIG. 3. Trajectories of a family of magnetization vectors during a 90°(X), 110°(Y) sequence representing pulse length missets between 0.8 and 1.0 times the nominal setting. Note the very small residual Z magnetization at the end of the sequence.

Compensation for Resonance Offset

It was demonstrated several years ago that a simple 90° pulse has a built-in compensation for resonance offset effects if judged only on the residual Z component of magnetization after the pulse (*3*). The reason is that the tilt of the effective field is largely compensated by the increase in H_{eff} as ΔH is increased. This can be illustrated by following a family of magnetization vectors representing a range of offsets ΔH and by plotting the loci of the tips of these vectors at various stages of their nutation about H_{eff} (Fig. 4). The curvature of the final locus is very small, so that it fits very close to the equator of the rotating reference frame. The phase errors are large and essentially a linear function of ΔH.

Should it be necessary to reduce these Z components even further, an improvement can be achieved with a composite two-pulse sequence. The two pulses act in opposite senses, and between them the nuclei are allowed to precess for a carefully chosen period τ, which introduces a curvature of the locus calculated to cancel the slight curvature of the final locus of Fig. 4. The time evolution is illustrated in Fig. 5. A 10°($-X$) pulse initiates the sequence and free precession for a period τ generates the locus $s - s'$. Then a 100°($+X$) pulse carries the family of vectors toward the XY plane, the final locus $t - t'$ lying very close to the equator. For a given offset ΔH, the amount of tilt of the effective field depends on the intensity of H_1, so the choice of τ for the best compensation depends on the value of H_1. For all of the pulse sequences described here, the radiofrequency field has been taken to be given by $\gamma H_1/2\pi =$ 10 kHz. With this condition good compensation is achieved with $\tau = 9$ μsec. The

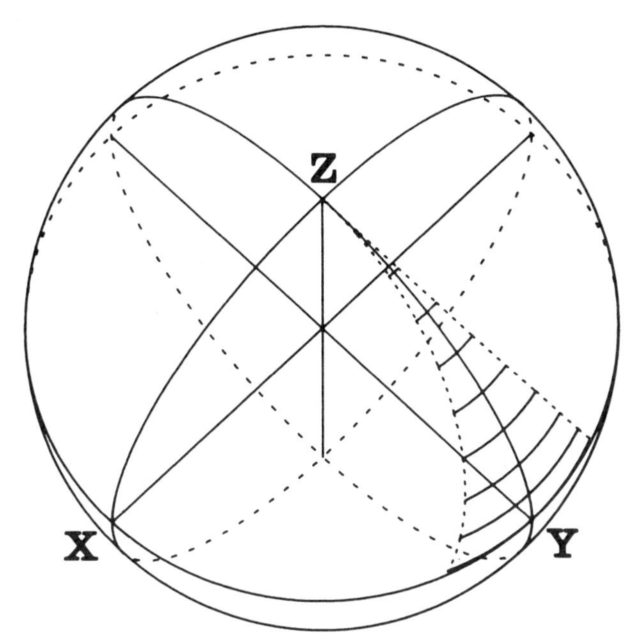

FIG. 4. The self-compensating effect of a conventional 90° pulse. A family of magnetization vectors representing resonance offsets $\Delta H/H_1$ from −0.33 to +0.33 has been considered, the locus of the tips of these vectors has been drawn at regular intervals during the pulse (full lines). Note that this locus coincides quite closely with the equatorial plane at the end of the pulse.

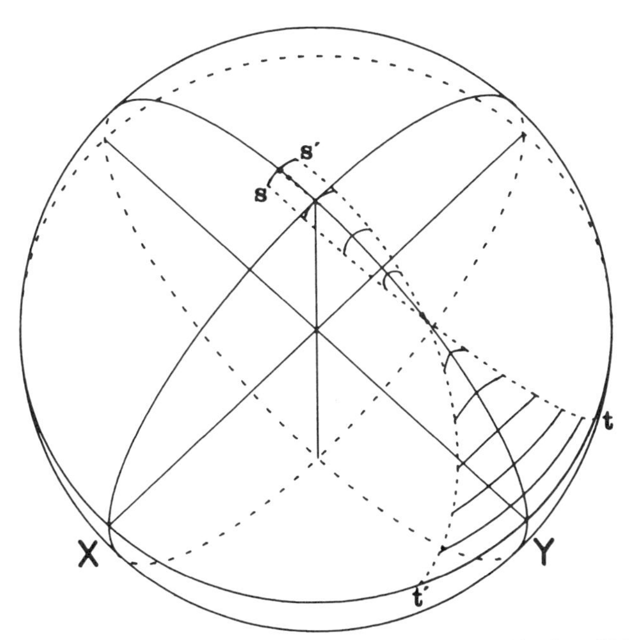

FIG. 5. Compensation of resonance offset effects with a composite $10°(-X)$, τ, $100°(+X)$ sequence. The period of free precession τ is chosen so as to minimize the residual Z magnetization after the pulse. With $\gamma H_1/2\pi = 10$ kHz, $\tau = 9$ μsec. As in Fig. 4, the full lines represent loci of the tips of a family of magnetization vectors corresponding to $\Delta H/H_1$ from −0.33 to +0.33.

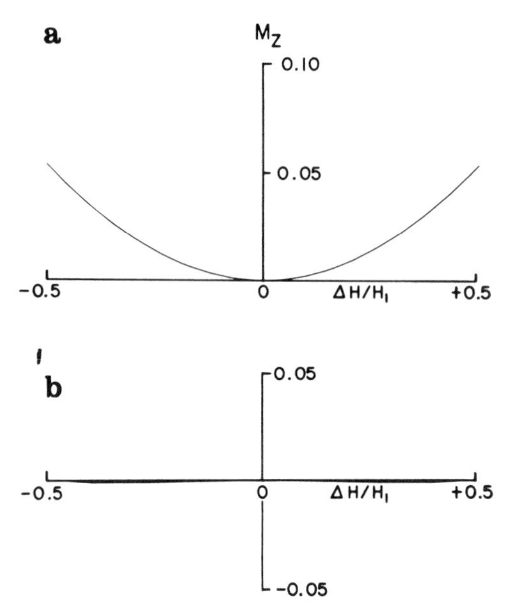

FIG. 6. Compensation for resonance offset effects with a 10°(−X), τ, 100°(+X) composite pulse. The residual Z magnetization after a conventional 90°(+X) pulse (a) is compared with that after the composite sequence (b).

improvement in the residual Z component after the pulse is shown in Fig. 6 as a function of the offset parameter $\Delta H/H_1$. For the conventional single 90° pulse (Fig. 6a) there is an approximately parabolic dependence whereas with the composite 10°(−X), τ, 100°(+X) sequence (Fig. 6b) the curve is very much flatter.

"*Spin Knotting*"

The two-pulse sequence of the previous section provides the clue for a composite sequence which compensates for both Z-component errors and phase errors at the same time. This has important applications in spin-locking experiments (4–6) where either phase errors or residual Z components after the first 90° pulse preclude the proper alignment of the magnetization vectors along the spin-locking field. The result is that not all magnetization components relax with the expected time constant $T_{1\rho}$ but precess about the spin-locking field undergoing the nutational motion first described by Torrey (*7*).

The time evolution of the magnetization vectors during this three-pulse sequence is most clearly visualized if the variable parameters (pulse lengths and interpulse intervals) are not at first optimized. The algebraic sum of the three nominal flip angles is maintained at +90°, so that an isochromat at exact resonance is carried from the Z axis to the Y axis. There are thus four independent variable parameters.

The three pulses carry the spins clockwise, counterclockwise, and clockwise about the X axis. The first pulse is 20°(+X) and is short enough that tilt effects are almost negligible. It is followed by free precession for a period of 30 μsec (Fig. 7). A family

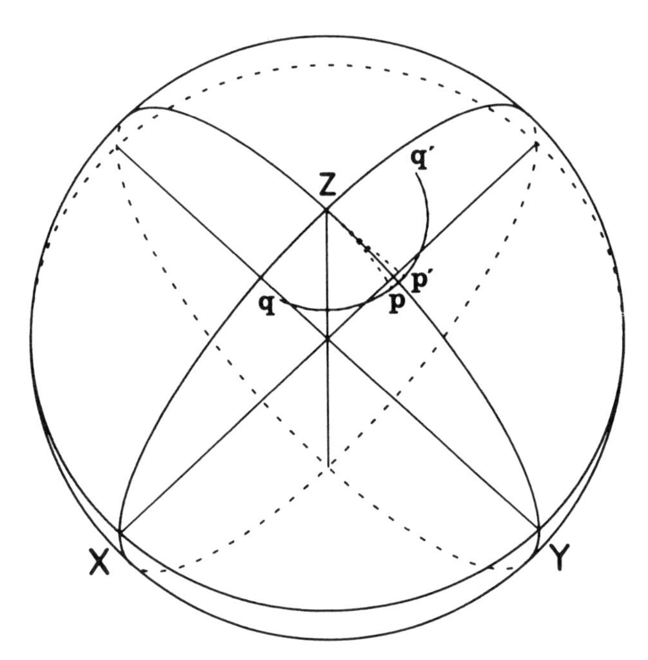

FIG. 7. "Spin-knotting" pulse sequence. A 20°(+X) pulse carries a family of magnetization vectors forward until the tips of the vectors lie along the locus $p-p'$. A period of 30 μsec free precession allows this locus to expand along $q-q'$. A range of offsets $\Delta H/H_1 = -0.5$ to $+0.5$ has been considered.

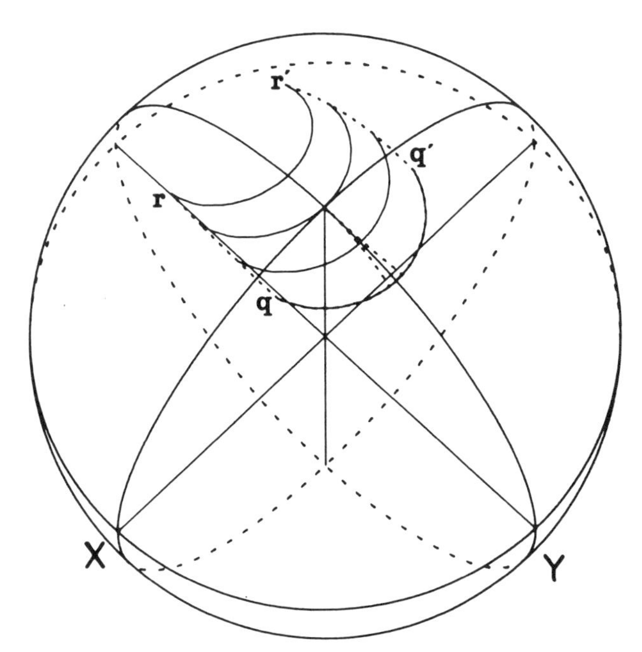

FIG. 8. "Spin-knotting" pulse sequence. The second pulse, 30°(−X), carries the locus $q-q'$ back to $r-r'$.

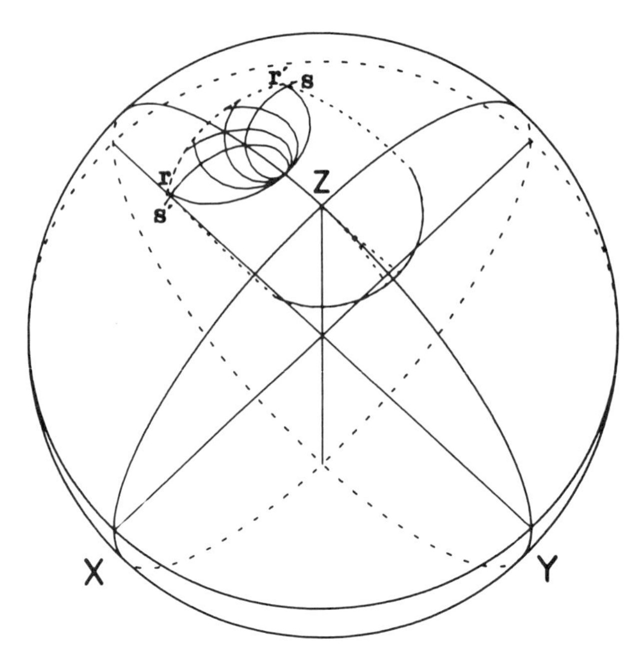

FIG. 9. "Spin-knotting" pulse sequence. The second period of free precession, 30 μsec in duration, allows the locus $r-r'$ to fold in upon itself, forming a loop $s-s'$ where the relative positions of "high-field" and "low-field" vectors have been reversed.

of isochromats is considered covering the range $\Delta H/H_1$ from -0.5 to $+0.5$, and gives rise to a locus $q-q'$. The second pulse, $30°(-X)$, carries this arc back over the pole $(+Z)$ leaving a locus $r-r'$ (Fig. 8). There is then a further 30-μsec period of free precession, during which this arc folds in upon itself, reversing the relative positions of clockwise and counterclockwise precessing vectors and giving the locus $s-s'$ (Fig. 9). Finally the third pulse, $100°(+X)$, carries this looped locus forward over the pole, converting it into a bow in the process $(t-t')$ (Fig. 10). A clearer view of the final locus $t-t'$ can be seen in Fig. 11 from a different angle.

This "knot" may be tightened by suitable adjustment of the parameters. Shortening the first pulse while lengthening the second improves the compensation. It may be noted that then the relative amplitudes of the three pulses follow roughly the form of one-half of a sinc curve, the form of excitation which would be expected to give uniform excitation in the limit of a linear response by the spins. Trial-and-error simulations indicate that a near-optimum sequence is

$$10°(+X), \quad 40\ \mu\text{sec}, \quad 60°(-X), \quad 11\ \mu\text{sec}, \quad 140°(+X).$$

Figure 12 shows the final locus to be a tight loop at the end of this sequence. With this sequence the residual Z component of magnetization after the composite pulse is quite small for a range of offset parameters from -0.3 to $+0.3$ as seen in Fig. 13 and the phase errors are similarly compensated (Fig. 14).

The compensation of phase errors (dispersion-mode components) has been tested experimentally in Fig. 15, which shows the offset dependence of a single carbon-13

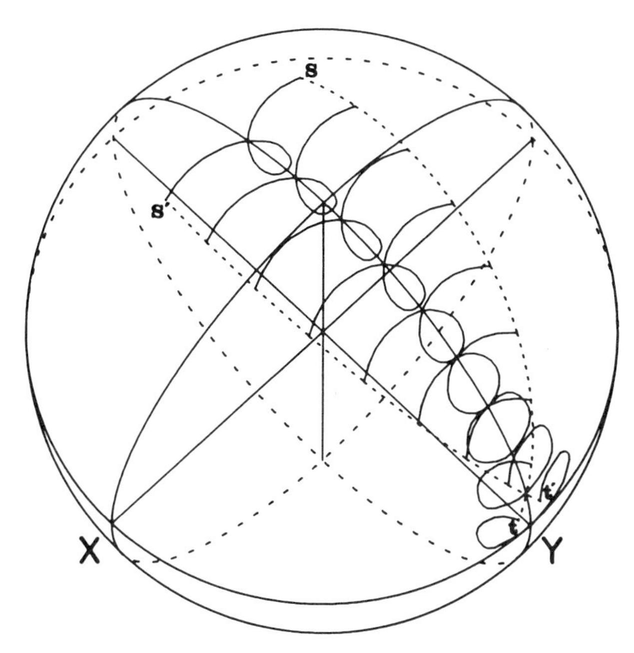

FIG. 10. "Spin-knotting" pulse sequence. The third pulse 100°(+*X*) carries the looped locus $s-s'$ forward toward the $+Y$ axis, leaving it in the form of a bow $t-t'$.

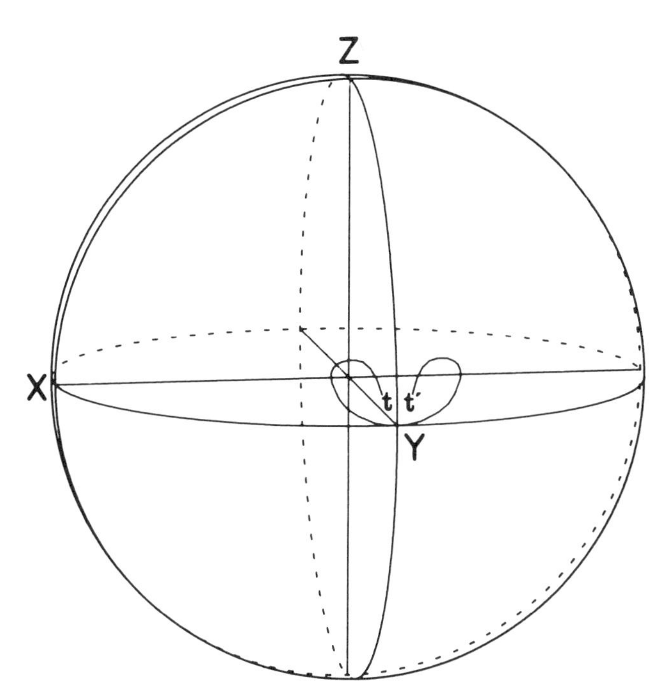

FIG. 11. The final locus of the tips of the magnetization vectors after a "spin-knotting" sequence, viewed from the $+Y$ axis.

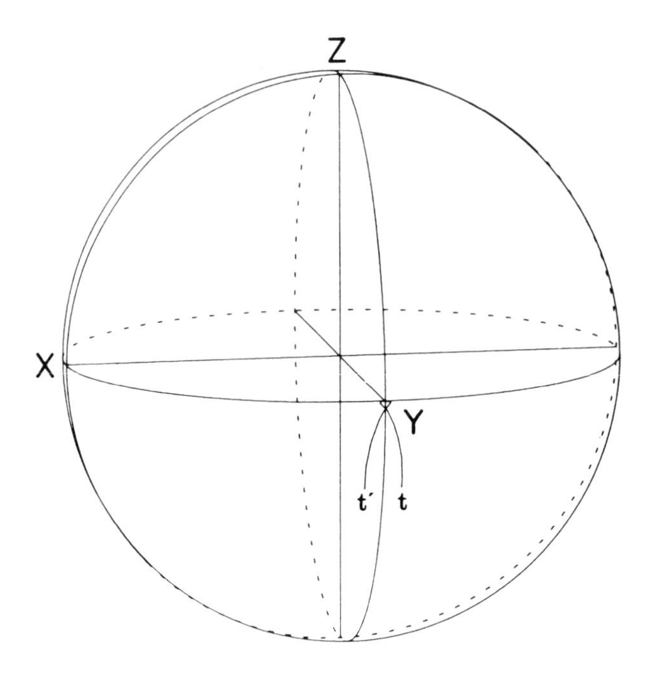

FIG. 12. The final locus after an optimized "spin-knotting" sequence, 10°(+X), 40 μsec, 60°(−X), 11 μsec, 140°(+X). Note that the "bow" of Fig. 11 has been tightened so that vectors representing small offsets are grouped close to the +Y axis.

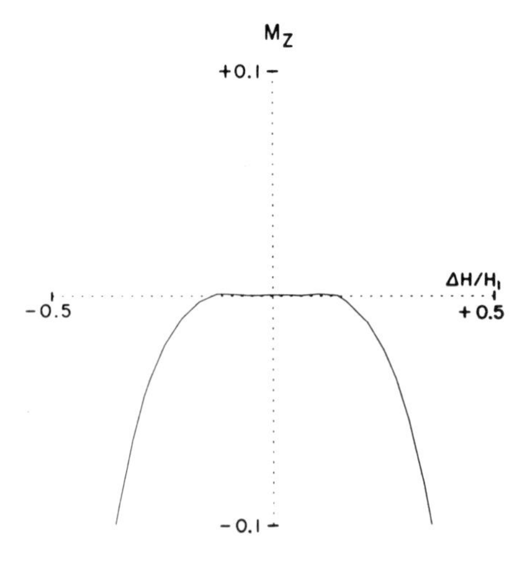

FIG. 13. The compensation of resonance offset effects achieved with the optimized "spin-knotting" sequence, bringing the Z magnetization after the composite pulse close to zero for small offsets.

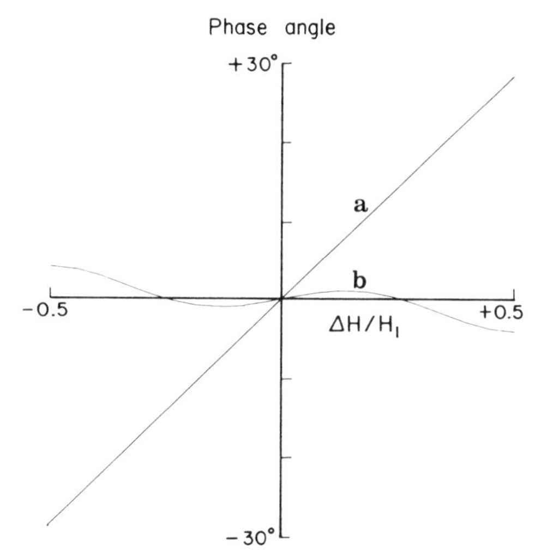

FIG. 14. The compensation of resonance offset effects achieved with the optimized "spin-knotting" sequence. The phase errors after a conventional 90°(+X) pulse (a) are compared with those at the end of the composite sequence (b).

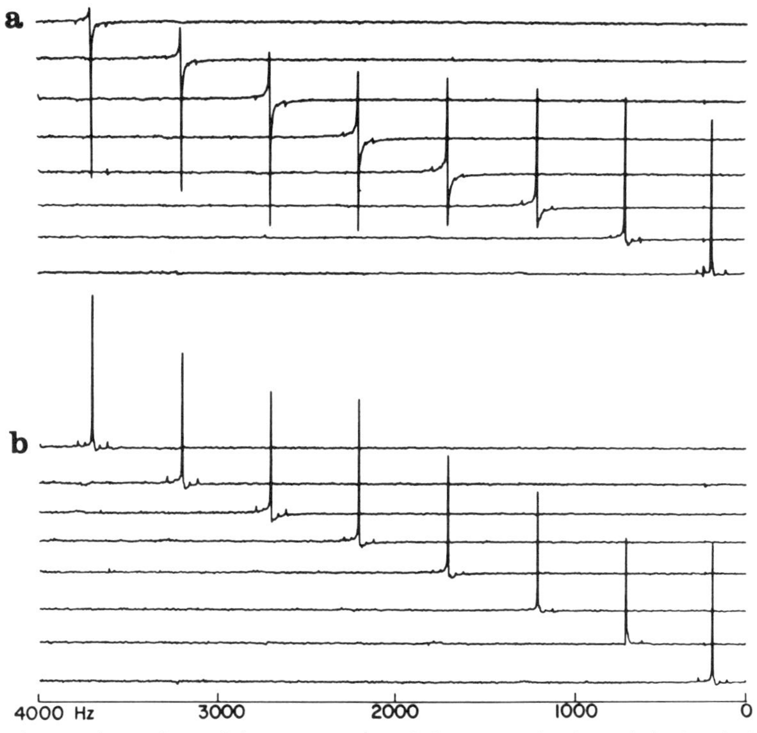

FIG. 15. An experimental test of the compensation of phase errors by the optimized "spin-knotting" sequence. (a) After a conventional 90°(+X) pulse, a signal which is correctly phased close to resonance rapidly acquires phase errors as it is moved away from resonance. (b) After the composite sequence, these phase errors are effectively compensated.

signal after a conventional 90° pulse (Fig. 15a) compared with the same series of signals after the near optimum spin-knotting sequence (Fig. 15b). Note the much better phasing in the latter series.

Spin-Locking Experiments

Of the possible applications of the composite sequence described in the previous section, the spin-locking experiment is probably the most important since it relies on all isochromats being aligned along the appropriate effective field H'_{eff}, irrespective of the offset ΔH. Failure to meet this condition changes the rate of decay and introduces oscillatory components onto the signal when it is followed as a function of the spin-locking time. The experiment can be considered to be divided into two parts, the preparation (alignment of isochromats along $+Y$) and the spin-locking period. For the experiment proposed here, H_1 is approximately twice as strong during the preparation period as during spin locking; hence the effective field during preparation (H_{eff}) is different from that during spin locking (H'_{eff}).

Suppose for the sake of this section that essentially perfect alignment has been achieved along the Y axis (for isochromats in a certain offset range) by a "spin-knotting" composite pulse. There remains the problem of turning each isochromat so as to align it along its characteristic H'_{eff}, since the direction of H'_{eff} varies with offset ΔH. Fortunately there is a simple trick to accomplish this. A 180°(Y) pulse (Fig. 16) carries each isochromat in a different arc, depending on θ and on H_{eff}, slightly overshooting the YZ plane. A very short period of free precession corrects this overshoot, leaving the isochromats fanned out along an arc in the YZ plane. Now if the radiofrequency field intensity H_1 is suitably reduced during the spin-locking period, each new H'_{eff} can be made to correspond quite closely to the direction of the appropriate isochromat. The choice of reduction factor is a mild compromise, depending on the offset range to be covered, but 0.48 is found to be suitable for offset parameters up to ± 0.5. For offsets small enough that $\tan\theta \simeq \theta$, the reduction factor is just 0.5. To date, this appears to be the most promising method of preparing nuclear spin magnetization vectors at different resonance offsets so that they may be effectively "locked" by a continuous radiofrequency field in order to measure the transverse relaxation times.

An appreciation of the improvement to be achieved can be obtained by following the trajectories of typical magnetization vectors by computer simulation. Four vectors, representing offsets $\Delta H/H_1 = \pm 0.33$ and ± 0.17 were allowed to evolve according to a spin-knotting sequence and then a 180°(Y) pulse. After free precession for 3 μsec, a simulated radiofrequency field was applied along the $+Y$ direction reduced in intensity by a factor 0.48. Figure 17 maps out the spiral paths of the four vectors as they precess about their appropriate effective fields. These are to be compared with the precession of much higher amplitude for a vector with offset parameter 0.33 prepared with a conventional 90°(X) pulse and the usual spin-locking field applied along the $+Y$ axis (Fig. 18). This trajectory along the surface of a wide-angle cone resembles the nutational motion described by Torrey (*7*) as much as the "forced precession" described by Solomon (*5, 6*) and hence involves a complicated combination of spin–spin and spin–lattice relaxation.

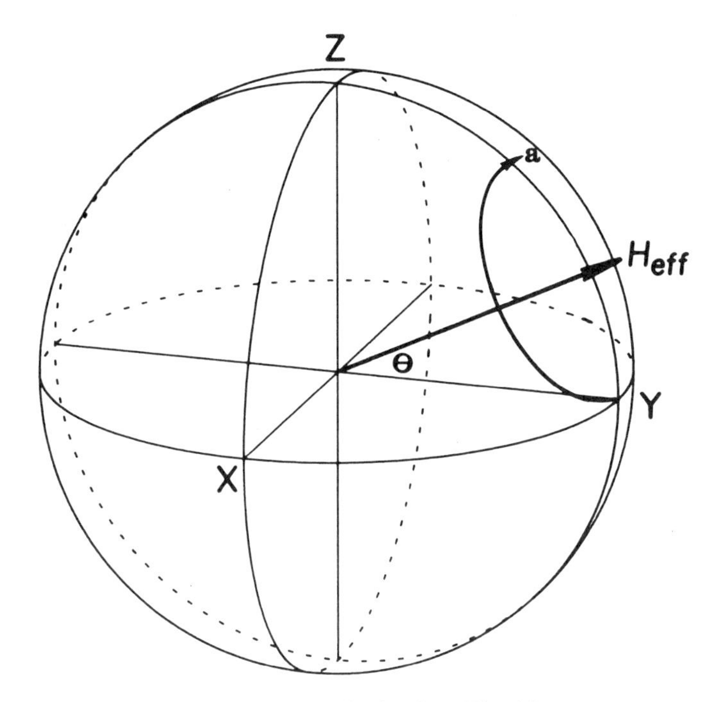

FIG. 16. Preparation for spin locking along the effective field H'_{eff}. All vectors are assumed to have been aligned along $+Y$ by a "spin-knotting" sequence. A 180°($+Y$) pulse then carries each vector along a different arc according to the appropriate offset ΔH, slightly overshooting the ZY plane (a). A very short period of free precession cancels this overshoot. The radiofrequency field intensity is then reduced to about half its normal value, so that the new effective field H'_{eff} is aligned along the appropriate magnetization vector, the angle θ being approximately doubled.

COMPOSITE 180° PULSES

Pulses of 180° are used for two main purposes, to turn magnetization vectors lying in the XY plane into mirror image positions with respect to the H_1 axis, or to take magnetization vectors from the $+Z$ to the $-Z$ axis in order to create a population inversion and an interchange of spin states. It has not yet been possible to devise a composite sequence suitable for this first application, the 180° refocusing pulse. Fortunately the well-known Meiboom–Gill modification (*8*) or the related phase-alternation scheme (*9*) satisfactorily compensates the effects of radiofrequency field inhomogeneity and resonance offset at the time of the even-numbered echoes.

Composite 180° pulses designed for population inversion are considered here. A brief account of a 90°(X), 180°(Y), 90°(X) sequence was presented in an earlier note (*10*). This composite pulse compensates for spatial inhomogeneity and resonance offset effects. Indeed it is readily seen that the compensation for pulse length error would be exact if the 180°(Y) pulse were perfect; that is, a θ°(X), 180°(Y), θ°(X) sequence is equivalent to a single 180° pulse for all values of θ. If θ approaches 90°, the excursions during the 180°(Y) pulse are small, and the effects of imperfections in the 180°(Y) pulse are minimized. Hence for 90°(X), 180°(Y), 90°(X) sequences,

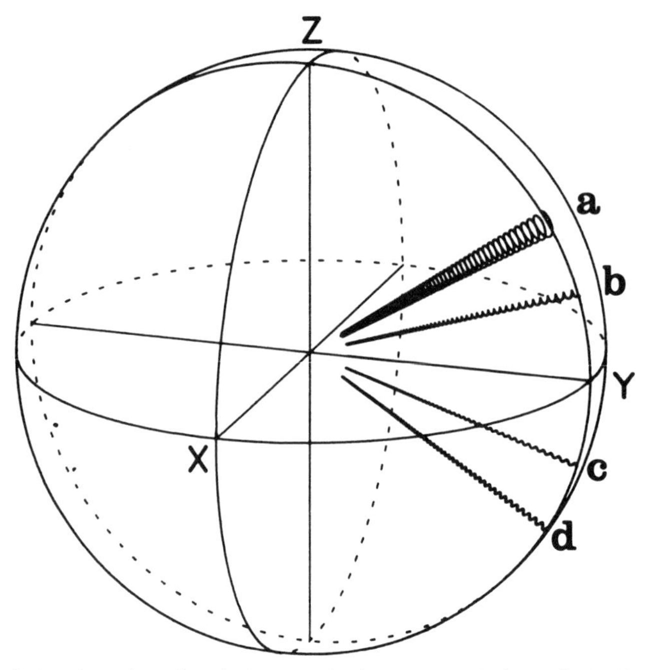

FIG. 17. The spiral trajectories of typical magnetization vectors during spin-locking experiments where the preparation involved the "spin-knotting" sequence, a 180°(+Y) pulse, a short period of precession, and a reduced spin-locking field 0.48 H_1, according to the prescription of Fig. 16. The four trajectories have been calculated for offset parameters $\Delta H/H_1$ equal to (a) +0.33, (b) +0.17, (c) −0.17, (d) −0.33.

small discrepancies in flip angle are particularly well compensated. Naturally this high degree of compensation is not maintained for sample regions where H_1 is very weak, but such regions contribute very little signal.

The degree of compensation can be illustrated by plotting contours of Z magnetization after the pulse as a function of pulse length error (H_1/H_1°) and offset parameter ($\Delta H/H_1$). It will be noted that all these contour diagrams are symmetrical with respect to offsets above and below the transmitter frequency. In the case of a three-pulse sequence the reasons for this property are not immediately obvious, but lie deep within the symmetry properties of the elements of the three rotation matrices. In the Appendix it is shown that the resultant Z magnetization possesses such symmetry after any three-pulse sequence of the form $\alpha(X)$, $\beta(Y)$, $\alpha(X)$ only if the first and last pulses are identical in length and flip axis. No such symmetry exists either in the final values of the X and Y components of magnetization or in the intermediate trajectories of the magnetization vectors.

Figure 19 shows that a conventional 180°(X) pulse has very little tolerance of pulse length error or resonance offset; note the small area within which the Z magnetization after the pulse is greater than 99% of M_0. This should be compared with the corresponding contour diagrams for the composite 90°(X), 180°(Y), 90°(X) sequence illustrated in Fig. 20, where pulse length errors four times larger can be accommodated within the same 99% contour. Note that it is only near the top of this

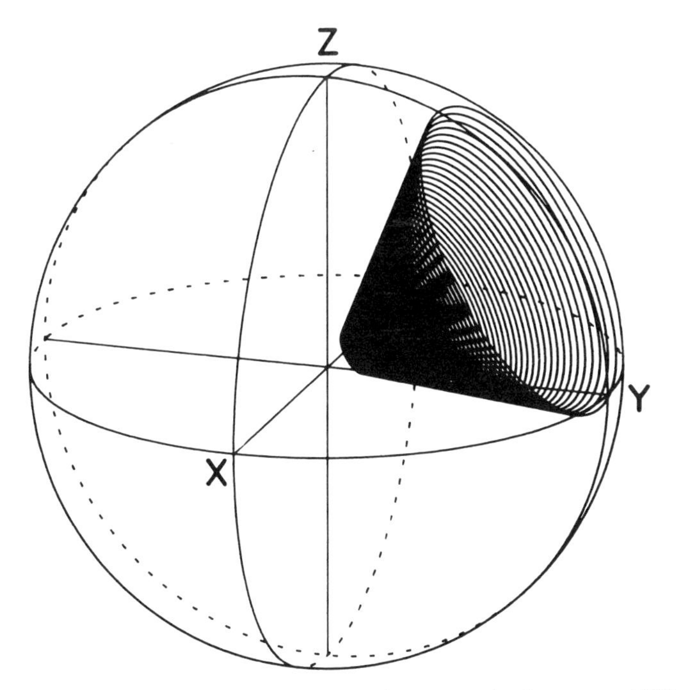

FIG. 18. The corresponding spiral trajectory of a typical magnetization vector ($\Delta H/H_1 = +0.33z$ during a spin-locking experiment where the preparation involved a conventional 90°(+X) pulse. These large excursions cause large errors in the derived relaxation time.

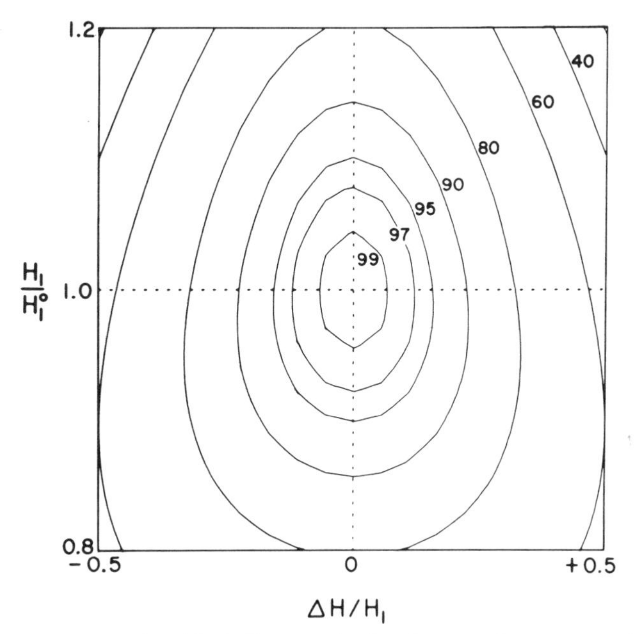

FIG. 19. The contours of residual Z magnetization (expressed as a percentage of M_0) after a conventional 180°(+X) pulse, as a function of pulse length misset H_1/H_1^0 and resonance offset $\Delta H/H_1$.

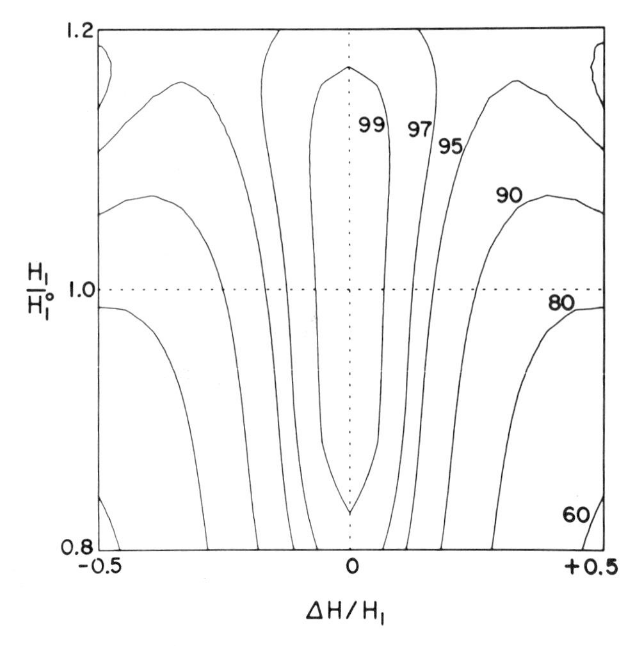

FIG. 20. Contours of residual Z magnetization after a composite 90°(+X), 180°(+Y), 90°(+X) sequence. Note that in comparison with the contours of Fig. 19, this diagram shows a higher tolerance of pulse length errors, and therefore of spatial inhomogeneity of the radiofrequency field.

diagram that there is much compensation for resonance offset effects. This is the clue that better compensation for offset can be achieved by deliberately increasing the pulse lengths beyond their nominal settings of 90 and 180°. It is not necessary to increase the two pulses in proportion, and in fact good results are obtained by increasing only the second pulse to give the composite sequence

$$90°(X),\ 240°(Y),\ 90°(X).$$

The trajectories of vectors representing a range of offset from $\Delta H/H_1 = 0.4$ to $\Delta H/H_1 = 0.6$ are followed in Fig. 21; the increased length of the second pulse is seen to compensate for the fact that it operates about a tilted axis. This family of vectors terminates quite close to the $-Z$ axis, although a conventional 180°(X) pulse would have taken these vectors far from the YZ plane. Figure 22 shows the contours of Z magnetization after the 90°(X), 240°(Y), 90°(X) sequence. There is a wide range of offset parameters which give only a small error in M_Z. In fact there is a roughly T-shaped area where the Z magnetization remains above 97% of its ideal value. Compensation for spatial inhomogeneity of H_1 is therefore good, for it is the *low* values of H_1 which require compensation, representing sample regions remote from the center of the transmitter coils. Figure 23 illustrates how the increased length of the second pulse compensates for a range of weaker-than-normal H_1 fields arising from spatial inhomogeneity, turning all the magnetization vectors down toward the $-Z$ axis.

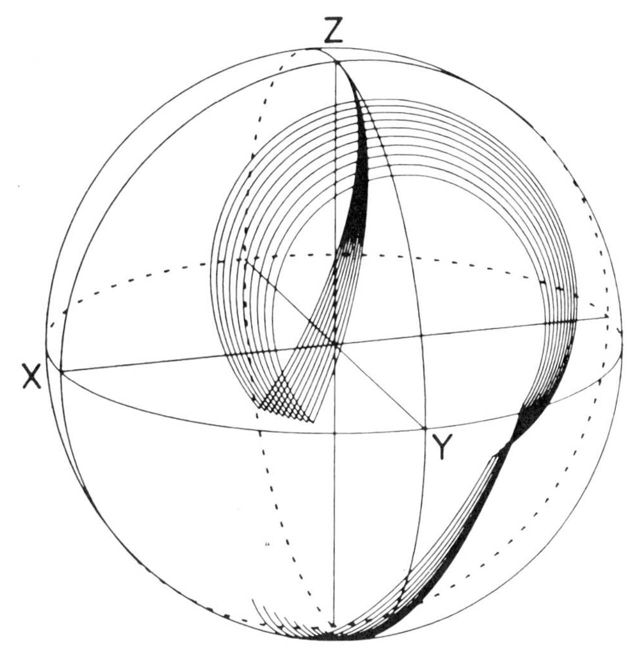

FIG. 21. Trajectories of a family of magnetization vectors (representing offset parameters $\Delta H/H_1$ between 0.4 and 0.6) during a composite pulse sequence 90°(+X), 240°(+Y), 90°(+X). Note that these trajectories terminate much closer to the $-Z$ axis than those after a single 180°(+X) pulse.

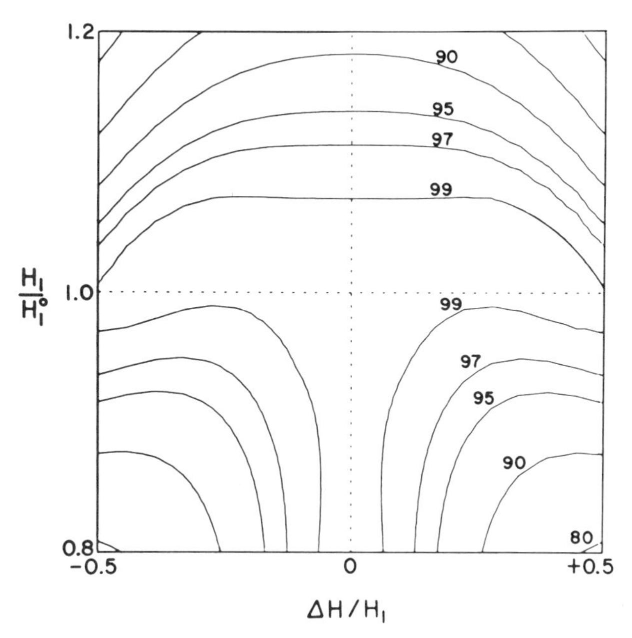

FIG. 22. Contours of residual Z magnetization after a composite 90°(+X), 240°(+Y), 90°(+X) sequence. Note the high tolerance of resonance offset $\Delta H/H_1$ as well as pulse length misset below the nominal value. In a large T-shaped region M_Z is held within 1% of M_0.

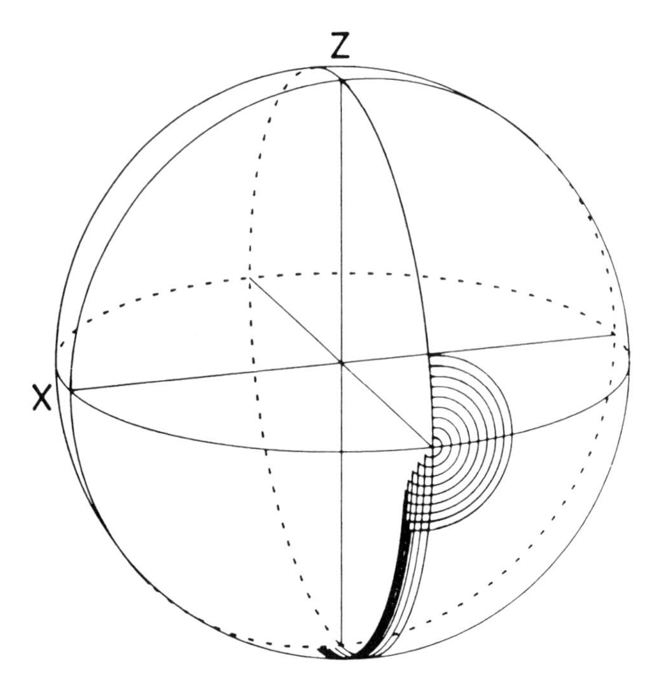

FIG. 23. Compensation of pulse length errors in the composite 90°(+X), 240°(+Y), 90°(+X) sequence. A family of magnetization trajectories has been traced out for pulse lengths between 0.8 and 1.0 times the nominal value. Note how the termini are grouped close to the $-Z$ axis.

An experimental test of the population inversion obtained with the composite pulse sequence can be made very simply by monitoring the Z magnetization with a subsequent 90° pulse, alternating the phase to cancel transverse magnetization (*11*), and allowing no time for spin–lattice relaxation. Figure 24 shows the equilibrium Z magnetization monitored by a single 90° pulse (a), the imperfect inversion achieved after a conventional 180° pulse (b), and the essentially complete inversion achieved after a composite 90°(X), 200°(Y), 90°(X) sequence (c). Spatial inhomogeneity effects were minimized by the use of a small spherical bulb sample, in order to test the effect of resonance offset, $\Delta H/H_1 = 0.4$.

A preliminary report (*10*) on composite 180° pulses utilized the null method for determining spin–lattice relaxation times as a test of the efficiency of population inversion. Pulse imperfections lead to imperfect inversion and a systematic underestimate of the spin–lattice relaxation time if calculated from the time at which the signal passes through the null. Thus if the true spin–lattice relaxation time is known, the timing of the null is a sensitive test of pulse imperfections. The original experiments (*10*) used a 90°(X), 180°(Y), 90°(X) sequence, which, as has been shown above, is not optimized for compensation of the effects of resonance offset. The 90°(X), 240°(Y), 90°(X) composite sequence gives better results.

An experimental test of the effectiveness of this composite sequence was made on a small spherical sample of methyl iodide by following the spin–lattice relaxation of the singlet carbon-13 signal, decoupled from protons. First a careful determination

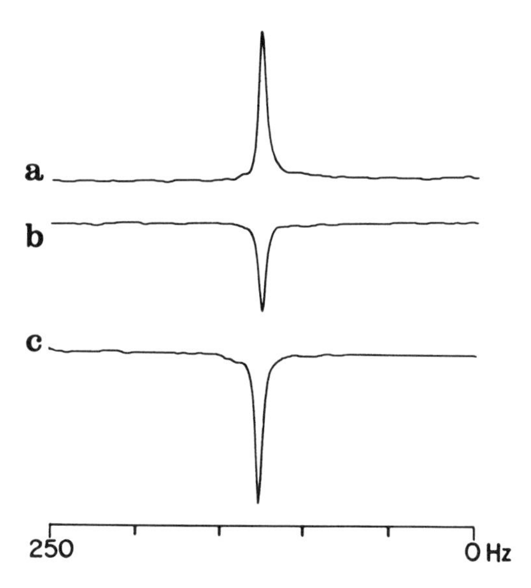

FIG. 24. Experimental test of the composite 90°(+X), 200°(+Y), 90°(+X) pulse sequence. The signals were all observed by means of a conventional 90°(+X) "read" pulse. The initial conditions were (a) Boltzmann equilibrium, (b) population inversion by a conventional 180°(+X) pulse, (c) population inversion by the composite sequence. A small bulb sample was used, and the offset parameter was $\Delta H/H_1 = 0.4$.

of the spin–lattice relaxation time by inversion–recovery was made, using a semi-logarithmic plot of the difference signal $S_\infty - S_t$. This method is known to be insensitive to pulse imperfections provided that the phase of the monitoring 90° pulse is alternated (*11*). A result of $T_1 = 13.4 \pm 0.1$ sec was obtained.

With this knowledge of the "true" T_1 it was possible to use the null point (determined from two measurements which bracketed the null) as a measure of pulse imperfections. Table 1 illustrates the dependence of the systematic error on the resonance offset for the case of a conventional single 180°(X) pulse, compared with a

TABLE 1

APPARENT SPIN–LATTICE RELAXATION TIMES DETERMINED BY THE NULL METHOD: EFFECT OF RESONANCE OFFSET

$\gamma \Delta H/2\pi$ (kHz)	$\Delta H/H_1$	Conventional pulse T_1 (sec)	Composite pulse T_1 (sec)
0	0.00	13.3 (−1)[a]	13.4 (<1)
1	0.09	13.0 (−4)	13.4 (<1)
2	0.18	12.0 (−11)	13.4 (<1)
3	0.27	10.8 (−20)	13.4 (<1)
4	0.36	9.7 (−28)	13.1 (−2)

[a] Systematic error (%) with respect to the value $T_1 = 13.4$ sec measured by careful inversion–recovery experiments and least-squares fitting to an exponential.

TABLE 2

APPARENT SPIN-LATTICE RELAXATION TIMES DETERMINED BY THE NULL METHOD: EFFECT OF PULSEWIDTH MISSET

t_p/t_p°	α (°)	Conventional pulse T_1 (sec)	Composite pulse T_1 (sec)
1.0	180	13.3 (−1)[a]	13.4 (<1)
0.9	162	12.8 (−5)	13.4 (<1)
0.8	144	12.1 (−10)	13.3 (−1)
0.7	126	11.0 (−19)	13.3 (−1)
0.6	108	6.2 (−54)	12.1 (−10)
0.5	90	—[b]	10.2 (−24)

[a] Systematic error (%) with respect to the value $T_1 = 13.4$ sec measured by careful inversion–recovery experiments and least-squares fitting to an exponential.

[b] Signal not inverted; null not observed.

composite 90°(X), 240°(Y), 90°(X) sequence. The apparent T_1 values showed large errors when the conventional pulse was used, increasing with the offset parameter $\Delta H/H_1$, while the composite sequence greatly reduced these errors, holding them below 1% for most of the offset range.

The effect of spatial inhomogeneity was then simulated by deliberately missetting the flip angles α used for population inversion. Initially the conventional pulse was 180°(X) and the composite pulse 90°(X), 240°(Y), 90°(X). Then all four pulse widths were reduced in the same proportion in order to follow the effects of incomplete inversion, the monitoring pulse being kept fixed throughout at 90°(X). This introduced large systematic errors into the apparent spin–lattice relaxation times determined from the null point if the conventional single pulse was used, but much smaller errors when the composite sequence was employed (Table 2). Even when the three pulses were set at 70% of their nominal widths, the composite sequence held the error within 1%. It may be inferred from these results that the corresponding errors due to spatial inhomogeneity effects should be similarly reduced (Table 3).

Good compensation may therefore be achieved for resonance offset effects or for radiofrequency inhomogeneity, but when both types of imperfection occur simultaneously, the contour diagram (Fig. 22) indicates that compensation is less effective. For high-resolution spectra with lines spread over a wide frequency range, it appears likely that the offset effect will be the more serious.

EXPERIMENTAL

The practical tests of various composite pulse sequences were carried out on a Varian CFT-20 spectrometer using carbon-13 signals from enriched methyl iodide. Only minor modifications are necessary to implement this technique, principally the provision of a second transmitter channel shifted in phase by 90° (*12*). The remaining changes are accomplished by machine code program modifications, one of the chief

TABLE 3
COMPOSITE PULSE SEQUENCES

Composite sequence	Equivalent	Compensation	Application
$90°(X), 90°(Y)$ or $90°(X), 110°(Y)$	$90°(X)$	H_1 inhomogeneity	Saturation[a]
$10°(-X), \tau, 100°(+X)$[b]	$90°(X)$	ΔH	Saturation[a]
$10°(+X), \tau_1, 60(-X), \tau_2, 140°(+X)$	$90°(X)$	ΔH	Saturation[a]
$10°(+X), \tau_1, 60°(-X), \tau_2, 140°(+X), 180°(+Y), \tau_3$	$90°(X)$	ΔH	Preparation for spin-locking
$90°(X), 180°(Y), 90°(X)$ or $90°(X), 240°(Y), 90°(X)$	$180°(X)$	ΔH or H_1 inhomogeneity	Population inversion
$90°(+X), \theta°(+Y), 90°(-X)$	$\theta°(+Z)$	None	Rotation about Z

[a] For a saturation–recovery or a progressive saturation experiment to measure T_1.
[b] τ represents a period of free precession.

difficulties on this instrument being to ensure that the interpulse intervals were sufficiently short when free precession between pulses had to be avoided. In practice the minimum interpulse interval attainable was 11 μsec. More sophisticated pulse programmers would be useful in this regard.

A sensitive and rapid experimental check of the accuracy of the 90° phase shift is provided by the composite $90°(X)$, $180°(Y)$, $90°(X)$ sequence applied to a resonance close to the transmitter frequency ($\Delta H \ll H_1$) on a small spherical sample where a spatial inhomogeneity of H_1 can be neglected. The three pulse lengths must be properly calibrated for this test. If the phase shift is correct and the second pulse is applied about the Y axis, then the detected signal should be a null, the magnetization vector being aligned along $-Z$. If there is a small deviation δ from the 90° phase shift, the second pulse carries the magnetization vector away from the YZ plane to a position where it subtends an angle 2δ with respect to the $+Y$ axis. After the third pulse this is translated into a residual (negative) dispersion-mode signal of amplitude $M_0 \sin 2\delta \simeq 2M_0\delta$ for small angles δ. Thus the amplitude and sign of this residual signal can be used as a test for the amount of the sense of the dephasing of the $180°(Y)$ pulse.

Pulsewidth calibration was carried out by searching for the condition which gave a null signal after a complete rotation, according to Eq. [3]. The pulsewidths for 90 and 180° pulses were reduced in proportion, since the "droop" in the intensity of H_1 over these time periods had been adjusted to be negligible. Pulse droop is a feature designed to protect the output transistors in the transmitter circuit; an alternative scheme consisting of a period of constant amplitude H_1 followed by a delayed droop was incorporated as a modification in the spectrometer used.

DISCUSSION

The concept of a composite pulse is not really new; a "2–1–4" sequence was used by Redfield (*13*) to achieve frequency selectivity, and quite complicated pulse sequences are commonplace in solid-state NMR (*14*). However, their use in high-resolution NMR has been extremely limited. The present work emphasizes the use of composite pulses to compensate the two common pulse imperfections, H_1 inhomogeneity and resonance offset effects. However, this is unlikely to be the only application, because the principle appears to be quite general. Whereas a conventional single pulse is described by two parameters, the flip angle α and the axis along which H_1 is directed, a composite pulse involves many more independently variable parameters (including periods of free precession between pulses) and its applications are correspondingly more flexible. The adjustment of these variable parameters can be carried out either by trial-and-error (aided by computer simulation), as in the "spin-knotting" sequence described above, or through a computer program to optimize these parameters according to some desired experimental criterion. It is nevertheless a good rule to keep the number of elements in the sequence to the minimum required to solve the problem.

One example will serve to illustrate that quite different applications are conceivable. Suppose that it becomes necessary to contrive a rotation of magnetization vectors by θ degrees about the Z axis of the rotating frame. For a high-resolution spectrum with several resonance lines, a simple free precession would not achieve this. Provided that the pulses can be considered ideal, then a $90°(+X)$, $\theta°(+Y)$, $90°(-X)$ composite sequence can be shown to be equivalent to $\theta°(+Z)$. This is readily appreciated by considering the behavior of the three magnetization components separately; M_Z remains unchanged while M_X and M_Y are rotated through $\theta°$ clockwise about the $+Z$ axis. There are several possible extensions of this principle.

Compensated 90° pulses should prove useful in spin–lattice relaxation experiments carried out by the saturation–recovery or progressive saturation techniques. Unfortunately, H_1 inhomogeneity and resonance offset effects cannot yet be compensated simultaneously, so a decision must be made about the relative importance of the two types of pulse imperfection. If spatial inhomogeneity of H_1 is the dominant factor in a particular experimental situation, then the $90°(X)$, $90°(Y)$ sequence (or an optimized variation) is indicated; if resonance offset effects are more critical, then the "spin-knotting" sequence is preferable.

Spin-echo methods of determining spin–spin relaxation times are often complicated by the modulation generated by homonuclear coupling (*15*), so the spin-locking experiment can be a simpler method of obtaining this relaxation information. However, there are practical limitations to the application of very strong spin-locking radiofrequency fields for periods of the order of seconds. Consequently, for high-resolution spectra with many lines, the tilt of the effective field H'_{eff} must be taken into account. It is suggested that the "spin-knotting" sequence might be used to prepare the nuclear magnetization vectors, leaving them suitably aligned along $+Y$, and then a 180° pulse would fan them out into an arc in the YZ plane, at which point a reduced H_1 field would be applied such that each individual isochromat is locked along its appropriate effective field H'_{eff}. Where radiofrequency heating is a severe problem, *three* levels of radiofrequency intensity could be employed—the full intensity for the short pulses in the preparation sequence, a significantly reduced H_1 for the $180°(Y)$ pulse, followed by a further twofold reduction for the spin-locking period.

The composite 180° pulse sequence can compensate for either H_1 inhomogeneity or resonance offset effects, although the correction is less effective where there is both an abnormally low flip angle and a large resonance offset. Once essentially complete population inversion can be guaranteed, the null method of determining spin–lattice relaxation times has much to recommend it. The method is fast because it requires no determination of the asymptotic signal S_∞, and the null point can be rapidly located by bracketing, using a linear interpolation (a very good approximation in this short section of the exponential recovery curve). It is a particularly effective method when there is already a good estimate of T_1 available, or when small changes in T_1 are being studied.

When composite 180° pulses are used for the "fast inversion–recovery" method (*16*), improved accuracy in the derived relaxation times is to be expected. In this mode, the waiting period between experiments is deliberately shortened so as to be comparable with some of the spin–lattice relaxation times under investigation; it therefore involves steady-state conditions. The danger with this technique is that pulse imperfections may have cumulative effects in the steady-state regime, introducing systematic errors in the observed T_1 values (*17*). For this application both the 180 and 90° pulses would need to be compensated for pulse imperfections.

Other applications which require accurate 180° pulses are to be found in the field of two-dimensional Fourier transform NMR spectroscopy. One such experiment generates spin echoes modulated by heteronuclear spin coupling by applying 180° pulses to the coupled heteronucleus (*18*). If these 180° pulses are imperfect then spurious responses appear in the two-dimensional spectrum. A similar problem arises in two-dimensional Fourier transform experiments designed to correlate the chemical shifts of two nuclear species, for example, protons and carbon-13 (*19, 20*); artifacts appear in the spectra if the 180° pulses used for "decoupling" are imperfect. There is a related experiment in conventional one-dimensional NMR where 180° pulses are used to flip the spin states of protons and thus effectively decouple or scale down the splittings in the carbon-13 spectrum (*21*). In this application the effects of pulse imperfections are cumulative, making compensation particularly important.

APPENDIX

Symmetry Properties of an $\alpha(X)$, $\beta(Y)$, $\alpha(X)$ Composite Pulse

The trajectories and final positions of nuclear magnetization vectors under the influence of a three-pulse sequence may be analyzed by examination of the three rotation matrices. During a pulse, an off-resonance isochromat is rotated through an angle α about a vector $\mathbf{r}$ tilted away from the XY plane by an angle θ given by Eq. [2]. The flip angle α is given by $\gamma H_{\text{eff}} t_{\text{p}}$, where H_{eff} is defined by Eq. [1]. Then the normalized rotation vector is

$$\mathbf{r}_X = (K, 0, S) \qquad [4]$$

for pulses about the X axis, and

$$\mathbf{r}_Y = (0, K, S) \qquad [5]$$

for pulses about the Y axis, where

$$K = H_1/H_{\text{eff}} \tag{6}$$

and

$$S = \Delta H/H_{\text{eff}}. \tag{7}$$

Rotation operators are then formed by straightforward vector algebra. For a pulse applied about the X axis,

$$\hat{R}_X(\alpha) = \begin{pmatrix} K^2(1-\cos\alpha)+\cos\alpha & S\sin\alpha & K(1-\cos\alpha)S \\ -S\sin\alpha & \cos\alpha & K\sin\alpha \\ K(1-\cos\alpha)S & -K\sin\alpha & S^2(1-\cos\alpha)+\cos\alpha \end{pmatrix} \tag{8}$$

and where H_1 is applied along the Y axis,

$$\hat{R}_Y(\alpha) = \begin{pmatrix} \cos\alpha & S\sin\alpha & -K\sin\alpha \\ -S\sin\alpha & K^2(1-\cos\alpha)+\cos\alpha & K(1-\cos\alpha)S \\ K\sin\alpha & K(1-\cos\alpha)S & K^2(1-\cos\alpha)+\cos\alpha \end{pmatrix}. \tag{9}$$

Each element possesses a symmetry with respect to the off-resonance parameter determined by the effect of changing the sign of S. For example, the element R_X^{31} clearly changes sign as S changes sign, and is thus antisymmetric, denoted $\bar{R}_X^{31}$.

The effect of a composite pulse $\alpha(X)$, $\beta(Y)$, $\alpha(X)$ on a unit magnetization vector **k** (initially along the Z axis) can now be calculated. The effect of the first pulse can be written

$$\hat{R}_{X1}k = \begin{pmatrix} R_{X1}^{13} \\ \bar{R}_{X1}^{23} \\ R_{X1}^{33} \end{pmatrix}. \tag{10}$$

Application of the second pulse gives

$$\hat{R}_Y\hat{R}_{X1}k = \begin{pmatrix} R_Y^{11}\bar{R}_{X1}^{13} + \bar{R}_Y^{12}R_{X1}^{23} + R_Y^{13}R_{X1}^{33} \\ \bar{R}_Y^{21}\bar{R}_{X1}^{13} + R_Y^{22}R_{X1}^{23} + \bar{R}_Y^{23}R_{X1}^{33} \\ R_Y^{31}\bar{R}_{X1}^{13} + \bar{R}_Y^{32}R_{X1}^{23} + R_Y^{33}R_{X1}^{33} \end{pmatrix}. \tag{11}$$

From Eq. [10] it is possible to see that in general the position of a magnetization vector at this point is completely asymmetric in the off-resonance parameter, each component consisting of a mixture of symmetric and antisymmetric terms. Nevertheless, under certain conditions, the application of a third pulse can restore symmetry to the final Z component. This third pulse, $\hat{R}_{X2}$, operating on the vector of Eq. [11] gives for the final Z component

$$\begin{aligned} M_Z = {} & \bar{R}_{X2}^{31}(R_Y^{11}\bar{R}_{X1}^{13} + \bar{R}_Y^{12}R_{X1}^{23} + R_Y^{13}R_{X1}^{33}) \\ & + R_{X2}^{32}(\bar{R}_Y^{21}\bar{R}_{X1}^{13} + R_Y^{22}R_{X1}^{23} + \bar{R}_Y^{23}R_{X1}^{33}) \\ & + R_{X2}^{33}(R_Y^{31}\bar{R}_{X1}^{13} + \bar{R}_Y^{32}R_{X1}^{23} + R_Y^{33}R_{X1}^{33}). \end{aligned} \tag{12}$$

The terms $M_Z(a)$ which are antisymmetric in S thus become

$$M_Z(a) = \bar{R}^{31}_{X2} R^{11}_{Y} R^{33}_{X1} + R^{32}_{X2} \bar{R}^{23}_{Y} R^{33}_{X1} + R^{33}_{X2} R^{31}_{Y} \bar{R}^{13}_{X1} + R^{33}_{X2} \bar{R}^{32}_{Y} R^{23}_{X1}. \quad [13]$$

Examination of the form of the matrices [8] and [9] reveals that $M_Z(a)$ vanishes identically if the first and third pulses are identical. Under these conditions the resultant Z magnetization contains only terms in $M_Z(s)$ which are unchanged on changing the sign of S. Hence the contour maps drawn above are symmetrical with respect to positive and negative offsets ΔH. On the other hand, it is easy to show that the X and Y components display no such symmetry. Moreover the symmetry of the Z components is destroyed if the third pulse is applied along the $-X$ axis (simulated by interchanging $-K$ for K in Eq. [8]).

These symmetry properties of M_Z are important because they permit a doubling of the effective range of allowed transmitter offsets when a quadrature detection scheme is employed for signal detection. The analysis is useful in that it allows a certain insight into the properties of pulse sequences without explicit multiplication of many large rotation matrices. Beyond this, where exact knowledge of the trajectories of the magnetization vectors is required, computer simulation becomes a practical necessity.

ACKNOWLEDGMENTS

This work was made possible by an equipment grant and research studentships (S.P.K. and M.H.L.) from the Science Research Council.

REFERENCES

1. H. Y. Carr and E. M. Purcell, *Phys. Rev.* **94,** 630 (1954).
2. J. L. Markley, W. J. Horsley and M. P. Klein, *J. Chem. Phys.* **55,** 3604 (1971).
3. R. Freeman and H. D. W. Hill, *J. Chem. Phys.* **54,** 3367 (1971).
4. A. G. Redfield, *Phys. Rev.* **98,** 1787 (1955).
5. I. Solomon, *Compt. Rend.* **248,** 92 (1959).
6. I. Solomon, *Compt. Rend.* **249,** 1631 (1959).
7. H. C. Torrey, *Phys. Rev.* **76,** 1059 (1949).
8. S. Meiboom and D. Gill, *Rev. Sci. Instrum.* **29,** 688 (1958).
9. R. Freeman and S. Wittekoek, *J. Magn. Reson.* **1,** 238 (1969).
10. M. H. Levitt and R. Freeman, *J. Magn. Reson.* **33,** 473 (1979).
11. D. E. Demco, P. van Hecke, and J. S. Waugh, *J. Magn. Reson.* **16,** 467 (1974).
12. G. Bodenhausen, R. Freeman, G. A. Morris, R. Niedermeyer, and D. L. Turner, *J. Magn. Reson.* **25,** 559 (1977).
13. A. G. Redfield, 17th Experimental N.M.R. Conference, Pittsburgh, Pa., April 1976.
14. J. S. Waugh, L. M. Huber, and U. Haeberlen, *Phys. Rev. Lett.* **20,** 180 (1968).
15. R. Freeman and H. D. W. Hill, "Dynamic N.M.R. Spectroscopy" (F. A. Cotton and L. M. Jackman, Ed.), Chap. 5, Academic Press, New York, 1975.
16. D. Canet, G. C. Levy, and I. R. Peat, *J. Magn. Reson.* **18,** 199 (1975).
17. H. Hanssum, W. Maurer, and H. Ruterjans, *J. Magn. Reson.* **31,** 231 (1978).
18. G. Bodenhausen, R. Freeman, G. A. Morris, and D. L. Turner, *J. Magn. Reson.* **28,** 17 (1977).
19. A. A. Maudsley, A. Kumar, and R. R. Ernst, *J. Magn. Reson.* **28,** 463 (1977).
20. G. Bodenhausen and R. Freeman, *J. Magn. Reson.* **28,** 471 (1977).
21. R. Freeman, S. P. Kempsell, and M. H. Levitt, *J. Magn. Reson.* **35,** 447 (1979).

JOURNAL OF MAGNETIC RESONANCE **39**, 297–308

An NMR Investigation into the Range of the Surface Effect on the Rotation of Water Molecules

D. E. WOESSNER

Mobil Research & Development Corporation, Field Research Laboratory, P. O. Box 900, Dallas, Texas 75221

Received October 12, 1979

It is well known that the properties of water in the vicinity of a surface or interface are different from those of pure bulk water. An important question is the thickness of the water layer with properties different from bulk water, i.e., the range of the effect. Different physical properties might have different ranges. The NMR relaxation time T_1 reflects the molecular motions in the megahertz frequency range. The deuteron T_1 of D_2O depends only on the molecular rotational properties, whereas the proton T_1 of H_2O depends on both the translational and rotational properties. In order to determine the range of the surface effect on the rotational properties of water, the room temperature deuteron T_1 was measured as a function of water layer thickness for a carefully prepared series of hectorite–D_2O samples. This clay was chosen because it has very low concentration of paramagnetic impurities which can complicate the experiment. The results are in complete agreement with the simple picture that at room temperature the hectorite surface influences and slows down the rotation of only the first layer or two of water molecules.

INTRODUCTION

It is well known that the properties of water in the vicinity of a surface or interface are different from those of pure bulk liquid water at the same temperature. The surface or interface perturbs the vicinal water through its effects on the arrangement and energetics of these water molecules. The perturbations can arise from specific ion–dipole forces, dispersion forces, hydrophobic effects on water structure, etc. Because of the nonspherical electrical and bonding properties of the water molecule, the perturbations are expected to be anisotropic with respect to an axis attached to a water molecule. Also, beyond the first molecular layer away from the surface, the predominant changes in the water properties arise from perturbations of intermolecular interactions in the water. An important question is the range over which the presence of the surface or interface affects the properties of the water. The range may be different for different properties. For example, the range for a cooperative property such as the freezing temperature may be different from the range for a dynamic property such as a given percentage change in the rate of molecular rotational motion.

It is beyond the scope of this paper to review the vast literature concerning the effects of surfaces or interfaces on various properties of water. Drost-Hansen (*1*) has

0022-2364/80/080297-12 $02.00/0

published a comprehensive review of experiments relating to the properties of vicinal water and concludes that vicinal water structures may apparently be propagated over considerable distances from the surface into the bulk liquid. In a recent review of the dynamics of water in heterogeneous systems, Packer (*2*) concludes that the effects of the nonaqueous components on the water extend only to a distance of the order of one or two molecular diameters.

The effect of a surface in inducing nuclear magnetic resonance (NMR) line splittings and enhanced relaxation rates in water is well documented (*3*). Indeed, both Drost-Hansen (*1*) and Packer (*2*) cite numerous NMR investigations. NMR line splittings are sensitive to time- (and space-) averaged molecular preferential orientation induced by the surface. NMR spin–lattice relaxation time (T_1) values are sensitive to the rates and anisotropy of molecular motions induced by the surface. Hence, NMR provides a powerful means of investigating the effects of surfaces on many properties of water.

The range of the surface effects on NMR-sensitive parameters can be investigated by varying the quantity of water available to a unit surface area over a range which includes a significant amount of bulk water (i.e., water whose NMR properties are identical to those of bulk water). The deuteron T_1 value is determined by intramolecular interactions while the proton T_1 in H_2O is dependent on both intra- and intermolecular interactions. Hence deuteron NMR measurements are ideal (as compared to proton NMR measurements) for investigating the range of surface effects on aqueous molecular rotational motions as well as preferential orientation.

The clay–water system is ideal (*4, 5*) for such concentration dependence NMR studies, especially if the clay is of the montmorillonite type (i.e., a smectite). The fundamental particle is an aluminosilicate sheet approximately ten angstroms thick and hundreds or thousands of angstroms in length and width. Nearly all of the surface area is on the faces of these sheets. The faces are made of oxygen atoms arranged in a hexagonal pattern, and exchangeable cations such as sodium are present. It is possible to prepare oriented samples in which the sheets are essentially parallel. The addition of water causes the "books" of clay sheets to expand uniformly. By varying the amount of water, it is possible to prepare samples with known thicknesses of water layers alternating with the parallel clay sheets.

If the water is deuterated, a doublet splitting in the deuteron NMR spectrum can be observed. This doublet splitting arises from the anisotropic orientation of water molecules with respect to the clay sheet. In a given oriented sample, one doublet splitting value is observed. In previous work (*4*), the deuteron doublet splittings were measured for a series of oriented sodium hectorite samples containing from 0.856 to 8.408 cm^3 of water per gram of dry clay. These values together with proton doublet splittings for samples containing as little as 0.3 cm^3 of water per gram of dry clay are in agreement with a very simple explanation. In a given oriented sample the rapid diffusion of water causes all of the water molecules to experience the same average preferential orientation because they all experience the various positions in the space between the clay sheets in a short time. The concentration dependence of the doublet splitting agrees, then, with the NMR-sensitive preferential orientation residing in the first one or two molecular layers. In other words, the NMR-sensitive preferential orientation is a short-range surface effect for this system.

It is of interest to determine for the same system the range of the surface effect on the rate of molecular rotations. In the present work, this was done by measuring the T_1 as a function of water content for a series of samples which were very carefully prepared so as to yield results of the highest quality.

CONTRIBUTIONS TO THE OBSERVED RELAXATION

It is well known that paramagnetic centers which can arise from the presence of iron atoms can have very strong relaxation effects. Clays typically contain some iron. However, in order to use relaxation data to study molecular motions, a correction must be made for this paramagnetic contribution to relaxation. The observed spin–lattice relaxation time T_1 is given by the following sum of relaxation contributions

$$\frac{1}{T_1}=\frac{1}{T_{1n}}+\frac{1}{T_{1p}}, \qquad [1]$$

where $1/T_{1p}$ is the contribution of the paramagnetic centers and $1/T_{1n}$ is the nonparamagnetic contribution. This equation is valid for both protons and deuterons. The quantity $1/T_{1n}$ contains the information on molecular motions. Fortunately, the paramagnetic effect for deuterons is different from that for protons so that $1/T_{1n}$ can be extracted from the experimental data. Theoretically (*6*),

$$\frac{1}{T_{1p}\mathrm{H}}=\left(\frac{\gamma(\mathrm{H})}{\gamma(\mathrm{D})}\right)^2\frac{1}{T_{1p}\mathrm{D}}, \qquad [2]$$

where $1/T_{1p}\mathrm{H}$ is the paramagnetic contribution to the proton relaxation, $1/T_{1p}\mathrm{D}$ is that for the deuteron relaxation, $\gamma(\mathrm{H})$ is the proton magnetogyric ratio, and $\gamma(\mathrm{D})$ is the deuteron magnetogyric ratio. The value of the theoretical ratio $(\gamma(\mathrm{H})/\gamma(\mathrm{D}))^2$ is 42.4374. In measurements on a D_2O+H_2O solution of $FeCl_3$, the theoretical ratio was experimentally verified within 1%. Hence, the theoretical ratio can be used to correct the data in order to obtain values of $1/T_{1n}\mathrm{D}$.

If the rotation of the water molecule is described by correlation times τ_{ci}, the deuteron nonparamagnetic relaxation contribution is given by the relationship (*5, 7, 8*)

$$\frac{1}{T_{1n}\mathrm{D}}=\frac{3}{40}\left(\frac{e^2qQ}{\hbar}\right)^2\sum_i K_i\left[\frac{\tau_{ci}}{1+\omega_0^2\tau_{ci}^2}+4\frac{\tau_{ci}}{1+4\omega_0^2\tau_{ci}^2}\right], \qquad [3]$$

where $e^2qQ/\hbar$ is 2π times the electric quadrupole coupling constant of the deuteron in the water molecule, $\sum_i K_i=1$, and ω_0 is 2π times the NMR frequency expressed in hertz. Early work (*9*) on the proton T_1 temperature dependence of low-water-content samples indicates that $\omega_0^2\tau_{ci}^2\ll 1$ at room temperature. This condition allows us to simplify Eq. [3] to

$$\frac{1}{T_{1n}\mathrm{D}}=\frac{3}{8}\left(\frac{e^2qQ}{\hbar}\right)^2\tau_{c,av}, \qquad [4]$$

where $\tau_{c,av}$ is the effective average of all the correlation times τ_{ci}. Molecular rotation determines the value of $\tau_{c,av}$ because $\tau_{c,av}$ is the average time required for a molecule to rotate 57.296° around a molecular axis. In the clay-water system, there are many different τ_{ci} values because the molecules at the surface rotate anisotropically and more slowly compared to molecules further away from the surface. A measurement of $1/T_{1n}$D enables us to estimate the average molecular rotational rate.

EXPERIMENTAL DETAILS AND RESULTS

Hectorite clay was chosen for these experiments because the exceptionally low iron content results in the smallest paramagnetic contribution to the deuteron $1/T_1$. It is desirable to have the smallest possible paramagnetic correction because of any possible uncertainties in the applicability of Eq. [2].

The samples were very carefully prepared to assure uniformity and accuracy of water contents. The hectorite was sodium ion-exchanged, washed, and centrifuged to select the fraction less than 0.1 μm in particle size. The clay gel was freeze-dried and portions were placed in weighed Pyrex sample vials. The freeze-drying process results in crystallites of parallel clay sheets. These samples were placed in an atmosphere of 51% relative humidity of H_2O for 52 days. Then, the sample vials together with the contents were weighed, deuterium oxide was added from a calibrated syringe, and the sample vials were sealed. The samples were allowed the equilibrate 37 days, and then the H_2O contents of the samples and a D_2O blank were carefully measured from pulsed NMR signal amplitudes. The consistency of the water analysis is shown by the resultant value of 0.11 gram of H_2O per gram of dry clay adsorbed by each sample during the residence in the 51% relative humidity atmosphere.

The NMR T_1 measurements were made with a classical pulsed NMR apparatus employing 180—90° pulse sequences and a boxcar integrator. The temperature was controlled at 25.0°C by means of a gas-flow cryostat. Proton T_1 measurements of the residual aqueous protons in the samples were made at 8.0- and 25.0-MHz NMR

TABLE 1

THE 25.0-MHz ROOM TEMPERATURE PROTON RELAXATION RATES IN THE D_2O–SODIUM HECTORITE SAMPLES

Sample	g/cm^3	$1/T_1$ (sec^{-1})
1	0.09172	3.80 ± 0.09
2	0.17771	7.39 ± 0.11
3	0.34840	14.45 ± 0.06
4	0.50848	21.34 ± 0.16
5	0.67167	27.93 ± 0.17
6	0.83555	35.18 ± 0.16
7	1.04299	43.80 ± 0.20
8	1.32529	56.00 ± 0.30
9	1.51883	64.80 ± 0.20
10	1.85726	79.20 ± 0.40
11	2.27482	96.10 ± 0.80
12	3.24363	135.40 ± 1.40

TABLE 2

DEUTERON RELAXATION RATES AT 25.0°C AND 15.3 MHz FOR D_2O–SODIUM HECTORITE SAMPLES

Sample	g/cm^3	$1/T_1D$ (sec^{-1})
Blank		2.2382 ± 0.0056
1	0.09172	2.5550 ± 0.0048
2	0.17771	2.8258 ± 0.0057
3	0.34840	3.3599 ± 0.0112
4	0.50848	3.9552 ± 0.0135
5	0.67167	4.4686 ± 0.0152
Blank		2.2294 ± 0.0074
6	0.83555	4.9874 ± 0.0134
7	1.04299	5.6064 ± 0.0128
8	1.32529	6.5018 ± 0.0148
9	1.51883	7.1278 ± 0.0331
10	1.85726	8.3458 ± 0.0233
11	2.27482	9.6660 ± 0.0438
12	3.24363	13.161 ± 0.124
12*	3.24363	12.743 ± 0.231
11*	2.27482	9.6991 ± 0.0880
1	0.09172	2.5504 ± 0.0052
Blank		2.2471 ± 0.0040

frequencies. The boxcar was gated such that the clay lattice hydroxyl protons did not interfere with the T_1 measurements on the aqueous protons. The T_1 values at the two different frequencies are nearly identical. The 25-MHz values shown in Table 1 are judged to be the more accurate. The quoted errors for these data and for the deuteron T_1 data are for the individual relaxation curves calculated for a 95% confidence index. Deuteron T_1 measurements were made at 8.0 and 15.3 MHz. As for the proton T_1 measurements, the T_1 values are nearly identical at both frequencies. Because of the greater signal-to-noise ratio, the 15.3-MHz data shown in Table 2 are the more accurate. Measurements of the D_2O blank were made several times as a check on the operation of the spectrometer. With the exception of the data marked by an asterisk, the NMR signal was measured by a boxcar gate initiated 40 μsec after the leading edge of the 8-μsec 90° pulse. Because of residual pulse transients and weak signals, samples 11 and 12 were also measured with the boxcar gate placed at a later time on the 90° pulse signal. These measurements are marked by an asterisk. The results for sample 11 are very close for both measurements; however, this is not the case for sample 12, where the results differ by 3%.

DATA ANALYSIS AND DISCUSSION

It is important to analyze the relaxation data in a manner appropriate to investigating the range of the surface effect on the rotational motions of the water molecules.

Conceptually, it is simpler to analyze the proton T_1 data. The proton $1/T_1$ values in Table 1 are much greater than the deuteron $1/T_1$ values in Table 2. However, in pure H_2O the proton $1/T_1$ value is an order of magnitude smaller than the deuteron $1/T_1$ in D_2O. Hence the greater proton $1/T_1$ values in these samples result from the relaxation effect of clay paramagnetic centers as given by Eq. [2]. Since the relaxation effect of these paramagnetic centers changes as the inverse sixth power of the distance between the proton and the paramagnetic center, the range of the effect is very short, about a molecular layer. At any instant, the T_1 of the protons of the surface water molecules is short while that of the other protons is very long. However, the rapid translational diffusion causes all of the aqueous protons to spend part of the time at the surface so that a single average $1/T_1$ is observed (*10, 11*). The value of $1/T_1$ is thus linearly related to the fraction of the water molecules at the surface at any one time. Since the number of molecules at the clay surface is directly proportional to the quantity of clay present, the fraction of molecules at the surface is directly proportional to g/cm^3, i.e., the number of grams of dry clay per cubic centimeter of water. As the observed $1/T_1$ is the weighted average (*10, 11*) $1/T_1$ for all the aqueous protons, these concepts can be formulated as

$$\frac{1}{T_1}=\frac{f_b}{T_{1b}}+\frac{f_s}{T_{1s}}, \qquad [5]$$

where f_b = fraction of bulk water, f_s = fraction of surface water, $T_{1b} = T_1$ of water in the bulk, and $T_{1s} = T_1$ of water while at the surface.

Of course, $f_b + f_s = 1$. If C is the number of cubic centimeters of water per gram of dry clay at any instant with a T_1 less than T_{1b},

$$f_s = C(g/cm^3), \qquad [6]$$

where g is the number of grams of dry clay in the sample and cm^3 is the number of cubic centimeters of water in the sample. Then, Eq. [5] can be rewritten as

$$\frac{1}{T_1}=\frac{1}{T_{1b}}+C\left(\frac{1}{T_{1s}}-\frac{1}{T_{1b}}\right)\left(\frac{g}{cm^3}\right). \qquad [7]$$

According to this equation, a plot of $1/T_1$ versus g/cm^3 is a straight line with an intercept of $1/T_{1b}$ at $g/cm^3 = 0$. Linear regression analysis of the 25-MHz proton data in Table 1 yields an intercept of $0.087\ sec^{-1}$, and a slope of $42.04 \pm 0.041\ cm^3\ g^{-1}\ sec^{-1}$. The error in the slope is calculated for a 95% confidence index. Since this value of the intercept is within experimental and statistical error of the measured proton $1/T_1$ of the blank, $0.11\ sec^{-1}$, Eq. [7] does appear to describe the proton data.

This form of Eq. [7] can also be used for a regression analysis of the observed deuteron relaxation data in Table 2. Figure 1 shows the observed deuteron $1/T_1$ values plotted versus g/cm^3. The data fit a straight line. The straight line represents the values predicted by the parameters obtained from linear regression analysis. In this analysis, the values for the D_2O blank are not included, the values for samples 1 and 11 are the average values, and the values for sample 12 are not included. These parameters thus obtained are an intercept of $2.249\ sec^{-1}$, and a slope of $3.255 \pm 0.044\ cm^3\ g^{-1}\ sec^{-1}$. The value for the intercept is very close to the average $1/T_1D$ value for the D_2O blank, $2.238\ sec^{-1}$. Table 3 contains the $1/T_1D$ values used in the

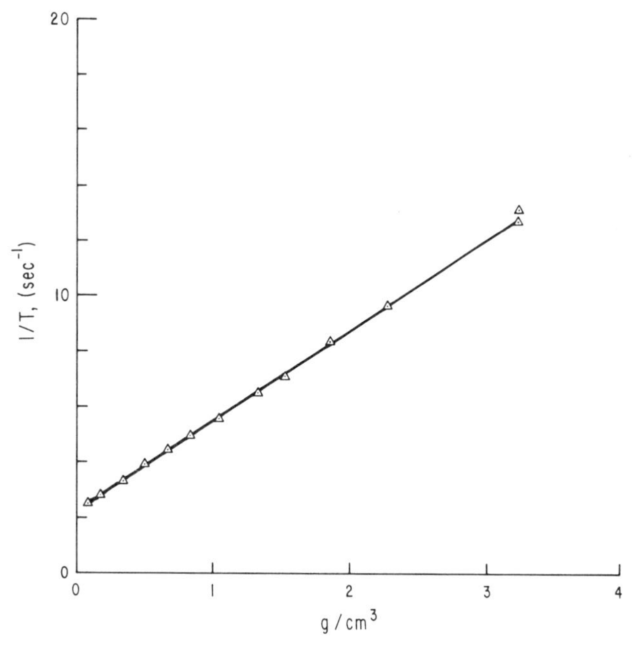

FIG. 1. The deuteron $1/T_1$ values plotted versus sodium hectorite concentration. The straight line is the concentration dependence predicted from the results of the linear regression analysis.

data analysis and also the difference between these values and the values predicted by the above values of slope and intercept. These differences are similar to the errors given in Table 2. Also, the value predicted for sample 12 is very close to one of the measured values.

TABLE 3

THE DEUTERON RELAXATION IN THE D_2O–SODIUM HECTORITE SAMPLES AND THE DEVIATIONS FROM THE LINEAR REGRESSION ANALYSIS

Sample	g/cm^3	$1/T_1D$ (sec^{-1})	$1/T_1D - 1/T_1(r)$ (sec^{-1})
1	0.09172	2.5527	0.0049
2	0.17771	2.8258	−0.0019
3	0.34840	3.3599	−0.0234
4	0.50848	3.9552	0.0509
5	0.67167	4.4686	0.0332
6	0.83555	4.9874	0.0186
7	1.04299	5.6064	−0.0375
8	1.32529	6.5018	−0.0610
9	1.51883	7.1278	−0.0649
10	1.85726	8.3458	0.0516
11	2.27482	9.6826	0.0294
12	3.24363	12.743	−0.0634
12	3.24363	13.161	0.3546

The object of this work is to examine the range of the surface effect on the rotational motions of the water molecules by using the nonparamagnetic part, $1/T_{1n}D$, of the deuteron relaxation. At this point, it is possible that the linear relationship of $1/T_1D$ versus g/cm^3, which would indicate a short-range surface effect, is an artifact of the paramagnetic contribution to $1/T_1D$. Accordingly, the $1/T_1D$ values in Table 3 are corrected by using Eqs. [1] and [2] together with the 25-MHz proton $1/T_1$ values in Table 1 to obtain the $1/T_{1n}D$ values given in Table 4. The results of linear regression analysis of the data from samples 1 through 11 using the form of Eq. [7] are an intercept of $2.256\ \text{sec}^{-1}$, and a slope $= 2.253 \pm 0.046\ \text{cm}^3\ \text{g}^{-1}\ \text{sec}^{-1}$. Again, the intercept is close to the $1/T_1D$ value for the blank. The $1/T_{1n}D$ values and the predictions of these slope and intercept values are shown in Fig. 2. The small differences between the experimental and predicted $1/T_{1n}D$ values are shown in Table 4. The excellent fit of the $1/T_{1n}D$ data to the form of Eq. [7] can be used in further considerations of the clay–water interaction.

In order to discuss the $1/T_{1n}D$ data in terms of the range of the surface effect on the rotational motions of the water molecules, $1/T_{1n}D$ must be formulated appropriately. In accordance with Eqs. [3] and [4] and rapid-exchange theory (*12, 13*), $\tau_{c,av}$ is the effective average rotational correlation time over all of the water molecules in all the various locations in the sample with respect to the clay surface. It is possible to divide the water molecule locations into two types. In one type, the average correlation time is equal to τ_{cb}, the value in bulk water. In the other type, the effective average correlation time τ_{cs} is different from τ_{cb} because of the surface effect. The effect of adding the hectorite clay to the water is to cause the introduction of water locations with correlation times different from τ_{cb}. Since the observed $1/T_{1n}D$ is determined by the weighted effective average correlation time, the introduction of a very small amount of clay allows the $1/T_{1n}D$ to be given by the relationship (*12, 13*)

$$\frac{1}{T_{1n}D} = \frac{3}{8}\left(\frac{e^2qQ}{\hbar}\right)^2 (p_b\tau_{cb} + p_s\tau_{cs}), \quad [8]$$

where p_b is the fraction of water molecules with average correlation time equal to τ_{cb}, and p_s is the fraction of water molecules with effective average correlation time τ_{cs} different from τ_{cb}. In the range of small clay contents, the value of τ_{cs} is independent of the clay content value. If K is the number of cubic centimeters of water per gram of dry clay at any instant with average correlation time different from τ_{cb},

$$p_s = K(g/cm^3), \quad [9]$$

where g and cm^3 are defined as before. Then, since $p_b + p_s = 1$, Eq. [8] can be rewritten as

$$\frac{1}{T_{1n}D} = \frac{3}{8}\left(\frac{e^2qQ}{\hbar}\right)^2 \tau_{cb} + \frac{3}{8}\left(\frac{e^2qQ}{\hbar}\right)^2 K(\tau_{cs} - \tau_{cb})(g/cm^3). \quad [10]$$

As in the case of Eq. [7], a plot of $1/T_{1n}D$ versus g/cm^3 is a straight line with an intercept equal to the $1/T_1$ value of the blank. With increasing g/cm^3, the value of $1/T_{1n}D$ should increase with a constant slope until $p_b = 1$ or until τ_{cs} changes. The point at which the slope changes gives the g/cm^3 value which defines the range of the effect of the surface in affecting the molecular rotational motions.

TABLE 4

THE $1/T_{1n}D$ VALUES IN THE D_2O–SODIUM HECTORITE SAMPLES AND THE DEVIATIONS FROM THE LINEAR REGRESSION ANALYSIS

Sample	g/cm^3	$1/T_{1n}D$ (sec^{-1})	$1/T_{1n}D - 1/T_1(r)$ (sec^{-1})
1	0.09172	2.4631	0.0004
2	0.17771	2.6518	−0.0046
3	0.34840	3.0194	−0.0215
4	0.50848	3.4522	0.0508
5	0.67167	3.8104	0.0414
6	0.83555	4.1585	0.0203
7	1.04299	4.5738	−0.0316
8	1.32529	5.1822	−0.0591
9	1.51883	5.6018	−0.0754
10	1.85726	6.4798	0.0403
11	2.27482	7.4191	0.0390
12	3.24363	9.5529	−0.0094
12	3.24363	9.9709	0.4086

The excellent fit of the $1/T_{1n}D$ data to Eq. [10] can be used to show that several models of clay–water interaction are acceptable and that some others are not acceptable. If we assume that the density of sodium hectorite is 2.80 g/cm^3, that the thickness (*14*) of a hectorite sheet is 9.60×10^{-8} cm, and that the molar volume of

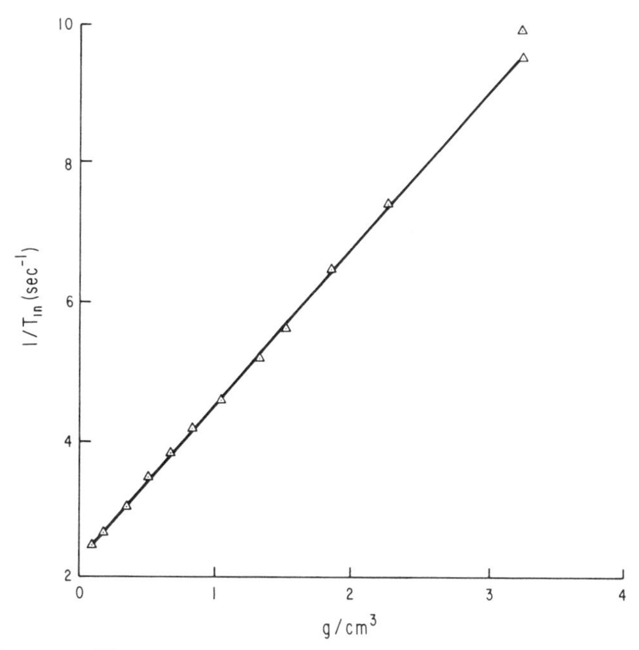

FIG. 2. The deuteron $1/T_{1n}$ values plotted versus sodium hectorite concentration. The straight line is the concentration dependence predicted from the results of the linear regression analysis.

water is 18.015 cm^3, the T_1 data cover the range from 2.7 to 108.3 molecular layers of water between consecutive clay sheets. The fit of Eq. [10] to the data over this entire range is consistent with the following picture: the effective average correlation time of the water in the molecular layers contacting a clay surface is much longer than that of pure water and the average correlation time of the water at least one molecular layer away from the surface is equal to that of pure D_2O. The range of the surface effect on the rotation of the water molecules thus is a molecular layer or two.

A longer range of significant slowing down of molecular rotation would cause a change in the slope of the plot of $1/T_{1n}D$ versus g/cm^3. On the other hand, an extremely long range would cause the intercept of the present data to deviate significantly from the $1/T_1$ of pure D_2O. These two points are illustrated below. For concreteness, a specific range equation is assumed. The above conclusion of short-range effects does not depend on the validity of this assumed equation.

More insight on the range parameter can be obtained from the following model, which has superficial attractiveness. The effective average correlation time for a clay–water system consisting of N layers of water molecules between clay sheets, each layer having the same number of water molecules, can be written generally as

$$\tau_{av} = \frac{1}{N}\sum_{i=1}^{N} \tau_i, \qquad [11]$$

where τ_i is the average correlation time of a molecule while it resides in layer i. Assume that the effect of the surface in slowing down molecular rotation (compared to pure water) decreases exponentially with distance of the molecular layer from the surface. If we replace the summation by an integral, we obtain for $1/T_{1n}D$

$$\frac{1}{T_{1n}D} = \frac{3}{8}\left(\frac{e^2qQ}{\hbar}\right)^2\left[\tau_{cb} + B\frac{\int_0^S \exp(-s/r)\,ds}{\int_0^S ds}\right], \qquad [12]$$

where B is a measure of the magnitude of the surface effect, r is an interaction distance parameter, s is the distance of the molecule from the surface, and S is the maximum distance of a molecular layer from the clay surface. Clearly, S is directly proportional to the water content so that $S = P(cm^3/g)$, where P is a proportionality constant. If we integrate Eq. [12], we obtain

$$\frac{1}{T_{1n}D} = \frac{3}{8}\left(\frac{e^2qQ}{\hbar}\right)^2\left\{\tau_{cb} + \frac{B}{P}\left(\frac{g}{cm^3}\right)r\left[1 - \exp\left(-\frac{S}{r}\right)\right]\right\}. \qquad [13]$$

Since S contains g/cm^3, a plot of $1/T_{1n}D$ versus g/cm^3 would exhibit a change in slope when g/cm^3 becomes great enough so that $s/r \approx 1$. Then, according to Fig. 2, the value of r is less than 1.4 molecular layers of water.

On the other hand, an exponential decay is very nearly linear for small values of the argument in the exponential function. Thus, the consequences of a very-long-range effect ($S/r \ll 1$) must be examined. In this case, Eq. [13] reduces to

$$\frac{1}{T_{1n}D} = \frac{3}{8}\left(\frac{e^2qQ}{\hbar}\right)^2\left\{\tau_{cb} + B\left(1 - \frac{1}{2}\frac{S}{r}\right)\right\}, \qquad [14]$$

so that the fractional change in $1/T_{1n}D$ is small compared to unity when S is changed by varying g/cm^3. Also, a significant long-range effect would yield an intercept (when plotting data for $S/r \ll 1$) which is significantly greater than the $1/T_1$ value of pure D_2O.

The use of Eq. [10] indicates that the $1/T_{1n}D$ data are consistent with a short-range clay surface effect. (The use of Eq. [12], which employs a specific distance dependency of the surface effect, serves merely to illustrate some of the concepts embodied in the application of Eq. [10].) The magnitude of this surface effect can be estimated by assuming that only the surface layer rotates at a different rate compared to the pure liquid water. Using the earlier assumptions of water density, hectorite platelet thickness, and hectorite density, we find that the value of K for Eq. [10] is 0.23116 cm^3 of water per gram of dry hectorite. Then, using the values slope = 2.253 $cm^3\ g^{-1}\ sec^{-1}$ and $1/T_1(D_2O) = 2.238\ sec^{-1}$, we obtain the ratio

$$\frac{1/T_1(C)}{1/T_1(D_2O)} = 5.35.$$

If Eq. [4] is applicable, the value of this ratio means that pure D_2O rotates 5.35 times as fast as the average rate for the surface monolayer. (Deuteron T_1 measurements made for the hectorite samples at 8.0 MHz are only a few percent shorter than the precise values given in Table 2. This observation, together with the proton T_1 data and Eq. [3], indicates that Eq. [4] is indeed applicable.) Since there may be a contribution of specific interacting groups to the slowing down of the rotation of the water molecules, these results show that the effect of the surface on slowing down the water molecules is unexpectedly small. The above result is similar to those reported (*15*) for diffusion of 0.25 cm^3 of water per gram of dry sodium montmorillonite. The diffusion coefficient of pure water is 3.3 times that of the clay water. The diffusion coefficient reflects the translational motion of the water, and hence it is another measure of the effect of the clay surface on molecular mobility.

CONCLUSIONS

These concentration dependence measurements of the deuteron $1/T_1$ measurements in the hectorite–water system agree with a short-range effect of the hectorite surface in affecting the rotational motions of the water molecules. The surface causes the nearby molecules to rotate approximately five times slower than those in bulk water. Since the hectorite surface consists of oxygen atoms, this relatively weak short-range effect may not be the case in systems in which the solid substrate contains strongly hydrophylic groups such as the hydroxyl group.

REFERENCES

1. W. Drost-Hansen, *Ind. Eng. Chem.* **61,** 10 (1969).
2. K. J. Packer, *Phil. Trans. R. Soc. London B* **278,** 59 (1977).
3. R. H. Walmsley and M. Shporer, *J. Chem. Phys.* **68,** 2584 (1978).
4. D. E. Woessner, *in* "Mass Spectrometry and NMR Spectroscopy in Pesticide Chemistry" (R. Haque and F. J. Biros, Eds.), p. 279, Plenum, New York, 1974.

5. D. E. WOESSNER, *Mol. Phys.* **34,** 899 (1977).
6. G. LAUKIEN AND F. NOACK, *Z. Phys.* **159,** 311 (1960).
7. D. E. WOESSNER, *J. Chem. Phys.* **37,** 647 (1962).
8. D. E. WOESSNER, *J. Chem. Phys.* **40,** 2341 (1964).
9. D. E. WOESSNER AND B. S. SNOWDEN, JR., *J. Colloid Interface Sci.* **30,** 54 (1969).
10. J. R. ZIMMERMAN AND W. E. BRITTIN, *J. Phys. Chem.* **61,** 1328 (1957).
11. C. F. HAZLEWOOD, D. C. CHANG, B. L. NICHOLS, AND D. E. WOESSNER, *Biophys. J.* **14,** 583 (1974).
12. D. BECKERT AND H. PFEIFER, *Ann. Phys.* **16,** 262 (1965).
13. H. G. HERTZ, *Ber. Bunsenges. Phys. Chem.* **71,** 979 (1967).
14. C. T. DEEDS AND H. VAN OLPHEN, *Clays Clay Miner.* **10,** 318 (1962).
15. W. D. KEMPER, D. E. L. MAASLAND, AND L. K. PORTER, *Soil Sci. Soc. Am. Proc.* **28,** 164 (1964).

JOURNAL OF MAGNETIC RESONANCE **43,** 502–507 (1981)

Composite Pulse Decoupling

Efficient decoupling over the entire band of proton resonance frequencies is an important factor in determining resolution and sensitivity of carbon-13 spectroscopy. Continuous monochromatic irradiation (*1*) does not achieve this goal, but several diverse modulation methods have been suggested, starting with coherent single-frequency modulation (*2*), noise modulation (*3*), square-wave phase modulation (*4*), and chirp frequency modulation (*5*). Unfortunately no general theory of wideband decoupling has yet emerged, and these experiments have been guided mainly by intuition or by trial and error.

Most present-day carbon-13 spectrometers employ the noise-decoupling method introduced by Ernst (*3*), where the second radiofrequency field B_2 is applied continuously but with phase inversions at pseudorandom intervals. The intensity of B_2 is normally set at its highest level compatible with the limitations on sample heating, leaving the frequency setting and the mean rate of phase inversion as adjustable parameters, the latter acting as a bandwidth control.

Broadband proton irradiation has two purposes—the establishment of a nuclear Overhauser enhancement in the period just prior to carbon-13 excitation, and decoupling during acquisition of the carbon-13 free-induction decay. The first goal only requires that the protons be saturated and is not particularly demanding; it is in the decoupling stage that shortcomings become apparent. Incomplete decoupling can leave small unresolved residual splittings which broaden the carbon-13 lines and reduce resolution and sensitivity. Noise modulation can be transmitted to the observed resonances, interfering with carbon-13 spin-echo measurements (*6*) and giving rise to noise sidebands in the flanks of the resonances, hindering the detection of carbon-13 satellite lines.

This communication presents some preliminary results which indicate that a new scheme for heteronuclear broadband decoupling is more effective than the techniques outlined above. Consider the case of weakly coupled spins I and S, where I represents the observed nuclear species (carbon-13) and S the strongly irradiated species (protons). The proposed method of broadband decoupling may be formulated on the basis of three principles:

(1) The S spins should undergo a coherent cyclic perturbation at a rate fast compared with the largest coupling constant J_{IS}.

(2) This cycle should consist of four elements, $R\ R\ R'\ R'$, where R is any perturbation known to invert longitudinal magnetization, and R' is the same perturbation with the radiofrequency pulse phase inverted.

(3) The operation of R and R' should be insensitive to the offset parameter $\Delta B/B_2$ over the frequency range of interest.

The innovation lies mainly in principle (2). The theoretical justification for this idea is quite complex and only an outline can be presented here. The Hamiltonian for two weakly coupled spins I and S is expressed as

0022–2364/81/060502-06$02.00/0

$$\mathcal{H}(t) = \mathcal{H}_I + \mathcal{H}_S(t), \qquad [1]$$

where

$$\mathcal{H}_I = -\Delta\omega_I I_Z + 2\pi J I_Z S_Z, \qquad \mathcal{H}_S(t) = \mathcal{H}_{rf}(t) - \Delta\omega_S S_Z.$$

In this equation $\mathcal{H}_S(t)$ represents a "stirring" of the S spins, made up of a sequence of rotations about tilted effective fields in the rotating frame. The theory can be easily generalized for the case of several coupled I and S spins.

The Liouville operator $L(t)$ for the system described by this Hamiltonian is given in general by

$$L(t) = \hat{T} \exp\left[-i \int_0^t \mathcal{H}(t')dt'\right], \qquad [2]$$

where $\hat{T}$ is the Dyson time-ordering operator. This expression can be shown to be equivalent to

$$L(t) = L_S(t)\tilde{L}_I(t), \qquad [3]$$

where

$$L_S(t) = \hat{T} \exp\left[-i \int_0^t \mathcal{H}_S(t')dt'\right], \qquad \tilde{L}_I(t) = \hat{T} \exp\left[-i \int_0^t \tilde{\mathcal{H}}_I(t')dt'\right] \qquad [4]$$

and

$$\tilde{\mathcal{H}}_I(t) = L_S^{-1}(t)\mathcal{H}_I L_S(t). \qquad [5]$$

Equations [3] to [5] have the following physical interpretation: $\tilde{\mathcal{H}}_I(t)$ represents $\mathcal{H}_I$ transformed into a second rotating frame defined by $L_S(t)$. $L_S(t)$ rotates S-spin operators about a sequence of effective fields $B_{eff}(t)$, the resultant of ΔB and B_2. At $t = 0$ the second rotating frame is coincident with the conventional rotating frame but rotates about B_{eff} at an angular frequency γB_{eff} radians per second. Since the phase of B_2 is switched during the cycle, the rotating frame similarly switches its axis of rotation. Only the part of $\mathcal{H}_I$ involving coupling with the S spins is affected by the transformation. In Eq. [3], $\tilde{L}_I(t)$ describes the evolution of the system in this second rotating frame, while $L_S(t)$ transforms this back into the conventional reference frame.

Now if a cycle of total duration τ can be chosen so that the tilted rotations return the S-spin operators to their starting position, then

$$L(\tau) = \tilde{L}_I(\tau). \qquad [6]$$

The term $\tilde{L}_I(\tau)$ may be evaluated by average Hamiltonian theory (*7*):

$$\tilde{L}_I(\tau) = \exp[-i\tau(\bar{\mathcal{H}}_I^{(0)} + \bar{\mathcal{H}}_I^{(1)} + \bar{\mathcal{H}}_I^{(2)} + \cdots)] \qquad [7]$$

with

$$\bar{\mathcal{H}}_I^{(0)} = \tau^{-1} \int_0^\tau L_S^{-1}(t)\mathcal{H}_I L_S(t)dt. \qquad [8]$$

The cycle must be rapid enough to satisfy the condition $2\pi J\tau \ll 1$, and since the higher-order correction terms $\bar{\mathcal{H}}_I^{(n)}$ have magnitudes of the order $(2\pi J\tau)^{n+1}/[(n+1)!\tau]$, it is permissible to retain only the zero-order term as a first approximation. Note that it was the choice of the decomposition of the Hamiltonian in Eq. [1] which caused the higher-order correction terms to be small; their frequency dependence is thus unimportant.

The transformed Hamiltonian can be written in general as

$$L_S^{-1}(t)\mathcal{H}_I L_S(t) = -\Delta\omega_I I_Z + 2\pi J I_Z[a(t)S_X + b(t)S_Y + c(t)S_Z]. \quad [9]$$

It is sufficient to calculate the behavior of the system over the complete cycle τ to define it for all times. This average is

$$\overline{\mathcal{H}}_I^{(0)} = -\Delta\omega_I I_Z + 2\pi J I_Z(\bar{a}S_X + \bar{b}S_Y + \bar{c}S_Z). \quad [10]$$

Providing that interest lies only in I-spin observables, this can be shown to be equivalent to

$$\overline{\mathcal{H}}_I^{(0)} = -\Delta\omega_I I_Z + 2\pi J I_Z S_Z(\bar{a}^2 + \bar{b}^2 + \bar{c}^2)^{1/2}, \quad [11]$$

the second term representing the residual IS interaction. This is the term which must vanish if the decoupling is to be effective, and it is clear that each of the averages $\bar{a}$, $\bar{b}$, and $\bar{c}$ must vanish. This criterion can be satisfied by the sequence of four perturbations, $R\ R\ R'\ R'$, provided that R accurately inverts longitudinal magnetization ($RI_ZR^{-1} = -I_Z$) over the frequency range of interest. The perturbation R' is similar except that it employs pulses shifted in phase π radians.

$$R' = R_{-Z}(\pi)RR_Z(\pi), \qquad \text{where} \qquad R_Z(\pi) = \exp(i\pi I_Z). \quad [12]$$

If each of the four elements of the cycle has a duration $d = \tau/4$, the average Hamiltonian may be written, for the first element,

$$\bar{h}_I^{(0)} = -\Delta\omega_I I_Z + 2\pi J I_Z d^{-1}\int_0^d [a(t)S_X + b(t)S_Y + c(t)S_Z]dt. \quad [13]$$

Over the full cycle of length τ

$$\overline{\mathcal{H}}_I^{(0)} = (1/4)\{\bar{h}_I^{(0)} + R^{-1}\bar{h}_I^{(0)}R + R^{-1}R^{-1}R_{-Z}(\pi)\bar{h}_I^{(0)}R_Z(\pi)RR + R^{-1}R^{-1}(R')^{-1}R_{-Z}(\pi)\bar{h}_I^{(0)}R_Z(\pi)R'RR\}. \quad [14]$$

Using the property $R^2 = 1$, it can be shown that the combination of the first term with the third, and the second with the fourth, cancels the mean values $\bar{a}$ and $\bar{b}$, while the combinations of first with the second, and the third with the fourth, cancels $\bar{c}$. Thus

$$\overline{\mathcal{H}}_I^{(0)} = -\Delta\omega_I I_Z, \quad [15]$$

which is the decoupled I-spin Hamiltonian. Note that in general a perturbation $R\ R'$ or a perturbation $R\ R$ would not alone guarantee that the spin-coupling interaction vanishes. More complicated sequences may be devised which partially cancel higher-order terms of the average Hamiltonian (*8*).

This prediction that a four-step cycle is essential throws an interesting light on the decoupling method proposed by Basus *et al.* (*5*), where "chirp modulation" causes S-spin inversion by adiabatic rapid passage. These authors observed that the method worked much better when the radiofrequency phase was inverted on the third and fourth sweeps of a four-step cycle. Unfortunately with this cycle it is difficult to meet the adiabatic conditions and still complete a full cycle in a time short compared with $(2\pi J)^{-1}$ sec.

The proposed criteria for broadband decoupling may be satisfied by several

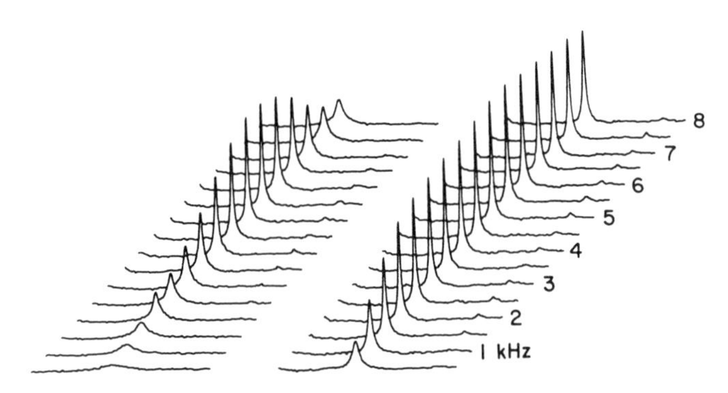

FIG. 1. A comparison of conventional noise decoupling (left) with composite pulse decoupling (right). The spectra are of carbon-13 in doped enriched methyl iodide with a spectral width of 125 Hz. The decoupler frequency has been offset in 500-Hz steps over a range of almost 8 kHz.

different pulse sequences with varying degrees of success. One that appears to be particularly promising may be written $R\ R\ R'\ R'$, where

$$R = R_X(\pi/2)R_Y(4\pi/3)R_X(\pi/2) \quad \text{and} \quad R' = R_{-X}(\pi/2)R_{-Y}(4\pi/3)R_{-X}(\pi/2). \quad [16]$$

The basic element of this sequence, R, may be recognized as a composite pulse sequence that has been previously described (*9*) and which behaves like a π pulse that is insensitive to resonance offset effects.

The proposed scheme for broadband "composite pulse decoupling" has been tested by studying the offset dependence of the carbon-13 spectrum of methyl iodide recorded on a Varian CFT-20 spectrometer. A comparison was made with conventional noise decoupling with a 2-kHz "bandwidth", usually found to be the optimum choice of the four bandwidths available. The decoupler field strength was measured as $\gamma B_2/2\pi = 6.3$ kHz. The observed signal peak heights were measured with respect to that observed with a coherent decoupler close to resonance for the methyl protons. Programming considerations dictated that the establishment of the nuclear Overhauser enhancement was achieved by noise irradiation in both series of experiments; this should have no effect on the comparison of the two methods.

Figure 1 shows a comparison of the spectra obtained by conventional noise decoupling (left) and composite pulse decoupling (right), while the normalized peaks heights are compared in Fig. 2. Clearly the proposed new method is effective over a much wider band. A further comparison was made on a sample of ethyl benzene with the decoupler level deliberately turned down to half ($\gamma B_2/2\pi = 3.1$ kHz) in both experiments. The signal-to-noise ratio is improved by almost a factor of 2 with composite pulse decoupling (Fig. 3).

These results are only preliminary, and further work is planned on the theory and practical aspects of the experiment. An alternative pulse sequence that satisfies the criteria for broadband decoupling may be written $R\ R\ R'\ R'$, where

$$R = R_X(3\pi/2) - t_p - R_X(3\pi/2) \quad \text{and} \quad R' = R_{-X}(3\pi/2) - t_p - R_{-X}(3\pi/2). \quad [17]$$

The interval t_p is set equal to $2/(\gamma B_2)$, allowing each isochromat to precess through

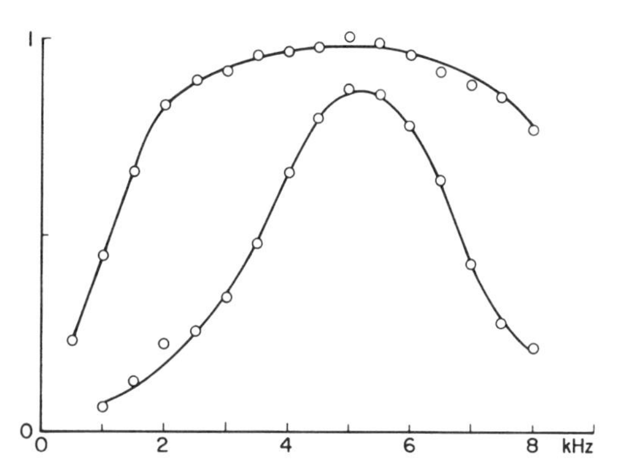

FIG. 2. A plot of the normalized peak heights observed in Fig. 1 against decoupler offset for conventional noise decoupling (lower curve) and composite pulse decoupling (upper curve). The full lines have no theoretical significance.

approximately 2θ radians (where $\tan \theta = \Delta B/B_2$), whatever the offset ΔB, provided that θ is not too large. This sequence is found to be better than noise decoupling but less effective than sequence [16]. It is also more sensitive to the correct setting of the pulse parameters, but has the practical advantage that no $\pi/2$ phase shifts of the radiofrequency B_2 are required, making it easier to implement on some spectrometers.

Sequences of the kind described above should prove their utility quite generally for heteronuclear broadband decoupling, and should be particularly appropriate for electrically conducting solutions or for samples that are temperature sensitive. For the decoupling of scalar interactions to fluorine, or dipolar interactions in a liquid crystal matrix, this method promises to work better than conventional irradiation techniques.

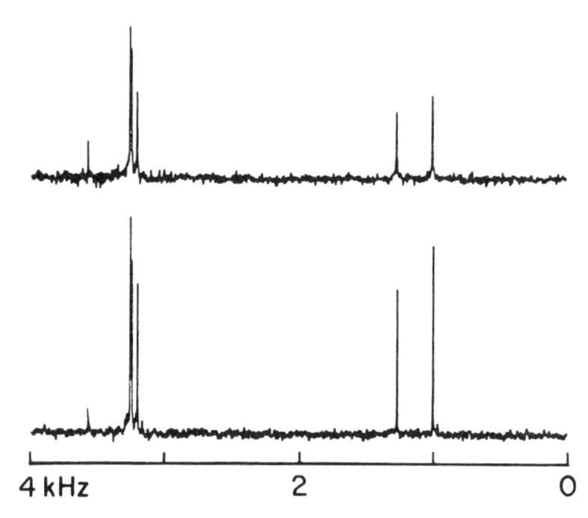

FIG. 3. Conventional noise decoupling (upper trace) compared with composite pulse decoupling (lower trace) applied to carbon-13 in a sample of ethyl benzene. The decoupler field intensity was deliberately reduced to one-half its normal setting for both experiments.

ACKNOWLEDGMENTS

This work was made possible by an equipment grant from the Science Research Council and was supported by a Research Studentship (M.H.L.). The authors are greatly indebted to Professor J. S. Waugh who presented a detailed exposition of average Hamiltonian theory during a recent NATO Summer School on NMR in Sardinia. Part of this work is the subject of a patent application by the National Research and Development Corporation of Great Britain.

REFERENCES

1. F. Bloch, *Phys. Rev.* **93,** 944 (1954).
2. W. A. Anderson and F. A. Nelson, *J. Chem. Phys.* **39,** 183 (1963).
3. R. R. Ernst, *J. Chem. Phys.* **45,** 3845 (1966).
4. J. B. Grutzner and R. E. Santini, *J. Magn. Reson.* **19,** 173 (1975).
5. V. J. Basus, P. D. Ellis, H. D. W. Hill, and J. S. Waugh, *J. Magn. Reson.* **35,** 19 (1979).
6. R. Freeman and H. D. W. Hill, *J. Chem. Phys.* **54,** 3367 (1971).
7. U. Haeberlen and J. S. Waugh, *Phys. Rev.* **175,** 453 (1968).
8. P. Mansfield, *J. Phys. C.* **4,** 1444 (1971).
9. R. Freeman, S. P. Kempsell, and M. H. Levitt, *J. Magn. Reson.* **38,** 453 (1980); M. H. Levitt and R. Freeman, *J. Magn. Reson.,* **43,** 65 (1981).

Malcolm H. Levitt*
Ray Freeman

Physical Chemistry Laboratory
Oxford University
Oxford, England

Received March 24, 1981

* Present address: Isotope Department, Weizmann Institute, Rehovot, Israel.

JOURNAL OF MAGNETIC RESONANCE **44**, 409–412 (1981)

Composite *Z* Pulses

With the advent of Fourier transform methods, experimental innovations in high-resolution NMR spectroscopy often involve the invention of new sequences of radiofrequency pulses. Each pulse in the sequence has a specified intensity, width, timing, and phase, all normally controlled by a small digital computer. The majority of these experiments can be carried out by selecting a radiofrequency phase which directs B_1 along one of the four directions $+X$, $+Y$, $-X$, $-Y$ of the rotating reference frame, but certain applications, for example, multiple-quantum spectroscopy, require phase shifts of less than $\pi/2$ rad. The computer-controlled hardware for this can be quite complicated (*1*). The present communication describes a new method of producing an effect equivalent to a radiofrequency phase shift of arbitrary magnitude without modification of the spectrometer. In fact, it is the precession phase of a nuclear magnetization vector or multiple-quantum coherence that is affected.

The method uses the concept of a "composite pulse," a combination of several radiofrequency pulses, usually with negligibly short intervals between the components, calculated to have an overall effect similar to that of a single pulse. Earlier applications of composite pulses (*2–4*) concentrated mainly on the compensation of pulse imperfections, principally the effects of resonance offset and spatial inhomogeneity of the radiofrequency field B_1 across the sample. However, a composite pulse may also be used to engineer a rotation about some normally inaccessible axis, an idea which has been much used in solid-state NMR experiments (*5*).

For the purposes of this treatment the radiofrequency pulses are assumed to be much shorter than the relaxation times, and much shorter than the inverse of all resonance offsets. Spatial inhomogeneity of the B_1 field is neglected. A pulse which rotates nuclear spin magnetization through α rad about the $+X$ axis of the rotating frame may be represented by the rotation operator

$$R_{+X}(\alpha) = \exp(i\alpha I_X), \qquad [1]$$

where $\alpha = \gamma B_1 t$, with B_1 the radiofrequency field intensity, and t the pulse duration. A rotation of θ rad about the $+Z$ axis of the rotating frame employs a combination of three radiofrequency pulses (*3*)

$$R_{+X}(\pi/2)R_{-Y}(\theta)R_{-X}(\pi/2) = R_{+Z}(\theta). \qquad [2]$$

The accepted convention here is that the order written is the reverse of the chronological order of the rotations; thus the first rotation is through $\pi/2$ rad about the *negative* X axis. Equation [2] can be proved by expanding the exponential of the second rotation operator as a power series and by introducing identities of the form $R_{+X}(\pi/2)R_{-X}(\pi/2) = 1$:

$$R_{+X}(\pi/2)R_{-Y}(\theta)R_{-X}(\pi/2) = \exp[-i\theta R_{+X}(\pi/2)I_Y R_{-X}(\pi/2)] = \exp(i\theta I_Z). \qquad [3]$$

0022-2364/81/080409-04$02.00/0

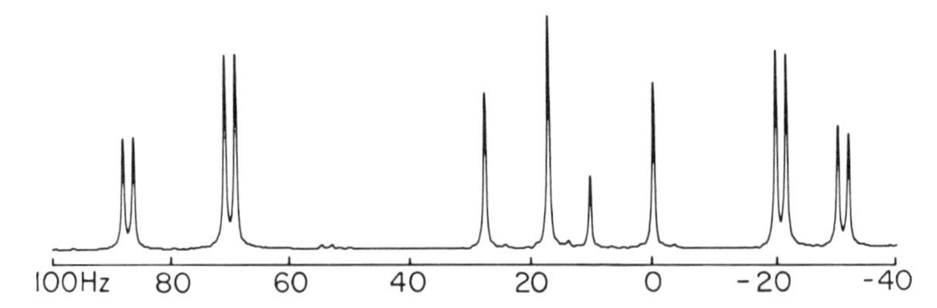

FIG. 1. The conventional high-resolution spectrum of acrylic acid at 200 MHz. Twelve allowed single-quantum transitions are observed and can be grouped into three "first-order" quartets.

The result can also be checked by considering the effect of the three successive rotations on the magnetization components M_X, M_Y, and M_Z separately.

For conventional nuclear magnetization vectors obeying the Bloch equations, the operator $R_{+Z}(\theta)$ is equivalent to a shift of precession phase through θ rad in a clockwise sense; for multiple-quantum coherence of order n it corresponds to a phase shift of $n\theta$ rad (*6*, *7*). This has already been exploited in double-quantum spectroscopy in order to discriminate between positive and negative double-quantum precession frequencies; a composite pulse nominally represented as $R_{+Z}(\pi/4)$ was used to achieve the equivalent of quadrature phase detection for the double-quantum signal (*8*).

The technique is tested here by using an $R_{+Z}(\pi/3)$ pulse to separate the various orders of multiple-quantum coherence in a three-spin system. This idea of incrementing the phases of multiple-quantum coherences at each increment of an evolution period t_1 in order to shift the frequencies of the various orders in a differential manner was introduced by Pines (*10*) and by Bodenhausen (*11*). The frequency shift is $n\theta\Delta F_1/\pi$, where ΔF_1 is the spectral width. Consequently for

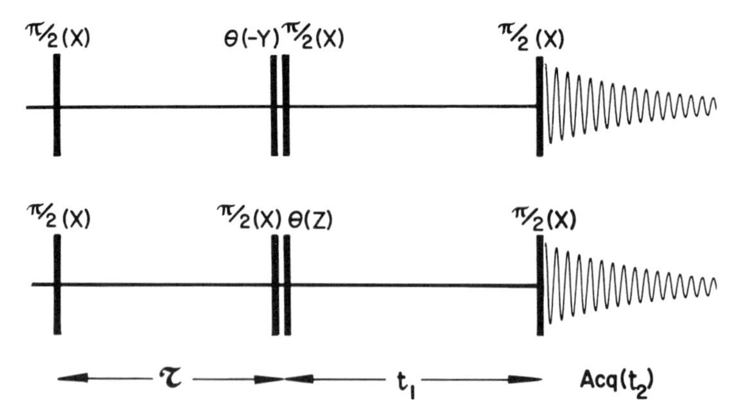

FIG. 2. The pulse sequence used for excitation, evolution, and detection of multiple-quantum coherences. The interval τ was fixed at 0.04 sec, comparable with the inverse of the proton–proton coupling constants. The bottom trace shows the effective sequence, with a rotation of θ rad about the $+Z$ axis at the beginning of the evolution period. The top trace shows the sequence used in practice, in which a three-component composite pulse has been consolidated with a $\pi/2$ pulse.

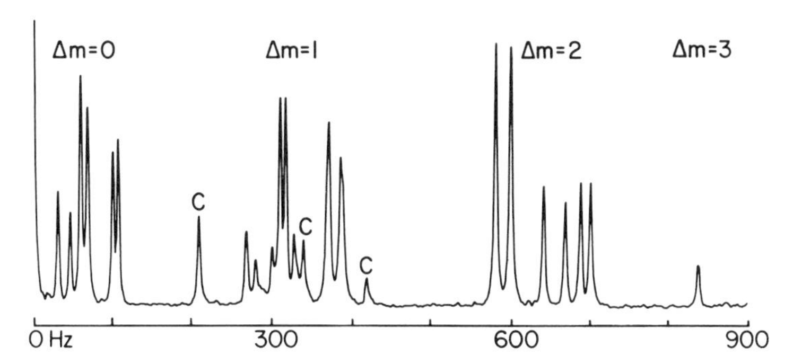

FIG. 3. Multiple-quantum signals from acrylic acid separated into their appropriate orders by the method described in the text. The three combination lines with $\Delta m = 1$ are marked C. The triple-quantum signal was aliased from above 900 Hz. There is a strong artifact at zero frequency.

a three-spin system, where the highest order of multiple-quantum coherence is 3, a rotation angle $\theta = \pi/3$ rad is the largest that would be appropriate.

The olefinic protons of acrylic acid provide an example of an AMX spin system at 200 MHz. The spectrum recorded on a Varian XL-200 spectrometer shows 12 allowed single-quantum transitions (Fig. 1). Two $\pi/2$ pulses separated by a suitable time interval τ (in practice 0.04 sec) generate all orders of multiple-quantum coherence; these are allowed to evolve for a variable period t_1 and then reconverted into detectable single-quantum signals by a third $\pi/2$ pulse (Fig. 2). Fourier transformation with respect to t_1 and t_2 generates a two-dimensional spectrum which may be projected onto the F_1 axis to give the multiple-quantum spectrum (*6*, *11*).

Figure 2 (bottom) shows the basic excitation and detection pulse sequence with the "*Z* pulse" inserted at the beginning of the evolution period. Since this composite pulse follows immediately after a $\pi/2$ pulse about the X axis, it is possible to consolidate the four pulses into two:

$$\pi/2(+X)\pi/2(-X)\theta(-Y)\pi/2(+X) \equiv \theta(-Y)\pi/2(+X). \qquad [4]$$

Here the pulses are written in chronological order, in contrast to the ordering of the rotation operators of Eq. [2]. This consolidation accounts for the pulse sequence actually used in practice (Fig. 2, top). For each increment in t_1, there is an incremental rotation about the $+Z$ axis of $\pi/3$ radians. While zero-quantum signals are unaffected by this, single-quantum signals are shifted to the right by one-third of the spectral width, double-quantum signals by two-thirds, and the triple-quantum signal by the entire spectral width. Choice of a suitably large spectral width (in practice 900 Hz) allows a clear separation of the various orders (Fig. 3). Note that in this spectrum the formally forbidden single-quantum "combination lines" make their appearance (marked C); they correspond to two protons flipping in one sense while the third flips in the opposite sense. Each order of multiple-quantum coherence generates a spectrum "referenced" with respect to 0, 300, 600, or 900 Hz, reflecting the transmitter offsets in Fig. 1. The triple-quantum transition is aliased from a frequency above 900 Hz. Because of the simple method of excitation of multiple-quantum coherences, the relative intensities in this spectrum are essentially arbitrary. Nevertheless, all the possible multiple-quantum signals are observed. An iterative analysis (*12*) of the conventional spectrum of Fig. 1 yielded accurate estimates of the energies of the eight energy levels of this spin system, which permitted the prediction of all the multiple-quantum frequencies. These were consistent with the observed frequencies derived from Fig. 3 within ± 0.3 Hz.

These results demonstrate that composite pulses are not restricted to the task of compensating pulse imperfections, but can have wider implications. In principle it should be possible to combine the function of rotation about an arbitrary axis with the compensatory properties of composite pulses, although the question of pulse imperfections has been disregarded in the present treatment.

ACKNOWLEDGMENTS

This work was made possible by an equipment grant from the Science Research Council and by two Research Studentships (M.H.L. and T.A.F.).

REFERENCES

1. G. Bodenhausen, *J. Magn. Reson.* **34,** 357 (1979).
2. M. H. Levitt and R. Freeman, *J. Magn. Reson.* **33,** 473 (1979).
3. R. Freeman, S. P. Kempsell, and M. H. Levitt, *J. Magn. Reson.* **38,** 453 (1980).
4. M. H. Levitt and R. Freeman, *J. Magn. Reson.* **43,** 65 (1981).
5. J. S. Waugh, L. M. Huber, and U. Haeberlen, *Phys. Rev. Lett.* **20,** 180 (1968).
6. A. Wokaun and R. R. Ernst, *Chem. Phys. Lett.* **52,** 407 (1977).
7. S. Vega and A. Pines, *J. Chem. Phys.* **66,** 5624 (1977).
8. A. Bax, R. Freeman, T. A. Frenkiel, and M. H. Levitt, *J. Magn. Reson.* **43,** 478 (1981).
9. A. Wokaun and R. R. Ernst, *Mol. Phys.* **36,** 317 (1978).
10. G. Drobny, A. Pines, S. Sinton, D. T. Weitekamp, and D. Wemmer, *Faraday Symp. Chem. Soc.* **13,** 49 (1979).
11. G. Bodenhausen, R. L. Vold, and R. R. Vold, *J. Magn. Reson.* **37,** 93 (1980).
12. A. A. Bothner-By and S. M. Castellano, "Computer Programs for Chemistry" (D. F. DeTar, Ed.), Benjamin, New York, 1968.

Ray Freeman
Thomas A. Frenkiel
Malcolm H. Levitt*

Physical Chemistry Laboratory
Oxford University
Oxford, England

Received May 14, 1981

* Present address: Isotope Department, Weizmann Institute, Rehovot, Israel.

JOURNAL OF MAGNETIC RESONANCE **44**, 542–561 (1981)

Investigation of Complex Networks of Spin–Spin Coupling by Two-Dimensional NMR

AD BAX* AND RAY FREEMAN

Physical Chemistry Laboratory, Oxford University, Oxford OX1 3QZ, England

Received April 9, 1981

Proton–proton couplings in a sample of 9-hydroxytricyclodecan-2,5-dione were studied by the two-dimensional Fourier transform method proposed by Jeener. Various techniques were tested for reshaping the two-dimensional responses to remove the long dispersion-mode tails, and to emphasize cross-peaks at the expense of diagonal peaks. Operation with a mixing pulse of relatively small flip angle allows relative signs of coupling constants to be determined by inspection of the two-dimensional spectrum. Weak long-range couplings, normally hidden within the linewidth, may be detected, although their magnitudes can only be estimated very approximately. A simple modification of the pulse sequence permits broadband decoupling in one frequency dimension, giving proton spectra without any J splittings.

INTRODUCTION

The very first two-dimensional Fourier transform NMR experiment was carried out by Jeener (*1*) and later analyzed in detail by Aue *et al.* (*2*). The concept was soon realized to be very general, and was extended to two-dimensional J spectroscopy (*3–5*), the correlation of proton and carbon-13 chemical shifts (*6–8*), investigations of chemical exchange (*9–11*), and the indirect detection of multiple-quantum transitions (*12–15*). These extensions were so successful that they diverted attention away from the original experiment in the form described by Jeener (*1*), where a simple sequence of two 90° pulses was applied to a proton spin system. One possible factor contributing to this neglect is that the experiment cannot be described in terms of a simple physical picture based on precessing magnetization vectors, but requires a density matrix treatment, which is quite cumbersome for systems of several coupled spins.

Nevertheless, even without a rigorous analysis, this experiment can be very useful for studying the pattern of proton–proton couplings in molecules of intermediate complexity (*16–18*). The present work investigates various methods of simplifying the resulting two-dimensional spectrum so that spin couplings are more readily assigned, and suggests modifications which render the technique more sensitive to the effects of long-range coupling. It is demonstrated that one such modification permits the relative signs of homonuclear spin–spin coupling constants to be obtained by inspection. Another variation generates a proton spectrum

* Present address: Department of Applied Physics, Delft University of Technology, Delft, The Netherlands.

0022–2364/81/090542–20$02.00/0

from which all J splittings have been removed (provided that the coupling is first order). This "broadband decoupling" of protons compares favorably with the established method, in which a 45° projection of a two-dimensional J spectrum is calculated (3). It is interesting to note the analogy between two-dimensional Fourier transform NMR and several different types of double irradiation experiment.

DESCRIPTION OF THE EXPERIMENT

In its simple form Jeener's experiment employs the radiofrequency pulse sequence

$$(T_W-90°-t_1-90°-t_2-)_n$$

to a system of homonuclear coupled spins, usually protons. The spectrometer receiver is activated only during the detection period t_2, acquiring a series of signals $S(t_2)$ as the sequence is repeated n times with the evolution period t_1 incremented in small steps Δt_1 over a suitable time range $n\Delta t_1$ of the order of the spin–spin relaxation time T_2. Between each sequence a period T_W is allowed for spin–lattice relaxation. Information about the evolution of the nuclear spin magnetization vectors during t_1 is coded into the modulation of the detected signals $S(t_2)$ and a data matrix $S(t_1, t_2)$ is built up and stored on a disk.

Two-dimensional Fourier transformation of $S(t_1, t_2)$ generates a two-dimensional spectrum $S(F_1, F_2)$ which contains three distinct categories of signal (2). Any nuclear magnetization that was aligned along the Z axis just before the second 90° pulse (the "mixing" pulse) gives rise to a series of lines along the F_2 axis essentially identical with the conventional proton spectrum. These "axial peaks" are of little interest for the purpose of studying spin–spin coupling, so it is convenient to suppress them by alternating the phase of the second radiofrequency pulse (18).

Two other types of signal appear in the two-dimensional spectrum (Fig. 1). The first are peaks which fall on or near the principal diagonal ($F_1 = F_2$); these are the resonances grouped around the coordinates (δ_A, δ_A) or around (δ_B, δ_B) of Fig. 1. They arise from magnetization which either maintains the same resonance frequency during t_1 and t_2 (these peaks fall exactly on the diagonal) or is transferred to another resonance in the same spin multiplet ("parallel" transitions). When the mixing pulse is 90° these peaks all have the same phase, the two-dimensional dispersion; that is to say, sections through such a response parallel to the F_1 or F_2 axes have a pure dispersion-mode profile.

It is the third type of signal on which attention is focused. These "cross-peaks" lie at a distance from the principal diagonal at coordinates determined by the chemical shifts of two coupled nuclei, and thus serve to identify coupled resonances. Cross-peaks arise from magnetization that was precessing in the transverse plane during t_1 and which is transferred to a second nuclear spin by the mixing pulse, thus changing its precession frequency by an amount comparable with the chemical shift difference $\delta_A - \delta_B$ between the two coupled sites. Thus in first-order spectra, cross-peaks lie so much further from the principal diagonal than any diagonal peaks that they are readily distinguished. Cross-peaks are naturally grouped into "two-dimensional spin multiplets," in the simple case illustrated in Fig. 1 a square pattern centered at (δ_A, δ_B) or at (δ_B, δ_A), with a splitting J in both dimensions.

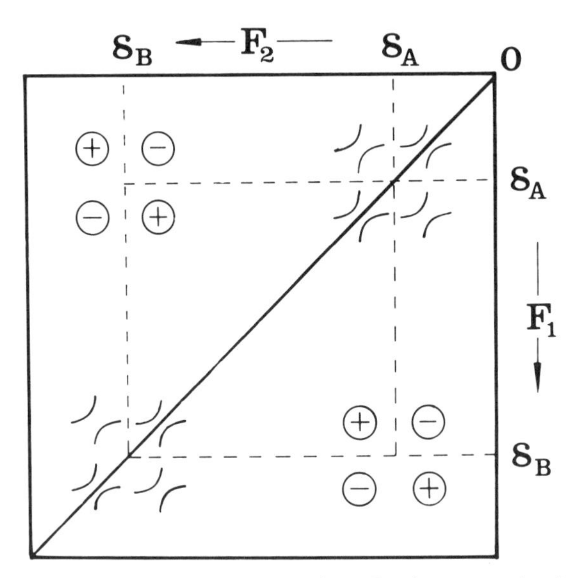

FIG. 1. Schematic diagram of the Jeener spectrum of an AB spin system, showing the diagonal peaks which have a dispersion shape in both dimensions, and the cross-peaks (circles) which have an absorption shape in both dimensions but which alternate in sense (+, −). Diagonal peaks are centered at the coordinates (δ_A, δ_A) and (δ_B, δ_B) while cross-peaks are centered at (δ_A, δ_B) and (δ_B, δ_A). Cross-peaks indicate spin coupling between A and B.

Each individual component is a two-dimensional absorption, and the intensities alternate as indicated in Fig. 1. This has the important consequence that if the multiplet components are incompletely resolved, mutual cancellation occurs, and in the limit the cross-peaks vanish. It is therefore essential to ensure that the digitization of the multiplet is sufficiently fine in both dimensions that mutual cancellation is avoided. When a wide range of chemical shifts has to be covered, this entails a large data matrix, which is one of the principal limitations on the method. The antiphase nature of the multiplet components may be put to good use in order to emphasize the cross-peaks in comparison with diagonal peaks. In the time domain, the vectors which give rise to a cross-peak always start out in antiphase and only give a strong resultant if their time development is permitted for a suitably long period. Consequently time-domain weighting functions which increase with time and reach a maximum after an interval of the order of $(2J)^{-1}$ favor cross-peaks at the expense of diagonal peaks, and shaping functions designed to achieve this are described in detail below. Another consequence of the antiphase nature is that when the absolute-value mode is employed, the dispersion-mode tails of cross-peaks tend to cancel, whereas the tails of diagonal peaks are much more pronounced because the dispersion contributions reinforce.

PRACTICAL ILLUSTRATION

Many of the modifications and refinements of the Jeener experiment can only be illustrated by reference to actual two-dimensional spectra. These were obtained

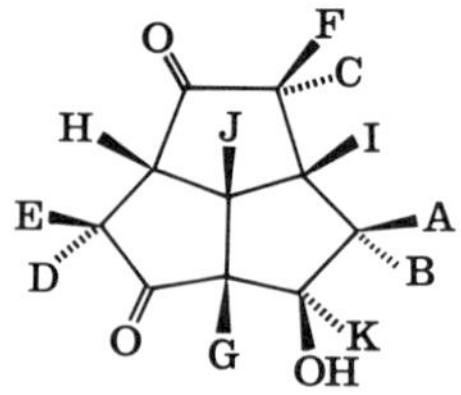

FIG. 2. The compound used for most of the two-dimensional Fourier transform experiments: one of the stereoisomers of 9-hydroxytricyclodecan-2,5-dione.

mainly from the compound sketched in Fig. 2, a tricyclodecane derivative chosen because of the wealth of proton–proton couplings that can be detected. Spectra were recorded on a Varian XL-200 spectrometer with software modifications for two-dimensional Fourier transformation. The phase of the second radiofrequency pulse (the "mixing" pulse) was cycled in 90° steps as described earlier (*18*) in order to discriminate the signs of the F_1 frequencies and thus reduce the size of the data matrix, and also to suppress the unwanted axial peaks. More sophisticated modifications of this pulse sequence are described in the relevant sections below.

Lineshapes

Two-dimensional spectra commonly use the absolute-value mode of presentation. This has the unfortunate result that the two-dimensional lineshape has marked ridges running through the center of the response parallel to the F_1 and F_2 axes; that is to say, the intensity contours form a four-pointed star rather than an ellipse (*19*, *20*). This effect has been discussed at some length for a two-dimensional Lorentzian response, but it is less well known that the two-dimensional Gaussian also has ridges if it is recorded in the absolute-value mode.

For the reasons discussed above, it is the diagonal peaks which have the most prominent tails, whereas cross-peaks are much narrower at the base, as can be appreciated from Fig. 3. Interference between the ridges of two different diagonal peaks can give rise to a spurious peak in the two-dimensional spectrum which might be mistaken for a true cross-peak. There is an example of this effect at the bottom left of Fig. 3. The ridges from strong diagonal peaks may make it difficult to pick out cross-peaks that lie near the principal diagonal. It is therefore important to eliminate dispersion-mode components from the two-dimensional lineshape, generating responses with circular intensity contours. This can be achieved by multiplying the time-domain signals $S(t_1)$ and $S(t_2)$ by one of the shaping functions described below.

A second reason for reshaping the time-domain signals is that this is a powerful method of enhancing the intensities of cross-peaks relative to the diagonal peaks. Because the components of a spin multiplet making up a cross-peak are in antiphase and can cause mutual cancellation, a resolution enhancement function increases the intensity of cross-peaks relative to diagonal peaks. Viewed in the time domain, the multiplet components may be represented as vectors which start out in antiphase and require a time of the order to $(2J)^{-1}$ to precess relative to one

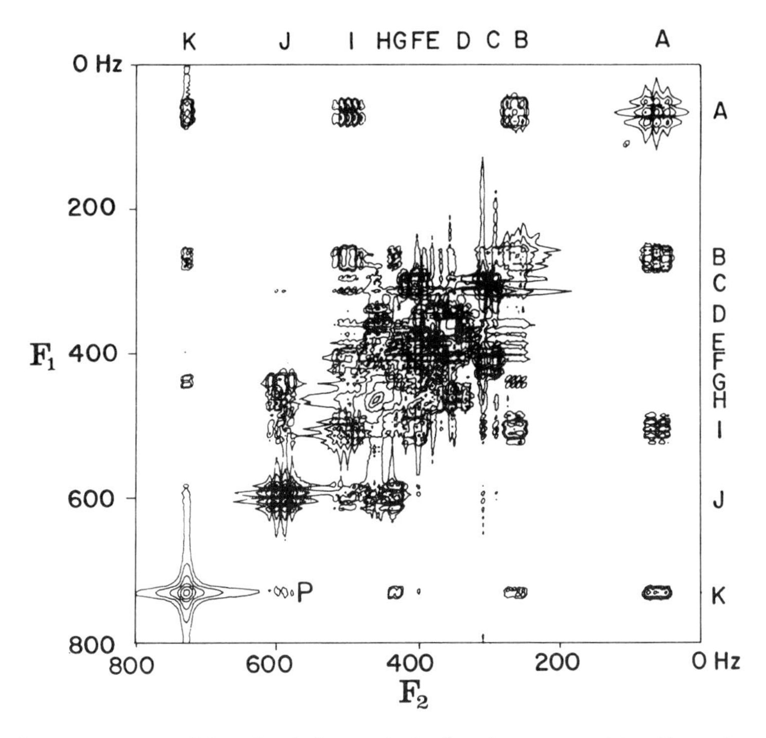

FIG. 3. Jeener spectrum of the tricyclodecane derivative shown as an intensity contour plot. The time-domain signals have been subjected to a shaping function which transforms an exponential decay into a Gaussian decay, giving Gaussian profiles in the frequency domain. Since an absolute-value mode has been recorded, the diagonal peaks show long tails in the F_1 and F_2 dimensions, but the cross-peaks, having equal numbers of antiphase multiplet components, do not show these tails. At the bottom left, marked P, is a spurious peak generated by the overlap of the tails of two diagonal peaks.

another to induce a detectable signal. A third reason for imposing a resolution enhancement function on the time-domain signals is that it can be approximately matched to the shape of the coherence transfer echo, described in more detail below.

One resolution enhancement procedure which has been used successfully is convolution difference (*21*) employing a multiplying function of the form

$$f(t) = 1 - \exp(-t/T_C), \qquad \text{where} \qquad T_C \simeq 1/(2J). \qquad [1]$$

This method was applied in both time dimensions to produce the multiple-trace plot of the Jeener spectrum of the tricyclodecane derivative (Fig. 4). Note the absence of ridges in this absolute-value-mode spectrum. Convolution difference first broadens the lines (causing mutual cancellation of antiphase signals) and then subtracts the unbroadened line. In this way the cross-peaks appear at almost full intensity while the diagonal peaks, where such cancellation cannot occur, are reduced in intensity.

There is another shaping function that may be used which has been specifically designed to eliminate dispersion-mode components altogether. It does so by

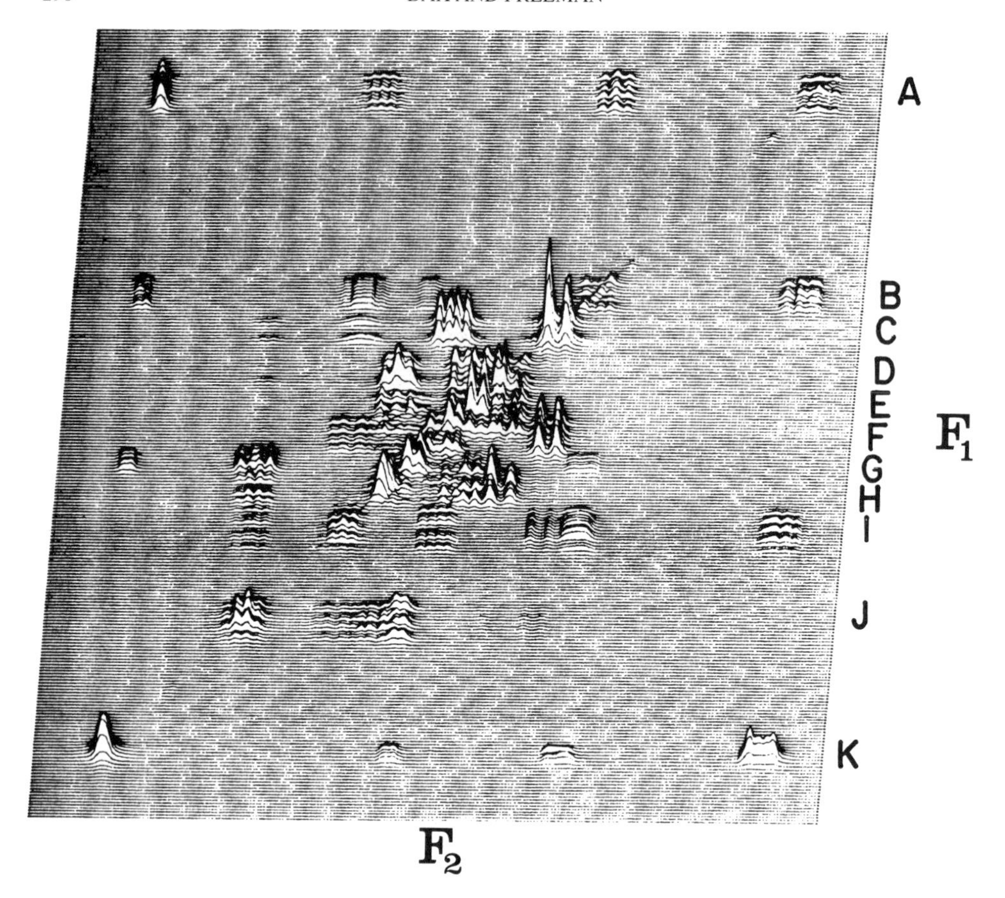

FIG. 4. The multiple-trace plot corresponding to the contour plot of Fig. 3. A convolution-difference resolution-enhancement function has been used in both dimensions in order to reduce the intensity of the diagonal peaks in relation to the cross-peaks and to reduce the tails on the diagonal peaks. The absolute-value mode has been used.

imposing exact symmetry on the envelope of the time-domain signal about its midpoint $t = T/2$, where T is the acquisition period. The resulting signal has been called a "pseudoecho," since it has the property of a true spin echo (when $T_2 \gg T_2^*$) that the dispersion-mode contributions in each half are equal and opposite in sign (*22*). Fourier transformation of a whole echo (*23*) or pseudoecho therefore produces a frequency-domain signal which is in the pure absorption mode. However, the pseudoecho differs from the true echo in that the various frequency components are not in phase at the center $t = T/2$ but at $t = 0$. If ϕ is the phase of a typical component of the pseudoecho at $t = T/2$, then the sine and cosine transforms contain that component in the proportion $\sin \phi$ and $\cos \phi$. A straightforward calculation of the sine transform would therefore show gross intensity anomalies; however, the square root of the sum of the squares of the sine and cosine transforms can be recorded, and this contains only the pure absorption mode. Any shaping function can be used which gives the pseudoecho a symmetrical envelope.

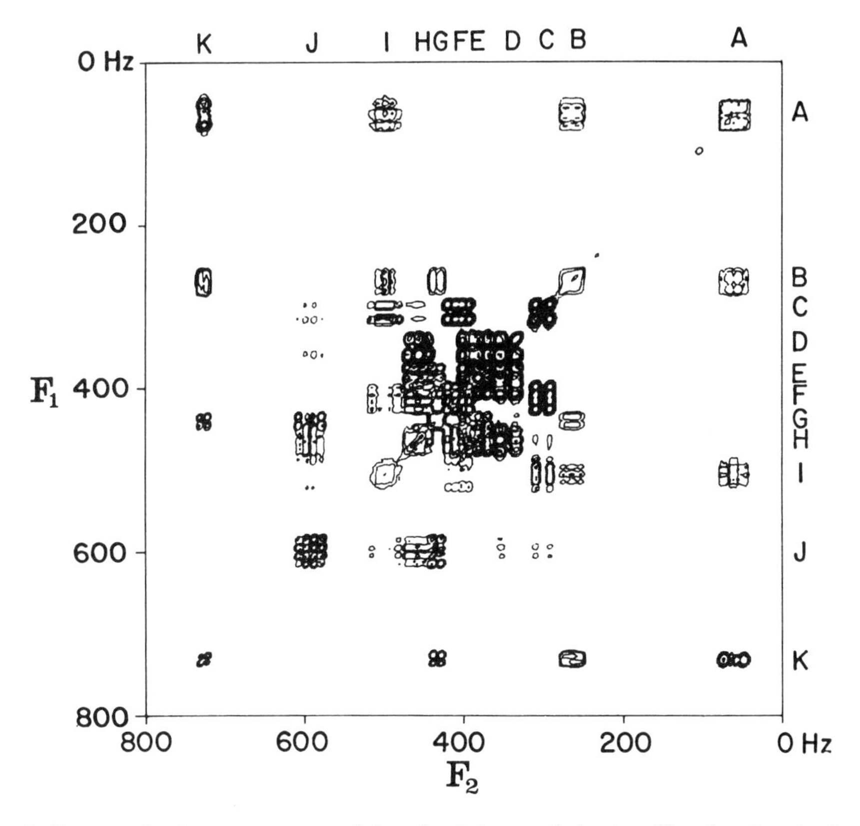

FIG. 5. Contour plot Jeener spectrum of the tricyclodecane derivative. The time-domain signals were multiplied by a shaping function which converts them into "pseudoechoes" with a symmetrical Gaussian envelope. This eliminates dispersion-mode components and leads to a basically circular contour for an isolated line. This spectrum is to be compared with that of Fig. 3. The cross-peaks indicate the spin couplings listed in the left-hand column of Table 1.

For two-dimensional spectra, a Gaussian envelope has advantages. The linewidth is chosen to avoid step-function discontinuities in the time domain, giving a pure Gaussian lineshape in the frequency domain. The same function is used in both t_1 and t_2, so that the two-dimensional spectrum has circular intensity contours. Figure 5 shows a Jeener spectrum where the time-domain data have been processed in this manner (see also Table 1).

Apart from enhancing the relative intensity of the cross-peaks, the pseudoecho method has the important property that it can be "tailored" to favor coupling constants of a given magnitude. Optimum transfer of magnetization from one nucleus A to another nucleus X occurs when $\sin(\pi J_{AX} t_1)\exp(-t_1/T_{2A})$ is a maximum, while optimum detection of the transferred magnetization occurs when $\sin(\pi J_{AX} t_2)\exp(-t_2/T_{2X})$ is a maximum. (In these expressions, the spin–spin relaxation times T_{2A} and T_{2X} have been used rather than the instrumental decay constant T_2^* for reasons discussed in the following section.) This matching of the shaping function to a particular order of magnitude of coupling constant is very useful for emphasizing long-range coupling constants. If they are so weak that

TABLE 1

COUPLINGS OBSERVED BETWEEN PROTONS IN THE TRICYCLODECANE DERIVATIVE SKETCHED IN FIG. 2

Normal couplings	Very weak couplings
AB, AK, AI	AC, AJ
BA, BG, BI, BK	BJ
CF, CH, CI, CJ	CA
DE, DH, DJ	DG
ED, EG, EH	
FC, FH, FI	FK
GB, GE, GJ, GK	GD
HC, HD, HE, HF, HJ	
IA, IB, IC, IF, IJ	IK
JC, JD, JG, JH, JI	JA, JB, JK
KA, KB, KG	KF, KI, KJ

$\pi J_{AX} \ll (T_{2A})^{-1}$, $(T_{2X})^{-1}$ then the optimum choice for the centers of the pseudo-echoes is $t_1 = T_{2A}$, $t_2 = T_{2X}$. This permits the observation of cross-peaks resulting from spin coupling which is so weak as to be unresolved in the conventional spectrum.

Normally the observation of cross-peaks resulting from very weak couplings would require the storage of a very large data matrix because of the requirement for very fine digitization. This requirement can be relaxed with pseudoecho shaping of the time domain signals by the introduction of a significant "dead time" Δ sec between $t_1 = 0$ or $t_2 = 0$ and the beginning of the corresponding pseudo-echoes. No data are acquired during the interval Δ. In this way long-range couplings may be investigated without increasing the data storage requirements. In general it is convenient to record the Jeener spectrum under two or more sets of conditions in order to discriminate between cross-peaks from large couplings and those from weak couplings. When the conditions are optimized for weak long-range couplings, then the cross-peaks arising from large couplings can be weak or even absent if there is an integral number of beats in the interval $T/2$.

The experiment is performed with the same increments and time ranges for t_1 and t_2 and the same shaping functions, with the result that the two-dimensional spectrum is exactly symmetrical with respect to F_1 and F_2 unless there are spurious responses. This symmetry can be a useful aid in resolving ambiguities, since if a cross-peak is "real" it must have a "twin" symmetrically placed with respect to the principal diagonal.

Coherence Transfer Echoes

The transient signal after the first radiofrequency pulse, although not normally observed in the spectrometer, decays as a function of t_1 through natural spin–spin relaxation (T_2) and the effects of spatial inhomogeneity of the magnetic field (T_2^*). The same is true of the detection period, except that, if the coherence transfer echo (*24*) is detected, field inhomogeneity effects are refocused to a good approximation. Furthermore, since magnetization transfer takes place within a given molecule, the macroscopic distribution of B_0 fields has no effect on the process, and the only decay which influences the amplitude of the transferred signal is that resulting from spin–spin relaxation. This produces the surprising result that the amplitude of cross-peaks is dependent on T_2 but essentially independent of T_2^*. Aue *et al.* (*2*) commented on this refocusing effect in the Jeener experiment, which gives rise to very narrow line profiles measured at right angles to the principal axis. There is a close analogy with the spin-tickling double-resonance experiment (*25*), where the intramolecular nature of the coupling allows B_0 inhomogeneity effects to be compensated.

This was why, in the section above, the pseudoechoes were centered at times $t_1 = T_2$ and $t_2 = T_2$ rather than $t_1 = T_2^*$ and $t_2 = T_2^*$, in order to optimize the magnetization transferred through weak long-range couplings. It has been demonstrated that even very weak couplings, an order of magnitude smaller than the natural linewidth, can give detectable cross-peaks. In some spectrometers the instrumental linewidth might be an order of magnitude broader than the natural width. This raises the possibility of detecting the presence of weak couplings many times too small to appear as splittings in the conventional spectrum. However, it would only be possible to get a very rough estimate of the magnitude of such a coupling from the Jeener experiment.

The selection of the coherence transfer echo is achieved by cycling the phase of the second radiofrequency pulse and the reference phase of the receiver in 90° steps, picking out the component of magnetization which precesses in opposite senses during t_1 and t_2 while rejecting the other rotating component, sometimes called the "antiecho" (*26*).

With a mixing pulse of less than 90°, detection of the coherence transfer echo emphasizes cross-peaks at the expense of diagonal peaks. This is important when the flip angle of the mixing pulse is kept small in order to restrict transfer to directly connected transitions, (see following section), because then a larger proportion of the coherence remains associated with the original nucleus, giving rise to relatively strong diagonal peaks. Discrimination against diagonal peaks is particularly important when there are true cross-peaks near the principal diagonal, as in spin systems with strong coupling ($J \sim \delta$). An example is provided by the cross-peaks of resonances D and E in the spectrum of Fig. 5, where the nuclei are strongly coupled.

Flip Angle Effects

When the second pulse of the sequence (the "mixing" pulse) is 90° then magnetization is transferred to all other coupled spins, but it has been shown theoretically (*2*) that if the mixing pulse has a small flip angle then magnetization is transferred predominantly to connected transitions, those which share a common energy level. This considerably simplifies the resulting two-dimensional spectrum. Furthermore, the transferred magnetization is distributed over a smaller number of cross-peaks and sensitivity can be improved, provided that the flip angle is not too small. In practice a 45° pulse is a good compromise.

This flip angle effect can be visualized by considering the mixing pulse to be

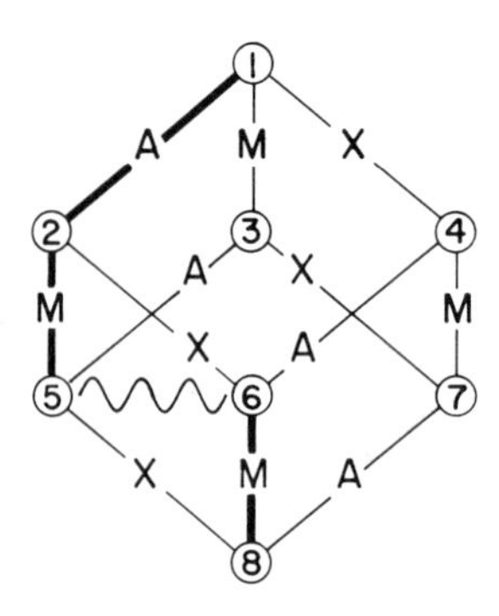

FIG. 6. The energy-level diagram appropriate to the AMX system. An argument based on the concept of pulse cascades has been used to calculate the relative efficiencies of magnetization transfer from $A(12)$ to $M(25)$ and to $M(68)$. The latter transfer involves the forbidden zero-quantum transition (*56*).

decomposed into a cascade (*27*) of selective pulses, each affecting only a single transition. Consider the three-spin AMX system with the energy level diagram shown in Fig. 6. The initial 90° pulse is not decomposed into a cascade, but simply creates precessing transverse magnetization in all the allowed transitions. The mixing pulse of flip angle α may be represented as a cascade of 12 selective pulses:

$$\alpha(12)\alpha(46)\alpha(35)\alpha(78)$$
$$\alpha(25)\alpha(13)\alpha(47)\alpha(68)$$
$$\alpha(26)\alpha(58)\alpha(14)\alpha(37)$$

Note that the pulses affecting the A spins must all be grouped together, and similarly the pulses affecting the M spins form a separate group. However, the order of the three groups and the order within any group can be permuted.

Attention may be focused on the fate of one typical transverse magnetization, $A_{XY}(12)$, and although this precesses as a function t_1, its amplitude R_0 is all that needs to be considered here. Consider, first of all, magnetization transfer to a typical connected transition $M_{XY}(25)$. Not all the elements in the pulse cascade are involved, and it is sufficient to calculate the effect of the cascade:

$$\alpha(12)\alpha(25)\alpha(26)\alpha(58).$$

The first element, $\alpha(12)$, converts transverse magnetization $A_{XY}(12)$ into longitudinal magnetization $A_Z(12)$ of amplitude $R_0 \sin \alpha$. The second element, $\alpha(25)$, converts one-half of this into transverse magnetization $M_{XY}(25)$ of amplitude $(1/2)R_0 \sin^2 \alpha$. The remaining two pulses withdraw magnetization from $M_{XY}(25)$ (creating multiple-quantum coherence) leaving an amplitude $(1/2)R_0 \sin^2 \alpha \cos^2 (\alpha/2)$.

A different sequence of events governs the transfer to a nonconnected transition, $M_{XY}(68)$, although the same four elements of the pulse cascade are applicable. Transverse magnetization $A_{XY}(12)$ is converted into $M_{XY}(25)$ by the first two pulses with amplitude $(1/2)R_0 \sin^2 \alpha$. But the third element, $\alpha(26)$, creates zero-quantum coherence ZQT(56) with an amplitude $(1/2)R_0 \sin^2 \alpha \sin (\alpha/2)$ and the last element, $\alpha(58)$, reconverts this into observable transverse magnetization $M_{XY}(68)$ with an amplitude $(1/2)R_0 \sin^2 \alpha \sin^2 (\alpha/2)$. A reversal of the sense of the last two elements of the cascade would have given the same result but by way of the double-quantum coherence DQT(28). It is a general rule that transfer to nonconnected transitions involves double- or zero-quantum coherence as an intermediate, whereas directly connected transitions acquire magnetization by conversion of the longitudinal magnetization of a connected transition.

As a result, the ratio of the intensities of connected cross-peaks to those of nonconnected cross-peaks is given by $\cot^2 (\alpha/2)$. Thus for a mixing pulse of 45° the connected transitions are 5.8 times stronger. This is the setting used in many of the experiments described below when good discrimination between connected and nonconnected transitions is required, as in the determination of relative signs of coupling constants. When sensitivity is an important consideration, a mixing pulse of 60° delivers higher intensities in the cross-peaks while still retaining a factor of 3 in favor of the connected transitions.

Similar pulse cascade arguments can be used to predict the relative intensities of parallel transitions, where magnetization is transferred between component lines of the same spin multiplet. These lines are situated just off the principal diagonal within a distance of the order of J. The general rule is that parallel transitions lose intensity compared with connected transitions as the flip angle is reduced below 90°, but that certain parallel transitions lose intensity faster than others, because they involve more stages of magnetization transfer in succession. In practice it can be very useful to reduce the relative intensity of parallel transitions so as to leave the region close to the principal diagonal clear in order to search for cross-peaks. Figure 7 shows an example of a spectrum where most of the parallel transitions have been reduced below the threshold of the lowest-intensity contour. These conclusions, based on the concept of pulse cascades, are consistent with the results derived more formally by Aue *et al.* (*2*).

Long-Range Couplings

The tricyclodecane derivative sketched in Fig. 2 provides a rich field for investigation of long-range proton–proton couplings. The introduction of a relatively long delay Δ before the beginning of data acquisition enhances the relative intensity of cross-peaks arising from long-range coupling. In practice Δ was set at 0.3 sec. When the coupling constants are much smaller than the natural linewidths, the optimum setting for the peak of the pseudoecho is near points $t_1 = T_2$ and $t_2 = T_2$, where the signal components from long range coupling are maximum. The cross-peak amplitude is then of the order $\sin^2 (\pi J T_2) \exp(-2)$ times smaller than a cross-peak from a large spin coupling constant with conventional data acquisition. Sensitivity, however, is still a critical consideration, entailing a compromise setting of 55° for the flip angle of the mixing pulse.

The intensity contour plot of Fig. 7 was obtained under the conditions described above. It shows six new cross-peaks that were not observed in the spectrum of Fig. 5; they arise from weak long-range couplings. Note that all of these cross-peaks show multiplet structure resulting from large couplings to other protons, the long-range splittings themselves are not of course resolved in this diagram. In addition, some of the cross-peaks resulting from large couplings do not appear

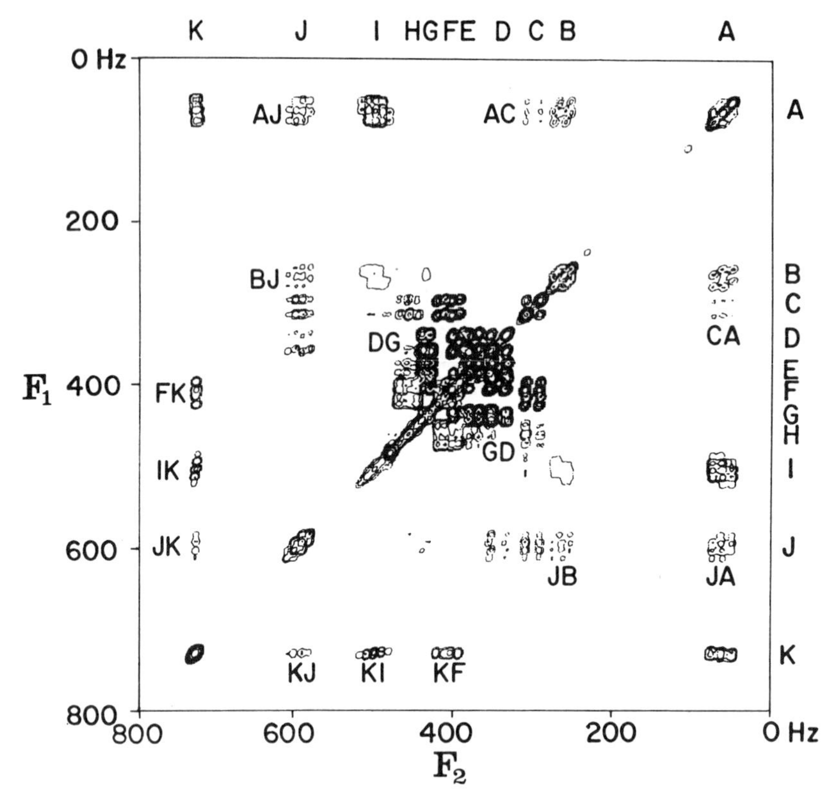

FIG. 7. Jeener spectrum of the tricyclodecane derivative obtained under conditions calculated to emphasize long-range couplings. This involves the use of a reduced flip angle (54.6°) for the mixing pulse, delays $\Delta = 0.3$ sec before data acquisition in the t_1 and t_2 dimensions, and "pseudoecho" shaping of the time-domain signals. The new cross-peaks are indicated and the corresponding couplings are listed in the right-hand column of Table 1.

in Fig. 7 as they have low intensity because of the pseudoecho shaping of the time-domain signals.

An estimate of the approximate magnitudes of these long-range coupling constants would require several experiments with different settings for the peak of the pseudoecho. In a later section, a method of comparing the approximate magnitudes of couplings within the same spectrum is presented. In general the Jeener experiment is quite effective at detecting the presence of weak couplings, but not at estimating their magnitudes.

Relative Signs of Coupling Constants

There is an interesting consequence of using small flip angles to restrict the transfer of magnetization to connected transitions. In a system of three or more coupled spins, if the appropriate splittings can be resolved, the relative signs of the coupling constants may be ascertained by inspection of the Jeener spectrum.

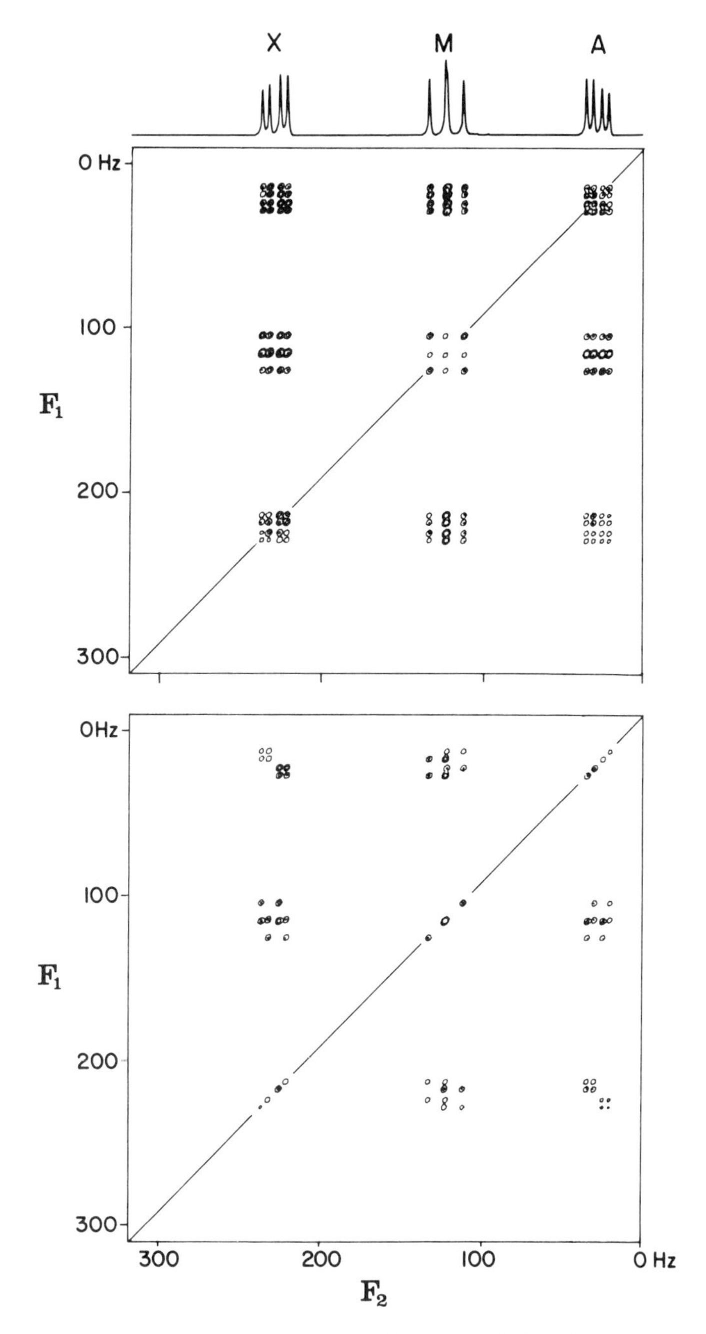

FIG. 8. Contour plots of the Jeener spectra of 2,3-dibromopropionic acid, an AMX spin system. The upper spectrum was obtained with a 90° mixing pulse and each multiplet contains 16 component lines, grouped in four squares of the kind shown in Fig. 1. The lower spectrum was obtained with a 45° mixing pulse so that only directly connected transitions are intense enough to show contours. Each cross-peak is now made up of two square patterns, and the sign of the slope of the line joining their centers gives the relative signs of the appropriate coupling constants.

The basis of the method is the same as that of the selective-decoupling experiments (*28*, *29*). Consider the AMX system of 2,3-dibromopropionic acid, where conventional double-resonance and double-quantum experiments have established that the two vicinal couplings have signs opposite to that of the geminal coupling (*30, 31*). When the flip angle of the mixing pulse is small, then AX cross-peaks involving a simultaneous flip of the M nucleus are of vanishingly small intensity; on the other hand, for a flip angle of 90°, all 16 multiplet components are observed with comparable intensity. The Jeener spectrum obtained with a small flip angle thus resembles a selective double-resonance experiment, where A and X are decoupled only in molecules with one of the spin states of the M nucleus, say $M(\alpha)$ but not $M(\beta)$. The local magnetic fields at the sites of the A and X nuclei resulting from the couplings to the M spin must be in the same sense if $J_{AM} \cdot J_{MX} > 0$. Thus the corresponding displacements of the AX cross-peak are in the same sense in the F_1 and F_2 dimensions, resulting in a tilt of the multiplet in the same sense as the principal diagonal. On the other hand, if these couplings have opposite signs, the AX multiplets are tilted in the opposite sense, making an angle greater than 45° with respect to the principal diagonal. If either J_{AM} or J_{MX} is vanishingly small, then the AX multiplet lies vertically on one side of the diagram and horizontally on the other, and no sign determination is possible.

Figure 8 shows the contour diagram for 2,3-dibromopropionic acid, first for a mixing pulse flip angle of 90°, showing each cross-peak made up of 16 components, then for a flip angle of 45°, where each cross-peak has been reduced to 8 components. (In the spectrum obtained at low flip angle, the "parallel" transitions—where magnetization has been transferred within a single spin multiplet—are absent, leaving only diagonal peaks which fall exactly on the principal diagonal.) Each cross-peak is made up of two groups of four lines which form an exact square (compare Fig. 1). It is the relative orientation of these two squares which determines the relative signs of the coupling constant, each cross-peak relating the signs of two coupling constants. In the case of 2,3-dibromoproprionic acid, J_{AX} and J_{MX} have like signs, opposite to the sign of J_{AM}, the geminal coupling.

Once the principle of relative sign determination has been established in a "textbook" case like that of dibromopropionic acid, it is readily applied to more complex spectra simply by noting the tilt of the pattern of lines in a cross-peak. Figure 9

TABLE 2

RELATIVE SIGNS OF CERTAIN COUPLING CONSTANTS IN THE TRICYCLODECANE DERIVATIVE DETERMINED BY INSPECTION OF THE JEENER SPECTRUM IN FIG. 9

J(AB) and J(BK)	Opposite signs
J(AB) and J(BI)	Opposite signs
J(AI) and J(BI)	Like signs
J(AK) and J(AB)	Opposite signs
J(AB) and J(AI)	Opposite signs
J(CF) and J(FI)	Opposite signs
J(DH) and J(EH)	Like signs

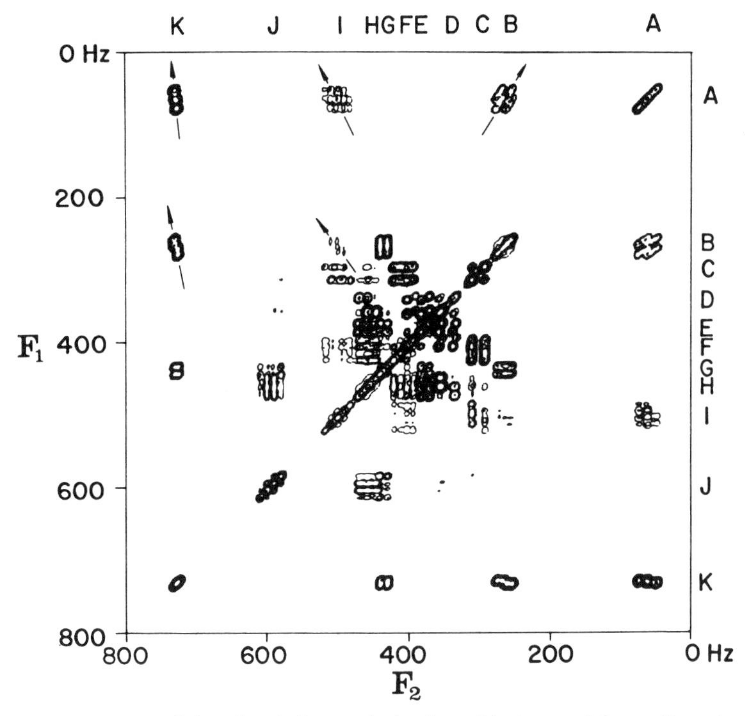

FIG. 9. Jeener spectrum of the tricyclodecane derivative with the conditions adjusted to determine the relative signs of certain coupling constants. The mixing pulse was reduced to 45° and a "pseudo-echo" shaping function used in both time dimensions, but with no dead time ($\Delta = 0$). The arrows indicate the approximate "tilt" of the cross-peaks. The relative signs determined from this spectrum are set out in Table 2.

shows a contour plot of the spectrum of the tricyclodecane derivative obtained with a mixing pulse of flip angle 45°. The tilt of several cross-peaks is indicated by the arrows. Arrows with a positive slope indicate like signs, those with a negative slope opposite signs; a vertical or horizontal slope means that one of the two couplings involved is vanishingly small. (See Table 2.)

The simplicity of the Jeener spectrum obtained with small flip angles for the mixing pulse stems from the fact that each cross-peak is basically from an AX or AB spin system; all other splittings are "passive" and simply create several independent AX or AB subspectra. In a more general case, groups of two or three spins could be equivalent, leading to A_2X, AA′X spectra, for example.

Broadband Decoupling in the F_1 Dimension

Throughout the interval between radiofrequency pulses nuclear magnetization vectors corresponding to the components of a given spin multiplet diverge continuously at a rate determined by J. Suppose that this interval were to be fixed at a value t_d seconds, and a 180° refocusing pulse introduced after a variable delay $t_1/2$ after the initial 90° pulse. Then, although the amplitude of the transferred

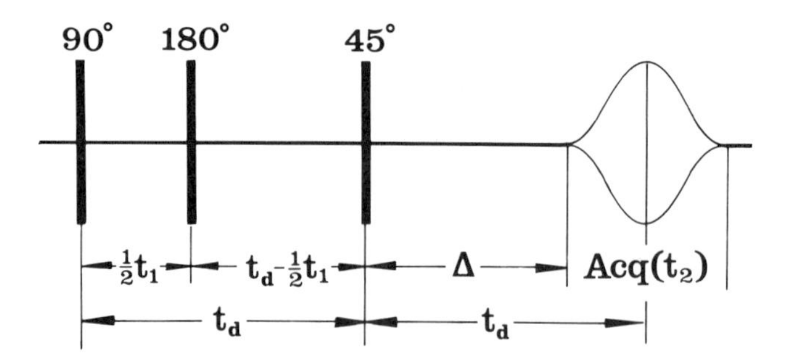

FIG. 10. The pulse sequence used to obtain Jeener spectra that are decoupled in the F_1 dimension. The delay t_d is fixed, and only t_1 and t_2 are varied. A "pseudoecho" shaping function is indicated in the t_2 dimension. A 45° mixing pulse is used in order to limit the responses to directly connected transitions.

magnetization would depend on Jt_d, there would be no J modulation of the signal as t_1 was varied. On the other hand, the effect of the chemical shift would change, going from a maximum at $t_1 = 0$ to zero when $t_1 = t_d/2$ (exact refocusing) and back to a maximum again at $t_1 = t_d$.

This is the basis of a method of broadband decoupling in the F_1 dimension (*3*). It relies on the assumption that the coupling is first order, otherwise spurious responses are excited; in an AB spin system, for example, there is a spurious response at the mean chemical shift frequency (*32*). By collapsing all spin multiplet structure in the F_1 dimension onto the appropriate chemical shift frequency, this greatly simplifies the Jeener spectrum, while at the same time concentrating the intensity of the multiplet components. The only information that is lost in this process is that discussed in the previous section—the relative signs of the coupling constants.

A pulse sequence and acquisition scheme (*33*) is set out in Fig. 10. In this application it is important to employ a mixing pulse of small flip angle (45°) to avoid mutual cancellation of antiphase components within a given cross-peak. The usual phase cycling of the 45° pulse and the receiver reference phase ensures that the coherence transfer echo is detected. In addition, the 180° pulse is carefully calibrated and phase alternated along the $\pm Y$ axes of the rotating frame in order to avoid spurious responses from pulse imperfections (*18*). In this application it is advantageous to increase the digitization along the F_1 axis in order to take advantage of the improvement in resolution. Figure 11 shows the effect of this broadband decoupling scheme on the Jeener spectrum of the tricyclodecane derivative.

Projections and Cross Sections

The technique provides an attractive alternative to the established method of obtaining decoupled proton spectra by 45° projection of two-dimensional J spectra (*3*). The Jeener spectrum with F_1 decoupling is simply projected onto the F_1 axis. Figure 11 shows such a projection together with an indication of the spurious lines

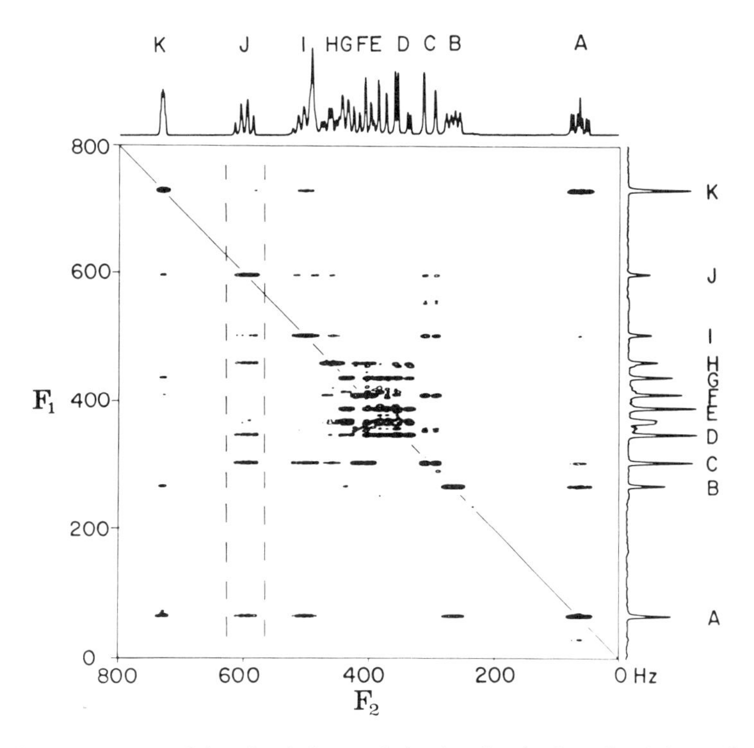

FIG. 11. Jeener spectrum of the tricyclodecane derivative showing broadband decoupling in the F_1 dimension through the use of the pulse sequence shown in Fig. 10. The spectrum running along the top of the diagram is the fully coupled spectrum; that shown in the right margin is the projection of the two-dimensional spectrum. In this mode the principal diagonal is reversed. A band of signals associated with proton J has been picked out of the data matrix and projected onto the F_1 axis in order to obtain the spectrum shown in Fig. 12.

arising from strong coupling effects. Operation with a redefined evolution period t_1 has resulted in reversing the sense of the principal diagonal in this spectrum.

A cross section through this Jeener spectrum at an appropriate point provides a clear indication of the pattern of couplings to a given proton. For example, if it is required to know which protons are coupled to proton J, a vertical section is taken so as to cut the principal diagonal at the chemical shift of J. In practice it is rather better to extract a vertical band of frequencies from the two-dimensional matrix, as indicated by the dashed lines in Fig. 11; the band is then projected horizontally onto the F_1 axis. This has the effect of summing all the individual component lines in a given cross-peak. The result is a one-dimensional spectrum (Fig. 12) which indicates the cross-peaks associated with proton J, each cross-peak appearing now as a single line. Not only does this show that proton J is coupled to A, B, C, D, H, I, and K, but it also gives an approximate estimate of the relative magnitudes of the long-range couplings, for the acquisition parameters were optimized for couplings of the order of the natural linewidth. Large coupling constants give anomalous intensities under these conditions, for example, the cross-peak resulting from coupling between J and G is absent in this trace. On the other

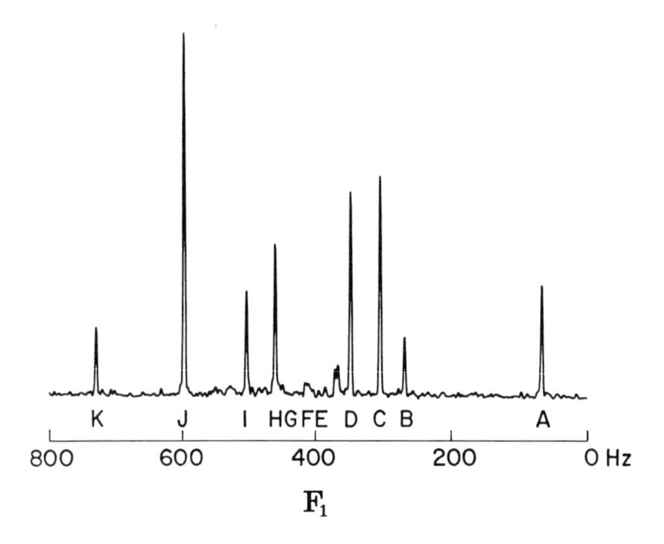

FIG. 12. The projection onto the F_1 axis of a narrow band of signals extracted from the data matrix of Fig. 11, representing all the cross-peaks associated with proton J. This indicates that J is coupled to protons A, B, C, D, H, I, and K, and some indication of the relative strengths of these couplings can be obtained provided that they are weak. An artifact appears between the chemical shift frequencies of protons D and E because these protons are strongly coupled.

hand, a weak spurious response is observed at the mean chemical shift of protons D and E, attributable to strong coupling effects.

The pattern of couplings in the tricyclodecane derivative is set out in Table 1. The column "normal couplings" was obtained under acquisition conditions adjusted for relatively strong couplings, with no dead time Δ before acquisition, using mainly the information from Fig. 5. The column "very weak couplings" was obtained mainly from Fig. 7, where the conditions had been suitably optimized for couplings of the same order as the natural linewidth. In several cases the two sets of results confirmed each other; in other cases they were complementary.

DISCUSSION

The simple two-dimensional experiment of Jeener (*1*), although relatively neglected in the past, proves itself well suited to the task of analyzing complex networks of proton–proton spin couplings. In practice it is necessary to modify the operating conditions in various ways in order to simplify the two-dimensional spectra and their interpretation. Reshaping of the two-dimensional responses turns out to be important, and several techniques are used to suppress axial peaks and to reduce the relative intensities of diagonal peaks, in order that attention can be focused on the cross-peaks which carry the coupling information. Intensity contour plots have proved to be the most convenient method of displaying Jeener spectra.

The present work emphasizes the advantage to be gained by operating with a reduced flip angle for the mixing pulse–typically 45° rather than 90°. It was recognized at an early stage (*2*) that this restricts magnetization transfer predominantly to directly connected transitions, but this fact does not seem to have been exploited in chemical applications. In multispin systems there is an interesting side effect of this property. Relative signs of coupling constants are obtained simply by inspection of the "tilt" of two-dimensional spin multiplets, a method closely analogous to the selective decoupling technique (*28*, *29*).

Weak long-range couplings, normally hidden within the instrumental linewidth, can be detected unequivocally if suitable changes are made in the method of acquiring the two-dimensional data matrix $S(t_1, t_2)$. This involves the introduction of a "dead-time" Δ, and the shaping of the data in the time domain in order to produce a "pseudoecho" which has the property that dispersion-mode contributions to the lineshape are suppressed.

Finally, a simple change in the pulse sequence permits "broadband decoupling" in the F_1 dimension, further simplifying the two-dimensional spectrum. Projection onto the F_1 axis then gives a proton spectrum with all scalar splittings removed; indeed partial spectra may be extracted showing only the resonances of protons coupled to a chosen proton. In the proton spectroscopy of large molecules, such a simplification might well prove crucial to the analysis.

ACKNOWLEDGMENTS

This work was made possible by an equipment grant from the Science Research Council, and a Stipend from the Delft University Fund (A.B). The tricyclodecane derivative was kindly provided by Mr. P. L. Beckwith.

REFERENCES

1. J. Jeener, Ampere International Summer School, Basko Polje, Yugoslavia, 1971.
2. W. P. Aue, E. Bartholdi, and R. R. Ernst, *J. Chem. Phys.* **64,** 2229 (1976).
3. W. P. Aue, J. Karhan, and R. R. Ernst, *J. Chem. Phys.* **64,** 4226 (1976).
4. E. L. Hahn and D. E. Maxwell, *Phys. Rev.* **88,** 1070 (1952).
5. R. Freeman and H. D. W. Hill, *J. Chem. Phys.* **54,** 301 (1971).
6. A. A. Maudsley and R. R. Ernst, *Chem. Phys. Lett.* **50,** 368 (1977).
7. A. A. Maudsley, L. Müller, and R. R. Ernst, *J. Magn. Reson.* **28,** 463 (1977).
8. G. Bodenhausen and R. Freeman, *J. Magn. Reson.* **28,** 471 (1977).
9. S. Forsén and R. A. Hoffman, *J. Chem. Phys.* **39,** 2892 (1963).
10. J. Jeener, B. H. Meier, P. Bachmann, and R. R. Ernst, *J. Chem. Phys.* **71,** 4546 (1979).
11. B. H. Meier and R. R. Ernst, *J. Am. Chem. Soc.* **101,** 6441 (1979).
12. H. Hatanaka, T. Terao, and T. Hashi, *J. Phys. Soc. Jpn.* **39,** 835 (1975).
13. A. Wokaun and R. R. Ernst, *Chem. Phys. Lett.* **52,** 407 (1977).
14. S. Vega and A. Pines, *J. Chem. Phys.* **66,** 5624 (1977).
15. A. Wokaun and R. R. Ernst, *Mol. Phys.* **36,** 317 (1978).
16. K. Nagayama, K. Wüthrich, and R. R. Ernst, *Biochem. Biophys. Res. Commun.* **90,** 305 (1979).
17. K. Nagayama, A. Kumar, K. Wüthrich, and R. R. Ernst, *J. Magn. Reson.* **40,** 321 (1980).
18. A. Bax, R. Freeman, and G. A. Morris, *J. Magn. Reson.* **42,** 164 (1981).
19. R. R. Ernst, *Chimia* **29,** 179 (1975).
20. G. Bodenhausen, R. Freeman, R. Niedermeyer, and D. L. Turner, *J. Magn. Reson.* **26,** 133 (1977).
21. I. D. Campbell, C. M. Dobson, R. J. P. Williams, and A. V. Xavier, *J. Magn. Reson.* **11,** 172 (1973).
22. A. Bax, R. Freeman, and G. A. Morris, *J. Magn. Reson.* **43,** 333 (1981).

23. A. Bax, A. F. Mehlkopf, and J. Smidt, *J. Magn. Reson.* **35,** 373 (1979).
24. A. A. Maudsley, A. Wokaun, and R. R. Ernst, *Chem. Phys. Lett.* **55,** 9 (1978).
25. R. Freeman and W. A. Anderson, *J. Chem. Phys.* **37,** 2053 (1962).
26. R. Freeman and G. A. Morris, *Bull. Magn. Reson.* **1,** 5 (1979).
27. G. Bodenhausen and R. Freeman, *J. Magn. Reson.* **36,** 221 (1979).
28. D. F. Evans and J. P. Maher, *Proc. Chem. Soc. London*, 208 (1961).
29. R. Freeman and D. M. Whiffen, *Mol. Phys.* **4,** 321 (1961).
30. R. Freeman, K. A. McLauchlan, J. I. Musher, and K. G. R. Pachler, *Mol. Phys.* **5,** 321 (1962).
31. K. A. McLauchlan and D. H. Whiffen, *Proc. Chem. Soc. London*, 144 (1962).
32. G. Bodenhausen, R. Freeman, G. A. Morris, and D. L. Turner, *J. Magn. Reson.* **31,** 75 (1978).
33. A. Bax, A. F. Mehlkopf, and J. Smidt, *J. Magn. Reson.* **35,** 167 (1979).

JOURNAL OF MAGNETIC RESONANCE 45, 188–191 (1981)

Proton Imaging for *in Vivo* Blood Flow and Oxygen Consumption Measurements

K. R. THULBORN,* J. C. WATERTON,† AND G. K. RADDA*

*Department of Biochemistry, University of Oxford, South Parks Road, Oxford OX1 3QU, England, and †ICI Corporate Laboratory, P.O. Box 11, The Heath Runcorn WA7 4QE, England

Received July 20, 1981

Proton imaging (zeugmatography) of animals and humans *in vivo* relies on the variation of the volume fraction and longitudinal and transverse relaxation times (T_1 and T_2, respectively) of tissue water among the different tissues and their different physiological states. We have previously reported that a ^{1}H NMR parameter, namely, the blood water transverse relaxation time (T_2) at high field, is dependent on the oxygenation state of hemoglobin in whole blood (*1*). In this communication, we demonstrate that spatially resolved values of T_2 and flow rate can be obtained from a simple imaging experiment allowing the determination of arteriovenous differences. We also demonstrate that the local availability of oxygen in tissues *in vivo* is reflected in T_2 values.

One of the simplest methods of obtaining spatial information from the NMR experiment is by the application of a linear z field gradient during acquisition. We have initially used the coil arrangement previously described in which a copper foil Helmholtz coil tuned to the proton frequency (182.4 MHz) is positioned in the center of a pair of z field gradient coils (*2*). A Teflon two-compartment phantom filled with water was located within the most homogeneous region of the B_1 field. Figure 1 shows that the application of a z gradient readily resolves the single proton line (Fig. 1a) into a one-dimensional image of the phantom (Fig. 1b). The size of the image in the frequency domain corresponds to the physical dimensions of the phantom (1 mm = 1800 Hz). The shape of the image is dependent on the proton density, B_1 homogeneity, and deviations of the z gradient from ideality (*3*). Artifacts visible at the ends of the image are largely due to B_1 inhomogeneity at the ends of the phantom.

In Fig. 2, water was replaced by blood of two different oxygenation states in separate compartments. A modified Carr–Purcell–Meiboom–Gill pulse sequence,

$$(\text{Delay}-90°-(t-180°-t)_n-\text{Acq(field gradient on)and FT}),$$

with the z gradient on only during acquisition was used. By plotting the signal from each compartment against n, T_2 values of 37 and 94 msec were determined for the two blood samples of lower and higher oxygenation states, respectively. In order to eliminate artifacts arising from pulse imperfections, the signal intensities were integrated only at those points in the image corresponding to positions where

0022–2364/81/100188-04$02.00/0

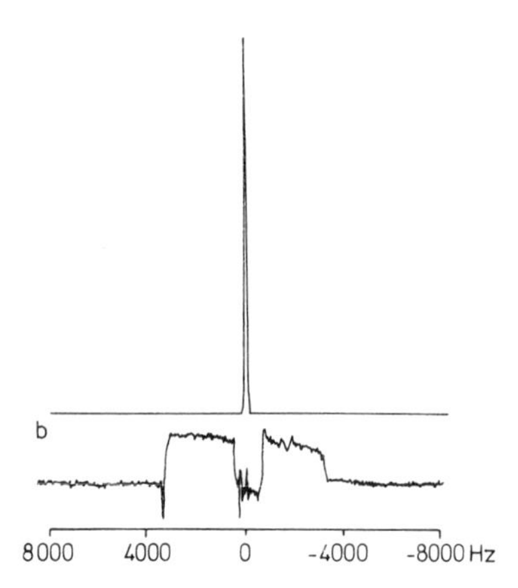

FIG. 1. ^{1}H NMR spectrum of water in a two-compartment phantom using the pulse sequence (a) 90°–Acq(field gradient off) and (b) 90°–Acq(field gradient on). Vertical scale of (a) is reduced ninefold relative to (b).

the rf field, B_1, was such that $\pi(\gamma B_1)^{-1}$ was closest to the 180° pulse width over most of the image. Within experimental error, these values were in agreement with those measured on separate blood samples by the method described previously (*1*) to within ±5% even though only 20 data points were used in this experiment (see Fig. 2) compared to 212 points used for the separate samples.

An application of such one-dimensional images would be in the improvement of the method reported previously (*1*, *4*) for the measurement of oxygen utilization in organs of laboratory animals. Although that procedure had good temporal resolution and was quantitative, the requirement for a surgically implanted by-pass to carry blood through the rf coil from and back to the blood vessel of interest limited its application to tissues supplied by relatively few blood vessels of large

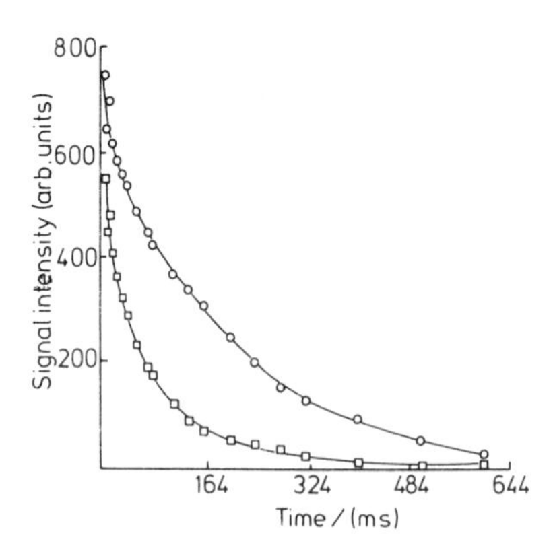

FIG. 2. Spatially resolved T_2 experiment using the apparatus of Fig. 1 in which water was replaced by rat blood of 35 and 60% oxygenation state in compartments x (□) and y (○), respectively. The temperature was 25°C and the delay (2τ) between refocusing pulses was 2 msec. Only the region of the image with a true 180° pulse was integrated for the echo intensity used in the calculation of T_2.

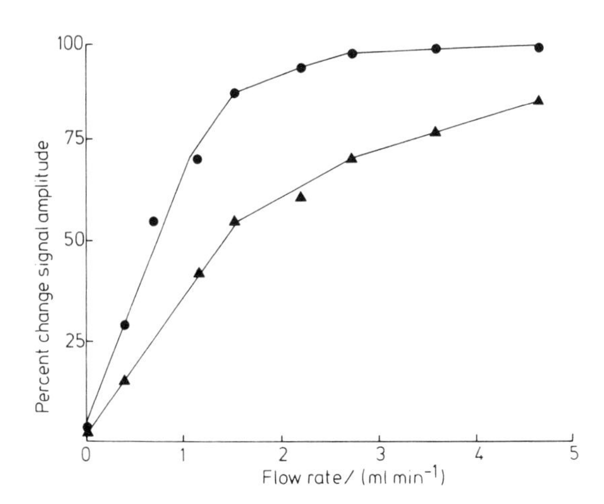

FIG. 3. Calibration curve for percentage of signal intensity recovered from saturation after a delay of 100 msec (●) and 50 msec (▲) against flow rate using a surface coil (1 mm diameter) placed on a glass capillary tube (1 mm diameter). A saturation recovery pulse sequence; (gradient–90°–gradient–delay–90°–Acq(gradient on)) was used. Only the region of the image having a true 90° pulse was integrated for the analysis.

diameter. The increased resistance to flow resulting from the bypass and its volume exchange time must also be considered. That method did not allow simultaneous measurement of arterial and venous oxygenation states without the implementation of a complicated circuit of taps.

The simple modification of using a z gradient to provide spatial resolution would allow the measurement of flow and oxygenation state of arterial and venous blood simultaneously while still requiring cannulation of two blood vessels to provide bypasses through the rf coil.

The above methodology provides information about *in vivo* physiology but requires surgery. A less invasive approach is desirable. Surface coils can be used noninvasively (*5*) but for multiple-pulse experiments, B_1 field inhomogeneity makes quantitative measurements difficult. However, by the use of imaging methods analogous to that described above, the region of homogeneous B_1 field can be selected, allowing quantitative determination of relaxation times and blood flow.

Figure 3 presents a flow calibration curve measured with a surface coil (1 mm diameter) using the above modification to the saturation recovery pulse sequence for flow measurement (*2*). The measurement was made in the region of highest B_1 field at the center of the coil which had a well-defined and measurable 90° pulse. The sample capillary had a bore size less than the coil diameter to allow the problem to be reduced to one dimension in the z direction. By reducing the diameter of the surface coil, lower flow rates than previously possible with Helmholtz coils can be measured *in vivo*. The surface coil can be placed directly on the mobilized blood vessel eliminating the need for cannulation.

As we do not yet have three-dimensional images it has not been possible to determine T_2 values for whole tissue. However, it is of interest to know if T_2 values of whole tissue rather than only blood reflect oxygenation levels. An indication that this is the case comes from an examination of the change in proton linewidth on induction of ischaemia during which oxygenation levels fall drastically. A small surface coil was placed directly onto a rat kidney *in vivo* through a loin incision (comparable to that used for renal biopsy). The proton linewidth of tissue water increased from 41 to 110 Hz on induction of ischaemia. Although the linewidth in such an experiment has a major contribution from B_0 inhomogeneity, the increase in linewidth is a measure of the decrease in T_2. Although this measurement was not made on a single blood vessel, the vascular volume of the kidney (15%) may be sufficiently large to account for the observed change. Such results have important clinical implications for distinguishing lesions resulting from inadequate oxygen delivery to renal tissue when imaging systems at high field become available. More detailed biophysical studies will be reported elsewhere.

ACKNOWLEDGMENTS

This work was supported by the British Heart Foundation, the Science Research Council, and the National Research and Development Corporation.

REFERENCES

1. K. R. Thulborn, J. C. Waterton, P. Styles, and G. K. Radda, *Biochem. Soc. Trans.* **9,** 233 (1981).
2. G. K. Radda, P. Styles, K. R. Thulborn, and J. C. Waterton, *J. Magn. Reson.* **42,** 488 (1981).
3. D. I. Hoult, *Progr. NMR Spectrosc.* **12,** 41 (1978).
4. K. R. Thulborn and G. K. Radda, *J. Cerebral Blood Flow Metabolism* **1,** S82 (1981).
5. O. C. Morse and J. R. Singer, *Science* **170,** 440 (1970).

JOURNAL OF MAGNETIC RESONANCE **47,** 328–330 (1982)

COMMUNICATIONS

Broadband Heteronuclear Decoupling

MALCOLM H. LEVITT

Isotope Department, Weizmann Institute, Rehovot, Israel

AND

RAY FREEMAN AND THOMAS FRENKIEL

Physical Chemistry Laboratory, Oxford University, Oxford, OX1 3QZ, England

Received November 19, 1981

In a recent communication (*1*) we described a new method of heteronuclear broadband decoupling based on a repeated sequence of inversions of one nuclear species S (usually protons) while observing the other species I (usually carbon-13). The inversion of the S spins may be achieved by means of a composite radiofrequency pulse (*2–4*) designed to be insensitive to decoupler offset, or by adiabatic rapid passage through all the S spin resonances. Several possible inversion pulses have been tried; one quite promising example has $R = 90°(X)\ 240°(Y)\ 90°(X)$, used in conjunction with the same sequence with all radiofrequency phases reversed, $\bar{R}$. By grouping the elements R and $\bar{R}$ into a cycle $R\ R\ \bar{R}\ \bar{R}$ which is repeated at a rate fast compared with the largest spin coupling constant J_{IS}, efficient decoupling is achieved over a wide band of S spin frequencies. Justification for this choice of cycle comes from average Hamiltonian theory (*5*). The cycle (denoted MLEV-4) is continuously repeated with minimum intervals between pulses, and the sampling of the carbon-13 free-induction decay is synchronized with the end of each cycle. The preliminary experiments (*1*) on a Varian CFT-20 spectrometer indicated an effective proton decoupling range of ±3 kHz, significantly wider than the results achieved by pseudo-random noise decoupling (*6*).

We now report an extension of this decoupling sequence with far better performance. Further development of the theory and more extensive experiments both confirm the superiority of the sequence MLEV-16,

$$R\ R\ \bar{R}\ \bar{R} \quad \bar{R}\ R\ R\ \bar{R} \quad \bar{R}\ \bar{R}\ R\ R \quad R\ \bar{R}\ \bar{R}\ R,$$

which provides further compensation for residual imperfections of the composite inversion pulse R. The requirement for fast cycling applies only to the subcycles of the type $R\ R\ \bar{R}\ \bar{R}$ but not to the complete cycle MLEV-16. A complete theoretical treatment, which is related to that used for multipulse experiments in solid-state

0022-2364/82/050328-03$02.00/0

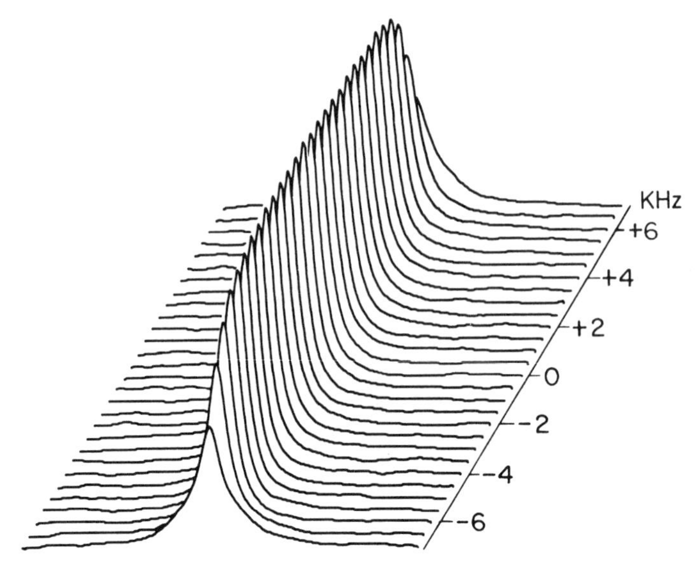

FIG. 1. Broadband proton decoupling of the carbon-13 resonance of dimethyl carbonate by the MLEV-16 sequence, as a function of proton resonance offset. The spectral width displayed is 40 Hz. Peak heights near the center of the proton offset range are essentially equal to that obtained by coherent on-resonance decoupling.

NMR (*7–9*) is presented elsewhere (*10*) together with experimental comparisons of the technique with the methods of noise decoupling (*6*), square-wave phase modulation (*11*), and chirp frequency modulation (*12*).

Figure 1 shows 50-MHz carbon-13 signals observed on a Varian XL-200 spectrometer as a function of proton resonance offset, obtained with the MLEV-16 cycle. The intensity of the irradiation field is given by $\gamma_H B_2/2\pi = 8.3$ kHz and a

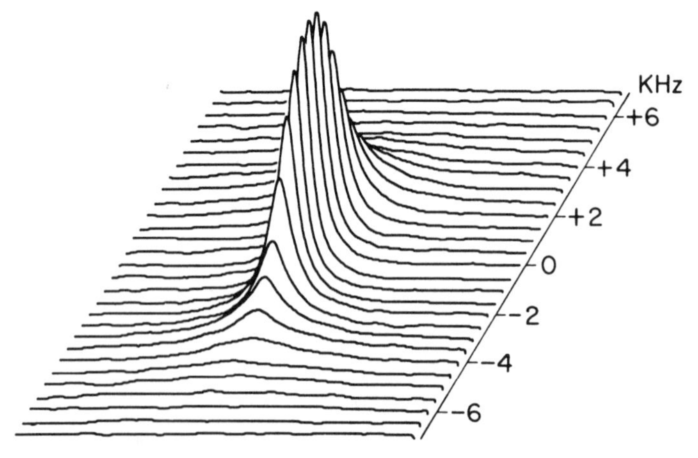

FIG. 2. Broadband decoupling by square-wave phase modulation. The other experimental parameters are identical with those used for Fig. 1.

sensitivity enhancement function has been used, corresponding to a full linewidth of 2 Hz. By way of comparison, Fig. 2 shows the results obtained with square-wave phase modulation, found to be rather more effective than noise modulation under the experimental conditions used. Note that the proposed MLEV-16 sequence operates over a significantly broader band, the peak heights remaining within 5% of that observed for coherent on-resonance decoupling, over a range of more than ±5 kHz. Such an "effective bandwidth" would be acceptable in a hypothetical high-field spectrometer operating at 1000 MHz for protons. Provided that $\gamma_H B_2/2\pi \gg J_{CH}$, this bandwidth decreases in almost direct proportion as B_2 is reduced; tests with a fivefold reduction in the B_2 intensity showed satisfactory decoupling over a ±1-kHz range, which is still quite adequate for the present spectrometer. This is a reduction in radiofrequency power by a factor of 25.

The principle behind these techniques appears to be rather general, and further extensions of the cycle are under investigation. The method can be applied not only to broadband decoupling, but also to the scaling of heteronuclear interactions by a fixed factor. The reader is directed to Ref. (*10*) for a fuller description of the theory and practical aspects of the method.

ACKNOWLEDGMENTS

This work was made possible by an equipment grant from the Science Research Council, which also provided a postdoctoral Fellowship (M.H.L.) and a Research Studentship (T.F.). The authors are pleased to acknowledge their initiation into average Hamiltonian theory by Professor J. S. Waugh at a recent NATO Summer School in Sardinia.

REFERENCES

1. M. H. Levitt and R. Freeman, *J. Magn. Reson.* **43,** 502 (1981).
2. M. H. Levitt and R. Freeman, *J. Magn. Reson.* **33,** 473 (1979).
3. R. Freeman, S. P. Kempsell, and M. H. Levitt, *J. Magn. Reson.* **38,** 453 (1980).
4. M. H. Levitt and R. Freeman, *J. Magn. Reson.* **43,** 65 (1981).
5. U. Haeberlen and J. S. Waugh, *Phys. Rev.* **175,** 453 (1968).
6. R. R. Ernst, *J. Chem. Phys.* **45,** 3845 (1966).
7. J. S. Waugh, L. Huber, and U. Haeberlen, *Phys. Rev. Lett.* **20,** 180 (1968).
8. P. Mansfield, *Phys. Lett. A* **32,** 485 (1970).
9. U. Haeberlen, "High Resolution NMR in Solids," Advances in Magnetic Resonance, Suppl. 1, Academic Press, New York, 1976.
10. M. H. Levitt, R. Freeman, and T. Frenkiel, *Proc. Roy. Soc. London*, in press.
11. J. B. Grutzner and R. E. Santini, *J. Magn. Reson.* **19,** 173 (1975).
12. V. J. Basus, P. D. Ellis, H. D. W. Hill, and J. S. Waugh, *J. Magn. Reson.* **35,** 19 (1979).

JOURNAL OF MAGNETIC RESONANCE **47**, 351–357 (1982)

The Relevance of *J* Cross-Peaks in Two-Dimensional NOE Experiments of Macromolecules

S. MACURA,*·†·‡ K. WÜTHRICH,† AND R. R. ERNST*

**Laboratorium für Physikalische Chemie und †Institut für Molekularbiologie und Biophysik, Eidgenössische Technische Hochschule, 8092 Zürich, Switzerland*

Received December 31, 1981

Two-dimensional (2D) exchange spectroscopy (*1*, *2*) is on the point of becoming a major technique for the elucidation of the spatial structure of biological macromolecules through the nuclear Overhauser effect (NOE) (*3–8*) and for the investigation of chemical exchange processes (*9*, *10*). In particular, two-dimensional NOE spectroscopy has already been well established as a powerful tool for investigating proteins in solution (*3*, *4*, *6–8*).

The nuclear Overhauser effect and chemical exchange cause an incoherent exchange of magnetization between different sites, either by cross-relaxation or by an actual exchange of nuclei, respectively. In 2D spectroscopy these processes are traced out by "frequency-labeling" the various magnetization components before exchange and by sorting out the labeled components after exchange (*1*). In addition to the cross-peaks caused by this incoherent exchange, so-called *J* cross-peaks may arise as a consequence of coherent transfer of magnetization between *J*-coupled resonances (*11*, *12*). Although the *J* cross-peaks contain valuable information on the coupling network, which is exploited in 2D-correlated spectroscopy (*13–15*), they are undesired in exchange spectra. They may be misinterpreted as additional exchange cross-peaks or may falsify the measured intensities in the case of coincidence with exchange cross-peaks. Suitable techniques for their suppression have been described recently (*11*, *12*).

In this communication, we would like to quantitatively evaluate the relevance of *J* cross-peaks in NOESY spectra of proteins by means of experimental measurements and by simple model calculations.

Zero-quantum coherence precessing during the mixing period of duration τ_m is primarily responsible for the occurrence of *J* cross-peaks. Double-quantum coherence may also contribute. However, by proper phase-cycling (*11*), it is possible to eliminate *J* cross-peak contributions originating from double-quantum coherence. The interference effects from transverse magnetization present during the mixing time are usually suppressed by a simple phase alternation technique. Higher-order multiple-quantum coherence is seldom of practical significance for the creation of *J* cross-peaks.

‡ Permanent address: Institute of Physical Chemistry, Faculty of Science, 1100 Beograd, Yugoslavia.

0022-2364/82/050351-07$02.00/0

The intensities of J cross-peaks are fast oscillatory functions of the mixing time τ_m. For zero-quantum transitions, the oscillation frequencies are equal to the difference between the Larmor frequencies of coupled spins, for double-quantum transitions they are equal to the sum of the Larmor frequencies. The zero-quantum and double-quantum J cross-peak contributions vanish with the decay rates $1/T_2^{(0)*}$ and $1/T_2^{(2)*}$, respectively (*11*).

Before describing experimental results, we develop criteria to judge the relevance of J cross-peaks based on a simple model calculation. The buildup of the NOE peak intensities during the mixing time τ_m is governed, in general, by a master equation of the form

$$\dot{\mathbf{m}}(t) = \mathbf{R}\mathbf{m}(t), \qquad [1]$$

where the off-diagonal elements of the relaxation matrix $\mathbf{R}$ are the cross-relaxation rate constants R_{kl}, measuring the magnetization transfer from spin l to spin k, which are directly related to internuclear distances and to molecular correlation times (*2*). It is well known that the solutions of Eq. [1] are superpositions of n exponential functions (n = dimension of $\mathbf{R}$). The NOE buildup is thus nonexponential (*6*). The n time constants (eigenvalues of $\mathbf{R}$) are usually complicated functions of all rate constants R_{kl} and are not easily interpretable. For a simple analysis of the experimental results it is necessary to operate in the linear regime at very short mixing times τ_m, where the NOE cross-peak intensity I_{BA} is directly proportional to the corresponding cross-relaxation rate constant R_{BA}

$$I_{BA} \propto m_A(0) R_{BA} \tau_m \, . \qquad [2]$$

To determine the time limits of the initial rate approximation, let us assume a nuclear spin A which cross-relaxes to $(n - 1)$ additional spins B, C $\cdots$ with $(n - 1)$ identical cross-relaxation rates $R_{BA} \equiv R_{CA} \equiv \cdots$ and with identical leakage rates R_L (*2*). The time evolution of the A and B magnetization is then, for $m_B(0) = m_C(0) = \cdots = 0$, given by the expressions

$$m_A(\tau_m) = m_A(0)\left[\frac{1}{n} + \left(1 - \frac{1}{n}\right)\exp\{-nR_{BA}\tau_m\}\right]\exp\{-R_L\tau_m\}, \qquad [3]$$

$$m_B(\tau_m) = m_A(0)\frac{1}{n}[1 - \exp\{-nR_{BA}\tau_m\}]\exp\{-R_L\tau_m\}. \qquad [4]$$

It is essential to note that the cross-relaxation rate constant R_{BA} is enhanced by the number of nuclei, n. We neglect in the following estimate the leakage relaxation. To apply the linear initial rate approximation with a maximum deviation of, say, 10% from linearity, we have to impose an upper limit for the mixing time τ_m. A straightforward evaluation of Eq. [4] leads to the maximum permitted mixing time

$$\tau_{m\,max} = 0.21[nR_{BA}]^{-1}. \qquad [5]$$

For a reliable measurement of NOE cross-peaks in the linear regime, the J cross-peak amplitude, which exponentially decays with increasing mixing time τ_m, must not exceed 10% of the linearly increasing NOE peak intensity. If we impose that this requirement is fulfilled for $\tau_m > (1/3)\tau_{m\,max}$, we can establish the following conditions for the multiple-quantum decay rates $1/T_2^{(0)*}$ and $1/T_2^{(2)*}$:

$$
\begin{aligned}
n = 2: &\quad 1/T_2^{(0)*}, \quad 1/T_2^{(2)*} > 160R_{BA}, \\
n = 4: &\quad 1/T_2^{(0)*}, \quad 1/T_2^{(2)*} > 320R_{BA}.
\end{aligned} \qquad [6]
$$

With these conditions fulfilled, two-thirds of the linear regime, i.e., $(1/3)\tau_{m\,max} < \tau_m < \tau_{m\,max}$, can be exploited for accurate NOE measurements. Equation [6] makes clear that J cross-peaks can be disregarded without special elimination

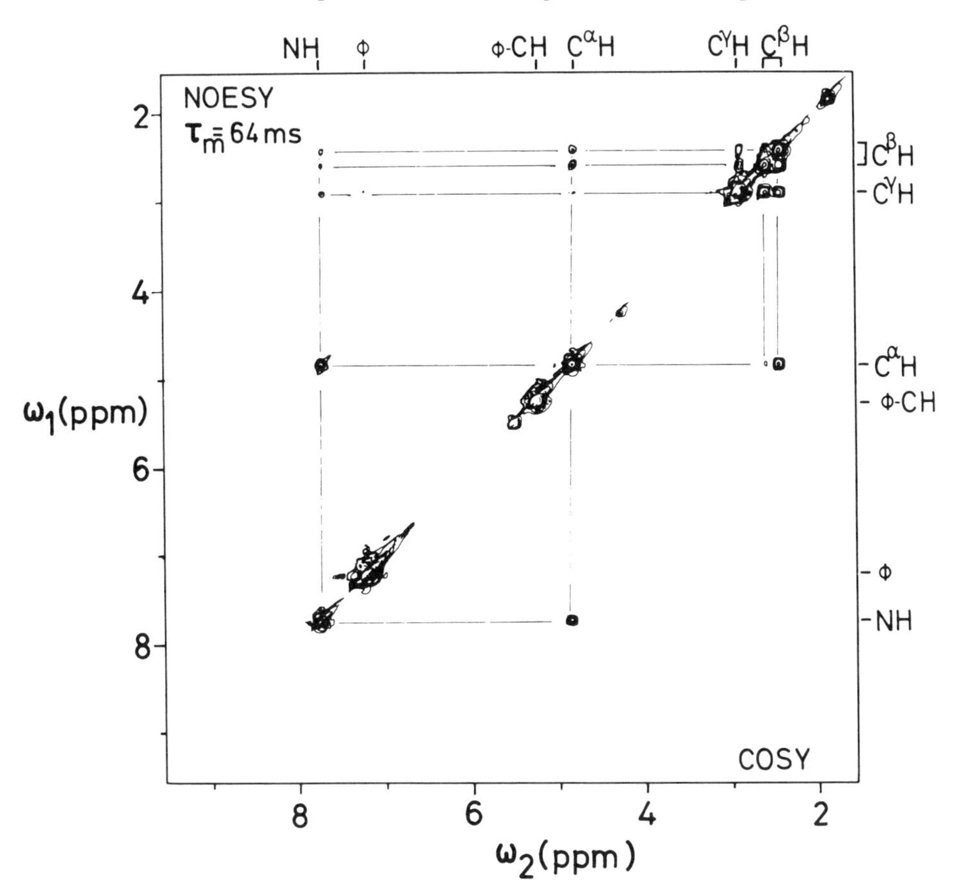

FIG. 1. Combined ^{1}H NOESY (2D nuclear Overhauser enhancement spectroscopy)–COSY (2D correlated spectroscopy) plot for a 2% solution of PBLG (poly-γ-benzyl-L-glutamate) in a mixed solvent of 30% TFA (trifluoroacetic acid) and 70% $CDCl_3$. The spectra were recorded at 360 MHz, the temperature was 24°C, and the NOESY spectrum was obtained with a mixing time of 64 msec. The individual spectra were cut along the diagonal into two triangles, and the combined plot was assembled using the upper-left triangle of NOESY and the lower-right triangle of COSY. For the present investigation this combined plot is particularly convenient since the presence of cross-peaks in symmetrical locations with respect to the diagonal indicates where both coherent and incoherent transfer of magnetization might contribute to the intensity of the NOESY peaks (see text). The chemical shifts of the proton resonances of PBLG are indicated at the top and at the right of the plot, where ϕ stands for the aromatic protons and ϕ-CH for the methylene protons of the benzyl group.

technique only when the multiple-quantum decay rates exceed the cross-relaxation rates by at least two orders of magnitude.

For an experimental check of the relevance of J cross-peaks, we measured the zero-quantum and double-quantum coherence decay rates and the cross- and leakage-relaxation rates in a 2% w/v solution of the synthetic polypeptide poly-γ-benzyl-L-glutamate (PBLG, DP = 1,100, MW = 240,000) in a 30% (v/v) trifluoroacetic acid/deuterochloroform mixture. It is known that under such conditions PBLG is in a random coil form and exhibits rapid internal motions (*16–19*). Despite its large molecular weight of approximately 200,000 it shows features similar to small globular proteins, with correlation times for the backbone skeleton of approximately 1×10^{-8} sec (which is in the spin diffusion limit, i.e., $\omega\tau_c \gg 1$, at proton resonance frequencies $\gtrsim$ 360 MHz).

Figure 1 shows at the same time the ^{1}H–^{1}H connectivities in PBLG observed with NOESY ($[90°\text{–}t_1\text{–}90°\text{–}\tau_m\text{–}90°\text{–}t_2]_n$ pulse sequence) (*3*), and with correlated spectroscopy (COSY, using the sequence $[90°\text{–}t_1\text{–}90°\text{–}t_2]_n$) (*8*, *13*, *15*). Since both NOESY and COSY connectivities are manifested redundantly by pairs of cross-peaks in symmetrical locations with respect to the diagonal (*16*, *17*), the triangular NOESY/COSY combination of Fig. 1 contains the information from both experiments in a single, square plot (*8*). The observed peaks can be interpreted in terms of a single building block of PBLG,

$$-\underset{\underset{O}{\|}}{C}-\overset{\overset{|}{NH}}{\overset{|}{C^{\alpha}}}H-C^{\beta}H_2-C^{\gamma}H_2-\overset{\overset{O}{\|}}{C}-O-CH_2-C_6H_5 \quad .$$

The chemical shifts for the individual protons are indicated in Fig. 1. It is seen that within the resolution of our experiments, the five aromatic protons, the two benzyl methylene protons, and the two γ-methylene protons, respectively, have identical chemical shifts. A priori, one expects both COSY *and* NOESY connectivities between geminal and vicinal protons, i.e., the occurrence of a pair of cross-peaks at symmetrical locations in a combined NOESY/COSY plot. This is verified by Fig. 1: Symmetrical cross-peaks connect NH with $C^{\alpha}H$, $C^{\alpha}H$ with both β-methylene protons, $C^{\beta}H$ with $C^{\beta}H$, and $C^{\gamma}H$ with both β-methylene positions. For all these combinations the intensity of the NOESY peak may contain, at the same time, contributions from J coupling and dipole–dipole coupling. On the other hand, the NOESY cross-peaks linking NH with $C^{\beta}H$ or $C^{\gamma}H$, and $C^{\alpha}H$ with $C^{\gamma}H$ must come entirely from dipolar cross-relaxation, since there is no COSY peak in the symmetrical location.

For the measurement of the zero- and double-quantum coherence decay rates, the J cross-peak contributions were separated along the ω_1 axis by a method analogous to the mixing time incrementation technique described earlier (*12*). We employed the scheme of Fig. 2, where an additional 180° refocusing pulse has been inserted into the mixing period. Its position is systematically shifted together with the t_1 incrementation. This leads in the t_1 domain to a characteristic modulation of zero-quantum and double-quantum coherence by the difference and sum of the two involved Larmor frequencies, respectively, and causes, after Fourier transfor-

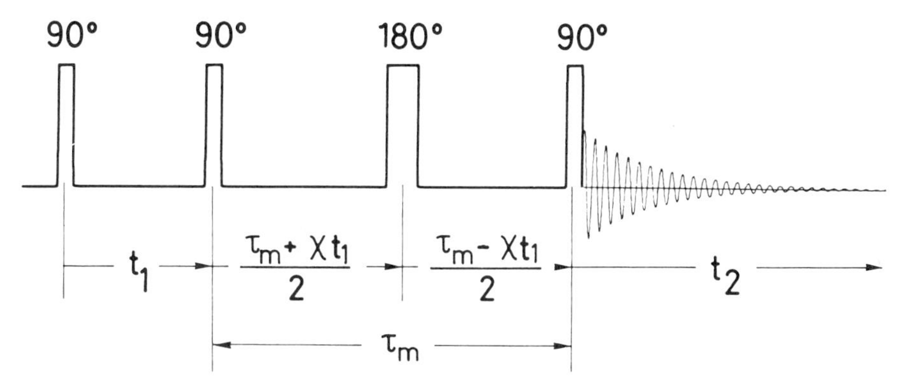

FIG. 2. Basic 2D exchange experiment (*1*, *3*) modified for separation of NOE and J cross-peaks. The position of the refocusing 180° pulse is systematically shifted in proportion to the evolution time t_1. The position of the 180° pulse affects exclusively the J cross-peaks but leaves the NOE cross-peaks invariant.

mation, shifts of the J cross-peak contributions by, for example, $\pm\chi(\omega_{\alpha CH}-\omega_{NH})$ and $\pm\chi(\omega_{\alpha CH} + \omega_{NH})$ (*12*).

We would like to mention at this point that the use of a refocusing 180° pulse during the mixing period (Fig. 2) leads to efficient J cross-peak elimination techniques. Thus, it is possible to eliminate the multiple-quantum J cross-peak contributions, separated by the technique described above, by symmetrization (*20*, *21*) in analogy to elimination by the mixing time incrementation technique (*12*). Alternatively, it is also feasible to eliminate J cross-peaks by a random variation of the position of the 180° pulse, in analogy to a previously proposed method using random variation of τ_m (*11*). The employment of shifted 180° refocusing pulses for separation or suppression of J cross-peak components has the advantage that the mixing time τ_m remains constant throughout the entire 2D experiment.

From a set of spectra in which the diagonal peak, the NOE cross-peak, and the zero- and double-quantum J cross-peak contributions were separated by the experiment of Fig. 2, we obtained the buildup and decay curves plotted in Fig. 3. From these the four parameters

$$\begin{aligned} R_{BA} &= 1.16 \pm 0.05 \text{ sec}^{-1}, \\ R_L &= 1.29 \pm 0.2 \text{ sec}^{-1}, \\ 1/T_2^{(0)*} &= 44 \pm 6 \text{ sec}^{-1}, \\ 1/T_2^{(2)*} &= 63 \pm 8 \text{ sec}^{-1}. \end{aligned} \qquad [7]$$

have been deduced (assuming $n = 2$ in Eq. [3]). The order of magnitude of these values can be considered typical for ^1H–^1H interactions in polypeptide segments with effective rotational correlation times of the order 1×10^{-8} sec, e.g., for large proteins in "random coil" form or for small proteins in globular form.

These experimental results show that in protein NOESY spectra recorded with mixing times τ_m shorter than approximately 100 msec the stringent condition for the neglect of J cross-peaks, Eq. [6], is usually not fulfilled. Hence, for accurate measurements of NOEs or even for a qualitative interpretation of NOESY spectra

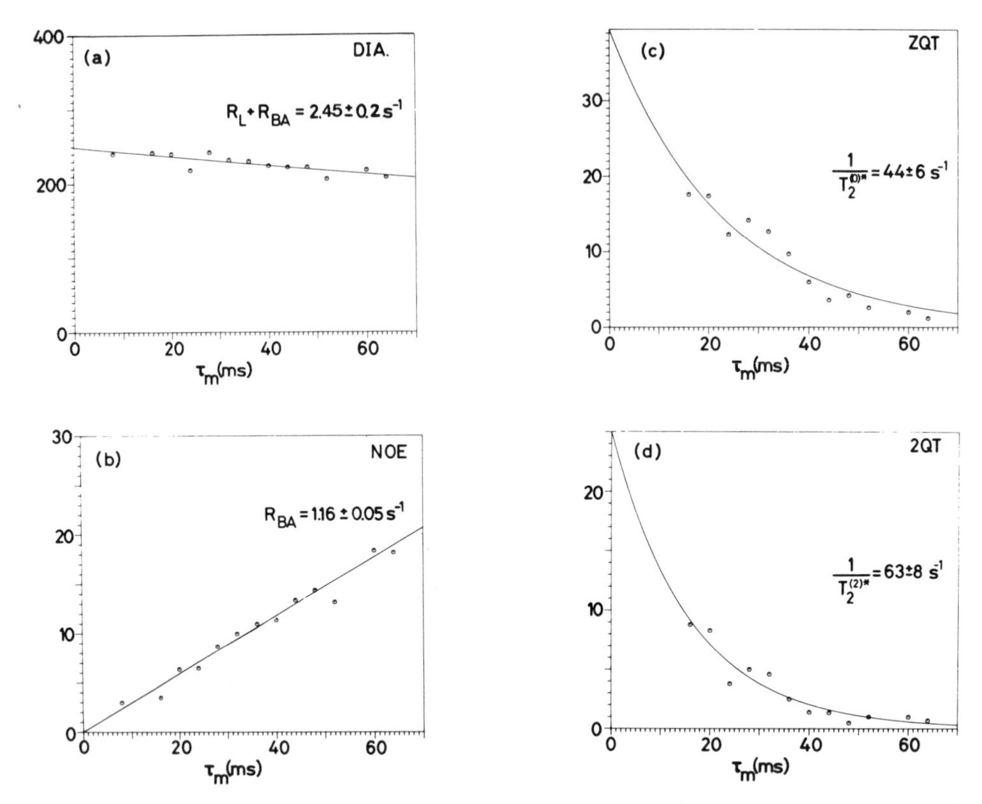

FIG. 3. Decay of diagonal and cross-peak contributions as functions of the mixing time τ_m for PBLG (Fig. 1). (a) Amplitude of diagonal NH peak. (b) Amplitude of NH-C$^\alpha$H NOE cross-peak. (c) Amplitude of zero-quantum contribution to diagonal NH peak. (d) Amplitude of double-quantum contribution to diagonal NH peak. The peak amplitudes have been obtained from cross sections taken at $\omega_2 = \omega_{NH}$ through a set of NOESY spectra recorded with the scheme of Fig. 2 at 360 MHz and 24°C using τ_m values between 8 and 64 msec and $\chi = 0.237$. Diagonal and NOE cross-peaks were located at frequencies $\omega_1 = \omega_{NH}$ and $\omega_1 = \omega_{C^\alpha H}$, respectively, while the J cross-peak contributions were shifted from those two positions by $\pm\chi\omega_{ZQT}$ and $\pm\chi\omega_{2QT}$, respectively. The rates given in (a) and (b) have been obtained by linear fits of the initial slopes, those in (c) and (d) from exponential least-squares fits.

in terms of the initial rate approximation, J cross-peaks must be eliminated by special techniques, e.g., by those described in Refs. (*11*) and (*12*) and those mentioned above, which use the experimental scheme of Fig. 2.

Finally, we would like to discuss briefly the origin of the measured decay rates for PBLG. The cross-relaxation between NH(A) and C$^\alpha$H(B) protons with the rate R_{BA} is determined by the dipolar interaction and is in the spin diffusion limit given by (*2*)

$$R_{BA} = q_{BA}\tau_{c\,BA}, \qquad q_{BA} = \frac{1}{10}\left(\gamma^2 \hbar r_{BA}^{-3} \frac{\mu_0}{4\pi}\right)^2. \qquad [8]$$

With an estimate for the average proton–proton distance of $r_{BA} \sim 3.0$ Å, we calculate a correlation time $\tau_{c\,BA} = 1.4 \times 10^{-8}$ sec for the motion of the backbone skeleton. The leakage relaxation rate R_L is governed primarily by interaction with additional 3–4 neighboring protons. A quantitative evaluation would involve numerous unknown parameters describing internal motion and shall not be attempted. The zero- and multiple-quantum relaxation rates are also affected by the dipolar interactions with the neighboring protons (*11*). However, the rather fast decay of multiple-quantum coherence cannot be accounted for by this mechanism alone, taking into account the rather low leakage relaxation rate R_L. For an explanation, it should be recalled that the multiple-quantum transitions are also split by J couplings to further nuclei (*22, 23*). The zero-quantum transition of nuclei A and B, for example, is split because of the interactions to a nucleus K by the difference of the respective coupling constants, $J_{AK} - J_{BK}$. The expected complicated coupling network of the NH and $C^{\alpha}H$ protons can explain the observed fast multiple-quantum decay rates which are comparable to the normal single-quantum decay rate $1/T_2^* = 46\ \text{sec}^{-1}$ of the NH proton.

In conclusion, it has been found that even in the presence of fast damping of the multiple-quantum coherence by J coupling to additional nuclei, J cross-peaks are not negligible in NOE measurements when initial rates in the linear regime are exploited. Suppression of coherent transfer effects is therefore generally necessary for reliable measurements of NOE peak intensities in NOESY spectra.

ACKNOWLEDGMENTS

This research has been supported by a research grant by the Swiss Federal Institute of Technology. Computational support by Mr. M. Linder is acknowledged.

REFERENCES

1. J. Jeener, B. H. Meier, P. Bachmann, and R. R. Ernst, *J. Chem. Phys.* **71**, 4546 (1979).
2. S. Macura and R. R. Ernst, *Mol. Phys.* **41**, 95 (1980).
3. Anil Kumar, R. R. Ernst, and K. Wüthrich, *Biochem. Biophys. Res. Commun.* **95**, 1 (1980).
4. Anil Kumar, G. Wagner, R. R. Ernst, and K. Wüthrich, *Biochem. Biophys. Res. Commun.* **96**, 1156 (1980).
5. S. Macura and R. R. Ernst, *Period. Biol.* **83**, 87 (1981).
6. Anil Kumar, G. Wagner, R. R. Ernst, and K. Wüthrich, *J. Am. Chem. Soc.* **103**, 3654 (1981).
7. C. Bösch, Anil Kumar, R. Baumann, R. R. Ernst, and K. Wüthrich, *J. Magn. Reson.* **42**, 159 (1981).
8. G. Wagner, Anil Kumar, and K. Wüthrich, *Eur. J. Biochem.* **114**, 375 (1981).
9. B. H. Meier and R. R. Ernst, *J. Am. Chem. Soc.* **101**, 6441 (1979).
10. Y. Huang, S. Macura, and R. R. Ernst, *J. Am. Chem. Soc.* **103**, 5327 (1981).
11. S. Macura, Y. Huang, D. Suter, and R. R. Ernst, *J. Magn. Reson.* **43**, 259 (1981).
12. S. Macura, K. Wüthrich, and R. R. Ernst, *J. Magn. Reson.* **46**, 269 (1982).
13. W. P. Aue, E. Bartholdi, and R. R. Ernst, *J. Chem. Phys.* **64**, 2229 (1976).
14. K. Nagayama, K. Wüthrich, and R. R. Ernst, *Biochem. Biophys. Res. Commun.* **90**, 305 (1979).
15. K. Nagayama, Anil Kumar, K. Wüthrich, and R. R. Ernst, *J. Magn. Reson.* **40**, 321 (1980).
16. D. I. Marlborough, K. G. Orreland, and N. H. Rydon, *Chem. Commun.* **21**, 518 (1965).
17. E. M. Bradbury, C. Crane-Robinson, H. Goldman, and H. W. E. Rattle, *Nature (London)* **217**, 812 (1968).
18. K. Nagayama and A. Wada, *Chem. Phys. Lett.* **16**, 50 (1972).
19. K. Nagayama and A. Wada, *Biopolymers* **14**, 2489 (1975).
20. R. Baumann, Anil Kumar, R. R. Ernst, and K. Wüthrich, *J. Magn. Reson.* **44**, 76 (1981).
21. R. Baumann, G. Wider, R. R. Ernst, and K. Wüthrich, *J. Magn. Reson.* **44**, 402 (1981).
22. G. Pouzard, S. Sukumar, and L. D. Hall, *J. Am. Chem. Soc.* **103**, 4209 (1981).
23. L. Braunschweiler, G. Bodenhausen, and R. R. Ernst, in preparation.

JOURNAL OF MAGNETIC RESONANCE **47**, 515–521 (1982)

COMMUNICATIONS

The Loop-Gap Resonator: A New Microwave Lumped Circuit ESR Sample Structure

W. FRONCISZ* AND JAMES S. HYDE

National Biomedical ESR Center, Department of Radiology, Medical College of Wisconsin, 8701 Watertown Plank Road, Milwaukee, Wisconsin 53226

Received February 23, 1982

Hardy and Whitehead (*1*), apparently working from an NMR perspective, have described a novel sample resonator for the range of 200 to 2000 MHz. This type of resonator was described as early as 1940 (*2*) in the context of magnetron designs and more recently in the literature of heavy-ion particle accelerators (*3*). We describe here the design and use of structures of this type for ESR spectroscopy between 1 and 10 GHz. As will be apparent, it offers some remarkable advantages. We have previously published data at 3.8 GHz (*4*).

Figure 1 shows one of our designs and also defines the notation. Various names have been used for this structure: we introduce here the name "loop-gap." The loop is the inductive element, *a*, surrounding the sample, while the gaps, *b*, are capacitive elements. We find it convenient to employ nouns describing the two key elements of the resonator as a compound name for the structure itself.

Previous workers have used single-gap resonators. We have discovered that the number of gaps can be increased and that the equivalent circuit of the resonator has the capacities of the individual gaps connected in series. This is a lumped circuit resonator. The dimensions of the structure should be small compared with one-quarter wavelength, $\lambda/4$. Capacitive and inductive elements are separated in space, and to a fair approximation the rf magnetic field exists inside the loop and the rf electric field inside the gaps. The classical expressions for the inductance and capacitance, assuming n identical gaps, are

$$C = \frac{\epsilon W Z}{tn}, \qquad L = \frac{\mu_0 \pi r^2}{Z} \qquad [1]$$

and the resonant frequency is given by

$$2\pi\nu = \frac{1}{(LC)^{1/2}} = \frac{1}{r} n^{1/2} \left(\frac{t}{W}\right)^{1/2} \left(\frac{1}{\pi \epsilon \mu_0}\right)^{1/2}, \qquad [2]$$

which is independent of length Z. These equations are only approximations. An advantage of using larger values of n is that larger values of r become possible, permitting ESR spectroscopy on larger samples at higher frequency. This is one of the points of this communication. We find that the use of multiple gaps together with variation of gap dimensions t and W and dielectric constant ϵ permits one to

* On leave from the Department of Biophysics, Institute of Molecular Biology, Jagiellonian University, 31-001, Krakow, Poland.

0022-2364/82/060515-07$02.00/0

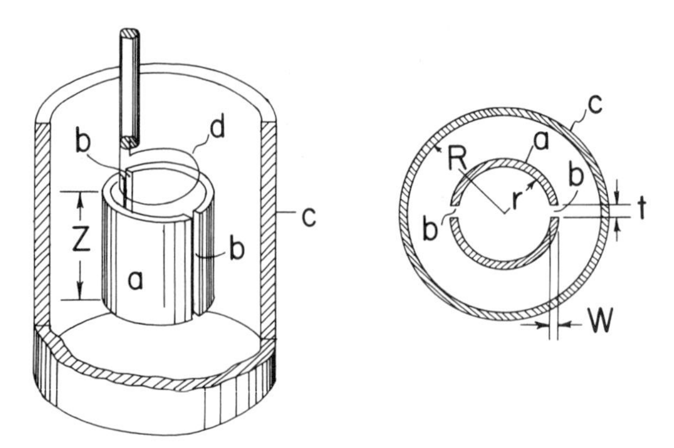

FIG. 1. The loop-gap resonator showing the principal components (a, loop; b, gaps; c, shield; d, inductive coupler) and the critical dimensions (Z, resonator length; r, resonator radius; R, shield radius; t, gap separation; W, gap width). The sample is inserted into the loop a through the coupler, d. The microwave magnetic field in the loop is parallel to the axis of the loop.

adjust the capacity C such that resonance can be achieved for almost any inductance L (or radius r) as long as dimensions remain small compared with $\lambda/4$.

When the dimensions begin to approach $\lambda/4$, the resonator may radiate, thereby decreasing the Q. Radiation can be suppressed by a cylindrical shield, c, of radius R, which should be less than the cutoff wavelength for the lowest excited propagation mode in cylindrical waveguide. We, as well as Hardy and Whitehead, have found an inductive coupler d convenient, achieving match to the resonator by a mechanical displacement of the separation of the coupling loop and the resonator loop.

The shield and the fringing fields of the gaps shift the frequency and cause a breakdown in Eq. [2]. The semiempirical equation

$$\nu = \frac{1}{2\pi}\left(1 + \frac{r^2}{R^2 - (r+W)^2}\right)^{1/2}\left(\frac{nt}{\pi W \epsilon \mu_0}\right)^{1/2}\frac{1}{r}\left(\frac{1}{1 + 2.5(t/W)}\right)^{1/2} \qquad [3]$$

has been satisfactory in calculating the frequency within 10% between 1 and 10 GHz over a very wide variation of t, W, r, R, and Z. The first term in parentheses was introduced by Hardy and Whitehead to describe the effect of the shield. The last term describes the effect of fringing fields of the gaps.

The Q of the resonator likewise can be calculated, neglecting effects of the shield and electric field losses in the gaps,

$$Q \simeq \frac{r}{\delta} \propto r\nu^{1/2}, \qquad [4]$$

where δ is the skin depth.

Hardy and Whitehead give an expression for the Q including shield effects

$$Q_L = \frac{r}{\delta}\left[1 + \frac{r^2}{R^2 - (r+W)^2}\right] \Big/ \left[1 + \left(1 + \frac{W}{r} + \frac{R}{r}\right)\left(\frac{r^2}{R^2 - (r+W)^2}\right)^2\right]. \qquad [5]$$

To this should be added the conductor losses of the capacitor formed by the gaps, using Caulton's semiempirical equation (*5*) as modified by us to take into account the fringing fields

$$Q_C = 1.7 \times 10^5 t[\nu^{3/2}\epsilon W^2(1 + 2.5(t/W))]^{-1} \quad [6]$$

where t and W in meters, ν in hertz and ϵ for vacuum 8.85×10^{-12} F/m, and therefore

$$\frac{1}{Q} = \frac{1}{Q_L} + \frac{1}{Q_C}. \quad [7]$$

The Q is predicted to be independent of resonator length Z. The peak rf field in a matched resonator varies as

$$B_1 \propto \left(\frac{QP_0}{\nu r^2 Z}\right)^{1/2} \quad [8]$$

or approximately,

$$B_1 \propto P_0^{1/2}\nu^{-1/4}(rZ)^{-1/2}, \quad [9]$$

where P_0 is the incident power. The peak rf field intensity has been measured in several different structures at several frequencies using the technique of perturbing metal spheres (*6*), in combination with ESR techniques, and the results are shown in Table 1.

Most of our experiments have been carried out with resonators fabricated from the machinable ceramic Macor (Corning Glass), and silver-plated in a two-step process: chemical deposition followed by electroplating. The shield is silver-plated fiberglass-epoxy. Experiments at X band were carried out using a standard Varian spectrometer. Since the loop-gap resonator is of dimensions that are similar to the Varian Q-band cavity, it was convenient to use the Varian Q-band field modulation and Dewar assemblies both at X and at S bands. The bridge used for S band is

TABLE 1
MICROWAVE CHARACTERISTICS OF LOOP-GAP RESONATORS

No.	r_0 (mm)	t (mm)	W (mm)	Z (mm)	ν (GHz)	Q_0	Λ[a]	B_1[b]	Resonator
1	0.5	0.1	2.7	2.5	8.8	600	8.2	9	Macor[c,d]
2	0.6	0.1	1.9	10	9.15	810	4.9	5.4	Macor[c,d]
3	0.6	0.1	1.9	5	9.5	700	6.1	6.7	Macor[c,d]
4	2.0	0.15	0.4	10	9.09	1400	2.4	2.6	Macor[c,e]
5	2.4	0.25	2.4	10	3.8	1300	—	3	Macor[d]
6	2.5	0.15	0.7	10	9.8	1200	—	2	Copper[f]
7	3.9	0.15	1.7	10	2.3	1200	—	—	Brass[d,g]
8	6.1	1.6	0.3	20	3.2	2500	—	—	Brass[e,g]

[a] In comparison with Varian TE_{102} multipurpose cavity.
[b] Gauss (rotating frame) at 1-W incident power.
[c] Macor (Corning) machinable ceramic, silver-plated, in Rexolite holder.
[d] One gap.
[e] Two gaps, supported on a quartz tube.
[f] Four gaps, supported on a quartz tube.
[g] Silver-plated.

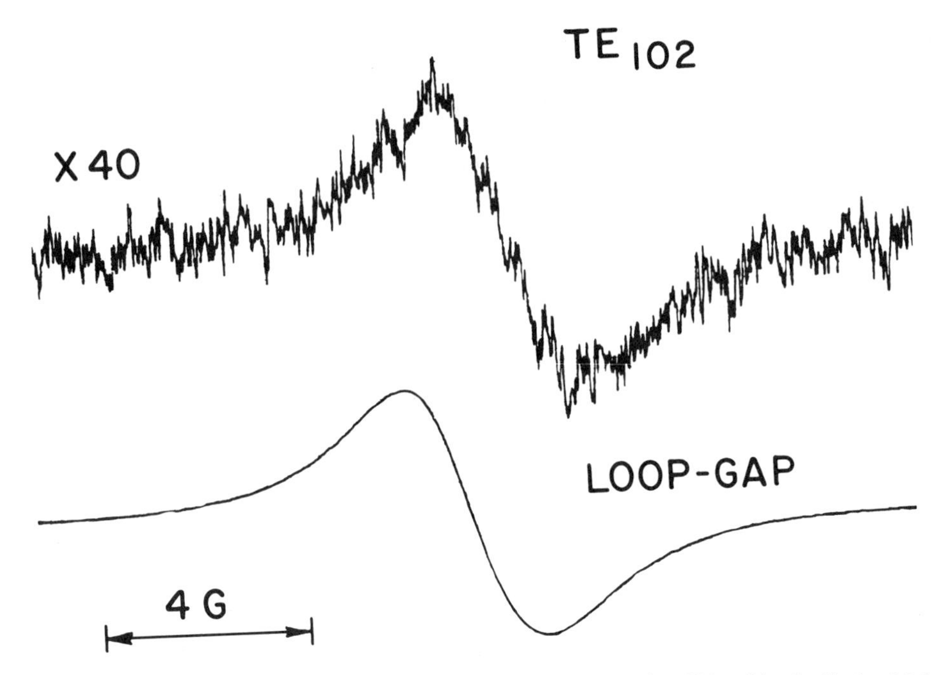

FIG. 2. Comparison of spectra of a point sample of DPPH measured at X band in the Varian Multipurpose (rectangular TE_{102}) cavity and in a loop-gap resonator of dimensions $r = 0.6$ mm, $Z = 5$ mm. The incident power was held constant at a low nonsaturating level. The loop-gap resonator yields 37 times greater signal, indicating that $\Lambda = 6.1$.

self-built, of the reference arm type, and employs a mechanically tunable octave bandwidth (2 to 4 GHz) transistor oscillator (Engelman CC-24).

We have found it convenient in comparing the loop-gap resonator with cavity resonators to introduce the parameter Λ, which is the ratio of peak magnetic field intensities at the sample positions in the two structures for constant incident magnetic field. In one loop-gap resonator that has been extensively used by us at X band, resonator 1 in Table 1, $\Lambda = 8.2$ in comparison with a Varian TE_{102} multipurpose cavity. Thus, 67 times higher incident power is required in the multipurpose cavity to achieve the same rf field intensity. For a small sample exhibiting no microwave power saturation, it can be shown that the relative signal intensity will vary as Λ^2. Figure 2 illustrates nearly a 40 times increase in signal intensity using a small sample of DPPH in resonator 3.

In order to examine the Varian weak pitch standard sample at X band, it is necessary that $n \geq 2$, yielding a loop diameter of 4 mm at 9.5 GHz. Figure 3a shows the resulting spectrum using resonator 4 in Table 1. The signal-to-noise ratio employing the usual extrapolation procedures to 200 mW is 1450:1. Figure 3b shows the same sample in resonator 5 at 3.8 GHz measured at 2 mW of incident power. With cavity resonators at 3.8 GHz, we have been unable to detect a signal from this sample.

Of particular interest is ESR of biological samples in water. Because to a first

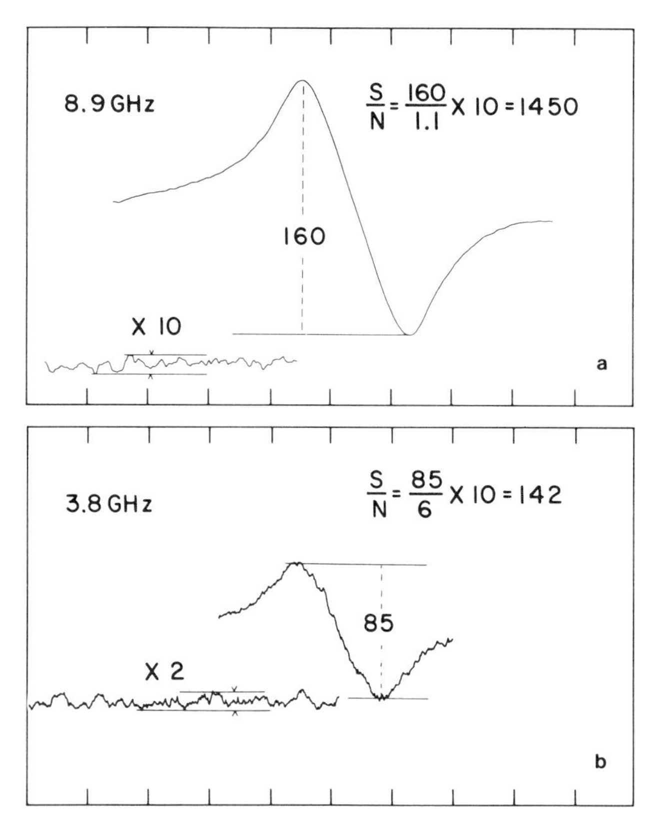

FIG. 3. (a) Sensitivity test using Varian weak pitch standard sample in a two-gap resonator at X band, $r = 2.0$ mm, $Z = 10$ mm. The spectrum was run at 12.5-mW incident power and the signal height extrapolated to the value expected at 200 mW assuming no microwave power saturation. (b) Sensitivity test using Varian weak pitch in an S-band loop-gap resonator, $r = 2.1$, $Z = 10$ mm. The spectrum was run at 2-mW incident power and the signal height extrapolated to 32 mW, which is the maximum power available in our S-band bridge. The signal to noise ratio would be 350:1 at 200 mW.

approximation the electric field exists only in the gaps and not in the loop, the loop-gap resonator would appear to be a particularly attractive structure. For saturable samples of unlimited size (such as spin labels), optimizing the amount of sample in both cavity and loop-gap resonator, we have achieved an improvement with the loop-gap resonator of $\Lambda/2$, while for unsaturable samples (such as aqueous Mn^{2+}), the ratio is $\Lambda^2/2$, factors of 3 and 20, respectively, for resonator 3.

If the amount of aqueous sample is limited, the advantages of the loop-gap resonator become more pronounced. In Fig. 4, resonator 2 ($\Lambda = 4.9$) is compared with the Varian TE_{102} cavity by examining a 0.4-mm-i.d. aqueous sample of a spin label in both structures. The effective length of the cavity is about 1 cm and thus the sample volume is about the same in the comparison. In order for the rf field to be the same, incident power must be 13 db higher when using the cavity than when using the loop-gap resonator. With 13-db lower power, the loop-gap resonator yields about five times higher signal intensity.

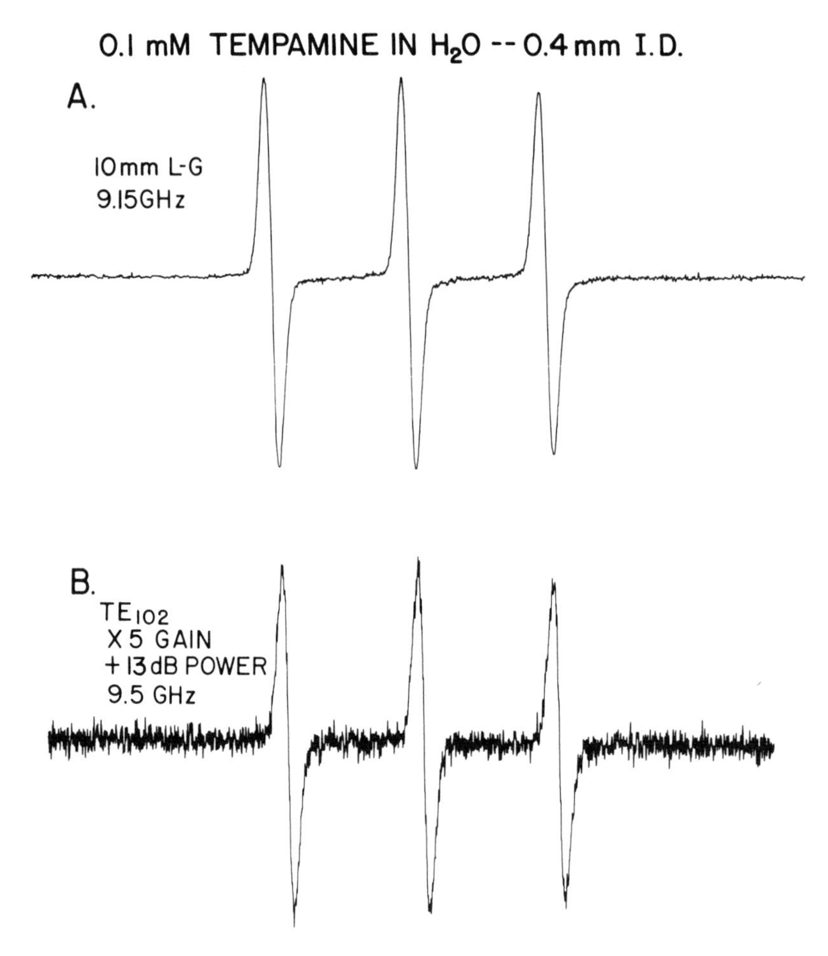

FIG. 4. Sensitivity comparison examining an aqueous spin label sample in a TE_{102} cavity and a loop-gap resonator with the same effective lengths at the same value of B_1 at the sample.

Resonant cavities tend to become increasingly disadvantageous at low microwave frequencies because (a) they are large and cumbersome, (b) the values of Q increase as $\nu^{-1/2}$, resulting in increasingly serious demodulation of microwave oscillator noise, and (c) the peak rf energy density falls approximately as $\nu^{-5/2}$, making it increasingly difficult to saturate samples. The loop-gap resonator overcomes these problems. The advantages at X band are less, but still pronounced. Thus weak pitch is several hundred times improved at S band and about 6 to 7 times improved at X band. The low Q of the resonator lessens the susceptibility of the structure not only to demodulation of source noise but also to microphonics and vibrations. The high rf field intensity for the available input power is attractive for pulse EPR experiments and cw saturation studies on transition metal ions.

ACKNOWLEDGMENTS

This work was supported by Grants PCM-7823206 and PCM-8118976 from the National Science Foundation and Grants GM-27665 and RR-01008 from the National Institutes of Health.

REFERENCES

1. W. N. HARDY AND L. A. WHITEHEAD, *Rev. Sci. Instrum.* **52,** 213 (1981).

2. G. B. COLLINS *in* "Microwave Magnetrons," pp. 49–50, 59–62, McGraw-Hill, New York, 1948.

3. (*a*) K. W. SHEPARD, J. E. MERCEREAU, AND G. J. DICK, *IEEE Trans. Nucl. Sci.* **NS-22,** 1179 (1975). (*b*) J. R. DELAYEN, G. J. DICK, AND J. E. MERCEREAU, *IEEE Trans. Nucl. Sci.* **NS-26,** 3664 (1979). (*c*) J. R. DELAYEN, G. J. DICK, AND J. E. MERCEREAU, *IEE Trans. Magn.* **MAG-17,** 939 (1981).

4. J. S. HYDE AND W. FRONCISZ, *in* "Proceedings, National Electronics Conference 1981," Vol. XXXV, pp. 602–606.

5. M. CAULTON *in* "Advances in Microwaves" (L. Young and H. Sobol, Eds.), pp. 143–202, Academic Press, New York, 1974.

6. J. H. FREED, D. S. LENIART, AND J. S. HYDE, *J. Chem. Phys.* **47,** 2762 (1967).

JOURNAL OF MAGNETIC RESONANCE **48**, 286–292 (1982)

A Two-Dimensional Nuclear Overhauser Experiment with Pure Absorption Phase in Four Quadrants*

D. J. STATES

The Chemical Laboratories, Harvard University, Cambridge, Massachusetts 02138

AND

R. A. HABERKORN AND D. J. RUBEN

Francis Bitter National Magnet Laboratory, Massachusetts Institute of Technology, Cambridge, Massachusetts 02139

Received December 29, 1981

A method is described for obtaining pure absorption phase spectra in four quadrants in a two-dimensional nuclear magnetic resonance spin exchange experiment. It is shown that phase correction results in a substantial increase in resolution and discrimination while maintaining a signal-to-noise ratio comparable to that of the usual magnitude spectrum. Experimental results are presented for the application of the method to a biological macromolecule, the bovine pancreatic trypsin inhibitor.

The use of two-dimensional NMR methods to observe nuclear Overhauser and chemical exchange phenomena have generated considerable excitement (*1–3*). These techniques are able to resolve many individual spin exchange effects even in complex systems such as biological macromolecules, and they permit rapid and systematic data collection. The quantitative interpretation of this data has been limited by the application of absolute value and power transforms to obtain suitable peak shapes (*4*). These nonlinear transforms generate positive definite spectra from data that would otherwise have both positive and negative peaks (*5*) but also result in differential scaling, peak broadening, and cross terms with overlapping peaks and baseline offsets.

In this communication we describe a technique for phasing two-dimensional spectra based on the use of separate quadrature in the two dimensions. We will show that this method can be applied with only minor modifications in the experimental design of the usual two-dimensional spin exchange experiment, and we will present experimental results obtained with the method. Finally we will discuss the impact of these methods on signal to noise ratio and resolution.

BASIS FOR TWO-DIMENSIONAL PHASING IN FOUR QUADRANTS

A two-dimensional exchange experiment can be described by considering an oscillator with frequency ω_1 during the labeling period and a frequency ω_2 after

* This work was supported in part by grants from the National Institutes of Health.

0022-2364/82/080286-07$02.00/0

some mixing period. The observed amplitude of this oscillator will then be a function of both the time t_1, used to label the spins and the time t_2, at which it is observed during the acquisition period. The two-dimensional spectrum is usually obtained by taking Fourier transforms in both dimensions

$$F(\omega_1, \omega_2) = \int\int e^{i\omega_1 t_1} e^{i\omega_2 t_2} f(t_1, t_2) dt_1 dt_2 ,$$

where $f(t_1, t_2)$ is the oscillation function in the time domain and $F(\omega_1, \omega_2)$ is its Fourier transform. Assuming an exponentially decaying oscillation we obtain the expression

$$F(\omega_1, \omega_2) = \frac{1/T_2 + i(\omega_1 - \omega_{01})}{1/T_2^2 + (\omega_1 - \omega_{01})^2} \frac{1/T_2 + i(\omega_2 - \omega_{02})}{1/T_2^2 + (\omega_2 - \omega_{02})^2} .$$

It is easily seen that both the real and imaginary parts of this expression are biphasic. Note that for each oscillator $f(t_1, t_2)$ can be separated into the product $f_1(t_1)*f_2(t_2)$. This permits the integrals to be separated. By accumulating the real and imaginary parts of each dimension independently it is possible to calculate the product of the real part in one dimension and the real part in the other dimension to obtain

$$G(\omega_1, \omega_2) = \frac{1/T_2}{1/T_2^2 + (\omega_1 - \omega_{10})^2} \frac{1/T_2}{1/T_2^2 + (\omega_2 - \omega_{20})^2} .$$

This latter function is linear in the peak amplitude and is positive definite. It decays rapidly away from ω_{01}, ω_{02}, as an absorption phase Lorentzian in both dimensions.

Pure absorption phase spectra from a single-quadrant two-dimensional spin exchange experiment have been reported (*6*). In a four-quadrant experiment it is not possible to achieve a pure absorption phase two-dimensional spectrum using analytical transformations of the data (*5, 7*). The transformation outlined above generates a pure absorption phase spectrum in four quadrants using a nonanalytical transformation. It is based on extracting the real part of a complex function, a transformation that is linear to addition and real multiplication, but is not analytical. Having used this nonanalytical transformation, the projection cross-section theorem (*8*) does not apply, and there is no theoretical objection to a positive definite four-quadrant spectrum.

Pure phase spectroscopy can be performed using the usual two-dimensional NMR pulse sequence (Fig. 1). The real part of the t_1 dimension is obtained by taking the

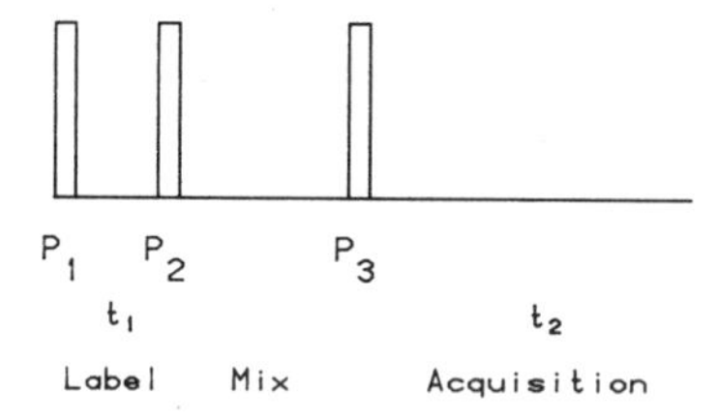

FIG. 1. The pulse sequence used in two-dimensional spin exchange spectroscopy. P_1 and P_2 are two $\pi/2$ pulses used to label spins in the t_1 dimension. The spins are allowed to mix and a final detection pulse P_3 is applied to excite a free-induction decay during t_2.

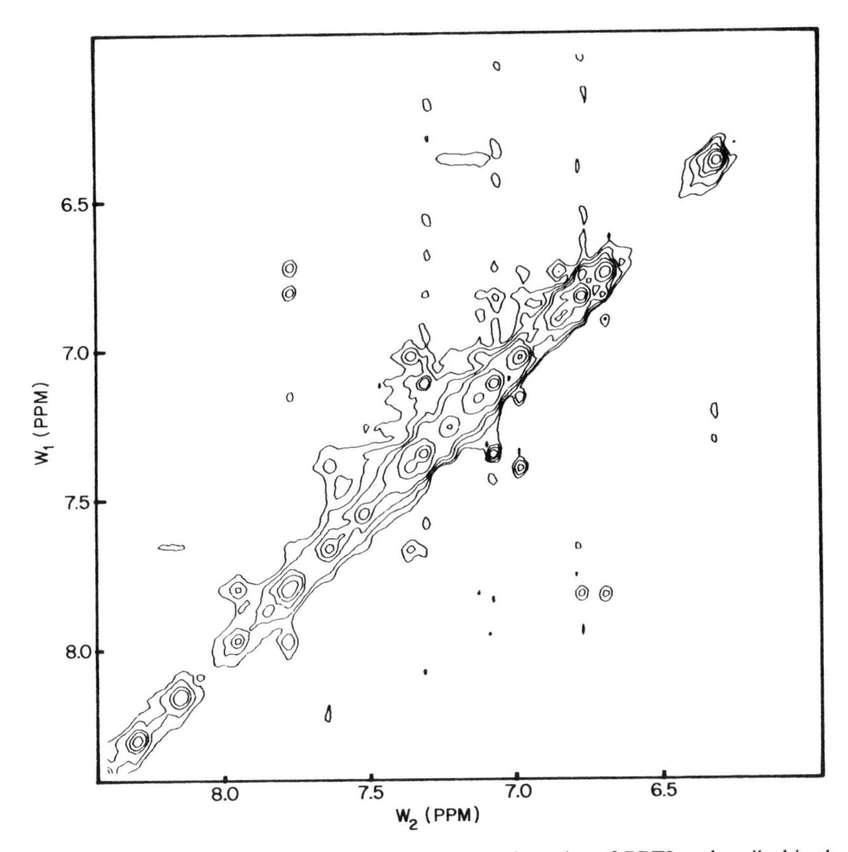

FIG. 2. The pure absorption phase spectrum of the aromatic region of BPTI as described in the text. The displayed section is 200 points square, extracted from the full 1024 point square spectrum. The spectrum was normalized so that the maximum peak (a dioxane marker) has an amplitude of 10,000. Contours are drawn at 15, 30, 60, 100, 250, 500, and 1000.

difference of the free-induction decays accumulated with the phase of P_2 at 0 and π with respect to P_1. The imaginary part of the t_1 dimension is the difference of the free-induction decays accumulated with P_2 at $+\pi/2$ and $-\pi/2$ with respect to P_1. These free-induction decays (a pair for each t_1) are stored and Fourier-transformed separately. They are all phase-corrected using the values appropriate for $t_1 = 0$. After transforming in the t_2 dimension, the complex time domain data in the t_1 dimension is assembled from the extracted real parts of the two spectra for each value of t_1. This complex matrix is transposed and Fourier-transformed to obtain the final phased two-dimensional spectrum. Phase correction can be applied in the second dimension, although in practice only a small linear phase correction is needed to compensate for the finite lengths of P_1 and P_2.

Figure 2 illustrates the application of this experiment to a macromolecular system, the bovine pancreatic trypsin inhibitor. Data were collected at 25°C pD at 3.8 using the 500-MHz spectrometer at the Francis Bitter National Magnet Laboratory. A mixing period of 100 msec was used without the application of any

TABLE 1

PULSE PHASES USED TO ACCUMULATE THE REAL AND IMAGINARY FREE-INDUCTION DECAYS FOR EACH t_1 VALUE

Pulse phases			
P_1	P_2	P_3	Accumulation
X	X	X	+ real
Y	Y	X	+ real
$\bar{X}$	$\bar{X}$	X	+ real
$\bar{Y}$	$\bar{Y}$	X	+ real
$\bar{X}$	X	X	− real
$\bar{Y}$	Y	X	− real
X	$\bar{X}$	X	− real
Y	$\bar{Y}$	X	− real
X	Y	X	+ imaginary
Y	$\bar{X}$	X	+ imaginary
$\bar{X}$	$\bar{Y}$	X	+ imaginary
$\bar{Y}$	X	X	+ imaginary
$\bar{X}$	Y	X	− imaginary
$\bar{Y}$	$\bar{X}$	X	− imaginary
X	$\bar{Y}$	X	− imaginary
Y	X	X	− imaginary

Note. Four-phase CYCLOPS rotation is applied to this set of pulse sequences resulting in a cycle of 64 accumulations.

homospoil pulse. Cyclic rotation of pulse phases and acquisition modes through the 64 possible phase combinations among the three pulses was used to cancel coherence effects between the labeling and acquisition pulses and to eliminate amplifier imbalance artifacts. The basic sequence of pulse phases is shown in Table 1. This subcycle of 16 phase settings was rotated through a four-phase CYCLOPS sequence (*9*, *10*) to give an overall cycle of 64 acquisitions; 1024 point free-induction decays were collected at 512 t_1 values with a spectral width of 6250 Hz. Gaussian line broadenings of 10 Hz were applied in both dimensions, and the data were zero-filled to 1024 points in the t_1 dimension prior to Fourier transformation.

This technique offers several practical advantages when compared to magnitude or power spectrum methods. Primary among these is the improved discrimination of the method. Figure 3 shows an absolute value spectrum calculated from the data presented in Fig. 2 with the same line broadenings and the same contour levels. It is much easier to recognize peaks in the phased spectrum and to distinguish them from the noise present in the ω_1 dimension and from overlapping tails of larger peaks. In the pure phase spectrum the noise is seen to be randomly phased, but in the magnitude spectrum it is positive definite like the signal of interest. Linewidths in the phased spectrum are narrower than those in the magnitude spectrum minimizing the need to apply resolution enhancements with their associated arti-

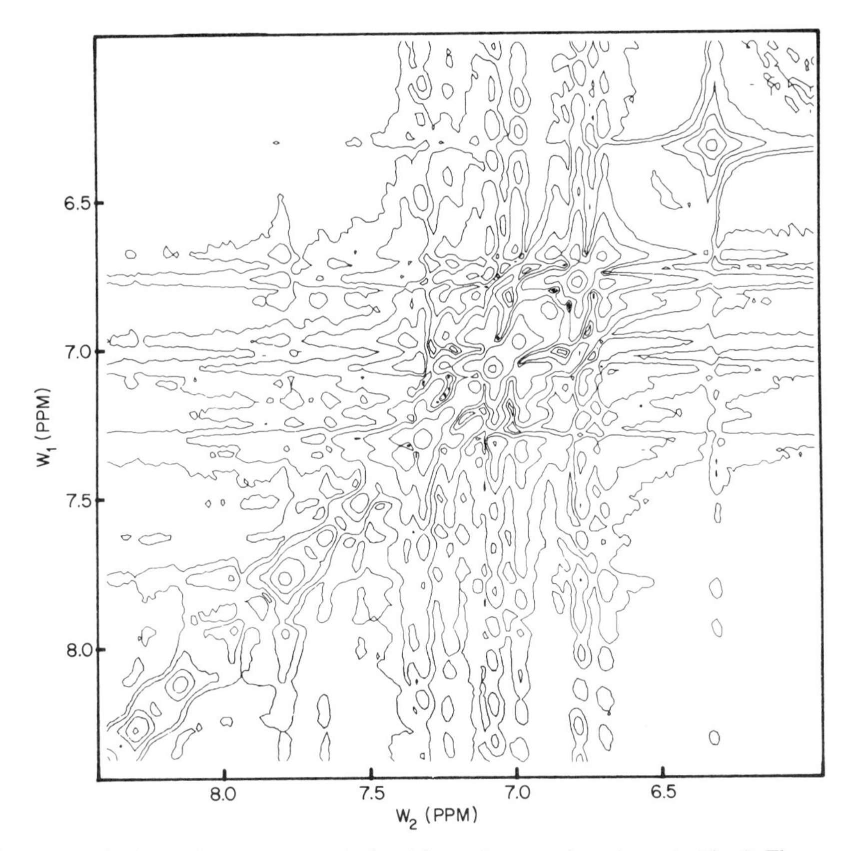

FIG. 3. An absolute value spectrum calculated from the same data shown in Fig. 2. The same normalization was applied, and the same contour levels are drawn.

facts, and resulting in an effective increase in signal-to-noise ratio. For Lorentzian lines the peak widths at half-height are $1/T_2$ an improvement of the square root of three over the magnitude spectrum width at half-height. Even more dramatic resolution improvements occur when one considers widths at lower heights because the limiting behavior of Lorentzian peaks is a decay with $(\omega - \omega_0)^{-2}$ while magnitude peaks decay as $(\omega - \omega_0)^{-1}$. The long tails present in magnitude spectra are particularly bothersome in samples with large dynamic ranges where they obscure significant regions of the spectrum. Finally, in crowded regions of the spectrum prominent overlap effects are seen in the magnitude spectrum which are absent in the phased spectrum. The amplitude in the pure phase spectrum is a linear function of the resonance amplitude without cross effects between overlapping peaks. For these reasons, pure phase spectra should greatly simplify quantitative data analysis.

The sensitivity of two-dimensional spectroscopy has been examined in detail previously (*11*). The effect of phasing on the sensitivity depends on how sensitivity is defined. A pure phase two-dimensional NMR experiment results in four linearly independent spectra; four-quadrant and one-quadrant experiments are equivalent (assuming that quadrature phase data acquisition is used in both). Each component will contain noise with a standard deviation of σ. All four components of the noise would appear in the power or magnitude spectrum

$$A(\omega_1\omega_2) = [(R(\omega_1)R(\omega_2) - I(\omega_1)I(\omega_2))^2 + (R(\omega_1)I(\omega_1) + I(\omega_1)R(\omega_2))^2]^{1/2}.$$

As Aue *et al.* note, there is a systematic shift of the data in a magnitude spectrum resulting from the average noise power

$$\langle N \rangle = \int dN_1 dN_2 (N_1^2 + N_2^2)^{1/2} f_1(N_1) f_2(N_2),$$

where f_1 and f_2 are the noise distribution functions for the two parts of the absolute value spectrum. Because each is composed of the sum of two of the four independent parts of the phased two-dimensional spectrum each will have a standard deviation of $2^{1/2}\sigma$. Assuming a Gaussian distribution for the noise the integral can be evaluated:

$$f(N_i) = \frac{1}{2\sigma(\pi)^{-1/2}} \exp(-N_i^2/4\sigma^2),$$

$$\langle N \rangle = \pi^{1/2}\sigma,$$

$$\langle N^2 \rangle = \langle N_1^2 \rangle + \langle N_2^2 \rangle$$

$$= 4\sigma.$$

The variance of the absolute value spectrum is therefore

$$\sigma_{\text{abs}}^2 = \langle N^2 \rangle - \langle N \rangle^2$$

$$= (4 - \pi)\sigma^2.$$

In the two-dimensional experiment the magnitude spectrum has a lower noise variance by a factor of 0.86. In a one-dimensional experiment there are only two components so the noise variance in a magnitude spectrum is lower than the phased spectrum by a factor of $2 - \pi/2$ or 0.42. Defining the noise level as the square root of the variance, the phased spectrum has a lower signal-to-noise ratio by a factor of 0.93 in two-dimensional spectroscopy (or 0.65 in one-dimensional spectra). It should be noted that the noise distribution in the magnitude spectrum has a significant third moment (skew) while the noise in a pure phase spectrum is symmetric. For practical purposes the magnitude and pure phase spectra appear to be essentially equivalent in signal-to-noise ratio. The rms error in the spectrum resulting from the addition of white noise of amplitude σ to each component of the data during acquisition is improved by a factor of 2 for phased spectra compared to magnitude spectra because there is no systematic offset of the data.

In summary, a method is presented for obtaining absorption phase two-dimensional NMR data. Experimental results are presented and several advantages of the method are discussed. These experiments should provide unique insight into the solution structure and dynamics of biological macromolecules.

ACKNOWLEDGMENTS

We wish to thank Dr. Walter Aue for his comments and Drs. Christopher Dobson and Martin Karplus for their support and encouragement. David States was supported by a Medical Scientists Training Program predoctoral training grant from the National Institutes of Health. The NMR Facility at MIT is supported by Grant No. RR00995 from the Division of Research Resources of the NIH and by the National Science Foundation under Contract No. C-670.

REFERENCES

1. J. Jeener, B. H. Meier, P. Bachmann, and R. R. Ernst, *J. Chem. Phys.* **71,** 4546 (1979).
2. R. Freeman and G. A. Morris, *Bull. Magn. Reson.* **1,** 5 (1979).
3. K. Wuthrich, K. Nagayama, and R. R. Ernst, *Trends Biochem. Sci.* **4,** N178 (1979).
4. A. Kumar, G. Wagner, R. R. Ernst, and K. Wuthrich, *J. Am. Chem. Soc.* **103,** 3651 (1981).
5. W. P. Aue, E. Bartholdi, and R. R. Ernst, *J. Chem. Phys.* **64,** 2229 (1976).
6. S. Macura and R. R. Ernst, *Mol. Phys.* **41,** 95 (1980).
7. K. Nagayama, P. Bachmann, K. Wuthrich, and R. R. Ernst, *J. Magn. Reson.* **31,** 75 (1978).
8. R. N. Bracewell, *Aust. J. Phys.* **9,** 198 (1956).
9. D. I. Hoult and R. E. Richards, *Proc. Roy. Soc. London Ser. A* **344,** 311 (1975).
10. E. O. Stejskal and J. Schaefer, *J. Magn. Reson.* **14,** 160 (1974).
11. W. P. Aue, P. Bachmann, A. Wokaun, and R. R. Ernst, *J. Magn. Reson.* **29,** 523 (1978).

JOURNAL OF MAGNETIC RESONANCE **49**, 203–211 (1982)

EPR Zeugmatography with Modulated Magnetic Field Gradient

THOMAS HERRLING, NORBERT KLIMES,* WOLFGANG KARTHE,†
UWE EWERT, AND BERND EBERT*

*Centre of Scientific Instruments, Academy of Sciences of GDR, 1199 Berlin, Rudower Chaussee 6, GDR, *Central Institute of Molecular Biology, Academy of Sciences of GDR, 1115 Berlin, Lindenberger Weg 70, GDR, and †Department of Physics, Friedrich Schiller University, 6900 Jena, Max-Wien-Platz 1, GDR*

Received January 4, 1982

A method for the investigation of the spatial distribution of paramagnetic centers in a sample by means of EPR is described. Using a modulated magnetic field gradient additional to the normal condition of EPR spectroscopy it is possible to detect paramagnetic centers in a small region of the sample. The method is theoretically based on the effect of modulation broadening. Resolution enhancement is achieved by recording the first or higher derivatives of the absorption line. Some experimental results are presented to show the capability of the method.

INTRODUCTION

NMR zeugmatography has become a new method for tomographic investigation of the human body with a resolution of about 10 mm (*1–4*). Recently an EPR method with a stationary field gradient was described for investigations of objects smaller than 1 cm^3 with a resolution of 10 μm (*5*). A wide field of application for such a method can be expected. In biology and in medicine various questions can be studied, e.g., the distribution of natural paramagnetic centers in tissue slices, the distribution pattern of probe molecules (spin probes) in membranes, and diffusion profiles of spin-labeled drugs in tissues. The investigation of the distribution and the diffusion of paramagnetic centers in the amorphous solid state, especially in high polymers and catalysts, represents a further field of application in chemistry. In the field of solid-state physics the possibility exists of measuring the distribution and diffusion of paramagnetic donor and acceptor centers, as well as other local defects in semiconductors, and paramagnetic centers adsorbed on interfaces.

The paramagnetic centers in EPR give broader lines than in NMR. The universality of the method demands also the measurement of distribution of paramagnetic species with a multiple-line spectrum of differing linewidths and lineshapes. For the method with stationary field gradient the convolution integral of the real distribution function and of the lineshape function of paramagnetic species is recorded. Deconvolution of a multiple-line spectrum of differing linewidths and lineshapes has not been solved generally up to now. Therefore we have looked for another method to overcome this problem.

The application of a time-dependent field gradient seems to be a proper way to

0022-2364/82/110203-09$02.00/0

solve these problems. In NMR zeugmatography low-power time-dependent field gradients were used by Hinshaw (*6*). By scanning the time-independent zero plane of the field gradient across the sample, the real distribution function can be measured without difficult mathematical deconvolution.

METHOD

For simplicity we shall restrict the explanation of the method to the one-dimensional projection of distribution of paramagnetic centers. The generalization to the three-dimensional case is straightforward. The specimen is placed into a uniform static magnetic field **B**. The field vector of **B** is directed parallel to the z axis. The sample absorbs power from an irradiated microwave field of constant angular frequency ω_0 if the resonance condition

$$\omega_0 = \gamma_{\text{eff}} \cdot \mathbf{B}_0 \quad [1]$$

is fulfilled, where γ_{eff} is the effective gyromagnetic ratio which is different for each resonance line of the paramagnetic species and $\mathbf{B}_0$ is the magnetic resonance field corresponding to the paramagnetic species and ω_0.

The spatial resolution is obtained using an additional inhomogeneous field $\hat{B}_m(z)$ in the z direction. The inhomogeneous field $\hat{B}_m(z)$ is modulated periodically in time, so that the resulting field $B_z(z, t)$ at the sample is the sum of the static field **B** and the time- and space-dependent field $B_m(z, t)$:

$$B_z(z, t) = \mathbf{B} + B_m(z, t) = \mathbf{B} + \hat{B}_m(z) \cdot f(t + nT), \quad [2]$$

where $\hat{B}_m(z)$ is the peak amplitude of the modulation field in the z plane (plane perpendicular to the z axis).

If we assume that the inhomogeneous field has a constant gradient $G_z = d\hat{B}_m(z)/dz$ then

$$\hat{B}_m(z) = G_z(z - z_0). \quad [3]$$

In accordance with Eqs. [1], [2], and [3], a time-independent resonance condition is fulfilled only in the zero-field plane z_0 ($z = z_0$), where $\hat{B}_m(z_0) = 0$ and $B_z(z_0, t) = \mathbf{B}$. In this way a time-independent spectrum $Y(z_0, \mathbf{B})$ is recorded corresponding to the paramagnetic centers in the z_0 plane. For every other z plane, a modulated spectrum $Y(z, B_z)$ is obtained, which arises from paramagnetic centers in these modulated planes. The width of an observed z plane is much smaller than the resolution distance Δz. In the simplest case the modulation function is described by $f(t) = \cos \omega_m t$ with $\omega_m \ll 1/T_2$.

The contour integration of the time-dependent signal $Y(z, B_z)$ leads to the spectra shown in Fig. 1a. The lineshape of the modulation broadened absorption curve Y^* is determined by the equation

$$Y^*(z, B_z) = y_m(z) \cdot a_0(B_z), \quad [4]$$

where a_0 is the first Fourier amplitude of the modulated absorption line (*7, 8*). For a Lorentzian line the coefficient a_0 is given by

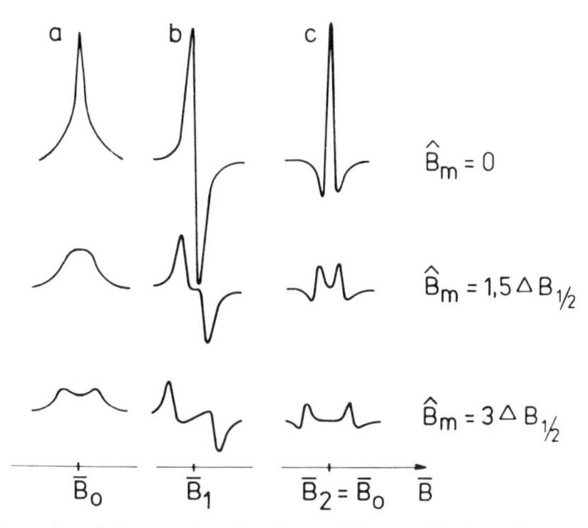

FIG. 1. Modulation-broadened Lorentzian lineshape. The amplitudes of the modulation-broadened absorption line (a), the first derivative (b), and the second derivative (c) are represented as a function of the modulation amplitude $\hat{B}_m$ with $\hat{B}_n \ll \Delta B_{1/2}$.

$$a_0 = \frac{\omega_m}{\pi} \int_{-\pi/\omega_m}^{\pi/\omega_m} \frac{dt}{((1/2)\Delta B_{1/2})^2 + (B_\delta + \hat{B}_m(z) \cos \omega_m t)^2}, \qquad [5]$$

where $B_\delta = (\mathbf{B} - \mathbf{B}_0)$ and $\Delta B_{1/2}$ is the linewidth. The number of spins $N(z)$ in the observed z plane is described by $y_m(z)$, which is the amplitude of the absorption curve at $\mathbf{B} = \mathbf{B}_0$.

In EPR it is usual to record the first derivative of the absorption line. In this case a second homogeneous modulation field $B_n(t) = \hat{B}_n \cos \omega_n t$ is superimposed on the static magnetic field $\mathbf{B}$ with $\omega_m \ll \omega_n \ll 1/T_2$. A time- and space-dependent magnetic field

$$B_z(z, t) = \mathbf{B} + \hat{B}_n \cos \omega_n t + \hat{B}_m(z) \cos \omega_m t \qquad [6]$$

acts now across the sample. The signal amplitude originating from the z_0 plane is only modulated with a time-dependent field of the frequency ω_n while the other planes are modulated with two time-dependent fields. The double modulation with ω_n and ω_m and the narrow band detection at $p\omega_n$ including the formation of the arithmetical average (by means of an RC filter) yield the spectra shown in Figs. 1b and c. Their lineshape is determined by

$$Y^{*(p)}(z, B_z) = \frac{d^p Y^*(z, B_z)}{d\mathbf{B}^p} = y_m(z) a_p(B_z), \qquad [7]$$

where p is the ordinal number of the recorded derivative. For a Lorentzian line a_p is given by

$$a_p = \frac{\omega_m}{\pi} \int_{-\pi/\omega_m}^{\pi/\omega_m} \frac{\cos p\omega_n t}{((1/2)\Delta B_{1/2})^2 + (B\delta + \hat{B}_n \cos \omega_n t + \hat{B}_m(z) \cos \omega_m t)^2}\, dt. \qquad [8]$$

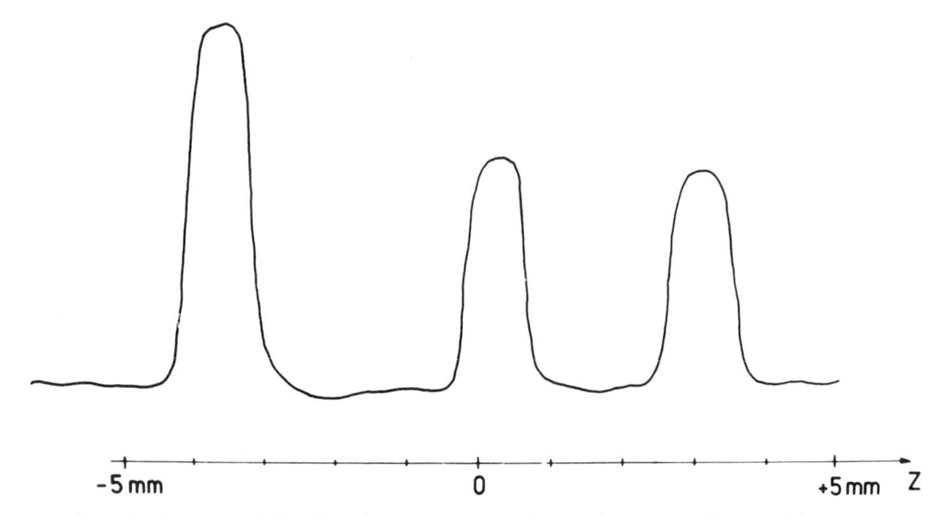

FIG. 2. Projection of the distribution of paramagnetic centers in three paper strips containing a different spin concentration of DPPH. All the strips have a geometrical width of about 1 mm and they are separated by two different distances of 3 and 2 mm. The gradient strength was $G_z = 0.015$ T/cm.

The amplitude a_p, describing the quantity of the pth harmonic of ω_n, decreases with increasing value of p (*8*).

PROJECTION FORMATION

The projection of spatial distribution of paramagnetic centers is achieved by shifting the z_0 plane across the sample during the time the static magnetic field is constant. The signal amplitude $S_p(z_0, \mathbf{B}_p)$ according to the position of the z_0 plane and at the fixed field $\mathbf{B} = \mathbf{B}_p$ ($\mathbf{B}_p$ is the magnetic field of the maximum signal amplitude of the pth derivative) can be written for a Lorentzian line as

$$S_p(z_0, \mathbf{B}_p) = \int_z y_m(z) \cdot a_p(z, \mathbf{B}_p)dz. \qquad [9]$$

Figure 2 illustrates the distribution of spin concentration in a sample consisting of three strips of filter paper on which various concentrations of the stable free radical α, α-diphenyl-β-picryl-hydrazyl (DPPH) were adsorbed. The prepared strips have different distances along the z axis.

The second possibility for application of this method is given by sweeping the static magnetic field $\mathbf{B}$ for a constant position z_0. In this case the observed spectrum can be described by

$$S_p(z_0, \mathbf{B}) = \int_z y_m(z)a_p(z, \mathbf{B})dz. \qquad [10]$$

In Fig. 3 the spectra ($p = 1$) are presented for the three strips according to the example in Fig. 2. At the tails of the DPPH lines the modulation broadening parts influenced by the other strips can be seen.

The method is also practicable for selective measurement of the distribution S_{pi} of only one center type i in a sample with j center types of various distributions and of different $\gamma_{\mathrm{eff}\,j}$. It is necessary to reach sufficient resolution that the spectra Y_j recorded in the homogeneous magnetic field do not overlap. The spatial distribution for the type i is described by

$$S_{pi}(z_0, \mathbf{B}_{pi}) = \sum_{k=1}^{j} \int_z y_{mk}(z, \mathbf{B}_{0k}) \cdot a_p(z, \mathbf{B}_{pi}) dz \qquad \text{for}$$

$$k = 1 \cdots i \cdots j \qquad \text{and} \qquad B_{\delta ik} = (\mathbf{B}_{pi} - \mathbf{B}_{0k}). \qquad [11]$$

This possibility includes the measurement of the distribution function of a species with a multiple-line spectrum.

Figure 4 shows the spectrum of a geometrical arrangement of a cylindrical DPPH sample (single-line spectrum) which was placed into an other cylindrical sample containing Mn^{2+} (six-line spectrum). The distribution of Mn^{2+} concentration (solid line) and DPPH concentration (dotted line) is presented in Fig. 5. The measurements were performed with a conventional EPR spectrometer (ERS 220, produced by the Centre of Scientific Instruments of the Academy of Sciences of GDR) and

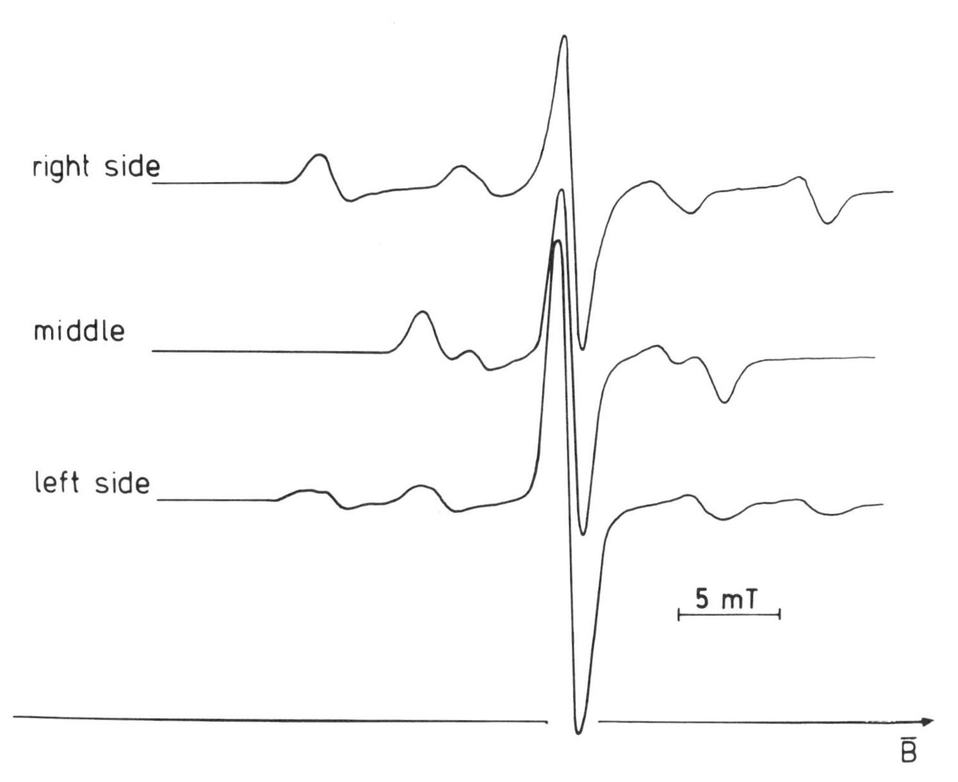

FIG. 3. Three DPPH spectra (first derivative) obtained from three selected planes in the sample used in Fig. 2. These planes were localized in the centers of the DPPH strips (left side, middle, right side).

an experimental zeugmatography arrangement consisting of two high-power gradient coils and a driving unit for shifting the z_0 plane across the sample.

RESOLUTION AND SENSITIVITY

It is shown in Figs. 1b and c that the spatial resolution increases with measurement of higher derivatives. The spectra in Fig. 1 are normalized to the same maximum signal amplitude (for the case of $\hat{B}_m = 0$ only) to provide a good comparison.

To obtain the best resolution and sensitivity it is necessary to chose the field $\mathbf{B} = \mathbf{B}_p$. For a Lorentzian line and $p = 1$ we obtain

$$\mathbf{B}_1 = \mathbf{B}_0 \pm \frac{\Delta B_{1/2}}{2(3)^{1/2}}. \quad [12]$$

The modulation amplitude $\hat{B}_n$ broadens the resonant line (*8*) so that Eq. [12] becomes

$$\mathbf{B}_1 = \mathbf{B}_0 \pm \frac{\Delta B_{1/2}}{2(3)^{1/2}} \left\{ \left(\frac{2(3)^{1/2}\hat{B}_n}{\Delta B_{1/2}} \right)^2 + 5 - 2\left[4 + \left(\frac{2(3)^{1/2}\hat{B}_n}{\Delta B_{1/2}} \right)^2 \right]^{1/2} \right\}^{1/2}. \quad [13]$$

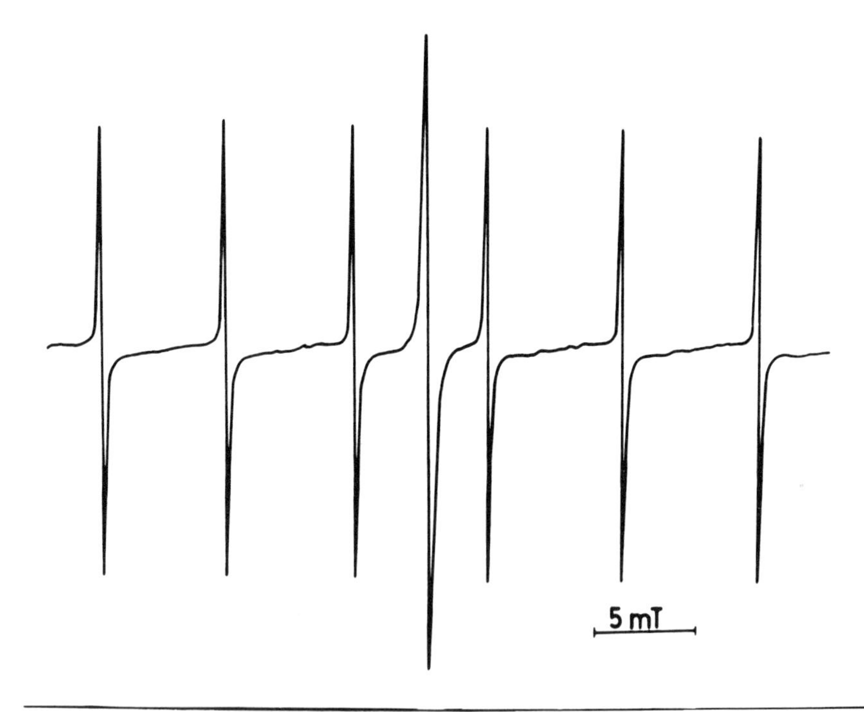

FIG. 4. Complex spectrum of a 2.8-mm-i.d. tube of Mn^{2+} (MgO) powder (six-line spectrum) and a second tube of DPPH powder (single-line spectrum) with an inner diameter of 0.8 mm. The smaller tube was localized inside the bigger one.

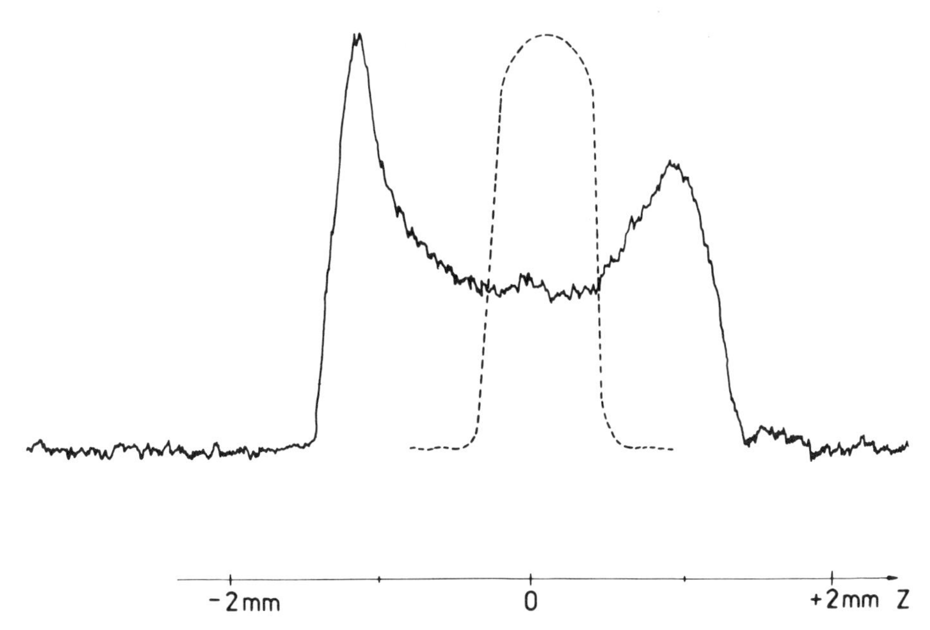

FIG. 5. Projection of the distribution of Mn^{2+} (solid line) and DPPH (dotted line) for the sample configuration in Fig. 4. The projection of Mn^{2+} was recorded with a higher amplification. The measurements were performed with a gradient strength of $G_z = 0.025$ T/cm.

The static magnetic field $\mathbf{B}_2 = \mathbf{B}_0$ has to be selected for the second derivative ($p = 2$).

The resolution of the method is described by the distance Δz, which is calculated on the basis of the Rayleigh criterion. If the absorption line has a Lorentzian shape we obtain for the first derivative ($p = 1$) and $\hat{B}_n \ll \Delta B_{1/2}$ a resolution

$$\Delta z \approx \frac{2}{3} \frac{\Delta B_{1/2}}{G_z} . \qquad [14]$$

Considering the influence of the amplitude $\hat{B}_n$ the resolution parameter is changed to

$$\Delta z = \frac{2\Delta B_{1/2}}{3G_z} \left\{ \left(\frac{2(3)^{1/2}\hat{B}_n}{\Delta B_{1/2}} \right)^2 + 5 - 2\left[4 + \left(\frac{2(3)^{1/2}\hat{B}_n}{\Delta B_{1/2}} \right)^2 \right]^{1/2} \right\}^{1/2} . \qquad [15]$$

The influence of $\hat{B}_n$ at the resolution Δz and at the amplitude a_1 is demonstrated in Fig. 6. It is shown that the best resolution Δz is obtained for small values of $\hat{B}_n$. If $\hat{B}_n \ll \Delta B_{1/2}$ we can get a resolution of $\Delta z \approx 5$ μm for $G_z = 0.15$ T/cm and for a Lorentzian line with $\Delta B_{1/2} = 0.1$ mT.

The amplitude a_1 (sensitivity) can be increased for an appropriate choice of $\hat{B}_n$ with decreasing the resolution Δz. The best sensitivity is obtained for $\hat{B}_n = \Delta B_{1/2}$. Corresponding to Eq. [9] the signal amplitude $S_p(z_0, \mathbf{B}_p)$ depends on the ordinal number of p and on the number of spins $N(\Delta z)$ in the region $z_0 \pm \Delta z/2$, which can be increased by increasing Δz.

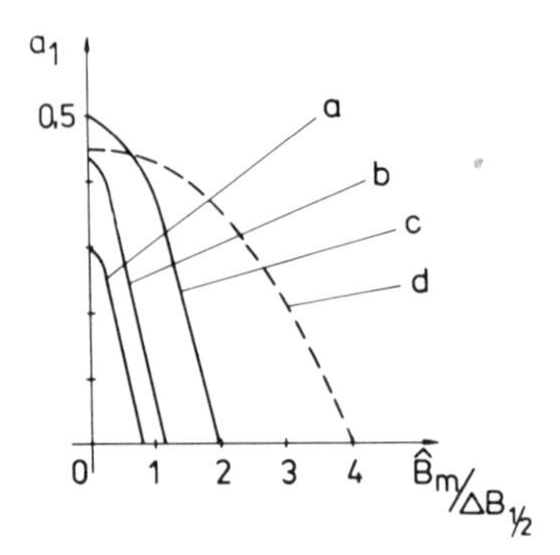

FIG. 6. Dependence of the first Fourier amplitude a_1 of a Lorentzian line on the normalized modulation amplitude $\hat{B}_m/\Delta B_{1/2}$. Curves are drawn for various amplitudes $\hat{B}_n$. Letters indicated (a) $\hat{B}_n = (1/4)\Delta B_{1/2}$, (b) $\hat{B}_n = (1/2)\Delta B_{1/2}$, (c) $\hat{B}_n = \Delta B_{1/2}$, and (d) $\hat{B}_n = 2\Delta B_{1/2}$.

DISCUSSION

The present method for EPR zeugmatography accomplishes the measurement of distribution of paramagnetic centers in a more direct way without use of the difficult mathematical deconvolution and simulation programs as is necessary in the case of a stationary field gradient. Various problems can now be solved, such as the distribution measurement of multiple-line spectra with differing shapes and widths, which are typical for various paramagnetic species (for instance, different lineshapes of spin labels). In addition this method enables the selective recording of the distribution of different types of paramagnetic centers in a complex sample. The third possibility is a spectral investigation of only a small sample region.

The main features for the practical application of the method are given by the resolution of the zeugmatography apparatus and the sensitivity of the EPR spectrometer. The best sensitivity and resolution are obtained by recording only one-dimensional projections for different gradient directions relative to the sample orientation, and reconstructing the image using a computer algorithm (*9–12*). Assuming we have a slice of 0.2 mm thickness in the x direction and dimensions of 10 mm in the y and z directions, the two-dimensional imaging of spin concentration in the y–z plane with a resolution of $\Delta z \approx 5\ \mu m$ requires a detectable concentration in the observed volume ($5 \times 200 \times 10^4\ \mu m$) of 3×10^{10} spins/0.1 mT (sensitivity of the spectrometer) to 10^{14} spins/0.1 mT (limit of dipole–dipole broadening). This concentration range is sufficient in our opinion to solve various problems as mentioned in the first section.

ACKNOWLEDGMENTS

We are indepted to W. Schneider from the Centre of Scientific Instruments for helpful discussions and support of this work.

REFERENCES

1. P. MANSFIELD AND I. L. PYKETT, *J. Magn. Reson.* **29,** 355 (1978).

2. E. R. Andrew, P. A. Bottomley, W. S. Hinshaw, G. N. Holland, W. S. Moore, and C. Simaroj, *Phys. Med. Biol.* **22,** 971 (1977).
3. J. M. S. Hutchison, W. A. Edelstein, and G. Johnson, *J. Phys. E* **13,** 947 (1980).
4. R. Damadian, M. Goldsmith, and M. Minkoff, *Physiol. Chem. Phys.* **9,** 97 (1977).
5. W. Karthe and E. Wehrsdorfer, *J. Magn. Reson.* **33,** 107 (1979).
6. W. Hinshaw, *Phys. Lett. A* **48,** 87 (1974).
7. H. Wahlquist, *J. Chem. Phys.* **35,** 1708 (1961).
8. C. P. Poole, "Electron Spin Resonance," pp. 398–413, Wiley–Interscience, New York, 1967.
9. P. Brunner and R. R. Ernst, *J. Magn. Reson.* **33,** 83 (1979).
10. P. C. Lauterbur, *Nature (London)* **242,** 190 (1973).
11. A. M. Cormack, *Phys. Med. Biol.* **18,** 195 (1973).
12. C. M. Lai and P. C. Lauterbur, *Phys. Med. Biol.* **26,** No. 5 (1981).

JOURNAL OF MAGNETIC RESONANCE 50, 30–49 (1982)

Theory of Broadband Spin Decoupling*

J. S. WAUGH

Massachusetts Institute of Technology, Cambridge, Massachusetts 02139

Received February 23, 1982

The NMR spectrum of a spin S coupled to another spin I which is undergoing an arbitrary time-dependent irradiation is analyzed. The efficacies of various schemes for spin decoupling are compared using the average Hamiltonian method, and the limitations of this method are pointed out. A procedure for exact numerical analysis is outlined. This analysis leads to a very simple criterion, namely: decoupling will be effective over a bandwidth δ_0 if the net precessional angle ϕ of a spin is insensitive to offset δ for $0 < \delta < \delta_0$. The application of this criterion in quantitative form permits us to exhibit decoupling sequences which are somewhat superior to any previously described.

INTRODUCTION

The technique of spin decoupling, first proposed by Bloch (*1*), has evolved into a universal and essential part of high-resolution NMR spectroscopy. The general idea is that the behavior of one spin or group of spins S is observed while an irradiation of relatively high strength H_2 is applied in the neighborhood of the Larmor frequency of another spin or group of spins I. The object is to arrange this irradiation so that splittings of the S resonance produced by spin–spin couplings J_{IS} are maximally suppressed.

When the decoupling irradiation is of constant strength and frequency, and when the resonances of I and S are well separated, the phenomenon is very easy to understand and analyze (*2, 3*) and the conditions for adequate decoupling are easily worked out. To obtain essentially complete decoupling the amplitude $\omega_2 = \gamma_I H_2$ of the irradiating field must be much larger than the displacement δ of the irradiating frequency from the I-spin resonance. The I spin resonant frequencies are usually not known a priori except for the fact that they may be presumed to lie within a certain range characteristic of I spin chemical shifts. Then the intensity of irradiation must be adjusted to exceed the largest possible such displacement δ_0 by a considerable factor. This procedure is inherently inefficient in terms of the radiofrequency power which must be deposited in the probe to ensure the desired degree of decoupling. Such inefficiency is offensive not only aesthetically but in a practical sense: The power required, particularly to maintain the necessary field strengths in large and perhaps lossy samples, results in intolerable heating of the probe and sample. Hence

* This work was supported in part by the National Science Foundation and in part by the National Institutes of Health.

0022-2364/82/130030-20$02.00/0

there is a great deal of interest in finding more elaborate schemes which will adequately decouple I spins having shifts within an agreed-upon range, with the minimum expenditure of power. The first such scheme to be widely employed involved the modulation of the irradiating field with random or pseudorandom noise having a suitably limited bandwidth (*4*). Others have employed modulation or sweeping of the irradiating frequency (*5*), phase reversals with or without sawtooth (chirp) frequency modulations (*5*, *6*), pulses (*7*, *8*), etc. Irradiation strategies in modern commercial spectrometers even have a tendency to pass into the realm of trade secrecy.

Unfortunately these newer methods have been nearly without exception unsupported by theory. Such rationales as exist are for the most part based on an intuition which is so seductive as to be nearly irresistible, but nevertheless completely erroneous: attention is focused on the Fourier spectrum (indeed often only the power spectrum) of the irradiating waveform. One seeks to pack the I-spin spectral range densely and uniformly with modulation sidebands, and imagines that those closest to the resonance of a particular I spin will somehow decouple it, much as cw irradiation does. It is further taken as obvious that for any *given* irradiation scheme an increase in transmitter power will lead to better decoupling. The simple example of decoupling with 180° pulses (*7*, *8*) serves to demolish these views: it is well known that this method works best when the pulses are (a) narrow and (b) closely spaced in time. The spectrum of such a pulse train (a) extends arbitrarily far beyond the I-spin spectral range and (b) has its components *widely* spaced in frequency. Moreover consider what happens when the transmitter power is increased fourfold: the Fourier spectrum is unchanged except for a common factor of 2 on all amplitudes, but the pulses are now 360° pulses and do not decouple at all (*9*).

In this paper we attempt to develop a sound theoretical basis for understanding the decoupling problem. We will greatly simplify our task by making several assumptions, all of which are appropriate to the situations of greatest interest, i.e., NMR of dilute species (^{13}C, ^{15}N, etc.) coupled to protons (indeed we shall often refer to the I spins as protons and the S spins as carbons where this shorthand avoids circumlocution):

1. I and S have widely separated Larmor frequencies so that irradiation near ω_{0I} has no direct effect on S.
2. The irradiating fields are strong in the sense that they dominate the J coupling insofar as the motion of the I spins is concerned.
3. I–I couplings are relatively weak: $J_{II} \ll J_{IS}$, and S–S couplings are absent.

These restrictions allow us to reduce the problem to that of a single proton coupled to a single carbon, which may be assumed at exact resonance. Then, in a frame of reference rotating with respect to the carbon at ω_{0S} and with respect to the proton at a carrier frequency chosen at the center of the assumed proton spectral range,

$$\mathscr{H} = -\omega_{2x}(t)I_x - \omega_{2y}(t)I_y - \delta(t)I_z + JI_zS_z = -\omega_e(t)\cdot\mathbf{I} + JI_zS_z\,. \qquad [1]$$

Here ω_e is an effective field made up of an offset δ of the proton resonance ω_{0I} from the carrier and an rf "field" of amplitude $\omega_2 = (\omega_{2x}^2 + \omega_{2y}^2)^{1/2}$ and phase $\tan^{-1} \times (\omega_{2y}/\omega_{2x})$. The value J is the coupling constant, of either electron-coupled or dipolar origin.[1] The time dependence in [1] allows for any desired modulation of amplitude, phase, and Zeeman field. Frequency modulation must be represented as a combination of field and phase modulation.

Given a desired time dependence, our task is to calculate the carbon-free induction decay $f(t)$

$$f(t) = \frac{\langle S_x(t) \rangle}{\langle S_x(0) \rangle}, \qquad [2]$$

$$\langle S_x(t) \rangle = \mathrm{Tr}\,\{S_x U(t, 0) S_x U^{-1}(t, 0)\}, \qquad [3]$$

$$U(t, 0) = \mathcal{T} \exp\{-i \int_0^t \mathcal{H}(t')dt'\}, \qquad [4]$$

where $\mathcal{T}$ denotes time ordering.

We will find it convenient to represent the time variation as a sequence of "states," during each of which ω_e is constant: as a matter of fact most currently used decoupling schemes, being digitally generated, are of this type. Continuous variations can be simulated when needed by a sequence of short steps. In addition we will assume, in correspondence to most but not all common schemes, that

4. The irradiation sequence is periodic, with repetition period t_r.

The FID [2] is always sampled regularly with spacing t_s: we should in principle compute $f(nt_s)$, $n = 1, 2, \cdots$ and Fourier transform the resulting array to obtain the spectrum. In practice t_s and t_r are usually derived from independent clocks. However, the theory is enormously simplified by assuming that

5. $t_r = t_s$ and moreover the irradiation has a fixed phase relation to the sampling cycle.

We will abide by this highly restrictive assumption for computational convenience, but will presently argue that it does not seriously affect the results.

Our analysis will be begun in terms of coherent averaging theory (*10*), with various decoupling schemes introduced at appropriate points in the development. This approach has great advantages in simple cases, but will be shown to lose its appeal in more complicated ones. As the difficulties become clear we will abandon this approach in favor of a direct integration of the equations of motion. The main result will be the establishment of an extremely simple criterion by which the effectiveness of any proposed decoupling scheme can be analyzed. Sequences will be exhibited which are predicted to perform considerably better than any in common use. However, we will not be able to go very far in inverting the theory, i.e., *deriving* an optimal decoupling scheme, even assuming that the meaning of "optimal" could be easily agreed upon.

Finally it must be noted that real NMR spectrometers are subject to nonidealities which affect the quality of decoupling, in particular inhomogeneities in Zeeman and

[1] The physical origin of the coupling in [1] cannot be determined from the NMR spectrum to which [1] leads: truncation gives [1] exactly the same mathematical form whether the coupling was isotropic and electron-coupled ("scalar"), anisotropic electron-coupled [$J = J(\theta, \phi)$] or dipolar [$J = \gamma_{\pm}\gamma_s \hbar r^{-3} P_2 \times (\cos\theta)$]. Accordingly the sometimes-used terms "scalar decoupling" and "dipolar decoupling" are misleading and should be replaced by "weak" and "strong" decoupling, respectively.

rf fields, which differ from one spectrometer to another in a manner not often controlled or even known by the experimenter. This circumstance makes it very difficult to compare published results on various decoupling methods. The role of these nonidealities, particularly of the inhomogeneity of H_2, will be discussed qualitatively.

AVERAGE HAMILTONIAN METHOD

General Features

Under assumption (2) above, it is productive to factor the propagator [4] as

$$U(t, 0) = R(t)U_J(t), \qquad [5]$$

where R gives the time development arising from applied fields in the absence of J coupling, and U_J isolates the effects of J coupling in modified form

$$R(t) = \mathcal{T} \exp\left\{i \int_0^t \omega_e(t') \cdot \mathbf{I} dt'\right\}, \qquad [6]$$

$$U_J(t) = \mathcal{T} \exp\left\{-iJS_z \int_0^t \tilde{I}_z(t')dt'\right\}, \qquad [7]$$

$$\tilde{I}_z(t') = R^{-1}(t')I_zR(t'). \qquad [8]$$

In the average Hamiltonian method (*10*) one uses Magnus' theorem (*11*) to express [7] in the form

$$U_J(t) = e^{-i\bar{\mathscr{H}}t}; \qquad \bar{\mathscr{H}} = \sum_{n=0}^{\infty} \bar{\mathscr{H}}(n), \qquad [9]$$

where, for $\omega_e \gg J$, $\bar{\mathscr{H}}$ is well approximated by a few terms:

$$\bar{\mathscr{H}}^{(0)} = JS_z\left(\frac{1}{t}\right) \int_0^t \tilde{I}_z(t')dt', \qquad [10]$$

$$\bar{\mathscr{H}}^{(1)} = J^2S_z^2\left(\frac{-i}{2t}\right) \int_0^t \int_0^{t'} [\tilde{I}_z(t'), \tilde{I}_z(t'')]dt'dt''. \qquad [11]$$

The series converges well if $Jt \ll 1$. In addition, if $S = 1/2$ (^{13}C, ^{15}N, etc.), all odd-order terms can be omitted because $S_z^2 = 1/4$, a c number which cannot affect the S-spin FID.

Now [6] represents a pure rotation in I space, so we can write

$$\tilde{I}_z(t) = \mathbf{I} \cdot \mathbf{m}(t), \qquad [12]$$

where $\mathbf{m}$ is a unit vector whose components are the direction cosines of $\mathbf{I}$. Then [10] becomes

$$\bar{\mathscr{H}}^{(0)} = JS_z\mathbf{I} \cdot \bar{\mathbf{m}} \qquad [13]$$

with $\bar{\mathbf{m}}$ the (vector) average of $\mathbf{m}(t)$. Since $\mathbf{m}(t)$ is a unit vector, $|\bar{\mathbf{m}}| \leqslant 1$.

Equation [13] already foreshadows a general result: the principal effect of decoupling is to scale down all values of J, irrespective of their individual magnitudes. Scaling by moderate values can be useful in making assignments: then it is important that the scaling factor λ be nearly independent of offset δ. This situation has been most thoroughly analyzed by Aue (*12*). For our purposes we are not so much interested in making the scaling factor constant as in making it very small.

Sequence of States

For each state k of a sequence an average Hamiltonian $\bar{\mathscr{H}}_k$ can be constructed as above with $\omega_e(t')$ in [6] constant. Then the total propagator [5] up to the first signal-sampling time $t_s = \sum_{k=1}^{N} t_k$ is

$$U(t_s, 0) = R_N[\![\bar{\mathscr{H}}_N t_N]\!]R_{N-1}[\![\bar{\mathscr{H}}_{N-1} t_{N-1}]\!] \cdots R_1[\![\bar{\mathscr{H}}_1 t_1]\!]. \tag{14}$$

Here we have used the notation

$$U_{Jk}(t_k, 0) = e^{-i\bar{\mathscr{H}}_k t_k} = [\![\bar{\mathscr{H}}_k t_k]\!]. \tag{15}$$

Equation [14] can be rearranged to a form which collects all the values of R on the left,

$$U(t_s, 0) = R_s[\![\tilde{\mathscr{H}}_N t_N]\!][\![\tilde{\mathscr{H}}_{N-1} t_{N-1}]\!] \cdots [\![\tilde{\mathscr{H}}_1 t_1]\!], \tag{16}$$

where

$$R_s = R_N R_{N-1} \cdots R_1 , \tag{17}$$

$$\tilde{\mathscr{H}}_k = R_1^{-1} R_2^{-1} \cdots R_{k-1}^{-1} \bar{\mathscr{H}}_k R_{k-1} \cdots R_2 R_1 . \tag{18}$$

Each of the $\bar{\mathscr{H}}_k$ in [14] can be characterized by an $\bar{\mathbf{m}}_k$ in the fashion of [13]. Recall that each of these corresponds to an average over the history of a unit vector $\mathbf{m}_k$, which in each case began as $\mathbf{m}_k(0) = \mathbf{k}$. The $\tilde{\mathscr{H}}_k$ in [16] can similarly be associated with a set of vectors $\tilde{\mathbf{m}}_k$, which are similar to the $\bar{\mathbf{m}}_k$ except that in each case the history of the vector being averaged begins at the terminus of the trajectory of $\mathbf{m}_{k-1}$. Clearly the magnitudes $|\tilde{\mathbf{m}}_k|$ and $|\bar{\mathbf{m}}_k|$ are the same. According to [16] the $\tilde{\mathscr{H}}_k$ can be associated with coupling Hamiltonians with $J_{k\,\mathrm{eff}} = |\bar{\mathbf{m}}|_k J \leqslant J$. Now just as the condition for using $\bar{\mathscr{H}}^{(0)}$ is $Jt_k \ll 1$, we can further combine the $\tilde{\mathscr{H}}_k$ of [16] into an average Hamiltonian $\bar{\mathscr{H}}_s$ if $J_{\mathrm{eff}} t_s \ll 1$, where J_{eff} is a representative value of $J_{\mathrm{eff}\,k}$. That is, we do not have to require that $Jt_s \ll 1$. The result is

$$U(t_s, 0) = R_s e^{-i\bar{\mathscr{H}}_s t_s}, \tag{19}$$

$$\bar{\mathscr{H}}_s^{(0)} = \sum_{k=1}^{N} \tilde{\mathscr{H}}_k (t_k/t_s). \tag{20}$$

The sequence is said to be *cyclic over the time* t_s if $R_s = 1$, i.e., the irradiation would have restored the spin system to its initial state if J coupling were absent. States are sometimes cyclic over the intervals t_k: If this is true for all states, Eqs. [14] and [16] are identical and the overall sequence is also cyclic.

Equation [20] can also be expressed in the form

$$\bar{\mathscr{H}}_s^{(0)} = JS_z \mathbf{I} \cdot \bar{\mathbf{m}}_s , \tag{21}$$

$$\bar{\mathbf{m}}_s = \sum_k \tilde{\mathbf{m}}_k (t_k/t_s). \tag{22}$$

Evidently the magnitude of $\bar{\mathbf{m}}_s$ is less than the average of the magnitudes of the $\tilde{\mathbf{m}}_k$ or $\bar{\mathbf{m}}_k$, by the same reasoning which showed $\lambda_k = |\bar{\mathbf{m}}_k| \leqslant 1$. The result is that we can define an overall coupling magnitude which is smaller than the (weighted) average of the $J_{\text{eff}\,k}$ mentioned above.

Some Conclusions

In the previous section we concentrated on finding the state of the system at the first instant t_s at which the FID is sampled. Now note that, in inserting [19] into [3], R_s can be dropped whether the sequence is cyclic or not. This is because of the invariance of the trace to unitary transformation of its argument, and the fact that R_s commutes with S_x. A somewhat similar argument shows that [21] can be replaced by

$$\bar{\mathscr{H}}_s^{(0)} = J_{\text{eff}} S_z I_z \,, \tag{23}$$

with $J_{\text{eff}} = \lambda J$. To see this, consider an I-spin rotation W such that $W(\mathbf{I} \cdot \mathbf{m}_s)W^{-1} = \lambda I_z$; W commutes with all components of $\mathbf{S}$ and can be carried through the argument of [3], whereupon the invariance of the trace shows that $f(t_s)$ is unchanged.

The result of these arguments is that $f(t_s)$ has exactly the value which would have been obtained in an unirradiated system with $J = J_{\text{eff}}$: the effect of decoupling in the context of average Hamiltonian theory is apparently simply to scale down the J splitting.

If the irradiation is cyclic over t_s, nothing more need be said. If not, the conclusion just drawn must be used cautiously. Suppose we wish to calculate $f(2t_s)$, using an irradiation which is periodic but not cyclic modulo t_s. We have

$$U(2t_s, 0) = R_s[\![\bar{\mathscr{H}}_s t_s]\!] R_s[\![\bar{\mathscr{H}}_s t_s]\!] = R_s^2[\![R_s^{-1}\bar{\mathscr{H}}_s R_s t_s]\!][\![\bar{\mathscr{H}}_s t_s]\!]. \tag{24}$$

As before, R_s^2 can be dropped. However, the effective Hamiltonian operating over the interval $(0, 2t_s)$ is *not* $\bar{\mathscr{H}}_s$, but rather the average of the two bracketed quantities in [24]. It leads to an apparent

$$J_{\text{eff}}(2t_s) = J_{\text{eff}}(t_s) \cdot \frac{1}{2}\left| \bar{\mathbf{m}}_s + R_s^{-1} \bar{\mathbf{m}}_s R_s \right| \leqslant J_{\text{eff}}(t_s). \tag{25}$$

The uncritical use of average Hamiltonian theory for such a case always results in an *underestimate* of the quality of decoupling.

The fact that successively sampled points do not all "correspond" to the same J_{eff} means that the effect of decoupling is not simply to scale the coupling constant unless the sequence is cyclic over t_s. The true situation will be made clear later.

Decoupling with Delta Pulses

A simple example will serve both to illustrate the use of the machinery just described and to illuminate the effect of relaxing restriction (5) made in the Introduction. We imagine the spin system to develop freely with *no* irradiation, except that a pulse of width t_w is applied along the x axis at time τ and a similar one of opposite phase is applied at a later time t_r. (The second pulse is introduced only to make the sequence cyclic.) In the limit $\omega_2 \longrightarrow \infty$, $t_w \longrightarrow 0$, $\omega_2 t_w = \theta$ we can write for the development up to some time $t_r > \tau$

$$U(t_r, 0) = e^{-i\theta I_x}[\![\mathscr{H}_J(t_r - \tau)]\!]e^{i\theta I_x}[\![\mathscr{H}_J\tau]\!], \quad [26]$$

with $\mathscr{H}_J$ the unperturbed coupling Hamiltonian. According to [16] the sequence is cyclic over the interval t_r and

$$U(t_r, 0) = [\![\tilde{\mathscr{H}}(t_r - \tau)]\!][\![\mathscr{H}_J\tau]\!], \quad [27]$$

where

$$\tilde{\mathscr{H}} = e^{-i\theta I_x}\mathscr{H}_J e^{i\theta I_x}$$

$$= JS_z(I_z \cos\theta - I_y \sin\theta). \quad [28]$$

Writing $\tau/t = \alpha$ we easily find

$$\bar{\mathscr{H}}_r^{(0)} = JS_z\{[\alpha + (1 - \alpha)\cos\theta]I_z - [(1 - \alpha)\sin\theta]I_y\}. \quad [29]$$

If $f(t)$ is sampled synchronously with the exciting sequence, $t_s = t_r$, then we know that the effect is simply to reduce J by the factor

$$\lambda = [1 - 2\alpha(1 - \alpha)(1 - \cos\theta)]^{1/2}, \quad [30]$$

which can be adjusted for $0 < \lambda < 1$ (*8*).

The minimum value of λ is zero, which requires both $\tau = t_s/2$ and $\theta = (2n + 1)\pi$ —the same conditions, incidentally, which give a perfect proton echo at $t = t_s$.

We now inquire what happens if $t_s \neq t_r$, in particular if the signal is sampled several or many times during t_r, as often occurs in practice. The sequence being discussed is simple enough that the continuous function $f(t)$ can be calculated exactly. If $f(t) = 1$ at $t = 0$, then for $t < \tau$ we of course have the unperturbed free precession signal caused by the J coupling:

$$f(t < \tau) = \cos\frac{J\tau}{2}. \quad [31]$$

For $\tau < t < t_r$ an elementary calculation gives (for $I = S = 1/2$)

$$U(t, 0)S_xU^{-1}(t, 0) = S_x \cos\frac{J\tau}{2}\cos\frac{J(t - \tau)}{2}$$

$$+ 2S_y\left(I_z \sin\frac{J\tau}{2}\cos\frac{J(t - \tau)}{2} - I_y \cos\frac{J\tau}{2}\sin\frac{J(t - \tau)}{2}\right), \quad [32]$$

whence

$$f(\tau < t < t_r) = \cos\frac{J\tau}{2}\cos\frac{J(t - \tau)}{2} = f(\tau)\cos\frac{J(t - \tau)}{2}. \quad [33]$$

Immediately following the final pulse of one sequence, I_z and I_y are interchanged in [32] but $f(t)$ is not changed. At $t = t_r$ we know $f(t_r) = \cos((\lambda J)t_r/2)$, λ being given by [30] with $\theta = \pi/2$. Thereafter the signal repeats itself with this new amplitude:

$$f(nt_r + t) = \cos\frac{(\lambda J)nt_r}{2} f(t < t_r), \quad [34]$$

where $f(t < t_r)$ can be regarded as a modulation of frequency $2\pi/t_r$ imposed on the free precession signal expected from the earlier analysis. Under conditions typically employed in practice, and indeed required for the convergence of the average Hamiltonian, one has $Jt_r \ll 1$. Accordingly the modulation indices associated with the amplitude and phase modulations are small, and the sidebands are very weak. When this condition is satisfied we are justified in analyzing the decoupling problem in terms of the highly restrictive stroboscopic conditions imposed in the Introduction.

CW Decoupling

The idealized sequence described in the preceding section has the property of being cyclic over a time t_r which, even if not the same as the sampling interval t_s, is at least fixed by the pulse programmer, is known to the operator, and can be considered a constant. Continuous-wave (cw) decoupling does not have this property. To bring out this fact in preparation for the following section we describe the appropriate average Hamiltonian theory, although an exact analysis in closed form was long ago carried out by Anderson and Freeman (*3*).

We are dealing with an irradiation having only one state, with ω_{2x}, ω_{2y}, and ω_{2z} constant, so from [6]

$$R(t) = e^{i\omega_e t \mathbf{n}\cdot\mathbf{I}}, \qquad [35]$$

where $\boldsymbol{\omega}_e = \omega_e \mathbf{n}$, and $\mathbf{n}$ is a unit vector. From [8] we then have

$$I_z(t') = JS_z \mathbf{I}\cdot\mathbf{m}(t'), \qquad [36]$$

where

$$m_x = n_y \sin\omega_e t + n_x n_z(1 - \cos\omega_e t),$$

$$m_y = -n_x \sin\omega_e t + n_y n_z(1 - \cos\omega_e t),$$

$$m_z = n_z^2 + (1 - n_z^2)\cos\omega_e t. \qquad [37]$$

The propagator to an arbitrary time is then, from [10],

$$U(t, 0) = e^{i\omega_e t\mathbf{n}\cdot\mathbf{I}} e^{-i\bar{\mathscr{H}}^{(0)}t},$$

$$\bar{\mathscr{H}}^{(0)} = JS_z\left\{n_z(\mathbf{n}\cdot\mathbf{I}) + \frac{1}{\omega_e t}[(n_y I_x - n_x I_y)(\cos\omega_e t - 1) + (I_z - n_z(\mathbf{n}\cdot\mathbf{I}))\sin\omega_e t\right\}. \qquad [38]$$

The irradiation is cyclic over $t_c = 2\pi/\omega_e$. If we could sample $f(t)$ at the appropriate times we would have

$$\bar{\mathscr{H}}^{(0)}(t_c) = JS_z n_z(\mathbf{n}\cdot\mathbf{I}) \qquad [39]$$

or a scaling factor

$$\lambda_{cw} = n_z = \frac{\delta}{(\omega_2^2 + \delta^2)^{1/2}}. \qquad [40]$$

This result is in fact correct, and is widely used to measure ω_2 by means of off-resonance decoupling (*4*). Yet $f(t)$ is certainly not sampled at nt_c in practice: one does not usually even know the value of ω_e, and if one did it would be different for protons having different chemical shifts, making the requisite synchronization impossible. The success of [40] can be accounted for along the lines of the preceding section: the actual signal contains modulation sidebands at $\pm\omega_e$, but these have intensities of the order of $(J/\omega_e)^2$ and can be neglected under ordinary decoupling conditions. (They *cannot* be ignored under weak irradiation or "tickling" conditions: they correspond when decoupling is strong to the weak satellite resonances which appear in an exact analysis (*3*).)

Before leaving this subject we record, for the sake of completeness, the result which corresponds to [40] when the average Hamiltonian expansion is carried through sixth order in $\bar{\mathscr{H}}$:

$$\lambda_{cw} = n_z\left\{1 - \frac{J^2}{8\omega_e^2} + \frac{9J^4 + 24J^2\delta^2}{64\omega_e^4} - \frac{5J^6 + 80J^3\delta^3}{128\omega_e^6}\right\}. \quad [41]$$

For $J/2\pi = 200$ Hz, $\omega_1/2\pi = 4$ kHz the correction amounts to less than one part in 3000.

Phase-Alternated Decoupling

Grutzner and Santini (*6*) (GS) have described a scheme, now widely used, in which the phase of an otherwise cw decoupling irradiation is reversed at intervals τ. In the two states we label the directions of the effective field by

$$\mathbf{n}_1 = \omega_2\mathbf{i} + \delta\mathbf{k}; \qquad \mathbf{n}_2 = -\omega_2\mathbf{i} + \delta\mathbf{k}. \quad [42]$$

Applying the machinery developed earlier we have

$$U(2\tau, 0) = e^{i\omega_e\tau\mathbf{n}_2\cdot\mathbf{I}}e^{i\omega_e\tau\mathbf{n}_1\cdot\mathbf{I}}[[\tilde{\mathscr{H}}_2\tau]][[\tilde{\mathscr{H}}_1\tau]]. \quad [43]$$

The irradiation is not cyclic unless $\omega_e\tau = 2n\pi$: the proton then precesses an integer number of times about the effective field in each state. If this condition is satisfied we have

$$\tilde{\mathscr{H}}_1 = JS_zn_z(\mathbf{n}_1\cdot\mathbf{I}); \qquad \tilde{\mathscr{H}}_2 = JS_zn_z(\mathbf{n}_2\cdot\mathbf{I}) \quad [44]$$

and

$$\bar{\mathscr{H}}^{(0)} = JS_zI_zn_z^2 \quad [45]$$

leading to a scaling factor

$$\lambda_{GS} = n_z^2 = \frac{\delta^2}{\omega_2^2 + \delta^2}, \quad [46]$$

which is superior to cw decoupling: both give perfect decoupling for $\delta = 0$ but the GS scheme degrades only quadratically rather than linearly with δ.

In practice τ is fixed by the experimenter and is usually quite long: $\omega_2\tau \gg 1$. The result is that as ω_e is increased by going further off resonance, the cyclic condition will recur at several values of δ, at which [45] is correct. At other offsets, however, the simple result is wrong. There are two effects: (1) The extra rotation operators must be inserted at the beginning of [42], and (2) the effective Hamiltonians of [43] must be converted to expressions of the form of [38]. The result is complicated, and illustrates the fact that the average Hamiltonian method loses its attractive simplicity in noncyclic situations. The one thing we can be sure of, on the basis of the discussion [24], is that the decoupling will be *better* than predicted by [45]. We will presently see in fact that it becomes perfect in the vicinity of $\omega_e \tau = (n + 1/2) \cdot 2\pi$!

A Different Interaction Representation

The trouble mentioned in the last paragraph can be avoided if the system Hamiltonian is divided in a different way, as has been done by Haeberlen (*13*). We rewrite [5] as

$$U(t, 0) = R'(t) U_{J\delta}(t), \quad [47]$$

$$R(t) = \mathcal{T} \exp\left\{ i \int_0^t (\omega_{2x}(t') I_x + \omega_{2y}(t') I_y) dt' \right\}, \quad [48]$$

$$U_{J\delta}(t) = \mathcal{T} \exp\left\{ -i(JS_z + \delta) \int_0^t I_z(t') dt' \right\}. \quad [49]$$

Then a sequence which is cyclic at $\delta = 0$ is always so. However, the convergence of the Magnus series is impaired in two respects:

1. Convergence requires $\delta/\omega_2 \ll 1$ rather than $J/\omega_e \ll 1$, which will not be well satisfied for large offsets.
2. The odd-order terms in $\bar{\mathcal{H}}$, in particular $\bar{\mathcal{H}}^{(1)}$, no longer can be discarded since they contain terms linear in S_z.

For these reasons the alternative approach is not desirable for analyzing the problem of broadband decoupling.

Composite Pulse Decoupling

Levitt and Freeman (*14*) (LF) recently devised a superior decoupling method based on an elegant use of the average Hamiltonian method. The essential points of their argument will be outlined to illustrate both the power and the defects of this approach.

First recall that the average Hamiltonian for any state or sequence of states is always of the form [13] or [21], i.e., depends on a linear combination of the components of **I**. Suppose a sequence A can be found for which R_A (cf. Eq. [19]) has the form $R_A = \exp(i\pi I_x)$. The simple δ-pulses described earlier have this property. However, real pulses of finite width possess a z component of ω_e which becomes substantial for large offsets δ. Levitt and Freeman envision a more complicated sequence for which R_A does not possess this defect even for large δ. In fact they had previously found such sequences in another connection (15), e.g.,

$$R_A = e^{(i\pi/2)I_x} e^{(2i\pi/3)I_y} e^{(i\pi/2)I_x}. \quad [50]$$

The average Hamiltonian appropriate to sequence A is

$$\bar{\mathcal{H}} = JS_z \mathbf{I} \cdot \mathbf{m}_A . \quad [51]$$

If now the same sequence is repeated we have, using $R_A^2 = 1$,

$$\bar{\mathscr{H}}_{AA} = \frac{1}{2}[R_A\bar{\mathscr{H}}_A R_A^{-1} + \bar{\mathscr{H}}_A] = JS_zI_xm_{Ax}\,. \quad [52]$$

Now Levitt and Freeman consider a sequence which is identical to A except that the rf phases are reversed for every state:

$$R_{Bk} = e^{i\pi I_z}R_{Ak}e^{-i\pi I_z}. \quad [53]$$

The average Hamiltonian for this sequence is easily seen to be

$$\bar{\mathscr{H}}_B = JS_z\mathbf{I}\cdot\mathbf{m}_B\,,$$

where $\mathbf{m}_B$ and $\mathbf{m}_A$ have the same z components but opposite x and y components. Thus

$$\bar{\mathscr{H}}_{BB} = JS_zI_xm_{Bx} = -JS_zI_xm_{Ax}\,. \quad [54]$$

If these sequences are now combined, one has

$$\bar{\mathscr{H}}_{AABB} = 0, \quad [55]$$

i.e., perfect decoupling. Indeed Levitt and Freeman have demonstrated that such a twelve-state sequence based on [50] performs extraordinarily well. As a result of their analysis they propose that an optimal (periodic) decoupling sequence must be (a) made up of four parts, $AABB$, where (b) B is identical to A except for reversal of all rf phases, and (c) both A and B invert longitudinal magnetization of a free spin, independent of offset δ. We shall later show that this criterion can be replaced by a less restrictive and more general one.

The LF analysis, elegant as it may be, is incomplete in the sense that it is based on the existence of ideal composite pulses R_A and R_B which perfectly invert z magnetization. At large enough offsets any sequence fails this test. When this occurs, the simple symmetries of the LF analysis are lost, and moreover the sequence becomes noncyclic in the same offensive manner as the GS sequence. Then one must inevitably resort to numerical analysis. That being the case, there is no particular advantage associated with the average Hamiltonian method. We now proceed to describe an exact means of analyzing the decoupling problem.

EXACT ANALYSIS

General Method

Anticipating a numerical analysis, we insist that the irradiation consist of a sequence of states during which [1] is independent of time. For state k we imagine writing [1] in the explicit matrix representation $|m_s m_I\rangle$. Since $[\mathscr{H}, S_z] = 0$ this is factored into a pair of 2×2 matrices, which we will call $\mathscr{H}_{k+}$ and $\mathscr{H}_{k-}$, corresponding to $m_s = +1/2$ and $-1/2$, respectively. (We restrict ourselves to $I = S = 1/2$.) Each can be thought of as the Hamiltonian for a single proton subject to an effective external field ω_{e+} or ω_{e-}: these differ only in their z components, $\delta + J/2$ or $\delta - J/2$. This being the case, the propagators derived from $\mathscr{H}_{\pm}$ are pure rotations through angles $\phi^{(k)}{}_{\pm}$:

$$R_{\pm}^{(k)} = e^{it_k \omega_{e\pm}^{(k)} \cdot \mathbf{I}} = e^{i\phi^{(k)}{}_{\pm}\mathbf{n}^{(k)}{}_{\pm}\cdot\mathbf{I}}. \qquad [56]$$

In the 2×2 representation of the rotation group these have the form

$$R_{\pm}^{(k)} = \begin{pmatrix} a_{\pm}^{(k)} & b_{\pm}^{(k)} \\ -b_{\pm}^{(k)*} & a_{\pm}^{(k)*} \end{pmatrix}, \qquad [57]$$

where the Cayley–Klein parameters (*13*) a and b are easily written in terms of the given properties of the state a appearing in (57):

$$a_{\pm}^{(k)} = \cos\frac{\phi_{\pm}^{(k)}}{2} + in_{z\pm}^{(k)} \sin\frac{\phi_{\pm}^{(k)}}{2},$$

$$b_{\pm}^{(k)} = (n_{y\pm}^{(k)} + in_{x\pm}^{(k)}) \sin\frac{\phi_{\pm}^{(k)}}{2}. \qquad [58]$$

The state propagators are now concatenated for successive states by explicit matrix multiplication out to the time t_s where the first measurement is desired. Denoting the concatenated 2×2 propagators by $R_{\pm}$ we construct the 4×4 matrix

$$U(t_s, 0) = \left(\begin{array}{c|c} R_+ & \\ \hline & R_- \end{array}\right) \qquad [59]$$

and evaluate $f(t)$ from [3]. One easily finds

$$f(t_s) = 2\,\mathrm{Tr}\,(R_+R_-^{\dagger}) = 4\,\mathrm{Re}[a_+a_-^* + b_+b_-^*]., \qquad [60]$$

In the case of a perfectly arbitrary decoupling sequence and unconstrained sampling interval, the above concatenation would then be continued to $t = 2t_s$ and $f(2t_s)$ again numerically found from [60]. The Fourier transform of a series of such points would give the S-spin spectrum exactly. The computing time involved would be rather large. Moreover (a) most decoupling sequences in use are periodic, and (b) we have seen earlier that sampling synchronously with the sequence does not make any substantial difference. Therefore we set $t_s = t_r$. Then the first point $f(t_r)$ contains all the information present in the full FID and the spectrum can be extracted as follows: because of the group property of rotations, the concatenated $R_{\pm}(t_r)$ have the form of [57] and [58]. From the numerical values of its elements, the parameters of the overall rotation can be extracted:

$$\phi_{\pm} = 2\cos^{-1}[\mathrm{Re}(a_{\pm})],$$

$$n_{z\pm} = \mathrm{Im}(a_{\pm})/\sin\phi_{\pm}, \qquad \text{etc.} \qquad [61]$$

Expressing [60] in terms of these,

$$f(t_r) = \frac{1}{2}(1 + \mathbf{n}_+ \cdot \mathbf{n}_-)\cos\left(\frac{\phi_+ - \phi_-}{2}\right) + \frac{1}{2}(1 - \mathbf{n}_+ \cdot \mathbf{n}_-)\cos\left(\frac{\phi_+ + \phi_-}{2}\right). \qquad [62]$$

The Fourier transform of $f(nt_r)$ consists of δ functions at frequencies $\pm\,\omega_a$, $\pm\omega_b$

$$\omega_a = \frac{\phi_+ - \phi_-}{2t_r}, \qquad \omega_b = \frac{\phi_+ + \phi_-}{2t_r}, \tag{63}$$

with intensities given by the corresponding coefficients in [62].

Note that these results depend only on the sequence being *periodic* modulo t_r: it need not be cyclic. The periodic condition is satisfied by most decoupling schemes in use, whereas as we have seen the cyclic condition is not.

Another Average Hamiltonian

A unitary operator such as $U(t_r, 0)$ can always be represented as a complex exponential function of a hermitian operator. We choose it in the form

$$U(t_r, 0) = e^{-i\mathfrak{H}t_r}, \tag{64}$$

where $\mathfrak{H}$ represents a fictitious time-independent Hamiltonian which would have caused the same change in quantum state over time t_r as the actual time-dependent sequence of states. It is easy to show that

$$\mathfrak{H} = \frac{1}{2t_r}(\boldsymbol{\phi}_+ + \boldsymbol{\phi}_-)\cdot\mathbf{I} + \frac{1}{t_r}S_z(\boldsymbol{\phi}_+ - \boldsymbol{\phi}_-)\cdot\mathbf{I}. \tag{65}$$

This Hamiltonian has not had the portion of [1] containing external fields factored out, and is therefore a very different entity from the average Hamiltonians discussed earlier. If [65] is diagonalized, and S_x regarded as a first-order perturbation causing transitions, the resulting absorption frequencies and intensities are the same as derived in the preceding section. The two terms of [65] resemble a Zeeman effect and a J coupling, respectively, where the quantization axes of $\mathbf{I}$ and $\mathbf{S}$ have been made nonparallel by the decoupling. The result is that the selection rules for S-spin transitions are $\Delta m_S = \pm 1$, but the usual condition $\Delta m_I = 0$ is relaxed as a result of mixing of states, leading to satellite resonances.

The resulting four line spectrum is the characteristic of any IS system (with $I = S = 1/2$) subject to a periodic perturbation and sampled synchronously at nt_r. The simplest example is provided by cw irradiation where t_r is arbitrary. For strong decoupling ($\omega_e \gg J$), ω_{e+} and ω_{e-} are nearly the same at all offsets. Thus $\mathbf{n}_+ \approx \mathbf{n}_-$; the first term of [66] dominates and the splitting $2\Omega_1 \approx \omega_{e+} - \omega_{e-}$. This splitting vanishes at $\delta = 0$: its dependence on δ resonance reproduces [40] to lowest order in J/ω_2 as expected. The present calculation, being exact, also gives the correct result for spin tickling: for $\omega_2 \longrightarrow 0$, $\mathbf{n}_1 \longrightarrow -\mathbf{n}_2$, $\phi_+ = \phi_- = Jt_r/2$ and $2\omega_a = J$. In general, the present theory leading to [65] provides a much simpler and more compact alternative to the previously published exact theory of cw irradiation (*3*).

Interpretation

The quantities $\boldsymbol{\phi}_+$ and $\boldsymbol{\phi}_-$ in [62] are simply the total rotations accumulated by a single proton under the influence of a succession of effective fields such that ω_{e+} and ω_{e-} are the same in every state except that the z component of each is augmented (ω_{e+}) or decreased (ω_{e-}) by $J/2$ for every state. Clearly $J/2$ simply plays the role of an additional offset: $\phi_{\pm} = \phi(\delta \pm J/2)$. Accordingly [66] can be evaluated merely by computing the offset-dependent rotation $\phi(\delta)$, $\mathbf{n}(\delta)$ of a free spin. The criterion for good decoupling is just that $\phi(\delta)$ change very little over $\Delta\delta = J$.

Moreover ϕ and $\mathbf{n}$ are continuous functions of δ, and may usually be assumed to vary slowly on the scale of J if the irradiating fields are strong. This means $\mathbf{n}(\delta + J/2)\cdot\mathbf{n}(\delta - J/2) \approx 1$ and the second term of [62] can be neglected. The spectrum then consists of just one doublet with splitting

$$J_{\text{eff}}(\delta) = 2\omega_{\text{a}} = \frac{\phi(\delta + J/2) - \phi(\delta - J/2)}{t_{\text{r}}}. \qquad [66]$$

In the limit of small J the result is simply a scaling:

$$\lambda = \frac{1}{t_{\text{r}}}\frac{\partial\phi}{\partial\delta} = \frac{\partial\Omega}{\partial\delta}. \qquad [67]$$

Here Ω is a fictitious proton precessional frequency which would cause a phase accumulation ϕ during t_{r}.

Equations [62] and [67] are the principal result of this paper. They constitute an exceedingly simple criterion by which the efficacy of arbitrary decoupling sequences can be tested. Qualitatively speaking, we require that *the magnitude of the net rotation of a free spin must be insensitive to offset* for given t_{r}, over the range of offsets of interest. This criterion is far less restrictive than the one proposed by Levitt and Freeman (*11*): it makes no reference to the number, properties, or relationship of the parts of which the sequence is composed. Moreover [67] is a *quantitative* criterion by means of which the derivations from ideality of a sequence can be easily calculated. Of course the specific sequence advocated by Levitt and Freeman satisfies both criteria in their qualitative sense: since their R_A and R_B invert z magnetization over a wide range of δ, one has $\phi \approx 0$ over that range.

We note in passing that the decoupling predicted is independent of the sampling phase, i.e., the point in a given sequence at which the FID is initiated and the signal samples taken. This is because, while the rotations making up a sequence do not commute, the *magnitude* ϕ of the resulting rotation is invariant to a cyclic permutation of its parts.

The criterion just stated could have been anticipated intuitively: while in the context of the present problem one tends to think of J coupling as an effect of I on S, of course the coupling is reflexive. If in any irradiation the motion of I is the same whether S has spin up or spin down one can say that I is unaware of the presence of S, i.e., they are decoupled.

Connection with Average Hamiltonian Theory

We can at this point shed additional light on our earlier assertion that the average Hamiltonian theory in general gives an overestimate of the residual coupling. We have established that, even for a noncyclic irradiation, that theory gives an accurate result for the first sampled point $f(t_{\text{s}})$. To interpret that point as resulting from an effective J doublet amounts to fitting it to the function cos $(J_{\text{eff}}t_{\text{s}}/2)$. But according to [62] the true $f(t)$ is a sum of two cosine terms, the *slower* of which we would like

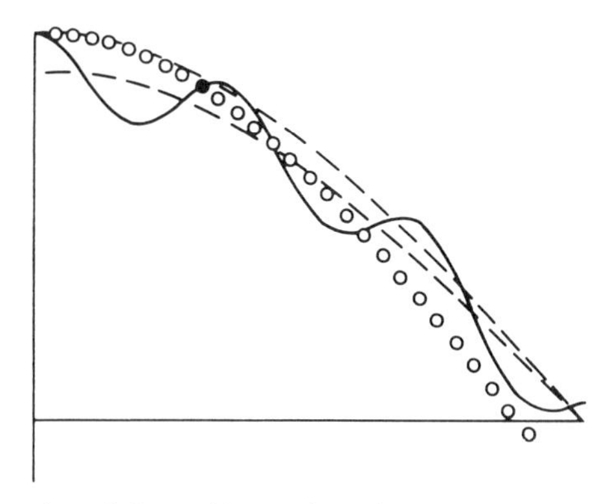

FIG. 1. Schematic illustration of the problems with naive average Hamiltonian theory for noncyclic sequences. The solid curve represents the true FID, corresponding to a spectrum with one strong and one weak doublet. The lower dashed curve is the part which gives the dominant doublet with small splitting. In simple average Hamiltonian theory the first sampled point (solid circle) would be forced into a fit to a single cosine wave (open circles), which has a shorter period than the upper dashed curve, which is a renormalized version of the lower one.

to interpret as corresponding to the effective coupling. In Fig. 1 we show schematically what happens: The solid curve represents the true FID and the solid circle the first sampled point. The open circles represent the FID which would be conjectured from average Hamiltonian theory. It can be seen to have a higher frequency than the upper dashed curve, which represents the dominant lower-frequency component of the true $f(t)$, normalized to unity. The conclusion becomes general if the first sampling point is taken early enough (as is done in practice but not shown in Fig. 1) to satisfy the Nyquist criterion for both components.

A more formal approach to the same comparison would be based on the observation that the rotation of the I spin embodied in [22] is a very good approximation to the rotation $R_+R_-^\dagger$ of the exact theory. To extend the average Hamiltonian result to nt_s would be tantamount to assuming

$$f(nt_s) = 2\ \mathrm{Tr}\ (R_+R_-^\dagger)^n,$$

whereas the correct result is

$$f(nt_s) = 2\ \mathrm{Tr}\ \{(R_+)^N(R_-^\dagger)^N\}.$$

EXAMPLES

Experimental

We now compare the expected performances of several decoupling schemes according to the theory developed in this paper. We have standardized all sequences to a nominal effective rf field strength of 5 kHz, i.e.,

$$\omega_{\text{eff}}^2 = \int_0^{t_r} \omega_2^2 dt = (2\pi \times 5000)^2 \qquad [68]$$

to equalize rf heating effects. The J-scaling factor, λ, is computed for offsets up to

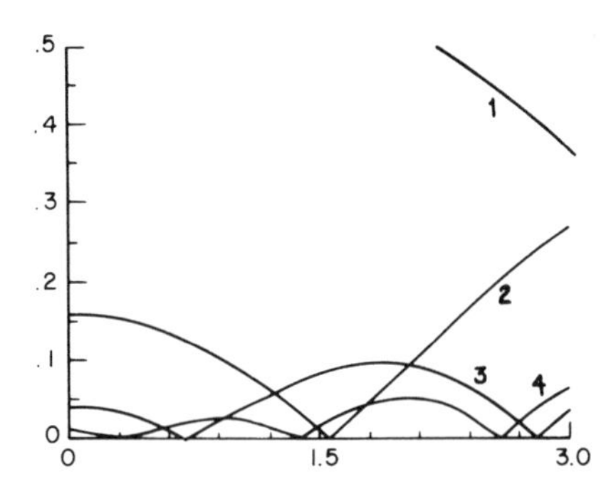

FIG. 2. Predicted scaling factor vs offset for decoupling with phase-alternated 180° pulses with effective rf field corresponding to $\omega_2/2\pi = 5$ kHz. From top to bottom the curves correspond to duty factors 1, 1/4, 1/16, and 1/64.

3 kHz, i.e., over a 6-kHz proton spectral range. In some cases we suggest the effects of rf field inhomogeneity by recalculating with all rf fields reduced. In all cases we assume $t_s = t_r$. The scaling properties of the theory are such that the values of λ are unchanged by $\omega_{\text{eff}} \rightarrow k\omega_{\text{eff}}$, $\delta_{\max} \rightarrow k\delta_{\max}$, $t_r \rightarrow k^{-1}t_r$. In some cases λ_{cw} is plotted for comparison.

Calculations were done on a Hewlett–Packard HP85 computer using programs obtained from the French Resource for NMR, primarily the program RFRMN2. This program accepts rf field amplitudes and phases and magnetic field shifts for a sequence of up to 50 states and computes the rotation angles ϕ and scaling factors λ as functions of δ, using both average Hamiltonian and exact methods. Computational speed was increased by using the real 3×3 representation of rotation operators rather than the complex 2×2 representation implied by the text.

Decoupling with 180° Pulses

As remarked earlier, a train of 180° δ pulses provides perfect decoupling, but with pulses of finite width the sequence becomes noncyclic for $\delta \neq 0$. Consider a sequence of length t_r containing two 180° pulses of opposite carrier phase and width t_w, i.e., having duty factor $x = 2t_w/t_r$. If [68] is to be obeyed while keeping $\omega_2 t_w = \pi$, t_r must be increased as the duty factor is decreased. We have calculated by way of example the results expected for values of x between 1 ($t_w = 0.1$ msec, $t_r = 0.2$ msec) and 1/64 ($t_w = 12.5$ μsec, $t_r = 1.6$ msec), with the results shown in Fig. 2. The topmost curve, for $x = 1$, shows extremely poor decoupling. Successively smaller values of x give rapidly improving results. However, for $x = 1/64$ we already have $H_2 \approx 10$ gauss (for protons) and further improvement much beyond that point would probably encounter difficulties with voltage breakdown of the probe in practice.

The progressive improvement as x is decreased in Fig. 2 is of course reversed if 360° pulses are used instead.

Phase-Alternated and Noise Decoupling

An important parameter characterizing the GS sequence (*6*) is $N(0) = \omega_2\tau/2\pi$, the number of full precessions of the proton during each half of the sequence at

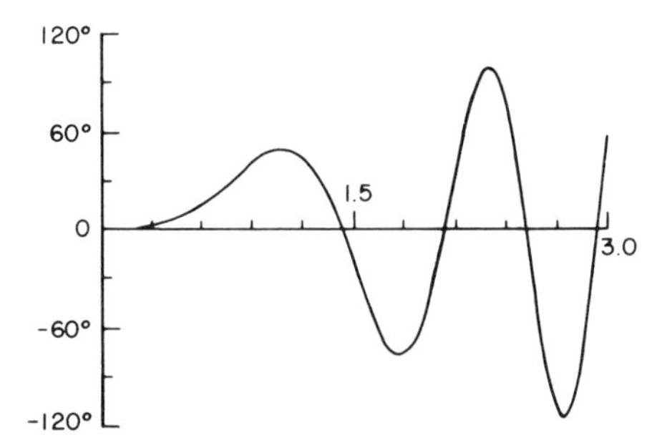

FIG. 3. Magnitude ϕ of the overall rotation in time t_r for a Grutzner–Santini sequence with $\omega_2/2\pi = 5$ kHz, $t_r = 10$ msec. At the zeroes of $\phi(\delta)$ the rotation is about the $\pm z$ axis of the rotating frame, when the sequence is cyclic. Near the extrema it is about the $\pm y$ axes. The slope of this curve corresponds to the predicted scaling factor $\lambda(\delta)$.

exact resonance. (The case $x = 1$ in the preceding section is an example with $N(0) = 1/2$.) The GS method has been found to perform best, however, with rather large values of $N(0)$. Here we choose $N(0) = 25$, i.e., $\omega_2/2\pi = 5$ kHz, $\tau = 5$ msec. As the offset δ is increased, $N(\delta)$ increases also, passing through successive integer values for which the sequence is cyclic and therefore the overall rotation $\phi = 0$. At intervening values of δ, ϕ has maxima and minima in the neighborhood of half-integer values of N.

A plot of $\phi(\delta)$ is presented in Fig. 3. Figure 4 shows the J-scaling factor λ, from exact calculation (solid curve) and average Hamiltonian theory (open circles). As expected, these agree at integer N. Near half-integer values, however, the effective coupling passes through zero. "Noise decoupling" as conventionally understood (*4*) is similar, except that after every interval τ the phase is either reversed or not, depending on the toss of a coin. Over long intervals the probability of occupations of each of the two states approaches 1/2, and the simplest cyclic average Hamiltonian theory, as in [44], would predict the same performance as the GS sequence. In practice, however, noise decoupling is usually practiced with $N(0)$ much smaller—usually less than 1. The pronounced zeroes which the exact theory predicts for the

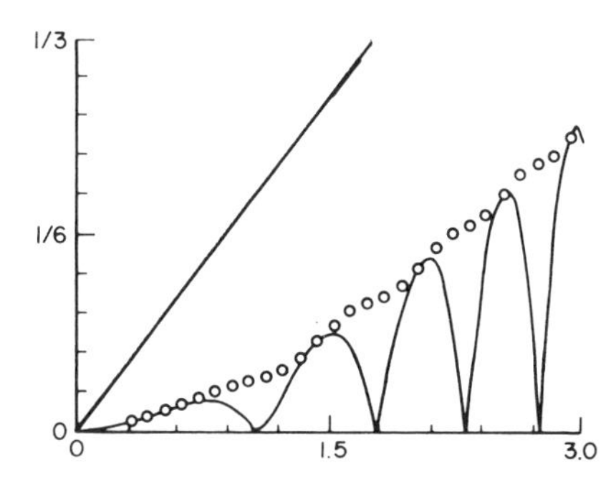

FIG. 4. Scaling factor vs offset, $\lambda(\delta)$, for the Grutzner–Santini sequence. The solid curve is the exact result; the open circles represent the prediction of simple average Hamiltonian theory. The two agree near $\phi(\delta) = 0$, where the sequence is cyclic. The straight line shows the result of cw decoupling for comparison.

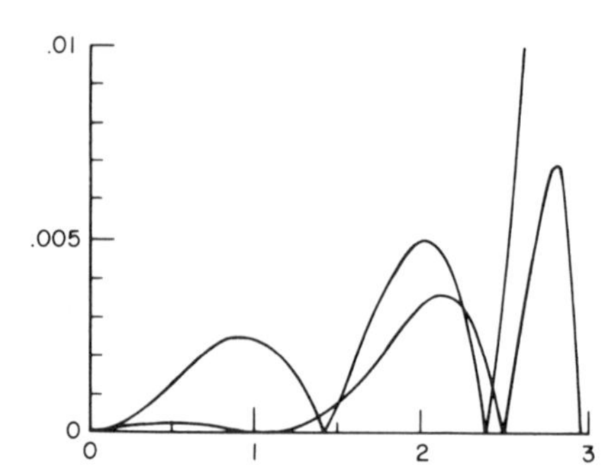

FIG. 5. Performance of two "composite pulse" sequences. The upper curve is appropriate to the Levitt–Freeman sequence, while the lower is for a sequence which is the same except that one of the 90° pulses has been shortened by 6%, destroying the symmetry of the sequence.

GS method are thus expected to be suppressed both because $N(\delta)$ is small and because it is only statistically defined.

Composite Pulse Sequences

The analysis of Levitt and Freeman predicts perfect decoupling at all offsets, *assuming* that their composite pulses correspond to perfect 180° flips. Indeed it performs spectacularly well, as shown by their experimental tests and by explicit calculation (see Fig. 5). However, we earlier established a criterion for decoupling efficiency which is much broader than theirs. Accordingly it would not be surprising if one could find sequences which perform still better by departing from their criterion while obeying the more general one. We content ourselves with exhibiting in Fig. 5 one such case, in which the LF sequence has been altered by reducing *any one* of the 90° pulses by 6% in length, thereby destroying its symmetry and rendering it noncyclic for $\delta = 0$ ($\phi = 5.4°$). In Fig. 6 we also provide evidence suggesting that the modified sequence is superior to the original one in its resistance to the effects of rf field imhomogeneity.

This illustration is not to be taken as advocating the use of the particular nonsymmetric sequence chosen, but only as indicating that there is room within our

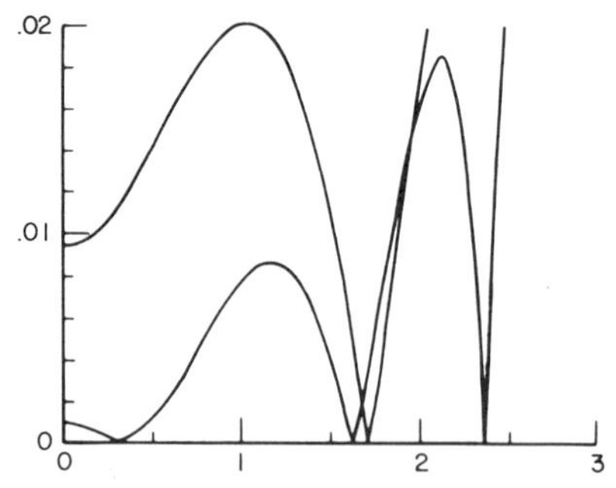

FIG. 6. Same as Fig. 5 except that the rf field strength has been reduced by 10% to indicate sensitivity of the sequences to inhomogeneity of H_1.

more general criterion for finding a variety of improved decoupling schemes. In that connection, one must at some point decide what one desires in the tradeoff between good decoupling over a wide spectral range and better at more uniform decoupling over a narrower range. For example, we have found schemes which, by comparison with Fig. 6, provide an order of magnitude better decoupling for $\delta \leqslant 1$ kHz but much worse decoupling for $\delta \gtrsim 1$ kHz. Further exploration of these matters is outside the scope of the present paper.

The Role of Phase Reversal

It sometimes happens that the performance of a sequence R is improved by combining it with a sequence R', which is the same but with all rf phases reversed. The GS and LF methods are cases in point. If we denote the rotation angle of R or R' by ϕ, it is clear that the overall rotation $\Phi \leqslant 2\phi$, because the components of the rotation axes $\mathbf{n}$, $\mathbf{n}'$ in the xy plane cancel. It might then be thought that the decoupling would necessarily be better, but recall that it is not $\Phi(\delta)$ but $\partial\Phi/\partial\delta$ that governs. Indeed a doubling of the noncyclic sequence of the preceding section *degrades* its performance slightly.

In certain cases it is clear that doublings with phase reversal have only a very special effect: This is true whenever the component sequence R has already been subjected to this artifice: $R = AA'$. Then $RR' = (AA')(A'A)$, and since $\phi(\delta)$ is invariant to cyclic permutations this is equivalent to $AAA'A'$, i.e., to a doubling in length of A and A'. For example, a phase-reversed doubling of the GS sequence leaves its effects the same except for a compression of the horizontal scale of Fig. 4.

CONCLUSION

We have shown that average Hamiltonian theory can provide insight into the decoupling problem, but that its results must be interpreted cautiously when, as usually occurs, one is dealing with noncyclic irradiations. As an alternative, we have devised an exact (though numerical) procedure which leads to an extremely simple test of any arbitrary decoupling scheme.

This result has a defect which must not be minimized: it provides no insight or procedure by which better decoupling schemes can be devised. Of course a sort of blind search could be made by linear programming or related techniques, but it is not clear how successful it might be: the problem is highly nonlinear, and its "natural" variables (strengths, phases, timing, ordering, etc., of pulses) are far from being orthogonal in their effects. It would be very much worthwhile seeking a deeper theory which has predictive rather than retrodictive value.

The reader is also reminded that we have clad the present theory in a straitjacket of restrictive assumptions: sequences are assumed to be periodic, with one sampling point per period, and the "satellite" lines in the spectrum predicted by [62] have been largely ignored. We have argued very qualitatively that these restrictions are not critical. However, we have found sporadic and local instances where they apparently are.

Finally, we note that we have made no comparisons with real experiments. Such comparisons are difficult for two reasons: (1) Published results tend to be based on intensities of almost-collapsed multiplets rather than on actual measurements of residual splittings, and the data needed to convert from one criterion to the other are not given. (2) Nonidealities of the experiments (rf inhomogeneity, timing aberations, phase transients, etc., are generally not known or reported. It is disquieting (to the author) that there seem to be cases in which the experimental performance of certain decoupling schemes is *better* than can be accounted for by idealized theory. Thus there is clearly further work to be done on the problem, both theoretically and experimentally.

ACKNOWLEDGMENTS

The author benefited from conversations with R. J. Silbey, A. Pines, and D. Weitekamp.

REFERENCES

1. F. BLOCH, Phys. Rev. **93,** 944 (1954).
2. R. FREEMAN AND D. H. WHIFFEN, *Proc. Phys. Soc. London* **79,** 794 (1962).
3. W. A. ANDERSON AND R. FREEMAN, *J. Chem. Phys.* **37,** 85 (1962).
4. R. R. ERNST, *J. Chem. Phys.* **45,** 3845 (1966).
5. V. J. BASUS, P. D. ELLIS, H. D. W. HILL, AND J. S. WAUGH, *J. Magn. Reson.* **35,** 19 (1979).
6. J. B. GRUTZNER AND A. E. SANTINI, *J. Magn. Reson.* **19,** 178 (1975).
7. M. MEHRING, A. PINES, W.-K. RHIM, AND J. S. WAUGH, *J. Chem. Phys.* **54,** 3239 (1971).
8. R. FREEMAN, S. P. KEMPSELL, AND M. H. LEVITT, *J. Magn. Reson.* **35,** 447 (1979).
9. A. PINES AND J. D. ELLETT, JR., *J. Am. Chem. Soc.* **95,** 4437 (1973).
10. U. HAEBERLEN AND J. S. WAUGH, *Phys. Rev.* **175,** 453 (1968).
11. W. MAGNUS, *Commun. Pure Appl. Math.* **7,** 649 (1954).
12. W. P. AUE, Ph.D thesis, Eidgenössische Technische Hochschule, Zürich, 1979.
13. U. HAEBERLEN, "High Resolution NMR in Solids," pp. 82–87, Academic Press, New York, 1976.
14. M. H. LEVITT AND R. FREEMAN, *J. Magn. Reson.* **43,** 502 (1981).
15. M. H. LEVITT AND R. FREEMAN, *J. Magn. Reson.* **43,** 65 (1981).

JOURNAL OF MAGNETIC RESONANCE **50**, 95–110 (1982)

Symmetrical Composite Pulse Sequences for NMR Population Inversion. II. Compensation of Resonance Offset

MALCOLM H. LEVITT*

Department of Isotope Research, Weizmann Institute of Science, Rehovot 76100, Israel

Received April 22, 1982

A five-pulse sequence $R_x(90°)R_y(200°)R_{-y}(80°)R_y(200°)R_x(90°)$ is suggested for population inversion of all resonances in a spectrum whose width may be nearly twice as large as the radiofrequency field strength. The population inversion is very fast and is easier to implement than comparable methods such as adiabatic rapid passage. The composite pulse is derived from previous three-pulse sequences with the aid of mathematical and geometrical arguments, and computer simulation. Experimental results are shown and are compared with computer simulations, and with results for previous composite pulses.

In Part I of this paper it was demonstrated that composite pulse sequences can be devised which provide an accurate population inversion even in the presence of an inhomogeneous radiofrequency field or when the pulse lengths have been wrongly estimated (*1*). These pulse sequences are expected to be of potential importance to the NMR of systems where the inhomogeneity of the radiofrequency field cannot be avoided, such as solutions of high electrolytic conductivity, or biological samples of large dimensions. A mathematical approach led to some general guidelines for the construction of such self-compensating pulse sequences, including the use of "anti-tangential trajectories," in which magnetization vectors reverse exactly their momentary direction of nutation in passing from one pulse element to the next at some point in the composite sequence, and the use of symmetry between trajectories in the northern and southern hemispheres of the rotating reference frame. Two pulse sequences were suggested which embody both of these features and accordingly provide an extremely good compensation of pulse length errors.

In this part a related pulse sequence will be suggested which offers a good population inversion even in the presence of severe off-resonance effects. By using this composite pulse all resonances in a spectrum of width almost twice the radiofrequency field strength can be inverted. Also this inversion can be achieved in a small fraction of a millisecond at usual irradiation field strengths. To produce a comparable wide-band population inversion by previous methods, such as adiabatic fast passage (*2*), would require special hardware and take an order of magnitude longer.

The range of applications of such a sequence should be more far reaching than that anticipated for those proposed in Part I. Resonance offset effects are a result

* Present Address: E.T.H., Zürich, Laboratorium für Physikalische Chemie, Switzerland.

0022-2364/82/130095-16$02.00/0

of real practical limitations on the strength of the radiofrequency field as compared with the spread in resonance frequencies of the irradiated isotope. In contrast to the problem of radiofrequency inhomogeneity, which may be solved in principle by careful coil design (at least for nonconducting samples), there are unavoidable limitations on the acceptable radiofrequency power input. The production of pulses of very high peak power requires large expensive radiofrequency amplifiers and is invariably associated with problems of instability and arcing. If the duty cycle of radiofrequency irradiation is high then simple thermal damage of the sample is a risk. Even mild power inputs can set up convection currents within liquid samples, often making shimming impossible. This is particularly troublesome with samples of appreciable electrical conductivity which absorb the radiofrequency energy effectively. All of these problems are becoming more pressing because of the continuing increase in static magnetic field strengths as the technology of superconducting magnets advances.

An application with particularly critical power requirements is heteronuclear spin decoupling which requires continuous irradiation of the abundant isotopic species at the same time as signals from a rare isotope are observed. Composite pulses already have been demonstrated to be of considerable importance in this technique (*3–5*). The spin states of the abundant isotope are repetitively inverted by a sequence of composite pulses which must also take into account the coherent properties of the spin evolution. Average Hamiltonian theory (*6*) has been used to elucidate rules for the optimum design of composite pulse decoupling sequences (*5*). It seems likely that the power requirements of heteronuclear decoupling will be reduced by an order of magnitude by using composite pulses which provide accurate, rapid, wide-band population inversion. The possible application of the new sequence to composite pulse decoupling methods will be discussed later.

THEORY

The compensation of radiofrequency inhomogeneity effects, as treated in Part I, is much simpler than the compensation of resonance offset effects. For a pulse sequence applied with an inhomogeneous radiofrequency field, magnetization vectors originating from different parts of the sample volume nutate about rotating-frame effective fields which differ only in their strength and not in their direction (providing the pulse is applied close to resonance). In contrast, when there is a distribution of resonance frequencies of width comparable with the irradiation field strength, the effective fields experienced by the various transitions differ both in magnitude and in direction. For a transition irradiated far off resonance, the rotating-frame effective field is tilted appreciably away from the equatorial plane and is also enhanced in magnitude. This condition applies to all pulses in a sequence, so that not only the rotation angles but even the geometrical relationship between the nutation axes of adjacent pulses differs from resonance to resonance in a wide spectrum. This makes it very difficult to ensure that the pulse sequence produces the same net desirable effect on a wide range of transitions. To add to our difficulties, the degree of off-resonance effects may be in practice quite large so that first-order approximations are insufficient. Nevertheless we will start by finding a pulse sequence which compensates off-resonance effects to first order and then try to improve it so as to compensate for at least some of the higher-order effects. Because of the complicated nature of the problem we will eventually have to be content with finding a rough solution by approximate analysis, followed by computer-aided optimization of the variable parameters.

The convention for pulse operators will be the same as that used in Part I. Consider a pulse in which a field of magnitude ω_1^0 rad sec^{-1}, rotating at a radiofrequency ω rad sec^{-1}, is applied to an ensemble of isolated spins located in a strong magnetic field applied along the z axis of a coordinate system in which they have a natural precession frequency ω_0 rad sec^{-1}. Including an arbitrary phase angle ϕ for the radiofrequency field, the laboratory frame Hamiltonian $\mathscr{H}^{\mathrm{L}}$ may be expressed (*7*)

$$\mathscr{H}^{\mathrm{L}} = -\omega_0 I_z - \omega_1^0 e^{i(\omega t-\phi)I_z} I_x e^{-i(\omega t-\phi)I_z}. \quad [1]$$

Analogous expressions can be written for more general transitions in terms of single-transition operators (*8*, *9*).

Transforming into the frame rotating at frequency ω the Hamiltonian becomes

$$\mathscr{H} = -\Delta\omega I_z - \omega_1^0 e^{-i\phi I_z} I_x e^{i\phi I_z},$$

in which we have the resonance offset $\Delta\omega = \omega_0 - \omega$ in angular frequency units. The same Hamiltonian can be rewritten in terms of the tilted effective field

$$\mathscr{H} = -\omega_e e^{-i\phi I_z} e^{i\theta I_y} I_x e^{-i\theta I_y} e^{i\phi I_z}.$$

This represents a rotating-frame effective field tilted away from the equatorial plane by an angle

$$\theta = \tan^{-1}(\Delta\omega/\omega_1^0) \text{ radians}$$

and of augmented magnitude

$$\omega_e = [(\omega_1^0)^2 + (\Delta\omega)^2]^{1/2} = \omega_1^0 \sec\theta \qquad \text{rad sec}^{-1}.$$

For offsets $\Delta\omega$ of positive sign (irradiation at a frequency lower than the transition frequency), the effective field is tilted above the xy plane. For offsets of negative sign the effective field is tilted below that plane.

Typically the radiofrequency field is applied for a time t intended to provide some desired flip angle α^0 under ideal conditions

$$\alpha^0 = \omega_1^0 t.$$

In the presence of off-resonance effects the pulse causes an evolution of the density matrix σ of the form

$$\sigma(t) = R_\phi(\alpha^0)^{-1}\sigma(0)R_\phi(\alpha^0), \quad [2]$$

where

$$R_\phi(\alpha^0) = e^{-i\phi I_z} e^{i\theta I_y} e^{-i\alpha(\theta) I_x} e^{-i\theta I_y} e^{i\phi I_z} \quad [3]$$

and

$$\alpha(\theta) = \omega_e t = \alpha^0 \sec\theta.$$

Relaxation and other effects such as coupling interactions have been neglected, this implying that the following approach is only valid for sequences completed in times much smaller than the relaxation times, or the inverse of interaction strengths.

The operator $R_\phi(\alpha^0)$ is the rotation operator for a pulse of phase ϕ and nominal flip angle α^0. With the definition given above, the operator for a sequence of pulses can conveniently be written as the product of the individual pulse operators, from left to right in chronological order. As usual we will denote these operators for the common cases $\phi = 0$, $\pi/2$, π, and $3\pi/2$ by $R_x(\alpha^0)$, $R_y(\alpha^0)$, $R_{-x}(\alpha^0)$, and $R_{-y}(\alpha^0)$, respectively. Note that in the presence of off-resonance effects, the inverse of the operator $R_\phi(\alpha^0)$ is not equal to $R_{\pi+\phi}(\alpha^0)$.

Equation [3] shows clearly the twofold dependence of the rotation operator on the resonance offset. The direction of the effective field, as expressed by the operators $\exp(\pm i\theta I_y)$, is dependent on the offset as well as the nutation angle $\alpha(\theta)$. As the latter effect is the smaller at low offsets we will often tend to ignore it as a first-order approximation.

Three-Pulse Sequences

We will now use such rotation operators to analyze three-pulse sequences of the form $R_x(\pi/2)R_y(\alpha^0)R_x(\pi/2)$, which already have been demonstrated to give a good population inversion compensated for off-resonance effects (*10–12*). We will employ a notation in which $\sigma(+n;\,\theta)$ represents the density matrix after the nth pulse in a given sequence, expressed as function of off-resonance tilt angle θ. The initial density matrix $\sigma(0)$ is assumed to be in thermal equilibrium and will be expressed in reduced form

$$\sigma(0) = I_z\,.$$

Consider first the action of the initial $R_x(\pi/2)$ pulse on the equilibrium z magnetization. The pulse is represented by the rotation operator

$$R_x(\pi/2) = e^{i\theta I_y}e^{-i\alpha(\theta)I_x}e^{-i\theta I_y},$$

in which here

$$\alpha(\theta) = (\pi/2)\sec\theta.$$

Applying this to the initial density matrix according to Eq. [2] leads to

$$\sigma(+1;\,\theta) = R_x(\pi/2)^{-1}\sigma(0)R_x(\pi/2) \quad [4]$$

$$= I_x\sin\theta\cos\theta\,(1-\cos\alpha(\theta)) + I_y\sin\alpha(\theta)\cos\theta + I_z(\cos\alpha(\theta)\cos^2\theta + \sin^2\theta). \quad [5]$$

Discarding some second-order and higher terms in θ:

$$\sigma(+1;\,\theta) \cong I_x\left\{1 + \left(\frac{\pi}{4} - \frac{1}{2}\right)\theta^2\right\}\sin\theta + I_y\cos\theta + I_z\left(1 - \frac{\pi}{4}\right)\theta^2.$$

It can be seen that to a good approximation the magnetization vector after the first pulse lies close to the equatorial plane of the rotating frame and at an angle of θ radians from the y axis (vector **C** in Fig. 1a). The fact that the z component of this vector almost vanishes already has been noted (*11*, *13*).

In what follows we will assume that for all practical resonance offsets, the density

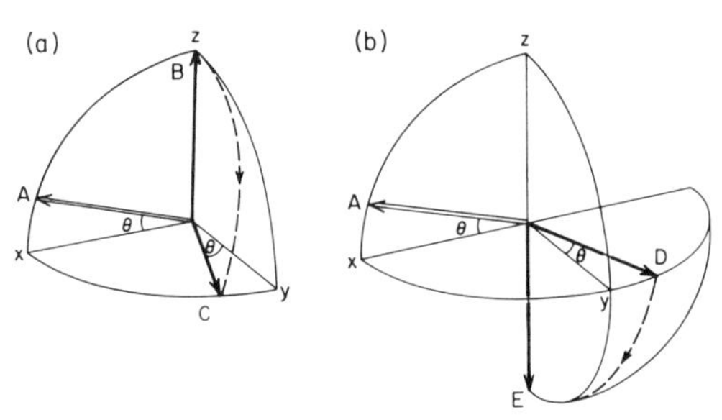

FIG. 1. (a) Three-dimensional projection of the trajectory of the tip of a typical magnetization vector during an off-resonance pulse. For clarity, only one octant of the sphere which is usually used to represent the rotating reference frame is shown. The off-resonance pulse $R_x(\pi/2)$ induces a rotation about the axis A, which is in the xz plane but tilted above the x axis by an angle θ radians. This rotation takes the vector from its starting position B along the z axis to a position C, close to the xy plane and subtending an angle also close to θ radians with the y axis. (b) Here the same pulse $R_x(\pi/2)$, producing a similar rotation about tilted axis A, takes a vector from the symmetrically related position D to a position E at the $-z$ axis. It is shown in the text that C and D are related by a formal rotation of π radians around the y axis.

matrix after the first pulse, $\sigma(+1; \theta)$, is given exactly by that represented by vector **C** in Fig. 1a. Thus we will assume that

$$\sigma(+1; \theta) \cong I_x \sin \theta + I_y \cos \theta.$$

This is a convenient assumption because it means that although the directions and magnitudes of the rotating-frame effective fields and the magnetization vectors are offset dependent, there remains a relatively simple geometrical relationship between them. This allows the use of straightforward geometrical arguments in calculating the effect of certain pulse sequences independently of offset.

We will not now treat the second pulse in the sequence but move immediately to the last pulse which produces a rotation identical to the first. However, this time we know that the trajectory of the magnetization vector should ideally terminate at the $-z$ axis. We will use this fact to deduce the desirable position of the magnetization vector *before* the last pulse by working backward. Thus for our three pulse sequence, ideally,

$$\sigma(+3; \theta) = R_x(\pi/2)^{-1}\sigma(+2; \theta)R_x(\pi/2) \cong -I_z\,,$$

with the solution for the density matrix before the last pulse

$$\sigma(+2; \theta) = -R_x(\pi/2)I_zR_x(\pi/2)^{-1}. \quad [6]$$

Now we can write a general property of rotations about any axis in the xz plane

$$R_x(\alpha^0) = e^{-i\pi I_y}R_x(\alpha^0)^{-1}e^{+i\pi I_y}.$$

This may be deduced from the form of the rotation operator

$$\begin{aligned} e^{-i\pi I_y}R_x(\alpha^0)^{-1}e^{+i\pi I_y} &= e^{-i\pi I_y}e^{i\theta I_y}e^{+i\alpha(\theta)I_x}e^{-i\theta I_y}e^{+i\pi I_y} \\ &= e^{i\theta I_y}e^{-i\alpha(\theta)I_x}e^{-i\theta I_y} \\ &= R_x(\alpha^0). \end{aligned}$$

Then solving for the density matrix before the last pulse, using Eqs. [4] and [6], we have

$$\sigma(+2;\,\theta) = -e^{-i\pi I_y}R_x(\pi/2)^{-1}e^{+i\pi I_y}I_z e^{-i\pi I_y}R_x(\pi/2)e^{+i\pi I_y}$$
$$= e^{-i\pi I_y}R_x(\pi/2)^{-1}I_z R_x(\pi/2)e^{+i\pi I_y}$$
$$= e^{-i\pi I_y}\sigma(+1;\,\theta)e^{+i\pi I_y}.$$

This clearly shows that ideally the positions of the magnetization vector before and after the second pulse in the sequence should be related by a mathematical rotation through π radians about the y axis. Thus if it assumed that $\sigma(+1;\,\theta)$ may be represented by vector **C** in Fig. 1a, then after the second pulse the density matrix should ideally be given by

$$\sigma(+2;\,\theta) \cong I_y \cos\theta - I_x \sin\theta$$

as represented by vector **D** in Fig 1b. As shown in the equations above, and demonstrated in Fig. 1b, population inversion can then be completed by applying another pulse $R_x(\pi/2)$ identical to the first, rotating the vector about the tilted axis A in Fig. 1.

The task remains of actually transforming vectors **C** into the symmetrically related vectors **D** independent of resonance offset. If resonance offset effects could be neglected, then clearly a pulse $R_y(\pi)$ would suffice. Indeed this is the basis of the compensation of radiofrequency inhomogeneity effects by the $R_x(\pi/2)R_y(\pi)R_x(\pi/2)$ sequence (*1*, *10*). However, in the present case rotations about the exact $+y$ axis are inaccessible since resonance offset effects cause a tilting of the rotation axis away from the xy plane. The rotation axis F which is made accessible by pulses $R_y(\alpha^0)$, is shown in Fig. 2b, together with the axis G for pulses $R_{-y}(\beta^0)$ which will be used later. These axes are to be compared with the desirable positions of the magnetization vectors before and after the central pulse, Fig. 2a. A positive sign of $\Delta\omega$ has been assumed in these diagrams. It is a remarkable and useful fact that the displacements of vectors **C** and **D**, and rotation axes F and G, away from the y axis are all identical, as shown more clearly in a view along the y axis, Fig. 2c.

We will now attempt to take magnetization vectors from positions C to positions D by a rotation about the axis F, implemented by a pulse $R_y(\alpha^0)$. The nominal flip angle α^0 which is required is given as a function of offset by the solution of the equations

$$R_y(\alpha^0)^{-1}(I_y \cos\theta + I_x \sin\theta)R_y(\alpha^0) = I_y \cos\theta - I_x \sin\theta,$$

with $R_y(\alpha^0)$ given by

$$R_y(\alpha^0) = e^{-i\theta I_x}e^{-i\alpha(\theta)I_y}e^{+i\theta I_x}.$$

Straightforward algebra shows that the rotation angle $\alpha(\theta)$ must obey the equation

$$\sin\alpha(\theta) + (1 - \cos\alpha(\theta))\cos\theta = 0, \qquad [7]$$

with the nominal flip angle α^0 differing from $\alpha(\theta)$ by a further factor $\cos\theta$.

If second-order terms in θ are neglected then the solution is

$$\alpha^0 \cong \alpha(\theta) \cong (2m + 1)\pi + \pi/2,$$

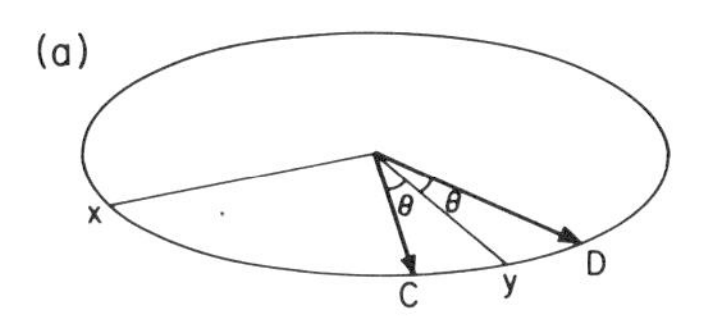

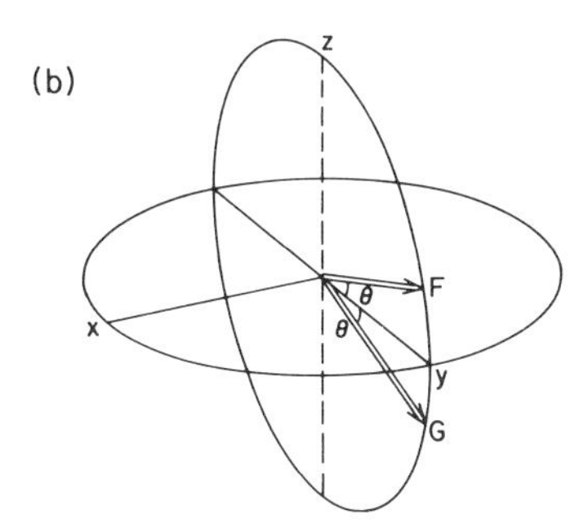

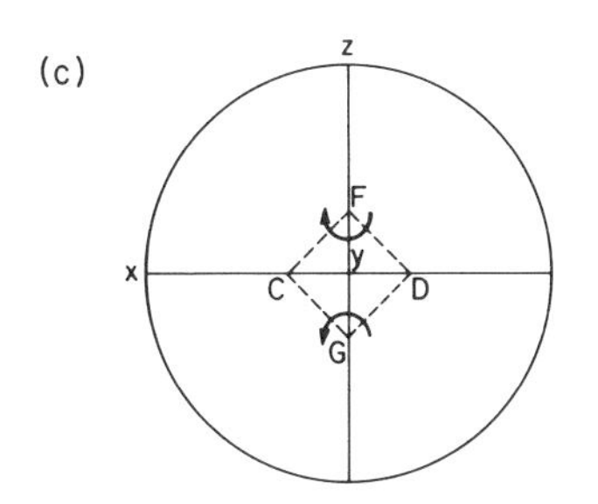

FIG. 2. Magnetization vectors and rotation axes involved in three- and five-pulse sequences. (a) The equatorial plane of the rotating frame with the symmetrically related positions C and D. Ideally the magnetization vector should be close to C after the first rotation, and close to D before the last rotation of the composite pulse. (b) Rotation axes accessible to pulses applied in the y or $-y$ directions, in the presence of off-resonance effects. $R_y(\alpha^0)$ pulses induce clockwise rotations about the tilted axis F, and $R_{-y}(\beta^0)$ pulses induce counterclockwise rotations about the tilted axis G. Both F and G lie in the yz plane, with F above and G below the equator. A positive sign of $\Delta\omega$ is assumed. (c) Ideal positions C and D, and off-resonance rotation axes F and G, viewed from along the y axis of the rotating frame. The sense of accessible rotations about F and G is shown. If the curvature of the unit sphere is neglected, the four vectors form a square of side $\theta 2^{1/2}$ units.

where m is an integer, and neglecting the false solutions $\alpha(\theta) = 2m\pi$. In a practical experiment only positive values of α^0 are accessible, which gives as the first solution $\alpha^0 = 3\pi/2$ radians, independent of offset θ. The same conclusion that to first order the sequence $R_x(\pi/2)R_y(3\pi/2)R_x(\pi/2)$ provides offset-insensitive population inversion was reached elsewhere by a different route (*12*).

A physical interpretation of this is given in Fig. 3. Here the region in the neighborhood of the y axis has been shown in expanded form, as viewed along the y axis. For very small offsets, all points C, D, and F are close to each other so that it is possible to neglect the curvature of the sphere in the vicinity and treat the problem by simple planar geometry. As was shown above, the displacements of

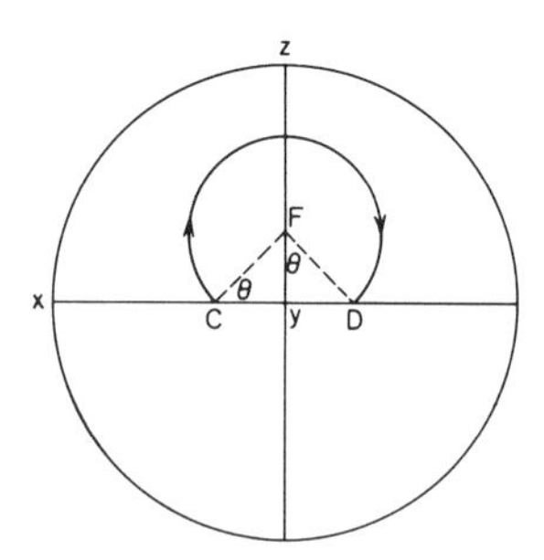

FIG. 3. The vector **C** can be converted into the symmetrically related vector **D** via a rotation through $3\pi/2$ radians about the tilted axis F. This corresponds to the central pulse of a $R_x(\pi/2)R_y(3\pi/2)R_x(\pi/2)$ sequence. Curvature of the sphere and the offset-dependence of the true rotation angle have been ignored, so that this picture is only accurate for very small resonance offsets.

vectors **C**, **D**, and **F** from the point of intersection of the y axis with the surface of the unit sphere are all directly proportional to the angle θ. It follows that the angle $C\hat{F}D$ is close to $\pi/2$ radians for small offsets, and that a clockwise rotation of $3\pi/2$ radians about the axis F, as produced by a pulse $R_y(3\pi/2)$, should take magnetization vectors from C to the desired position D. Because the relationship between the vectors remains constant for small tilt angles θ, the desired flip angle is independent of resonance offset, provided this is small.

However, a different answer is reached for a practical offset range. In this case the equation for the rotation angle, Eq. [7], must be solved exactly. The accurate solution is

$$\alpha(\theta) = (2m + 1)\pi + 2 \tan^{-1} (\cos \theta), \qquad [8]$$

where m is an integer. The desired rotation angle is no longer offset independent. The first solution of Eq. [8] is plotted as a function of absolute offset parameter $\tan \theta = \Delta\omega/\omega_1^0$ in Fig. 4 (curve (a)). Also shown is the nominal flip angle α^0, which should actually be used in a rotation $R_y(\alpha^0)$ so as to transform vector **C** into vector **D**. Because of the enhanced strength of the tilted effective field, this angle differs from $\alpha(\theta)$ by a further factor $\cos \theta$. It can be seen that the desirable values for both angles are always less than the first-order estimate 270 degrees and have a strong offset dependence. For example, for an absolute offset parameter $\tan \theta = 0.5$, vectors **C** and **D** are only connected by a rotation through an angle 263 degrees about the tilted y axis, requiring a pulse $R_y(\alpha^0)$ of nominal flip angle $\alpha^0 = 236$ degrees (1.31π radians). For a large offset $\tan \theta = 1.0$, a nominal flip angle as low as $\alpha^0 = 177$ degrees would be required (0.98π radians).

In practice a compromise solution must be reached for a given range of offsets. Computer simulation showed that a pulse sequence $R_x(\pi/2)R_y(\alpha^0)R_x(\pi/2)$ with $\alpha^0 = 240$ degrees (1.33π radians), accomplishes a good offset-insensitive population inversion over a range of offsets $-0.5 \leqslant \tan \theta \leqslant 0.5$. To compensate larger offsets the flip angle of the central pulse must be reduced as is apparent from Fig. 4. If the central pulse is shortened to $R_y(\pi)$ then good inversion is achieved around offsets $\tan \theta \cong 0$ or ± 0.9. However, population inversion is now unsatisfactory away from these values, as can be seen from the experimental results (Fig. 6).

Another conclusion to be drawn from Fig. 4 is that the most important effect

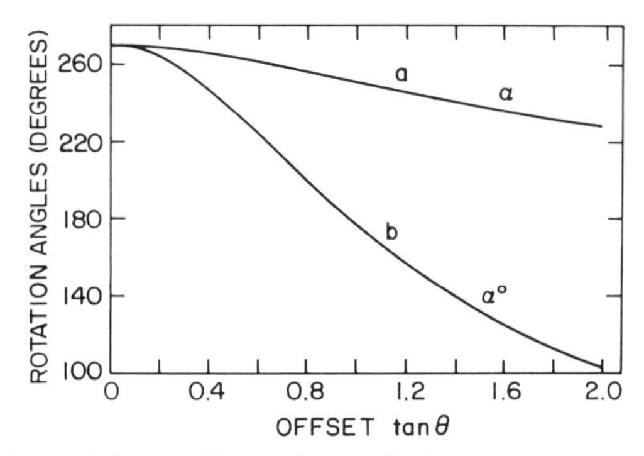

FIG. 4. Offset-dependence of the rotation angles required to convert vectors **C** into symmetrically related vectors **D** via rotations about tilted axes F. The resonance offset has been plotted in absolute units $\Delta\omega/\omega_1^0$ and the angles presented in degrees ($3\pi/2$ radians = 270°). (a) Offset-dependence of the true rotation angle α which is required, taking into account the increased effect of curvature of the sphere for large offsets. (b) The nominal flip angle α^0 of a pulse $R_y(\alpha^0)$ which would have to be applied in order to produce this true rotation angle α. This curve includes the effect of augmented effective field strength at large offsets.

preventing offset-insensitive transformation of vectors **C** into vectors **D** is the enhanced strength of the tilted effective field at appreciable offsets. This is responsible for the large divergence between curve a in Fig. 4, representing the desirable rotation angles around the tilted axis, and curve b, representing the required experimental flip angles. Accordingly, in trying to improve the three-pulse sequence, we will concentrate on achieving some measure of compensation of this effect. Thus we will continue to neglect the effect of nonplanar geometry on the rotation angles, since it is apparent from Fig. 4 that this has a comparatively minor influence. The assumption of planar geometry in the vicinity of the y axis makes visualization of the pulse sequence much easier. We will also continue to assume that after the first pulse all vectors end up exactly at positions C, and that if they are located at positions D before the last pulse, that final rotation will take them all precisely to the south pole, as required.

Five-Pulse Sequences

We will now demonstrate that by replacing the central pulse of a three-pulse sequence $R_x(\pi/2)R_y(\alpha^0)R_x(\pi/2)$ by a sequence $R_y(\alpha^0)R_{-y}(\beta^0)R_y(\alpha^0)$, the major effect responsible for imperfect performance of the three-pulse sequence, the offset dependence of the true rotation angles, can be much reduced. This leads to five-pulse sequences of the form $R_x(\pi/2)R_y(\alpha^0)R_{-y}(\beta^0)R_y(\alpha^0)R_x(\pi/2)$. We will derive approximate values for the nominal flip angles α^0 and β^0 by geometrical arguments to serve as a basis for optimization with the aid of computer simulation.

Consider again the region of the y axis, with the ideal positions of magnetization vectors after the first pulse (C), and before the last one (D), and the points F and G at which the tilted effective field axes for pulses $R_y(\alpha^0)$ and $R_{-y}(\beta^0)$ intersect the unit sphere (Fig. 5a). Neglecting deviations from planar geometry, these four points form a square of side $\theta 2^{1/2}$ units. Rotation arcs around the tilted axes F and G are

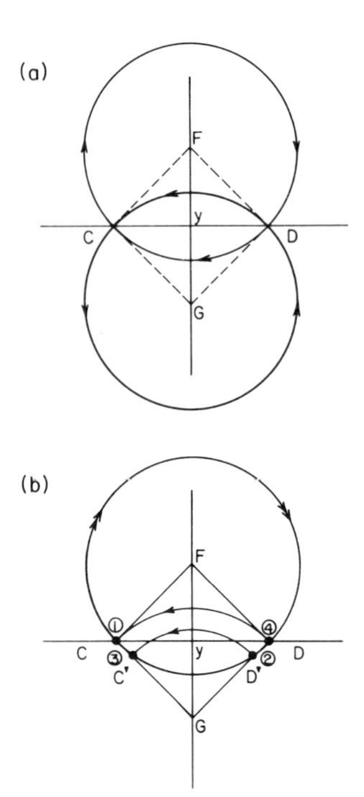

FIG. 5. Operation of the central pulses of a five-pulse sequence $R_x(\pi/2)R_y(3\pi/2)R_{-y}(\pi/2)R_y(3\pi/2)R_x(\pi/2)$. (a) Another view of the area around the y axis, with the rotation axes F and G, and the ideal positions of magnetization vectors, **C** and **D**. Pulses $R_y(\alpha^0)$ rotate vectors in a clockwise direction around F, pulses $R_{-y}(\beta^0)$ rotate vectors in an anticlockwise direction around G, assuming a positive sign of offset. The diagram shows that if tangents to the arcs are drawn at points C and D, they are mutually perpendicular and form radii of the other rotation arc. (b) A vector starting at C, and rotated around axis F by a pulse $R_y(3\pi/2)$, ends up near D. The principal error is the increased rotation angle, so that it ends up rather closer to D', being displaced from D along the radius DG, to a good approximation. A rotation $R_{-y}(\pi/2)$ acting about the axis G and assumed to be ideal transforms this into a symmetrically related position C'. A second enhanced rotation $R_y(3\pi/2)$ around F now places the vector much closer to D. The circled numbers should clarify the chronological sequence of positions occupied by the magnetization vector.

also drawn passing through points C and D. It must be remembered that for positive signs of $\Delta\omega$, only clockwise rotations about F are allowed (representing pulses $R_y(\alpha^0)$), and counterclockwise rotations around G (representing pulses $R_{-y}(\beta^0)$). It may be noted that as a consequence of the planar geometry, the tangents to the arcs around F and around G are mutually perpendicular where the arcs intersect at points C and D. This will be of some use.

Now consider a vector placed at point C by the initial $R_x(\pi/2)$ pulse. As was seen above, a pulse $R_y(3\pi/2)$ will take the tip of the vector in an arc around the axis F, ending up close to position D, if high-order effects are omitted. As was seen before, the principle error in this process is due to the enhanced rotation angle around the tilted effective field F. This error can be treated in a similar way to the distribution of rotation angles caused by radiofrequency inhomogeneity (*1*). As a first-order approximation, the deviation from the ideal position D at this time is along a tangent to the arc and hence is in the direction DG. We assume that the vector ends up close to point D' in Fig. 5b.

The expected size of the deviation DD' gives an estimate of the nonideality of the sequence as a whole. This deviation should be proportional to both the error in the rotation angle and the arc length of the trajectory. Accordingly it is of approximate magnitude

$$\epsilon \cong (3\pi/2)(\sec\theta - 1)\theta 2^{1/2}.$$

We will follow this by a pulse $R_{-y}(\pi/2)$, which produces a counterclockwise rotation about the axis G. As a first approximation we will assume that the rotation angle about the tilted axis G is not offset dependent and is always equal to $\pi/2$ radians. As can be seen from Fig. 5b, such a rotation takes the tip of the magnetization vector from point D' to point C', close again to the starting position C but displaced so as to be symmetrically related to D'. Symmetry is preserved because the error vector $\mathbf{DD'}$, which is along a tangent to the arc around F, is also directed along the radius DG. Hence the vector $\mathbf{CC'}$ is also along a radius CG and is again at a tangent to the arc around F. Thus the vector is now in such a position that if a rotation $R_y(3\pi/2)$ identical to the first is applied, the errors compensate each other and the vector ends up much closer to D than after the first rotation. Hence if errors due to deviations from nonplanar geometry and due to increased effective field strength during the central pulse are ignored, the sequence $R_y(3\pi/2)R_{-y}(\pi/2)R_y(3\pi/2)$ takes vectors from ideal positions C to ideal positions D, with compensation of errors in the outer rotations.

If the offset dependence of the rotation angle during the central pulse is now taken into account, the symmetry of Fig. 5b is broken somewhat, and compensation is incomplete. However, the errors produced should only be about a third of what they were before. This is because the arc length of the trajectory during the central $R_{-y}(\pi/2)$ rotation is only a third of the arc length of the trajectories during the $R_y(3\pi/2)$ rotations.

If it is wished to analyze this problem to higher than first order, there is a whole multitude of error terms which must be taken into account. Instead of trying to analyze these terms the first-order solution $R_x(\pi/2)R_y(3\pi/2)R_{-y}(\pi/2)R_y(3\pi/2)R_x(\pi/2)$ was simply used as a basis for optimizing the lengths of the central pulses with the aid of accurate computer calculations of the motions of magnetization vectors. As in the three-pulse sequence, it was found advisable to reduce the flip angles of the central pulse below the first-order recommendation. The modified sequence $R_x(\pi/2)R_y(\alpha^0)R_{-y}(\beta^0)R_y(\alpha^0)R_x(\pi/2)$ with $\alpha^0 = 1.12\pi$ radians (200 degrees) and $\beta^0 = 0.44\pi$ radians (80 degrees) was found to be a near-optimum solution, and was predicted to provide an accurate population inversion for all offsets in the range $-0.9 \leqslant \tan\theta \leqslant 0.9$.

Positive and Negative Resonance Offsets

As all the pulse sequences described in this paper are completely symmetrical in time, the population inversion they produce is independent of the sign of the resonance offset, as was proved previously (1). Note that this conclusion remains true even when the sequence is performing very imperfectly.

EXPERIMENTAL

Mr. Tom Frenkiel of the Physical Chemistry Laboratory, Oxford, has kindly made available some experimental results which verify the conclusions of the preceding discussion. The degree of population inversion achieved by various composite pulses was measured as a function of offset in the course of evaluating suitable sequences for composite pulse decoupling (*3–5*). For this reason the test resonance was the proton signal from dimethyl carbonate, which was excited and detected through the decoupler coil of an XL-200 spectrometer. The use of the decoupler coil was responsible for radiofrequency inhomogeneity effects which may be discerned in the experimental results.

The degree of population inversion was measured as described in the previous publication (*1*). The method used was to measure the partially inverted longitudinal magnetization produced by a composite pulse applied to an equilibrium spin system. Equilibrium was established by an adequate delay ($\geq 5T_1$), and a composite pulse was applied. The longitudinal magnetization left by the composite pulse was then measured by applying almost immediately a single $R_x(\pi/2)$ "read" pulse. The precessing transverse magnetization so produced induces a signal whose amplitude, after certain precautions to be discussed below, gives a direct measure of the longitudinal magnetization left by the composite pulse.

The frequency of the "read" pulse was set to be always exactly on resonance, while the offset frequency of the composite pulse was varied in a series of experiments between 0 and 15 kHz. (The insensitivity of the population inversion to the sign of the resonance offset was checked for some individual cases.) To allow the phase lock to stabilize after the frequency jump, a delay of 1 msec was left between the composite pulse and the read pulse. This was sufficiently short in comparison with the spin–lattice relaxation time, which was measured as 260 msec. Within the composite pulse itself it is particularly important for proper offset compensation that all delays are minimized to prevent unwanted free precession. In our experiments the delays were less than 0.5 μsec.

A phase-cycling scheme was used to suppress any contributions from transverse magnetization remaining after the composite pulse: The experiment was repeated with the phases of all components of the composite pulse inverted, but the phase of the "read" pulse and the receiver reference signal unchanged, and the signals added together. The resultant signal is proportional to longitudinal magnetization after the composite pulse alone. In principle this phase-cycling is usually unnecessary if a sufficient number of transients are averaged and if the composite pulse and the "read" pulse are applied at different frequencies so that there is no coherent relationship between their phases. However, since the composite pulse was sometimes also exactly on resonance, it was convenient to retain the phase-cycling scheme for all measurements.

The experimental results for a single conventional $R_x(\pi)$ pulse and three different composite pulses are shown in Fig. 6. The longitudinal magnetizations were all normalized by a comparison with the signal produced by a "read" pulse alone. The offset frequencies were also converted into absolute terms by division by the radiofrequency field strength of 9.3 kHz, determined by off-resonance decoupling.

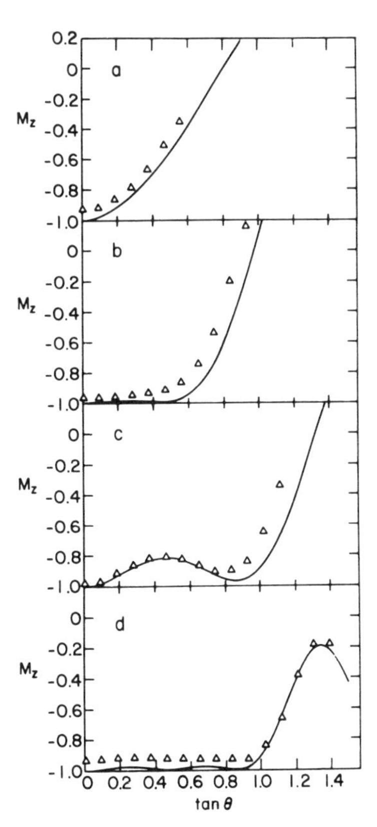

FIG. 6. Experimental test of the offset dependence of the population inversion produced by various pulse sequences, by the method described in the text. Partially inverted z magnetizations (normalized to -1.0 for full inversion) are plotted against absolute resonance offset ($\tan\theta = \Delta\omega/\omega_1^0$). Experimental points (triangles) are compared with computer-simulated curves (unbroken lines). (a) Single $R_x(\pi)$ pulse. (b) Composite pulse $R_x(\pi/2)R_y(4\pi/3)R_x(\pi/2)$. (c) Composite pulse $R_x(\pi/2)R_y(\pi)R_x(\pi/2)$. (d) Composite pulse $R_x(\pi/2)R_y(\alpha^0)R_{-y}(\beta^0)R_y(\alpha^0)R_x(\pi/2)$, with α^0 and β^0 set to the values recommended by computer-aided optimization, $\alpha^0 = 200$ degrees, $\beta^0 = 80$ degrees.

In all cases the partially inverted longitudinal magnetizations were compared with the results of accurate computer simulations of the motion of magnetization vectors under the sequence of tilted effective fields (smooth curves).

Experimental and computer-simulated results for a single conventional $R_x(\pi)$ pulse are compared in Fig. 6a. It is obvious from these results that the conventional method is very sensitive to offset effects. The small divergence between experimental and computer-simulated results may be attributed to radiofrequency inhomogeneity in the decoupler coil which was used for these measurements.

Results for composite pulses of the form $R_x(\pi/2)R_y(\alpha^0)R_x(\pi/2)$ are shown in Figs. 6b and c. For Fig. 6b, the flip angle α^0 was set to be $4\pi/3$ radians (240 degrees), so as to achieve compensation of offset effects in the range $-0.5 \leqslant \Delta\omega/\omega_1^0 \leqslant 0.5$. The experimental results verify this compensation and account for the success of this pulse in such applications as heteronuclear decoupling (*3*). When the resonance offset is small, there is less divergence between the experimental points and the theoretical curve as for the conventional pulse, Fig. 6a. This is to be expected since close to resonance, this sequence is also expected to compensate radiofrequency inhomogeneity effects (*11*).

If α^0 is reduced to π radians (180 degrees), compensation is improved at large offsets, as predicted by the above treatment and demonstrated in Fig. 6c. However, population inversion becomes unsatisfactory at intermediate offsets around $\Delta\omega/\omega_1^0 \cong 0.45$. This is because the central pulse is now too short to accomplish transformation between the symmetrically related ideal positions *C* and *D* of Fig. 3. Further from resonance, the effects of curvature of the sphere and augmented effective field strength assist this process. Over a wide range of offsets, agreement between experimental and theoretical curves is now found to be excellent. This is presumably because the $R_x(\pi/2)R_y(\pi)R_x(\pi/2)$ sequence accomplishes a good compensation of radiofrequency inhomogeneity (*10*).

In Fig. 6d the results for a five-pulse sequence of the form $R_x(\pi/2)R_y(\alpha^0) \times R_{-y}(\beta^0)R_y(\alpha^0)R_x(\pi/2)$ are shown, with α^0 and β^0 equal to the recommended values $\alpha = 200$ degrees, $\beta^0 = 80$ degrees. It can be seen that compensation of resonance offset effects is now excellent, the longitudinal magnetization after inversion being held below -0.9 units over an offset range $-0.95 \leq \Delta\omega/\omega_1^0 \leq 0.95$. By way of comparison, the radiofrequency field strength would have to be increased by a factor of at least 5, and the peak power by a factor of 25, in order to reproduce this range with a conventional single pulse. For this sequence, there is again a small divergence between the experimental results and the simulated curve, which may once more be attributed to radiofrequency inhomogeneity. Relaxation during the composite pulse may also be a factor here, since the pulse now occupies 200 μsec.

DISCUSSION

The five-pulse sequence proposed here should find wide application whenever spin inversion is required and off-resonance effects are a problem. An obvious example is the measurement of spin-lattice relaxation times in wide spectra such as those of ^{13}C, where an accurate population inversion can allow the safe use of rapid methods such as location of the null-point in the inversion-recovery curve (*10*). Spin-spin relaxation measurements should also benefit, especially when multiple refocusing must be used (*14*), and the conventional Meiboom-Gill compensation scheme (*15*) breaks down because of large pulse imperfections or the presence of spin-spin coupling (*12*, *16*). In this application it must always be remembered that the population inversion induced by a composite pulse may be accompanied by an indeterminate phase shift, which must be removed by always employing an even number of echoes (*12*). Two-dimensional spectroscopy (*17–19*) may also benefit from the use of composite pulses. Although most of the artifacts caused by imperfect pulses in two-dimensional experiments may be removed by phase-cycling schemes (*20*), this is not always the case (*21*), and composite pulses can then be useful (*22*). Even for those artifacts which may be removed by phase-cycling, the use of composite pulses should prevent the unwanted signals from sapping the intensity of the desirable peaks. Related problems can be observed in one-dimensional sequences such as INEPT (*23*) and treated in the same fashion (*24*).

A particularly promising application for offset-insensitive composite pulses is heteronuclear spin decoupling (*3–5*). However, in this case it may be that little is to be gained by using the five-pulse sequence as against the three-pulse sequence $R_x(\pi/2)R_y(\pi)R_x(\pi/2)$, whose utility already has been demonstrated in this field (*5*). As can be seen by comparison of Figs. 6(c) and d, the range of offset compensation of these two sequences is rather similar, the only major difference being the imperfect performance of the three-pulse sequence at intermediate offset values. However, this imperfection is not particularly important in composite pulse decoupling because long decoupling sequences can be constructed which compensate such errors very effectively (*4*, *5*). The five-pulse sequence also suffers the disadvantage of being almost twice as long as the three-pulse sequence, which tends to introduce problems in very low-power work (*5*). Accordingly the five-pulse sequence is not expected to be of much use in decoupling schemes. One possibility is that further adjustment of flip angles α^0 and β^0 may allow even more stretching of the usable offset range, again at the expense of accuracy of the population inversion. However, since the sequences used at present for composite pulse decoupling seem more than capable of achieving good low-power decoupling of protons over any conceivable present and future offset range, such adjustments seem hardly worthwhile.

In summary it can be said that by using the proper composite pulse it is now possible to achieve an excellent NMR population inversion even when the radiofrequency homogeneity is very poor or when it is impossible to avoid considerable offset effects. However, a combination of these two problems together, such as might occur in biomedical imaging (*25–26*), cannot yet be handled. This might be a direction for further research.

ACKNOWLEDGMENTS

The author would like to thank the Science Research Council of Great Britain for the provision of a NATO Overseas Fellowship, and Shimon Vega for his help and encouragement.

REFERENCES

1. M. H. Levitt, *J. Magn. Reson.* **48,** 234 (1982).
2. A. Abragam, "The Principles of Nuclear Magnetism," p. 65, Oxford Univ. Press (Clarendon), Oxford, 1961.
3. M. H. Levitt and R. Freeman, *J. Magn. Reson.* **43,** 502 (1981).
4. M. H. Levitt, T. A. Frenkiel, and R. Freeman, *J. Magn. Reson.* **47,** 328 (1982).
5. M. H. Levitt, T. A. Frenkiel, and R. Freeman, *J. Magn. Reson.*, in press.
6. U. Haeberlen and J. S. Waugh, *Phys. Rev.* **175,** 453 (1968).
7. C. P. Slichter, "Principles of Magnetic Resonance," 2nd ed., Chap. 2, Springer-Verlag, New York, 1978.
8. S. Vega, *J. Chem. Phys.* **68,** 5518 (1978).
9. A. Wokaun and R. R. Ernst, *J. Chem. Phys.* **67,** 1752 (1977).
10. M. H. Levitt and R. Freeman, *J. Magn. Reson.* **33,** 473 (1979).
11. R. Freeman, S. P. Kempsell, and M. H. Levitt, *J. Magn. Reson.* **38,** 453 (1980).
12. M. H. Levitt and R. Freeman, *J. Magn. Reson.* **43,** 65 (1981).
13. R. Freeman and H. D. W. Hill, *J. Chem. Phys.* **54,** 3367 (1971).
14. H. Y. Carr and E. M. Purcell, *Phys. Rev.* **94,** 630 (1954).

15. S. Meiboom and D. Gill, *Rev. Sci. Instrum.* **29,** 688 (1958).
16. R. Freeman and H. D. W. Hill, *in* "Dynamic NMR Spectroscopy" (L. M. Jackman and F. A. Cotton, Eds.), Chap. 5, Academic Press, New York, 1975.
17. W. P. Aue, E. Bartholdi, and R. R. Ernst, *J. Chem. Phys.* **64,** 2229 (1976).
18. G. Bodenhausen, R. Freeman, R. Niedermeyer, and D. L. Turner, *J. Magn. Reson.* **26,** 133 (1977).
19. R. Freeman and G. A. Morris, *Bull. Magn. Reson.* **1,** 5 (1979).
20. G. Bodenhausen, R. Freeman, and D. L. Turner, *J. Magn. Reson.* **27,** 511 (1977).
21. G. Bodenhausen and D. L. Turner, *J. Magn. Reson.* **41,** 200 (1980).
22. R. Freeman and J. Keeler, *J. Magn. Reson.* **43,** 484 (1981).
23. G. A. Morris and R. Freeman, *J. Am. Chem. Soc.* **101,** 760 (1979).
24. D. M. Thomas, M. R. Bendall, D. T. Pegg, D. M. Doddrell, and J. Field, *J. Magn. Reson.* **42,** 298 (1981).
25. P. C. Lauterbur, *Nature (London)* **242,** 190 (1973).
26. Various Authors, *Phil. Trans. Roy. Soc. London B* **289,** 379–559 (1980).

JOURNAL OF MAGNETIC RESONANCE **51**, 169–173 (1983)

Simplification of NMR Spectra by Filtration through Multiple-Quantum Coherence

A. J. SHAKA AND RAY FREEMAN

Physical Chemistry Laboratory, Oxford University, Oxford, England

Received September 21, 1982

One of the most useful applications from the pioneering work on pulse excitation and detection of multiple-quantum coherence (*1–9*) has been the simplification of high-resolution NMR spectra by selecting one desired feature while eliminating the rest. For example, Bodenhausen and Dobson (*10*) have made this the basis of methods to suppress singlet or doublet proton resonances, while Bax *et al.* (*11*) have used double-quantum coherence to single out signals from coupled pairs of carbon-13 spins in natural abundance while rejecting the much more intense signals from isolated carbon-13 nuclei.

Multiple-quantum coherence has a characteristic dependence on the relative phases of the radiofrequency pulses used to excite and reconvert it into observable magnetization. For instance, a *p*-quantum coherence is *p* times as sensitive to radiofrequency phase shifts as single-quantum coherence (*5*, *6*), and this can be made the basis of methods for separating the various orders of coherence (*5*, *12*, *13*). When a

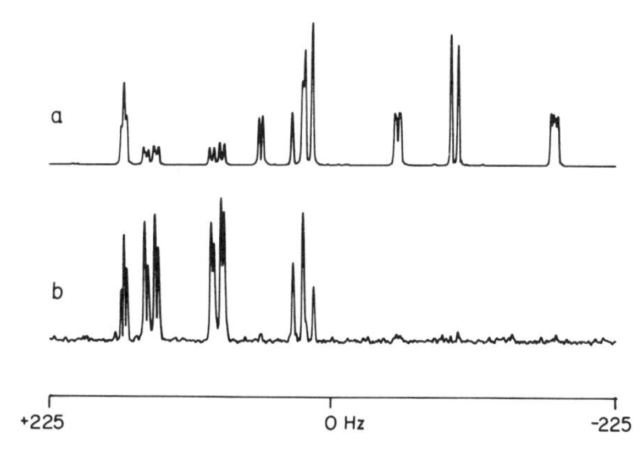

FIG. 1. (a) The 200-MHz proton NMR spectrum of a mixture of 2,3-dibromothiophene, 2-furoic acid and 1-bromo-3-nitrobenzene in acetone-d_6 displayed in the absolute-value mode. Symmetric Gaussian pseudoecho weighting has been used. The relaxation times have been shortened by adding a small amount of $Cr(acac)_3$. (b) The absolute-value spectrum of the same sample after filtering through four-quantum coherence. Signals from the two-spin and three-spin systems have been reduced by a large factor. Differential relaxation has reduced the relative intensity of the low-field multiplet. The slight changes in lineshape are due to the antiphase relationship between component lines.

0022-2364/83/010169-05$03.00/0

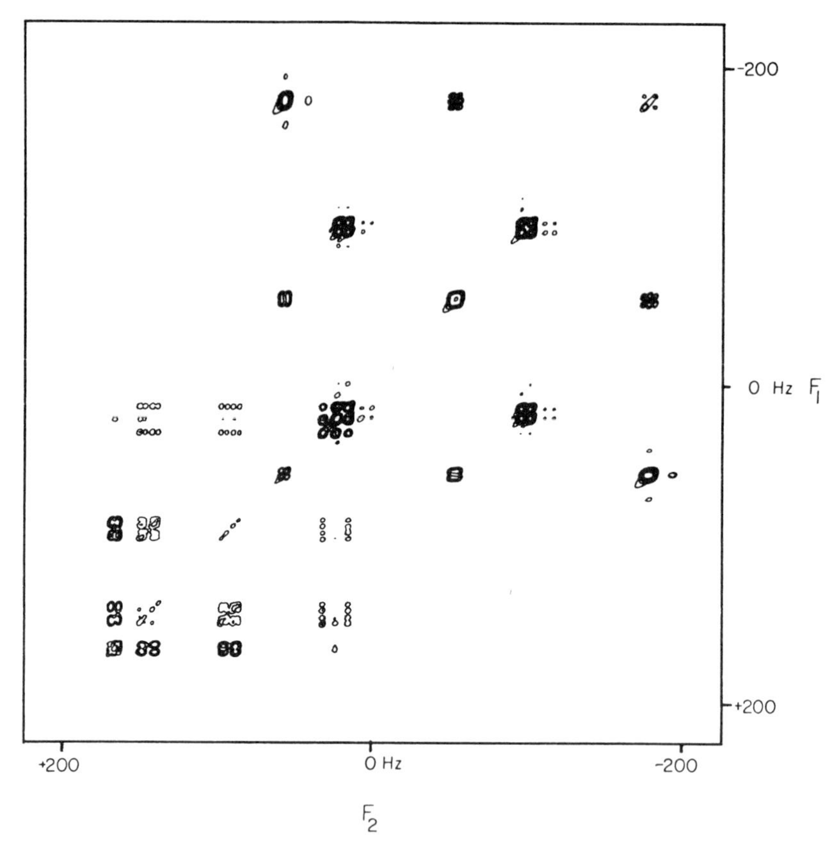

FIG. 2. Two-dimensional homonuclear correlation spectrum of the mixture displayed as a contour plot. Cross peaks symmetrically disposed about the slope 1 diagonal indicate spin–spin coupling. Spinning sidebands are apparent on some of the most intense peaks from the AX and AMX spin systems.

spectrum is a complicated superposition of several subspectra from different spin systems, considerable simplification is possible by generation of multiple-quantum coherence followed by immediate reconversion into observable magnetization. The simplification is achieved by selective detection of a particular order of coherence.

A system of $(n - 1)$ coupled spin-½ nuclei cannot give rise to coherences of order greater than $(n - 1)$, so that selective detection of signals derived from n-quantum coherence will eliminate the entire contribution of the $(n - 1)$ spin system from the observed spectrum. Only spin systems consisting of n or more coupled nuclei may contribute to the final spectrum, via a transient state of n-quantum coherence. While large values of n may offer the greatest simplification, there is a loss in signal intensity due to competing excitation of the more numerous lower order coherences so that sensitivity falls off rapidly with increasing order. For proton NMR this loss of sensitivity may well be an acceptable penalty to pay for the simplification achieved.

We illustrate the filtering technique by studying the 200-MHz proton spectrum of a mixture of 2,3-dibromothiophene (AX), 2-furoic acid (AMX) and 1-bromo-3-nitrobenzene (AMQX). These proton spectra show some overlap (Fig. 1a). Several methods of exciting multiple-quantum coherence are known; for simplicity we choose the two-pulse scheme, 90°–τ–90°, followed essentially immediately by a reconversion pulse:

$$90°(\phi)–\tau–90°(\phi)–\Delta–90°(0)–\text{Acquisition } (\psi).$$

In practice the delay Δ was set to 10 μsec to ensure a clean shift in radiofrequency phase. The radiofrequency phase ϕ was incremented in eight steps of $\pi/4$ radians from 0 to $7\pi/4$ radians (*14*), while the receiver reference phase ψ was alternated between 0 and π radians, selecting signals derived from four-quantum coherence (*5*). In addition the usual "CYCLOPS" phase cycling (*15*) was superimposed on the basic cycle.

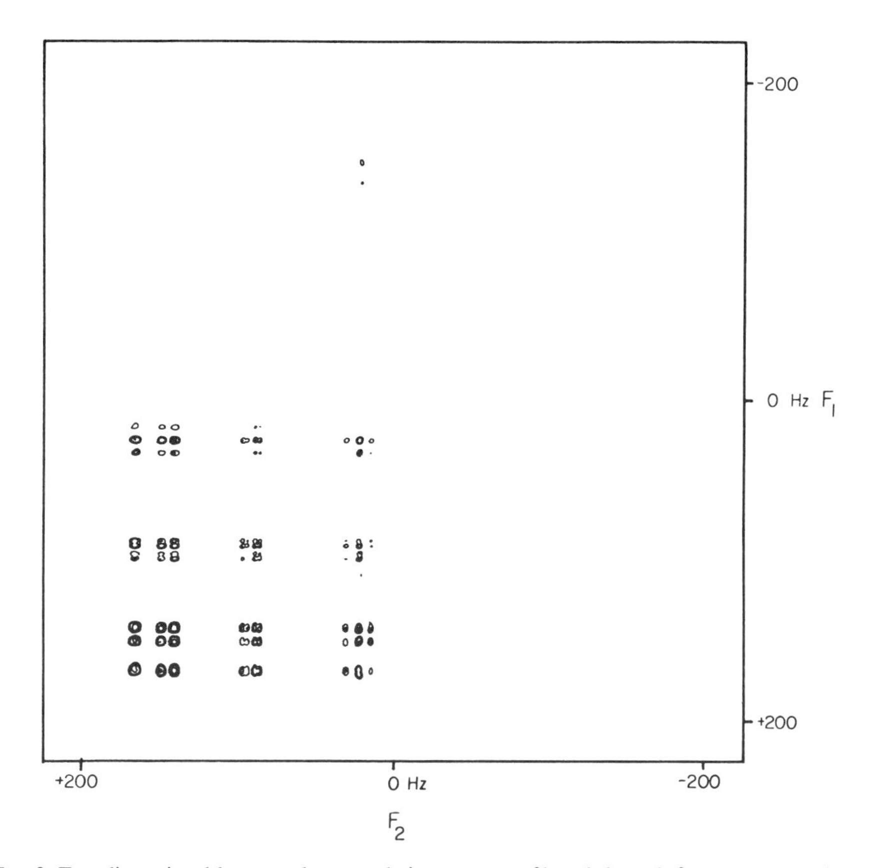

FIG. 3. Two-dimensional homonuclear correlation spectrum filtered through four-quantum coherence. Addition of a fixed delay of 0.3 sec to both the evolution and detection periods has ensured adequate signal intensity. The intense peaks from the two-spin and three-spin systems are eliminated, leaving the correlation spectrum of the weaker four-spin system.

The amount of multiple-quantum coherence formed by this sequence is a function of the interval τ and the coupling constants J_{ij} and offsets ω_i of the spins involved. For an even number n of weakly coupled inequivalent spins the total contribution to the n-quantum coherence from a spin k is proportional to $\cos(\omega_k\tau)\prod_{i\neq k}^{n}\sin(\pi J_{ik}\tau)$; if τ is inadvertently set near to $(2m+1)\pi/2\omega_k$ the rapidly oscillating first factor is small, giving poor sensitivity. This problem can be avoided by repeating the entire experiment with the modified sequence:

$$90°(\phi)\text{–}\tau\text{–}90°(\phi+\pi/2)\text{–}\Delta\text{–}90°(\pi/2)\text{–Acquisition }(\psi+\pi)$$

which gives a contribution proportional to $\sin(\omega_k\tau)\prod_{i\neq k}^{n}\sin(\pi J_{ik}\tau)$. (For an odd number of spins the sine and cosine terms are interchanged.) The two experiments, combined in quadrature, give a contribution proportional to $\exp(i\omega_k\tau)$, which has unit magnitude. This has the additional effect of selecting the coherence transfer echo (*1*, *16*) which may be further emphasized by symmetrical Gaussian pseudoecho weighting (*17*), giving Gaussian absorption lineshapes when the spectrum is displayed in the absolute-value mode. The absolute-value display is most convenient owing to the antiphase nature of peaks derived from multiple-quantum coherence. Application of the method with τ set to 0.3 sec gives adequate generation of four-quantum coherence and good suppression of the AX and AMX spectra, revealing the AMQX spectrum of 1-bromo-3-nitrobenzene (Fig. 1b).

The requirement that a particular value of τ be chosen is completely avoided in a two-dimensional version of the experiment, in which τ becomes the variable evolution period t_1. The sequence is then the original two-dimensional experiment described by Jeener (*3*, *18*) simplified by filtration through multiple-quantum coherence. Since small values of τ result in only weak excitation of multiple-quantum coherence, the two-dimensional experiment can be improved by the introduction of a fixed delay at the end of the evolution period and at the beginning of the detection period (*19*). Again selecting the four-quantum coherence by means of the phase-cycling, the two-dimensional spectrum of the mixture (Fig. 2) can be filtered to give the two-dimensional spectrum of 1-bromo-3-nitrobenzene (Fig. 3) with good suppression of the signals from the more intense two-spin and three-spin systems.

ACKNOWLEDGMENTS

This work was made possible by an equipment grant from the Science and Engineering Research Council and a Rhodes Scholarship (A.J.S.). Tom Frenkiel and James Keeler designed and constructed the phase shifter and provided useful comments.

REFERENCES

1. H. Hatanaka, T. Terao, and T. Hashi, *J. Phys. Soc. Jap.* **39,** 835 (1975).
2. H. Hatanaka and T. Hashi, *J. Phys. Soc. Jap.* **39,** 1139 (1975).
3. W. P. Aue, E. Bartholdi, and R. R. Ernst, *J. Chem. Phys.* **64,** 2229 (1976).
4. S. Vega, T. W. Shattuck, and A. Pines, *Phys. Rev. Lett.* **37,** 43 (1976).
5. A. Wokaun and R. R. Ernst, *Chem. Phys. Lett.* **52,** 407 (1977).
6. S. Vega and A. Pines. *J. Chem. Phys.* **66,** 5624 (1977).
7. M. E. Stoll, A. G. Vega, and R. W. Vaughan, *J. Chem. Phys.* **67,** 2029 (1977).
8. A. Wokaun and R. R. Ernst, *Mol. Phys.* **36,** 317 (1978).

9. G. Bodenhausen, *Prog. NMR Spectrosc.* **14,** 137 (1980).
10. G. Bodenhausen and C. M. Dobson, *J. Magn. Reson.* **44,** 212 (1981).
11. A. Bax, R. Freeman, and S. P. Kempsell, *J. Am. Chem. Soc.* **102,** 4849 (1980).
12. G. Drobny, A. Pines, S. Sinton, D. P. Weitekamp, and D. Wemmer, *Symp. Faraday Soc.* **13,** 49 (1979).
13. G. Bodenhausen, R. L. Vold, and R. R. Vold, *J. Magn. Reson.* **37,** 93 (1980).
14. T. Frenkiel and J. Keeler, *J. Magn. Reson.* **50,** 479 (1982).
15. D. I. Hoult and R. E. Richards, *Proc. R. Soc. London Series A* **344,** 311 (1975).
16. A. A. Maudsley, A. Wokaun, and R. R. Ernst, *Chem. Phys. Lett.* **55,** 9 (1978).
17. A. Bax, R. Freeman, and G. A. Morris, *J. Magn. Reson.* **43,** 333 (1981).
18. J. Jeener, Ampère International Summer School, Basko Polje, Yugoslavia, 1971.
19. A. Bax and R. Freeman, *J. Magn. Reson.* **44,** 542 (1981).

JOURNAL OF MAGNETIC RESONANCE **52**, 147–152 (1983)

COMMUNICATIONS

Correlation of Isotropic Shifts and Chemical Shift Anisotropies by Two-Dimensional Fourier-Transform Magic-Angle Hopping NMR Spectroscopy

AD BAX, NIKOLAUS M. SZEVERENYI, AND GARY E. MACIEL*

Department of Chemistry, Colorado State University, Fort Collins, Colorado 80523

Received November 29, 1982

During the 1960s Andrew and others examined the rapid spinning of a sample about an axis that makes an angle of 54° 44′ with the direction of the static magnetic field ($\mathbf{H}_0$) in order to remove broadening effects in the NMR spectra of solids (1–3). It was much later when Schaefer and Stejskal (*4*) applied this approach, magic-angle spinning (MAS), to remove broadening due to chemical shift anisotropy (CSA) in ^{13}C NMR, combining this approach with high-power ^{1}H decoupling and cross polarization (CP). The resulting levels of resolution and sensitivity obtained with this combination have made the ^{13}C CP–MAS experiment the most widely applied solid state NMR experiment in recent years.

As powerful, versatile, and popular as the ^{13}C CP–MAS experiment has become, there remain some characteristics that limit its usefulness in certain types of applications. Technological problems persist in techniques for spinning the sample rapidly, problems that are intensified by the scaling of CSA with increasing magnitude of the static field ($\boldsymbol{H}_0$), although recent advances show great promise for alleviating these problems (*5, 6*). Another limitation of the usual CP–MAS ^{13}C experiment is that it eliminates the potentially useful information embodied in the CSA pattern, i.e., independent values of the three principal elements of the shielding tensor, σ_{11}, σ_{22}, and σ_{33}. Only the trace, actually $(\sigma_{11} + \sigma_{22} + \sigma_{33})/3$, of the shielding tensor survives under MAS. Techniques have been proposed for retrieving CSA information from a MAS experiment (*7–11*); although each of these techniques has merits, each suffers from disadvantages.

Introduced here is a two-dimensional (2-D) Fourier transform (FT) technique which presents the isotropic average chemical shift, $\sigma_i = (\sigma_{11} + \sigma_{22} + \sigma_{33})/3$, in one frequency dimension (F_1) and the static CSA powder pattern along the other frequency axis (F_2). The experiment is carried out using discrete "hops" between evolution segments, rather than continuous sample spinning, and no spinning sidebands are produced. As the detection occurs on a static sample, the signal decays more rapidly than in a normal MAS experiment, and sensitivity suffers correspondingly. Nevertheless, the experiment shows considerable promise, not only for the CSA results it is capable of

* To whom correspondence should be addressed.

0022-2364/83 $3.00

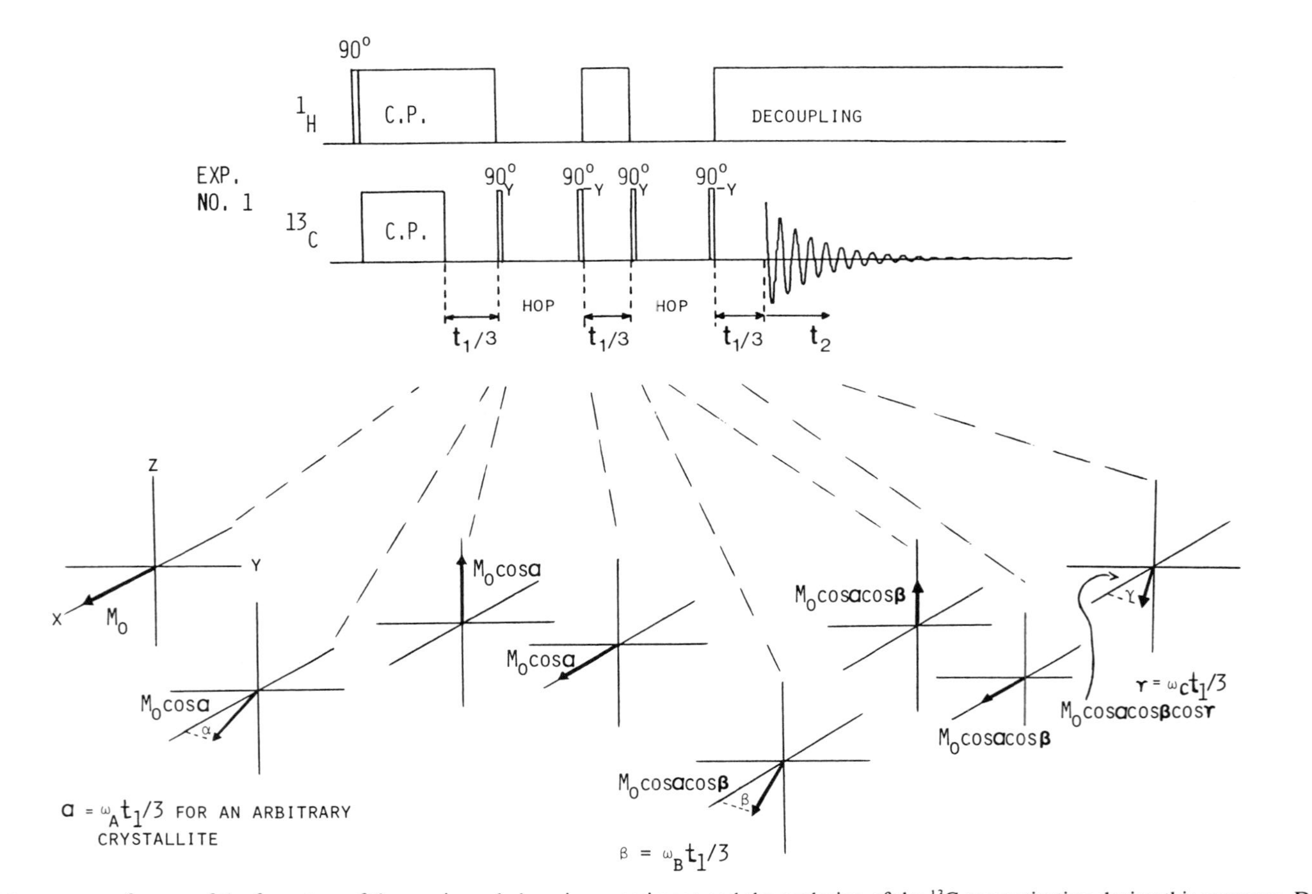

FIG. 1. Pulse sequence for one of the four steps of the magic-angle hopping experiment and the evolution of the ^{13}C magnetization during this sequence. During the time intervals indicated with "hop" the sample is rotated by 120° about the magic-angle axis. During the "hop" period, proton decoupling is switched off, and transverse magnetization defocuses rapidly. The results of three similar experiments with the different ^{13}C rf phases given in Table 1 are combined as described in the text.

providing in its present form, but also for the class of new types of future experiments for which it can serve as a prototype.

The experiments were carried out at 25.0 MHz (^{13}C) on a home-built spectrometer, employing a Nicolet 1180 data system and a wide-bore 3.7 T Nalorac magnet operating at 2.3 T. The probe contains a mechanical device for rotating the sample (about 1.5 cm^3 volume) in a Kel-F cylinder of 1.1 cm inside diameter) in discrete 120° jumps about an axis oriented at the magic angle relative to $\mathbf{H}_0$. Each jump is executed in about 150 ms, driven by a computer-controlled stepping motor mounted at the top of the probe and coupled to the sample by a suitable connecting-rod/gear arrangement. The ^{13}C rf field was 48 G and the ^1H field 12 G.

A simple form of the magic-angle hopping experiment is shown in the diagrams of Fig. 1. The experiment begins with a conventional CP sequence. This is followed by a $t_1/3$ evolution period during which the ^{13}C resonance frequency of a given carbon in an arbitrary crystallite in the static powder is ω_A. After $t_1/3$ the x component of transverse ^{13}C magnetization is $M_0 \cos \alpha$, where $\alpha = \omega_A t_1/3$. This component is then stored along z by a 90°_y pulse. At this point the ^1H decoupler is turned off and the sample is quickly rotated by 120° about the magic-angle axis. A 90°_{-y} pulse brings the stored z magnetization back along x, where evolution for a time $t_1/3$ is allowed to proceed under the new ^{13}C resonance frequency, ω_B. As shown in the vector diagram, after a similar sequence of events involving evolution under a third resonance frequency ω_C, the x component of ^{13}C magnetization at the end of the third $t_1/3$ period is $M_0 \cos \alpha \cos \beta \cos \gamma$ (where $\beta = \omega_B t_1/3$, $\gamma = \omega_C t_1/3$). At this point, when data acquisition in the t_2 domain begins under evolution of the ^{13}C magnetization according to ω_C, the total xy magnetization, $M_1(t_1)$, can be represented as $M_0 \cos \alpha \cos \beta \exp(i\omega_C t_1/3)$.

Three similar experiments are carried out which differ from the one shown in Fig. 1 only in the combinations of rf phases employed in the 90° ^{13}C pulses. These combinations are summarized in Table 1. For the four hopping experiments thereby specified, the ^{13}C xy magnetizations at the beginning of the t_2 period are

$$M_1(t_1) = M_0 \cos (\omega_A t_1/3) \cos (\omega_B t_1/3) \exp(i\omega_C t_1/3)$$

$$M_2(t_1) = iM_0 \sin (\omega_A t_1/3) \cos (\omega_B t_1/3) \exp(i\omega_C t_1/3)$$

$$M_3(t_1) = iM_0 \cos (\omega_A t_1/3) \sin (\omega_B t_1/3) \exp(i\omega_C t_1/3)$$

$$M_4(t_1) = -M_0 \sin (\omega_A t_1/3) \sin (\omega_B t_1/3) \exp(i\omega_C t_1/3).$$

The sum of these four $M(t_1)$ values is

$$M_\Sigma(t_1) = M_0 \exp(i\omega_A t_1/3) \exp(i\omega_B t_1/3) \exp(i\omega_C t_1/3)$$
$$= M_0 \exp(i\{\omega_A + \omega_B + \omega_C\} t_1/3). \quad [1]$$

For this case, in which ω_A, ω_B, and ω_C correspond to chemical shifts of a crystal at three orientations related to each other by 120° rotations about a magic-angle axis, it can be shown (*7, 10, 11*) that the sum in brackets on the right side of Eq. [1] corresponds to the trace of the chemical tensor (Tr $\sigma = \sigma_{11} + \sigma_{22} + \sigma_{33}$). Hence

$$M_\Sigma(t_1) = M_0 \exp(i\omega_i t) \quad [2]$$

TABLE 1

PHASES OF THE ^{13}C rf PULSES AND OF THE RECEIVER IN THE SEQUENCE OF FIG. 1

Experiment no.	CP segment[a]	Hopping segment[b] ϕ_1	ϕ_2	ϕ_3	ϕ_4	Acquisition[c]
1	x	y	$-y$	y	$-y$	+
2	x	$-x$	$-y$	y	x	+
3	x	y	$-y$	x	x	−
4	x	$-x$	$-y$	$-x$	$-y$	−

[a] Phase of the ^{13}C pulse in the CP segment.
[b] Phases of the rf in the four 90° ^{13}C pulses in the sequence of Fig. 1.
[c] Phase of the ^{13}C receiver during data acquisition.

where ω_i is the resonance frequency corresponding to the isotropic shielding, $(\mathrm{Tr}\ \sigma)/3$.

Taking this ω_i modulation of the ^{13}C magnetization at the beginning of the t_2 period into account, the time dependence of M_Σ during data acquisition can be written

$$M_\Sigma(t_1, t_2) = M_0 \exp(i\omega_i t_1) \exp(i\omega_C t_2) \exp(-[t_1 + t_2]/T_2) \quad [3]$$

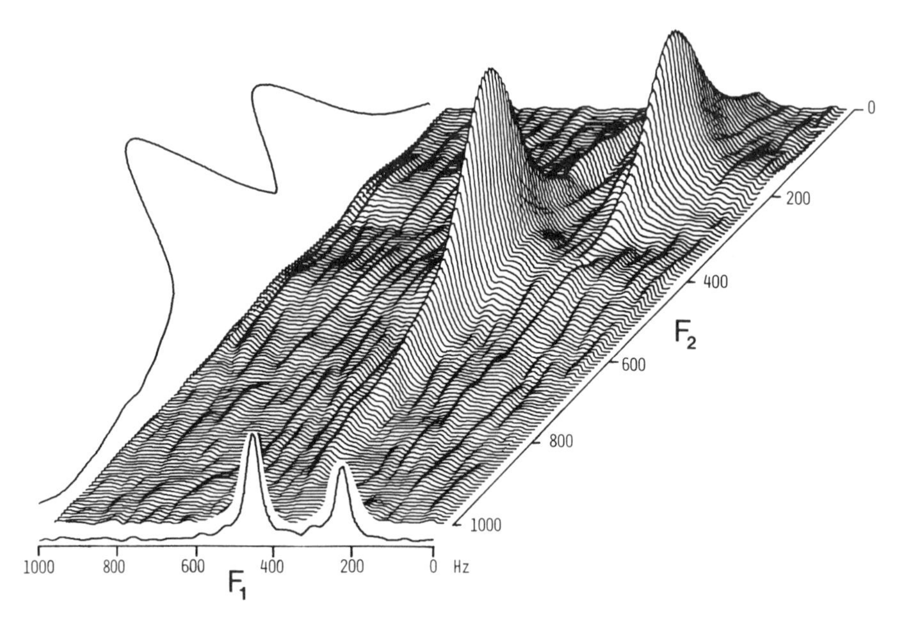

FIG. 2. Two-dimensional magic-angle hopping ^{13}C spectrum of adamantane in the absolute-value-mode presentation, plus the projections of this absolute-value-mode spectrum on both the F_1 and F_2 axes. Four acquisitions were performed per t_1 value with a delay time between experiments equal to 3 sec, and 64 values of t_1 were used. The t_1 increment was equal to 1 msec. The total measuring time was approximately 15 min.

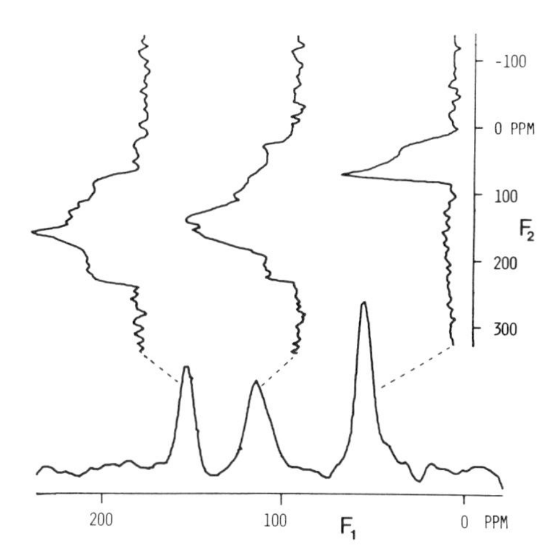

FIG. 3. Magic-angle hopping ^{13}C spectrum of *p*-dimethoxybenzene obtained from a projection of the absolute-value-mode two-dimensional spectrum onto the F_1 axis, and the absorption-mode cross sections through the two-dimensional spectrum showing the static powder patterns for the different ^{13}C chemical shifts. The two proton-bearing aromatic-carbon signals are not resolved in this experiment and the corresponding powder patterns are overlapping in the cross section shown. Six hundred acquisitions were performed for each value of t_1, with a delay time between experiments equal to 3.5 sec, and 24 values of t_1 were used. The t_1 increment was equal to 75 μsec. The total measuring time was approximately 16 hr.

where the term in T_2 accounts for transverse relaxation and ^{13}C–^{13}C dipolar broadening has been neglected. Further, it has been assumed that ^{13}C spin–lattice relaxation is negligible during the hopping periods. The factor $\exp(i\omega_C t_2)$ in Eq. [3] describes the evolution of the transverse ^{13}C magnetization of the *static* sample. Hence, Fourier transformation in the t_2 domain yields a frequency domain (F_2) in which the CSA powder pattern is manifested. Fourier transformation in the t_1 domain, which carries modulation at ω_i, yields the isotropic chemical shift in the F_1 domain.

Figure 2 shows the 2-D FT spectrum obtained on adamantane, using the sequence shown in Fig. 1. Figure 3 shows results on *p*-dimethoxybenzene. The projection along F_1 (horizontal axis) shows the isotropic shift spectrum. The cross sections parallel to F_2 (vertical axis) show the chemical shift anisotropy patterns for the various ^{13}C sites.

These preliminary results show the promise for obtaining powder patterns for individual peaks of complex molecules for which a straightforward nonspinning approach would yield only broad bands of inextricably overlapping powder patterns. This and related experiments are under extensive study.

ACKNOWLEDGMENTS

The authors are grateful for partial support of this research by grants from the U.S. Geological Survey and the Colorado State University Experiment Station.

REFERENCES

1. E. R. ANDREW, A. BRADBURY, AND R. G. EADES, *Nature (London)* **182,** 1659 (1958).
2. E. R. ANDREW, *Phil. Trans. R. Soc. London Ser. A* **299,** 505 (1981).

3. I. J. LOWE, *Phys. Rev. Lett.* **2,** 285 (1959).
4. J. SCHAEFER AND E. O. STEJAKAL, *J. Am. Chem. Soc.* **95,** 1031 (1976).
5. W. T. DIXON, *J. Magn. Reson.* **44,** 226 (1981).
6. W. T. DIXON, *J. Chem. Phys.* **77,** 1800 (1982).
7. M. M. MARICQ AND J. S. WAUGH, *J. Chem. Phys.* **70,** 3300 (1979).
8. E. LIPPMAA, M. ALLA, AND T. TUHESR, "Proceedings of the 19th Congress Ampere, Heidelberg, 1976," p. 113.
9. W. P. AUE, D. J. RUBEN, AND R. G. GRIFFIN, *J. Magn. Reson.* **43,** 472 (1981).
10. Y. YARIM-AGAEV, P. N. TUTUNJIAN, AND J. S. WAUGH, *J. Magn. Reson.* **47,** 51 (1982).
11. A. BAX, N. M. SZEVERENYI, AND G. E. MACIEL, *J. Magn. Reson.*, in press.

JOURNAL OF MAGNETIC RESONANCE 52, 335–338 (1983)

An Improved Sequence for Broadband Decoupling: WALTZ-16

A. J. SHAKA, JAMES KEELER, TOM FRENKIEL, AND RAY FREEMAN

Physical Chemistry Laboratory, Oxford University, Oxford, England

Received December 14, 1982

Progressive improvements in broadband decoupling performance have recently been achieved with the pulse sequences known as MLEV-4, MLEV-16, MLEV-64, etc. (*1*–*5*). Applied to carbon-13 spectroscopy, such sequences permit operation with a much lower radiofrequency power. A common feature of these and related experiments (*6*) is that their effectiveness can be improved by combining different versions of the primitive cycle into extended "supercycles" in which some of the residual pulse imperfections are compensated in a manner reminiscent of the folklore of solid state NMR. The original treatment of these experiments was based on average Hamiltonian theory (*7*) which, although it provides insight into the mechanism of error compensation, can be rather cumbersome in its application (*5*). An elegant new theory has recently been proposed (*6, 8*) which represents the effects of the proton irradiation sequence by means of a train of spin rotation operators, the overall effect at the end of the cycle being calculated by explicit matrix multiplication. The offset dependence of this proton response then determines the residual splitting of the carbon-13 resonance and hence the effectiveness of the decoupling. A particular virtue of this treatment is that it provides a simple mechanism for testing new decoupling sequences by computer simulation, and it acts as a guide to the intuitive approach.

The principal criteria for decoupling performance are (a) wide effective proton bandwidth for a given power dissipation, (b) residual splittings of carbon-13 small compared with the line width, (c) insensitivity to pulse length error or B_2 inhomogeneity, (d) insensitivity to errors in the radiofrequency phase shifts, (e) negligible sidebands due to sampling within the decoupling cycle (*4, 5*), and (f) programming simplicity.

Prime importance is attached to criterion (a). In what follows the bandwidth has arbitrarily been set as the proton offset range for which the carbon-13 peak remains above 80% of its maximum. In practice this depends rather critically on the choice of sensitivity enhancement function, since line broadening obscures the effect of small residual splittings. In many routine applications of carbon-13 spectroscopy there is little point in considering linewidths narrower than 1 or 2 Hz because of the coarse digitization in the frequency domain. On the other hand, some applications demand much higher resolution, and then careful attention must be paid to the magnitude of the residual splittings.

With these criteria in mind, a search was made for better decoupling sequences, using computer simulation to predict the residual splitting as a function of proton

0022-2364/83 $3.00

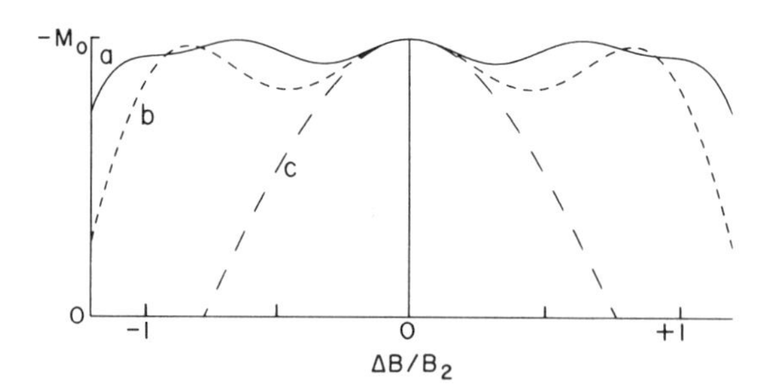

FIG. 1. The efficiency of certain pulse sequences for inversion of Z magnetization, plotted as a function of resonance offset. (a) 90°(X) 180°($-X$) 270°(X). (b) 90°(X) 180°(Y) 90°(X). (c) 180°(X).

resonance offset (*6, 8*). The culmination of this search was the new spin inversion sequence

$$R = 90°(X)\ 180°(-X)\ 270°(X) = 1\bar{2}3$$

where the shorthand notation (*6*) refers to the number of 90° pulses and the bar denotes phase inversion. The effectiveness of this element for inverting proton Z magnetization may be judged from Fig. 1, where its offset dependence is compared with that of the 90°(X) 180°(Y) 90°(X) element and a simple 180° pulse. It is a general rule that a wide bandwidth for spin inversion goes hand in hand with a wide effective decoupler bandwidth.

The primitive cycle is WALTZ-4:

$$R\,R\,\bar{R}\,\bar{R} = 1\bar{2}3\ 1\bar{2}3\ \bar{1}2\bar{3}\ \bar{1}2\bar{3}.$$

One way to proceed would be to permute one of the spin inversion elements of this cycle as in earlier experiments (*2–5*) and to combine these two cycles with their phase-inverted counterparts. A better strategy (*6*) is to permute a 90° pulse since this is known (*9, 10*) to have a built-in compensation for offset effects if judged on its ability to take a vector from $+Z$ to the XY plane of the rotating reference frame. The net effect of an imperfect $R\,R\,\bar{R}\,\bar{R}$ cycle is a small-angle rotation about an axis close to the Z axis. Cyclic permutation of a 90° pulse converts this into an equal rotation about an axis near the Y axis, which is then canceled by an equal and opposite rotation generated by a subsequent phase-inverted sequence. In this manner, cyclic permutation of the first pulse of WALTZ-4 followed by a phase-inverted cycle generates the supercycle WALTZ-8:

$$K\,\bar{K}\,\bar{K}\,K = \bar{2}4\bar{2}3\bar{1}\ 2\bar{4}2\bar{3}1\ 2\bar{4}2\bar{3}1\ \bar{2}4\bar{2}3\bar{1}.$$

Here adjacent pulses of the same phase have been combined. Note that this sequence has the same general form as the starting cycle $R\,R\,\bar{R}\,\bar{R}$. Sequences of this kind can be represented by a single overall rotation about an axis very close to $+Z$, and are therefore suitable for further expansion by cyclic permutation of a 90° pulse.

There are several possible modes of expansion, not all equivalent in performance. The best procedure at this stage is a cyclic permutation of a 90° pulse from the end to the beginning of the sequence, matched by a subsequent phase-inverted cycle, giving WALTZ-16:

$$Q\bar{Q}\bar{Q}Q = \bar{3}4\bar{2}3\bar{1}2\bar{4}2\bar{3}\ 3\bar{4}2\bar{3}1\bar{2}4\bar{2}3\ 3\bar{4}2\bar{3}1\bar{2}4\bar{2}3\ \bar{3}4\bar{2}3\bar{1}2\bar{4}2\bar{3}.$$

Note that K or Q can be regarded as very efficient composite inversion pulses; they have very flat offset-dependence curves.

The effectiveness of WALTZ-16 for broadband decoupling was simulated by calculating the carbon-13 spectrum which consists of a pair of strong lines separated by a very small residual splitting, flanked by two very weak satellites (*6*). This was convoluted with a Lorentzian broadening function (full width 0.25 Hz) and the result displayed for a series of proton resonance offsets (Fig. 2). A bandwidth equal to $2B_2$ is predicted. At the extremes of the range, a residual splitting of approximately 0.2 Hz is observed. A key feature of this sequence is that it tolerates appreciable errors in the 180° radiofrequency phase shift; the simulations of Fig. 2 are not changed perceptibly by the introduction of phase errors as large as ±5°. This remarkable insensitivity to phase error can be shown to be a general property of sequences made up of pulses with alternating radiofrequency phase. Similarly a 5% increase in all nominal pulse flip angles does not significantly alter the results of Fig. 2. However, since the B_2 intensity distribution is necessarily skewed towards low values, a reduction in all nominal pulse lengths curtails the effective bandwidth. Since there is some experimental error involved in calibrating B_2, it is good practice to set the nominal flip angles a few percent high, a precaution adopted in the experiments described below.

The WALTZ-16 sequence was verified experimentally by studying the spectrum of formic acid (J_{CH} = 221 Hz) on a Varian XL-200 spectrometer. The sample was mildly doped, giving a carbon-13 spin–lattice relaxation time of 5 sec. The instrumental line width was approximately 0.19 Hz and the sensitivity enhancement function increased this to 0.25 Hz. The radiofrequency level of the proton decoupler was calibrated by measuring the splittings in a series of coherent off-resonance decoupling experiments and was adjusted to the condition $\gamma B_2/2\pi$ = 2 kHz. There were no significant delays between radiofrequency pulses or between cycles. For programming simplicity the carbon-13 acquisition was synchronized with the decoupler cycling. Figure 3 shows the carbon-13 resonance as a function of the proton resonance offset, incremented in 200 Hz steps with exact resonance in the center. Even though the resolution is almost an order of magnitude higher than in earlier experiments (*1–5*)

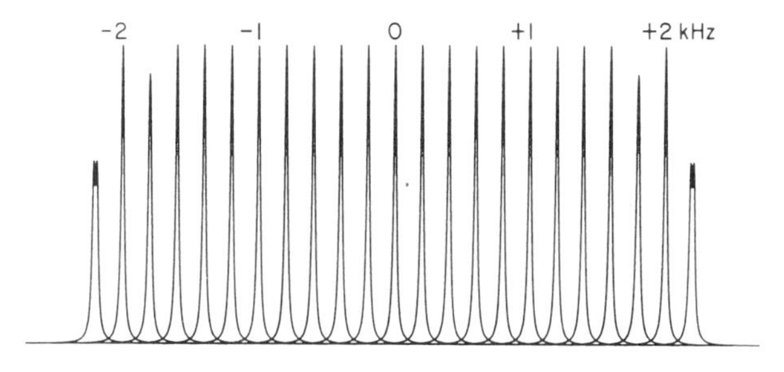

FIG. 2. Computer simulation of carbon-13 resonances decoupled with the WALTZ-16 sequence with $\gamma B_2/2\pi$ = 2 kHz and J = 220 Hz. The Lorentzian line broadening function has a full width of 0.25 Hz. The proton resonance offset has been incremented in steps of 200 Hz.

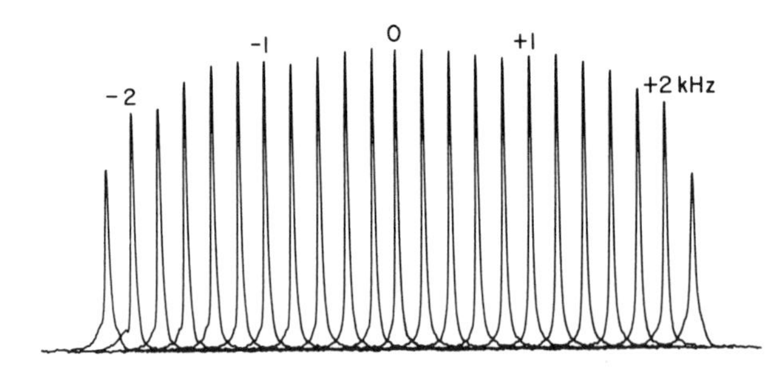

FIG. 3. Experimental carbon-13 spectra from formic acid decoupled by the WALTZ-16 sequence with $\gamma B_2/2\pi = 2$ kHz. Line broadening and offset increments as in Fig. 2. An effective bandwidth of approximately 4 kHz is obtained.

the effective bandwidth was equal to $2B_2$. It is believed that the discrepancies between the simulations of Fig. 2 and the experimental spectra of Fig. 3 arise mainly from the spatial inhomogeneity of the decoupler field. Some parts of the sample experience quite low B_2 fields and their effective offsets $\Delta B/B_2$ are therefore abnormally high, well outside the range of effective decoupling. The simulations assume a perfectly uniform B_2 field.

Further experiments are in hand to test this sequence and its variations, but it is already clear that it is superior to its predecessors with regard to bandwidth, residual splitting, and insensitivity to phase shift errors.

ACKNOWLEDGMENTS

This work was made possible by an equipment grant from the Science and Engineering Research Council and by a research studentship (T.F.), a Domus senior scholarship from Merton College (J.K.), a Rhodes scholarship and an N.C.A.A. postgraduate fellowship (A.J.S.). The authors are indebted to Professor J. S. Waugh for making available two manuscripts (*6, 8*) prior to publication. The inventor of the acronym WALTZ (Wideband, Alternating-phase, Low-power Technique for Zero-residual-splitting) wishes to remain anonymous.

REFERENCES

1. M. H. LEVITT AND R. FREEMAN, *J. Magn. Reson.* **43,** 502 (1981).
2. M. H. LEVITT, R. FREEMAN, AND T. FRENKIEL, *J. Magn. Reson.* **47,** 328 (1982).
3. M. H. LEVITT, R. FREEMAN, AND T. FRENKIEL, *J. Magn. Reson.* **50,** 157 (1982).
4. R. FREEMAN, T. FRENKIEL, AND M. H. LEVITT, *J. Magn. Reson.* **50,** 345 (1982).
5. M. H. LEVITT, R. FREEMAN, AND T. FRENKIEL, "Advances in Magnetic Resonance" (J. S. Waugh, Ed.), Academic Press, New York, in press.
6. J. S. WAUGH, *J. Magn. Reson.* **49,** 517 (1982).
7. U. HAEBERLEN AND J. S. WAUGH, *Phys. Rev.* **175,** 453 (1968).
8. J. S. WAUGH, *J. Magn. Reson.* **50,** 30 (1982).
9. R. FREEMAN AND H. D. W. HILL, *J. Chem. Phys.* **54,** 3367 (1971).
10. M. H. LEVITT, *J. Magn. Reson.* **50,** 95 (1982).

JOURNAL OF MAGNETIC RESONANCE **53**, 517–520 (1983)

Broadband Homonuclear Decoupling in Heteronuclear Shift Correlation NMR Spectroscopy

AD BAX

Department of Chemistry, Colorado State University, Fort Collins, Colorado 80523

Received March 7, 1983

Recently, Garbow, Weitekamp, and Pines introduced a new approach to homonuclear broadband decoupling (*1*). Independently, a closely related experiment has been proposed for semiselective heteronuclear *J* spectroscopy (*2*). The two types of experiments have the common feature that all protons which are not directly attached to a ^{13}C nucleus experience an effect as if a nonselective 180° pulse were applied, while protons bonded to a ^{13}C nucleus are not affected. This type of behavior can be used conveniently in many of the heteronuclear types of two-dimensional experiments (*3–9*). Here, its use for sensitivity and resolution enhancement in heteronuclear spectroscopy will be demonstrated. It is mentioned here that other approaches to homonuclear broadband decoupling in heteronuclear shift correlation spectroscopy have been proposed by Frenkiel (*10*), but those experiments rely on a different principle (*11*), and are of a less general nature.

The experimental scheme is sketched in Fig. 1. The theory of the conventional heteronuclear decoupled shift correlation experiment (*3*) has been described elsewhere (*3–6*) and will be repeated here. The difference between the conventional heteronuclear decoupled shift correlation experiment and the scheme of Fig. 1 is that the single 180° ^{13}C pulse at the center of the evolution period has now been replaced by a $90°_x(^1\mathrm{H})$–$1/(2J)$–$180°_x(^1\mathrm{H})$, $180°_x(^{13}\mathrm{C})$–$1/(2J)$–$90°_{-x}(^1\mathrm{H})$ sequence, where *J* is the magnitude of the heteronuclear coupling constant in hertz. It has been shown (*1, 2*) that this sequence has the effect of a single 180° pulse for those protons that are not attached to a ^{13}C nucleus, while other protons (those attached to ^{13}C) are essentially

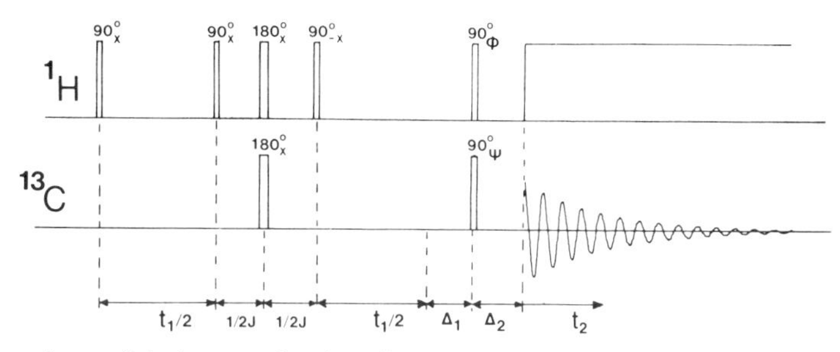

FIG. 1. Pulse scheme of the homonuclear broadband decoupled shift correlation experiment. The delays Δ_1 and Δ_2 are set to the same values as in the conventional shift correlation experiment (*3, 6*). The phases ϕ and ψ are cycled according to Table 1.

0022-2364/83 $3.00

unaffected. If a proton, A, directly coupled to a ^{13}C, X, is also coupled to a proton, M, which is not attached to ^{13}C and is initially, for example in the $m = 1/2$ spin state, the pulse sequence inverts the state of M to $m = -1/2$, and does not affect the (transverse) A magnetization. If the sequence is applied at the center of the evolution period, there will be no overall effect of coupling between A and M at the end of the evolution period, because the A nucleus is coupled for a time $t_1/2$ to the M nucleus in the $m = 1/2$ state and for another time $t_1/2$ to the M nucleus in the $m = -1/2$ state. Therefore, the position of the A magnetization vector in the transverse plane depends only on the chemical shift, δ_A, and the length of the evolution period t_1. In the usual way (*3–5*), this magnetization of proton A is then transferred to its directly coupled ^{13}C nucleus, X. Therefore, the observed ^{13}C signal will be modulated by the chemical shift, δ_A, only. If broadband heteronuclear proton decoupling is employed during

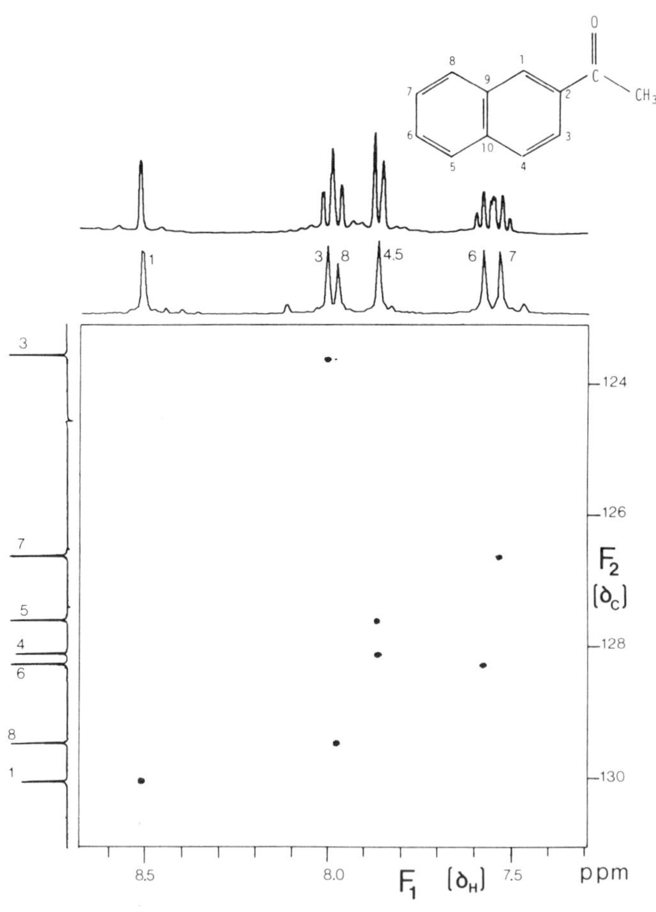

FIG. 2. Homonuclear broadband decoupled absolute-value-mode heteronuclear shift correlation spectrum of the aromatic resonances in 2-acetonaphtalene (inset). A 192 × 512 data matrix was acquired, and four experiments were performed for each value of t_1, with the phases of the rf pulses as given in Table 1. Along the F_2 axis, the corresponding conventional ^{13}C spectrum is shown. Along the F_1 axis, the projection of the absolute value mode 2D spectrum on this axis, and the conventional proton spectrum are shown.

acquisition (t_2), a two-dimensional Fourier transformation will give a single resonance at $(F_1, F_2) = (\delta_A, \delta_X)$. No homonuclear proton splittings will appear in the F_1 dimension.

Experiments were performed on a NT-360 spectrometer, controlled by a 1180 computer and a 293A′ pulse programmer. A 1 M concentration of 2-acetonaphtalene in acetone-d_6 in a 12 mm sample tube was used. Figure 2 shows a proton–proton decoupled 2D shift correlation spectrum of the protonated aromatic region. A 192 × 512 data matrix was acquired and four experiments were performed for each value of t_1. The total measuring time was approximately 40 min. The phases of the final proton and ^{13}C pulses in the scheme of Fig. 1 were cycled in the four steps to allow distinguishing between positive and negative proton modulation frequencies and for suppression of axial peaks at $F_1 = 0$ Hz (*5, 6*). The cycling of the rf pulses actually employed here is given in Table 1, and all data are co-added with identical receiver phase (the only option with the NT-360 used). Along the F_2 axis, the conventional ^1H-decoupled ^{13}C spectrum is shown. Along the F_1 axis, an absolute value mode projection of the 2D spectrum on this axis is shown, representing the homonuclear broadband decoupled proton spectrum. For comparison, the conventional proton spectrum, recorded on the same sample in the same 12 mm sample tube, is also shown along this axis. Absorption mode cross sections parallel to the F_1 axis through the data matrix of Fig. 2 taken at the various ^{13}C F_2 frequencies are shown in Fig. 3. Those cross sections demonstrate the excellent proton resolution that can be obtained. In this case, the main limitation for high resolution in the F_1 dimension is the limited acquisition time in this dimension (384 msec). Gaussian multiplication was used in both dimensions to avoid truncation. Small mirror images with respect to the proton transmitter frequency (midpoint of the traces) can be seen in Fig. 3; these images are attributable to imperfect 90° phase shifts in the decoupler channel.

In cases where high F_1 resolution is required and ^1H–^1H splittings in the conventional experiment would be resolved, the sensitivity of the new, homonuclear decoupled version should be better, because the sensitivity is proportional to the reciprocal of the number of resonances over which signal energy is being distributed (*12*). On the other hand, one loses information about the proton–proton couplings, which sometimes could be of practical use if the proton spectrum shows severe overlap. It has been found experimentally (*6*), that in selecting the delays $1/(2J)$ at the center of the

TABLE 1

THE rf PHASES OF THE FINAL PROTON PULSE ϕ AND OF THE FINAL ^{13}C PULSE ψ IN THE FOUR STEPS OF THE EXPERIMENT OF FIG. 1[a]

ϕ	ψ
x	x
y	$-y$
$-x$	$-x$
$-y$	y

[a] All data are co-added with identical receiver phase.

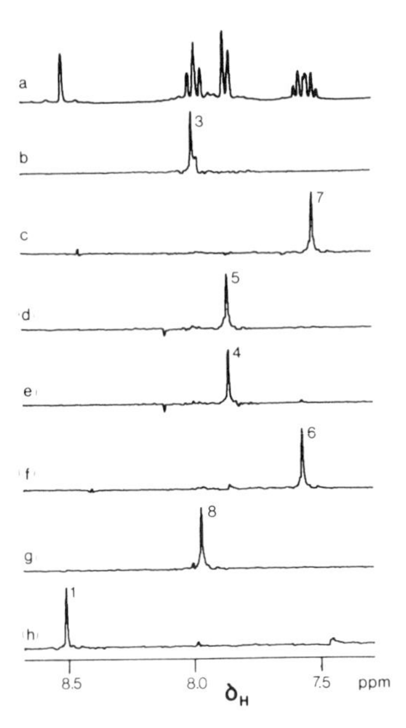

FIG. 3. (a) The conventional proton spectrum of the aromatic resonances in 2-acetonaphtalene, and (b–h) absorption mode cross sections parallel to the F_1 axis through the data matrix of Fig. 2 for carbons C1, C8, C6, C4, C5, C7, and C3, respectively.

evolution period, the value of J for the direct heteronuclear coupling has to be accurate to within approximately 10%. It also has been found that the homonuclear proton multiplet width has to be a factor of five or more smaller than the value of J. Garbow *et al.* (*1*) reported that other sequences for selectively inverting spin states of protons that are not bonded to a ^{13}C are less sensitive to the limitations mentioned above, and their applicability in heteronuclear spectroscopy is under investigation.

ACKNOWLEDGMENT

The author is indebted to Professor G. E. Maciel for many valuable comments in preparing the manuscript and further gratefully acknowledges partial support of this work from the U.S. Department of Energy (Laramie Energy Technical Center) and use of the Colorado State University Regional NMR Center, supported by National Science Foundation Grant CHE-78-18581.

REFERENCES

1. J. R. GARBOW, D. P. WEITEKAMP, AND A. PINES, *Chem. Phys. Lett.* **93,** 504 (1982).
2. A. BAX, *J. Magn. Reson.*, in press.
3. A. A. MAUDSLEY, L. MULLER, AND R. R. ERNST, *J. Magn. Reson.* **28,** 463 (1977).
4. R. FREEMAN, *Proc. Roy. Soc. London, Ser. A* **373,** 149 (1980).
5. A. BAX AND G. A. MORRIS, *J. Magn. Reson.* **42,** 501 (1981).
6. A. BAX, "Topics in ^{13}C NMR" (G. C. Levy, Ed.), Vol. 4, Wiley Interscience, in press.
7. P. H. BOLTON AND G. BODENHAUSEN, *Chem. Phys. Lett.* **89,** 139 (1982).
8. R. FREEMAN AND G. A. MORRIS, *Bull. Magn. Reson.* **1,** 5 (1979).
9. L. MULLER, *J. Am. Chem. Soc.* **101,** 4481 (1979).
10. T. A. FRENKIEL, Part II thesis, Oxford University, 1980.
11. A. BAX, A. F. MEHLKOPF, AND J. SMIDT, *J. Magn. Reson.* **35,** 167 (1979).
12. W. P. AUE, P. BACHMANN, A. WOKAUN, AND R. R. ERNST, *J. Magn. Reson.* **29,** 523 (1979).

JOURNAL OF MAGNETIC RESONANCE **53**, 521–528 (1983)

Coherence Transfer by Isotropic Mixing: Application to Proton Correlation Spectroscopy

L. BRAUNSCHWEILER AND R. R. ERNST

Laboratorium für Physikalische Chemie, Eidgenössische Technische Hochschule, 8092 Zürich, Switzerland

Received April 11, 1983

In this communication we propose a new mechanism of coherence transfer useful for two-dimensional correlation spectroscopy in liquids. Cross-peaks are generated between all members of a coupled spin network leading to a most complete correlation diagram. The scheme has the advantage of producing net coherence transfer and can be arranged to create pure absorption mode spectra with peaks of positive intensity.

Coherence transfer processes are of central importance in modern pulse NMR. In fact many of the successful schemes of two-dimensional (2D) spectroscopy (*1–4*), of multiple quantum spectroscopy (*1, 5–8*), and of rare spin resonance in liquids (*9–14*) and solids (*14–16*) utilize the concept of coherence transfer for exploring nuclear spin systems and for enhancing sensitivity. Many of the known processes, however, lead only to differential coherence transfer and cause spectra with equal positive and negative intensities. In particular, traditional homonuclear 2D correlation spectroscopy (*1–4*) exhibits cross-peak multiplets invariably with zero integrated intensity, and transfer is restricted to directly spin-coupled nuclei.

In the context of heteronuclear coherence transfer in liquids, schemes have already been proposed for net magnetization transfer. It has been shown that refocusing after a single pulse transfer will lead to net coherence transfer (*10, 13*). It has also been demonstrated that net coherence can be transfered in the rotating frame either by Hartmann–Hahn matching of the applied rf fields (*12*) or by an adiabatic mixing process (*18*).

We demonstrate here that net magnetization and net coherence transfers are obtained whenever a mixing Hamiltonian $\bar{\mathscr{H}}_m$ is tailored which is dominated by the full isotropic coupling terms,

$$\bar{\mathscr{H}}_m = \mathscr{H}_J = \sum_{i<j}\sum 2\pi J_{ij}\mathbf{I}_i\mathbf{I}_j . \qquad [1]$$

The individual single spin operators, which under the weak coupling Hamiltonian precess independently in the sense of *single spin modes,* no longer evolve independently under the influence of $\bar{\mathscr{H}}_m$ but take part in so-called *collective spin modes.* This leads then to an oscillatory transfer of coherence throughout the coupled spin systems. The situation is to some extent analogous to a network of coupled mechanical oscillators where normal modes also involve collective motions of several oscillators.

0022-2364/83 $3.00

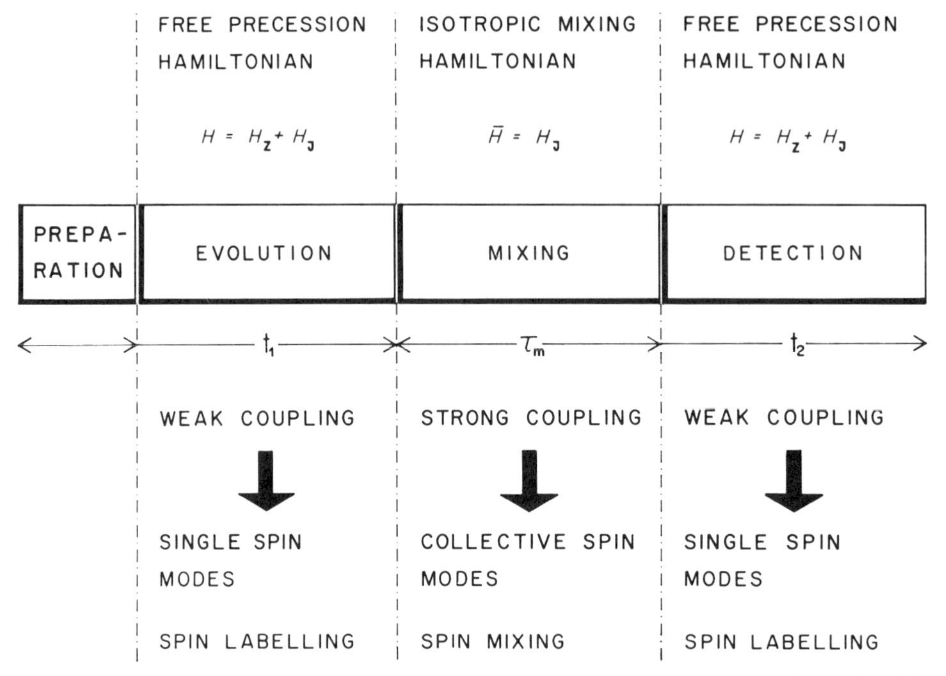

FIG. 1. Schematic representation of the two-dimensional correlation experiment with an isotropic mixing process.

The features of coherence transfer through isotropic coupling can be appreciated most easily by considering a two-spin-½ system with $\mathscr{H}_J = 2\pi J_{12}\mathbf{I}_1\mathbf{I}_2$. The orthogonal collective spin modes in this simple case correspond just to the sum and the difference of the single spin operators and to the sum and difference of product spin operators:

$$\Sigma_\alpha = \tfrac{1}{2}\{I_{1\alpha} + I_{2\alpha}\},$$
$$\Delta_\alpha = \tfrac{1}{2}\{I_{1\alpha} - I_{2\alpha}\},$$
$$\Sigma_{\alpha\beta} = \{I_{1\alpha}I_{2\beta} + I_{1\beta}I_{2\alpha}\},$$
$$\Delta_{\alpha\beta} = \{I_{1\alpha}I_{2\beta} - I_{1\beta}I_{2\alpha}\}, \qquad \alpha, \beta = x, y, z \tag{2}$$

with the following commutation relations:

$$[\mathscr{H}_J, \Sigma_\alpha] = 0$$
$$[\mathscr{H}_J, \Sigma_{\alpha\beta}] = 0$$
$$[\mathscr{H}_J, \Delta_\alpha] = i\Delta_{\beta\gamma}$$
$$[\mathscr{H}_J, \Delta_{\beta\gamma}] = -i\Delta_\alpha \tag{3}$$

where (α, β, γ) is a cyclic permutation of (x, y, z). This leads immediately to the following time evolution of the difference terms (*19*):

$$\Delta_\alpha \xrightarrow{\mathcal{H}_J \tau_m} \Delta_\alpha \cos(2\pi J_{12}\tau_m) + \Delta_{\beta\gamma} \sin(2\pi J_{12}\tau_m) \quad [4]$$

and

$$\Delta_{\beta\gamma} \xrightarrow{\mathcal{H}_J \tau_m} \Delta_{\beta\gamma} \cos(2\pi J_{12}\tau_m) - \Delta_\alpha \sin(2\pi J_{12}\tau_m),$$

while the sum terms remain invariant. For the evolution of the x component of spin 1, for example, we find then in more explicit notation

$$I_{1x} \xrightarrow{\mathcal{H}_J \tau_m} \tfrac{1}{2} I_{1x}\{1 + \cos(2\pi J_{12}\tau_m)\} + \tfrac{1}{2} I_{2x}\{1 - \cos(2\pi J_{12}\tau_m)\} + (I_{1y}I_{2z} - I_{1z}I_{2y}) \sin(2\pi J_{12}\tau_m). \quad [5]$$

Similar expressions apply also to the other components of in-phase (I_{1y}) and antiphase coherence ($2I_{1x}I_{2z}$, $2I_{1y}I_{2z}$) as well as to the polarization (I_{1z}). All of these components are periodically exchanged between spins 1 and 2 under the influence of $\mathcal{H}_J$. Equation

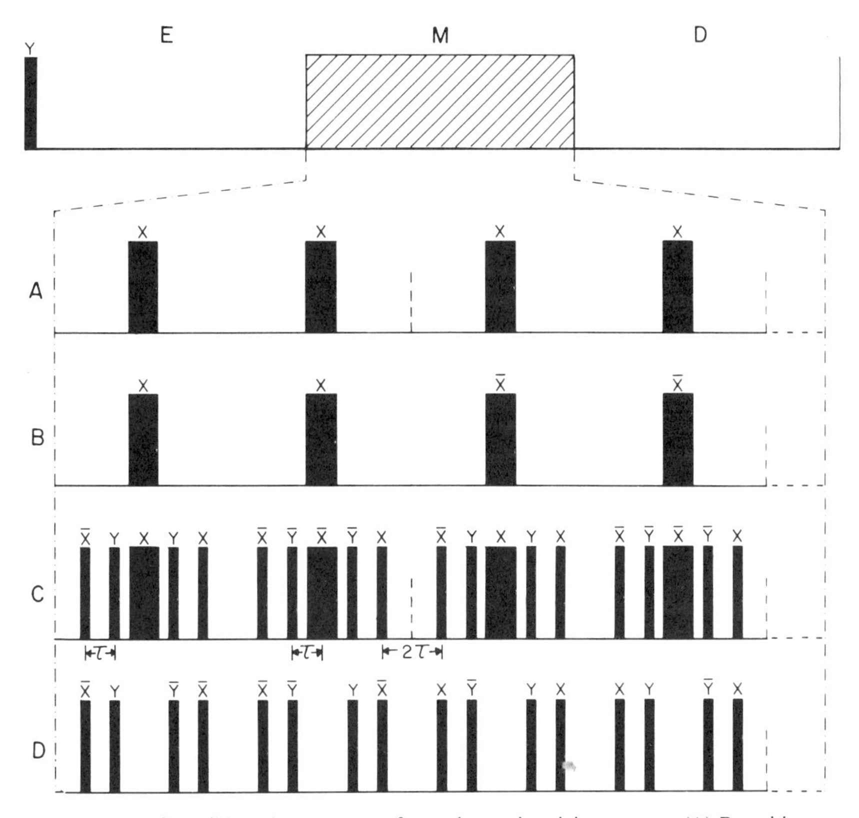

FIG. 2. Examples of possible pulse sequences for an isotropic mixing process. (A) Repetitive sequence of π_x pulses. (B) Repetitive sequence of phase-alternated π_x pulses. Radiofrequency field inhomogeneity is compensated by the phase alternation. (C) Cycle with the minimum number of 10 pulses leading to isotropic averaging of the Hamiltonian and having at the same time symmetrizing properties. The cycle consists of eight $\pi/2$ and two π pulses. (D) 16-pulse cycle leading to isotropic averaging of the Hamiltonian and having at the same time symmetrizing properties. The cycle consists exclusively of $\pi/2$ pulses equally distributed among the four possible phases.

[5] demonstrates that for $\tau_m = (2J_{12})^{-1}$ the transfer is complete. However, for intermediate values of τ_m additional antiphase components appear for both nuclear spins ($I_{1y}I_{2z}$ and $I_{1z}I_{2y}$).

In extended coupling networks, for example, in a linear chain of couplings, the oscillatory exchange proceeds through the entire network so that net magnetization may be transferred from one spin to another even without direct coupling. This leads then to multiple step relayed magnetization transfer (*20, 21*).

The outlined facts suggest a new type of a homonuclear (or heteronuclear) 2D correlation experiment illustrated in Fig. 1. After a preparation pulse, the spin system is allowed to evolve freely under the (assumed) weak coupling Hamiltonian in terms

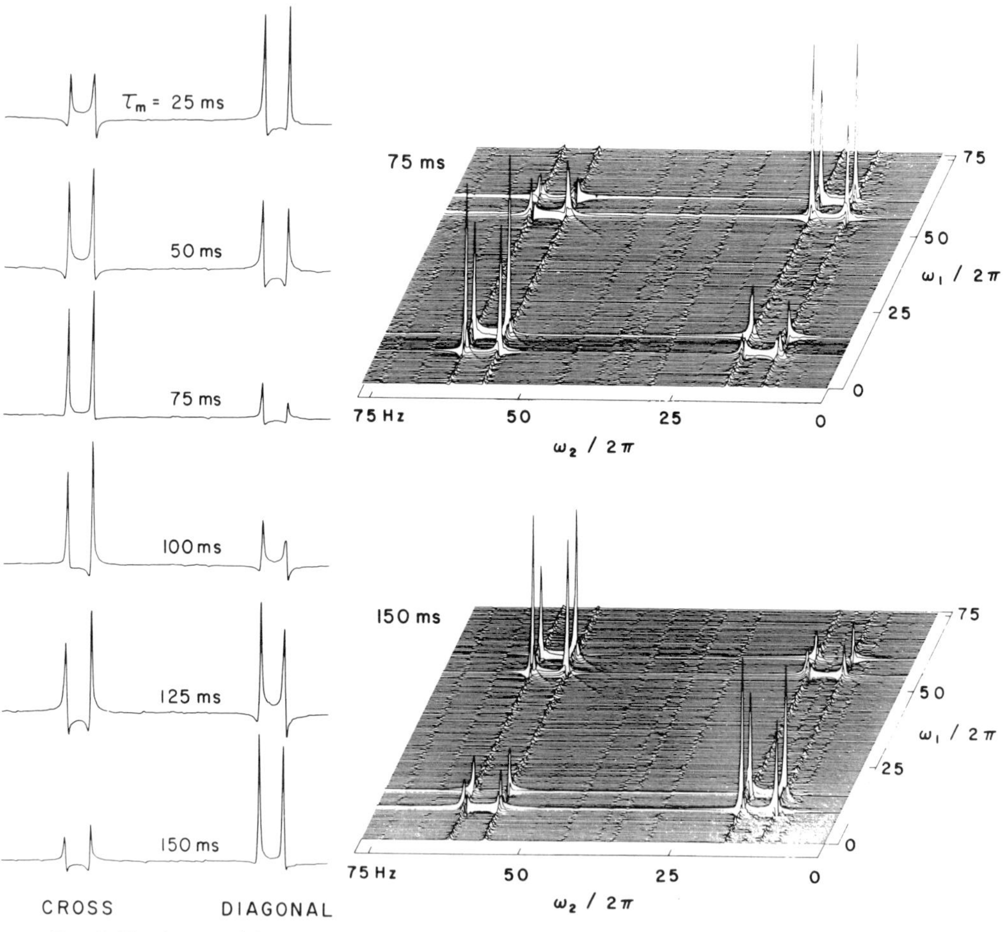

FIG. 3. Six phase-sensitive cross sections taken at $\omega_1/2\pi$ = 17 Hz through 2D correlated spectra of 2,3-dibromothiophene employing an isotropic mixing process of duration τ_m. For τ_m = 75 and 150 msec, the full phase-sensitive 2D spectra are shown. Pulse sequence A of Fig. 2 has been employed with a pulse repetition rate of 2 kHz. Free induction decays for 256 t_1 values have been processed. A two-experiment phase cycle ([X, X, +] and [$-X$, X, $-$]) has been employed for the single phase experiments at 90 MHz on a home-built instrument.

of single spin modes. For the mixing process the Hamiltonian is modified to the desired form of Eq. [1]. The coherence of the spins is periodically exchanged during the mixing time τ_m and is observed during the detection period again under the unperturbed Hamiltonian.

Many pulse sequences can be conceived to create the required mixing Hamiltonian consisting exclusively of the isotropic coupling terms of Eq. [1]. We indicate here the basic principles only. A more detailed analysis of the various sequences with regard to compensation of imperfections will be presented elsewhere. The task of the mixing pulse sequence is the efficient suppression of the chemical shift terms of the full Hamiltonian for energy matching to allow spin exchange. The scalar coupling terms remain invariant for any pulse sequence. All the sequences indicated in Fig. 2 lead, for a sufficiently fast pulse rate and sufficiently short pulses, to a pure isotropic coupling Hamiltonian.

For absolute value 2D spectra, the sequences of Fig. 2 gave comparable experimental results. For the recording of phase-sensitive spectra, however, sequence A is superior. It has been mentioned that isotropic mixing exchanges all in-phase as well as antiphase components. This actually precludes the recording of phase-sensitive spectra, and

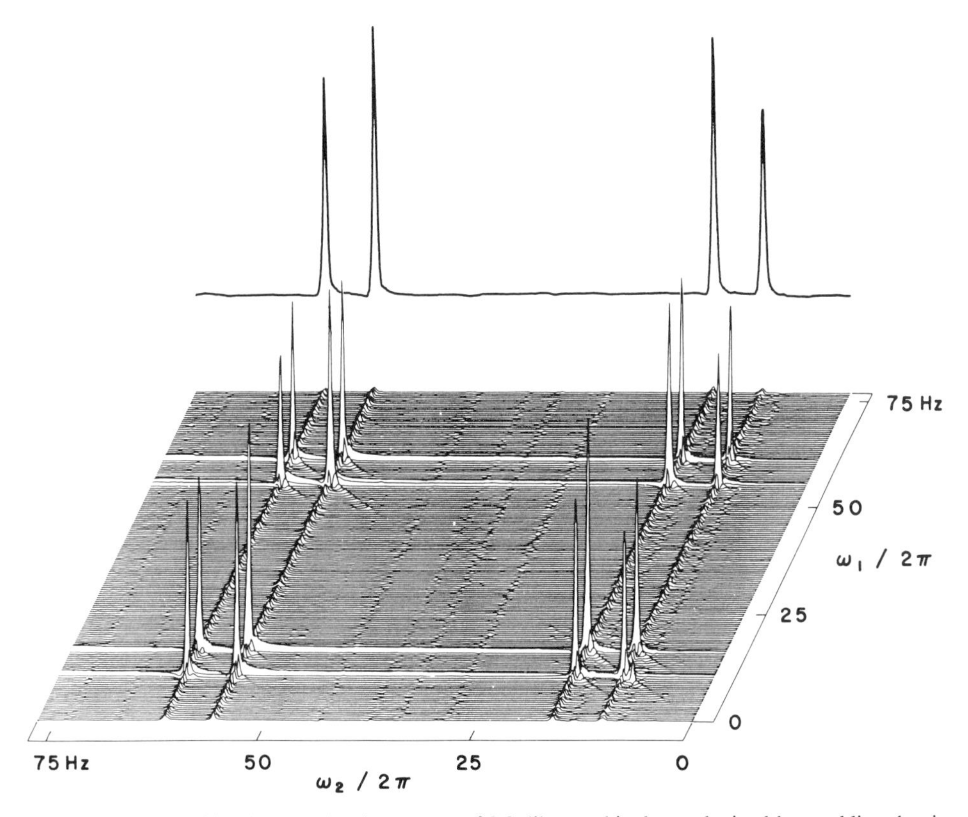

FIG. 4. Phase-sensitive 2D correlated spectrum of 2,3-dibromothiophene obtained by coadding the six spectra of Fig. 3 with mixing times between 25 and 150 msec. To better demonstrate the vanishing negative signal excursions, a cross section taken at $\omega_1/2\pi$ = 17 Hz is included.

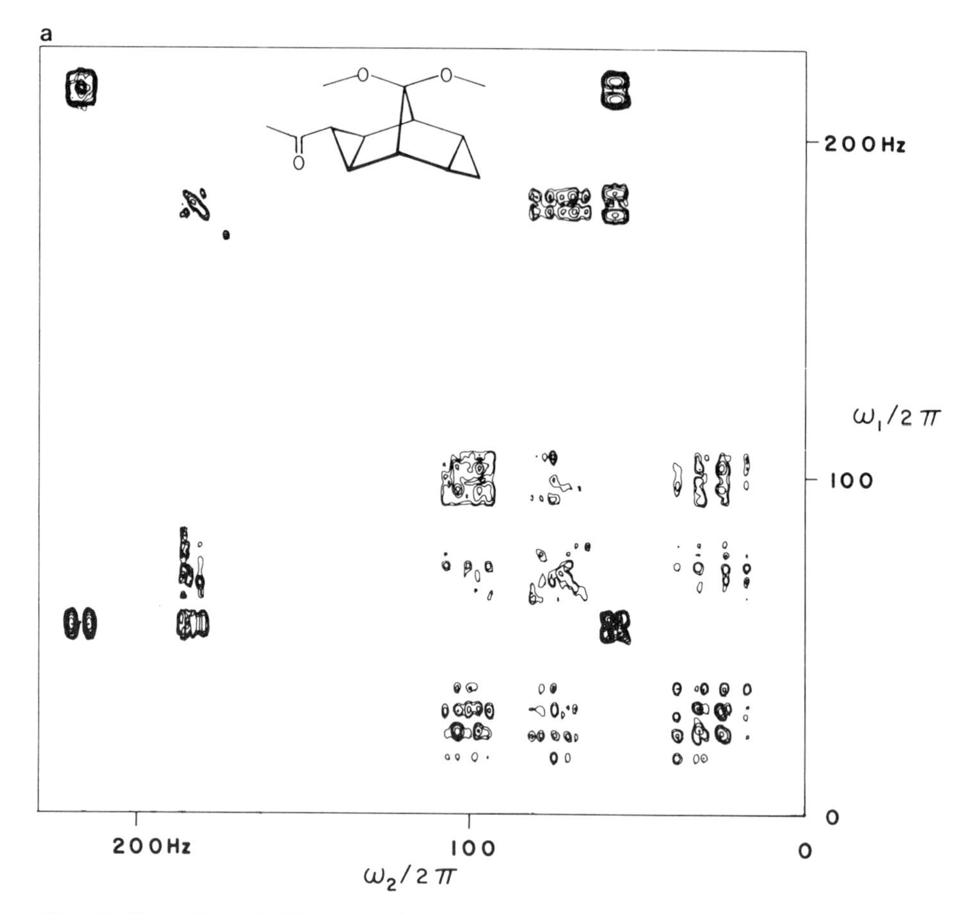

FIG. 5. Comparison of 2D correlated spectra of 3-acetyl-9,9-dimethoxy-3-*exo*-7-*endo*-tetracyclo-[3.3.1.0^{2,4}.0^{6,8}]-nonane without and with relayed peaks. (a) Contour plot of the absolute value spectrum recorded by a traditional correlation experiment (COSY) at 90 MHz. 512 free induction decays for different t_1 values, each with four repetitions, have been processed. (b) Contour plot of the phase-sensitive spectrum recorded with the TOCSY sequence A of Fig. 2. A total of 448 free induction decays for different t_1 values have been recorded whereby for each of them eight scans with mixing times between 25 and 200 msec have been coadded. The pulse sequence A of Fig. 2 has been employed. The contour levels have been set at 1, 1.3, 1.6, 2, 3, 5, 8, 16, 33, 58, 83, and 100% of the maximum peak amplitude.

special precautions must be taken to eliminate out-of-phase as well as antiphase components.

In this context it is important to note that sequence A is not rf inhomogeneity compensated and leads to a rapid decay of all components which do not commute with $F_x = I_{1x} + I_{2x}$. After extended mixing, the only remaining components in a two-spin system are I_{1x}, I_{2x}, $I_{1x}I_{2x}$, and $I_{1y}I_{2z} - I_{1z}I_{2y}$, and many undesired terms are eliminated. The features of the resulting 2D spectra for variable mixing time are demonstrated in Fig. 3 for the two-spin system of 2,3-dibromothiophene. It is interesting to note that for a mixing time $\tau_m = 75$ msec $= (2J_{12})^{-1}$ the diagonal peak multiplets are almost completely suppressed while the cross-peaks become maximum, indicating

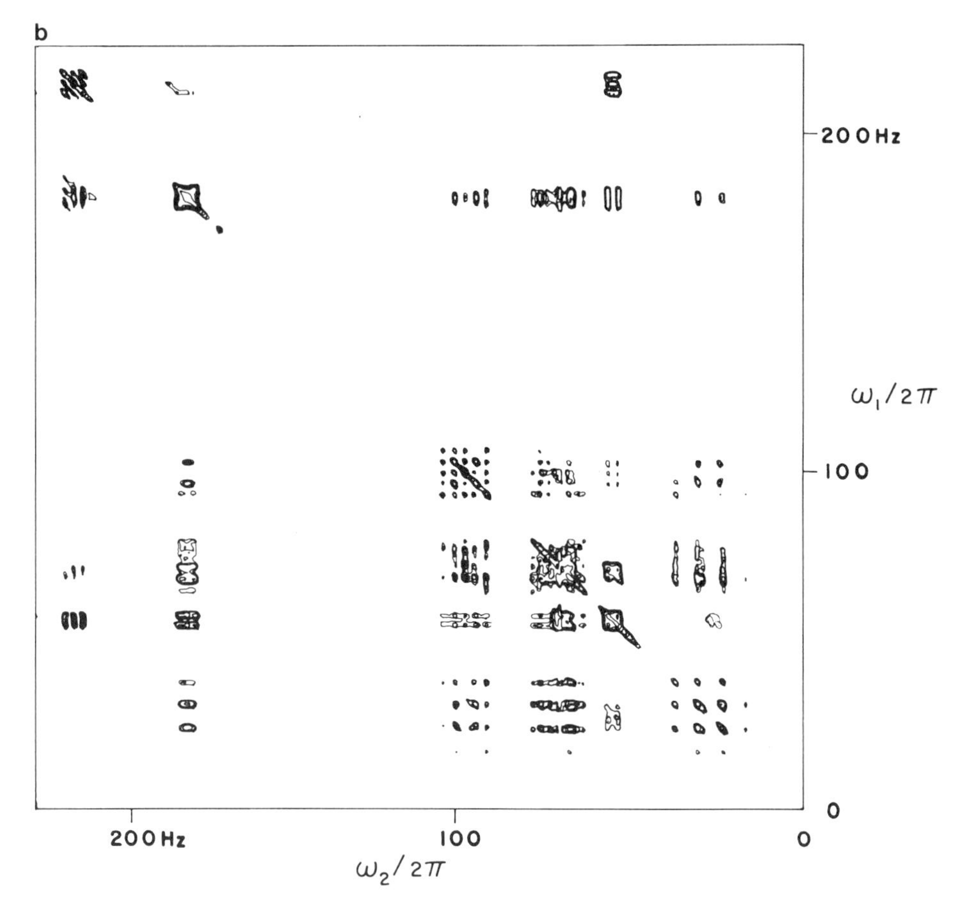

FIG. 5—*Continued.*

complete transfer. For $\tau_m = 150$ msec $= J_{12}^{-1}$ only diagonal peak multiplets remain.

It is apparent that the spectra contain in general in-phase signals which appear in absorption and antiphase signals (originating from $I_{1y}I_{2z} - I_{1z}I_{2y}$) which are in dispersion. The in-phase components remain positive while the antiphase components change sign as a function of τ_m. These undesired signals can thus be averaged to zero by coadding numerous signals with variable mixing time τ_m. The only term which survives after mixing with sequence A and variable mixing time is the operator $I_{1x} + I_{2x}$. It leads to pure absorption mode spectra. An example is shown in Fig. 4 for the same sample. This spectrum should be compared with a conventional correlation experiment (*1–4*) where diagonal and cross-peak multiplets are 90° out of phase and the cross-peaks are alternatingly positive and negative.

For a more extensive coupling network, all pairs of nuclei will show cross-peaks even without direct coupling. This feature may suggest the name "total correlation spectroscopy" (TOCSY) to this experiment. It comprises all orders of multiple step relayed coherence transfer. Figure 5 compares the conventional correlation spectrum (COSY) of 3-acetyl-9,9-dimethoxy-3-*exo*-7-*endo*-tetracyclo[3.3.1.$0^{2,4}$.$0^{6,8}$]-nonane with the corresponding TOCSY spectrum. Even without providing a detailed analysis at this place, it is obvious that numerous additional relayed correlation peaks appear in the TOCSY spectrum. Its resolution is higher because the phase-sensitive spectrum is free of dispersion mode signals, while the COSY spectrum is represented in absolute value mode to cope with the partially negative intensities.

Very recently Weitekamp *et al.* (*22*) have pointed out a similar possibility of isotropic mixing for heteronuclear transfer in solids using two synchronized pulse sequences applied to the two nuclear species.

Based on our results, we expect that isotropic mixing will become a valuable tool for obtaining 2D correlated spectra which include all possible relay peaks. Particularly when absorption mode spectra are desired, the inherent net magnetization transfer is very attractive.

ACKNOWLEDGMENTS

We acknowledge helpful discussions with Dr. G. Bodenhausen, H. Kogler, Dr. M. H. Levitt, and O. W. Sørensen. The sample used for Fig. 5 was provided by Professor H. Kessler and H. Kogler. This research has been supported in part by the Swiss National Science Foundation.

REFERENCES

1. W. P. Aue, E. Bartholdi, and R. R. Ernst, *J. Chem. Phys.* **64,** 2229 (1976).
2. K. Nagayama, Anil Kumar, K. Wüthrich, and R. R. Ernst, *J. Magn. Reson.* **40,** 321 (1980).
3. A. Bax and R. Freeman, *J. Magn. Reson.* **44,** 542 (1981).
4. A. Bax, "Two-Dimensional Nuclear Magnetic Resonance in Liquids," Delft Univ. Press, Dordrecht, 1982.
5. G. Drobny, A. Pines, S. Sinton, D. Weitekamp, and D. Wemmer, *Faraday Div. Chem. Soc. Symp.* **13,** 49 (1979).
6. A. Wokaun and R. R. Ernst, *Chem. Phys. Lett.* **52,** 407 (1977).
7. G. Bodenhausen, *Progr. NMR Spectrosc.* **14,** 137 (1981).
8. L. Braunschweiler, G. Bodenhausen, and R. R. Ernst, *Mol. Phys.* **48,** 535 (1983).
9. A. A. Maudsley and R. R. Ernst, *Chem. Phys. Lett.* **50,** 368 (1977).
10. A. A. Maudsley, L. Müller, and R. R. Ernst, *J. Magn. Reson.* **28,** 463 (1977).
11. G. A. Morris and R. Freeman, *J. Am. Chem. Soc.* **101,** 790 (1979).
12. L. Müller and R. R. Ernst, *Mol. Phys.* **38,** 963 (1979).
13. D. P. Burum and R. R. Ernst, *J. Magn. Reson.* **39,** 163 (1980).
14. D. P. Doddrell, D. T. Pegg, and M. R. Bendall, *J. Magn. Reson.* **48,** 323 (1982).
15. S. R. Hartmann and E. L. Hahn, *Phys. Rev.* **128,** 2042 (1962).
16. A. Pines, M. G. Gibby, and J. S. Waugh, *J. Chem. Phys.* **56,** 1776 (1972).
17. L. Müller, Anil Kumar, T. Baumann, and R. R. Ernst, *Phys. Rev. Lett.* **32,** 1402 (1974).
18. R. D. Bertrand, W. B. Moniz, A. N. Garroway, and G. C. Chingas, *J. Am. Chem. Soc.* **100,** 5227 (1978).
19. O. W. Sørensen, G. Eich, M. H. Levitt, G. Bodenhausen, and R. R. Ernst, submitted for publication.
20. G. Eich, G. Bodenhausen, and R. R. Ernst, *J. Am. Chem. Soc.* **104,** 3731 (1982).
21. P. H. Bolton and G. Bodenhausen, *Chem. Phys. Lett.* **89,** 139 (1982).
22. D. P. Weitekamp, J. R. Garbow, and A. Pines, *J. Chem. Phys.* **77,** 2870 (1982).

JOURNAL OF MAGNETIC RESONANCE 55, 494–497 (1983)

Chemical Shift Anisotropy in Powdered Solids Studied by 2D FT NMR with Flipping of the Spinning Axis

AD BAX, NIKOLAUS M. SZEVERENYI, AND GARY E. MACIEL*

Department of Chemistry, Colorado State University, Fort Collins, Colorado 80523

Received August 22, 1983

Information about chemical shift anisotropy is usually difficult to extract from the spectrum of a nonspinning powdered sample because of the usually extensive overlap of the powder patterns from the chemically different sites in the molecule. Several types of experiments have been proposed to facilitate the measurement of chemical shift anisotropy (*1–9*). All have the common feature that the sample is rotated about the magic-angle axis (*1–8*) or an axis very close to the magic angle (*9*), either by rapid sample spinning (*1–7, 9*) or by rotating the sample in three discrete steps (*8*).

We propose a new two-dimensional approach for obtaining the anisotropy information. In this new experiment, the spinning axis of the sample is flipped from 90 to 54.7° between the evolution and detection periods. The experiment appears to be widely applicable and has great promise for the study of complex samples.

The experimental scheme is set out in Fig. 1. Cross polarization of, in our case, ^{13}C nuclei is performed while the sample is spun about an axis that makes an angle of 90° with the static magnetic field. It can be shown (*10*) that the powder anisotropy pattern that obtains under these conditions is reversed and collapsed to half the width of the static nonspinning case, but keeps the same shape. At the end of the evolution period (t_1), the x component of the transverse ^{13}C magnetization is stored along the z axis, parallel to the static magnetic field, by means of a $90_y°$ ^{13}C pulse. The orientation of the spinning axis of the sample is then changed to the magic angle. The sample is spun fast compared with the width of the anisotropy patterns, so that spinning sidebands have negligible intensities. A final 90° ^{13}C pulse rotates the z-stored ^{13}C magnetization back into the transverse plane, where it precesses in the time domain, t_2, with the corresponding isotropic chemical shift frequencies.

Cycling of the phase of the first 90° ^{13}C pulse alternately along $+y$ and $-y$, together with adding and subtracting of the acquired data, is used to eliminate spurious signals. The detected isotropic spectrum, $S(t_1, F_2)$ obtained by Fourier transformation with respect to t_2, is modulated in amplitude with the frequencies existing during the evolution period, t_1. Hence, the powder anisotropy information and the isotropic chemical shifts will appear in the F_1 dimension. Because of the amplitude modulation, a pure 2D absorption spectrum can be obtained by calculating the cosine Fourier transform, $S^{cc}(F_1, F_2)$ (*11, 12*).

* To whom correspondence should be addressed.

0022-2364/83 $3.00

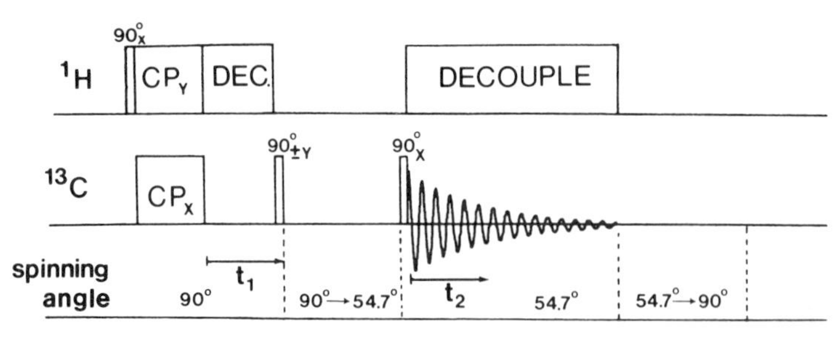

FIG. 1. Schematic representation of the two-dimensional experiment for mapping out the CSA pattern in the F_1 dimension. The angle between the sample spinning axis and the static magnetic field is changed from 90 to 54.7° between the evolution and detection periods. In the present case a sample spinning at 2.5 kHz undergoes this change of angle in 0.5 sec.

The method is demonstrated for a sample of polycrystalline *p*-dimethoxybenzene. Experiments were performed on a home-built spectrometer, equipped with a 2.35 T superconducting magnet and a Nicolet 293B pulse programmer. The sample was spun at 2.3 kHz in a turbine/air bearing spinning system obtained from Chemagnetics, Inc. (*13*). The orientation of the spinning axis was changed by means of a computer-controlled stepper motor from 90 to 54.7° in a time less than 1 sec. Further details about the experimental setup will appear in a forthcoming publication. The length of the evolution period was incremented in 64 steps of 83 μsec each, giving a spectral

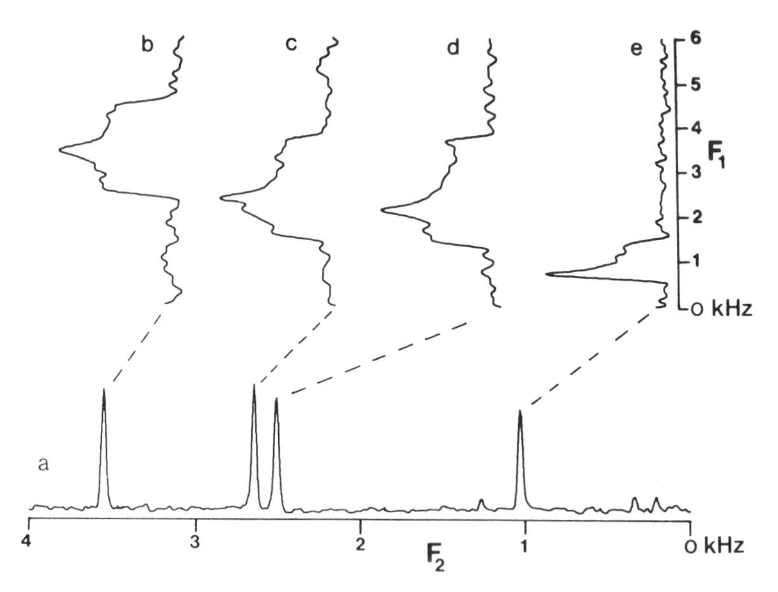

FIG. 2. (a) Phase-sensitive projection of the absorption-mode 2D spectrum of *p*-dimethoxybenzene, obtained with the sequence of Fig. 1, onto the F_2 axis, and (b–e) absorption mode cross sections through the 2D spectrum parallel to the F_1 axis, displaying the anisotropy patterns for the various sites. Because the ^{13}C transmitter frequency was placed at the high-shielding side of the spectrum, the anisotropy patterns have the conventional shape (not the mirror image). The patterns are half as wide as what would be obtained from the static, nonspinning case.

TABLE 1

MEASURED COMPONENTS, σ_{11}, σ_{22}, AND σ_{33}, OF THE ^{13}C CHEMICAL SHIFT SHIELDING TENSOR OF *p*-DIMETHOXYBENZENE GIVEN WITH RESPECT TO THE CORRESPONDING ISOTROPIC SHIELDING, σ_i (in ppm)

Site	σ_{11}	σ_{22}	σ_{33}	σ_i
$a(CH_3)$	−41 (−40)[a]	18 (15)	22 (24)	56
b or *c*[b]	−104 (−101)	24 (21)	82 (80)	113
b or *c*[b]	−95 (−96)	13 (17)	82 (79)	119
d	−82 (−82)	4 (7)	79 (75)	155

[a] Values in parentheses given by Maricq and Waugh (*1*).

[b] Unequivocal assignments of the *b* and *c* carbons have not yet been made.

width of 6.025 kHz in the F_1 dimension, and 24 transients were recorded for each value of t_1. The total measuring time was approximately 2.5 hr. The ^{13}C transmitter frequency was placed at the high-shielding side of the ^{13}C spectrum.

Figure 2 shows the isotropic shift spectrum obtained from a phase-sensitive projection (*14*) of the 2D spectrum onto the F_2 axis, and absorption-mode cross sections parallel to the F_1 axis through the 2D spectrum taken at the various isotropic F_2 shift frequencies, displaying the individual anisotropy powder patterns. Values for the ^{13}C chemical shift tensor components measured from the patterns shown in Fig. 2 are given in Table 1 and are in good agreement with those given by Maricq and Waugh (*1*).

Our new experiment yields the anisotropy information in a very straightforward way with hardly any distortion. The sensitivity of the experiment is rather good; but it is, of course, lower than that of the conventional CP/MAS experiment carried out in the same amount of time. Sensitivity decreases if very high resolution in the anisotropy pattern is needed, because in this case a long sampling time in the t_1 dimension is required and for long t_1 values only very little signal is acquired. The experiment is not very sensitive to the adjustment of experimental parameters such as pulse width, etc., and requires only a stable spinning device that remains stable when the orientation of the spinning axis is changed rapidly during the course of the experiment.

ACKNOWLEDGMENTS

The authors are grateful to Chemagnetics, Inc., for providing the sample-spinning system, and to the U.S. Geological Survey and U.S. Department of Energy (Laramie Energy Technology Center) for partial support of this research.

REFERENCES

1. M. M. Maricq and J. S. Waugh, *J. Chem. Phys.* **70,** 3300 (1979).
2. J. S. Waugh, M. M. Maricq, and R. Cantor, *J. Magn. Reson.* **29,** 183 (1978).
3. E. Lippmaa, M. Alla, and T. Tuherm, "Proceedings, 19th Congress Ampere, Heidelberg, 1976," p. 113.
4. J. Herzfeld and A. Berger, *J. Chem. Phys.* **73,** 6021 (1980).
5. W. P. Aue, D. J. Ruben, and R. G. Griffin, *J. Magn. Reson.* **43,** 472 (1981).
6. Y. Yarim-Agaev, P. N. Tutunjian, and J. S. Waugh, *J. Magn. Reson.* **47,** 51 (1982).
7. A. Bax, N. M. Szeverenyi, and G. E. Maciel, *J. Magn. Reson.* **51,** 400 (1983).
8. A. Bax, N. M. Szeverenyi, and G. E. Maciel, *J. Magn. Reson.* **52,** 147 (1983).
9. E. O. Stejskal, J. Schaefer, and R. A. McKay, *J. Magn. Reson.* **25,** 569 (1977).
10. M. Mehring, "High Resolution NMR in Solids," pp. 40–43, Springer, New York, 1983.
11. G. Bodenhausen, R. Freeman, R. Niedermeyer, and D. L. Turner, *J. Magn. Reson.* **26,** 133 (1977).
12. A. Bax, "Two-Dimensional Nuclear Magnetic Resonance in Liquids," pp. 26–30, Reidel, Boston, 1982.
13. Chemagnetics Inc., 208 N. Commerce Dr., Ft. Collins, Colo. 80524.
14. K. Nagayama, K. Wuthrich, P. Bachmann, and R. R. Ernst, *J. Magn. Reson.* **31,** 133 (1978).

JOURNAL OF MAGNETIC RESONANCE **56**, 207–234 (1984)

Homonuclear Two-Dimensional ^1H NMR of Proteins. Experimental Procedures

G. WIDER,*,‡ S. MACURA,*,†,§ ANIL KUMAR,*,†,‖ R. R. ERNST,† AND K. WÜTHRICH*

* *Institut für Molekularbiologie und Biophysik, and* †*Laboratorium für Physikalische Chemie, Eidgenössische Technische Hochschule, CH-8093 Zurich, Switzerland*

Received May 9, 1983; revised July 26, 1983

Experimental techniques used for homonuclear 2D ^1H NMR studies of proteins are described. A brief survey of the general strategy for structural studies of proteins by 2D NMR is included. The main part of the paper discusses guidelines for the selection of experimental techniques, the elimination of artifacts and unwanted peaks in protein 2D ^1H NMR spectra, suppression of the solvent line in H_2O solutions, experimental parameters, numerical data processing before and after Fourier transformation, and suitable presentations of complex 2D NMR spectra.

I. INTRODUCTION

During the past five years it was demonstrated that well-resolved, informative two-dimensional (2D) ^1H NMR spectra of proteins in solution can be recorded with commercially available high-resolution NMR equipment (*1–5*). The fundamental experimental schemes of the experiments which have most profitably been used for studies of proteins are shown in Fig. 1 (*1–3, 6–10*). With the use of 2D NMR nearly complete individual proton resonance assignments have been obtained for several small proteins (*11–19*) and 2D NMR investigations of static and dynamic aspects of the conformations of these proteins in solution are in progress (e.g., *20–24*). With the exception of some reports on conceptually new procedures (*8, 25–33*), our earlier publications contain little information on experimental details of the 2D NMR measurements. The present paper describes salient points from our practical experience gained when optimizing the experimental conditions for homonuclear 2D ^1H NMR with proteins.

While the scope of the present article is limited to technical aspects which have been of particular relevance for homonuclear ^1H experiments with proteins, much of this information should be of interest also for studies of other classes of large molecules. Several reviews are already available which survey the procedures used for the study of low molecular weight compounds (e.g., *34–36*).

‡ Present address: Spectrospin AG, Industriestrasse 26, CH-8117 Fällanden, Switzerland.
§ Present address: Institute of Physical Chemistry, Faculty of Science, 1100 Beograd, Yugoslavia.
‖ Present address: Department of Physics, Indian Institute of Science, Bangalore, India.

0022-2364/84 $3.00

II. STUDIES OF PROTEIN STRUCTURE AND CONFORMATION BY 2D NMR

This section presents a brief discussion of a 2D NMR strategy used for the elucidation of protein conformation. Three basic types of 2D experiments have proved valuable in this context (Fig. 1):

(A) 2D NOE spectroscopy (NOESY) (10, 3) delivers maps which demonstrate spatial proximity of nuclei based on their mutual cross-relaxation producing magnetization transfer manifested in nuclear Overhauser effects (NOE). The experiment requires an extensive mixing period separated from evolution and detection periods by two $\pi/2$ pulses (Figs. 1A and A′).

(B) 2D correlated spectroscopy (COSY) (7, 2, 9) produces correlation maps which display the connectivity of nuclei by spin–spin coupling. A COSY spectrum thus provides information on proximity of nuclei along the chemical bonds. The correlation is tested by a β pulse, in the present context usually of flip angle $\beta = \pi/2$, applied in the middle or at the end of the evolution period (Figs. 1B and B′). The variant of

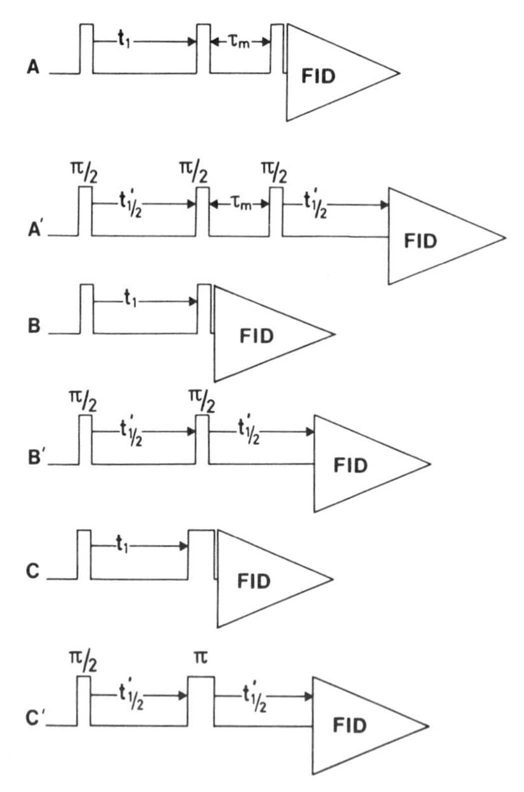

FIG. 1. Experimental schemes of the homonuclear 2D ^{1}H NMR experiments which have so far been used for the elucidation of protein conformation. (A, A′) 2D NOE spectroscopy (NOESY) without and with delayed accumulation, respectively (*3, 10*). (B, B′) 2D correlated spectroscopy; (B) without delayed accumulation (COSY) (*7–9*) and (B′) with delayed accumulation leading to 2D spin-echo correlated spectroscopy (SECSY) (*2, 8*). (C, C′) 2D *J*-resolved spectroscopy without and with delayed accumulation, respectively (*1, 6*). The evolution time is denoted by t_1 and τ_m is the mixing time.

Fig. 1B′ with the $\pi/2$ pulse in the middle of evolution is called 2D spin-echo correlated spectroscopy (SECSY).

(C) 2D J-resolved spectroscopy (6, 1) allows the separation of spin–spin multiplets and chemical shifts in two orthogonal frequency dimensions. This may lead to an efficient resolution of overlapping multiplets and allows their detailed analysis. The separation of multiplets and shifts is effected by a π pulse applied during the evolution period (Figs. 1C and C′).

For studies of protein structure we have so far primarily used a suitable combination of COSY or (SECSY) and NOESY spectra. 2D *J*-resolved spectroscopy, in addition, may be used for more refined studies of medium-size proteins where the multiplet structure is resolvable. The information contained in the 2D NMR spectra is first used for spectral assignments. In a first step the spin systems of the different amino acid types are identified uniquely or in groups on the basis of the *J*-connectivity patterns in SECSY or COSY spectra recorded in D_2O (*11, 14*). Additional support for these identifications may come from determination of the spin–spin coupling fine structure in the 2D *J*-resolved spectra (*11*). In a second step neighboring residues in the amino acid sequence are identified with NOESY experiments. As shown in Fig. 2 all the crucial ^1H–^1H connectivities involve the labile backbone amide protons (*12–19, 37*). It is therefore of critical importance that high-quality NOESY and COSY spectra of proteins can be recorded in H_2O solution (*26, 32*). In a third step the data on the identification of amino acid side-chain spin systems and sequential neighborhood relations are combined to obtain individual assignments in the primary structure (*13–18, 38*).

Once individual resonance assignments have been obtained, all the experiments of Fig. 1 may be useful for obtaining further information on the protein conformation. With regard to determination of three-dimensional polypeptide structures by NMR, NOESY plays a key role by providing a network of short ^1H–^1H distance

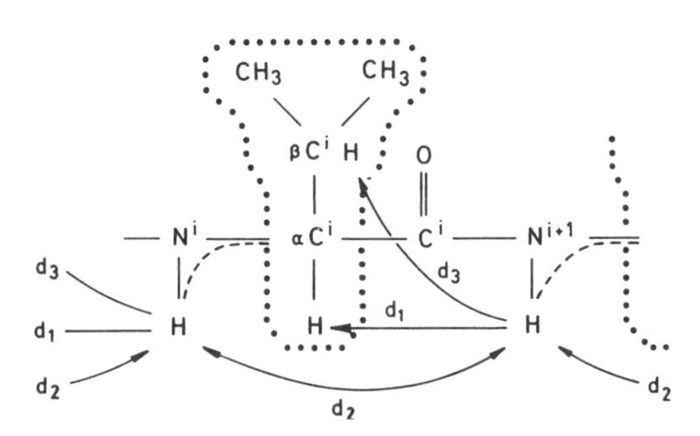

FIG. 2. Peptide segment ---Val---(NH)--- with indication of the *J* and NOE connectivities used to obtain individual resonance assignments for proteins. Dotted lines are drawn around the spin systems of the nonlabile hydrogen atoms in the individual amino acid residues which are identified from the *J* connectivities in SECSY or COSY spectra recorded in D_2O. The broken lines indicate the *J* connectivities between the backbone amide proton and the C^α proton of the individual residues, which are determined from COSY or SECSY spectra recorded in H_2O. The arrows indicate the through-space distances d_1, d_2, and d_3 used to identify sequentially neighboring residues from NOESY spectra recorded in H_2O (*37*). Note that d_2 is, in contrast to d_1 and d_3, symmetrical with respect to the direction of the polypeptide chain.

constraints. These can be used either for interactive model building (*39*) or as an input for a distance geometry algorithm (*40*) to develop molecular models which would be compatible with the experimental data. The problem tackled in the distance geometry approach is fundamentally the following (*40–42*): Given upper limits on the distances between N atoms by the network of covalent bonds and by the NOESY data, and lower distance limits by the van der Waals radii of the atoms, what are the possible conformations which are compatible with the ensemble of these constraints? Considering that the correlations between peak intensity in the NOESY spectra and proton–proton distances are presently at best semiquantitative (*23, 40*), it is important that the results of the distance geometry calculations do not depend critically on the accuracy of the distance measurements as long as a sufficiently large number of distance constraints is available (*40–42*).

For refinement of the structures obtained via the distance geometry approach, these results may be checked against other NMR parameters, such as chemical shifts, spin–spin coupling constants, and amide proton exchange rates (*43*). J-coupling constants may be obtained from 2D J spectra and/or from SECSY and COSY (*1, 2, 44, 45*). The relative signs of coupling constants become apparent in COSY spectra for flip angles $\beta < \pi/2$ (*7, 9*). Again, the data which involve backbone amide protons are of particular interest so that such studies must be based on NMR recordings in H_2O solution of the proteins.

2D NMR may further provide information on the internal mobility of protein molecules. For example, 2D J spectroscopy was used to investigate rotational mobility of amino acid side chains about the torsion angle χ^1, i.e., about the bond linking C^α and C^β (Fig. 2) (44), and COSY was employed to measure the exchange rates of individual labile protons in the polypeptide chain with the solvent (*21, 22*).

Before discussing the relevant practical aspects of the 2D experiments in the following sections, we present in Figs. 3 and 4 representative NOESY and COSY spectra of bull seminal inhibitor II A (BUSI II A) (*18*), a protein of molecular weight 6500. The spectra were recorded at 500 MHz on a Bruker WM 500 spectrometer, using the experiments of Fig. 1A for Fig. 3 and of Fig. 1B for Fig. 4. Quadrature detection was employed, with the carrier frequency at the low-field end of the spectrum. To eliminate experimental artifacts, a 16-step phase cycle was used for each value of t_1 (*2, 8*). The H_2O resonance was suppressed by selective, continuous irradiation at all times except during data acquisition (*26, 32*). A total of 512 measurements with t_1 values from 0.3 to 51 msec were recorded with 1024 data points in t_2. To end up with a 1024 $\times$ 1024-point frequency-domain data matrix for COSY and NOESY, which corresponds to the digital resolution given in the figure captions, the time-domain matrix was expanded to 4096 points in t_1, i.e., 2048 points each for the real and the imaginary part, and 4096 points in t_2 by "zero filling." Prior to Fourier transformation the time-domain data matrix was multiplied in the t_1 direction with a phase-shifted sine bell, $\sin(\pi(t + t_0)/t_s)$, and in the t_2 direction with a phase-shifted sine squared bell, $\sin^2(\pi(t + t_0)/t_s)$. The length of the window functions, t_s, was adjusted for the sine bells to reach zero at the last experimental data point in the t_1 or t_2 direction, respectively. The phase shifts, t_0/t_s, were 1/64 and 1/128 in the t_1 and t_2 directions, respectively. The spectra in Figs. 3 and 4 are shown in the absolute value representation.

The NOESY spectrum in Fig. 3A contains the diagonal peaks on the diagonal

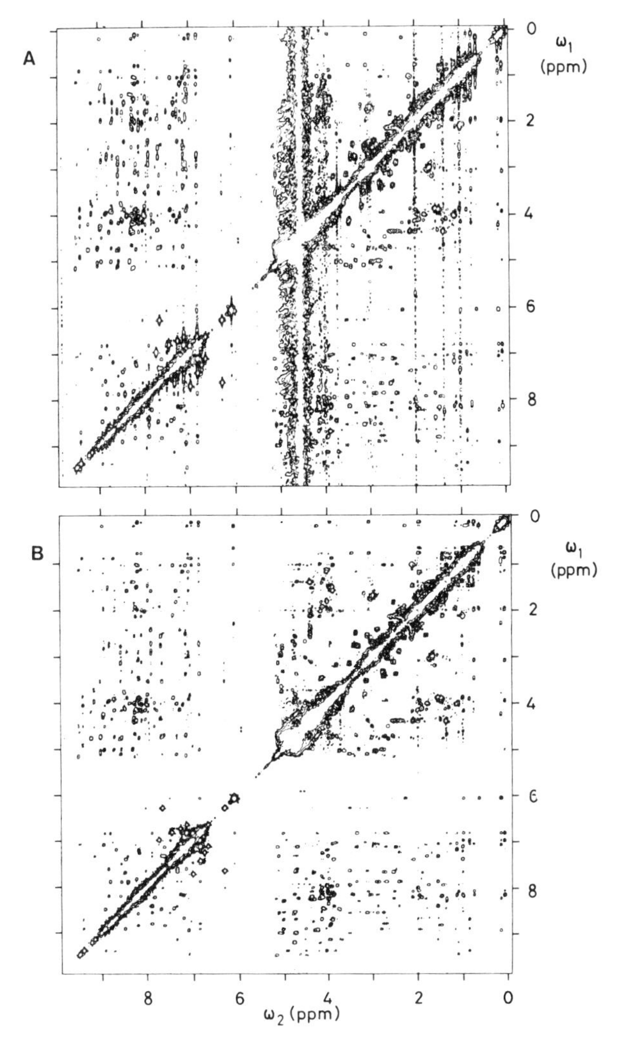

FIG. 3. Contour plot of a 500 MHz ^{1}H NOESY spectrum of a 0.016 *M* solution of bull seminal inhibitor II A (BUSI II A), which is a protein of molecular weight 6500, in H_2O, pH 4.9, T = 45°C. The digital resolution is 4.8 Hz/point. The spectrum was recorded in ca. 25 hr. and a mixing time of 200 msec was used. The strong vertical noise band at 4.6 ppm is at the chemical shift of the water resonance, which was suppressed by selective, continuous irradiation at all times except during the acquisition period t_2 (*26, 32*). (A) unsymmetrized spectrum; (B) spectrum after symmetrization (*29*).

from the lower left to the upper right corner. These peaks correspond to those in the conventional 1D ^{1}H NMR spectrum (*3*). In addition numerous well resolved cross-peaks occur in all the different spectral regions. These manifest NOEs between protons attached to the same amino acid residue, between sequentially neighboring residues (e.g., d_1, d_2, and d_3 in Fig. 2) and between residues which are far apart in the amino acid sequence. In NOESY spectra recorded with different mixing times, different relative intensities prevail between the diagonal spectrum and the cross-peaks, and

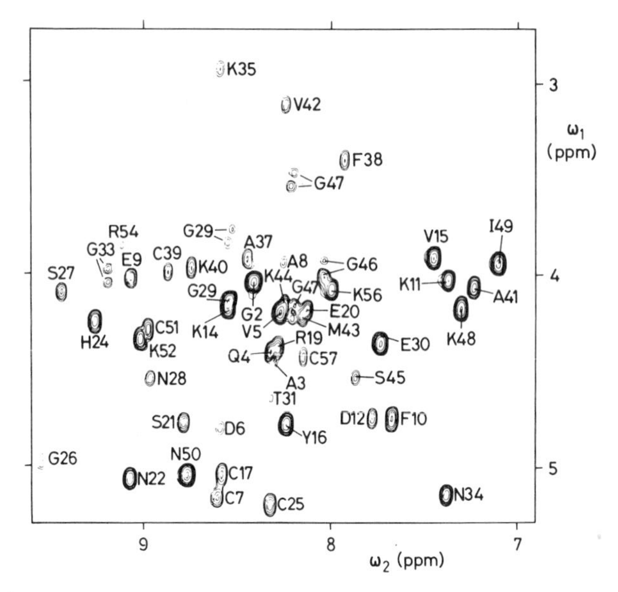

FIG. 4. NMR "fingerprint" of BUSI IIA in the spectral region (ω_1 = 2.8 to 5.2 ppm, ω_2 = 6.9 to 9.7 ppm) of a 500 MHz ^{1}H COSY spectrum recorded in a 0.012 *M* solution of the protein in H_2O, pH 4.9, at 45°C. The digital resolution is 4.8 Hz/point, the spectrum was recorded in ca. 17 hr. The water resonance was suppressed by selective, continuous irradiation at all times except during the acquisition period t_2. The assignments for the individual cross-peaks are indicated by the one-letter symbol for the amino acid residue and the position in the amino acid sequence. Since they were bleached out by the water irradiation (*26*) the cross-peaks of Thr 18, Tyr 32, His 53, and Gly 55 are missing and the Thr 31 peak is very weak in this spectrum. Furthermore, the cross-peak of Cys 36 is outside of this spectral region at (ω_1 = 2.44 ppm, ω_2 = 7.92 ppm) (*18*).

between different cross-peaks (*27*). As a rule of thumb for the structural interpretation of the NOESY spectra we assume that the distance between a pair of protons is shorter than ca. 5.0 Å when a NOESY cross-peak can be observed with a mixing time of 100 msec (*40*). The use of this information is further discussed below.

Figure 3A contains numerous vertical noise bands at the ω_2 positions of intense diagonal peaks. The most prominent artifacts are at the position of the solvent water resonance. These and additional artifacts can be efficiently suppressed by symmetrization of the spectrum (Fig. 3B) (*28, 29*). A symmetrized spectrum is usually more practical for the initial spectral analysis and is also more readily amenable for automated peak picking. Potential pitfalls of the technique are discussed in the course of this paper.

A COSY spectrum has the same general appearance as the NOESY spectrum in Fig. 3. It provides with a single instrument setting a complete set of all the scalar spin–spin coupling connections in the macromolecular structure. This includes long-range connectivities with small coupling constants which have in practice not been accessible to 1D ^{1}H NMR studies of proteins, e.g., those between the nonlabile imidazole ring protons of histidine (*15, 46*). The spectral region in Fig. 4 contains the nearly complete "NMR fingerprint" (*13*) of BUSI II A. The fingerprint of a protein in the $C^{\alpha}H$–NH region of the COSY spectrum contains one peak for each amino acid residue (Fig. 2) except for the prolines, which give no peak, and the glycines, which may give two peaks. When the amino acid sequence is known, inspection of the fingerprint region of the COSY spectrum then shows whether or not the resonances of all the amino acid residues in the protein can be observed and resolved in the ^1H 2D NMR spectra (*13*).

III. GUIDELINES FOR THE SELECTION OF THE EXPERIMENTAL TECHNIQUE

In setting up a 2D experiment it is at first necessary to select the most appropriate pulse scheme (Fig. 1). This section presents some guidelines.

Each of the three basic experiments shown in Fig. 1 occurs in two variants: Using the same pulse sequence, data acquisition can be started either immediately after the observation pulse ("direct acquisition 2D experiment") or delayed at the time of the expected spin echo ("delayed acquisition 2D experiment") (Fig. 1). In the case of correlated spectroscopy, which is in the following used to further illustrate the two different procedures, this leads to COSY and SECSY, respectively (*2, 8*). The two types of experiments are to some extent equivalent in the sense that a SECSY spectrum can be computed from a COSY data set while the opposite conversion is only approximately possible since the acquired set for a SECSY-type experiment lacks data in a triangular area of the (t_1, t_2) domain.

A SECSY-type experiment (with *P*-type peaks suppressed, see Section IVe) exhibits along ω_1 only differences of initial and final frequencies and spans a reduced frequency range, provided that all connected peaks are near in the frequency domain. This is always true in 2D *J*-resolved spectra of proteins and often true in protein correlation spectra, but happens only rarely in NOESY spectra where cross-peaks appear also between resonance lines which are remote in the frequency domain. SECSY-type experiments are therefore useful particularly for 2D *J*-resolved and correlation spectroscopy.

The smaller frequency range to be covered in ω_1 allows one to save acquisition time and/or permits higher resolution in ω_1. The frequency range to be covered in a delayed acquisition experiment should be at least $\pm(\Delta\omega)_{max}/2$, while in the SECSY form of correlated spectroscopy (Fig. 1B′) $(\Delta\omega)_{max}$ is the maximum frequency difference of *J*-connected peaks. In 2D *J*-resolved spectroscopy (Fig. 1C′) $(\Delta\omega)_{max}$ is the largest spread of multiplet components in the spectrum (*47*).

In an echo-type experiment, like SECSY, it is normally impossible to achieve pure absorption and mixed lineshapes are obtained (*2, 8*). For pure absorption spectra, COSY has to be used with $\beta = \pi/2$. However, it should be noted that even with COSY-type experiments a reduced frequency range in ω_1 of $\pm(\Delta\omega)_{max}/2$ is allowed. The resulting foldover can be corrected mathematically with the FOCSY procedure (foldover-corrected spectroscopy) (*8, 48*) and leads then to a representation of difference frequencies along ω_1 which is reminiscent of SECSY-type spectra but can, if required, be recorded in pure absorption. Details on the FOCSY procedure are presented in Section VI below. From an information–theoretical standpoint FOCSY appears to be the preferred technique. In practice, a combination of SECSY and COSY spectra of the same protein molecule has proven to be quite efficient for assignment and distinction of incompletely resolved peaks.

IV. ELIMINATION OF ARTIFACTS AND "UNWANTED" PEAKS

(a) Elimination of Axial Peaks

In addition to the cross-peaks of interest and the diagonal peaks (Fig. 3) resonances known as "axial peaks" (*7*) may appear along the $\omega_1 = 0$ line in the two-dimensional spectra. These peaks originate from longitudinal magnetization that builds up during the time period immediately before the detection pulse. Such peaks do not contain information of interest and may be eliminated. In a COSY experiment a 180° phase shift of the second rf pulse (mixing pulse) inverts the signs of the axial peaks whereas the signs of the diagonal and cross-peaks are not affected. Addition of the signals from the following two experiments will therefore eliminate axial peaks:

$$\begin{array}{ccc} x & x & + \\ x & -x & +. \end{array} \quad [1]$$

where x stands for a $90°_x$-pulse, $-x$ for a $90°_{-x}$ pulse, and $\pm$ for addition or subtraction of the signals in the computer memory.

In a NOESY experiment, a 180° phase shift of the second pulse inverts the sign of the exchanging z magnetization. The longitudinal magnetization which builds up during the mixing time and leads to the axial peaks, however, does not change sign. Subtraction of the signals from the following two experiments will thus eliminate axial peaks:

$$\begin{array}{cccc} x & x & x & + \\ x & -x & x & -. \end{array} \quad [2]$$

(b) Single and Quadrature Mode Detection

In 2D NMR spectroscopy of large molecules, performance time and signal to noise (S/N) ratio are always limiting factors. It is therefore mandatory to use quadrature phase detection in the ω_2 direction in order to gain the available factor of 2 in time or $\sqrt{2}$ in S/N. In quadrature detection the well-known CYCLOPS sequence (*49*) should be employed to correct for inaccuracies in the two detector channels and for mismatch of the two audiofrequency filters. Using quadrature phase detection, it is possible to distinguish between positive and negative ω_2 frequencies (Fig. 5) and to set the carrier frequency in the center of the spectrum for reducing the bandwidth to be covered. This will, however, also cause positive and negative ω_1 frequencies to occur (Fig. 5) and quadrature detection in ω_1 will also be necessary. To eliminate image frequencies, two experiments have to be performed for each t_1 value with the rf phases of the mixing pulses shifted by $\pi/2$ between the two experiments. Because of the large differences between the intensities of diagonal and cross-peaks in COSY and NOESY spectra, the image suppression must be very good to prevent the appearance of a disturbing "antidiagonal" in the resulting 2D spectrum (Fig. 6). Experience shows that this happens frequently in protein spectra. To prevent this, the carrier frequency is often set at one end of the spectrum, but quadrature detection in ω_2 is nevertheless employed to gain the factor $\sqrt{2}$ in S/N. However, image suppression in ω_1 is then unnecessary.

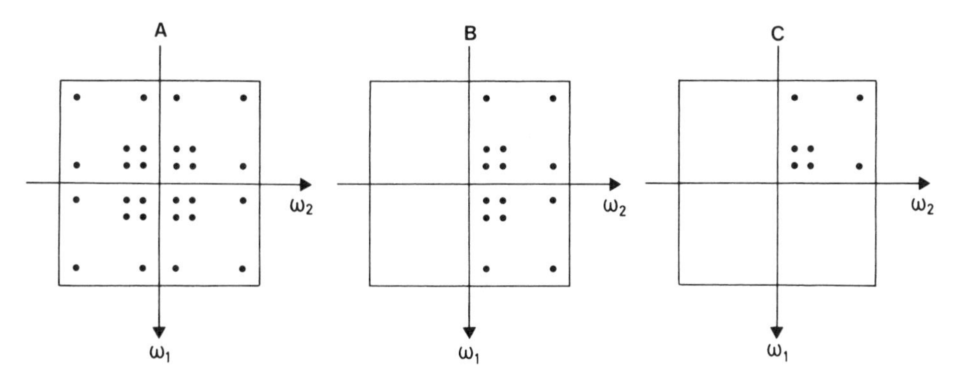

FIG. 5. Scheme illustrating single-channel and quadrature detection in 2D spectroscopy. (A) With single-channel detection the four quadrants in a 2D spectrum cannot be separated. (B) With quadrature detection positive and negative ω_2 frequencies can be distinguished. (C) By linear combination of two different measurements with different phases of the rf pulses (see text) P and N peaks can be distinguished and may be selectively suppressed.

(c) Separation of P-type and N-type Peaks

In this section, we discuss in some detail the forementioned problem of quadrature detection in ω_1. The application of a mixing pulse of flip angle β between evolution and detection periods (Figs. 1A–C) or in the middle of the evolution period (Figs. 1A′–C′) induces transfer of coherence between different transitions as well as between components corresponding to apparently opposite sense of precession during the evolution period t_1. This leads to pairs of peaks in the 2D spectrum at mirror image frequencies $\pm\omega_1$. Peaks which correspond to precession in the same sense in t_1 and

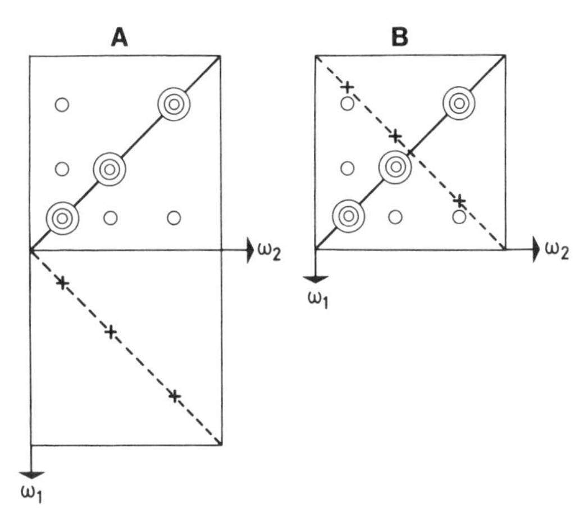

FIG. 6. Suppression of P peaks. (A) Scheme of a NOESY or COSY spectrum (Figs. 1A and B). The carrier frequency is chosen outside of the spectrum. Quadrature detection along ω_1 is assumed, with suppression of the P peaks. Because of incomplete suppression spurious diagonal P peaks (+) may be retained. (B) When full 2D quadrature detection is applied the spurious peaks are folded into the spectral region of interest and may obscure the spectrum of the N peaks.

t_2 are called P peaks (*2*) or "antiecho peaks," while peaks with precession in opposite sense during t_1 and t_2 are identified as N peaks (*2*) or "echo peaks." The two types of peaks have different character in the presence of magnetic field inhomogeneity; N-type peaks tend to be narrower than P-type peaks due to refocusing caused by the opposite sense of precession. They have also different flip-angle dependence. P-type peaks dominate for small flip angle and disappear for $\beta = \pi$ while N-type peaks have vanishing amplitude for $\beta = 0$ and become maximum for $\beta = \pi$. For $\beta = \pi/2$, the two kinds have equal amplitude.

Complete COSY and SECSY spectra for an AMX spin system are shown in schematic form in Figs. 7 and 8. It is obvious that only the N-type peaks in a SECSY spectrum exhibit the advantage of a restricted ω_1 frequency range. There is thus a strong motivation to devise experimental schemes by which the P-type peaks can be suppressed. Another reason for separation is the severe overlap of the two classes of peaks whenever the rf carrier is set in the center of the spectrum.

Suppression of one of the two classes of peaks is possible by a set of experiments with a suitable phase cycle. For a COSY experiment the following two experiments have to be combined

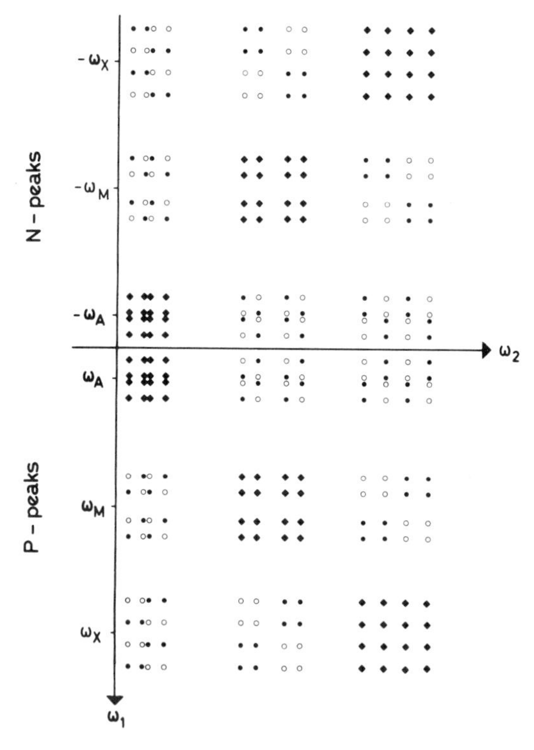

FIG. 7. Schematic COSY spectrum of an AMX spin system; weak coupling is assumed. The coupling constants are $J_{AM} = 5$ Hz, $J_{AX} - 7$ Hz, and $J_{MX} = -14$ Hz. Positive absorptive (●), negative absorptive (○), and dispersive (♦) resonance lines are distinguished. All absolute-value intensities are equal. Note that this AMX system, which is important in protein ^{1}H NMR spectra (*44*), contains different groups of cross-peaks with identical signs.

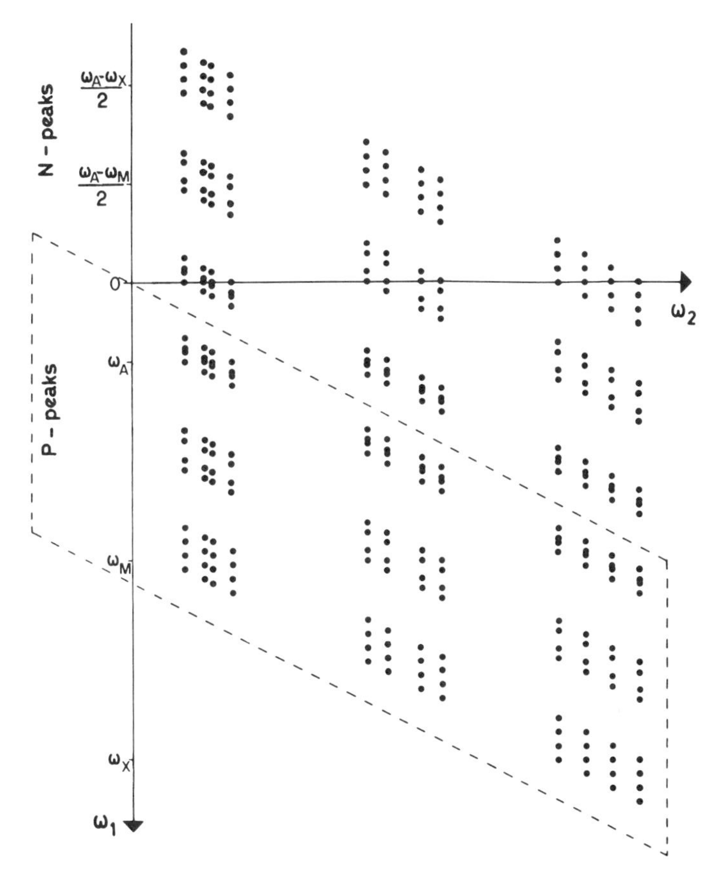

FIG. 8. Schematic SECSY spectrum of an AMX spin system. The coupling constants are $J_{AM} = 5$ Hz, $J_{AX} = 7$ Hz, and $J_{MX} = -14$ Hz and weak coupling is assumed. All peaks in a SECSY spectrum consist of a superposition of an absorptive and a dispersive contribution. Cross-peak multiplets are in antiphase such that severe overlap leads to cancellations. SECSY spectra are normally plotted in absolute-value mode.

$$\begin{array}{cccc} & & N & P \\ x & x & + & + \\ x & y & - & + \end{array} \tag{3}$$

Here x and y indicate the phases of the applied rf pulses while $\pm$ refers to addition or subtraction of the signals to obtain either N- or P-type peaks. Combined with the suppression cycle for axial peaks of Eq. [1], the following phase cycle is obtained:

$$\begin{array}{cccc} & & N & P \\ x & x & + & + \\ x & y & - & + \\ x & -x & + & + \\ x & -y & - & + \end{array} \tag{4}$$

which is also called EXORCYCLE (*50*). An experimental example of an AMX system is shown in Fig. 9. Figure 9A gives the complete SECSY spectrum without separation of *N*- and *P*-type peaks while in Fig. 9B the *P*-type peaks have been suppressed by the cycle of Eq. [4].

For NOESY experiments cycle [4] can be used as well. The two mixing pulses use the same phase as the second pulse in cycle [4]. For quadrature detection in both directions the phase cycles must be combined with CYCLOPS (*49*) to compensate for phase and amplitude errors. In the notation used here CYCLOPS is written as

$$\begin{array}{ccc} x & +A & +B \\ y & -B & +A \\ -x & -A & -B \\ -y & +B & -A. \end{array} \qquad [5]$$

The signals A and B of the two detectors are added (+) or subtracted (−) into two memory regions in the computer. The two memory regions are indicated by the second and third column in [5]. The combination of phase cycles [4] and [5] leads to a cycle with 16 phases for both COSY and for NOESY (Tables 1 and 2). The application of the full scheme is necessary when the carrier frequency is set in the middle of the spectrum, otherwise partial schemes may profitably be employed (see footnotes to Tables 1 and 2).

(d) Setting the Carrier Frequency

In a polypeptide or protein 2D spectrum many of the cross-peaks have only a few percent of the intensity of the diagonal peaks. Mirror image diagonal peaks in quadrature detection (Fig. 6) can therefore be quite disturbing even when the quadrature balance is well adjusted. For this reason it is often advisable in COSY and NOESY experiments to set the carrier frequency to one end of the spectrum even when quadrature detection is available. An additional advantage is that incompletely suppressed axial peaks do then not appear in the central region of the 2D spectrum.

In SECSY experiments, where the P peaks must always be suppressed, frequency

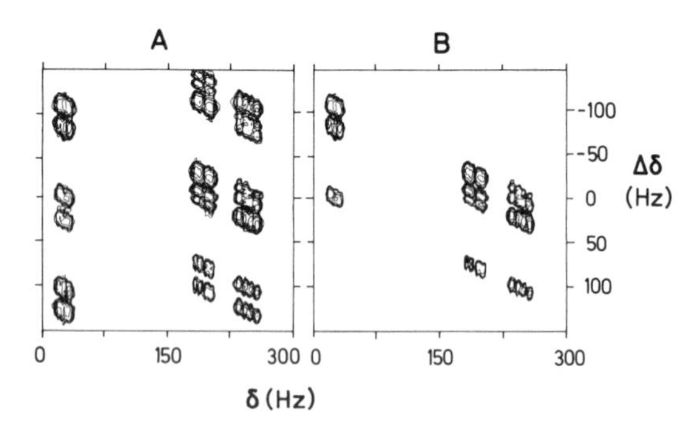

FIG. 9. SECSY spectrum of a 0.04 M tyrosine solution in D_2O, pD = 11.5, T = 30°C. The spectral region is shown which contains the resonances of the $C^{\alpha}H–C^{\beta}H_2$ AMX spin system. (A) N- and P-type peaks are present. Some of the P peaks are folded in the ω_1 direction. (B) Only the N-type peaks are present, the P peaks have been suppressed.

TABLE 1

PHASE CYCLE FOR HOMONUCLEAR 2D CORRELATED SPECTROSCOPY (COSY/SECSY) ALLOWING FULL QUADRATURE DETECTION (AXIAL AND *P* PEAKS ARE SUPPRESSED)[a]

Phases		Computer memory	
1st Pulse	2nd Pulse	Region I	Region II
x	x	$+A$	$+B$
x	$-x$	$+A$	$+B$
y	y	$-B$	$+A$
y	$-y$	$-B$	$+A$
$-x$	$-x$	$-A$	$-B$
$-x$	x	$-A$	$-B$
$-y$	$-y$	$+B$	$-A$
$-y$	y	$+B$	$-A$
x	y	$-A$	$-B$
x	$-y$	$-A$	$-B$
y	$-x$	$+B$	$-A$
y	x	$+B$	$-A$
$-x$	$-y$	$+A$	$+B$
$-x$	y	$+A$	$+B$
$-y$	x	$-B$	$+A$
$-y$	$-x$	$-B$	$+A$

[a] The schemes [1], [3] and [5] are the basis for this phase cycle. Application of the full sequence of 16 phases is needed for recording of absolute-value spectra when the carrier is in the center of the spectrum. When the carrier frequency is set to one end of the spectrum only the first block of 8 phases is profitably employed. The data can then be transformed in either the absolute-value mode or the phase sensitive mode. Use of the full scheme of 16 phases results in mixing of absorptive and dispersive lineshapes and phase-sensitive spectra can only be obtained when the recordings with the first and second cycle of 8 phases are stored in separate memory locations. This allows one to obtain phase-sensitive spectra from recordings with a centrally located carrier frequency (*76–78*).

folding of spurious *P*-type peaks cannot be avoided and there is no particular reason not to set the carrier frequency in the center of the spectrum. In 2D *J*-resolved spectroscopy, it is convenient to set the carrier frequency in the center of the spectrum. *P*-type peaks do not occur in this case provided that the 180° pulse is perfect.

(e) Suppression of J Cross-Peaks and Transverse Magnetization in NOESY

For the proper functioning of a NOESY experiment exclusively frequency-labeled longitudinal magnetization should exist during the mixing period. Additional com-

TABLE 2

PHASE CYCLE FOR 2D NOE SPECTROSCOPY (NOESY) ALLOWING FULL QUADRATURE DETECTION (AXIAL AND *P* PEAKS ARE SUPPRESSED) AND REMOVING TRANSVERSE MAGNETIZATION[a]

Phases			Computer memory	
1st Pulse	2nd Pulse	3rd Pulse	Region I	Region II
x	x	x	$+A$	$+B$
x	$-x$	x	$-A$	$-B$
x	x	y	$-B$	$+A$
x	$-x$	y	$+B$	$-A$
x	x	$-x$	$-A$	$-B$
x	$-x$	$-x$	$+A$	$+B$
x	x	$-y$	$+B$	$-A$
x	$-x$	$-y$	$-B$	$+A$
x	y	y	$-A$	$-B$
x	$-y$	y	$+A$	$+B$
x	y	$-x$	$+B$	$-A$
x	$-y$	$-x$	$-B$	$+A$
x	y	$-y$	$+A$	$+B$
x	$-y$	$-y$	$-A$	$-B$
x	y	x	$-B$	$+A$
x	$-y$	x	$+B$	$-A$

[a] The phase cycles discussed in Section IVc are the basis of this table. See footnote to Table 1.

ponents, like transverse magnetization and multiple-quantum coherence, will lead to artifacts in the NOESY spectrum in the form of incorrect intensities and phases, and in the appearance of *J* cross-peaks (*10, 51*). The elimination of all unwanted components, except for zero-quantum coherence, is quite easy and can be achieved either by the application of a field gradient pulse during the mixing period (*10*) or much better by a suitable phase cycle (*51*). Phase cycle [2], for example, eliminates all odd order multiple-quantum transitions. The phase cycle of Table 2 suppresses in addition effects from double-quantum coherence. Zero-quantum coherence can, however, not be removed by phase cycling. Unfortunately, it contributes significantly to the *J* cross-peaks in polypeptide spectra (*31*).

The elimination of zero-quantum coherence can utilize its rapid oscillation during the mixing time (*30, 52*). A random modulation (*52*) of the mixing time, τ_m (Fig. 1A), within a selected interval τ_m^{var} averages zero-quantum coherence nearly to zero provided that τ_m^{var} extends over several oscillation periods (determined by chemical shift differences). NOE cross-peaks are only insignificantly affected by such a variation of τ_m. The insertion of a 180° pulse with variable position in the mixing period has a similar effect as the variation of the mixing time (*31*). The random variation of the mixing time (or the 180° pulse position) must cover time intervals longer than one period of the lowest zero-quantum transition frequency

$$\tau_m^{var} > 1/\omega_{ZQT}^{min}. \qquad [6]$$

ω_{ZQT}^{min} is equal to the smallest chemical shift difference $\Delta\omega$ between coupled nuclei. In practice this means that τ_m^{var} must be of the order of 2–10 msec to cover the low-frequency range from 100–500 Hz.

Instead of a random variation of the mixing time τ_m or of the position of the 180° pulse, a systematic incrementation of the mixing time correlated with the increase of t_1 can also be employed (*30, 31*)

$$\tau_m = \tau^{\circ}{}_m + \frac{k}{N}\tau_m^{var}; \qquad k = 0, 1, \ldots, N. \qquad [7]$$

This leads to a shift displacement of the J cross-peaks in the ω_1 direction, leaving exclusively the NOE cross-peaks in place. By a symmetrization of the 2D spectrum (*28, 29*) it is then possible to eliminate the J cross-peaks efficiently.

(f) Suppression of the Solvent Line in H_2O Solutions of Proteins

In ^1H NMR experiments with H_2O solutions of proteins, which are of crucial importance in structural studies (see Section II), one faces severe problems with regard to detecting weak resonance signals in the presence of a strong solvent line. Since for proteins the experimental conditions can usually be selected such that magnetization transfer with the solvent by exchange of the amide protons is inefficient (*43, 53*), this problem can be circumvented by selective saturation of the solvent resonance (*26*). Among different water suppression schemes that have been used (*12–19, 26, 32*), the experiment shown in Fig. 10 appears to be the best choice. A continuous, selective rf field is applied to the H_2O line at all times except during the periods t_1 and t_2, which have to be chosen sufficiently short so that no significant recovery of the water magnetization occurs. The effect of the solvent suppression is illustrated in Fig. 11. Without suppression of the water line, the signal to noise ratio is such that (with the possible exception of two or three particularly strong peaks) the positions of the cross-peaks cannot be determined (compare Fig. 11A with Figs. 11B and C).

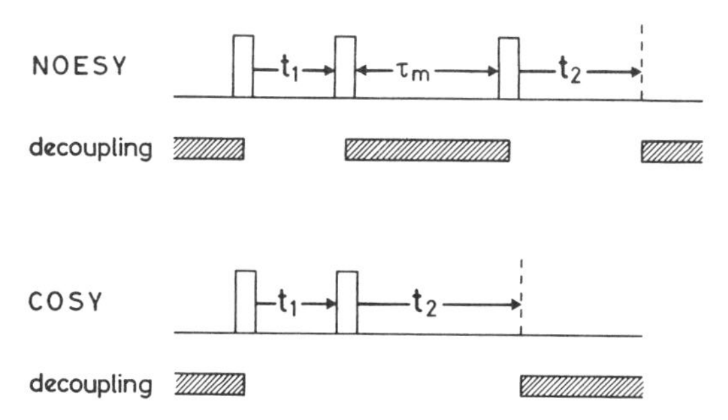

FIG. 10. Experimental scheme for the H_2O suppression in COSY and NOESY experiments. The H_2O resonance is continuously irradiated at all times except during t_1 and t_2. Compared to the procedure used for the experiments of Figs. 3 and 4, where the decoupler was on also during t_1, this scheme has the advantage that no Bloch–Siegert shifts will distort the spectrum even at high amplitudes of the decoupler field.

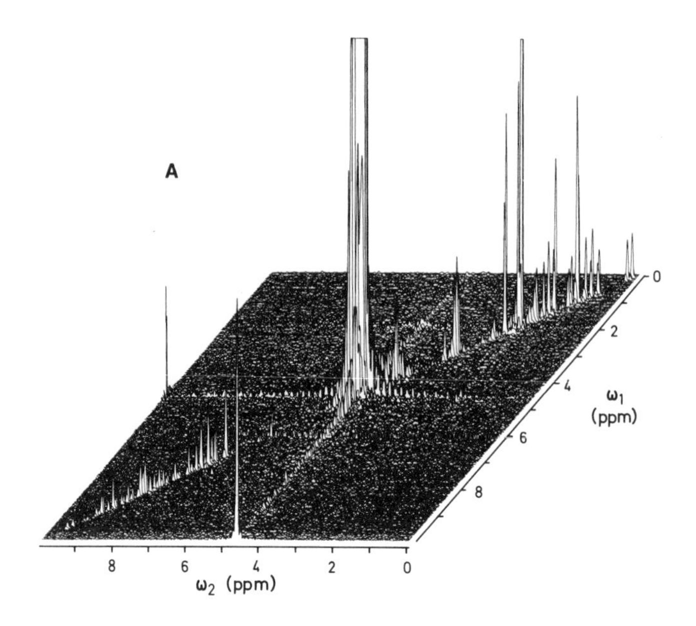

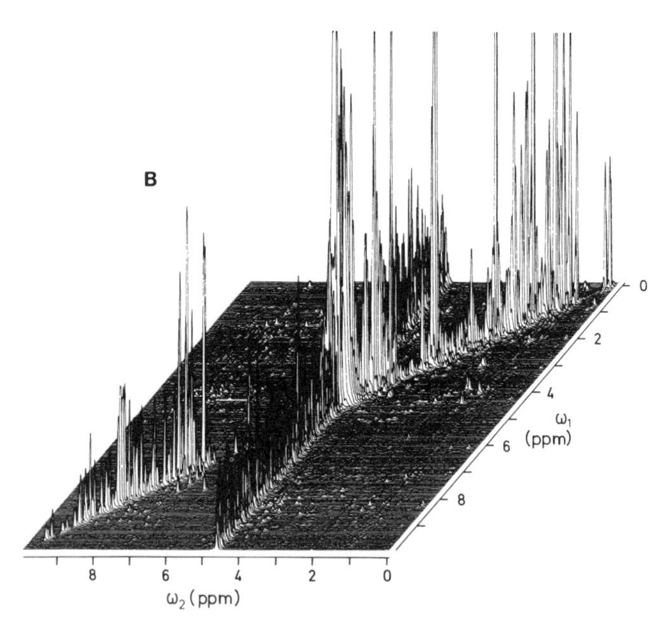

FIG. 11. Stacked plot presentation of three 500 MHz ^{1}H NOESY spectra of the BUSI II A solution used in Fig. 3. (A) Spectrum recorded without water suppression. (B) Spectrum recorded with water suppression (same data as in Fig. 3A). (C) Spectrum B after symmetrization (same data as in Fig. 3B). For all three spectra identical experimental parameters were used with the exception of the solvent suppression and symmetrization mentioned above.

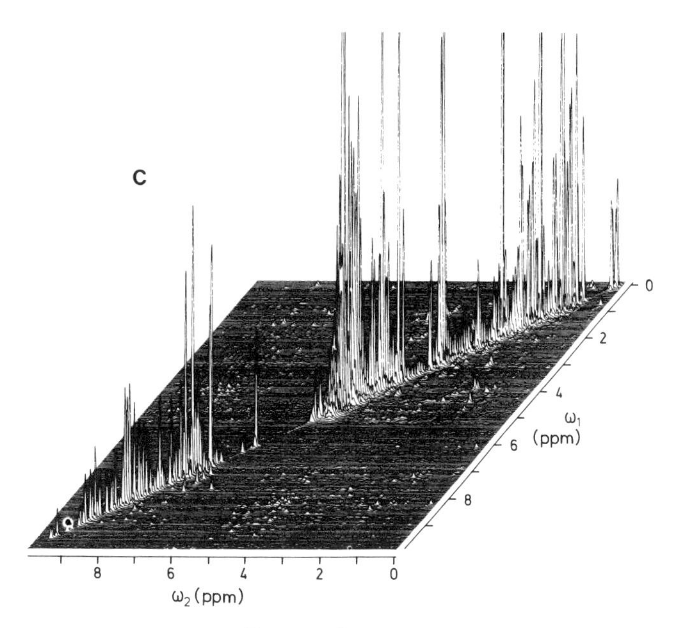

FIG. 11—*Continued.*

Solvent suppression normally leaves some residual perturbations in the spectrum. Thus a strong ridge parallel to the ω_1 axis at the water resonance originates from incomplete saturation of the solvent line, which may differ slightly from experiment to experiment (Fig. 3A and 11B). Furthermore, protein resonances near the water line may also be saturated, and as a consequence cross-peaks with these lines may disappear from the spectrum (*26*). These difficulties can be minimized by optimization of the field homogeneity, since the better the field homogeneity the less rf power is needed for effective solvent suppression. As a result a weaker ridge is obtained and a minimal number of cross-peaks is bleached out. With regard to the latter effect it is also possible to exploit the large temperature coefficient of the chemical shift of water (*43*). By a suitable choice of the temperature the water line can often be placed in an otherwise empty region of the spectrum, or a complete set of cross-peaks can be obtained by two complementary measurements at different temperatures (*26*).

Besides the experiment of Fig. 10, a suppression scheme which we recently used (e.g., Figs. 3, 4, and 11B and C) consists of continuous irradiation at all times except during t_2 (*32*). Fundamentally this provides for a more complete solvent suppression, but there is the disadvantage that unless a weak rf field can be applied, the 2D NMR spectra may be distorted by Bloch–Siegert shifts (*32*). A numerical correction of the Bloch–Siegert shifts in COSY and NOESY spectra has recently been described (*33*).

The well-known Redfield solvent suppression technique (*54*), offers an alternative but more delicate procedure. It has the significant advantage that cross-peaks to resonances coinciding with the solvent line are not suppressed and even more important that protons which are chemically exchanging with the solvent can also be observed. This is crucial when rapid exchange of labile protons is studied by a 2D chemical exchange experiment (*55*). In work with biopolymers, this type of suppression scheme may be needed for experiments with proteins at high temperature and high pH, and quite generally for studies of nucleic acids (*43, 54*).

V. SELECTION OF PERFORMANCE PARAMETERS

A 2D experiment differs from a 1D experiment in that experiments for a large number X_1 of equidistant t_1 values, with an increment Δt_1 have to be performed. The increment Δt_1 is determined by the maximum frequency range $\omega_{1_{max}}$ to be covered through the sampling theorem,

$$\Delta t_1 \leqslant 1/(2\omega_{1_{max}}). \quad [8]$$

In direct acquisition 2D experiments (Figs. 1A–C), the spectral ranges in ω_1 and ω_2 are the same. In a SECSY experiment $\omega_{1_{max}}$ is equal to half the largest frequency difference between coupled nuclei. In protein spectra, alanine and threonine usually set the upper limit, with roughly 4 ppm chemical shift difference between coupled protons. In delayed acquisition 2D *J*-resolved spectra, a value $\omega_{1_{max}}$ of ± 20 to ± 30 Hz is normally sufficient if strong coupling effects are neglected (*47*). For FOCSY the same considerations apply as for the corresponding delayed acquisition experiments.

The number X_1 of t_1 values determines the resolution in ω_1 and obviously at the same time the total performance time. X_1 also influences the *S/N* ratio when the sampling process is extended too far into the t_1 decay of the signal. In any case, apodization in t_1 is equally important as in t_2. Table 3 shows some parameters which are routinely applied on a 500 MHz spectrometer for measuring protein 2D ^{1}H NMR spectra. For 10 m*M* protein solutions in 5 mm sample tubes, spectra with acceptable *S/N* ratio are obtained in about 10 to 24 hr. The maximum value of X_2 is usually determined by the disk storage capacity. The relaxation delay can be correspondingly shorter when more t_2 data points are acquired.

Before starting a 2D experiment a precise calibration of the 90 and 180° pulse

TABLE 3

SAMPLES OF PARAMETERS USED FOR RECORDINGS OF RELATIVELY "LOW RESOLUTION" 2D ^{1}H NMR EXPERIMENTS AT 500 MHz[a]

2D Experiment	$\pm\omega_{1_{max}}/2\pi$ (Hz)	$\omega_{2_{max}}/2\pi$ (Hz)	X_1[b]	X_2[b]	Hz/pt[b] $\omega_1/2\pi$	Hz/pt[b] $\omega_2/2\pi$
2D *J* (Fig. 1C′)	20	4,000	64	4096	0.6	2
SECSY (Fig. 1B′)	1000	4,000	512	3072	4	2.6
COSY[c] (Fig. 1B)	5000	10,000	512	2048	20	10
NOESY[c] (Fig. 1A)	5000	10,000	512	2048	20	10

[a] The values given are useful for small diamagnetic proteins with molecular weight up to ca. 15,000 (for bigger proteins the choice of parameters will in general have to be reconsidered). With these parameters and a relaxation delay of 1.5 sec a spectrum with good *S/N* ratio of a 10 m*M* protein solution can be recorded in ca. 10 to 24 hr.

[b] X_1, number of t_1 values; X_2, number of sampling points in each FID; Hz/pt, resolution in ω_1 and ω_2 without zero filling.

[c] For COSY and NOESY it is assumed that the carrier frequency is set to one end of the spectrum.

angles is needed. Whenever possible the calibration should be made with the sample to be measured, because the pulse angle may depend on experimental parameters such as the solvent, the salt concentration, and the sample temperature.

Special care has to be taken to properly adjust the receiver gain in 2D experiments to avoid overflow because the first experiment with $t_1 = 0$ does not necessarily produce the largest signal. Consider, for example, a COSY experiment with two $90°_x$ pulses. For $t_1 = 0$ the two 90° pulses act like a 180° rotation which generates no signal at all. For a setup experiment, it is advisable to apply a single 90° pulse to the sample to be studied and use the resulting signal amplitude as a measure for the expected signal maximum.

Normally, many scans have to be averaged for each t_1 value. Here the question arises whether it is advantageous to cycle first through all t_1 values before the next repetition or whether all experiments for one t_1 value should be performed in sequence. To our experience there is no advantage to cycle several times through all t_1 values. Such a procedure would have the disadvantage that in case of a system failure (e.g., lost lock) toward the end of the experiment the entire experiment may be destroyed, whereas in a consecutive experiment only the samples for longer t_1 would be lost, resulting merely in a slightly reduced resolution.

A special comment concerns the choice of the mixing time, τ_m, in NOESY experiments (Fig. 1A and A′). The τ_m-dependence of NOESY was investigated for the basic pancreatic trypsin inhibitor (BPTI) (*27*) and for micelle-bound glucagon (*23*). For BPTI a plot of the peak intensities vs τ_m (Fig. 5 of Ref. (*27*)), which is representative for the τ_m dependence of NOESY spectra of small proteins, showed a monotonous decrease of the diagonal peak intensity. Two different types of behavior were observed for different cross-peaks. Some cross-peaks showed rapid growth at short τ_m values, reached a maximum intensity after 150 to 200 msec and subsequently decayed at still longer mixing times. For other cross-peaks the intensity started to increase after a lag period of up to ca. 100 msec, indicating that the magnetization transfer was dominated by "spin diffusion" (*27*). For conformational studies of proteins it is therefore advisable to collect several NOESY spectra with different mixing times, so that spin-diffusion processes can be identified and separated from "direct" NOEs, which provide more reliable distance information. In practice, mixing times between 30 and 200 msec have been used (*12–19, 23, 27*). On principal grounds measurements with very short mixing times are of prime interest for measurements of ^1H–^1H distances. However, the interpretation of such spectra is often limited by the appearance of intense bands of "t_1 noise" (Fig. 3A) and depends crucially on efficient *J*-peak suppression (see Section IVe). We found it helpful to record a spectrum with a long mixing time of ca. 200 msec, where the location of the cross-peaks can be accurately determined, and to use this spectrum as a guide for the interpretation of the data obtained with short mixing times (*23*).

VI. DATA PROCESSING

(a) Data Manipulation before 2D Fourier Transformation

In two-dimensional spectroscopy filtering procedures have to be applied in both frequency domains. Normally the Fourier transformation is executed in two steps, first with respect to t_2 and then with respect to t_1. Filtering in ω_2 can be performed by multiplication of the original $s(t_1, t_2)$ data with a suitable weighting function $h_2(t_2)$, while filtering in ω_1 is applied most conveniently prior to the second Fourier transformation by multiplication of the $S(t_1, \omega_2)$ data with a weighting function $h_1(t_1)$.

(i) Baseline correction. Baseline correction by subtracting a constant or a linearly increasing function from the original data is a standard operation in 1D spectroscopy. In 2D spectroscopy, it must be used with care. In particular, the same correction should be applied to all FIDs in $s(t_1, t_2)$ or $S(t_1, \omega_2)$. If individual baseline corrections are used, differing from FID to FID, artifacts may arise. In addition, signals at $\omega_1 = 0$ or $\omega_2 = 0$ may be lost or suppressed by application of a baseline correction. As an illustration Fig. 12 shows the suppression of an axial peak by the use of a baseline correction in the t_1 domain.

(ii) Resolution enhancement. One-dimensional and two-dimensional protein ^{1}H NMR spectra invariably win when moderate digital resolution enhancement is applied. Resolution enhancement is indispensible when absolute-value spectra are displayed to reduce the broad signal tails originating from the dispersive parts. As in 1D spectroscopy digital filtering is performed by multiplication of the time-domain data by a suitable filtering function (*56, 57*). The multiplication in the two time domains is usually performed independently, possibly with different filtering functions. All suitable resolution enhancement functions $h(t)$ start for $t = 0$ at a low value, increase to a maximum and fall off towards the end, preferentially to zero. The initial low values suppress broad signal components, the intermediate maximum is responsible for the enhancement of narrow lines, and the ultimate fall-off avoids truncation errors by apodization and maintains an acceptable S/N ratio by attenuating the noisier parts of the FID. Various suitable weighting functions can be found in the literature (*43, 56–58*).

A good compromise between resolution enhancement and sensitivity can be achieved by applying a Lorentz–Gauss transformation (*56*) which is theoretically justified for

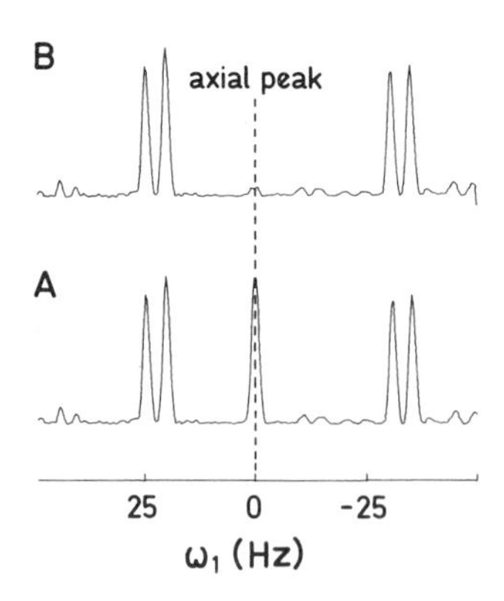

FIG. 12. Suppression of axial peaks at $\omega_1 = 0$ by application of a baseline correction. The same cross section through a ^{1}H COSY spectrum of 2,3,4-trichlorophenol is shown in the two traces. (A) No baseline correction was used and there appears an intense axial peak. (B) After the baseline correction the axial peak is almost completely suppressed. The carrier frequency has been set within the recorded spectral range.

Lorentzian lines, i.e., for exponential FIDs. Another concept is based on the idea of a "pseudo-echo" (*59*). It allows the exact suppression of dispersive parts and leads to similar weighting functions. A further, empirical approach utilizes sine-bell functions (*58*). The sine bell or in an improved version a shifted sine bell (*60*) is very easily generated and applied. It requires specification of a single parameter only, the time shift. Another function with the same virtue is the squared sine-bell function which approaches the Lorentz–Gauss transformation. In 2D ^{1}H NMR of proteins, the use of shifted sine-bell and shifted squared sine-bell functions was found convenient and efficient (e.g., Figs. 3, 4, and 11 and Refs. (*13–19*)) and no need for a more profoundly justified weighting function became apparent.

The influence of different digital filtering functions on a protein 2D ^{1}H NMR spectrum is illustrated in Fig. 13. The three spectra were obtained from the same data set by applying different filtering functions. The result of a cosine apodization, which eliminates truncation errors but does not provide resolution enhancement, is shown in Fig. 13A. This spectrum is hardly suitable for analysis. The best result was obtained with the squared sine-bell filter (Fig. 13B), which adopts zero value for $t = 0$ (Fig. 13B), whereas the Lorentz–Gauss transform for the chosen parameter set is less efficient in suppressing broad lines (Fig. 13C). Comparison of Figs. 13B and C affords an illustration of how critical the selection of the filter functions can be, in particular when absolute value presentations are used.

(iii) Zero filling. In two-dimensional protein spectroscopy, the available performance time invariably sets an upper limit to the number of t_1 values which can be measured. To improve the visual appearance of the resulting spectra, zero filling (*61*), particularly in the t_1 domain, is required in almost all cases. Experience with protein ^{1}H NMR spectra showed that N data points can be supplemented by up to $3N$ zero values with a noticeable improvement of resolution, further zero filling is ineffective. In Fig. 14 spectra with different zero filling prior to the Fourier transformation illustrate this point.

(b) Data Manipulation after 2D Fourier Transformation

(i) Triangular multiplication and symmetrization. Ideal COSY and NOESY spectra are symmetrical with respect to the main diagonal $\omega_1 = \omega_2$. In practice, however, noise and instrumental artifacts destroy the symmetry (Figs. 3A and 11B). By a suitable combination of the upper left and the lower right triangle, it is possible to improve the significance of a 2D spectrum (Figs. 3 and 11). One possibility is calculating the geometric mean (*28*) of upper and lower triangles ("triangular multiplication"). In another procedure of "triangular symmetrization," the lower one of each pair of symmetrically located values is retained (*29*). A slight improvement of the S/N ratio can at the same time also be achieved with these procedures (*28, 29*). Furthermore, triangular symmetrization is important for J cross-peak elimination from NOESY spectra after application of the τ_m incrementation method (*30*) (see Section IVe).

We should note that symmetrization may lead to erroneous peaks when the original spectrum contains strong ridges extending parallel to one of the two axes. The crossing of two ridges in the symmetrization process will feign the presence of two symmetrically located peaks. Such artifactual peaks can, for example, be found in Figs. 3B and 11C

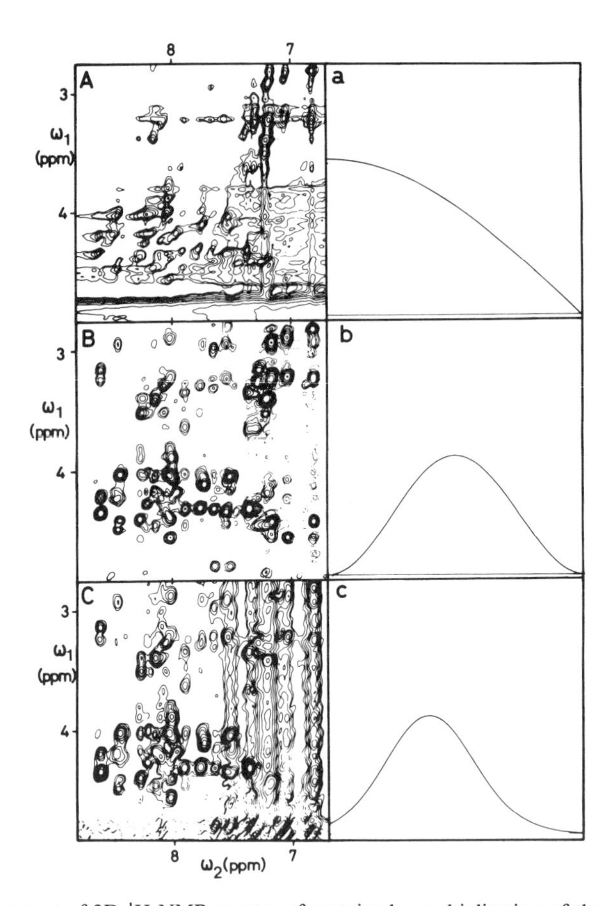

FIG. 13. Improvement of 2D ^{1}H NMR spectra of proteins by multiplication of the time-domain data with different resolution enhancement functions. (A) to (C) show the region (ω_1 = 2.8 − 4.9 ppm, ω_2 = 6.7 − 8.8 ppm) from a 500 MHz ^{1}H NOESY spectrum recorded with a mixing time of 200 msec at 37°C in a H_2O solution which contained the polypeptide hormone glucagon bound to micelles of perdeuterated dodecyl phosphocholine (0.015 *M* glucagon, 0.7 *M* [$^2H_{38}$]dodecylphosphocholine, 0.05 *M* phosphate buffer, pH 6.0). The H_2O resonance was suppressed by continuous irradiation at all times except during t_2. The glucagon-containing micelles have a molecular weight of ca. 17,000 (*14*) and correspondingly the ^{1}H NMR lines are quite broad. (A) The time-domain data matrix was multiplied with the cosine function a. (B) Multiplication with the sine squared function b. (C) Multiplication with the Gaussian function c.

for which the crossings of the strong water ridge with the weaker t_1-noise bands of the methyl groups are responsible. In such cases it is advisable to compare symmetrized and original spectra. It is obvious that symmetrization has also its limitation when the *S*/*N* ratio is below a certain limit. It is not possible to retrieve signals which are smaller than the noise level, and accidental coincidences of strong noise peaks can never be excluded.

(ii) Foldover-corrected spectroscopy (FOCSY). It has been mentioned in Section III that foldover correction of a conventional COSY spectrum can be used as an alternative to SECSY. It permits the same optimized usage of performance time, data storage, and resolution in the ω_1 domain. Foldover in ω_1 caused by a low sampling

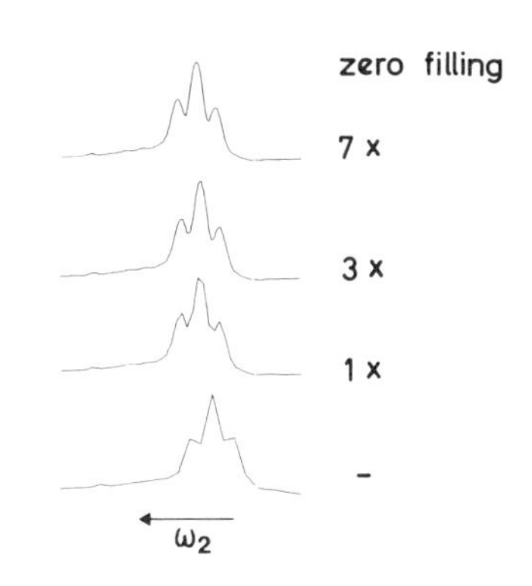

FIG. 14. Influence of "zero filling" in the time domain on the appearance of the frequency-domain spectrum. The numbers on the right indicate the extent of the zero filling relative to the number of experimental data points. There is no use to supplement N data points by more than $3N$ zeros.

rate in t_1, violating the sampling theorem, can be corrected whenever all cross-peaks are located within a range of $\pm\omega_N$ from the main diagonal (*8*). The rearrangement of the originally folded spectrum $S(\omega_1, \omega_2)$ into a foldover-corrected spectrum $S'(\omega_1', \omega_2')$ can be formulated as (*8*)

$$S'(\omega_1', \omega_2') = S(\omega_1, \omega_2) \qquad [10]$$

with

$$\omega_1' = (\omega_1 - \omega_2 + \omega_N) \bmod (2\omega_N) - \omega_N$$

$$\omega_2' = \omega_2$$

where ω_N stands for the Nyquist frequency. Figure 15 presents an example of a FOCSY-type spectrum. A 2D J-resolved spectrum of the basic pancreatic trypsin inhibitor (BPTI) was measured. The sampling in this experiment started immediately after the 180° pulse (Fig. 1C). In Fig. 15A a section of this spectrum from $\omega_2 = 4.19$ to 4.73 ppm is shown. The 1D spectrum extends along the diagonal because in contrast to the conventional echo-type J-resolved experiment (Fig. 1C′) both chemical shift and coupling information is present during the evolution period t_1 and the detection period t_2. The restricted ω_1 frequency range leads to multiple folding of the diagonal (Fig. 15A). With the foldover correction applied, the spectrum Fig. 15B results. Only six resonances between 4.25 and 4.45 ppm could not be properly relocated because their distance from the diagonal is larger than $\pm\omega_N$. In Fig. 15C the spectrum of Fig. 15B is presented after tilting (*25*). Figure 15D shows the same spectrum as Fig. 15C after symmetrization. In addition, the ω_1 frequencies were divided by two to remove the artifactual doubling of the frequency scale in the foldover correction calculation. This type of experiment has recently also been used by S. Macura and L. R. Brown (private communication).

VII. PRESENTATION OF 2D PROTEIN SPECTRA

The enormous information content of 2D spectra of proteins necessitates a careful selection of the presentation of the data. In the past, stacked plot representations (Fig.

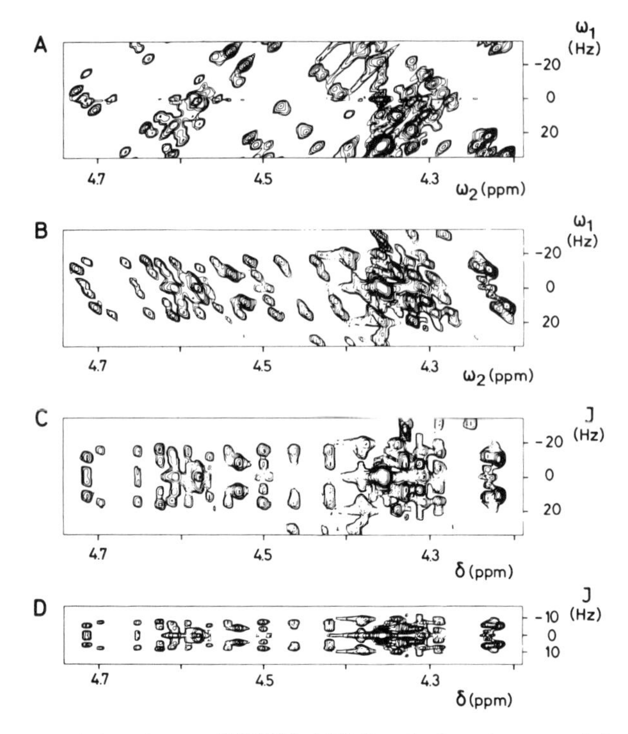

FIG. 15. Foldover-corrected spectroscopy (FOCSY). A 2D *J*-resolved spectrum recorded with the experiment of Fig. 1C in a 0.02 *M* solution of the basic pancreatic trypsin inhibitor (BPTI) in ^2H_2O, $T = 68°$, $p^2H = 4.7$ is shown. (A) Since the spectral width in the ω_1 direction is only 70 Hz, the spectrum is folded back many times. In this trace five fragments of the diagonal flanked by the multiplet components can be observed to rise at an angle of 45° relative to the ω_2 axis. (B) The same spectrum as in (A) after foldover correction; the data manipulation caused a doubling of the multiplet splittings. For six resonances between 4.25 ppm and 4.35 ppm the correction did not work because they have a larger distance from the diagonal than $\pm\omega_N$ (see text). (C) Same spectrum as in (B) after tilting. (D) Same spectrum as in (C) after symmetrization and reduction of the ω_1 frequencies by a factor 2.

11) and contour plots (Figs. 3, 4, 9, 13, 15) have been employed to represent complete 2D spectra. Often, the relevant information is confined to selected cross sections (Figs. 12 and 14) and sometimes one-dimensional projections of the 2D spectra are useful means of representation (*6, 25, 47*).

(a) Stacked Plots

For obtaining a quick visual impression of a 2D spectrum, stacked plots are quite instructive although for a quantitative evaluation of the data they are usually useless. Stacked plots do not require much preprocessing of the data and can be obtained with any sufficiently fast XY recorder. Problems of resolution may arise when the number of lines to be drawn exceeds 256. The utility of stacked plots is thus confined to spectra with a limited number of spectral elements in the ω_1 domain, particularly to 2D *J*-resolved spectroscopy and to low-resolution SECSY plots. In some cases it is of advantage to suppress hidden lines, in other cases the information loss by this cleanup procedure may be intolerable. Despite their disadvantages stacked plots are valid tools for "honest" representation of the original data. This contrasts somewhat to contour plots where essential features may often be hidden by an improper selection of contour levels.

(b) Contour Plots

For a quantitative evaluation of 2D spectra contour plots are invaluable. They are the most widely used representation of 2D data. The computation of the contours is a demanding task and requires a fair amount of computing time. Numerous procedures have been proposed in literature. The technique adopted in our software package uses a search procedure briefly described in Ref. (*10*).

The selection of the proper contour levels is of central importance for a pleasing and at the same time fair representation of the 2D data. It is recommended that for routine applications equidistant contour lines are selected. In special cases, a non-equidistant spacing may be applied. In all cases, the selected contour levels should clearly be stated in figure captions.

The representation of phase-sensitive spectra with positive and negative intensities is difficult in a contour plot and can most satisfactorily be solved by using a colour plot. For publications a representation of positive and negative intensities in two separate plots seems preferable. In reasonably well resolved spectra a superposition of broken contour lines representing negative peaks and solid contours depicting positive peaks may also be considered.

(c) Cross Sections

In many cases, selected cross sections contain all relevant information of 2D spectra (Figs. 12, 14). They should be used whenever accurate peak intensities or peak shapes must be represented (*25, 47*). They are also suitable in the case of very large dynamic ranges which are difficult to visualize in contour plots. In addition, negative signal intensities can be represented without problems in cross sections. Cross sections may become misleading when the resolution perpendicular to the cross section is insufficient and "cross-talk" between different cross sections cannot be avoided. Sometimes, it is useful to plot several adjacent sections through the same peak to visualize perpendicular tendencies.

(d) Projections

The utility of projections is limited to 2D *J*-resolved spectra where homonuclear broadband decoupled spectra can be obtained (*6, 25, 47*). They allow the "counting" of the individual protons and may be useful for the proper selection of the position of cross sections. Otherwise their utility in 2D protein NMR is somewhat limited.

VIII. CONCLUSIONS

This paper attempted an account of practical procedures which have proved useful in two-dimensional spectroscopy of small proteins in solution. Two-dimensional spectroscopy has become an indispensible tool in our laboratory for the elucidation of protein structure, which is also routinely used by biochemists and biophysicists without much previous knowledge of sophisticated NMR technology. Quite obviously, the procedures described are not exclusively applicable to proteins but would be equally well suited for the analysis of larger organic molecules in solution as well as for other biological macromolecules, such as nucleic acids (*62*).

We have limited this paper to the description of experimental procedures for the presently most widely used homonuclear 2D ^{1}H NMR techniques in protein research. Many more techniques are known which are potentially of interest for the study of large molecules. We would like to mention relayed magnetization transfer for tracing out remote connectivities in spin systems (*63, 64*). Similar information can be obtained from multiple-quantum spectra (*65, 66*). For the simplification of overcrowded spectra, multiple-quantum filters (*67, 68*) and diagonal peak suppression procedures (*69, 70*) can be employed. Heteronuclear correlation experiments (*71–75*) also look promising for protein NMR despite inherent sensitivity problems. Most of these techniques involve similar experimental requirements as those discussed in this paper. Recent experience has also shown that absorption mode spectra can be considerably more informative than absolute value displays due to the narrower lineshapes and the resulting increased spectral resolution (*7, 76–79*).

For future work with proteins and other macromolecules the 2D NMR experiments will be selected from the techniques described in detail in this paper as well as from the more recently introduced measurements which are briefly mentioned in this last section. Various different considerations will influence the choice. These will include that relatively low-resolution, absolute-value spectra using similar parameters to those in Table 3 can be obtained with relatively short times for accumulation and data manipulation, whereas, for example, the recording of highly resolved phase-sensitive spectra is much more time consuming (*78, 79*). One may thus, for example, decide to forego resolution for efficient data accumulation in the early phases of a structure determination and complement this data at a later stage with highly resolved presentations of selected spectral regions. In view of the very large amounts of data obtained, for example, from a single NOESY spectrum of a protein, it can be foreseen that the use of computers for bookkeeping and for the structural analysis will be an important factor in future developments. This will undoubtedly also lead to new demands on the quality of the experimental 2D NMR spectra.

ACKNOWLEDGMENTS

Financial support by special grants of the Eidgenössische Technische Hochschule, Zurich, and by the Schweizerischer Nationalfonds (Project 3.528.79) are gratefully acknowledged. The authors are grateful to Dr. G. Bodenhausen and Dr. M. P. Williamson for helpful suggestions with regard to the manuscript, to Mr. A. Eugster for recording the spectra of Fig. 11, to Mrs. Y. Hunziker for drawing some of the figures and to Mrs. I. Kowalski (Spectrospin AG, Fallanden) for typing the manuscript. The authors also acknowledge the support by Spectrospin AG, Fallanden.

REFERENCES

1. K. Nagayama, K. Wüthrich, P. Bachmann, and R. R. Ernst, *Biochem. Biophys. Res. Commun.* **78,** 99 (1977).

2. K. Nagayama, K. Wüthrich, and R. R. Ernst, *Biochem. Biophys. Res. Commun.* **90,** 305 (1979).

3. Anil Kumar, R. R. Ernst, and K. Wüthrich, *Biochem. Biophys. Res. Commun.* **95,** 1 (1980).

4. O. JARDETZKY, "NMR and Biochemistry" (S. J. Opella and P. Lu, eds.), pp. 141–167, Decker, New York, 1979.
5. K. WÜTHRICH, K. NAGAYAMA, AND R. R. ERNST, *Trends Biochem. Sci.* **4,** N178 (1979).
6. W. P. AUE, J. KARHAN, AND R. R. ERNST, *J. Chem. Phys.* **64,** 4226 (1976).
7. W. P. AUE, E. BARTHOLDI, AND R. R. ERNST, *J. Chem. Phys.* **64,** 2229 (1976).
8. K. NAGAYAMA, ANIL KUMAR, K. WÜTHRICH, AND R. R. ERNST, *J. Magn. Reson.* **40,** 321 (1980).
9. A. BAX AND R. FREEMAN, *J. Magn. Reson.* **44,** 542 (1981).
10. J. JEENER, B. H. MEIER, P. BACHMANN, AND R. R. ERNST, *J. Chem. Phys.* **71,** 4546 (1979).
11. K. NAGAYAMA AND K. WÜTHRICH, *Eur. J. Biochem.* **114,** 365 (1981).
12. G. WAGNER, ANIL KUMAR, AND K. WÜTHRICH, *Eur. J. Biochem.* **114,** 375 (1981).
13. G. WAGNER AND K. WÜTHRICH, *J. Mol. Biol.* **155,** 347 (1982).
14. G. WIDER, K. H. LEE, AND K. WÜTHRICH, *J. Mol. Biol.* **155,** 367 (1982).
15. A. S. ARSENIEV, G. WIDER, F. J. JOUBERT, AND K. WÜTHRICH, *J. Mol. Biol.* **159,** 323 (1982).
16. R. M. KELLER, R. BAUMANN, E. H. HUNZIKER-KWIK, F. J. JOUBERT, AND K. WÜTHRICH, *J. Mol. Biol.* **163,** 623 (1983).
17. R. V. HOSUR, G. WIDER, AND K. WÜTHRICH, *Eur. J. Biochem.* **130,** 497 (1983).
18. P. ŠTROP, G. WIDER, AND K. WÜTHRICH, *J. Mol. Biol.* **166,** 641 (1983).
19. P. ŠTROP, D. ČECHOVÀ, AND K. WÜTHRICH, *J. Mol. Biol.* **166,** 669 (1983).
20. K. WÜTHRICH, G. WIDER, G. WAGNER, AND W. BRAUN, *J. Mol. Biol.* **155,** 311 (1982).
21. G. WAGNER AND K. WÜTHRICH, *J. Mol. Biol.* **160,** 343 (1982).
22. K. WÜTHRICH AND G. WAGNER, Ciba Foundation Symposium No. 93 on Mobility and Function in Proteins and Nucleic Acids, Pitman, London, 1983.
23. W. BRAUN, G. WIDER, K. H. LEE, AND K. WÜTHRICH, *J. Mol. Biol.* **169,** 921 (1983).
24. G. WAGNER, A. PARDI, AND K. WÜTHRICH, *J. Am. Chem. Soc.* **105,** 5948 (1983).
25. K. NAGAYAMA, P. BACHMANN, K. WÜTHRICH, AND R. R. ERNST, *J. Magn. Reson.* **28,** 29 (1978).
26. ANIL KUMAR, G. WAGNER, R. R. ERNST, AND K. WÜTHRICH, *Biochem. Biophys. Res. Commun.* **96,** 1156 (1980).
27. ANIL KUMAR, G. WAGNER, R. R. ERNST, AND K. WÜTHRICH, *J. Am. Chem. Soc.* **103,** 3654 (1981).
28. R. BAUMANN, ANIL KUMAR, R. R. ERNST, AND K. WÜTHRICH, *J. Magn. Reson.* **44,** 76 (1981).
29. R. BAUMANN, G. WIDER, R. R. ERNST, AND K. WÜTHRICH, *J. Magn. Reson.* **44,** 402 (1981).
30. S. MACURA, K. WÜTHRICH, AND R. R. ERNST, *J. Magn. Reson.* **46,** 269 (1982).
31. S. MACURA, K. WÜTHRICH, AND R. R. ERNST, *J. Magn. Reson.* **47,** 351 (1982).
32. G. WIDER, R. V. HOSUR, AND K. WÜTHRICH, *J. Magn. Reson.* **52,** 130 (1983).
33. R. V. HOSUR, R. R. ERNST, AND K. WÜTHRICH, *J. Magn. Reson.* **54,** 142 (1983).
34. R. FREEMAN AND G. A. MORRIS, *Bull. Magn. Reson.* **1,** 5 (1979).
35. A. BAX, "Two-Dimensional Nuclear Magnetic Resonance in Liquids," Reidel, London, 1982.
36. W. E. HULL, "Two-Dimensional NMR," Bruker, Karlsruhe, 1982.
37. M. BILLETER, W. BRAUN, AND K. WÜTHRICH, *J. Mol. Biol.* **155,** 321 (1982).
38. K. WÜTHRICH, *Biopolymers* **22,** 131 (1983).
39. M. BILLETER, Diploma Thesis, ETH–Zurich. Computerunterstützter graphischer Modellbau von Polypeptidketten unter Berücksichtigung von NMR-Daten, 1980.
40. W. BRAUN, C. BÖSCH, L. R. BROWN, N. GO, AND K. WÜTHRICH, *Biochim. Biophys. Acta* **667,** 377 (1981).
41. G. M. CRIPPEN AND T. F. HAVEL, *Acta Crystallogr. Sect. A* **34,** 282 (1978).
42. T. F. HAVEL, G. M. CRIPPEN, AND I. D. KUNTZ, *Biopolymers* **18,** 73 (1979).
43. K. WÜTHRICH, "NMR in Biological Research: Peptides and Proteins," North-Holland, Amsterdam, 1976.
44. K. NAGAYAMA AND K. WÜTHRICH, *Eur. J. Biochem.* **115,** 653 (1981).
45. K. NAGAYAMA, *Advan. Biophys.* **14,** 139 (1981).
46. G. KING AND P. E. WRIGHT, *Biochem. Biophys. Res. Commun.* **106,** 559 (1982).
47. G. WIDER, R. BAUMANN, K. NAGAYAMA, R. R. ERNST, AND K. WÜTHRICH, *J. Magn. Reson.* **42,** 73 (1981).
48. L. MÜLLER, *J. Magn. Reson.* **36,** 301 (1979).
49. D. I. HOULT AND R. E. RICHARDS, *Proc. Roy. Soc. London Ser. A* **344,** 311 (1975).
50. G. BODENHAUSEN, R. FREEMAN, AND D. L. TURNER, *J. Magn. Reson.* **24,** 511 (1977).

51. S. Macura and R. R. Ernst, *Mol. Phys.* **41,** 95 (1980).
52. S. Macura, Y. Huang, D. Suter, and R. R. Ernst, *J. Magn. Reson.* **43,** 259 (1981).
53. K. Wüthrich and G. Wagner, *J. Mol. Biol.* **130,** 1 (1979).
54. A. G. Redfield, S. D. Kunz, and E. K. Ralph, *J. Magn. Reson.* **19,** 116 (1975).
55. J. D. Cutnell, *J. Am. Chem. Soc.* **104,** 362 (1982).
56. R. R. Ernst, *Advan. Magn. Reson.* **2,** 1 (1966).
57. J. C. Lindon and A. G. Ferrige, *Prog. NMR Spectrosc.* **14,** 27 (1980).
58. A. de Marco and K. Wüthrich, *J. Magn. Reson.* **24,** 201 (1976).
59. A. Bax, A. F. Mehlkopf, and J. Smidt, *J. Magn. Reson.* **35,** 373 (1979).
60. G. Wagner, K. Wüthrich, and H. Tschesche, *Eur. J. Biochem.* **86,** 67 (1978).
61. E. Bartholdi and R. R. Ernst, *J. Magn. Reson.* **11,** 9 (1973).
62. A. Pardi, R. Walker, H. Rapoport, G. Wider, and K. Wüthrich, *J. Am. Chem. Soc.* **105,** 1652 (1983).
63. G. Eich, G. Bodenhausen, and R. R. Ernst, *J. Am. Chem. Soc.* **104,** 3731 (1982).
64. G. Wagner, *J. Magn. Reson.* **55,** 151 (1983).
65. L. Braunschweiler, G. Bodenhausen, and R. R. Ernst, *Mol. Phys.* **48,** 535 (1983).
66. G. Wagner and E. R. P. Zuiderweg, *Biochem. Biophys. Res. Comm.* **113,** 854 (1983).
67. U. Piantini, O. W. Sørensen, and R. R. Ernst, *J. Am. Chem. Soc.* **104,** 6800 (1982).
68. T. H. Mareci and R. Freeman, *J. Magn. Reson.* **51,** 531 (1983).
69. G. Bodenhausen and R. R. Ernst, *Mol. Phys.* **47,** 319 (1982).
70. K. Nagayama, Y. Kobayaski, and Y. Kyogoku, *J. Magn. Reson.* **51,** 84 (1983).
71. A. A. Maudsley and R. R. Ernst, *Chem. Phys. Lett.* **50,** 368 (1977).
72. A. A. Maudsley, L. Müller, and R. R. Ernst, *J. Magn. Reson.* **28,** 463 (1977).
73. G. Bodenhausen and R. Freeman, *J. Magn. Reson.* **28,** 471 (1977).
74. R. Freeman and G. A. Morris, *J. Chem. Soc. Chem. Commun.,* 684 (1978).
75. T. M. Chan and J. L. Markley, *J. Am. Chem. Soc.* **104,** 4010 (1982).
76. P. Bachmann, W. P. Aue, L. Müller, and R. R. Ernst, *J. Magn. Reson.* **28,** 29 (1977).
77. D. J. States, R. A. Haberkorn, and D. J. Ruben, *J. Magn. Reson.* **48,** 286 (1982).
78. D. Marion and K. Wüthrich, *Biochem. Biophys. Res. Commun.* **113,** 967 (1983).
79. M. Williamson, D. Marion, and K. Wüthrich, *J. Mol. Biol.,* submitted for publication.

JOURNAL OF MAGNETIC RESONANCE 57, 404–426 (1984)

A Theoretical Study of Distance Determinations from NMR. Two-Dimensional Nuclear Overhauser Effect Spectra

JOE W. KEEPERS* AND THOMAS L. JAMES†

Department of Pharmaceutical Chemistry, University of California, San Francisco, California 94143

Received July 11, 1983; revised October 26, 1983

In principle, two-dimensional nuclear Overhauser effect NMR experiments can be used to establish internuclear distances and thus to determine molecular structure in solution. Theoretical calculations have been carried out for the 2D NOE experiment addressing the questions of (1) how sensitive 2D NOE cross-peak intensities are to spatial location of nuclear spins, and (2) how well the motional characteristics (and, therefore, spectral densities) of the molecule must be known in order to determine accurate distances. Theoretical values for cross-peak intensities obtained in 2D NOE spectra were calculated for a set of three-proton spins and a set of four-proton spins. The relaxation rate matrix whose elements establish the cross-peak intensities was diagonalized using a numerical procedure which enables accurate calculations of cross-peak intensities in large spin systems at any mixing time. Several calculations were performed; each calculation entailed a different set of circumstances. Specifically, the transformation of cross-peak intensities into proton–proton distances was examined for 21 configurations of the three-proton system and 7 configurations of the four-proton spin system. Additionally the influence of specific types of overall and internal motion on the calculated cross-peak intensities and the distances determined from them was tested. The conclusions from these calculations are as follows. With the experimental signal-to-noise ratio usually attained for biomolecular concentrations < 10 mM, interproton distances up to 5 Å should lead to detectable cross-peaks in the 2D NOE spectrum with mixing times within a few hundred milliseconds. Recording cross-peak intensities at several mixing times is important for obtaining accurate distances from them. Simple motional models assuming only a single isotropic motion with an effective correlation time may be used to obtain the spectral densities needed to calculate cross-peak intensities even though the actual motional dynamics of the molecule may be more complicated. Using the simplified models introduces approximately 10% error in the distance determination. Finally, it is shown how the cross-peak between a proton and each of its neighbors may increase the accuracy of the determination and help in locating the position of the proton. Our calculations indicate that distances with an accuracy of ± 0.5 Å should be attainable with knowledge of the overall molecular reorientation rate or individual proton relaxation times.

INTRODUCTION

Two-dimensional nuclear magnetic resonance spectroscopy provides a number of methods for obtaining previously inaccessible molecular structural information. Wagner and Wüthrich have elegantly demonstrated 2D correlated spectroscopy (COSY)

* Present address: Chemistry Department, Rutgers University, New Brunswick, N.J. 08903.
† To whom correspondence should be addressed.

0022-2364/84 $3.00

as a method for using *J*-coupled peaks to sequentially assign the proton resonances in bovine pancreatic trypsin inhibitor (BPTI) (*1*). Additionally, they used nuclear Overhauser enhancement spectroscopy (NOESY) to provide sequence information from through-space dipolar connectivities (*1*). For this purpose, the quantitative dependence of the cross-peak intensity upon the distance between two dipolar coupled protons need not be known; only the upper limit on the proton–proton distance needs to be determined.

It is well known that the constraints provided by the bonded structure alone are not sufficient to determine the three-dimensional structure of a macromolecule (*2*). Two-dimensional NOE spectra may provide valuable information in addition to that of the bonded structure by mapping specific through-space internuclear distances. In contrast to the constraints provided by the bonded structure alone, the constraints provided by the bonded structure and the through-space distances provided by 2D NOE experiments could be sufficient to completely determine the conformation of the three-dimensional structure.

Consequently, the precision and accuracy with which 2D NOE experiments may determine through-space proton–proton distances is of extreme importance for use in the conformational determination of macromolecular structures. How well these distances are determined by 2D NOE experiments is the question we wish to answer. Previously, Kumar *et al.* (*3*) outlined important experimental design criteria for 2D NOE experiments. In their work, they determined the essentiality of performing 2D NOE experiments at a number of mixing times. To interpret their cross-relaxation rates, they utilized mixing times during the initial magnetization buildup, circumventing the complications arising from spin diffusion to other nuclei (*3*).

It should be noted that previous reports have appeared dealing with the relationship between the location of nuclear spins and the time development of one-dimensional nuclear Overhauser effects in which selective peak irradiation was employed (*4–7*). Bothner-By and Noggle (*4*) have carried out calculations on a linear chain of nuclei assuming isotropic motion. Wagner and Wüthrich (*5*) used a three-spin system undergoing isotropic motion with approximations necessitating very short mixing times. Tropp (*6*) has examined the case of internal jump motions superimposed upon an overall tumbling motion with special emphasis upon a proton coupled to methyl protons undergoing a three-state jump. Dobson *et al.* (*7*) examined the possibility of using an approximate approach with only two strongly coupled protons.

We have chosen to examine the somewhat different problem of the two-dimensional NOE experiment utilizing an approach which is not limited by the number or geometry of interacting spins, by the range of mixing time, nor by specific motional characteristics. In the first section of this paper, we summarize the theory necessary to calculate cross-relaxation rates for a multispin system. We then describe the step-by-step procedure for computing the 2D NOE cross-peak intensities.

In the second section of this paper, we have considered relaxation and cross-relaxation in a three-proton system in an effort to examine the through-space distance dependence of the proton–proton cross-peak intensity. Our approach entails calculating the cross-peak intensities at several mixing times for several configurations of the three protons (see Fig. 1). In all of the calculations, the distance between protons H2 and H3 was held constant, while the distance between proton H1 and the other two was varied. In this way the calculated intensity at discrete distances was obtained as was the change in cross-peak intensity as a function of distance. By defining the experimentally discernible change in cross-peak intensity, the distance resolution for this system has been empirically determined.

In the third section we present calculations designed to assess the effect of internal motions on the cross-peak intensities. Protons H2 and H3 were allowed to rotate isotropically, while proton H1 tumbled both isotropically and with internal motion. We examined two cases for the internal motion. In the first case, the internal motion was due to free diffusion at a fixed distance from protons H2 and H3. In the second case, we examined several lattice models in which the distance from H1 to H2 and H3 changes with the occupancy of the lattice sites. In the third section we are assessing the effects of molecular dynamics more complicated than simple isotropic motion on the calculated cross-peak intensities and consequently on the distances determined from cross peaks if the experimenter assumes simple isotropic motion.

Finally, in the fourth section we use the isotropic rotational model to examine the cross-peak intensities within a four-spin system. We performed a series of calculations in which $\omega\tau_0 = 10$ and H1 occupied positions **h–n** while proton H4 occupied positions **a′–g′** in the lattice depicted in Fig. 1. Thus, for this series, the distance between H2 and H3 and between H1 and H4 is kept constant at 2.0 Å while the distance between H2 and H1,H4 and between H3 and H1,H4 varies with the position (see Fig. 1).

THEORY

The 2D NOE experiment employs the basic sequence $(\pi/2)_x$–t_1–$(\pi/2)_x$–t_m–$(\pi/2)_x$–t_2 where the first Fourier transform (yielding one dimension in the 2D spectrum) is over the signal acquisition period t_2 and the second Fourier transform (yielding the second dimension) is over t_1 as the pulse sequence is repeated for varying values of t_1. The mixing time t_m is the length of time z magnetization can be exchanged between components of a spin system via cross-relaxation.

The theory for calculating the cross-peak intensities in the 2D NOE spectrum was developed for a two-spin system by Macura and Ernst (*8*). Later, Bodenhausen and Ernst (*9*) outlined a matrix method for analyzing exchange processes affecting magnetic nuclei. The mathematics of chemical exchange and cross-relaxation are equivalent; thus extension of the work of Macura and Ernst is straightforward. We develop expressions for the cross-peak intensities in a multispin system in an effort to provide a unified treatment pertinent to the discussion here. The intensity of an auto- or cross-peak for any mixing time t_m may be calculated from the expression

$$\mathbf{a}(t_m) = \chi \exp(-\lambda t_m)\chi^{-1}. \quad [1]$$

In this expression $\mathbf{a}$ is a matrix whose elements a_{ij} give the cross-peak intensity for magnetic nuclei i and j, χ is the matrix containing eigenvectors of the rate matrix (given below) describing the relaxation behavior of the system of spins, and λ is a diagonal matrix giving the eigenvalues for that system and is the solution to the system of equations defining the rate matrix.

For a system of protons, the relaxation behavior may be written in the form

$$\Delta\mathbf{M}_z = \mathbf{R}\Delta\mathbf{M}_z \quad [2]$$

where $\Delta\mathbf{M}_z$ is the deviation of magnetization from thermal equilibrium. The relaxation rate matrix may be written as

$$\mathbf{R} = \begin{bmatrix} R_{ii} & R_{ij} & \cdot \\ R_{ji} & \cdot & \cdot \\ \cdot & \cdot & \cdot \\ \cdot & \cdot & \cdot \end{bmatrix}. \quad [3]$$

Within this matrix the R_{ij} element represents the cross-relaxation rate between spin i and spin j. The elements R_{ii} represent the direct relaxation contributions from all other sources to spin i. These terms may be written as functions of the zero-, single-, and double-quantum transition probabilities:

$$R_{ii} = \sum_j (W_0^{ij} + 2W_1^{ij} + W_2^{ij}) + R_{1i}$$

$$R_{ij} = (W_2^{ij} - W_0^{ij}). \quad [4]$$

In writing these equations, we have considered "unlike" protons only, and the element R_{1i} represents relaxation contributions from uncoupled sources to spin i. The transition probabilities may in turn be written as a function of the spectral densities:

$$W_1^{ij} = 1.5q_{ij}J(\omega_i)$$

$$W_0^{ij} = q_{ij}J(\omega_i - \omega_j)$$

$$W_2^{ij} = 6q_{ij}J(\omega_i + \omega_j) \quad [5]$$

The $J(\omega)$ are, of course, the spectral densities and contain motional information about the dipolar coupled nuclei within the system. As is well known, a spectral density is the Fourier transform of the autocorrelation function for the motion of the dipolar vector connecting the two coupled nuclei. As we (*10*) and others have shown before, the autocorrelation function may be constructed as simply or as complicated as needed to describe the motional dynamics of the system. If the distances between the dipolar coupled nuclei are constant (as they are in the case of a simply constructed autocorrelation function), the factor q is given by

$$q_{ij} = 0.1\gamma_i^2\gamma_j^2\hbar^2 r_{ij}^{-6}. \quad [6]$$

If the distance between the dipolar nuclei is not constant, however, then the distance fluctuations are included specifically in the calculation of the spectral densities (*10*). In this case,

$$q_{ij} = 0.1\gamma_i^2\gamma_j^2\hbar^2. \quad [7]$$

Thus, the first step in calculating the cross-peak intensities requires calculating the spectral densities using an appropriate motional model. Second, one constructs the appropriate rate matrix using Eqs. [2]–[4]. Finally, the rate matrix is diagonalized; the resulting eigenvalue matrix λ and the eigenvector matrix χ (and its inverse) may be used dirctly in Eq. [1] to calculate the auto- and cross-peak intensities as a function of mixing time. We have used numerical procedures for diagonalizing the rate matrix program (11). The results of the calculations are the intensities as a function of the mixing time. Each autointensity in our calculations is normalized to 1 and the cross-peaks are given as a fraction of this value. A test calculation using the method outlined was carried out with the calculations satisfactorily reproducing the experimental chloroform results of Macura and Ernst (*8*).

CALCULATIONS

Distance Determinations

Figure 1 shows the three-proton (and four-proton) system for which calculations were performed. From Fig. 1, it is clear that the H2–H3 distance was fixed at 2.0 Å in all calculations. The position of H1 was changed so that the cross-peak intensity for H1–H2 and H1–H3 proton pairs was calculated for each of the positions **a–u.** Table 1 lists the distances from H1 to H2 and H3 for each of these positions. We examined one of the more favorable cases for determining distances from 2D NOE spectra by setting $\omega_0\tau_0 = 10$. For H1 at positions **a–g** (see Fig. 1), the cross-peak intensities from H1 to H2 and H3 are equal (see Fig. 2), and we need only answer the question of discerning the distance dependence of the cross-peak intensity. For H1 at positions **c–g** the cross-peak magnetization increases throughout the range of

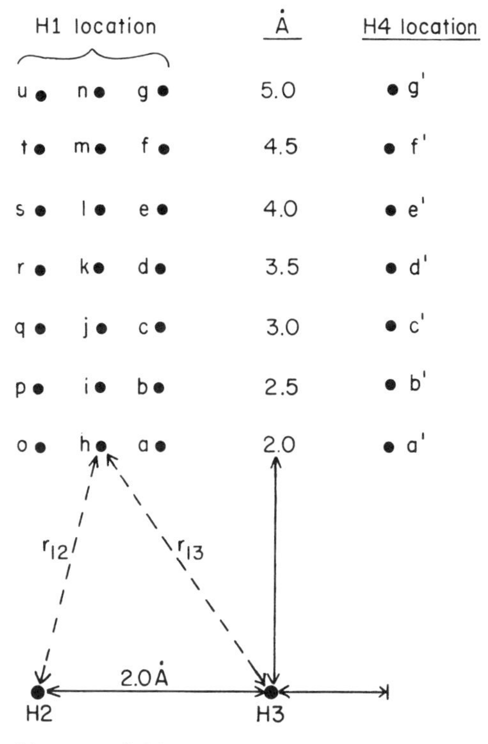

FIG. 1. The relative positions occupied by protons H1, H2, H3, and H4. Protons H2 and H3 remain fixed 2.0 Å apart. Proton H1 may occupy any of the positions **a–u,** depending on the calculation. For calculations involving four spins, H4 occupies one of the positions **a′–g′**. As an example, the internuclear distances r_{12} and r_{13} are shown if H1 is located at position **h.**

TABLE 1

INTERPROTON DISTANCES FROM H1 TO H2 AND H3 AT EACH OF THE POSITIONS IN FIG. 1

Position[a]	H1–H2 Distance (Å)	H1–H3 Distance (Å)
a	2.24	2.24
b	2.69	2.69
c	3.16	3.16
d	3.64	3.64
e	4.12	4.12
f	4.61	4.61
g	5.10	5.10
h	2.06	2.50
i	2.55	2.92
j	3.04	3.35
k	3.54	3.81
l	4.03	4.27
m	4.53	4.74
n	5.00	5.22
o	2.00	2.83
p	2.50	3.20
q	3.00	3.61
r	3.50	4.03
s	4.00	4.47
t	4.50	4.92
u	5.00	5.39

[a] See Fig. 1.

mixing times examined, while for positions **a** and **b** the curve increases to a maximum and then decreases. If we take 10% as the detectable lower limit for cross-peak intensity, only distances less than 3.64 Å are detectable. If the lower limit of detection is 1%, distances up to 5 Å would be detectable at mixing times less than 2 sec. Experimentally in our experience, cross-peak intensities slightly less than 1% of the auto-peak intensities have been discernible.

It is clear from Fig. 2A that the cross-peak intensities produced by the protons at positions **a–g** are significantly different. Using curves like that in Fig. 2A to derive a proton–proton distance from experimental cross-peak intensity is complicated by two considerations. At a given mixing time, significant differences in cross-peak intensities for positions **a–g** are recorded; however, a particular value of intensity for the cross-peak is obtained for many of the positions **a–d** but at different mixing times. Consequently, transforming cross-peak intensities into only a single distance from a single mixing time determination is impossible.

The cross-peak intensity at a fixed distance is a function of the molecular motion in the form of a correlation time τ_0. Variations in τ_0 (and $\omega_0\tau_0$) shift the cross-peak intensity-vs-mixing time curve (cf. Fig. 3), further obscuring the interpretation of a cross-peak intensity at a single mixing time. To establish the distance separating two

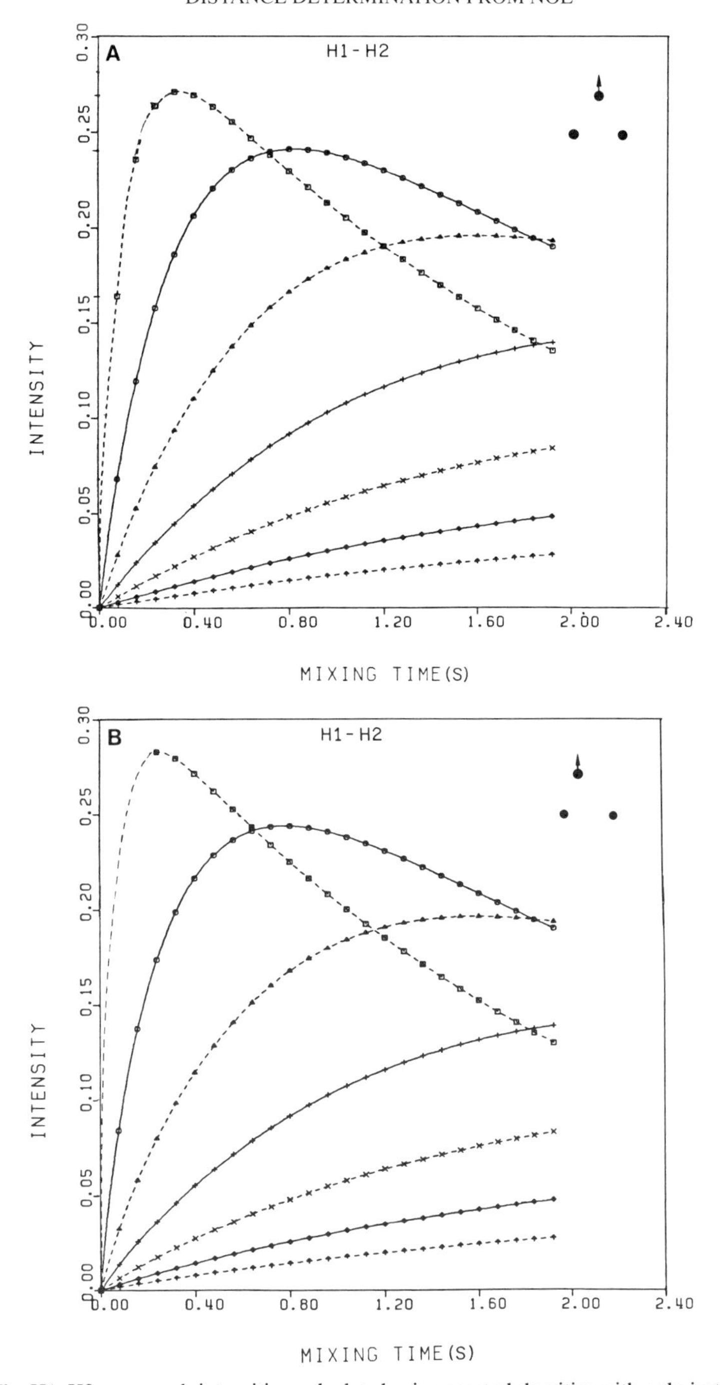

FIG. 2. The H1–H2 cross-peak intensities, calculated using spectral densities with only isotropic overall motion, as a function of mixing time t_m for a three-spin system. (A) H1–H2 cross-peak intensity (=H1–H3 intensity) is shown for H1 at positions **a** (□), **b** (○), **c** (△), **d** (+), **e** (×), **f** (◇), and **g** (↑); $\omega_0\tau_0 = 10$. (B) The H1–H2 cross-peak intensities for $\omega_0\tau_0 = 10$ with H1 at positions **h** (□), **i** (○), **j** (△), **k** (+), **l** (×), **m** (◇), and **n** (↑). (C) The H1–H2 cross-peak intensities for $\omega_0\tau_0 = 10$ with H1 at positions **o** (□), **p** (○), **q** (△), **r** (+), **s** (×), **t** (◇), and **u** (↑).

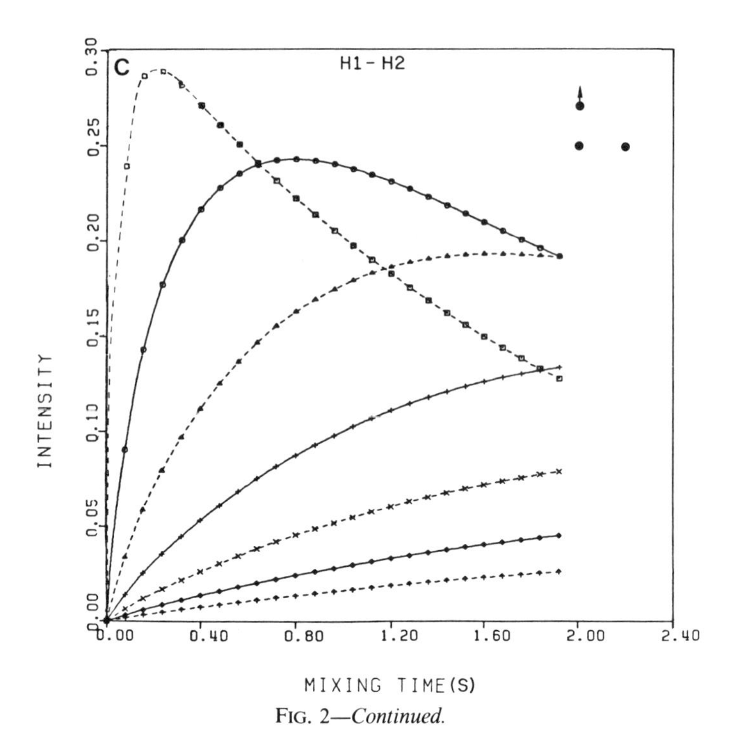

FIG. 2—*Continued.*

interacting protons, intensities at several mixing times need to be determined (3). In the absence of a priori knowledge of τ_0, a two-parameter fit for τ_0 and r_{ij} should be applied to the resulting curve.

The next question we attempted to answer was whether the cross-peak intensities with H1 in positions lateral to positions **a–g** can be distinguished. That is, can we distinguish positions **o** and **h** from one another and from **a**? Again we set $\omega_0\tau_0$ equal to 10. The cross-peak intensity curves for positions **h–n** and **o–u** are shown in Figs. 2 and 4. Figure 2B shows the H1–H2 cross-peak intensity, while Fig. 4A shows the H1–H3 cross-peak intensities for positions **h–n.** For positions **o–u,** Fig. 2C shows the H1–H2 cross-peak intensity, while Fig. 4B shows the H1–H3 cross-peak intensity. Comparing the H1–H2 cross-peak intensity curves in Fig. 2, only the curves generated by positions **a, h,** and **o** are distinguishable from one another. The curves for these three positions lose their distinguishability quickly as H1 is moved away. For example, the intensity difference at t_m = 320 msec between positions **a** and **o** is approximately 2%. At 80 msec, the intensity difference is about 8%. Clearly, only by establishing the entire curve between 0 and 320 msec could one hope to distinguish positions **a** and **o.** Some help in distinguishing position **a** from position **o** is provided by considering curves for both H1–H2 and H1–H3 for positions **a** and **o.** As the position of the proton changes from **a** to **o,** the increase in H1–H2 cross-peak intensity is accompanied by a simultaneous decrease in H1–H3 intensity. At a mixing time of 80 msec the difference in H1–H3 cross-peak intensities for positions **a** and **o** is about 7%. This difference is reduced to about 2% at 400 msec. As with the H1–H2 cross-peak intensity, establishing the entire curve for mixing times less than 400 msec would be necessary

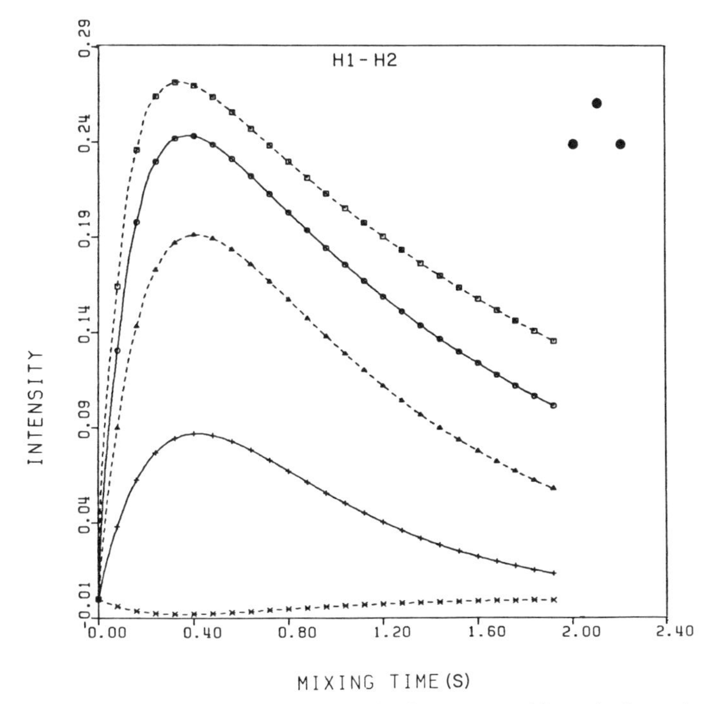

FIG. 3. The H1–H2 (or H1–H3) cross-peak intensity for H1 at position **a** is shown for various values of $\omega_0\tau_0$ assuming only isotropic motion: 10.0 (□), 7.5 (○), 5.0 (△), 2.5 (+), and 1.0 (×).

for distinguishing between positions from H1–H3 intensities. It is obvious, however, that using both H1–H2 and H1–H3 cross-peak intensities would provide a much better determination of the position of H1 relative to H2 and H3.

Internal Motion and Cross-Peak Intensity

It is evident from the discussion above that the motional properties of the molecule affect the cross-peak intensities via the spectral density terms. Our purpose here will be to test the sensitivity of the internuclear distances, derived from the 2D NOE cross-peak intensities, on the form of the spectral densities. Specifically, we wish to learn if distances can be determined to reasonable accuracy assuming simple motional models for the spectral densities when the true motion may be more complicated.

Again, we consider the three-proton spin system shown in Fig. 1. In this set of calculations, the H1–H2 and H1–H3 dipolar vectors are tumbling at a greater rate than the H2–H3 dipolar vector resulting from an internal motion we have imposed on the H1–H2 and H1–H3 vectors in addition to the isotropic motion affecting all three protons. Our intention here is to examine the direct effect of internal motion of the H1–H2 and H1–H3 dipolar vectors on the calculated cross-peak intensity.

The first model we tested for internal motion was the very simple free diffusion model. The angle between the internal rotation axis and the dipolar vector was set to 90°. In this calculation we set $\omega_0\tau_0$ and $\omega_0\tau_i$ both equal to 10. While τ_0 is the

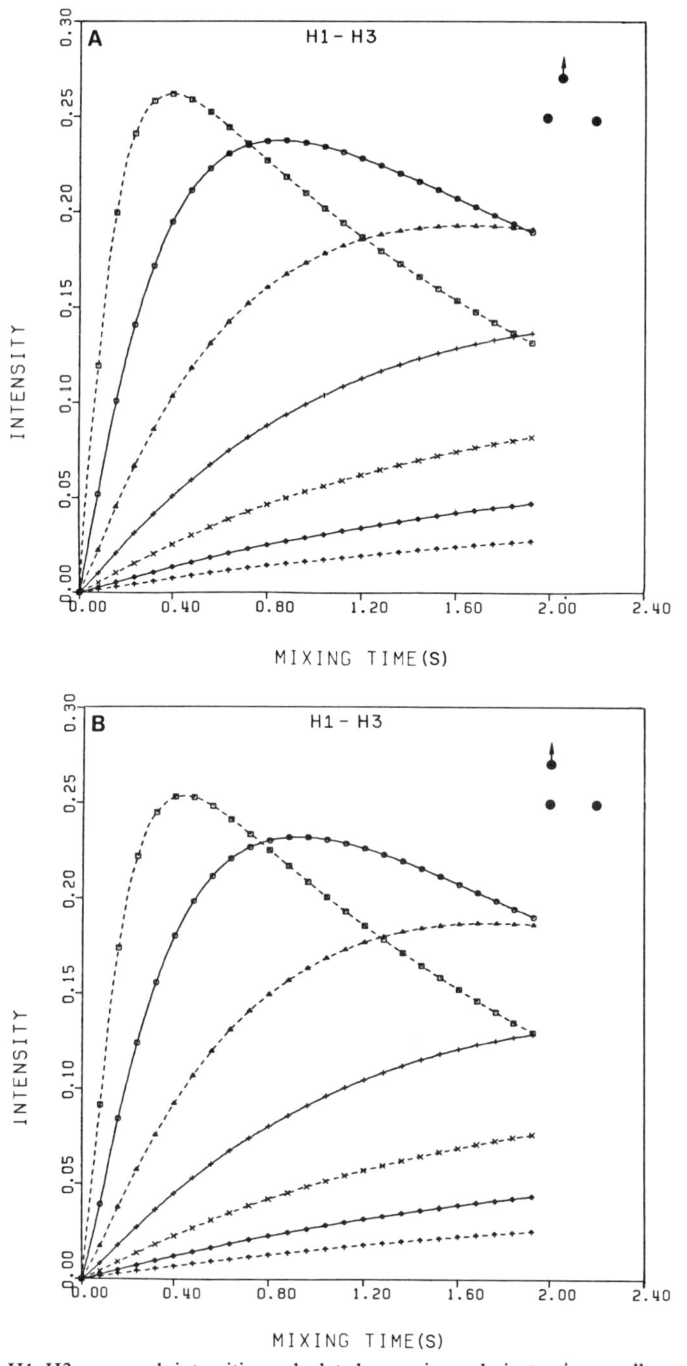

FIG. 4. The H1–H3 cross-peak intensities, calculated assuming only isotropic overall motion with $\omega_0\tau_0 = 10$, as a function of mixing time t_m for a three-spin system. (A) Intensities calculated with H1 at positions **h** (□), **i** (○), **j** (△), **k** (+), **l** (×), **m** (◊) and **n** (↑). (B) Intensities calculated with H1 at **o** (□), **p** (○), **q** (△), **r** (+), **s** (×), **t** (◊), and **u** (↑).

correlation time for overall isotropic tumbling motion, τ_i is the correlation time for the internal rotation.

The cross-peak intensities for positions **a–g** are shown in Fig. 5A for H1–H2 cross-peaks and Fig. 5B for H1–H3 cross-peaks; for positions **h–n,** Fig. 6A reports H1–H2 cross-peak intensities while Fig. 6B reports H1–H3 cross-peak intensities. Similarly, for positions **o–u,** Figure 7A reports the H1–H2 cross-peak intensities and Fig. 7B reports the H1–H3 cross-peak intensities. Performing the same comparisons as under Theory, the same general attributes of the curves and general utility in determining the distances between H1 and H2 or H3 are evident.

We compare, with this set of calculations, the distance predicted by the isotropic rotation model with the actual distance if the true motion is isotropic rotation plus free internal diffusion. To do this, the distance was obtained from the r^{-6} value required to give the same cross-peak intensity calculated with the isotropic motion spectral density (with $\omega_0\tau_0 = 10$) as the cross-peak intensity calculated with the two-correlation time model. The comparison of distances is shown in Fig. 8.

The general conclusion discernible from Fig. 8 is that the distances predicted by the isotropic motion model are somewhat larger than the distances predicted by the correct, isotropic plus free diffusion, model. For a specific mixing time, the isotropic plus free diffusion model predicts less cross-peak intensity than does the isotropic model. It is well-known that the cross-peak intensity decreases and can become negative as $\omega_0\tau_0$ decreases (see Fig. 3). The effect of the additional internal motion is to decrease the effective correlation time τ_{eff} compared with τ_0 in the isotropic model. The same decrease in cross-peak intensity may also be effected by increasing the distance separating the two protons, leading to the erroneous conclusion that the internuclear distances are longer than in actuality.

It is apparent from Fig. 3 that decreasing the isotropic tumbling time decreases peak intensities. In keeping with this observation, a considerable decrease in cross-peak intensities was found when calculations were carried out using the same internal rotation axis as above, but with $\omega_0\tau_0 = 10$ and $\omega_0\tau_i = 0.1$. In this situation, the effective correlation time to be used if one assumes only a single isotropic tumbling is approximately $\tau_{eff} = 0.25\tau_0$. From the appropriate curve in Fig. 3, it is seen that the maximum cross-peak intensity of about 0.09 is achieved at 400 msec mixing time. An intensity of 0.11 is obtained when the spectral density explicitly accounts for the internal motion. Such intensity errors of approximately 20% arising from use of a simple motional model with a single effective correlation time when the true motion includes internal rotational diffusion and overall tumbling lead to errors in distance estimation of less than 0.5 Å as seen by interpolating between the curves in any of Figs. 2–7.

In interpreting cross-peak intensities, just as in interpreting relaxation data, workers are likely to use a single model rather than several models from which to evaluate the proton–proton distances. The calculations and data presented in Fig. 8 give some indication of the errors which could be introduced by using an incorrect model for the motion. In general, the error is about 0.2–0.3 Å. At distances slightly greater than 5 Å, the cross-peak intensity predicted by the isotropic motion-only model is less than 1% of the diagonal peak intensity, a figure which we consider to be approximately the lower limit of detectability in solutions of proteins or nucleic acids.

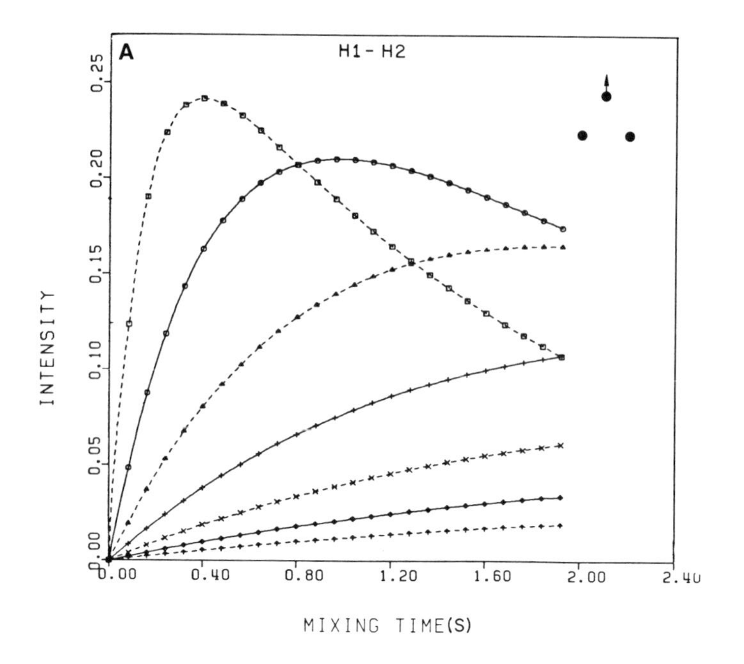

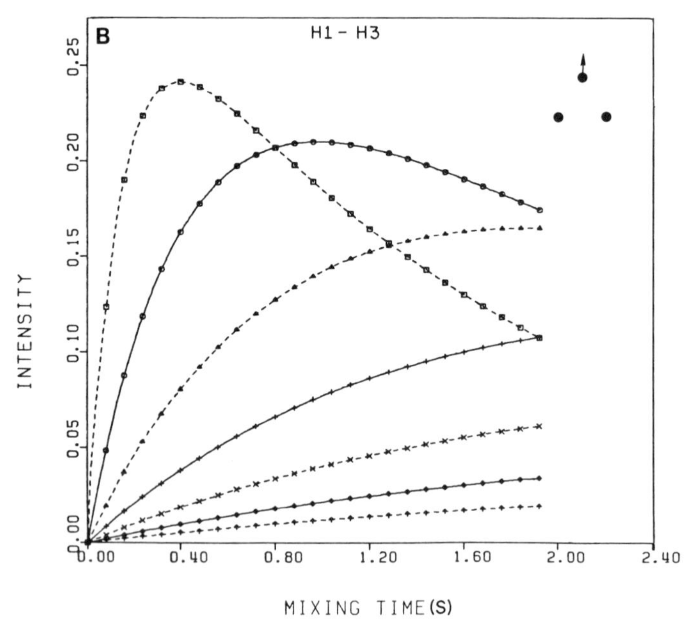

FIG. 5. The H1–H2 and H1–H3 cross-peak intensities, calculated using spectral densities with free internal diffusion superimposed on overall isotropic reorientation, as a function of mixing time with $\omega_0\tau_0 = 10$ and $\omega_0\tau_i = 10$. (A) H1–H2 cross-peak intensity is shown for H1 at positions **a** (□), **b** (○), **c** (△), **d** (+), **e** (×), **f** (◇), and **g** (↑). (B) H1–H3 cross-peak intensities for the same positions as in A.

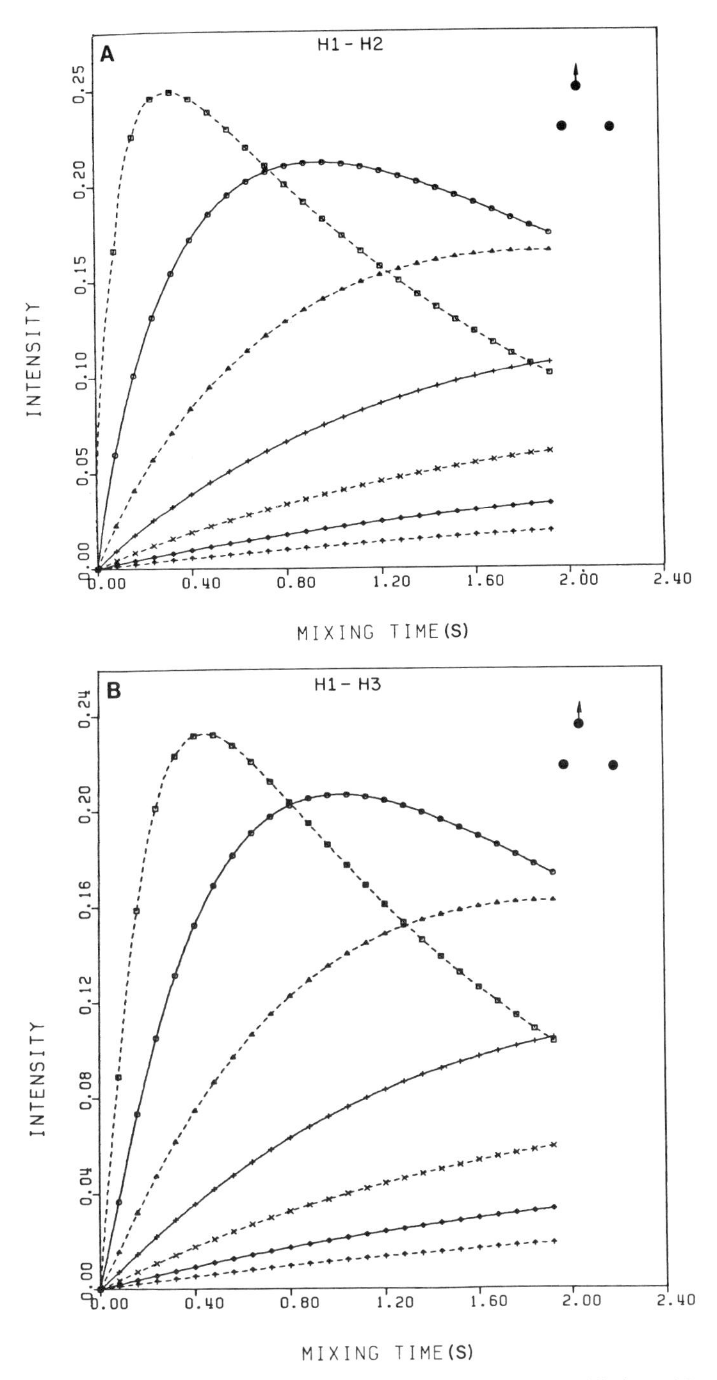

FIG. 6. H1–H2 and H1–H3 cross-peak intensities assuming free internal diffusion with $\omega_0\tau_i = 10$ superimposed upon overall isotropic tumbling with $\omega_0\tau_0 = 10$. The H1–H2 cross-peak intensities for H1 at positions **h** (□), **i** (○), **j** (△), **k** (+), **l** (×), **m** (◇), and **n** (↑). (B) The H1–H3 cross-peak intensities for the same positions as in A.

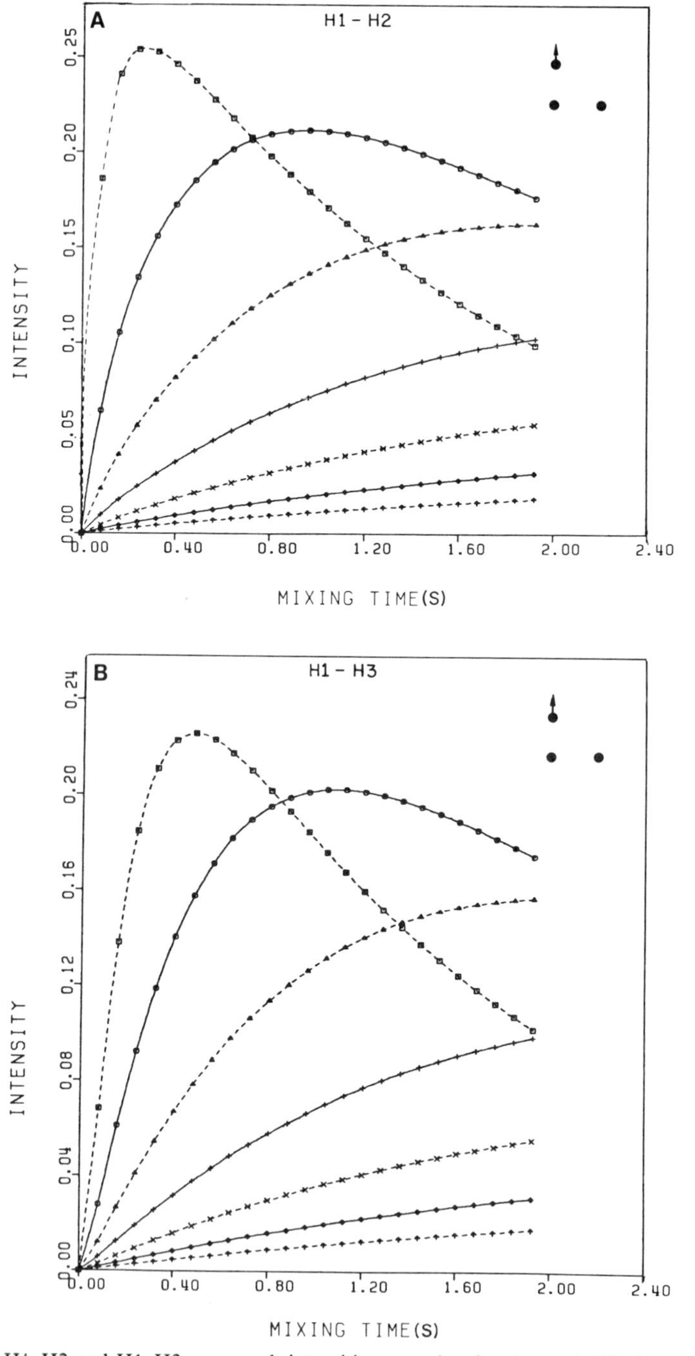

FIG. 7. (A) H1–H2 and H1–H3 cross-peak intensities assuming free internal diffusion with $\omega_0\tau_i = 10$ superimposed upon overall isotropic tumbling with $\omega_0\tau_0 = 10$. The H1–H2 cross-peak intensities for H1 at positions **o** (□), **p** (○), **q** (△), **r** (+), **s** (×), **t** (◇), and **u** (↑). (B) The H1–H3 cross-peak intensities for the same positions as in A.

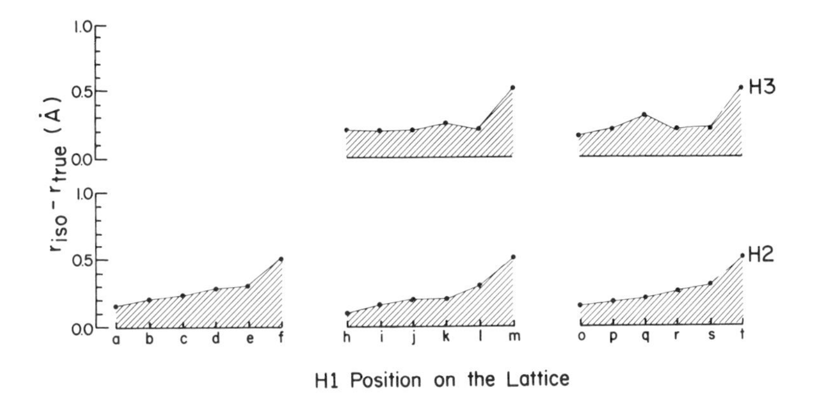

FIG. 8. Comparisons of the H1–H2 and H1–H3 distances derived from 2D NOE cross-peak intensities using the isotropic rotation model (r_{iso}) when the motion is correctly modeled with free internal diffusion superimposed upon the overall isotropic motion (r_{true}). The distances for the three-spin system were calculated for H1 at each of the positions depicted in Fig. 1. Details are given in the text.

Lattice Models for Internal Motion

With molecular motion, internuclear distances can change. In general, most motional studies have employed an "average" internuclear distance. As noted under Theory, we have explicitly accounted for the distance fluctuations by utilizing lattice models to describe the internal motions (*10*). In our opinion, a more accurate motional model includes explicitly the distance dependence in the spectral densities as proton H1 moves relative to H2 and H3. Again we wanted to examine the errors introduced in the distance determination by using the isotropic model to evaluate the cross-peak intensity information in a system in which the lattice model might be more appropriate.

Consequently, we calculated cross-peak intensities using spectral densities obtained from a model in which H1 rotates isotropically ($\omega_0\tau_0 = 10$) and also jumps between selected lattice sites ($\omega_0/k = 10$ where k is the elementary jump rate between sites). A subset of the positions **a–u** was chosen as the lattice of jump sites for each calculation. Protons H2 and H3 rotated with the same isotropic motion as H1 but without the internal motion lattice jumps.

In the first two calculations, we examined variations of a two-state lattice model. In Fig. 9, we compare an isotropic plus two-state lattice model with an isotropic rotation-only model. In this calculation, the lattice was constructed so that jumps between vertical lattice sites simultaneously increase or decrease the distance from H1 to H2 and H3. Figure 10 presents a similar comparison with the exception that the jump is between two lateral lattice positions and thus conversely affects the distance from H1 to protons H2 and H3.

In both Figs. 9 and 10, the isotropic rotation model, i.e., ignoring internal motions, predicts the H1–H2 or H1–H3 distance to be between the inner and outer limits of the lattice points until the minimum distance of separation is greater than 3.5 Å. Beyond this distance, the intensity is so low that the protons may be placed anywhere within the distances delimited by the lattice sites. This result is unexceptional, and

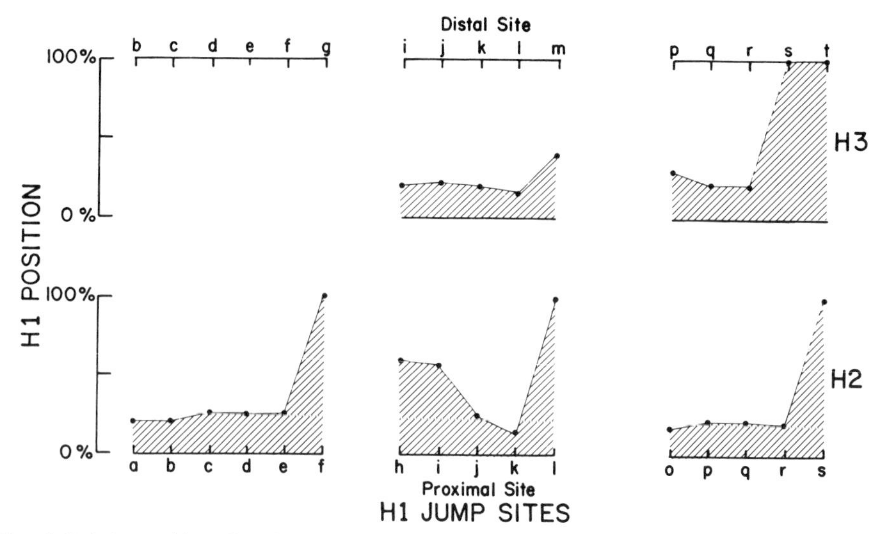

FIG. 9. Relative position of H1 in the three-spin system depicted in Fig. 1 based on calculated 2D NOE cross-peak intensities. The H1–H2 and H1–H3 distances determining the position were calculated assuming only isotropic reorientation when the true motion includes an additional two-site jump between various adjacent *vertical* points on the lattice of Fig. 1. The relative position is shown as a percentage of the distance between the two points on the lattice and is measured from the position closest to the H2–H3 vector.

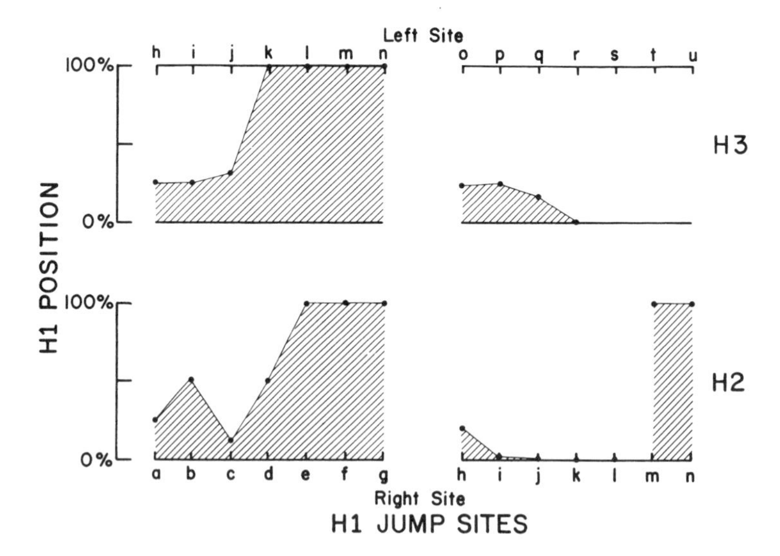

FIG. 10. Location of H1 in the three-spin system based on calculations of 2D NOE cross-peak intensities. The H1–H2 and H1–H3 distances establishing the relative position were calculated assuming only isotropic reorientation when the true motion includes a two-site jump between various adjacent *horizontal* points on the lattice of Fig. 1. The relative position is shown as a percentage of the distance between the two points with zero referring to the lattice point distal to H2.

TABLE 2

DISTANCES OBTAINED FROM THE ISOTROPIC ROTATION MODEL WHEN THE CORRECT MODEL IS A THREE-STATE LATTICE MODEL[a]

Jump sites on the lattice[b]	H1–H2 Distance (Å)	H1–H3 Distance (Å)
a, h, o	2.20	2.39
b, i, p	2.62	2.83
c, j, q	3.1	3.28
d, k, r	3.64	3.71
e, l, s	4.12	3.81
f, m, t	4.61–4.53	4.12–4.27
g, n, u	5.10	5.10–5.22

[a] Calculations assume $\omega_0\tau_0 = 10$ and $\omega_0/k = 10$, where k is the jump rate.
[b] See Fig. 1.

the only trend worth noting is the tendency of the isotropic model to place the H1 to H2 or H3 distance closer to the more proximal lattice points in Fig. 9. In Fig. 10, however, the trends for the H1–H2 distance and for the H1–H3 distance are different. The isotropic model places the H1–H2 distance closer to the more distal lattice points from H2 and the H1–H3 distance closer to the more proximal lattice points from H3 in the two-state jump model.

In Table 2, we report results using a three-state lattice internal motion model. For each calculation, a horizontal row of the lattice sites shown in Fig. 1 was chosen. Thus jumps between lattice sites conversely affect the distances from proton H1 to protons H2 and H3.

Inspecting Table 2, the distances predicted by the isotropic model are somewhat larger than the closest approach distances of H1 to either H2 or H3 (see Table 1). For H1–H2 distances, we were surprised to find that the distance predicted by the isotropic model fell between the middle and most distal positions included in a particular calculation, while the distance predicted by the isotropic model for the H1–H3 distance fell between the middle and most proximal lattice sites of the three positions. This result is similar to that found for the two-state lattice model.

In Table 3, we present the results of a four-state lattice model. The four lattice positions listed for each calculation represent four adjacent points in Fig. 1 where the H1 proton can be situated. As before, we have calculated the distance predicted by the isotropic motion model at the same cross-peak intensity predicted by the four-state lattice model for jumps among the four sites. It should be appreciated that there is not one discrete location of the H1 proton which, when used in the isotropic motion calculation, will yield the same cross-peak intensity as the lattice model provides. Rather, there will be a family of H1 locations which will suffice. Calculations for a couple of the possibilities were performed. If the four lattice points are considered to form a box, we calculated distances to H1, using the isotropic model, by interpolation along the edges of the box. In most cases, the intensities calculated from the four-

TABLE 3

DISTANCES OBTAINED FROM THE ISOTROPIC ROTATION MODEL WHEN THE CORRECT MODEL IS A FOUR-STATE LATTICE MODEL[a]

Jump sites on the lattice[b]	H1–H2 Distance[c] (Å)	H1–H3 Distance[c] (Å)
a, b, h, i	2.30/2.25	2.31/2.35
b, c, i, j	2.97/2.97	3.06/3.14
c, d, j, k	3.22/3.21	3.29/3.35
d, e, k, l	3.86/3.58	4.01/4.04
e, f, l, m	4.16/4.16	4.24/4.27
f, g, m, n	5.10/5.02	5.10/5.22
h, i, o, p	2.13/2.11	2.61/2.81
i, j, p, q	2.89/2.86	3.26/3.40
j, k, q, r	3.15/3.12	3.46/3.39
k, l, r, s	3.86/3.83	4.14/4.21
l, m, s, t	4.16/4.13	4.46/4.47
m, n, t, u	5.02/5.00	4.92/5.24

[a] Calculations assume $\omega_0\tau_0 = 10$ and $\omega_0/k = 10$, where k is the jump rate.

[b] See Fig. 1.

[c] The first entry corresponds to interpolation of the H1 position along the right vertical edge of the box describing the four lattice points, and the second entry corresponds to interpolation along the left edge.

state model were such that the distances were interpolated along vertical edges of the box (see Fig. 1) and these are the calculations reported. In Table 3, the first distance entry corresponds to interpolation along that edge of the box closer to the line bisecting the H2–H3 vector, while the second distance entry corresponds to interpolation along the vertical edge formed by the two most lateral points. Inspection of Table 3 reveals that although we did not perform interpolations along horizontal edges in Fig. 1, in some cases this interpolation would also be appropriate.

Finally, in Table 4 we review calculations in which the "true motion is a six-state lattice model. The same procedure that was used with the four-state lattice model for interpolation along edges was used. Again, more than one distance was obtained by interpolation along vertical edges of the rectangle because the intensity obtained by the lattice model was intermediate between those obtained using the isotropic model with discrete points.

It is efficacious to compare the results from the two-state jump (Figures 9 and 10) with the four-state lattice model (Table 3) with regard to the distance predicted by the isotropic model associated with it. In every case these distances are larger when the "true" motion includes a two-state internal jump than they are with a four-state internal jump. This effect indicates that the four-state lattice model allows a greater degree of internal motion and reduces the cross-peak intensity below that of the two-state lattice model at any particular mixing time. When the intensity produced by

TABLE 4

DISTANCES OBTAINED FROM THE ISOTROPIC ROTATION MODEL WHEN THE CORRECT MODEL IS A SIX-STATE LATTICE MODEL[a]

Jump sites on the lattice[b]	H1–H2 Distance[c] (Å)	H1–H3 Distance[c] (Å)
a, h, o, b, i, p	2.32/2.20/2.25	2.41/2.53/—
b, i, p, c, j, q	3.00/2.97/2.95	3.12/3.19/3.33
c, j, q, d, k, r	3.24/3.21/3.19	3.31/3.40/—
d, k, r, e, l, s	4.01/3.94/3.91	4.12/4.14/4.21
e, l, s, f, m, t	4.24/4.13/4.13	4.16/4.27/—
f, m, t, g, n, u	5.10/5.20/5.39	5.10/5.22/4.92

[a] Calculations assume $\omega_0\tau_0 = 10$ and $\omega_0/k = 10$, where k is the jump rate.

[b] See Fig. 1.

[c] The first entry corresponds to interpolation of the H1 position along the right vertical edge of the rectangle described by the six lattice positions, the second entry corresponds to interpolation along the bisector of the rectangle's long axis, and the third entry corresponds to interpolation along the left vertical edge of the rectangle.

motion correctly modeled as a two- or four-state lattice model is analyzed with the isotropic model, the effect is to lengthen the effective distance to a value somewhat greater than the distance of closest approach.

A Four-Spin System

There are several differences between the four-spin system and the similar three-spin system analyzed in the previous sections. While protons H2 and H3 occupy their usual position in Fig. 1 and H1 occupies positions **a–g,** H4 occupies the corresponding positions **a′–g′**, i.e., when H1 is at **b,** H4 is at **b′**. First, the H1–H2 and H1–H3 interactions are no longer equivalent interactions; consequently, the dependence of the cross-peak intensity on mixing time for the H1–H2 and H1–H3 interactions is different (see Fig. 11). As inspection of Figs. 11A and B shows, the differences are not large and decrease as the mixing time gets larger and as protons H1 and H4 are moved away from H2 and H3 through positions **a** to **g** and **a′** to **g′** simultaneously. The difference in interaction is not due to a different distance from H1 to H2 and H3, but is due to the presence of H4. In fact, two interactions are unique in this particular spin system: the H1–H3 dipolar interaction and the H2–H4 interaction. The H1–H2 and H3–H4 interactions are equivalent in this spin system, as are the H1–H4 and H2–H3 interactions.

We calculated cross-peak intensities using a four-spin system simply because it is the smallest spin system in which the distances may be varied such that the four spins formed a tightly coupled spin system or two loosely coupled spin systems of two tightly coupled spins.

Consider first the configuration of four spins in which protons H1, H4 occupy positions **a** and **a′**, respectively, while H2 and H3 occupy their usual positions shown in Fig. 1. We show the H1–H2 (or H3–H4) interactions in Fig. 11A. The H1–H4 interaction is shown in Fig. 11C, and the comparison clearly illustrates the gradual division of the four-spin system into two subsystems as H1 and H4 are moved further away from H2 and H3. At positions **a** and **a′**, the H1–H4 (or H2–H3) cross-peaks have intensities comparable to the H2–H1, H2–H4, H3–H1, or H3–H4 intensities. As H1 and H4 are moved to positions **g**, **g′**, respectively, the H1–H4 and H2–H3 cross-peak intensities increase substantially. Simultaneously, the H1–H2, H1–H3, H2–H4, and H3–H4 cross-peak intensities decrease substantially. This behavior is indicative of the increasing distance between the H1,H4 subsystem and the H2,H3 subsystem.

DISCUSSION

First, we wanted to demonstrate the ease with which cross-peak intensities may be calculated, and secondly, the need to perform such calculations as an aid in interpreting actual experimental cross-peak intensities. Perhaps the biggest barrier to performing this kind of calculation is the required numerical procedures we use for calculating the eigenvalues and eigenvectors needed for solution of Eq. [4]. The diagonalization routines are readily available from the source given in Ref. (*11*).

Our calculations using a four-spin system also clearly indicate the need for using such procedures rather than relying on approximations involving only the two spins which are closest to one another (*8*). Even for H1 and H4, at positions **g** and **g′**, respectively, this subsystem was not isolated from subsystem H2 and H3. This statement is true for any mixing time for positions **a–d** and **a′–d′** and is true for all of the positions **a–g** and **a′–g′** for mixing times greater than 400 msec. This observation is in accord with previously published advice to utilize only the shortest mixing times. If spin pairs are sufficiently isolated, any other spins can be ignored for short mixing times.

No attempt has been made here to select internuclear distances, correlation times or spin systems to reproduce any experimental results; that is not our purpose. It will be noted in Fig. 11C, however, that as more spins are brought together, the mixing time of the cross-peak intensity maximum occurs at shorter times and the intensity decreases to a lower level. This approaches the situation observed in proteins (e.g., (*3*)).

The second point we illustrated with our calculations is that relatively simple motional models may be used to generate the spectral densities used in Eq. [4] without introducing a large error. We gave a number of examples illustrating this point for a number of motional models of increasing complexity. While quite complicated motional models are sometimes required to fit relaxation data (*10*), expending such effort (and computer time) for calculating cross-peak intensities apparently is not required.

Related to this point is the observation that internal motions may actually diminish in influencing 2D NOE cross-peak intensities as internuclear distances increase. We can qualitatively understand this by considering the spin system of Fig. 12. In Fig. 12, we can imagine that both the proton and methyl group are in the same molecule

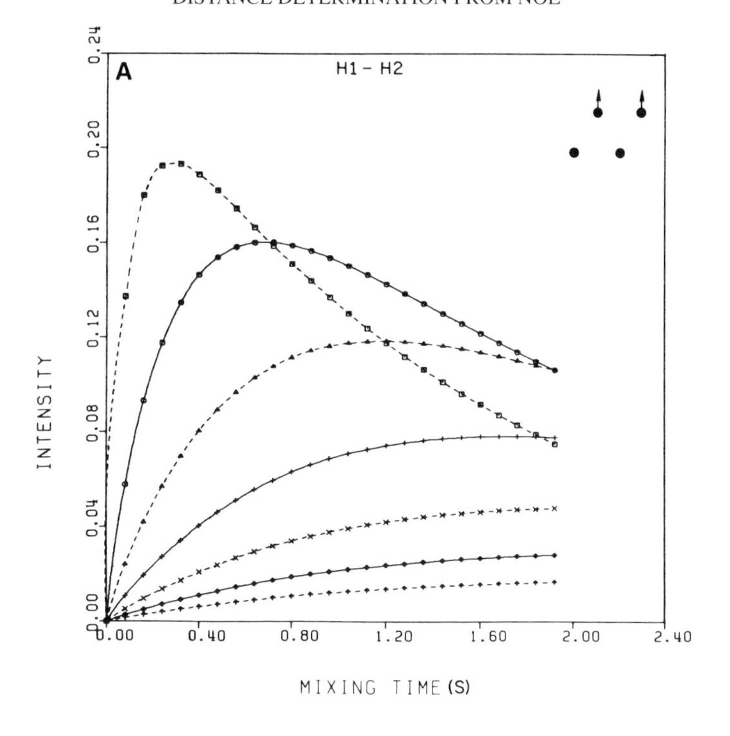

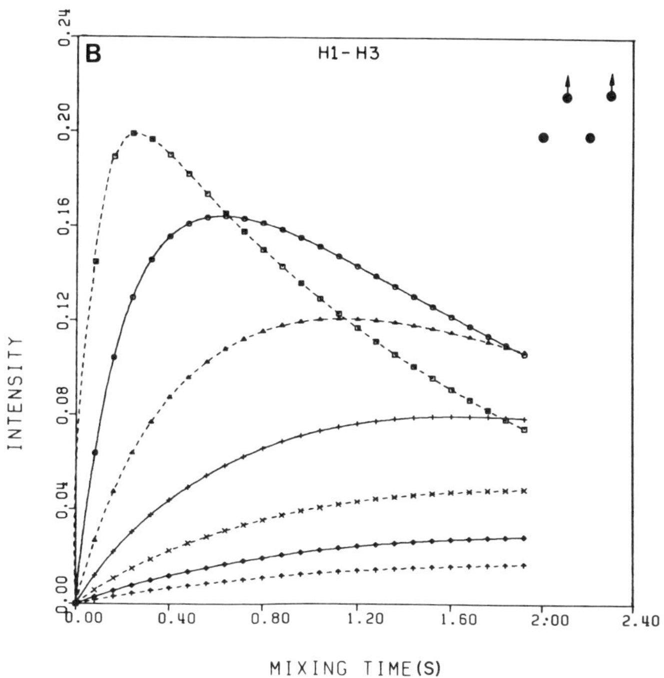

FIG. 11. 2D NOE cross-peak intensities as a function of mixing time for the four spins H1, H2, H3, and H4 in which $\omega_0\tau_0 = 10$. Protons H2 and H3 occupy their usual position in Fig. 1. Protons H1 and H4 occupy the positions **a,a′** (□); **b,b′** (○); **c,c′** (△); **d,d′** (+); **e,e′** (×); **f,f′** (◇); and **g,g′** (↑), respectively. (A) The H1–H2 (or equivalent H3–H4) cross-peak intensities. (B) The H1–H3 cross-peak intensities. (C) The H1–H4 (or equivalent H2–H3) cross-peak intensities.

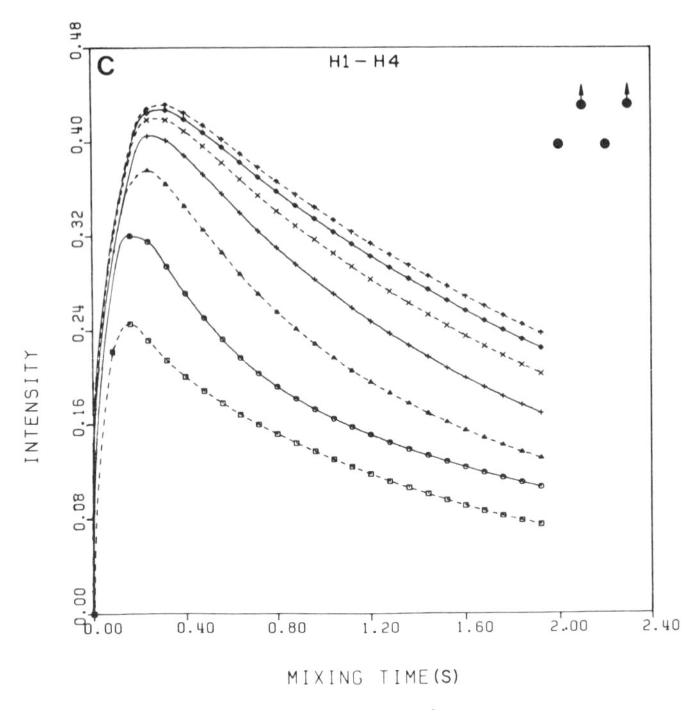

FIG. 11—*Continued.*

which is rotating isotropically with correlation time τ_0. Additionally, the methyl group undergoes internal motion with some correlation time τ_i. If the methyl carbon atom is ^{13}C, rotation of the methyl group will affect relaxation of that methyl carbon. Consider the relaxation and the cross-relaxation determining the 2D NOE peaks for a proton H at position A or B in Fig. 12. Rotation of the methyl group causes the dipolar vector connecting H to the proton(s) of the methyl group to rotate on the surface of a cone with half angle 22° (position A) or 11° (position B). Clearly, as the distance between H and the methyl protons increases, the angle through which the H–methyl H dipolar vectors fluctuate decreases. This effect, coupled with the direct loss in strength of the dipolar interaction due to the increasing distance, makes the internal motion very ineffective in the relaxation process. Consequently, at the normal

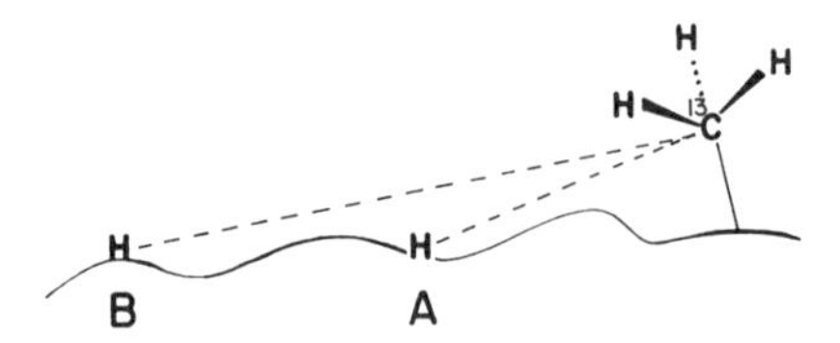

FIG. 12. A hypothetical spin system demonstrating the decreasing influence of internal motion as the distance separating protons increases. A proton at position A is 2.5 Å from the projection of the methyl protons on the axis connecting the methyl rotor. A proton at position B is 5.0 Å away.

distances separating protons on different residues within macromolecules, τ_0 will dominate the cross-relaxation process. This example illustrates the relatively small errors introduced by using spectral densities from an isotropic rotation model when the motion is more correctly modeled using spectral densities derived from more complicated molecular dynamics.

Finally, we want to point out the resolution we think our interpretative technique has. In the worst case, we assume that the intensity has been determined at a single mixing time. In this case, the experimenter obtains an upper limit to the internuclear distance because the intensity will be less than or equal to the maximum intensity for that proton, and the investigator must assume it is the maximum. If the cross-peak intensity curve is established by obtaining cross-peak intensities at several mixing times, then a more accurate range for the mixing time may be established. Our calculations show that ±0.5 Å should be obtainable if one knows the correlation times or individual proton relaxation rates accurately. At this resolution, refinement techniques (2) should prove useful in determining overall structures which are consistent with distances provided by 2D NMR experiments.

ACKNOWLEDGMENTS

This work was supported by Research Grants GM 25018 and CA27343 from the National Institutes of Health. T.L.J. also gratefully acknowledges receipt of Research Career Development Award AM 00291 from NIH.

REFERENCES

1. G. Wagner and K. Wüthrich, *J. Mol. Biol.* **155,** 347 (1982).
2. T. F. Havel, G. M. Crippen, and I. D. Kuntz, *Biopolymers* **18,** 73 (1979).
3. A. Kumar, G. Wagner, R. R. Ernst, and K. Wüthrich, *J. Am. Chem. Soc.* **103,** 3654 (1981).
4. A. A. Bothner-By and J. H. Noggle, *J. Am. Chem. Soc.* **101,** 5152 (1979).
5. G. Wagner and K. Wüthrich, *J. Magn. Reson.* **33,** 675 (1979).
6. J. Tropp, *J. Chem. Phys.* **72,** 6035 (1980).
7. C. M. Dobson, E. T. Olejniczak, F. M. Poulsen, and R. G. Ratcliffe, *J. Magn. Reson.* **48,** 97 (1982).
8. S. Macura and R. R. Ernst, *Mol. Phys.* **41,** 95 (1980).
9. G. Bodenhausen and R. R. Ernst, *J. Am. Chem. Soc.* **104,** 1304 (1982).
10. J. W. Keepers and T. L. James, *J. Am. Chem. Soc.* **104,** 929 (1982).
11. Quantum Chemistry Program Exchange, program 62, available from Indiana University.

JOURNAL OF MAGNETIC RESONANCE **58,** 370–388 (1984)

Selection of Coherence-Transfer Pathways in NMR Pulse Experiments

GEOFFREY BODENHAUSEN, HERBERT KOGLER,* AND R. R. ERNST

Laboratorium für Physikalische Chemie, Eidgenössische Technische Hochschule, 8092 Zurich, Switzerland

Received September 29, 1983; revised December 13, 1983

NMR pulse experiments are described in terms of pathways through various orders of coherence. A general procedure is indicated for the systematic design of phase cycles that select desirable coherence-transfer pathways.

INTRODUCTION

Some of the most successful new pulse techniques in high-resolution nuclear magnetic resonance rely on coherence-transfer processes. To name a few, we mention homonuclear and heteronuclear two-dimensional (2D) correlation spectroscopy (*1–7*), multiple-quantum spectroscopy (*8–12*), multiple-quantum filtering (*13–16*), spin pattern recognition (*17*), and various methods for the enhancement and simplification of 1D carbon-13 spectra (*18–21*). The development of these experiments has been accompanied by a proliferation of recipes for the elimination of unwanted signals and artifacts by phase-cycling techniques. Because many different, often rather intuitive approaches have been used, the common basis of these techniques is not always transparent.

In an effort to provide a unified picture, we describe pulse experiments in terms of pathways through various orders of coherence. This "coherence-transfer pathway" approach turns out to be useful to design novel experiments for specific purposes. The selection of the desired pathway is accomplished experimentally by means of phase-cycling procedures.

There are parallels between this work and a paper recently submitted by A. D. Bain (*22*), who also applied the concept of coherence-transfer pathways. In the present paper, we demonstrate its utility in homonuclear experiments involving several coherence-transfer steps, while Bain is concerned with the systematic description of phase cycles in connection with quadrature detection and heteronuclear coherence transfer (*22*).

COHERENCE AND COHERENCE-TRANSFER PATHWAYS

The concept of "coherence" is a generalization of the notion of transverse magnetization. Coherence can be associated with a transition between a pair of eigenstates

* Present address: Institut für Organische Chemie, Universität Frankfurt, D-6000 Frankfurt, West Germany.

0022-2364/84 $3.00

$|r\rangle$ and $|s\rangle$ with an arbitrary difference in magnetic quantum numbers $p_{rs} = M_r - M_s$. Transverse magnetization corresponds to a particular class of coherence associated with a change in quantum number $p = \pm 1$.

Formally, coherence can be conceived as a coherent superposition of two eigenstates (*23*)

$$\psi_{rs} = a_r|r\rangle + a_s|s\rangle. \quad [1]$$

Such a non-equilibrium state develops in time under the time-independent free precession Hamiltonian. In terms of the density operator σ, coherence between the states $|r\rangle$ and $|s\rangle$ is expressed by the existence of nonzero density matrix elements $\sigma_{rs} = |r\rangle\langle s|$ and $\sigma_{sr} = |s\rangle\langle r|$. These elements indicate a "transition in progress" between the two connected states.

In high-field NMR, each eigenstate $|r\rangle$ is characterized by a magnetic quantum number M_r and each coherence σ_{rs} by a magnetic quantum number difference $p_{rs} = M_r - M_s$ which we call "coherence order." Note that each transition is associated with two coherences σ_{rs} and σ_{sr} with coherence orders of opposite sign. The quantities M_r and p_{rs} are "good" quantum numbers,[1] and each coherence σ_{rs} conserves its quantum number p_{rs} in the course of free precession. Radiofrequency (rf) pulses, however, may induce a transfer between coherences σ_{rs} and σ_{tu}, a process that may change the coherence order.

It is often sufficient to classify the various terms of the density operator according to the coherence order p:

$$\sigma(t) = \sum_p \sigma^p(t). \quad [2]$$

For a system of K spins 1/2, p extends from $-K$ to K. This classification can be carried out explicitly if the density operator is expressed in matrix elements, or alternatively in a suitable set of base operators, such as irreducible tensor operators T_{lp} (*24*), products of shift operators (e.g., $I_k^+ I_l^+$) (*25*), or single-transition shift operators (e.g., $I^{+(rs)} = I_x^{(rs)} + iI_y^{(rs)}$). On the other hand, products of Cartesian operators (e.g., $I_{kx}I_{lx}$) (*25*) or Cartesian single-transition operators $I_x^{(rs)}$ (*26, 27*) are not particularly suitable for a classification according to p.

The characteristic properties of coherence of order p (or simply "p-quantum coherence") are demonstrated by the transformation under rotations about the z axis:

$$\exp\{-i\varphi F_z\}\sigma^p \exp\{i\varphi F_z\} = \sigma^p \exp\{-ip\varphi\} \quad [3]$$

where

$$F_z = \sum_{k=1}^{N} I_{kz}.$$

We found it convenient to represent the sequence of events in various experiments in a "coherence transfer map" such as shown in Fig. 1. Free precession proceeds within the levels of this map, while pulses may induce "transitions" between coherence

[1] This is related to the fact that the Hamiltonian has rotational symmetry. The eigenstate $|r\rangle$ transforms according to the irreducible representation M_r of the one-dimensional rotation group (*24*). Hence $|r\rangle\langle s|$ transforms according to the representation $p_{rs} = M_r - M_s$.

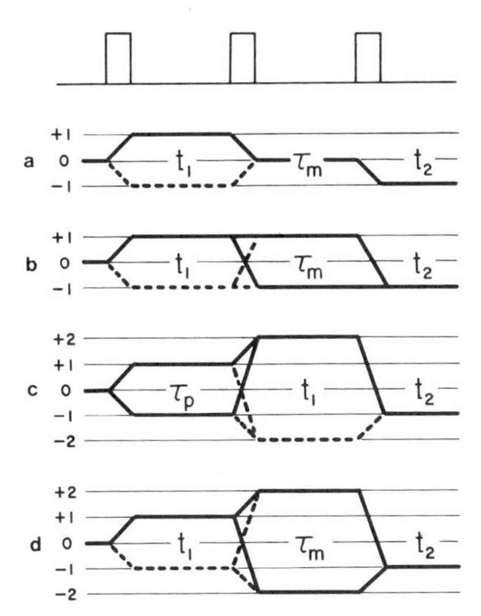

FIG. 1. Coherence-transfer maps (CT maps) for various 2D experiments involving three pulses. Solid lines indicate pathways that involve a single-order p in the evolution period. For a basic understanding of the experiments, these pathways suffice. If pure phase lineshapes are not essential (e.g., if composite lineshapes or absolute-value plots are acceptable), it is sufficient to select the pathways shown by solid lines. Mirror-image pathways with $-p$ in t_1 are indicated by dashed lines. For pure phase spectra (i.e., pure 2D absorption lineshapes), both solid and dashed pathways must be retained. Four experimental schemes are shown: (a) 2D exchange spectroscopy (NOESY), (b) relayed correlation spectroscopy (pathways shown for fixed mixing interval τ_m), (c) double-quantum spectroscopy and (d) 2D correlation spectroscopy with double-quantum filter ($\tau_m = 0$).

orders. The route of a particular component of coherence is referred to as a "coherence-transfer pathway." All coherence-transfer pathways of a pulse experiment start with $p = 0$ (thermal equilibrium) and must end with single-quantum coherence to be detectable. If we choose to observe the complex signal in the detection interval by quadrature detection,

$$s^+(t) = s_x(t) + is_y(t) = \mathrm{Tr}\{\sigma(t)F_x\} + i\,\mathrm{Tr}\{\sigma(t)F_y\} = \mathrm{Tr}\{\sigma(t)F^+\} \qquad [4]$$

where $F^+ = \sum_k I_k^+$, only density operator components proportional to I_k^- can contribute to the signal, and all pathways that do not lead to $p = -1$ can be disregarded. However, as noted by Bain (*22*), imperfect quadrature detection (i.e., imbalance of the two receiver channels) leads to partial observation of single-quantum coherence components with $p = +1$.

The examples in Fig. 1 show the coherence transfer pathways that are relevant to four well known techniques involving three consecutive coherence-transfer steps. Apart from the incrementation of the intervals in the course of the experimental sequence, these methods merely differ in the selection of coherence-transfer pathways. A single pathway suffices if absolute-value spectra or phase-sensitive spectra with composite ("phase-twisted") lineshapes are acceptable, while "mirror image" pathways (dashed lines in Fig. 1) must be retained simultaneously if pure phase spectra (i.e., pure 2D absorption lineshapes) are essential, as will be discussed below.

It is important to note that the order $p = 0$ may comprise Zeeman polarization (represented by density operator terms proportional to I_{kz}), longitudinal scalar or dipolar spin order (e.g., $I_{kz}I_{lz}$), and zero-quantum coherence (e.g., $I_k^+ I_l^-$). This is particularly relevant for 2D exchange spectroscopy (Fig. 1a) (*28, 29*). In the case of relayed magnetization transfer (Fig. 1b) (*30*) the delay τ_m can be kept constant, in which case the sign of the coherence order in τ_m is irrelevant and two pathways can be allowed simultaneously. It is also possible to vary τ_m in concert with t_1 (*31, 32*), in which case the pathway selection determines whether the ω_1 domain will contain sums or differences of chemical shifts. In multiple-quantum spectroscopy (Fig. 1c) (*8–12*), we have the option of observing both $+p$ and $-p$ coherences in the evolution period, or we may restrict the transfer as shown by solid lines. In correlation spectroscopy with multiple-quantum filters (Fig. 1d) (*14*), it is not necessary to select the sign of the coherence order in the τ_m interval, but one has the option of selecting only $p = +1$ coherence in the evolution period.

SELECTION OF COHERENCE-TRANSFER PATHWAYS

In experiments employing nonselective pulses, numerous coherence-transfer pathways can occur simultaneously. In principle, it is possible to use cascades of selective pulses to restrict the number of pathways, but it turns out that phase-shifted nonselective pulses provide a more flexible approach to the selection of desirable pathways.

Consider a complete pulse experiment with n coherence transfer processes expressed by the propagators $U_1, U_2, \cdots U_n$:

$$\sigma_0 \xrightarrow{U_1} \xrightarrow{U_2} \cdots \xrightarrow{U_n} \sigma(t). \qquad [5]$$

In the context of 2D spectroscopy, the first propagator typically represents the excitation process, while the last propagator corresponds to the conversion into observable magnetization. The intermediate propagators, which only occur in some experiments, induce coherence transfer between various orders. A propagator may represent a single pulse or a sequence of pulses, such as the composite sequence $(\pi/2)$–τ–(π)–τ–$(\pi/2)$ commonly used for multiple-quantum excitation (*25*). Each propagator U_i causes a transfer of a particular order of coherence $\sigma^p(t_i^-)$ into numerous different orders $\sigma^{p'}(t_i^+)$:

$$U_i \sigma^p(t_i^-) U^{-1} = \sum_{p'} \sigma^{p'}(t_i^+) \qquad [6]$$

where the arguments of the density operators refer to the state just before and immediately after the transformation by U_i. This leads to a "branching" or "fanning out" of the coherence-transfer pathways. After n consecutive coherence-transfer steps, each pathway can be characterized by a set of n values:

$$\Delta p_i = p'(t_i^+) - p(t_i^-) \qquad [7]$$

corresponding to the changes in coherence order under the propagators U_i. The complete pathway is therefore specified by a vector

$$\mathbf{\Delta p} = \{\Delta p_1, \Delta p_2, \ldots, \Delta p_n\}. \qquad [8]$$

Since all pathways must begin with $p = 0$ and are assumed to end with $p = -1$ to be observable (see Eq. [4]), the sum of the components of the vector $\mathbf{\Delta p}$ is fixed,

$$\sum_i \Delta p_i = -1. \qquad [9]$$

Thus if $(n - 1)$ values of Δp_i are specified by $(n - 1)$ independent phase cycles as discussed below, the entire vector $\mathbf{\Delta p}$ and hence the complete pathway are defined unambiguously. Because the rf phase shifts required for the pathway selection are often subject to systematic errors, it may however be advisable in practice to employ n independent phase cycles to select the desired Δp_i values under *all* n coherence transfer steps.

The key to the separation of coherence-transfer pathways is the use of propagators $U_i(\varphi_i)$ that are shifted in phase

$$U_i(\varphi_i) = \exp\{-i\varphi_i F_z\} U_i(0) \exp\{i\varphi_i F_z\}. \qquad [10]$$

If a particular propagator U is made up of a sequence of pulses, each constituent pulse must be incremented in phase. For example, the excitation sequence commonly used in double-quantum spectroscopy becomes $(\pi/2)_\varphi\text{–}\tau\text{–}(\pi)_\varphi\text{–}\tau\text{–}(\pi/2)_\varphi$.

Under a phase-shifted propagator $U_i(\varphi_i)$, Eq. [6] takes the form

$$U_i(\varphi_i)\sigma^p(t_i^-)U(\varphi_i)^{-1} = \sum_{p'} \sigma^{p'}(t_i^+) \exp\{-i\Delta p_i \varphi_i\}. \qquad [11]$$

Thus the phase shift of a coherence component that is transferred by the propagator U_i is given by $\Delta p_i \varphi_i$. In symbolic notation, Eq. [11] may be written

$$\sigma^p(t_i^-) \xrightarrow{U_i(\varphi_i)} \sum_{p'} \sigma^{p'}(t_i^+) \exp\{-i\Delta p_i \varphi_i\}. \qquad [12]$$

After n consecutive coherence-transfer steps, one obtains single-quantum coherence components ($p = -1$) with phases that reflect the pathways $\mathbf{\Delta p}$ and the propagator phases φ_i:

$$\begin{aligned} \sigma^{p=-1}(\varphi_1, \varphi_2, \ldots, \varphi_n, t) &= \sigma^{p=-1}(\varphi_1 = \varphi_2 = \cdots = \varphi_n = 0, t) \\ &\quad \times \exp\{-i(\Delta p_1 \varphi_1 + \Delta p_2 \varphi_2 + \cdots + \Delta p_n \varphi_n)\} \qquad [13a] \\ &= \sigma^{p=-1}(\boldsymbol{\varphi} = \mathbf{0}) \exp\{-i\mathbf{\Delta p}\boldsymbol{\varphi}\} \qquad [13b] \end{aligned}$$

with the vector notation for $\mathbf{\Delta p}$ in Eq. [8] and

$$\boldsymbol{\varphi} = \{\varphi_1, \varphi_2, \ldots, \varphi_n\}. \qquad [14]$$

The phase shifts of Eq. [13] also occur in the complex signal observed during the detection period (Eq. [4]). It is convenient to decompose the signal into contributions of individual pathways:

$$s(t) = \sum_{\mathbf{\Delta p}} s^{\mathbf{\Delta p}}(t). \qquad [15]$$

With a given vector of phase shifts $\boldsymbol{\varphi}$, the signal associated with a certain pathway carries the phase

$$s^{\mathbf{\Delta p}}(\boldsymbol{\varphi}, t) = s^{\mathbf{\Delta p}}(\mathbf{0}, t) \exp\{-i\mathbf{\Delta p}\boldsymbol{\varphi}\}. \qquad [16]$$

The characteristic phase shift in Eq. [16] can be used, following Wokaun and Ernst (9), to separate the different pathways under a particular propagator U_i by a Fourier analysis with respect to the rf phase of this propagator U_i.

To restrict the coherence transfer under U_i to a particular change Δp_i in coherence order, we may perform N_i experiments where the rf phase φ_i of the propagator is incremented systematically:

$$\varphi_i = k_i 2\pi/N_i, \qquad k_i = 0, 1, \ldots, N_i - 1. \qquad [17]$$

The N_i signals $s(\varphi_i, t)$ observed in the detection period are then combined according to a discrete Fourier analysis with respect to the phase φ_i,

$$s^{\Delta p_i}(t) = \frac{1}{N_i} \sum_{k_i=0}^{N_i-1} s(\varphi_i, t) \exp(i\Delta p_i \varphi_i). \qquad [18]$$

By this process, all coherence-transfer pathways are selected which undergo a change in coherence order Δp_i under the propagator U_i. However, by carrying out a series of N_i experiments, it is not possible to select a unique Δp_i, but rather a series of values $\Delta p_i \pm nN_i$ with $n = 0, 1, 2, \ldots$. This situation is reminiscent of aliasing in Fourier analysis and is a consequence of the sampling theorem. Clearly, if a unique Δp_i value must be selected from a range of r consecutive values, N_i must be chosen at least equal to r.

It is useful to exhibit the required selectivity of the phase cycle by listing all possible changes in coherence order, for example,

$$\Delta p_i\text{: } -3, -2, \mathbf{-1}, (0), (1), 2, 3, \qquad [19]$$

where the values of Δp_i that must be blocked are set in parentheses, while the desired value is set in boldface. The minimum number of experiments to be performed in this case would be $N_i = 3$. The examples discussed below will illustrate the resulting phase cycles.

In many experiments a more restrictive selection of pathways is desired than can be obtained by cycling the phase of a single propagator U_i. In such cases, a desired pathway with successive changes in coherence order $\Delta p_1, \Delta p_2, \ldots, \Delta p_n$ under the n propagators can be retained selectively by cycling the phases of each of the relevant propagators $U_1(\varphi_1), U_2(\varphi_2), \ldots, U_n(\varphi_n)$ separately:

$$\varphi_1 = k_1 2\pi/N_1, \ldots, \qquad \varphi_n = k_n 2\pi/N_n,$$

for

$$k_1 = 0, 1, \ldots, N_1 - 1, \ldots, \qquad k_n = 0, 1, \ldots, N_n - 1. \qquad [20]$$

A unique prescription for the phase cycle is obtained by incrementing k_1 through all N_1 steps before incrementing k_2. The total number of experiments to be performed is determined by the product $N = N_1 \cdot N_2 \cdot \ldots \cdot N_n$. To select the desired pathway characterized by the vector $\mathbf{\Delta p}$ in Eq. [8], the signals must be combined according to

$$s^{\Delta p}(t) = \frac{1}{N} \sum_{k_1=0}^{N_1-1} \sum_{k_2=0}^{N_2-1} \cdots \sum_{k_n=0}^{N_n-1} s(t) \exp\{+i\mathbf{\Delta p}\boldsymbol{\varphi}\} \qquad [21]$$

where

$$\mathbf{\Delta p}\boldsymbol{\varphi} = \Delta p_1 k_1 2\pi/N_1 + \Delta p_2 k_2 2\pi/N_2 + \cdots + \Delta p_n k_n 2\pi/N_n. \qquad [22]$$

The principle of the pathway selection becomes obvious by observing that the signal $s(t)$ consists of the contributions of all possible pathways $\Delta\mathbf{p}'$ (see Eqs. [15] and [16]):

$$s(t) = \sum_{\Delta\mathbf{p}'} s^{\Delta\mathbf{p}'}(\mathbf{0}, t) \exp\{-i\Delta\mathbf{p}'\boldsymbol{\varphi}\}. \quad [23]$$

Clearly, the discrete n-dimensional Fourier analysis in Eq. [21] leads to a nonvanishing signal for $\Delta\mathbf{p}' = \Delta\mathbf{p}$. However, since the selectivity under each propagator U_i is determined by the number N_i of phase increments, there are a manifold of pathways that survive the selection process, with

$$\Delta\mathbf{p} = \{\Delta p_1 \pm n_1 N_1, \Delta p_2 \pm n_2 N_2, \ldots, \Delta p_n \pm n_n N_n\}. \quad [24]$$

Because the maximum order of coherence $|p_{\max}| \leq K$ in a system with K spins 1/2, and because the amplitude of coherence transfer into very high orders is small, it is usually possible to retain a unique pathway by relatively small increment numbers N_i.

There are three different strategies to achieve the multiplication of the signals by the phase factors necessary for the Fourier analysis in Eq. [21]: (a) multiplication of the complex signals with complex phase factors (this can be achieved conveniently with routine phase-correction procedures); (b) phase-shifting of all pulses in the sequence through $\Sigma\ \Delta p_i \varphi_i$ and addition of the signals without weighting; (c) shifting of the phase of the receiver reference channel. Strategy (c) was adopted in the experimental examples discussed below. With the definition of the observable operator F^+ in Eq. [4], the reference phase must be $\varphi^{\text{ref}} = -\Sigma\ \Delta p_i \varphi_i$. The opposite phase shift must be applied if the observable operator is F^-, in which case the pathways terminate at the level $p = +1$.

The parameters involved in Eq. [21] are shown schematically in Fig. 2 for a hypothetical experiment involving three coherence-transfer steps. To select the desired pathway, the reference channel of the phase-sensitive detector (PSD) can be shifted in phase as indicated in the figure.

PURE 2D ABSORPTION LINESHAPES

In the evolution period t_1, there are always two coherences associated with each transition $|r\rangle \leftrightarrow |s\rangle$ that have opposite orders $p = M_r - M_s$ and $p' = -p$ and opposite frequencies. If the two coherence-transfer pathways $+p \rightarrow -1$ and $-p \rightarrow -1$ are both allowed, the counterrotating components lead, after complex Fourier transformation with respect to t_1, to signals that are symmetrically disposed about $\omega_1 = 0$. Each signal has a lineshape that consists of an admixture of 2D absorption and 2D dispersion components (*1*). This composite lineshape is often referred to as a "phase-twisted" lineshape (*33*).

Consider by way of example a 2D correlation spectrum (COSY) (*1–4*) obtained with the basic pulse sequence $(\pi/2)_x$–t_1–$(\beta)_\varphi$–t_2 and complex Fourier transformation in both dimensions. The schematic spectrum in Fig. 3a shows only diagonal peaks for clarity. The peaks that appear symmetrically with respect to $\omega_1 = 0$ (open and filled symbols) have been referred to as "*P*-type" and "*N*-type" signals (*3*) or "anti-echoes" and "echoes" (*4*). In terms of coherence transfer, these signals result from $p = 0 \rightarrow -1 \rightarrow -1$ and $p = 0 \rightarrow +1 \rightarrow -1$ pathways, respectively.

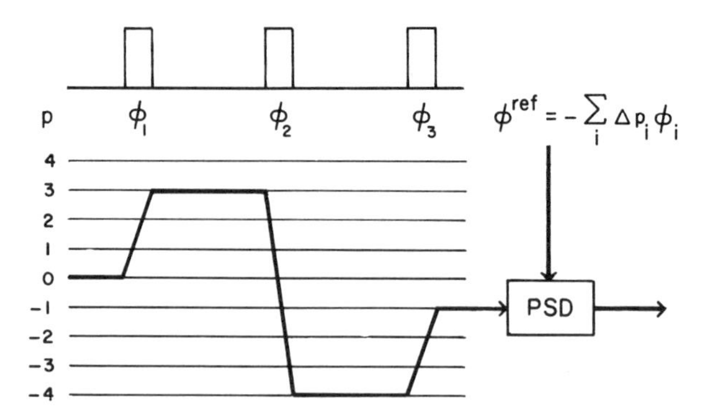

FIG. 2. The selection of a coherence transfer pathway, characterized in this hypothetical example by the changes in coherence order $\Delta p_1 = +3$, $\Delta p_2 = -7$, and $\Delta p_3 = +3$, can be achieved by cycling the phases of the three coherence transfer pulses and by shifting the phase of the reference channel of the phase-sensitive detector (PSD).

If the two frequency components at $\pm\omega_1$ have equal amplitude, a real Fourier transformation with respect to t_1 leads to a symmetrical superposition of the signals associated with mirror-image pathways $0 \rightarrow +p \rightarrow -1$ and $0 \rightarrow -p \rightarrow -1$. This superposition yields pure lineshapes, i.e., either pure 2D absorption or pure 2D dispersion (*33, 34*). Pure phase is obtained regardless of inhomogeneous broadening, which may however lead to different lineshapes and different peak heights of the two components.

Cross-peaks in COSY spectra and remote connectivity signals in double-quantum spectra (*11*) are symmetrical for mixing pulses with arbitrary β. Diagonal peaks in

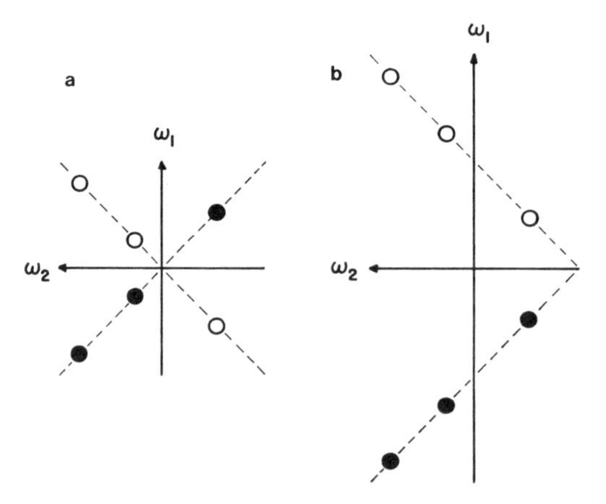

FIG. 3. (a) Schematic representation of the diagonal peaks in single-quantum correlation spectroscopy (COSY) with complex Fourier transformation in both dimensions. All signals have composite (phase-twisted) lineshapes; filled symbols correspond to $p = 0 \rightarrow +1 \rightarrow -1$ pathways ("*N* peaks"), open symbols are associated with $p = 0 \rightarrow -1 \rightarrow -1$ pathways ("*P* peaks"). If the amplitudes of the two classes are equal, pure phase spectra can be obtained by a real Fourier transformation with respect to t_1. If the carrier is positioned within the spectrum, the two classes of signal overlap. (b) The two types can be separated by incrementing the rf phase of the initial preparation pulses in concert with the evolution time (TPPI procedure). The principle is applicable to homo- and heteronuclear single- and multiple-quantum spectra.

COSY spectra of weakly coupled systems and direct connectivity signals in multiple-quantum spectra have symmetrical amplitudes only for $\beta = \pi/2$.

The overlap of the two classes of signal in Fig. 3a is undesirable and complicates the analysis. To separate the two classes, three strategies can be employed:

(a) If the carrier is positioned outside the spectrum, the two types of signals do not overlap, and both can be retained with a phase cycle with $2p$ steps. In the case of single-quantum correlation spectroscopy, the cycle boils down to a two-step phase alternation for the elimination of axial peaks.

(b) By means of a phase cycle with $N \geqslant 2p + 1$, it is possible to select one type of signal irrespective of the position of the carrier. The selection of a single pathway invariably leads to composite lineshapes. However, by properly combining the signals originating from the two pathways $0 \longrightarrow +p \longrightarrow -1$ and $0 \longrightarrow -p \longrightarrow -1$, it is possible to obtain pure phase spectra. Note that the two classes of signals can be extracted from the same set of N experiments with the Fourier analysis given in Eq. [18]. Separate complex Fourier transformations lead to two spectra analogous to Fig. 3a, one with filled symbols only, the other with open symbols. After reversal of the ω_1 axis in one of the spectra, the addition of the two matrices leads to pure phase lineshapes. In the case of single quantum spectroscopy, it is of course possible to use $N = 4 > 2p + 1$, and the procedure is equivalent to the linear combinations described by Bachmann *et al.* (*35*) and States *et al.* (*36*).

(c) Alternatively, it is possible to shift the signals in the ω_1 domain in such a way that the two classes of signal do not overlap even if the carrier is positioned within the spectrum. In the context of 1D spectroscopy, it has been shown that the effective receiver reference frequency can be shifted with respect to the transmitter carrier frequency by recording a free induction decay where the receiver reference phase is incremented for subsequent sampling points (*37*). The same idea can be incorporated in the ω_2 (*38*) and ω_1 domains (*39*) of 2D spectra. In the latter case, the experimental procedure closely resembles a method that has found widespread use in multiple-quantum NMR, known as "time-proportional phase incrementation" (TPPI) (*8, 40*). To obtain pure lineshapes, the rf phase of the excitation propagator is incremented in concert with t_1 according to

$$\varphi = \frac{t_1}{\Delta t_1}\frac{\pi}{2|p|} \qquad [25]$$

where $|p|$ is the order of multiple-quantum coherence evolving in t_1. The characteristic transformation of p-quantum coherence under rf phase shifts causes signals with opposite orders p and $p' = -p$ to shift in opposite directions by $(4\Delta t_1)^{-1}$. The case of single-quantum correlation spectroscopy (COSY) is shown schematically in Fig. 3b. If the amplitudes are symmetrical, a real Fourier transformation can be calculated with respect to t_1 to obtain pure phase spectra (*41*).

An advantage of strategies (b) and (c) in comparison to (a) is the possibility of setting the carrier in the center of the spectrum, which reduces rf power requirements. With regard to data storage requirements, strategy (b) with the selection of a single pathway is most economical. This method is sufficient for absolute-value displays.

Pure phase spectra obtained with strategies (b) and (c) require twice as much data storage space, while pure phase spectra obtained with method (a) demand a fourfold number of points in time domain (since twice the number of points must be recorded in t_2). Clearly, strategy (c) is of particular simplicity with regard to data handling. This TPPI procedure for generating pure phase spectra will be discussed in the examples in the following sections.

HOMONUCLEAR CORRELATION SPECTROSCOPY

In situations where pure 2D absorption is not essential, it is possible to simplify homonuclear correlation spectra obtained with the basic pulse sequence $(\pi/2)$–t_1–(β)–t_2 by retaining only $p = 0 \rightarrow +1 \rightarrow -1$ ("N peaks"). This approach yields spectra with minimum ω_1 bandwidth, similar to those of Fig. 3a but where the signals indicated by open symbols are eliminated by phase cycling. The required selectivity of the mixing process

$$\Delta p_2\text{: } -3, |\mathbf{-2}, (-1), (0)|, 1, 2, 3 \qquad [26]$$

can be achieved with a three-step cycle with $N_2 = 3$, with mixing pulse phases $\varphi_2 = 0, 2\pi/3$, and $4\pi/3$, and receiver phases $\varphi^{\text{ref}} = 0, 4\pi/3$, and $2\pi/3$. Aliasing in the resulting three-point Fourier transform leads to the selection of a periodic series of Δp values with identical behavior. The fundamental period is set between bars in Eq. [26]. The additional pathways that are retained ($\Delta p_2 = \cdots -5, 1, 4, \cdots$) are not relevant in this experiment.

To avoid the uncommon phase shifts of $2\pi/3$ and $4\pi/3$, it is of course allowed to select $N_2 = 4$,

$$\Delta p_2\text{: } -3, |\mathbf{-2}, (-1), (0), 1|, 2, 3 \qquad [27]$$

which leads to four experiments with mixing phases $\varphi_2 = 0, \pi/2, \pi$, and $3\pi/2$, and receiver phases $\varphi^{\text{ref}} = 0, \pi, 0, \pi$ (i.e., alternating addition and subtraction of the signals). This cycle is equivalent to "Exorcycle" (*42*) and has been used in standard SECSY spectroscopy (*2, 3*), in heteronuclear 2D correlation spectroscopy (*43, 44*), and in heteronuclear relayed magnetization transfer (*31, 32*). In cases where two consecutive coherence-transfer steps call for two selection cycles, significant time savings can be obtained by reducing the total number of complementary experiments $N_2 \cdot N_3$ from $4^2 = 16$ to $3^2 = 9$ by using three-step cycles.

To obtain 2D correlation spectra with pure phase with the TPPI procedure, both pathways $p = 0 \rightarrow +1 \rightarrow -1$ and $p = 0 \rightarrow -1 \rightarrow -1$ have to be retained. The required selectivity of the mixing process

$$\Delta p_2\text{: } \mathbf{-2}, (-1), \mathbf{0} \qquad [28]$$

can be achieved with a two-step cycle ($N_2 = 2$, $\varphi_k = 0, \pi$; $\varphi^{\text{ref}} = 0, 0$).

2D EXCHANGE SPECTROSCOPY (NOESY)

The coherence-transfer pathways that may occur in 2D exchange spectroscopy (*28, 29*) are shown in Fig. 1a. If pure 2D absorption lineshapes are not required, it is possible to select the pathway $p = 0 \rightarrow +1 \rightarrow 0 \rightarrow -1$ by cycling the phase of the first pulse to select

$$\Delta p_1\text{: } (-1), (0), \mathbf{1} \qquad [29]$$

while the third pulse must achieve the selection

$$\Delta p_3\text{: } (-p^{\max} - 1), \cdots, \mathbf{-1}, \cdots, (p^{\max} - 1) \qquad [30]$$

where p^{max} is the highest order of multiple-quantum coherence that can contribute significantly, which depends on the number of coupled nuclei. Except for the desired value $\Delta p = -1$, all Δp values in the interval given in Eq. [30], including the limits which are set in brackets, must be suppressed. This can be achieved with a cycle $N_1 \cdot N_3 = 3 \cdot (p^{max} + 1)$ steps.

If pure 2D absorption lineshapes are to be obtained, two pathways must be retained with $p = 0 \rightarrow \pm 1 \rightarrow 0 \rightarrow -1$. This task can be accomplished by eliminating longitudinal magnetization in the evolution period:

$$\Delta p_1: \mathbf{-1}, (0), \mathbf{+1} \qquad [31]$$

and by selecting the transfer $0 \longrightarrow -1$ by cycling the third pulse:

$$\Delta p_3: (-p^{max} - 1), \cdots, \mathbf{-1}, \cdots, (p^{max} - 1) \qquad [32]$$

which can be achieved with a cycle of $N_1 \cdot N_3 = 2 \cdot (p^{max} + 1)$. For $p^{max} = 3$, the cycling of the last pulse and the receiver phase ($\varphi^{ref} = k\pi/2$) corresponds to the well-known "Cyclops" sequence (*22, 45*).

The shortest possible phase cycles (irreducible cycles) are shown in Table 1 for systems without resolved couplings ($p^{max} = 1$) and in Table 2 for coupled systems with $p^{max} = 3$. The abbreviation TPPI (time proportional phase shift) indicates that the first pulse must be incremented in phase by $\Delta\varphi_1 = \pi t_1/(2\Delta t_1)$.

In practice, it may be advisable to use more extensive phase cycles, particularly if the rf phase shifts are subject to systematic errors. Thus Table 1a can be extended

TABLE 1

PHASE CYCLES FOR 2D EXCHANGE SPECTROSCOPY

(a) Spin system without resolved couplings ($p^{max} = 1$): Selection of $p = 0 \rightarrow +1 \rightarrow 0 \rightarrow -1$ pathway

$\Delta p_1 = +1$	Δp_2 = free	$\Delta p_3 = -1$	
φ_1 = TPPI	$\varphi_2 = 0$	$\varphi_3 = 0$	$\varphi^{ref} = 0$
$= 2\pi/3$ + TPPI	$= 0$	$= 0$	$= 4\pi/3$
$= 4\pi/3$ + TPPI	$= 0$	$= 0$	$= 2\pi/3$
= TPPI	$= 0$	$= \pi$	$= \pi$
$= 2\pi/3$ + TPPI	$= 0$	$= \pi$	$= \pi/3$
$= 4\pi/3$ + TPPI	$= 0$	$= \pi$	$= 2\pi/3$

(b) Spin system without resolved couplings ($p^{max} = 1$): Selection of $p = 0 \rightarrow \pm 1 \rightarrow 0 \rightarrow -1$ pathways to obtain pure 2D absorption lineshapes

$\Delta p_1 = \pm 1$	Δp_2 = free	$\Delta p_3 = \pm 1$	
φ_1 = TPPI	$\varphi_2 = 0$	$\varphi_3 = 0$	$\varphi^{ref} = 0$
$= \pi$ + TPPI	$= 0$	$= 0$	$= \pi$
= TPPI	$= 0$	$= \pi$	$= \pi$
$= \pi$ + TPPI	$= 0$	$= \pi$	$= 0$

TABLE 2

PHASE CYCLE FOR 2D EXCHANGE SPECTROSCOPY

Coupled spins with $p^{max} = 3$: Selection of $p = 0 \rightarrow \pm 1 \rightarrow 0 \rightarrow -1$ to obtain pure 2D absorption lineshapes			
$\Delta p_1 = \pm 1$	Δp_2 = free	$\Delta p_3 = -1$	
φ_1 = TPPI	$\varphi_2 = 0$	$\varphi_3 = 0$	$\varphi^{ref} = 0$
$= \pi$ + TPPI	$= 0$	$= 0$	$= \pi$
= TPPI	$= 0$	$= \pi/2$	$= \pi/2$
$= \pi$ + TPPI	$= 0$	$= \pi/2$	$= 3\pi/2$
= TPPI	$= 0$	$= \pi$	$= \pi$
$= \pi$ + TPPI	$= 0$	$= \pi$	$= 0$
= TPPI	$= 0$	$= 3\pi/2$	$= 3\pi/2$
$= \pi$ + TPPI	$= 0$	$= 3\pi/2$	$= \pi/2$

by specifying that Δp_2 must be -1 with $N_2 = 3(\varphi_2 = k_2 2\pi/3,\ k_2 = 0, 1, 2)$ with an additional shift of the receiver reference phase $\Delta\varphi^{ref} = k_2 2\pi/3$. Tables 1b and 2 can be extended by specifying that Δp_2 must be ± 1 with $N_2 = 2(\varphi_2 = k_2\pi,\ k_2 = 0, 1)$ with an additional shift of the reference phase $\Delta\varphi^{ref} = k_2\pi$. These additions are strictly speaking redundant, but may improve the degree of suppression in practical circumstances.

CORRELATION SPECTROSCOPY WITH MULTIPLE-QUANTUM FILTERS

Recently, a modification of homonuclear correlation spectroscopy has been proposed where coherence is transferred in two steps via multiple-quantum coherence in order to edit 2D spectra (*14, 15*).

The coherence transfer pathways that are relevant to double-quantum filtered correlation spectroscopy are shown in Fig. 1d. If pure 2D absorption lineshapes are not required, the transfer can be restricted to the two pathways $0 \rightarrow +1 \rightarrow +2 \rightarrow -1$ and $0 \rightarrow +1 \rightarrow -2 \rightarrow -1$ by selecting the following Δp values in the first and third pulse

$$\Delta p_1: (-1), (0), \mathbf{1} \quad [33]$$

$$\Delta p_3: \mathbf{-3}, (-2), (-1), (0), \mathbf{1} \quad [34]$$

which requires a cycle with $N_1 \cdot N_3 = 3 \cdot 4 = 12$ steps. To avoid uncommon phase shifts, we can of course select the desired pathway with a 16-step cycle.

In general if the filtration procedure is supposed to retain orders $\pm p$, and if pure phase lineshapes are not required, it is sufficient to select

$$\Delta p_1: (-1), (0), \mathbf{1} \quad [35]$$

$$\Delta p_3: \mathbf{-p-1}, \cdots, \mathbf{p-1}, \quad [36]$$

where all Δp_3 values in the interval between the desired values must be blocked, which requires a cycle with $N_1 \cdot N_3 = 3 \cdot 2p$ experiments.

To obtain pure 2D absorption lineshapes, the four pathways $0 \rightarrow +1 \rightarrow \pm p \rightarrow -1$ and $0 \rightarrow -1 \rightarrow \pm p \rightarrow -1$ must be allowed simultaneously. The pathway $0 \rightarrow 0 \rightarrow \pm p \rightarrow -1$ is impossible, since the second nonselective pulse cannot transform longitudinal polarization into multiple-quantum coherence. It is therefore possible to allow all pathways under the first two pulses, and to select the two pathways $+p \rightarrow -1$ and $-p \rightarrow -1$ under the last pulse:

$$\Delta p_3: \mathbf{-p-1}, \cdots, \mathbf{p-1}, \qquad [37]$$

where all values between the limits must be suppressed. This can be achieved with a cycle with $N_3 = 2p$ steps. In the case of double-quantum filtered correlation spectroscopy one obtains $N_3 = 4$ with rf phases $\varphi_3 = 0, \pi/2, \pi, 3\pi/2$ and receiver phases $\varphi^{\text{ref}} = 0, 3\pi/2, \pi, \pi/2$. This cycle corresponds to well-known procedures for double-quantum selection (*7, 9*).

The selection of the pathways $0 \rightarrow +1 \rightarrow \pm p \rightarrow -1$ has been tested experimentally by recording homonuclear proton 2D correlation spectra of thymidine. All three spectra in Fig. 4 were obtained with delayed acquisition, according to the scheme known as "spin-echo correlation spectroscopy" (SECSY) (*2, 3*). This representation relies on the suppression of $0 \rightarrow -1 \rightarrow \pm p \rightarrow -1$ signals (so-called "*P* peaks"), since this is a condition for reducing the ω_1 bandwidth (*3*). The conventional SECSY spectrum (Fig. 4a) can be simplified with a double-quantum filter (Fig. 4b) which in effect eliminates the responses of isolated spins. If a triple-quantum filter (Fig. 4c) is used, only signals stemming from subunits with at least three coupled protons survive, in accordance with coherence transfer selection rules (*11, 14*).

MULTIPLE-QUANTUM SPECTROSCOPY

In conventional two-dimensional *p*-quantum spectra (*8–11*), the coherences of order $+p$ and $-p$ lead to pairs of signals symmetrically disposed about $\omega_1 = 0$. If the mixing propagator consists of a single pulse with $\beta = \pi/2$, these signals have equal amplitudes (*11*). If both types of signals are retained, pure 2D absorption lineshapes can be obtained with the procedures described above. If, on the other hand, the $0 \rightarrow \pm 1 \rightarrow +p \rightarrow -1$ pathways are selected, the bandwidth in the ω_1 domain can be reduced, although at the expense of pure 2D absorption lineshapes. This selection has been achieved with *z* rotations (*46*), with phase shifts in increments of $\pi/4$ (*47*), and, for the special case of two-spin systems, by exploiting the dependence on the rotation angle of the rf pulse (*48*), and can also be achieved with field gradient pulses (*49*).

In the case of double-quantum spectroscopy, shown schematically in Fig. 1c, the phase of the mixing pulse (or of the sequence of pulses that constitute the mixing propagator) can be cycled in order to select the pathways that involve coherence of order $p = +2$ in t_1:

$$\Delta p_3: -4, \mathbf{-3}, (-2), (-1), (0), (1), 2, 3. \qquad [38]$$

Values with $\Delta p_3 < -3$ or $\Delta p_3 > 1$ are irrelevant if we assume that there are no coherences of order $|p| > 2$ in the evolution period. In this case, a five-step cycle with

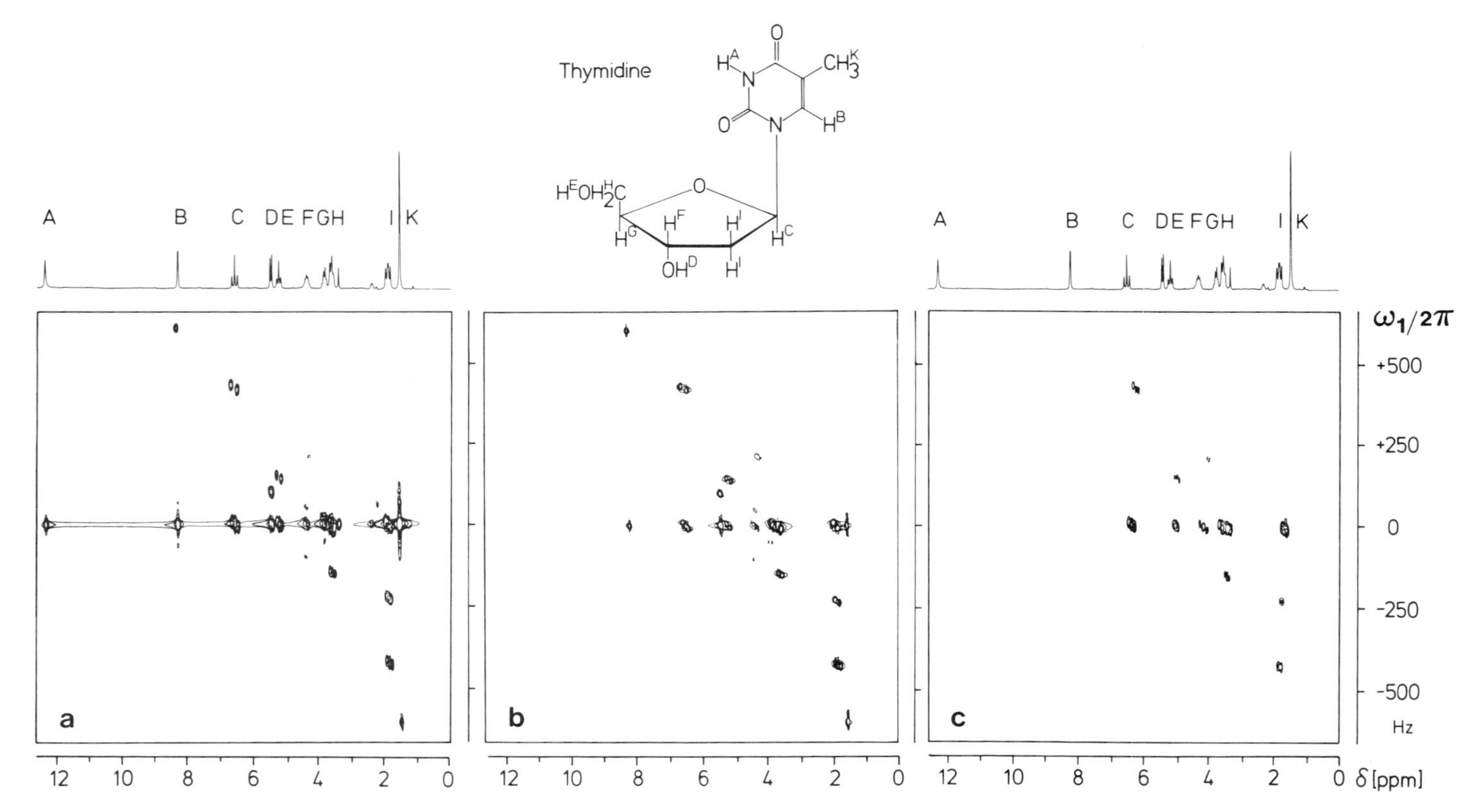

FIG. 4. Homonuclear 2D correlation spectra of thymidine. (a) Conventional spectrum obtained with delayed acquisition ("SECSY"). (b) Double-quantum filtered SECSY spectrum obtained with the sequence $(\pi/2)$–t_1–$(\pi/2)$–$(\pi/2)$–t_1–(acquisition) with selection of the pathways $p = 0 \rightarrow +1 \rightarrow \pm 2 \rightarrow -1$. Note the elimination of the singlet of proton A and the attenuation of the ridge at $\omega_1 = 0$. (c) Triple-quantum filtered SECSY spectrum obtained with the same sequence but with selection of $p = \pm 3$. Nuclei with less than two coupling partners are eliminated altogether (resonances *A*, *B*, *D*, and *K*), while a pair of coupled nuclei must have at least one additional common coupling partner to give rise to cross-peaks (e.g. cross-peaks between *C* and *I*, with common coupling partner *J*). The spectra were obtained at 90 MHz with a solution of 1 mol/liter at 300 K, 256 × 1024 data matrices, contour levels at 5, 8, 16, and 25% of the highest peak and without digital filtering.

TABLE 3

PHASE CYCLE FOR DOUBLE-QUANTUM SPECTROSCOPY

Selection of $p = 0 \rightarrow \pm 1 \rightarrow +2 \rightarrow -1$ pathway in systems with $p^{max} = 3$

Δp_1 = free	Δp_2 = free	$\Delta p_3 = -3$	
$\varphi_1 = 0$	$\varphi_2 = 0$	$\varphi_2 = 0$	$\varphi^{ref} = 0$
$= 0$	$= 0$	$= \pi/3$	$= \pi$
$= 0$	$= 0$	$= 2\pi/3$	$= 0$
$= 0$	$= 0$	$= \pi$	$= \pi$
$= 0$	$= 0$	$= 4\pi/3$	$= 0$
$= 0$	$= 0$	$= 5\pi/3$	$= \pi$

$N_3 = 5$ is sufficient. If the transfers $p = \pm 3 \rightarrow -1$ must be suppressed as well, we select $N_3 = 6$. By way of example, the six-step cycle is shown in Table 3; this cycle has been used for simplifying the experimental double-quantum spectrum discussed below.

Selective observation of the $0 \rightarrow \pm 1 \rightarrow +3 \rightarrow -1$ pathway in triple-quantum spectroscopy can be achieved with

$$\Delta p_3\text{: } \mathbf{-4}, (-3), (-2), (-1), (0), (1), (2), 3 \quad [39]$$

which can be realized with $N_3 \geqslant 7$. If $p = \pm 4 \rightarrow -1$ transfers are to be suppressed as well, it is necessary to use a cycle with $N_3 \geqslant 8$.

If pure 2D absorption lineshapes are required, the pathways involving the coherence orders $+p$ and $-p$ in the evolution period must be retained simultaneously. In the case of double-quantum spectroscopy, the selection

$$\Delta p_3\text{: } -4, \mathbf{-3}, (-2), (-1), (0), \mathbf{1}, 2, 3. \quad [40]$$

can be achieved with $N_3 = 4$. In general, simultaneous transfer is possible with $N_3 = 2p$ experiments. Such cycles have been used in many applications of multiple-quantum NMR (*9, 11*).

The selection of the $p = 0 \rightarrow \pm 1 \rightarrow +2 \rightarrow -1$ pathway in double-quantum spectra makes it possible to delay the beginning of data acquisition to a point in time $2t_1$ after initial excitation (*50–52*). This procedure causes the signals to shift in the ω_1 domain, as shown in Fig. 5, leading to a presentation of double-quantum spectra that closely resembles the familiar picture of single-quantum correlation spectra (COSY). The signals associated with directly connected pairs of nuclei are indicated by filled symbols. They are contained within a frequency band indicated by dotted lines (*53*). Signals associated with remote connectivity, which arise from double-quantum coherence involving two nuclei A and M that is transferred to a third nucleus X (*11*), are indicated by open symbols. These signals, which may fall outside the frequency band indicated by dotted lines, cannot occur in double-quantum spectra of two-spin systems, e.g., in INADEQUATE spectra of carbon-13 in natural abundance.

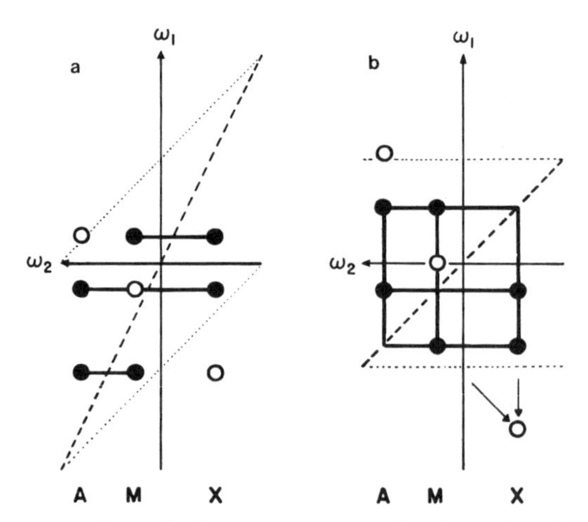

FIG. 5. Schematic representations of +2-quantum spectra of a three-spin AMX system obtained with selection of the pathways $0 \rightarrow \pm1 \rightarrow +2 \rightarrow -1$ (Fig. 1c). (a) Conventional form, (b) COSY-like representation obtained by delaying the acquisition to $2t_1$ after initial excitation. Filled and open symbols represent signals associated with directly and remotely connected nuclei (*11*). The former fall within a frequency band indicated by dotted lines. The virtual diagonals are indicated by dashed lines. In the COSY-like representation (b), the filled symbols appear at the same frequency coordinates as cross-peaks in conventional single-quantum COSY spectra. Additional information can be derived from the remote connectivity signals. For example, the signal at the bottom right stems from double-quantum coherence involving the nuclei A and M that is transferred into observable X magnetization. Its location is found by drawing a line through the two A, M cross-peaks (see arrows).

In this case the ω_1 bandwidth may be reduced by a factor of two without loss of information.

An experimental example of a double-quantum spectrum in COSY-like representation is shown in Fig. 6. The signals associated with remote connectivity do not have symmetrically related counterparts and can be eliminated by symmetrization. The remaining signals, which correspond to filled symbols in Fig. 5b, have the same frequency coordinates (and the same information content) as cross-peaks in single-quantum correlation spectra.

It should be noted that experimental methods involving delayed acquisition suffer from sensitivity losses due to transverse relaxation after the mixing pulse. The same COSY-like representation could be obtained with better sensitivity and pure phase lineshapes by a mathematical transformation of a double-quantum spectrum obtained without delayed acquisition, in analogy to the foldover correction procedure (FOCSY) (*3*).

CONCLUSIONS

Our recent experience has shown that coherence-transfer maps which portray the relevant coherence-transfer pathways are powerful tools for understanding and designing new pulse experiments. In several cases, the pulse sequence alone does not characterize the essential features of an experiment. It is rather the selection of specific

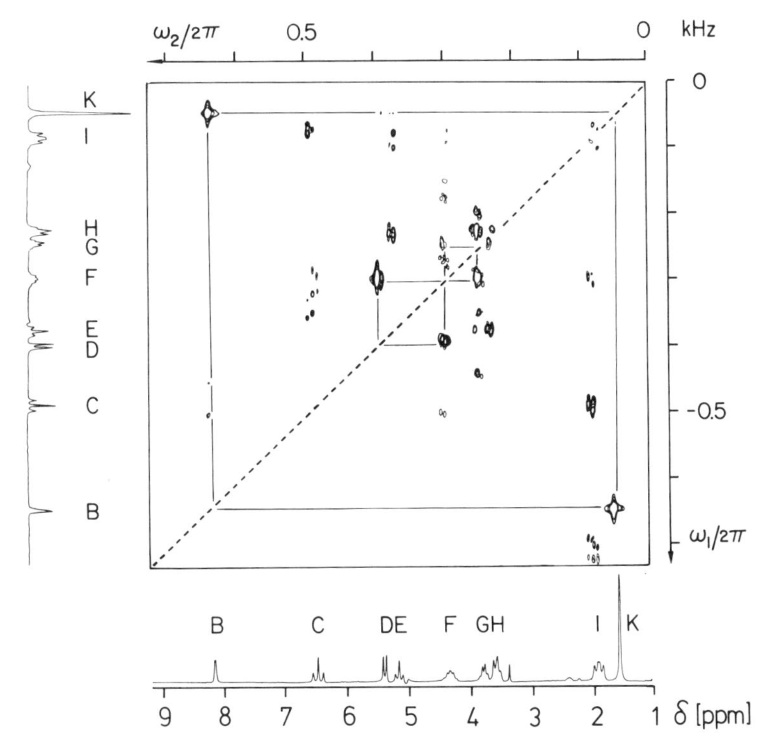

FIG. 6. Absolute-value double-quantum spectrum of thymidine (assignment as in Fig. 4), presented in COSY-like form with selection of the solid pathways in Fig. 1c and delayed acquisition, as shown schematically in Fig. 5b. The carrier was positioned at the high-field end of the spectrum. Symmetrical excitation and detection was used with the pulse sequence $(\pi/2)_x$–τ–$(\pi)_x$–τ–$(\pi/2)_x$–t_1–$(\pi/2)_x$–τ–$(\pi)_x$–τ–$(\pi/2)_y$–t_1–(acquisition) as described by Sørensen *et al.* (*54*). The phase of the detection sandwich was cycled in increments of $2\pi/6$ (60°). Same sample and conditions as in Fig. 4.

coherence-transfer pathways by an appropriate phase cycle which constitutes the essence of an experiment. For example, three-pulse experiments can be designed for relayed coherence transfer, for 2D exchange spectroscopy, for multiple-quantum filtering, and for multiple-quantum spectroscopy merely by selecting different coherence-transfer pathways.

At first sight it may appear unnecessary and artificial to distinguish the sign of the order of coherence. However, despite the Hermitian character of operators in quantum mechanics, it is indeed possible to trace out individual pathways violating Hermitian symmetry by combining results obtained from a phase-cycled sequence of experiments. In actual fact, it turns out that the distinction of the sign of coherence is of central importance for the design of optimized experiments.

In the practical examples described in this paper, phase cycles have been confined to individual pulses. It should however be noticed that the formalism is more general and also allows phase-cycling of entire groups of pulses as well as interlaced phase cycles of different hierarchy.

The discovery that many of the commonly used four-step phase cycles can in principle be replaced by shorter three-step phase cycles may serve as an example illuminating the power of coherence-transfer pathway considerations. It is likely that the same concepts can also help in the design of pulse experiments in electron spin resonance and in optical spectroscopy.

ACKNOWLEDGMENTS

This research has been supported by the Swiss National Science Foundation and by a grant of the Stiftung des Deutschen Volkes (H.K.). The authors are grateful to Dr. M. H. Levitt, Dr. E. R. P. Zuiderweg, Dr.G. Wagner, and O. W. Sørensen for helpful discussions and to Dr. A. D. Bain for providing a copy of Ref. (*22*) prior to publication. We are indebted to a referee for many constructive comments.

REFERENCES

1. W. P. Aue, E. Bartholdi, and R. R. Ernst, *J. Chem. Phys.* **64,** 2229 (1976).
2. K. Nagayama, K. Wüthrich, and R. R. Ernst, *Biochem. Biophys. Res. Commun.* **90,** 305 (1979).
3. K. Nagayama, Anil Kumar, K. Wüthrich, and R. R. Ernst, *J. Magn. Reson.* **40,** 321 (1980).
4. A. Bax and R. Freeman, *J. Magn. Reson.* **44,** 542 (1981).
5. A. A. Maudsley and R. R. Ernst, *Chem. Phys. Lett.* **50,** 368 (1977).
6. G. Bodenhausen and R. Freeman, *J. Magn. Reson.* **28,** 471 (1977).
7. A. Bax, "Two-Dimensional Nuclear Magnetic Resonance in Liquids," Delft Univ. Press, Dordrecht, 1982.
8. G. Drobny, A. Pines, S. Sinton, D. Weitekamp, and D. Wemmer, *Faraday Div. Chem. Soc. Symp.* **13,** 49 (1979).
9. A. Wokaun and R. R. Ernst, *Chem. Phys. Lett.* **52,** 407 (1977).
10. G. Bodenhausen, *Progr. Nucl. Magn. Reson. Spectrosc.* **14,** 137 (1981).
11. L. Braunschweiler, G. Bodenhausen, and R. R. Ernst, *Mol. Phys.* **48,** 535 (1983).
12. A. Bax, R. Freeman, and S. P. Kempsel, *J. Am. Chem. Soc.* **102,** 4849 (1980).
13. G. Bodenhausen and C. M. Dobson, *J. Magn. Reson.* **44,** 212 (1981).
14. U. Piantini, O. W. Sørensen, and R. R. Ernst, *J. Am. Chem. Soc.* **104,** 6800 (1982).
15. A. J. Shaka and R. Freeman, *J. Magn. Reson.* **51,** 169 (1983).
16. P. J. Hore, R. M. Scheek, and R. Kaptein, *J. Magn. Reson.* **52,** 339 (1983).
17. M. H. Levitt and R. R. Ernst, *Chem. Phys. Lett.* **100,** 119 (1983).
18. G. A. Morris and R. Freeman, *J. Am Chem. Soc.* **101,** 760 (1979).
19. D. P. Burum and R. R. Ernst, *J. Magn. Reson.* **39,** 163 (1980).
20. D. M. Doddrell, D. T. Pegg, and M. R. Bendall, *J. Magn. Reson.* **48,** 323 (1982).
21. O. W. Sørensen and R. R. Ernst, *J. Magn. Reson.* **51,** 477 (1983).
22. A. D. Bain, *J. Magn. Reson.* **56,** 418 (1984).
23. K. Blum, "Density Matrix Theory and Applications," Plenum, New York, 1981.
24. M. Hamermesh, "Group Theory and Its Applications to Physical Problems," Addison–Wesley, Reading, Mass., 1962.
25. O. W. Sørensen, G. W. Eich, M. H. Levitt, G. Bodenhausen, and R. R. Ernst, *Progr. Nucl. Magn. Reson. Spectrosc.* **16,** 163 (1983).
26. A. Wokaun and R. R. Ernst, *J. Chem. Phys.* **67,** 1752 (1977).
27. S. Vega, *J. Chem. Phys.* **68,** 5518 (1978).
28. J. Jeener, B. H. Meier, P. Bachmann, and R. R. Ernst, *J. Chem. Phys.* **71,** 4546 (1979).
29. S. Macura, Y. Huang, D. Suter, and R. R. Ernst, *J. Magn. Reson.* **43,** 259 (1981).
30. G. Eich, G. Bodenhausen, and R. R. Ernst, *J. Am. Chem. Soc.* **104,** 3731 (1982).
31. P. H. Bolton and G. Bodenhausen, *Chem. Phys. Lett.* **89,** 139 (1982).
32. O. W. Sørensen and R. R. Ernst, *J. Magn. Reson.* **55,** 338 (1983).

33. G. Bodenhausen, R. Freeman, R. Niedermeyer, and D. L. Turner, *J. Magn. Reson.* **26,** 133 (1977).
34. R. Freeman, S. P. Kempsell, and M. H. Levitt, *J. Magn. Reson.* **34,** 663 (1979).
35. P. Bachmann, W. P. Aue, L. Müller, and R. R. Ernst, *J. Magn. Reson.* **28,** 29 (1977).
36. D. J. States, R. A. Haberkorn, and D. J. Ruben, *J. Magn. Reson.* **48,** 286 (1982).
37. A. G. Redfield and S. D. Kunz, *J. Magn. Reson.* **19,** 250 (1975).
38. G. Bodenhausen, R. Freeman, G. A. Morris, R. Niedermeyer, and D. L. Turner, *J. Magn. Reson.* **25,** 559 (1977).
39. G. Bodenhausen and R. Freeman, *J. Magn. Reson.* **28,** 303 (1977).
40. G. Bodenhausen, R. L. Vold, and R. R. Vold, *J. Magn. Reson.* **37,** 93 (1980).
41. D. Marion and K. Wüthrich, *Biochem. Biophys. Res. Commun.* **113,** 967 (1983).
42. G. Bodenhausen, R. Freeman, and D. L. Turner, *J. Magn. Reson.* **27,** 511 (1977).
43. A. Bax and G. A. Morris, *J. Magn. Reson.* **42,** 501 (1981).
44. P. H. Bolton and G. Bodenhausen, *J. Magn. Reson.* **46,** 306 (1982).
45. D. I. Hoult and R. E. Richards, *Proc. Roy. Soc. London Ser. A* **344,** 311 (1975).
46. A. Bax, R. Freeman, T. A. Frenkiel, and M. H. Levitt, *J. Magn. Reson.* **43,** 478 (1981).
47. D. L. Turner, *Mol. Phys.* **44,** 1051 (1981).
48. T. H. Mareci and R. Freeman, *J. Magn. Reson.* **48,** 158 (1982).
49. A. Bax, P. G. de Jong, A. F. Mehlkopf, and J. Smidt, *Chem. Phys. Lett.* **69,** 567 (1980).
50. D. L. Turner, *J. Magn. Reson.* **49,** 175 (1982).
51. D. L. Turner, *J. Magn. Reson.* **53,** 259 (1983).
52. G. Wagner and E. R. P. Zuiderweg, *Biochem. Biophys. Res. Commun.* **113,** 854 (1983).
53. A. Bax and T. H. Mareci, *J. Magn. Reson.* **53,** 360 (1983).
54. O. W. Sørensen, H. Kogler, M. Rance, and R. R. Ernst, Sixth International Meeting on NMR Spectroscopy, Edinburgh, July 1983 (manuscript in preparation).

JOURNAL OF MAGNETIC RESONANCE **60**, 161–163 (1984)

Pattern Recognition in Two-Dimensional NMR Spectra

BEAT U. MEIER, GEOFFREY BODENHAUSEN, AND R. R. ERNST

Laboratorium für Physikalische Chemie, Eidgenössische Technische Hochschule, 8092 Zurich, Switzerland

Received May 30, 1984

The fully automated determination of molecular structure by spectroscopic means is an old dream of analytical chemistry (*1–10*). Many procedures proposed so far are based on comparison with extensive libraries of spectra or computer simulations rather than on direct analysis. The structure of NMR spectra suggests the feasibility of a deductive analysis, since the multiplet patterns allow one to trace out the connectivities of nuclear spins. Unfortunately, the assignment of multiplet patterns in conventional one-dimensional (1D) spectra is by no means unique, since a multiplet cannot be distinguished a priori from an accidental juxtaposition of chemical shifts. Pattern recognition (*11, 12*) in 1D spectra can therefore be misleading. On the other hand, two-dimensional NMR spectra are much less ambiguous. Their information content is sufficiently high to avoid pitfalls in pattern recognition procedures.

Of the many methods proposed, 2D correlation spectroscopy or "COSY" (*13*), appears particularly convenient for analyzing coupling networks. A cross-peak multiplet centered at $(\omega_1, \omega_2) = (\Omega_A, \Omega_X)$ indicates a resolved coupling between two nuclei with chemical shifts Ω_A and Ω_X. In a two-spin AX system, the cross-peak amplitude is distributed over four signals with alternating signs at $(\Omega_A \pm \pi J_{AX}, \Omega_X \pm \pi J_{AX})$, and four additional signals appear at mirror image positions with respect to the main diagonal. Diagonal multiplets are disregarded as they tend to suffer from severe overlap. The 16 coordinates of the eight cross-peaks of the AX system contain redundancy since they only carry information on two chemical shifts and one coupling constant. In AMX systems, each cross-peak multiplet is further split by couplings to the "passive" nucleus (*13, 14*), and the total number of cross-peaks increases from 8 to 96 for a mixing pulse with a rotation angle $\beta = \pi/2$. For N inequivalent spins with $I = 1/2$ with nonvanishing couplings, $N(N-1)2^{2(N-1)}$ cross-peaks, each with two coordinates, are available for the determination of $N(N+1)/2$ parameters. By using a small flip angle (e.g., $\beta = \pi/4$), coherence transfer occurs predominantly between connected transitions, and the amplitudes of the corresponding cross-peaks are favored. The redundancy of the dominant signals with respect to the number of unknown parameters remains high, since $N(N-1)2^N$ dominant cross-peaks appear, each with two coordinates. Pattern recognition is appropriate under these circumstances.

Our preliminary experience indicates that the *sign alternation* of cross-peak multiplet patterns in phase-sensitive 2D spectra is the most useful feature for

0022-2364/84 $3.00

recognizing patterns. The basic antiphase square pattern occurs both in simple AX systems and in larger systems with nondegenerate couplings (see Fig. 1b), while rectangular patterns may occur in systems with equivalent nuclei. To search for the basic eight-peak patterns (two symmetrically positioned squares), the computer systematically tests the environment of each matrix point for a square with the side length $2\pi J$ in the interval $J_{min} < J < J_{max}$. If the proper alternation of signs is found, a linear combination with addition and subtraction of the eight amplitudes is stored in a reduced 2D matrix at (Ω_A, Ω_X) and the value of the coupling constant J is deposited in a separate table.

More complicated spin systems can be identified by searching for regular patterns in the reduced 2D matrix, or by searching for specific patterns in the original data. For an AMX system, this procedure requires a three-dimensional search around each matrix point, since there are three unknown coupling constants. The success of a search procedure based on 8 + 8 predominant peaks (including the symmetry-related peaks), expected in an AMX spectrum for a small pulse rotation angle β, is shown in Fig. 1a.

The example of Fig. 1c demonstrates that the sign pattern search procedure is also applicable to 2D spectra with rather low signal-to-noise ratio. The probability W of finding a correct sign pattern accidentally in noise rapidly decreases with the number N_p of peaks in the desired pattern: $W = 2^{-N_p}$ ($\simeq 1.5 \cdot 10^{-5}$ for $N_p = 16$).

In combination with 2D correlation spectroscopy, the use of double-quantum filtering (*15, 16*) is useful to eliminate the dispersive part of the diagonal multiplets, which facilitates the recognition of cross-peaks near the diagonal. This method has been applied for Figs. 1b and c. The spectra in Fig. 1 were obtained with mixing

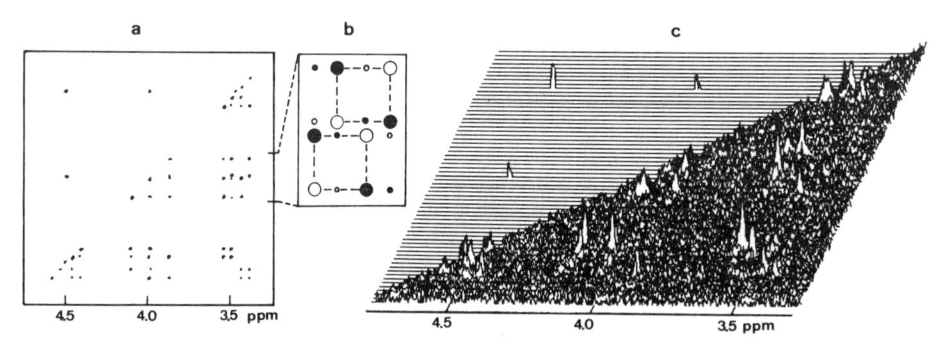

FIG. 1. (a) Lower triangle: experimental phase-sensitive double-quantum filtered ^{1}H correlation spectrum of 2,3-dibromopropanoic acid, recorded with the sequence $(\pi/2)-t_1-(\pi/4)(\pi/4)-t_2$ on a home-built 90 MHz spectrometer; the data were processed with an IBM CS 9000 computer. Positive and negative signals exceeding a threshold ($S > S_{th}$ and $S < -S_{th}$) are shown. Upper triangle: reduced spectrum obtained by 8 + 8 peak AMX-pattern recognition procedure described in text. (b) Schematic cross-peak pattern characteristic for three-spin systems. Filled and open symbols correspond to positive and negative signals, respectively; large symbols correspond to predominant peaks for small flip angles. (c) Pattern recognition in a noisy spectrum, obtained by adding computer-generated Gaussian noise to the experimental spectrum in (a). The smallest significant signal has an amplitude of 2.7 with respect to the rms amplitude of the random noise. All significant patterns are correctly identified (upper triangle) while the noise does not lead to false recognition of accidental multiplets in this case although no threshold has been set to discriminate noise.

pulses $\beta = \pi/4$ to prevent accidental cancellations of peaks that may occur for near-degenerate coupling constants when using $\beta = \pi/2$.

Many refinements and extensions of sign pattern recognition can be conceived. To increase the available information, it appears desirable to perform a search on two or more complementary experiments, for example by using different flip angles (e.g., $\beta = \pi/4$ and $\beta = 3\pi/4$), or to include multiple-quantum spectra, relayed magnetization transfer spectra, or *J*-resolved spectra.

ACKNOWLEDGMENTS

We are indebted to Dr. F. Baumann, W. Jaeggi, and L. Braunschweiler for valuable assistance and to IBM Instruments Inc. for putting an IBM CS 9000 computer at our disposal. This research has been supported in part by the Swiss National Science Foundation.

REFERENCES

1. P. C. JURS, B. R. KOWALSKI, T. L. ISENHOUR, AND C. N. REILLEY, *Anal. Chem.* **41,** 690 (1969); ibid **41,** 1949 (1969).
2. T. L. ISENHOUR AND P. C. JURS, *Anal. Chem.* **43,** 20A (1971).
3. B. R. KOWALSKI AND C. F. BENDER, *Anal. Chem.* **44,** 1405 (1972).
4. L. R. CRAWFORD AND J. D. MORRISON, *Anal. Chem.* **40,** 1464 (1968).
5. T. R. BRUNNER, R. C. WILLIAMS, C. L. WILKINS, AND P. J. MCCOMBIE, *Anal. Chem.* **46,** 1978 (1974).
6. P. C. JURS AND T. L. ISENHOUR, "Chemical Applications of Pattern Recognition," Wiley-Interscience, New York, 1975.
7. T. L. ISENHOUR, B. R. KOWALSKI, P. C. JURS, *CRC Crit. Rev. Anal. Chem.* **1974,** 4.
8. P. S. SHOENFELD AND J. R. DEVOE, *Anal. Chem.* **48,** 403R (1976).
9. K. VARMUZA, "Pattern Recognition in Chemistry," Lecture Notes in Chemistry, Vol. 21, Springer, Berlin, 1982.
10. R. RICHARDS, W. AMMANN, AND T. WIRTHLIN, *J. Magn. Reson.* **45,** 270 (1981).
11. K. FUKANAGA, "Introduction to Statistical Pattern Recognition," Academic Press, New York, 1972.
12. B. G. BATCHELOR, "Practical Approach to Pattern Classification," Plenum, London, 1974.
13. W. P. AUE, E. BARTHOLDI, AND R. R. ERNST, *J. Chem. Phys.* **64,** 2229 (1976).
14. A. BAX AND R. FREEMAN, *J. Magn. Reson.* **44,** 542 (1981).
15. U. PIANTINI, O. W. SØRENSEN, AND R. R. ERNST, *J. Am. Chem. Soc.* **104,** 6800 (1982).
16. M. RANCE, O. W. SØRENSEN, G. BODENHAUSEN, G. WAGNER, R. R. ERNST, AND K. WÜTHRICH, *Biochem. Biophys. Res. Commun.* **117,** 479 (1983).

JOURNAL OF MAGNETIC RESONANCE **61**, 465–481 (1985)

Retrieval of Frequencies, Amplitudes, Damping Factors, and Phases from Time-Domain Signals Using a Linear Least-Squares Procedure

H. BARKHUIJSEN, R. DE BEER, W. M. M. J. BOVÉE, AND D. VAN ORMONDT

Applied Physics Department, Delft University of Technology, P.O. Box 5046, 2600 GA Delft, The Netherlands

Received July 13, 1984

A new method for quantitative analysis of time-domain signals is reported. It amounts to fitting a function consisting of exponentially damped sinusoids with arbitrary phases to the data. By invoking the principle of linear prediction (LP) the fitting can be carried out by a linear least-squares (LS) procedure, and therefore needs no starting values. The LS procedure is based on singular value decomposition (SVD), which enables one to distinguish between signal and noise. The method, denoted by LPSVD, yields a list comprising the frequency, damping factor, amplitude, and phase of each retrieved sinusoid. In addition, LPSVD is insensitive to truncation at the beginning and/or the end of the signal, and in fact is capable to accurately reconstruct the missing part. Preprocessing of the data is not necessary. Finally, the method achieves higher resolution than fast Fourier transformation.

INTRODUCTION

Spectral analysis of sampled time-domain signals is usually accomplished by means of fast Fourier transformation (FFT) (*1, 2*). This method is computationally efficient and often produces satisfactory results. A fundamental limitation of FFT is, however, that the smallest observable splitting is approximately equal to the inverse of the duration of the signal (*1*). It follows that there is an increasing risk of peak overlap as the duration of the signal becomes shorter. This in turn hampers a quantitative analysis of the spectrum. (See, for instance, Refs. (*3*) and (*4*) for various ways of quantitative analysis in the frequency domain.) Enhancement of the resolution can be achieved by model (curve) fitting of the spectrum (*3*), which involves nonlinear least squares (LS) techniques. It should be noted that truncation of the time-domain signal may complicate the fitting procedure because it causes distortion of the spectrum.

The present paper is concerned with a new method of spectral analysis based on the work of Kumaresan and Tufts (*5*). The method, which yields higher resolution than FFT, involves fitting a model directly to the time-domain data and possesses the following important properties:

1. The model is devised in such a manner that a *linear* LS procedure can be used, thus eliminating the need for starting values and the iterative execution inherent to nonlinear techniques.

0022-2364/85 $3.00

2. The linear LS procedure used is based on singular value decomposition (SVD) (*5, 6*). This enhances the numerical stability of the mathematical process. In addition, analysis of the singular values enables one to distinguish between signal and noise.

3. The method yields a table of frequencies, damping factors, amplitudes, and phases of each sinusoidal component in the signal. This implies among other things that integration in the frequency domain to determine line intensities is not necessary.

4. Preprocessing of the data is not needed.

5. The phases of the sinusoidal components in the time-domain signal are not subject to restrictions. Thus, truncation of the signal, either at the start or at the end does not affect the procedure.

In the next section we treat the mathematics of the model, singular value decomposition, calculation of the model parameters, and implementation of the method, respectively. After this we give four applications to pulsed magnetic resonance experiments. First a simulated signal is analyzed for several values of the signal-to-noise ratio (SNR), then two signals from the field of electron spin echoes (ESE), and finally an *in vivo* ^{31}P NMR free induction decay.

Before proceeding, we mention that a fundamentally different approach to spectral analysis, based on maximizing the so-called configurational entropy, has very recently been applied to NMR (*7*).

THEORY

Mathematical Model

The mathematical model that is to be fitted to the data is set up in two steps. First, it is assumed that the signal, which is sampled at regular times $n\Delta t$ $(0 \leqslant n \leqslant N - 1)$, is made up of K exponentially damped sinusoids plus white noise. We write

$$x_n \equiv x(n\Delta t) = \sum_{k=1}^{K} c_k e^{-b_k n \Delta t} \cos(\omega_k n\Delta t + \varphi_k) + w_n, \qquad [1]$$

which involves $4K$ parameters, namely ω_k, b_k, c_k, and φ_k with $k = 1, \ldots, K$. The magnitude of K is unknown as yet; the noise is represented by w_n. Fitting Eq. [1] as it stands to the data leads to a nonlinear LS procedure, which first requires a set of starting values and then proceeds via a number of iterations. Various problems may arise in this process. The nonlinearity and concomitant complications can be circumvented by invoking the principle of linear prediction (LP) (*8*), also called autoregressive (AR) modeling (*1*). This constitutes the second step. The LP principle amounts to assuming that each data point can be expressed as a linear combination of M previous ones according to

$$x_n = a_1 x_{n-1} + a_2 x_{n-2} + \cdots + a_M x_{n-M}, \qquad [2]$$

in which n can have any value between M and $N - 1$, and the LP coefficients a_m $(m = 1, \ldots, M)$ are independent of n. The validity of Eq. [2] can easily be verified for a signal consisting of a single exponentially damped sinusoid without noise, i.e., for $x_n = \exp(-bn\Delta t)\cos(\omega n\Delta t + \varphi)$. After some algebraic manipulation one finds $a_1 = 2\exp(-b\Delta t)\cos(\omega\Delta t)$, and $a_2 = -\exp(-2b\Delta t)$, independent of the value of n. It can be shown (*1, 5*) that Eq. [2] also holds exactly for an arbitrary number, K, of exponentially decaying sinusoids without noise. In that case the required number of LP coefficients equals $2K$. The situation is less amenable to mathematical rigor when the signal contains noise (*1*). In practice, the number of LP coefficients is then increased to far beyond $2K$, usually to as high as $0.75N$ (*5*).

Equation [2] is now to be fitted to the data. This involves a standard linear LS procedure that needs no starting values and immediately yields the final solution. As already mentioned in the introduction, the method of Kumaresan and Tufts makes use of singular value decomposition. This will be treated in the next subsection.

Least-Squares Solution by Singular Value Decomposition

Equation [2] can be written for $N - M$ values of n. The resulting set of equations is represented by

$$\mathbf{X}\bar{\mathbf{a}} = \bar{\mathbf{x}}, \qquad [3]$$

in which $\mathbf{X}$ is a rectangular matrix (data matrix) of dimensions $(N - M) \times M$, and $\bar{\mathbf{a}}$ and $\bar{\mathbf{x}}$ are vectors of dimension M and $N - M$, respectively. $\mathbf{X}$ and $\bar{\mathbf{x}}$ are known, while $\bar{\mathbf{a}}$ is to be determined. The maximum rank of $\mathbf{X}$ equals the smallest of the numbers $N - M$ and M. Eq. [3] is solved by a linear LS procedure. To this end $\mathbf{X}$ is decomposed according to (*6*)

$$\mathbf{X} = \mathbf{U}\Lambda\tilde{\mathbf{V}}, \qquad [4]$$

in which Λ is a diagonal matrix having the same dimensions as $\mathbf{X}$. The diagonal entries $\lambda_i \equiv \Lambda_{ii}$ of Λ are the so-called singular values of $\mathbf{X}$, which are equal to the square roots of the eigenvalues of the nonnegative definite matrix $\mathbf{X}\tilde{\mathbf{X}}$ ($\tilde{}$ denotes transposition), in descending order. $\mathbf{U}$ and $\mathbf{V}$ are orthogonal matrices that reduce $\mathbf{X}\tilde{\mathbf{X}}$ and $\tilde{\mathbf{X}}\mathbf{X}$, respectively, to diagonal form. If the rank of $\mathbf{X}$ is l, l singular values are greater than zero while the remaining ones (numbering $N - M - l$ or $M - l$) equal zero. It can be proved that the data matrix of a signal comprising K noiseless exponentially damped sinusoids has rank $l = 2K$, implying that then only $2K$ singular values are nonzero (*5*). However, when noise is present, the remaining singular values become also nonzero. It is important to note that so long as the SNR is reasonably good it is possible to distinguish between those singular values that are related to the signal and those related to the noise by considering the relative magnitudes.

It has been shown that the unique minimum length solution of Eq. [3] is (*6*)

$$\bar{\mathbf{a}} = V\tilde{\Lambda}^{-1}\tilde{\mathbf{U}}\bar{\mathbf{x}}, \qquad [5]$$

in which the 'inverse' rectangular matrix $\tilde{\Lambda}^{-1}$ (the exponent -1 has only symbolical meaning) is an $M \times (N - M)$ dimensional diagonal matrix with diagonal entries λ_i^{-1} such that

$$\Lambda\tilde{\Lambda}^{-1} = \begin{vmatrix} \mathbf{E}_l & 0 \\ 0 & 0 \end{vmatrix}, \qquad [6]$$

$\mathbf{E}_l$ being the unit matrix of dimension $l \times l$. Here we have reached an important step in the LS procedure: since one is able to distinguish whether singular values are related to the signal or to the noise, it is possible to truncate the summation implied by Eq. [5]. Here truncation amounts to omitting the contribution of the "noise-related singular values" to the LP coefficients. Note that this has no effect on the number of LP coefficients. In addition, truncation has favorable consequences for the numerical stability of the remaining calculations below.

Having obtained the LP coefficients of Eq. [2], we must proceed by finding a relation between these and the actual parameters of interest, namely the frequencies, amplitudes, damping factors, and phases. This is the subject of the next subsection. However, before passing on to this we mention briefly a method used by Kumaresan and Tufts to reduce the contribution of the noise to the singular values (*5*). To understand the gist of the method one should realize that $\mathbf{X}\tilde{\mathbf{X}}/(N-M)$ approaches the autocorrelation matrix if N is sufficiently large. Furthermore we point out that the noise contribution to the autocorrelation matrix is equal to the mean square of w_n for each diagonal element, and zero for the other elements. A reasonable substitute for the mean square of the noise is the mean square of the noise-related eigenvalues of $\mathbf{X}\tilde{\mathbf{X}}/(N-M)$. It follows then that one can improve the signal-related eigenvalues of $\mathbf{X}\tilde{\mathbf{X}}$ by subtracting from each the arithmetic mean of the noise-related eigenvalues of $\mathbf{X}\tilde{\mathbf{X}}$.

Relation between Model Parameters and LP Coefficients

The relation to be given below between the model parameters in Eq. [1] and the LP coefficients in Eq. [2] can be understood by considering first a single noiseless exponentially damped sinusoid. We recall that for this case only two LP coefficients are needed, given by the relations $a_1 = 2\exp(-b\Delta t)\cos(\omega\Delta t)$ and $a_2 = -\exp(-2b\Delta t)$. When a_1 and a_2 are known, and b and ω unknown, one can write conversely

$$\exp[(-b \pm i\omega)\Delta t] = [a_1 \pm \sqrt{(a_1^2 + 4a_2)}]/2. \qquad [7]$$

Equation [7] shows that b and ω can be obtained from the LP coefficients by taking the roots of the polynomial

$$z^2 - a_1 z - a_2 = 0. \qquad [8]$$

Depicting a root as a vector in the complex plane, the damping factor can be obtained from the radius, and the frequency from the angle with the real axis. b is positive, zero, or negative when the root falls within, on, or outside the unit circle respectively. Thus, if one predicts backward (instead of forward as happens in Eq. [2]), the roots of an exponentially damped sinusoid fall outside the unit circle. The latter point will become important shortly. Only one of the two roots is needed, of course.

Analogous to this, it can be shown (*1, 5*) that in the case of K noiseless exponentially damped sinusoids the quantities $\exp[(-b_k \pm i\omega_k)\Delta t]$ with $k = 1, \ldots, K$ are equal to the roots of the polynomial

$$z^{2K} - a_1 z^{2K-1} - \cdots - a_{2K} = 0. \qquad [9]$$

When more LP coefficients than necessary for the noiseless case are taken ($M > 2K$), then the order of the polynomial increases accordingly. The extraneous roots resulting from this always fall within the unit circle (*5*). Therefore they can be distinguished from the signal roots if the latter are made to fall outside the unit circle by predicting backward. We point out that the numerical stability of the rooting procedure is improved substantially by the truncation mentioned in the previous subsection. The importance of this should be clear given the situation that the order of the polynomial can be as high as 750 in our applications.

At this point we have obtained the frequencies and damping factors of the sinusoids present in the signal. The amplitudes and phases are still unknown. This is not surprising if one realizes that the same LP equation applies everywhere in the data record and therefore cannot possibly contain time-dependent information such as amplitude and phase. The problem is solved by substituting the frequencies and damping factors into Eq. [1], and writing this equation down for all N data points. Each equation can be seen to depend linearly on the unknowns $c_k \cos \varphi_k$ and $c_k \sin \varphi_k$. Thus the amplitudes and phases can be determined by another linear LS procedure.

Implementation

In practice the procedure for fitting the mathematical model of Eqs. [1] and [2] to the data runs as follows (using FORTRAN 77 with double precision):

1. The $(N - M) \times M$ dimensional matrix $\mathbf{X}$ is set up from the data, in compliance with backward prediction. Most often $M = 0.75N$ is used; in the following this rule will be adhered to. With our present minicomputer adapted software the maximum dimensions of $\mathbf{X}$ are about 250 (rows) $\times$ 750 (columns).

2. Computation of the $(N - M) \times (N - M)$ dimensional nonnegative definite matrix $\mathbf{X}\tilde{\mathbf{X}}$, and subsequent diagonalization yielding the singular values (after taking the square roots) and the $(N - M) \times (N - M)$ dimensional eigenvector matrix $\mathbf{U}$. The $M \times M$ dimensional eigenvector matrix $\mathbf{V}$ need not be calculated by a similar procedure. Instead, one can use (*6*)

$$\mathbf{V} = \tilde{\mathbf{X}}\mathbf{U}\Lambda^{-1}, \qquad [10]$$

in which Λ^{-1} (the exponent -1 has only symbolical meaning, as in Eqs. [5] and [6]) is an $(N - M) \times M$ dimensional diagonal matrix with diagonal entries equal to λ_i^{-1}. The diagonalization subroutine was taken from Ref. (*9*), and rewritten for efficient use on a minicomputer (HP1000/45).

3. Perusal of the set of $N - M$ singular values leading to a reduction of the rank $N - M$ of $\mathbf{X}$ to an effective rank l. When the number of sinusoids is small compared to the maximum rank $N - M$, and the amplitudes are well above the noise, this step is straightforward and can be automated. Understandably, difficulties may arise when the number of sinusoids is high and/or several of them have amplitudes which are comparable to the level of the noise. This aspect is the subject of continued study (*10*). Practical examples are given in the next section.

4. Computation of the LP coefficients using a combination of Eqs. [5] and [10], and truncating the singular values at the number l.

5. Computation of the roots of the Mth order polynomial made up from the LP coefficients according to Eq. [9]. For this we have used a subroutine taken from Ref. (*11*), rewritten for efficient use on a minicomputer. From the M roots only the l ones with the largest radii are retained. Those lying on the real axis each represent a damped exponential function with zero frequency. Such components together make up the possible background signal (physical or instrumental) on which the oscillating part of the signal is superposed. The other roots can be divided into two groups which are each others' complex conjugate. Each complex conjugate pair of roots represents one exponentially damped sinusoid.

6. Calculation of the amplitudes and phases by a second linear LS procedure, using Eq. [1] for $n = 0, \ldots, N - 1$, and the frequencies and damping factors determined in the previous step. The maximum number of sinusoids with nonzero frequency is $l/2$.

APPLICATIONS

Simulated Signal

The first example concerns spectral analysis of a simulated time-domain signal. The advantage of analyzing a simulated instead of a measured signal is that the result can be compared with known theoretical parameters. The simulated signal, $x(n\Delta t)$, used in the present work comprises two exponentially damped sinusoids plus spectrometer noise. It was calculated according to Eq. [1] for $n = 0, \ldots, 511$ and $\Delta t = 100$ ns. The values of $\nu_k = \omega_k/2\pi$, b_k, c_k, and φ_k are listed in the two upper rows of Table 1. (In absence of noise the results of an LPSVD analysis approach the theoretical parameters with a high degree of accuracy. Therefore the numbers in the indicated rows can represent both.) The frequencies of the two components are 5% apart, while the amplitude of one component is significantly larger than that of the other, the ratio being 1:0.3 at $t = 0$. The SNR is to be related to the initial value of the amplitude of the larger component (*5*). Figure 1 shows the signal for SNR = 30 dB, and Fig. 2a the corresponding spectrum obtained with cosine FFT in conjunction with a rectangular window. For Fig. 2a the phases were temporarily put equal to zero to avoid distortion. Note that the two components are not resolved. Their resolution can possibly be achieved by curve fitting of the spectrum (*3*). Finally, the ratio of the damping factors was chosen to be 0.2:1, so as to amplify the difference in size when going from the time domain to the frequency domain. This is illustrated by the dotted line in Fig. 2a that represents the smaller component (on same scale as the total signal).

Next we apply LPSVD. Using only the first 256 data points ($N = 256$), we set the number of LP coefficients at $M = 192$, which fixes the maximum rank at $N - M = 64$. First we consider the singular values λ_i ($i = 1, \ldots, 64$), the first five of which are listed in Table 2 for SNR = ∞ and 30 dB. For SNR = ∞, only the first four singular values are seen to be nonzero (λ_5 in the table is still slightly greater than zero because $x(n\Delta t)$ has been calculated with single precision). If the SNR is finite, on the other hand, the other sixty singular values become also nonzero, but they remain smaller than those related to the signal, at least down to SNR = 30 dB. This is shown in Fig. 3, where a linear as well as a logarithmic scale

TABLE 1

LPSVD Analysis of the First 256 Data Points of a Simulated Signal with Various Amounts of Noise[a]

SNR	l	k	ν_k (MHz)	b_k (MHz)	c_k	φ_k (deg)
∞	4	1	1.000	0.100	1.00	20.0
		2	1.050	0.500	0.30	80.0
50	4	1	1.000	0.100	1.00	20.4
		2	1.052	0.510	0.29	78.5
40	4	1	1.000	0.100	1.00	21.1
		2	1.055	0.518	0.28	76.4
30	4	1	0.999	0.105	1.06	22.6
		2	1.058	0.279	0.17	91.2
30	4	1	0.999[b]	0.101[b]	1.00[b]	23.2[b]
		2	1.065[b]	0.529[b]	0.23[b]	68.5[b]
30	13	1	0.999	0.105	1.05	22.6
		2	1.058	0.299	0.18	89.2
30	20	1	0.999	0.104	1.04	22.7
		2	1.062	0.341	0.18	77.4
30	64	1	0.999	0.103	1.03	23.1
		2	1.066	0.376	0.18	75.0
30	4	1	1.000[c]	0.101[c]	1.00[c]	19.7[c]
		2	2.008[c]	0.483[c]	0.30[c]	77.0[c]

[a] See Fig. 1 for SNR = 30 dB. The parameters K, k, $\nu_k = \omega_k/2\pi$, b_k, c_k, and φ_k are defined through Eq. [1]. l is the effective rank ($l = 2K$), which equals the number of significant (i.e., signal-related) singular values. The results for SNR = ∞ are equal to the input parameters. If $l > 4$, the sinusoids with $k > 2$ can be ignored because the related amplitudes are of order 10^{-2} or less.

[b] Singular values corrected for noise.

[c] Frequency of small component changed from 1.05 to 2.00 MHz, so as to avoid overlap.

have been used to illustrate both the entire range and the transition from noise-related to signal-related singular values (see also Table 2). When considering Fig. 3 from right to left, the first significant change of λ_i is observed on passing from index $i = 5$ to 4. The smaller changes, occurring for $i > 5$, decrease steadily as the SNR increases. It follows that we find the correct number of components, $K = l/2 = 2$, for SNR $\geqslant$ 30 dB. For the sake of completeness the LPSVD analysis for SNR = 30 dB is executed also with larger effective ranks, namely $l = 13$, 20, and 64.

The results for SNR = ∞, 50, 40, and 30 dB have been compiled in Table 1. As already mentioned, the first two rows represent both the result of the analysis for SNR = ∞ and the original parameters. The results for finite SNR are to be compared with these numbers. Table 1 shows that the analysis is rather accurate down to SNR = 40 dB, especially when taking into consideration that FFT cannot resolve the two components, even for SNR = ∞. Only after lowering the SNR further, by 10 dB, do significant deviations develop, although the resolution of the

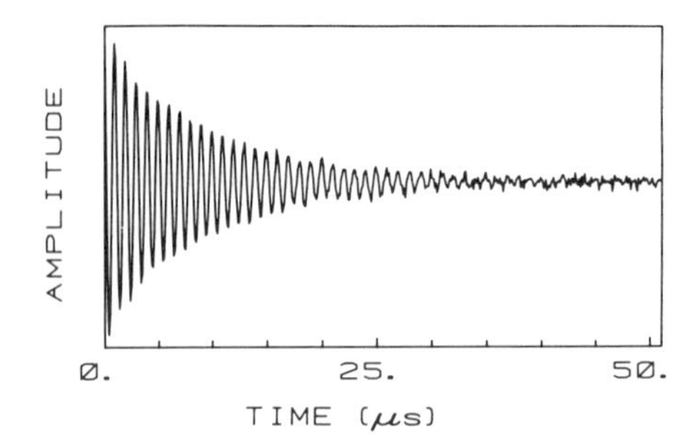

FIG. 1. Simulated time-domain signal comprising two exponentially damped sinusoids according to Eq. [1], for SNR = 30 dB. The frequencies (ν_k), damping factors (b_k), amplitudes (c_k), and phases (φ_k) are listed in the first two rows of Table 1 (for SNR = ∞). Only the first half of the signal is used for the LPSVD analysis, the result of which is given in Table 1.

two components is still established beyond doubt. At this noise level it seemed useful to try some alternative treatments within LPSVD. These will now be discussed briefly, turning first to the smaller component. It appears that the frequencies are reproduced satisfactorily in all cases. However, their precision tends to decrease somewhat when increasing l, or using noise corrected singular values. On the other hand, the latter two measures have a distinct positive effect on the amplitude and damping factor. As for the large component, this is reproduced rather well in all cases, but especially with noise-corrected singular values. Finally, the last two rows of the table show that the LPSVD analysis is very accurate (considering the SNR of the small component in Fig. 2a) when the frequencies of the components are placed sufficiently far apart.

A measure for the quality of fitting the model is obtained from the Fourier transform of the time domain residue. This has been calculated for 512 points from the results of the case SNR = 30 dB, l = 4, and uncorrected singular values (see Fig. 2b). It follows from the noisy appearance of the residue in the frequency domain that a near optimum fit has been obtained. Finally, we stress that the trends noted above for SNR = 30 dB apply only to a single noise realization. A more

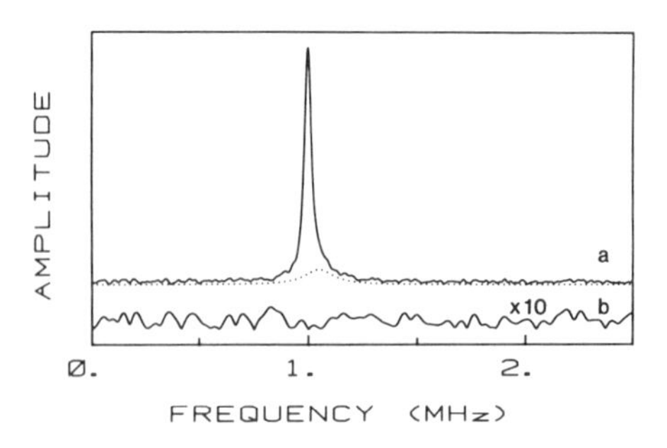

FIG. 2. (a) Cosine FFT, without explicit windowing, of the signal depicted in Fig. 1, the phases of the constituent sinusoids being temporarily set to zero (drawn curve). The two components are not resolved. The dotted curve represents the cosine FFT of the smaller of the two components. (b) Residue of the LPSVD fit in the frequency domain obtained with amplitude FFT and multiplied by 10.

TABLE 2

First Five Singular Values of the Data Matrix of a Simulated Signal Consisting of Two Damped Sinusoids Plus Noise (Parameters Given in First Two Rows of Table 1) for Two Values of the SNR[a]

SNR (dB)	λ_1	λ_2	λ_3	λ_4	λ_5
∞	20.998	20.456	0.73908	0.65618	0.30129×10^{-4}
30	20.999[b]	20.457[b]	0.78180[b]	0.71158[b]	0.49534

[a] Note the large difference between λ_5 for SNR = ∞ and 30 dB. The ratio between the first and last singular value, λ_1/λ_{64}, is 0.4×10^7 for SNR = ∞, and 135 for SNR = 30 dB.
[b] Average value for 12 noise realizations.

appropriate way to investigate the influence of noise is to use a great number of different noise realizations and study the ensuing scatter of the results (*5*). In this way it can be established that the variances of LPSVD results approaches the proper statistical Cramer–Rao bounds (*5*).

1D Electron Spin Echo

This example is concerned with three-pulse electron spin-echo (ESE) experiments leading to a stimulated echo. The magnitude of the echo is recorded as a function of the delay τ' between the first and third transmitter pulse, while the delay τ between the first and second pulse is kept constant. Hence the prefix 1D. As τ' is swept one observes that besides the usual exponential decay, the echo amplitude often exhibits also sinusoidal variations. The latter are caused by hyperfine interactions (HFI) between the paramagnetic center and surrounding nuclei. It follows that the envelope of the echo amplitude as a function of τ' contains information about the

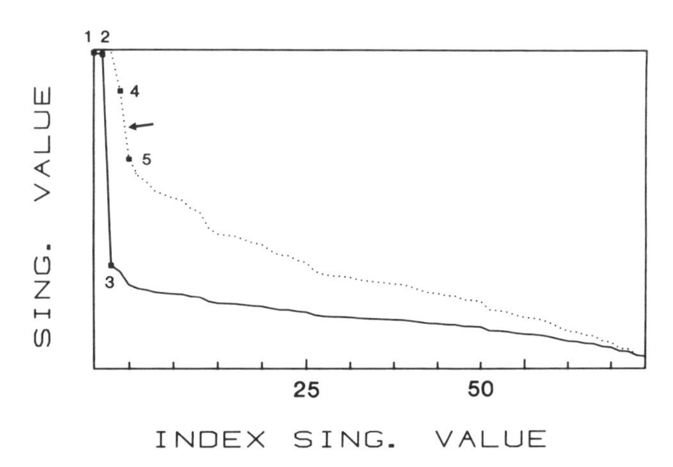

FIG. 3. Singular values, λ_i, related to the first half of the signal in Fig. 1, on a logarithmic scale (points connected by drawn line), and on a linear scale (points connected by dotted line). λ_1, λ_2, and λ_3 are explicitly indicated on the log plot, and λ_4 and λ_5 on the linear plot. It can be seen that when coming from the right, the first clear jump occurs between index i = 5 and 4.

HFI (*12–14*). Transformation to the frequency domain yields spectra similar to those obtained with the well-known method of electron nuclear double resonance (ENDOR) (*15–17*). The ESE method to measure HFI has been given the name ESE envelope modulation or ESEEM. In this work we consider ESEEM of an irradiated single crystal of potassium dihydrogen arsenate (KDA), KH_2AsO_4. Details about the measurements are given in a previous paper (*18*). The point of interest in the present case is the correct retrieval of the numerous sinusoids (from ^{39}K mostly, and some from ^{1}H) in the time-domain ESEEM signal which is depicted in part in Fig. 4 (to the right of the dotted line). Also, for some applications it is useful to reconstruct the missing initial part of the ESEEM signal (left of the dotted line in Fig. 4, see text below).

The number of (real) data points is 1024. At present this is the maximum number that can be processed with LPSVD on our minicomputer. As pointed out in the previous section, preprocessing is not necessary. Here this means that the decaying background on which the sinusoids are superposed need not be removed beforehand. In fact, the background signal shows up naturally in the analysis as a limited number (usually two or three) of damped exponentials with zero frequency. The singular values of the data matrix are plotted in Fig. 5. It can be seen that there is no sharp transition between noise-related and signal-related singular values when going from right to left. It turned out that such a condition does not pose a serious problem. In this case we have taken the effective rank l to be 110, as marked by an arrow. All roots were then found to lie outside the unit circle, in agreement with backward prediction. (This is not the case for all practical signals, however. If signal roots are to be taken from within the unit circle, those lying farthest from the origin are chosen. Admittedly, the latter represent sinusoids that increase with time, but their amplitudes are usually found to be small.) Next, we proceed by calculating the frequencies, damping factors, amplitudes, and phases, in the same way as in the previous example. These results are too numerous to be reproduced here in the form of a table, however. Therefore we devised a method to present the material graphically, that is, at least a major part of it. Briefly, the procedure is as follows.

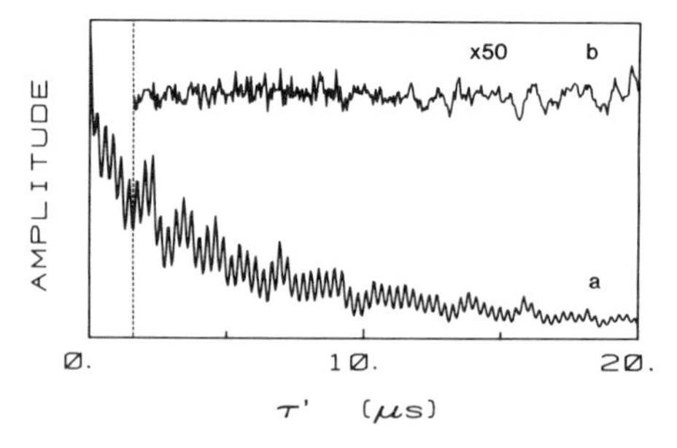

FIG. 4. (a) ESEEM signal from a single crystal of X-irradiated KH_2AsO_4, to the right of the dotted line at $\tau' = 1.6$ μs, and continuing beyond the frame up to $\tau' = 53$ μs. The part to the left of the dotted line could not be measured, and was reconstructed with the aid of LPSVD. Constructive interference of the latter (up to the top of the frame) is correctly reproduced at only 8 ns away from the time origin. (b) Residue of the LPSVD fit to the time domain signal, magnified by 50. Note the change in appearance as a function of τ'.

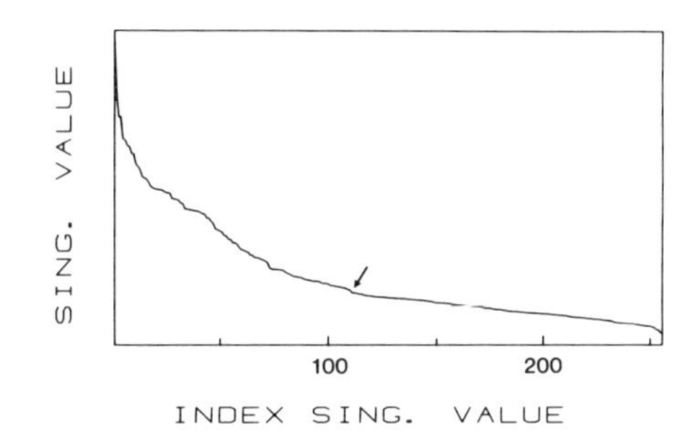

FIG. 5. Singular values λ_i related to the signal of Fig. 4. Note that there is no clear transition from noise-related singular values to signal-related singular values. The arrow at index $i = 110$ indicates the chosen position of the transition, which in turn fixes the effective rank l of the data matrix.

First, an undamped time domain signal $x'(n\Delta t)$, $n = 0, \ldots, N - 1$ is calculated from Eq. [1] and the LPSVD parameters ν_k, c_k, φ_k, $k = 1, \ldots, K$; $n = 0$ corresponds with $\tau' = 0$ in Fig. 4. Subsequently, $x'(n\Delta t)$ is transformed to the frequency domain by cosine FFT in conjunction with an appropriate window. The resulting spectrum exhibits a resolved peak at each frequency ν_k provided N is made sufficiently high. In addition, the size of each peak is proportional to the corresponding amplitude c_k, while the shape carries information about the related phase φ_k. Thus, a graphical display of the LPSVD parameters, except for the damping factors, can be obtained.

Before treating the results we digress briefly on the missing initial part of the ESEEM signal. Since the LPSVD parameters are now available, we are in a position to reconstruct these elusive data points and insert them in their proper place. This is shown in Fig. 4, the actual measurement starting only after the dotted line at $\tau' = 1.600$ μs. (This includes the extra time needed to avoid interference with unwanted echoes.) The quality of the reconstruction can be tested by considering the properties near $\tau' = 0$. According to ESEEM theory all sinusoids should interfere constructively at that time, leading to a sharp maximum. This pattern was indeed obtained, the maximum being as high as the upper frame of the picture, while the deviation on the time scale is only 8 ns. Adding the reconstructed initial part to the data enables one to apply cosine FFT without having to cope with distortion.

Figure 6 summarizes the results, showing the graphical display of the LPSVD parameters (save damping factors), the FFT of the time domain signal as depicted in Fig. 4, and the residue of the LPSVD fit (using the correct damping factors) in the frequency domain. The residue in the time domain is shown in Fig. 4. It follows from Fig. 6 that LPSVD has indeed brought about enhanced resolution. This is illustrated by the clear splitting of the peak at 3.469 MHz in the "LPSVD spectrum" (owing to a slight misorientation of the crystal), while the corresponding peak in the "FFT spectrum" is unsplit. It should be noted that the LPSVD spectrum reveals more splittings, in other regions of the spectrum. Next we consider the quality of the fit, bearing in mind that we are dealing with a solid sample, that will probably not produce Lorentzian lines. Despite the possible deficiencies of the model of Eq. [1], it appears that the SNR of the residue in the frequency domain (see Fig. 6) is low. This might be explained by assuming that the strongest part of the time domain

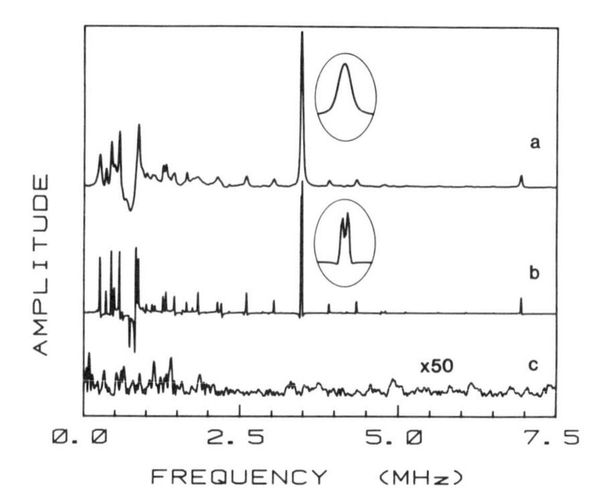

FIG. 6. (a) Cosine FFT of the time domain signal in Fig. 4, including the reconstructed initial part and the last part which falls outside the frame. Splitting of the (proton) peak near 3.5 MHz owing to misalignment of the crystal is not resolved; see insert on enlarged horizontal scale. (b) Graphical display in the frequency domain (via cosine FFT) of the LPSVD results minus the damping factors. The latter have been set to zero, resulting in a sharpened spectrum; see text. The splitting of the peak near 3.5 MHz is now clearly established as shown in the insert on enlarged horizontal scale. Other splittings are also revealed. The background time-domain signal has been removed before FFT in both a and b. (c) Residue of the complete (i.e., including the damping factors) LPSVD fit in the frequency domain (amplitude FFT), multiplied by 50. Note that modest deviations occur below 2 MHz, i.e., in the region of ^{39}K.

signal satisfies the model rather well, the remaining weaker part needing a different description. This conjecture is supported by the shape of the time-domain residue which is displayed in Fig. 4. It can be seen that the first part is random in nature, whereas the rest seems to be more deterministic. The latter observation is confirmed by FFT of separate parts of the time domain residue (not shown).

Summarizing, it has been shown that LPSVD can successfully retrieve a large number of sinusoids from a time domain signal, and this including a list with four relevant parameters for each. In comparison with FFT the resolution is clearly enhanced.

2D Electron Spin Echo

The sinusoidal variation of the echo decay as a function of τ', noted in the previous subsection, is also observed when varying τ. Sweeping then both τ' and τ, one obtains a 2D set of ESEEM data points indicated by $x(\tau, \tau')$. Transformation to the frequency domain has been shown to yield a 2D spectrum that reveals symmetry relations between peaks. This in turn facilitates the assignment of peaks to specific transitions between nuclear levels (*18–20*). A serious problem is that the decay in the τ space is usually one or two orders of magnitude faster than that in the τ' space. By consequence, the lines may become prohibitively wide in the τ space when using FFT. The situation can be improved significantly by applying the Burg autoregressive modeling algorithm in the τ space and FFT in the τ' space (18,

20). In the present example we shall apply LPSVD in *both* the τ and the τ' space, which results in further improvement.

The 2D-LPSVD procedure adopted here is as follows. First we put together an averaged time series $y(\tau')$ given by

$$y(\tau') = \sum_{\tau} g(\tau)x(\tau, \tau'), \qquad [11]$$

where $g(\tau)$ is a weighting factor proportional to $\sqrt{\sum_{\tau'} x(\tau, \tau')^2}$. The summations run over all values of τ and τ', respectively. The idea behind this step is that the frequencies and damping factors in the τ' space are independent of τ. (As for the frequencies, this is evident, while for the damping factors it was verified experimentally.) It follows that one can save much time by calculating these quantities via executing LPSVD just once, on $y(\tau')$. Next, the amplitudes and phases in the τ' space can be calculated for all values of τ, using the frequencies and damping factors determined from $y(\tau')$. At this point the analysis in the τ' space is completed, resulting in the quantities K, ω_k, b_k, $c_k(\tau)$, and $\varphi_k(\tau)$, for $k = 1, \ldots, K$. It turned out that the phases φ_k are almost independent of τ, so that the remaining analysis can be restricted to the amplitudes $c_k(\tau)$, $k = 1, \ldots, K$. The latter quantities constitute K time-domain signals, each of which is to be analyzed with LPSVD. Obviously it would be rather cumbersome to determine the effective rank by user interaction as is done so far. Because the number of frequencies contained in the $c_k(\tau)$ is as a rule rather limited (*18–20*), this step can easily be handled with a simple algorithm, however. Thus the K LPSVD analyses are executed completely automatically. After this one reaches the last step which consists of calculating the amplitudes and phases in the τ space. The results can be summarized in part through a 2D plot in the frequency domain, in which the peaks are indicated by circles whose radii are proportional to the respective amplitudes.

The data points were taken from a previous 2D measurement on F centers in a single crystal of KCl for B‖[110] (*20*). A typical data file consists of 2048×64 data points, the higher and lower number being related to the τ' and the τ space, respectively. With the present software this constitutes the largest possible data record for our mini computer. It turned out that the LPSVD method did not require all 2048 data points in the τ' space to attain good resolution. In fact, the first 512 data points already yielded satisfactory results. This implies that LPSVD enables us to change the strategy of 2D measurements. Henceforth we can take four times fewer data points in the τ' space, but four times more in the τ space. The result for 512×64 points is given in Fig. 7, for the region of ^{39}K. Peaks from shells III, V, and IX can clearly be identified.

NMR Free Induction Decay

Finally, we apply LPSVD to an *in vivo* ^{31}P free induction decay. The sample was an anaesthetized mouse, which limited the measuring time. Therefore prolonged time averaging to raise the SNR to a satisfactory degree was not possible. The number of data points N equals 768. Notwithstanding the use of the cyclops phase cycling technique, the first few data points were disturbed, owing to instrumental

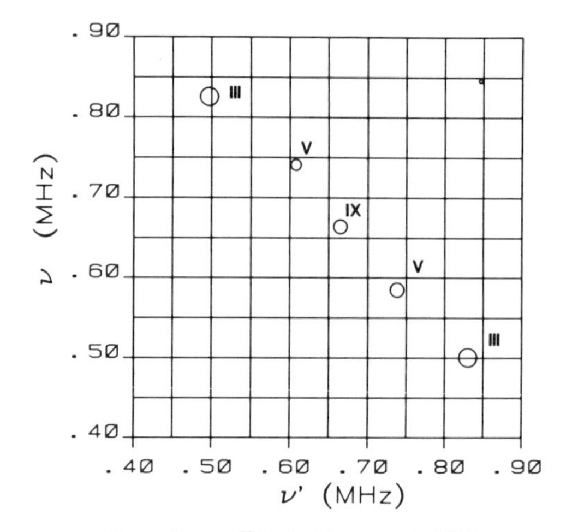

FIG. 7. 2D ESEEM spectrum of the region of ^{39}K for F centers in KCl, computed by LPSVD in both dimensions. The peak intensities are represented by the radii of the contours. The symmetry with respect to the bisector of the axes is well established. The difference in linewidth in the two dimensions has been effectively removed by ignoring the damping factors. The signals originate from shells III, V, and IX, around the F center.

effects (varying pulse breakthrough). When using FFT, one has to cope with this by applying some sort of time window. In any case, the quality of the spectrum deteriorates. This in turn causes additional problems in the subsequent quantitative analysis.

Since quadrature detection was used, the time-domain signal is complex. The theory in the previous section was restricted to real data, but extension to complex data requires only minor changes. It suffices to point out that the LP coefficients become complex and therefore carry twice as much information. As a result, the necessary number of coefficients in the LP equation for K complex noiseless damped sinusoids is reduced by a factor of two.

Applying LPSVD, we used only $N = 125$ data points; the first 3 were rejected because of the distortion mentioned earlier, the last 640 because of poor SNR. The number of prediction coefficients M was set at 88, which leaves $N - M = 37$ for the maximum rank. The singular values λ_i ($i = 1, \ldots, N - M$) calculated from the data matrix are shown graphically in Fig. 8 on a linear and a logarithmic scale. Scrutinizing the graph from right to left as before, the first really significant jump is seen to take place when going from $i = 10$ to $i = 9$ (indicated by an arrow). Thus, the effective rank l is set at 9. Since we are dealing with complex data, the number of sinusoids then also equals 9. Continuing the LPSVD procedure to the end, we obtain a set of nine frequencies, damping factors, amplitudes, and phases. These have been compiled in Table 3, after extrapolating the amplitudes and phases to the theoretical time origin. The numbers in the table contain all relevant information. However, for the sake of comparison one can calculate a synthetic time domain signal from the LPSVD data and subsequently Fourier transform this. The result is shown in Fig. 9, together with the FFT of the properly windowed original time

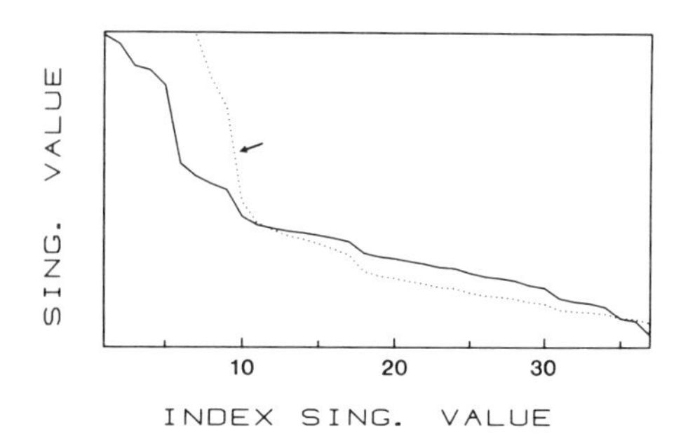

FIG. 8. Singular values λ_i related to an *in vivo* ^{31}P NMR free induction decay measurement; see spectrum in Fig. 9. The transition from noise-related to signal-related singular values takes place when going from index $i = 10$ to 9, as indicated by an arrow. The drawn line represents a logarithmic scale, the dotted line a linear scale.

domain signal. It can be seen that all essential features in the FFT spectrum are reproduced in the LPSVD spectrum. We note further that peak No. 8 of the LPSVD result rises barely above the noise level in the FFT spectrum. At present it is difficult to decide how much significance can be attached to this peak. The related phase is in agreement with a priori knowledge, but the damping factor is not, the latter being about three times below the value that corresponds with the inhomogeneity of the field. Application of the theory of subsets (*10*) should settle this type of problem.

In summary, we establish the fact that LPSVD has unambiguously retrieved at least eight peaks in the *in vivo* ^{31}P spectrum and has provided a table with frequencies and amplitudes (among other things) for each peak. No preprocessing of the data was needed.

TABLE 3

LPSVD Analysis of Data Points 4 through 128 of *in Vivo* ^{31}P NMR Free Induction Decay Comprising 768 Complex Data Points[a]

k	ν_k (kHz)	b_k (kHz)	c_k	φ_k (deg)
1	−1.571	0.387	2.32	21
2	−1.324	0.191	1.21	−11
3	−0.738	0.040	0.13	−11
4	−0.430	0.242	1.35	−13
5	−0.018	0.065	0.20	−24
6	0.178	0.472	3.40	6
7	0.454	0.181	0.37	−22
8	0.531	0.014	0.07	−26
9	1.235	0.311	1.57	−4

[a] The amplitudes and phases have been extrapolated to the time origin. The symbols are defined by the complex version of Eq. [1].

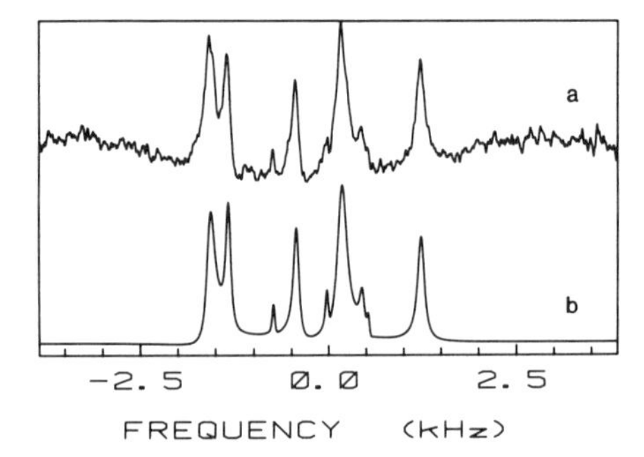

FIG. 9. (a) *In vivo* ^{31}P free induction NMR spectrum, computed by cosine FFT of 768 data points, using the optimal SNR filter with a time constant of 0.015 s, and putting the three initial data points equal to zero. (b) Graphical display of the LPSVD results extracted from data points 4 to 128. This was obtained by cosine FFT of 768 data points computed from the parameters listed in Table 3. Note that all essential features of spectrum a are reproduced by LPSVD.

CONCLUDING REMARKS

The examples in the previous section indicate that LPSVD can become a useful tool for processing time-domain magnetic resonance signals. The main feature of the method is that it alleviates the task of quantitative spectral analysis, and this also for components that cannot be resolved by FFT. Another attractive feature, not so much emphasized above, is that LPSVD enables one to accurately reconstruct missing parts of the time-domain signal.

A problem needing further study is the effect of inadequacy of the mathematical model. This problem arises especially in measurements on solid disordered systems which usually entail wide frequency distributions having no likeness to Lorentzian lines. Preliminary LPSVD results for ESE on solid disordered systems show that the fit can usually be made quite good using only a limited number of components (i.e., exponentially damped sinusoids). It follows that LPSVD can at least be used for partial reconstruction of a missing initial part of a signal (21), which in turn leads to reduction of the distortion of the spectrum as obtained by subsequent application of FFT. Interpretation of the actual values of LPSVD parameters obtained for solid disordered systems has not yet been attempted. Ultimately, fitting a time-domain model that contains the parameters of the spin Hamiltonian (see, for instance, Refs. (*22, 23*)) may yield the best results. This is a very complicated task however.

Finally, the reader will realize that there is a price to pay for the increased information provided by LPSVD. As with other parameter fitting procedures, the price is increased computing time. For example, analysis of 256 real data points with LPSVD takes about one minute on our HP 1000/45 minicomputer, while 1024 data points require about an hour. This is of course much more than needed for a single FFT, but we point out that interactive preprocessing prior to FFT also demands time. In addition, use of a mainframe computer reduces the quoted times considerably: the central computer (Amdahl) of our university requires only three minutes for analyzing 1024 real data points.

ACKNOWLEDGMENTS

This work was performed as part of the research program of the "Stichting voor Fundamenteel Onderzoek der Materie" (FOM) with financial support from the "Nederlandse Organisatie voor Zuiver Wetenschappelijk Onderzoek" (ZWO). The authors thank Professor H. Postma for stimulating remarks.

REFERENCES

1. S. M. KAY AND S. L. MARPLE, *Proc. IEEE* **69,** 1380 (1981).
2. "Fourier, Hadamard, and Hilbert Transforms in Chemistry" (A. G. Marshall, Ed.), Plenum, New York, 1982.
3. C. L. DUMOULIN AND G. C. LEVY, *Bull. Magn. Reson.* **6,** 47 (1984).
4. S. O. CHAN AND M. B. COMISAROW, *J. Magn. Reson.* **51,** 252 (1983); **54,** 201 (1983).
5. R. KUMARESAN AND D. W. TUFTS, *IEEE Trans.* **ASSP-30,** 833 (1982).
6. C. L. LAWSON AND R. J. HANSON, "Solving Least Squares Problems," Prentice-Hall, Englewood Cliffs, N.J., 1974.
7. S. SIBISI, *Nature (London)* **301,** 134 (1983); see also ibid. **311,** 446 (1984).
8. J. MAKHOUL, *in* "Modern Spectral Analysis" (D. G. Childers Ed.), pp. 99–118, IEEE Press, Wiley, New York, 1978.
9. Y. BEPPU AND I. NINOMIYA, *Comput. Phys. Commun.* **23,** 123 (1981).
10. R. R. HOCKING, *Technometrics* **25,** 21g (1983).
11. K. STEIGLITZ AND B. DICKINSON, *IEEE Trans.* **ASSP-30,** 984 (1982).
12. W. B. MIMS, *in* "Electron Paramagnetic Resonance" (S. Geschwind, Ed.), Chap. 4, Plenum, New York, 1972.
13. L. KEVAN, *in* "Time Domain Electron Spin Resonance" (L. Kevan and R. N. Schwartz, Eds.), Chap. 8, Wiley, New York, 1979.
14. J. R. NORRIS, M. C. THURNAUER, AND M. K. BOWMAN, *in* "Advances in Biological and Medical Physics" (J. H. Lawrence, J. W. Gofman, and T. L. Hayes, Eds.), Vol. 17, Academic Press, New York, 1980.
15. R. P. J. MERKS AND R. DE BEER, *J. Magn. Reson.* **37,** 305 (1980).
16. W. B. MIMS, *in* "Fourier, Hadamard, and Hilbert Transforms in Chemistry" (A. G. Marshall, Ed.), pp. 307–322, Plenum, New York, 1982.
17. P. A. NARAYANA AND L. KEVAN, *Magn. Reson. Rev.* **7,** 239 (1983).
18. H. BARKHUIJSEN, R. DE BEER, A. F. DEUTZ, AND D. VAN ORMONDT, *Solid State Commun.* **49,** 679 (1984).
19. R. P. J. MERKS AND R. DE BEER, *J. Phys. Chem.* **83,** 3319 (1979).
20. H. BARKHUIJSEN, R. DE BEER, E. L. DE WILD, AND D. VAN ORMONDT, *J. Magn. Reson.* **50,** 299 (1982).
21. W. B. MIMS, *J. Magn. Reson.* **59,** 291 (1984).
22. S. A. DIKANOV, YU. D. TSVETKOV, AND A. V. ASTASHKIN, *Chem. Phys. Lett.* **91,** 515 (1982).
23. M. ROMANELLI, M. NARAYANA, AND L. KEVAN, *J. Chem. Phys.* **80,** 4044 (1984).

JOURNAL OF MAGNETIC RESONANCE **64**, 81–93 (1985)

Stimulated Echo Imaging

J. FRAHM, K. D. MERBOLDT, W. HÄNICKE, AND A. HAASE

Max-Planck-Institut für biophysikalische Chemie, Postfach 2841, D-3400 Göttingen, Federal Republic of Germany

Received March 11, 1985

A new form of NMR imaging is described using stimulated echoes. The technique, dubbed STEAM (*st*imulated *e*cho *a*cquisition *m*ode) imaging, turns out to become a versatile tool for multipurpose NMR imaging. Stimulated echoes can be excited by a sequence of at least three rf pulses, which in the basic experiment have flip angles of 90° or less. Thus no selective or nonselective 180° pulses are needed, which eliminates a variety of problems associated with such pulses in conventional spin-echo NMR imaging. Further advantages of STEAM imaging are concerned with the functional flexibility of an imaging sequence comprising three pulses and three intervals and the possibility of "storing" information prepared during the first interval into the form of longitudinal magnetization during the second interval. In general, the applied rf power is considerably reduced as compared to spin-echo-based imaging sequences. Here the general principles of the technique are outlined and first applications to multislice imaging of directly neighboring slices are demonstrated. Subsequent papers will be concerned with modifications of the basic STEAM sequence which, for example, allow multiple chemical-shift-selective (CHESS) imaging, complete imaging of the spin–lattice relaxation behavior, diffusion imaging, and single-shot real-time imaging.

INTRODUCTION

Except for the very first attempts involving free induction decays, NMR imaging has been based on the acquisition of spin-echo (SE) NMR signals excited by rf pulse sequences of the form 90°–(180°–SE–)$_n$. These sequences provide the maximum signal available for reconstruction of an image and yield excellent anatomical and diagnostic details. Nevertheless, NMR experiments including 180° pulses are limited in various respects in particular when applied to the imaging of living subjects.

(i) The use of selective 180° pulses in combination with a slice selection gradient raises problems for proper refocusing of the excited spin moments, see, e.g., (*1*). A solution requires increase of the bandwidth of the 180° pulse but distorts magnetization immediately on both sides of the wanted slice. Moreover, in most cases a second "compensation" experiment is needed to eliminate magnetization components contributing from outside the slice. The first effect degrades the use of multislice imaging, while the second effect prolongs the measuring time by a factor of two.

(ii) SE-NMR sequences are entirely based on transverse magnetization which can be excited once and then must be used in a time short with respect to T_2 under the assumption of multiecho formation by means of a sequence of 180° pulses. For a more quantitative imaging, it is a disadvantage that the attenuation of the spin

0022-2364/85 $3.00

echoes results from a variety of processes which together destroy the phase coherence of the spin moments. *In vivo* the simultaneous presence of T_2 relaxation, diffusion, and flow makes a clear separation or discrimination of contributions almost impossible. Further, since no longitudinal magnetization is involved, spin echoes do not contain any T_1 information. In SE-NMR imaging, therefore, "T_1 contrast" has to be acquired in a rather time-consuming way by performing multiple imaging experiments with different repetition times or by means of recording multiple inversion–recovery images with different relaxation delays.

(iii) To provide a maximum of information within the measuring time of a single NMR image, SE techniques have been extended to multiecho and/or multislice versions using a large number of 180° pulses. These sequences, however, become a problem in high-field NMR imaging at field strengths of 1.5 T or higher because of the considerable increase of the rf power dissipation in the human body. A reduction of SE-NMR sequences to the safety guidelines in most cases might preclude a truly economic use of appropriate imaging systems.

Here we will present a new form of NMR imaging using stimulated echoes which is free from the above-mentioned drawbacks simply by avoiding the application of 180° pulses. Although the stimulated echo intensity is only half that of a spin echo taken at the same readout time (i.e., the time corresponding to the first and third intervals of the three-pulse sequence, see below), the advantages of STEAM imaging and some of its modified versions will definitely compensate for this partial loss in signal-to-noise. It is our belief that new pulse sequences which reduce the measuring times, which give access to new or more accurate information, and which thus provide improved specificity rather than optimized signal-to-noise, will become increasingly important in future applications of NMR imaging. In fact, STEAM imaging experiments will be described which often are without SE-NMR alternatives including simultaneous chemical-shift-selective (CHESS) imaging of multiline spectra (*2*), simultaneous imaging of complete spin–lattice relaxation curves (*3*), imaging of self-diffusion coefficients (*4*), and single-shot real-time imaging with a variable spatial resolution (*5*).

STIMULATED ECHOES

The basic rf pulse sequence that gives rise to a stimulated echo (STE) signal at $t_3 = t_1$ is

$$90°\text{–}t_1\text{–}90°\text{–}t_2\text{–}90°\text{–}t_3(\text{STE}) \qquad [1]$$

where the pulses may have arbitrary phase relations. It was first described by Hahn in 1950 (*6*) and was later used for NMR diffusion measurements (*7, 8*). More recently, STE signals have been observed with certain two-dimensional and multiple-quantum NMR experiments. Figure 1 depicts an experimental demonstration of the signals excited by sequence [1] in the presence of a magnetic field gradient. In general, five echoes may be observed by application of three pulses. Assuming $t_2 > t_1$ the primary echo signals are the SE signal at $t_2 = t_1$ created by the first two pulses and the STE signal at $t_3 = t_1$ which is due to the action of all three pulses. Secondary echoes occur as spin echoes excited by the third pulse which mirrors the primary SE signal to yield a second SE at $t_3 = t_2 - t_1$, the FID following the second

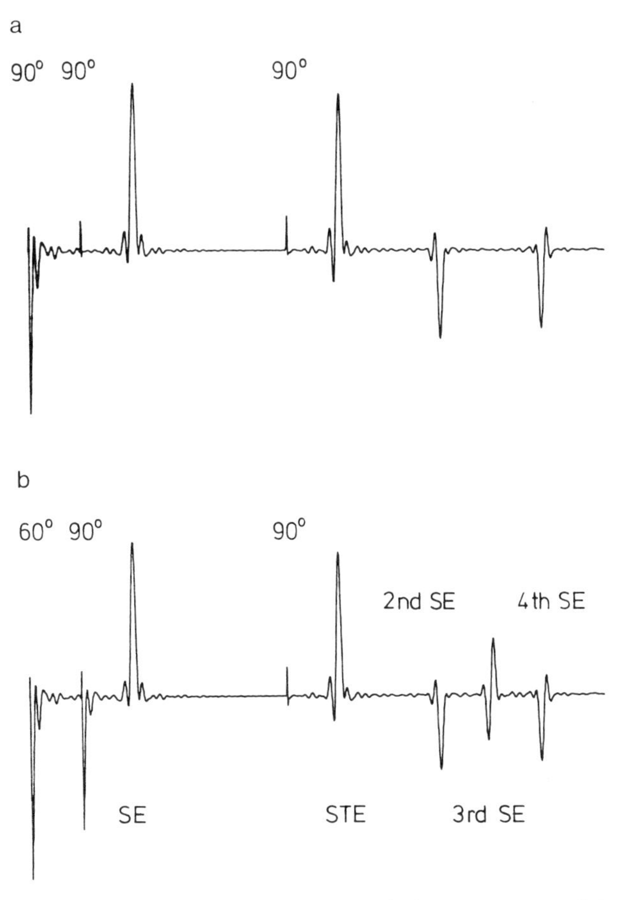

FIG. 1. Experimental demonstration of NMR signals excited by a sequence of three nonselective rf pulses in the presence of a gradient. SE, spin echo; STE, stimulated echo; 2nd SE, spin echo of the initial spin echo; 3rd SE, spin echo of the FID following the second pulse; and 4th SE, spin echo of the FID following the first pulse. (a) 90°–90°–90° pulse sequence. (b) 60°–90°–90° pulse sequence.

pulse to yield a third SE at $t_3 = t_2$, and the FID following the first pulse to yield a fourth SE at $t_3 = t_1 + t_2$. The third SE signal is seen only in Fig. 1b, where longitudinal magnetization has been left by the initial 60° pulse for excitation by the second pulse. Without an FID after the second pulse the corresponding spin echo after the third pulse is missing as demonstrated in Fig. 1a.

Transverse magnetizations following the initial $90°(x')$ pulse acquire phase information while precessing during the first interval of length t_1. At the time of a second $90°(x')$ pulse only those vector components that are aligned along the y' axis in the rotating frame of reference are transformed into (phase-encoded) longitudinal magnetizations. Assuming an equal distribution of transverse components within the x', y' plane prior to the second pulse, the affected $M_{y'}$ components represent half of the total magnetization and give rise to a stimulated echo at $t_3 = t_1$ after application of a third ("read") pulse. The remaining half of the total magnetization, i.e., the $M_{x'}$ components, refocus at $t_2 = t_1$ to form a conventional spin echo. Phase information may be related to spin–spin couplings or resonance offsets, i.e., chemical shifts in an NMR spectrum or a distribution of spatially encoded frequencies due to magnet inhomogeneities or deliberate applications of magnetic field gradients. In fact, the experimental conditions encountered in 1950 by Hahn directly corresponded to the situation in current state NMR imaging where homogeneous magnets are combined with magnetic field gradients of the order of 1 to 10 mT m^{-1}.

Another important feature of sequence [1] is that the flip angles may be reduced without much loss in intensity of the primary SE and STE signals. The effect is clearly demonstrated in Fig. 2 where the flip angles of the second and third pulses are varied from 90 to 10°, respectively. Experimentally, a 10° pulse in the third position still gave a signal intensity of almost 20% of the 90° result. If both the second and third pulses are reduced to 60° (45°), the corresponding STE amplitude is attenuated to about 85% (60%) with respect to that obtained by a 90° pulse. A variation of the flip angle of the first pulse parallels the well-known findings for the intensity of the conventional FID signal. On the other hand, a single 180° pulse in any of the three positions eliminates the stimulated echo.

In general, not only the STE but all signals resulting from application of a three-pulse sequence may be exploited for NMR imaging. At least the primary spin echo is well suited for imaging purposes and will be used in many STEAM imaging techniques in addition to the stimulated echo. In fact, the SE and STE signals may even be added to regain the same signal intensity as obtained with a conventional SE-NMR sequence, although it is often just the different use of both images that may increase the information available from a single STEAM imaging experiment. The outstanding properties, however, which make STE signals particularly attractive for NMR imaging, are due to (i) the "storage" effect of the intermediate interval t_2, where the phase-encoded magnetization relaxes rather slowly with the spin–lattice relaxation time T_1, so that phase information can be conserved for a long period and then be refocused as a STE signal; (ii) the possibility of a successive partial readout of STE signals by means of a series of "third" pulses which may be slice selective or chemical-shift selective or which may have small flip angles; and (iii) the possibility of discriminating between phase-encoded longitudinal magnetizations yielding a STE signal and normal (random) longitudinal magnetizations resulting in an FID after application of a third pulse. The storage, the portioning, and the compartmentation of phase-encoded ("prepared") magnetization offer a variety of ways for the design of improved or entirely new NMR imaging experiments to be discussed below as well as in subsequent papers.

BASIC SEQUENCES FOR NMR IMAGING USING STIMULATED ECHOES

Some basic pulse and gradient sequences for NMR imaging using stimulated echoes are shown in Fig. 3. In their most simple form STEAM sequences contain only a single slice-selective (90°) pulse and give rise to the observation of a single stimulated echo. The corresponding sequences shown in Figs. 3a–c demonstrate the great functional flexibility of the pulses as well as the variability of the gradient switches. Frequency-selective pulses of arbitrary pulse shapes are represented by rf

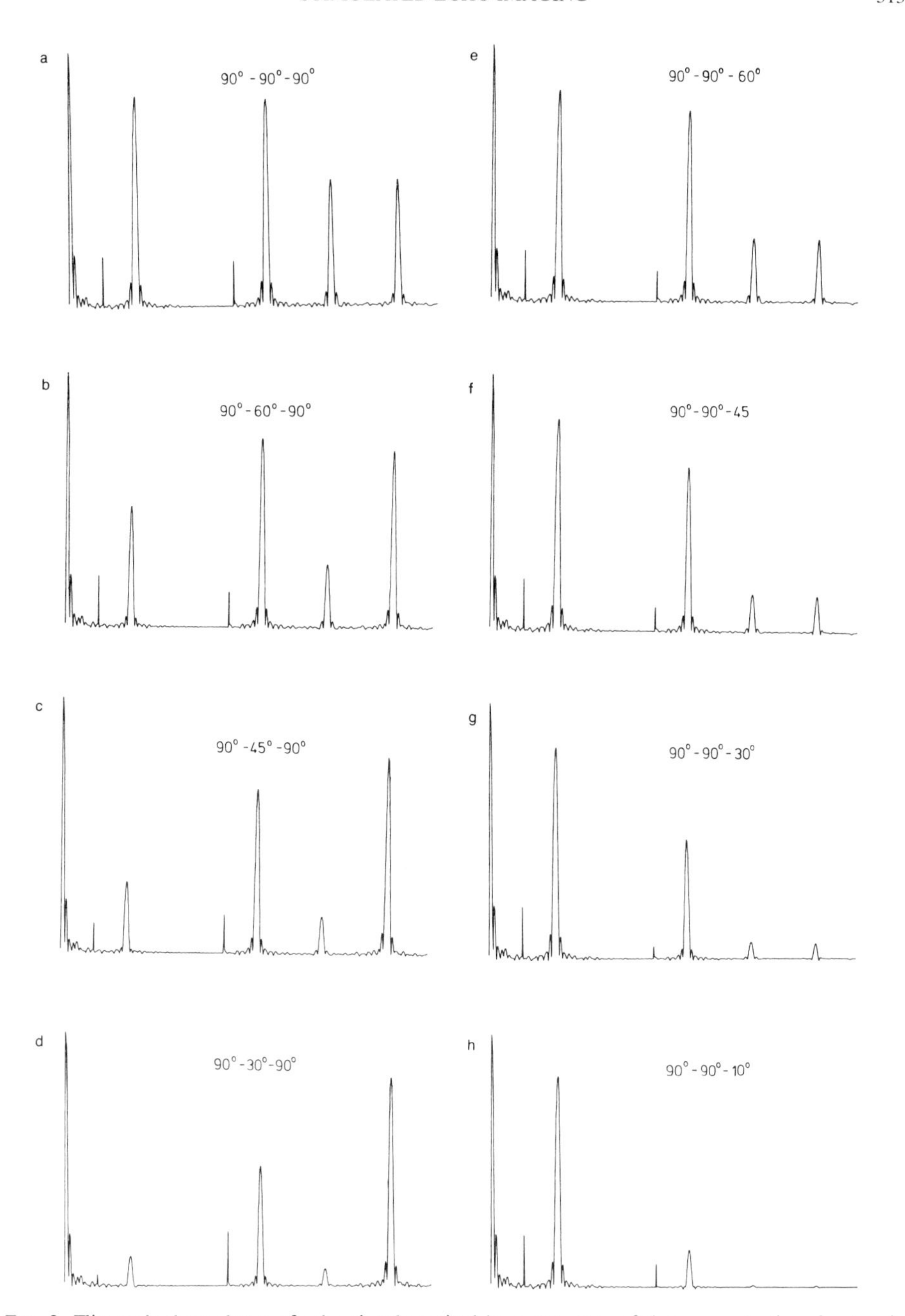

FIG. 2. Flip-angle dependence of echo signals excited by a sequence of three nonselective rf pulses in the presence of a gradient. Amplitudes correspond to magnitude values. (a) 90°–90°–90° pulse sequence. (b–d) Three-pulse sequence with flip angles of 60, 45, and 30° for the second pulse, respectively. (e–h) Three-pulse sequence with flip angles of 60, 45, 30, and 10° for the third pulse, respectively.

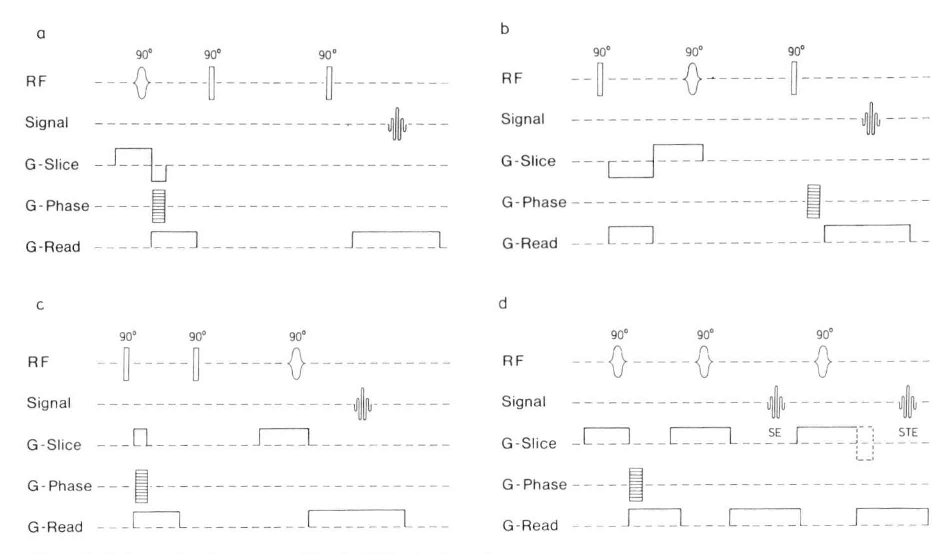

FIG. 3. Schematic diagrams of basic STEAM imaging sequences. (a–c) Sequences using a single slice-selective rf pulse in the first, second, or third position, respectively. (d) Sequence using three slice-selective rf pulses. Broken lines indicate refocusing parts of the slice gradient which are self-compensating and thus may be omitted.

pulses with a Gaussian pulse shape. They may be chemical-selective pulses or, in combination with a gradient, slice selective or so-called zoom pulses. Nonselective pulses of arbitrary shapes are represented by pulses with rectangular pulse shapes and are defined by the fact that their application excites the entire NMR spectrum. In many cases nonselective pulses may be advantageously replaced by frequency-selective pulses of various functions.

The three intervals of the STEAM sequence may be used to minimize situations in which different gradients have to be switched simultaneously. For example, proper refocusing of the slice-selection gradient may take place immediately after or before application of the respective rf pulse by inversion (compare Figs. 3a, b), or by using a defocusing gradient of the same sign in the first interval (compare Fig. 3c). It is worth noting that the diagrams shown in Figs. 3b, c represent new gradient schemes not applicable with SE-NMR imaging sequences because defocusing gradients are applied prior to the combined use of the slice-selection pulse and gradient. Figure 3d gives an example for a STEAM sequence comprising three slice-selective pulses with the simultaneous observation of the primary SE and STE signals. As is demonstrated by the broken lines, some parts of the various refocusing gradients may become self-compensating thereby simplifying the resulting gradient sequence.

As already indicated by the scheme shown in Fig. 3d slice-selective rf pulses within STEAM imaging sequences may be arbitrarily combined even independent of the directions of the gradients. Furthermore slice-selective pulses may have an identical length/bandwidth as is demonstrated by computer simulations shown in Figs. 4a–d. The diagrams refer to magnetization profiles as obtained by a numerical solution of the Bloch equations for slice selection by some typical STEAM imaging experiments. For details of the calculations see Ref. (*1*). The profiles of interest correspond to the $M_{y'}$ components of the transverse magnetization (solid lines). For comparison with the STEAM sequences, Fig. 4a depicts the case of a single slice-selective 90°(x') pulse with a Hamming pulse shape and refocusing by inversion of the gradient. Figures 4b, c contain pulse and gradient sequences with the slice-selective pulse in the third and second positions, respectively, while Fig. 4d shows the combination of two slice-selective pulses of identical length. In all cases similar slice profiles could have been obtained without inducing distortions of magnetizations close to the excited slice and thus without the need for a second "compensating" experiment.

The phase-encoding gradient as well as the read gradients are used in the same way as for conventional SE-NMR imaging. However, the position of the phase-encoding gradient within the STEAM sequence will select its action on the SE and/ or STE signals. If applied in the first interval as demonstrated in Fig. 3d, then both the SE and STE signals are similarily affected. In contrast, an application after the second pulse will affect only the SE signal, while an application after the third pulse will affect only the STE signal. Thus it becomes possible to use different phase-encoding gradients for independent imaging with the SE or STE signals, respectively. Appropriate experiments may be concerned with the simultaneous imaging of a plane with normal gradients and with magnified gradients producing a so-called zoom image. Such a technique is demonstrated in the pulse and gradient scheme shown in Fig. 5a. Another example may be the use of two different phase-encoding gradients for encoding spatial and flow information separately into the SE and STE signals.

MULTISLICE IMAGING USING STIMULATED ECHOES

Conventional multislice imaging techniques rely on repetitions of the entire pulse and gradient sequence with different irradiation frequencies. Although multislice SE-NMR imaging may be performed as a time-sharing experiment, the time needed for a single slice is of the order of 100 ms which leads to the recording of 10 slices with repetition times of about 1 to 1.5 s. STEAM imaging sequences provide a very efficient alternative: a general scheme for multislice imaging even of arbitrarily spaced slices is shown in Fig. 5b. The entire imaging volume is prepared by two

---

FIG. 4. Plane selection in STEAM imaging using selective 90°(x') pulses (Hamming profile) in combination with a z gradient of strength 10 mT m^{-1}. The pulse and gradient sequences are symbolically indicated at the top of each figure. The durations of the intervals t_1 and t_2 are of the order of 10 to 30 ms depending on the gradients which have to be applied within a complete imaging sequence. In the lower parts, computer-simulated magnetization profiles are shown which display the magnetizations M_z (dotted line), $M_{x'}$ (dashed line), and the observed magnetization $M_{y'}$ (solid line) as a function of the spatial coordinate z. The equilibrium magnetization M_0 is normalized to one. Relaxation effects (T_1 = 1 s, T_2 = 0.05 s) are included. (a) Application of a single selective 90° pulse with proper refocusing of the gradient. (b) STEAM imaging sequence with a single selective 90° pulse in the third position. (c) STEAM imaging sequence with a single selective 90° pulse in the second position using defocusing prior to application of the rf pulse. (d) STEAM imaging sequence with two selective 90° pulses of identical length (i.e., identical bandwidth) in the first and third positions.

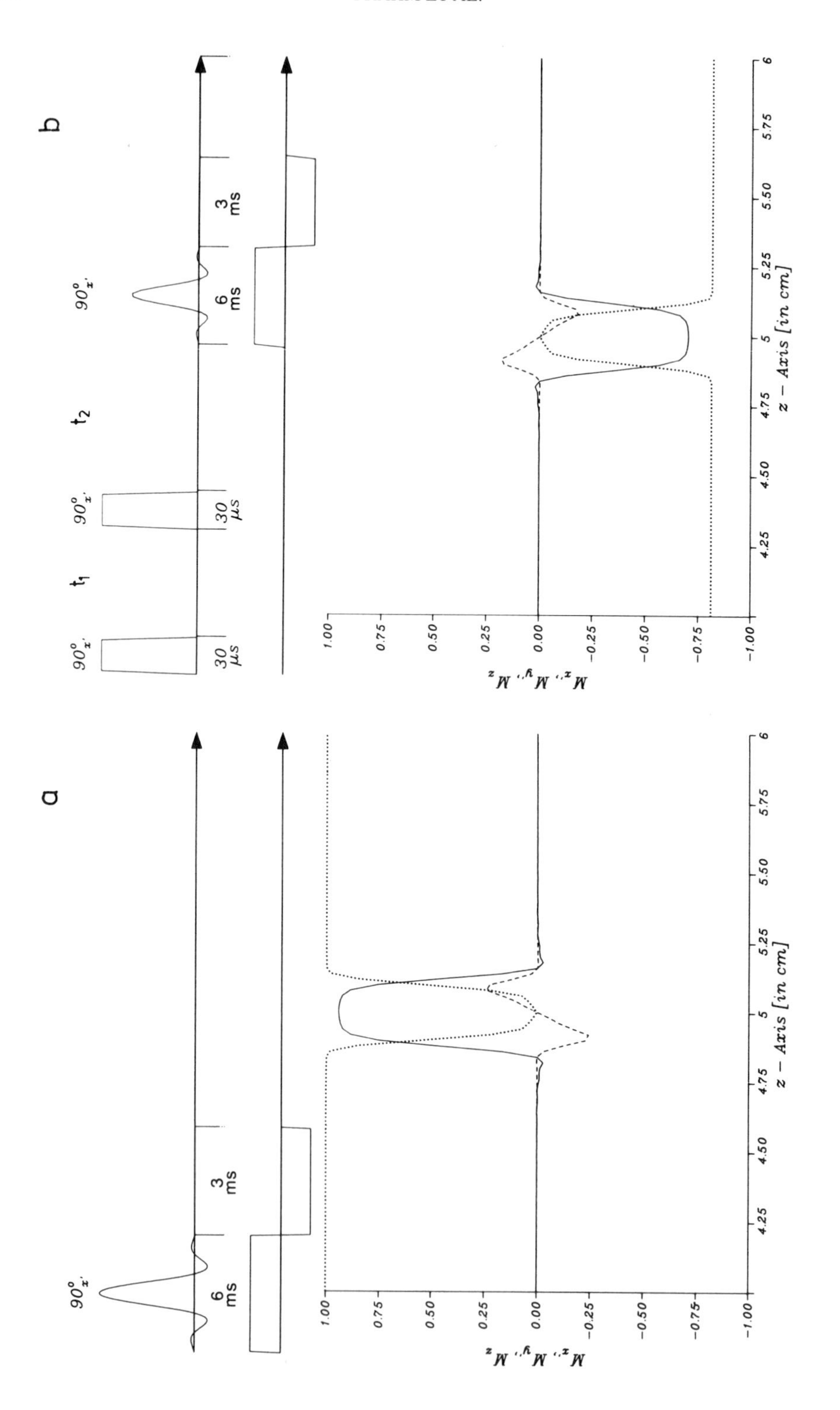
a
b
90°x'
t1
t2
30 µs
6 ms
3 ms
Mx', My', Mz
z − Axis [in cm]
1.00
0.75
0.50
0.25
0.00
−0.25
−0.50
−0.75
−1.00
4.25
4.50
4.75
5
5.25
5.50
5.75
6

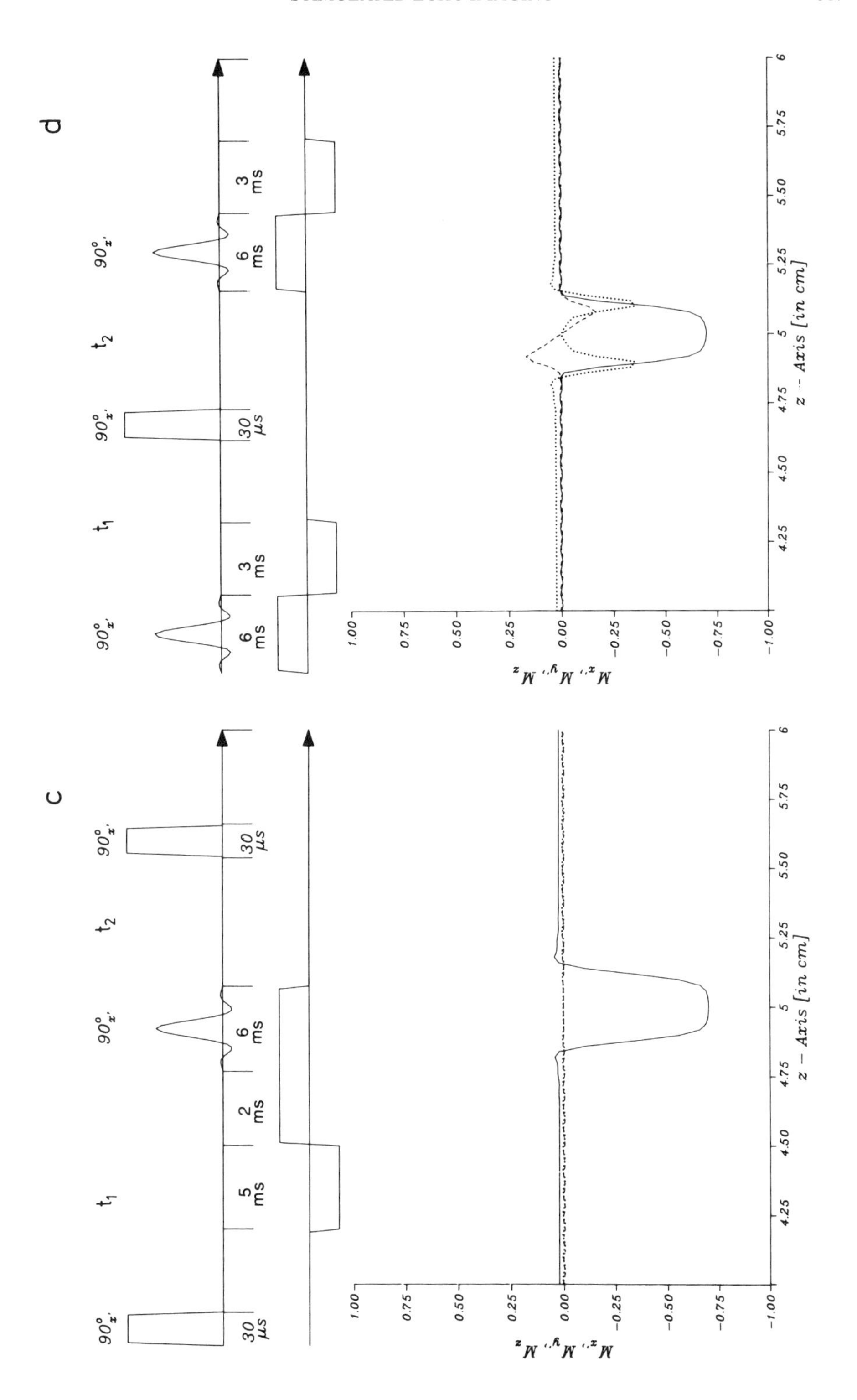

c
d
$90^o_{x'}$
t_1
t_2
30 μs
5 ms
2 ms
6 ms
3 ms
$M_{x'}, M_{y'}, M_z$
z - Axis [in cm]

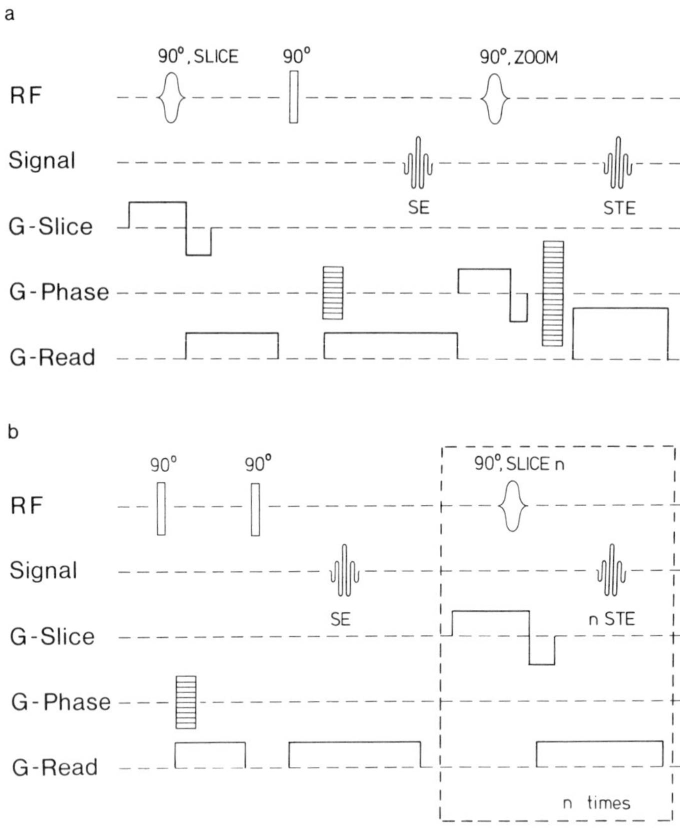

FIG. 5. Schematic diagrams of STEAM imaging sequences. (a) Simultaneous acquisition of an overview and zoom image using the SE and STE signals, respectively. (b) Multislice imaging using stimulated echoes.

leading nonselective rf pulses. They give rise to a SE signal in the second interval which, if wanted, may be used to create a so-called "transmission" image without slice selection. The final part of the sequence performs the slice selection and the acquisition of the corresponding STE signal. Only this part is repeated n times to create n cross-sectional images out of the entire excited volume. It should be noted that the irradiation frequency is selected prior to application of the slice selective pulses and is reset immediately after its termination to enable data readout with respect to the normal frequency, i.e., spatial position. The time needed for a single slice is of the order of 20 to 30 ms; the rf power corresponds to a single 90° pulse. Subsequent readout of different slices becomes possible due to the property of the STEAM imaging sequences of "storing" magnetization prepared by the first two pulses into longitudinal magnetization which then decreases rather slowly with T_1. Slice-selective "third" pulses induce only STE signals which belong to magnetization of the corresponding frequency interval. All other magnetization is unaffected and remains "stored."

The major advantages of multislice STEAM imaging are threefold: low rf power and reduced gradient switching, reduced measuring times or increased number of slices, and the benefit of avoiding 180° pulses which allows one to image directly

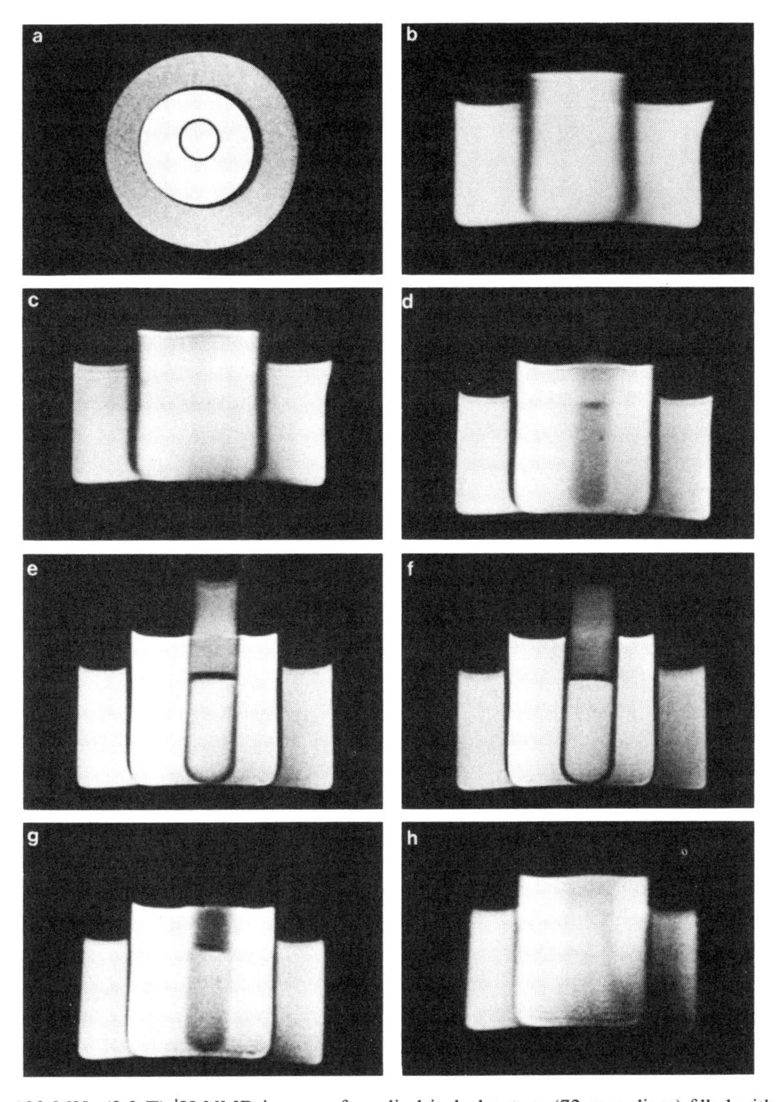

FIG. 6. 100 MHz (2.3 T) ^{1}H NMR images of a cylindrical phantom (72 mm diam) filled with olive oil containing a bottle (40 mm diam) of water and another tube (10 mm diam) filled with both water (on bottom) and oil (on top). All images are "raw" images which have not been subjected to a smoothing or filtering procedure. (a) Image of a horizontal slice of thickness 4 mm. (b–h) Transaxial multislice STEAM images of directly neighboring slices (spacing 4 mm, thickness 4 mm) selected out of a series of 10 images. Although the entire acquisition time, i.e., the time needed for recording a single set of "projections' of all 10 slices, was only 250 ms, the repetition time has been set to 1 s due to limitations of our present computer facilities.

neighboring slices. If desired, slice distances, slice thicknesses, and image sizes may be varied from slice to slice within the sequence. It should be emphasized that multislice STEAM imaging provides a solution to the problem of economic imaging in high magnetic fields where rf power absorption becomes a problem for multiple applications of SE-NMR imaging sequences. The reduced measuring time per slice improves the imaging efficiency insofar as faster repetition times become possible. Actual limitations might be due to slow data transfer rates and/or small computer memory sizes. NMR limitations are only determined by the spin–lattice relaxation times. However, the recording of 16 slices typically requires a measuring interval of about 0.4 s which is still short enough to account for signal losses because of water proton relaxation with a T_1 of the order of 1 s or higher in magnetic fields of about 2 T. On the other hand, a region scanned by 16 neighboring slices corresponds to a volume of 8 cm thickness if slices of only 5 mm thickness are chosen. Since proper computer equipment will allow repetition times of about 0.5 s for the above example, a set of 16 (overview) images, each based on 128 different projections, may be recorded in about 1 min.

Experimental results demonstrating multislice images of directly neighboring slices are summarized in Fig. 6. The images were obtained at 100 MHz proton NMR frequency using an imaging/spectroscopy system (Bruker BNT 100) based on a 2.3 T superconducting magnet with a 40 cm bore. The measuring time was about 4 min corresponding to the recording of 256 different phase-encoded projections with a repetition time of 1 s. Thus no averaging or compensation experiment was employed. The SE transmission image is not shown. Instead, Fig. 6a displays a cross-sectional slice perpendicular to the multislice images to illustrate the arrangement of the three concentric bottles: the outer one (inner diameter of 72 mm) has been filled with water, the middle one (40 mm diam) with olive oil, and the inner one (10 mm diam) contained both oil (on top) and water (on bottom). The chemical-shift artifact is clearly demonstrated in this plane. The thickness of the slices in all cases as well as the distance of the centers of the multislice images was 4 mm. This can be judged by comparing the number of multislice images affected by the inner tube (i.e., Figs. 6d–g) with Fig. 6a and the tube diameter of 10 mm.

CONCLUSIONS

A new principle for NMR imaging has been developed which is based on the acquisition of stimulated echo signals. The general characteristics of the technique are described and first experimental results are presented demonstrating an improved method for multislice imaging. STEAM imaging arises as a versatile tool for a variety of applications in NMR imaging. General advantages are concerned with improvements in security by reducing the applied rf power, in efficacy of imaging investigations by increasing the quality and quantity of information per time, and in parametric resolution as will be described in subsequent papers on chemical-shift-selective imaging, T_1 imaging, and diffusion imaging. Stimulated echo imaging will arise as a new tool for multipurpose investigations leading to a considerable gain in accuracy and specificity in NMR imaging.

ACKNOWLEDGMENTS

This work has been supported by the Bundesminister für Forschung und Technologie (BMFT) of the Federal Republic of Germany (Grant 01 VF 242). Computer simulations were performed using the facilities of the Gesellschaft für Wissenschaftliche Datenverarbeitung Göttingen (NAG software).

REFERENCES

1. J. FRAHM AND W. HÄNICKE, *J. Magn. Reson.* **60,** 320 (1984).
2. A. HAASE AND J. FRAHM, *J. Magn. Reson.* **64,** 94–102 (1985).
3. A. HAASE AND J. FRAHM, *J. Magn. Reson.*, in press.
4. K. D. MERBOLDT, W. HÄNICKE, AND J. FRAHM, *J. Magn. Reson.*, in press.
5. J. FRAHM, A. HAASE, D. MATTHAEI, K. D. MERBOLDT, AND W. HÄNICKE, *J. Magn. Reson.*, in press.
6. E. L. HAHN, *Phys. Rev.* **80,** 580 (1950).
7. D. E. WOESSNER, *J. Chem. Phys.* **34,** 2057 (1961).
8. J. E. TANNER, *J. Chem. Phys.* **52,** 2523 (1970).

JOURNAL OF MAGNETIC RESONANCE **66**, 530–535 (1986)

NMR Images of Solids

G. C. CHINGAS,* J. B. MILLER,† AND A. N. GARROWAY‡

Chemistry Division, Code 6120, Naval Research Laboratory, Washington, D.C. 20375-5000

Received September 19, 1985

A way of obtaining two-dimensional NMR images of solids using multiple-pulse line narrowing is demonstrated. The method is adaptable to any fast recovery solid-state spectrometer having 2D FT NMR software and provisions for applying magnetic gradients. The use of molecular mobility as an NMR image contrast mechanism is demonstrated.

In recent years, NMR imaging has become an important technique in biological and medical studies (*1*). Progress in applying this method to materials research, on the other hand, has been hampered by difficulties induced by broad NMR linewidths associated with the solid state. For the most frequently studied nuclei, the resonance broadening arises from spin–spin dipolar couplings. Although images have been obtained in certain favorable solids where molecular motions greatly reduce the dipolar coupling (*2*), in general some technique must be invoked to surmount this obstacle. Thus, rudimentary, one-dimensional solid images (*3, 4*) have been obtained by using coherent averaging (*5*) to suppress homonuclear dipolar couplings; by observing low-abundance nuclei with heteronuclear dipolar decoupling (*6*); by multiple-quantum, NMR (*7*); and by incrementing the field gradient during a fixed evolution time to phase encode a free induction decay (*8*).

The approach we take here is to create two-dimensional images of abundant nuclei (protons) by a variation of the multiple-pulse method (*3*). We use the technique of two-dimensional Fourier imaging (*9*) which has been altered with the MREV-8 line-narrowing sequence (*10*) to reduce the effective size of the dipolar coupling during the evolution (t_1) and detection (t_2) periods. Echoes are created in the t_1 and t_2 periods so that a simple magnitude calculation will provide a properly phased image (*11*). A further important feature is the storage of magnetization to suppress precession in the static field: this storage obviates the need for rapid gradient switching (*6*) and can provide a convenient contrast mechanism based on molecular mobilities. This sequence is intended to be a relatively easy to implement method for imaging abundant nuclei in organic solids.

As Fig. 1 shows, the experiment begins in the presence of a stabilized x gradient with the application of a $\pi/2$ prepulse. The magnetization now precesses about the effective field of the MREV-8 cycle for a time t_1. Between successive experiments in

* Present address: GE NMR Instruments, 255 Fourier Ave., Fremont, Cal. 94539.
† NRC–NRL postdoctoral fellow.
‡ To whom correspondence should be addressed.

0022-2364/86 $3.00

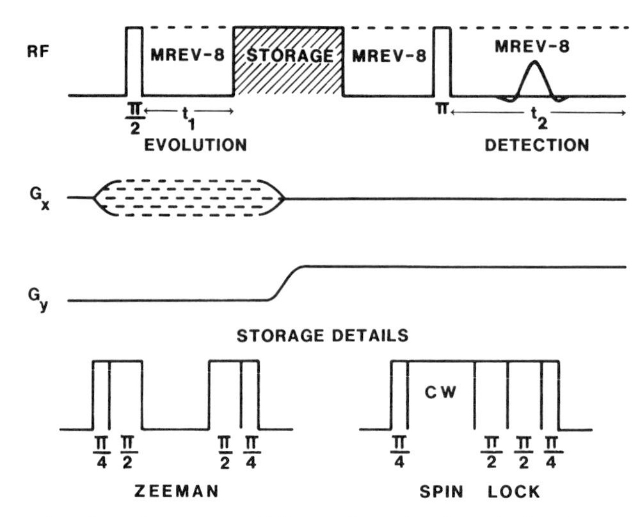

FIG. 1. Solid-state imaging pulse sequence. This two-dimensional Fourier imaging scheme generates spin echoes in both t_1 and t_2 domains. A refocus π pulse forms the t_2 echo actually acquired, while t_1 "echoes" are formed by incrementing G_x from negative to positive values so that maximum signal occurs in the center of the evolution data set where $G_x = 0$. The sequence also incorporates a storage interval which holds the relatively short-lived signal while the gradients are switched. The two types of storage used in these experiments are illustrated, and provide signal contrast as discussed in the text. The phases of the $\pi/2$ and cw rf pulses are cycled to reduce artifacts and alternately transport each signal phase across the storage period. The phases of the $\pi/4$ pulses are chosen to be orthogonal to the effective field of the MREV-8 cycle.

the 2D data set, t_1 is held constant while the magnitude of the gradient is incremented by ΔG from $-G_{max}$ to $+G_{max}$ effectively creating a t_1 echo. In this manner only evolution due to the gradient is observed in the t_1 dimension.

As indicated in Ref. (*8*), any two-dimensional sequence, even without dipolar decoupling, which holds the t_1 interval constant will remove dipolar broadening from the ω_1 domain. (Indeed such an approach could be improved by multiple-quantum imaging (*7*) which increases the rate of phase evolution during the t_1 interval.) However, without dipolar decoupling the rapid signal decay during t_1 reduces the signal-to-noise ratio in the final image. Such sensitivity considerations motivate our choice of a line-narrowing sequence to prolong the NMR signal during the t_1 evolution. The length of t_1 and the size of the gradient are chosen to insure adequate spatial resolution Δx and image window size L. The specific relations are

$$t_1 = \frac{\pi}{\gamma G_{max} \Delta x}, \quad [1]$$

and

$$\Delta G = \frac{2\pi}{\gamma t_1 L}. \quad [2]$$

The storage interval begins by aligning one of the two quadrature phases of the magnetization along the "storage field." This may be either the static field B_0 for Zeeman storage or the rotating frame field B_1 for spin-lock storage. During this time precession in the static field is suppressed, but T_1 or $T_{1\rho}$ decay processes will be active so signals from portions of the sample with short relaxation times will decay. Although decay during storage can be advantageously used to provide mobility contrast as shown below, the primary function of magnetization storage is to provide time for gradients and eddy currents to settle, obviating the need for rapid gradient switching. Unlike the case for liquid systems, this time may be appreciably longer than the signal lifetime under multiple-pulse conditions. At best, multiple-pulse techniques can prolong the signal lifetime only up to $T_{1\rho}$; in certain materials this may already be substantially shorter than the gradient settling time. Even for solids with long $T_{1\rho}$, lifetimes may still be reduced by inhomogeneous broadening mechanisms. Two of the most likely are chemical-shift anisotropy and the presence of a variety of chemically distinct species: e.g., aromatic and aliphatic species. On this basis, one would estimate an 8–10 ppm linewidth which, even at the relatively low proton Larmor frequency of 57 MHz, corresponds to a signal lifetime on the order of milliseconds. This time is shorter than the 5 ms required for the eddy currents and their fields to stabilize in our aluminum probe with 1/16 in. (1.6 mm) walls. Because only a single quadrature phase can be stored at a time, the other is lost. Both phases are recovered by phase cycling (*12*), but at a cost of twice the number of acquisitions.

At the end of the storage period when the gradients have stabilized, the magnetization is again set precessing about the MREV-8 effective field. Following a suitable delay the magnetization is refocused by a π pulse to generate an echo in the t_2 acquisition period.

Many of the steps discussed above involve phase cycling the pulses and acquisition, which is crucial for obtaining good image quality. In particular, phase cycling is necessary for the reconstruction of quadrature information during the storage period and for the removal of certain artifacts. We were concerned with four types of artifacts: (1) baseline offset associated with receiver ringing and magnetization spin-locked along the MREV-8 effective field; (2) signals generated by T_1 recovery during the storage period; (3) secondary echoes from mobile species; (4) quadrature "ghosting" due to the direction of the MREV-8 effective field.

Removal of this latter artifact requires some care. During the multiple-pulse sequence, the spins precess about an effective field which for most sequences is not colinear with B_0. Since ordinary rf phase shifting corresponds to a pure rotation about the B_0 axis, it is not always possible to remove multiple-pulse artifacts by straightforward rf phase shifting. However, as previously realized (*12*), MREV-8 is special in that the effective field is along the $(-1, 0, 1)$ direction (in the xz plane of the rotating frame). Hence, magnetization may be rotated from the xy plane to a plane perpendicular to the effective field and back by $\pi/4$ pulses with phases of $(0, -1, 0)$ and $(0, 1, 0)$ ($-y$ and y, respectively). In particular, such $\pi/4$ pulses are inserted at the beginning and end of the storage interval to compensate for such tilted effective fields. For the same reason, quadrature signals detected during the MREV-8 sequence are imbalanced (*13*); further phase cycling is required to equalize their average value. Compensation for all the above effects required 32 acquisitions to complete the phase cycling.

The experiments reported here were performed on a home-built spectrometer controlled by a Nicolet/GE data system. The spectrometer operates at a proton frequency of 57 MHz. Gradients are controlled using custom assembly language software drivers to increment the output of the digital-to-analog converters (DAC) in a Nicolet 293B pulse programmer. These outputs are amplified and used to drive the shim coils of an electromagnet. Coil currents of 400 mA generate 10 kHz/cm gradients.

The probe built for the multiple-pulse imaging experiments employs a $5\frac{1}{2}$ turn 18 mm inside diameter coil, and is based on the overcoupling scheme (*14*). The deviation in B_1 homogeneity is less than 2% throughout a cylindrical sample volume of 0.6 cm^3. The $\pi/2$ pulses are 4.5 μs long ($B_1 \approx$ 13 G) from a 100 W amplifier.

The cycle time for the MREV-8 sequence was 54 μs. The chemical-shift scaling (experimental = 1.80; theoretical = 1.86) was constant from -15 to $+15$ kHz resonance offset. No major change in the efficiency of dipolar decoupling was observed for adamantane over the frequency range used in these experiments.

With the pulse sequence of Fig. 1, we obtain two-dimensional proton images of a sample containing a neoprene rubber annulus and a 5 mm tube of adamantane. A photograph of the sample, end on, is shown in Fig. 2. A contour plot of the results from an experiment employing storage along B_0 is shown in Fig. 3a. Both the rubber annulus and the adamantane are well defined in the NMR image, viewed from the same direction as in the photograph. The spatial resolution of the two materials is the

FIG. 2. End-on view of sample used in imaging experiment, showing a 5 mm tube of adamantane surrounded by a 12 mm o.d. neoprene hose section inside a 15 mm glass tube.

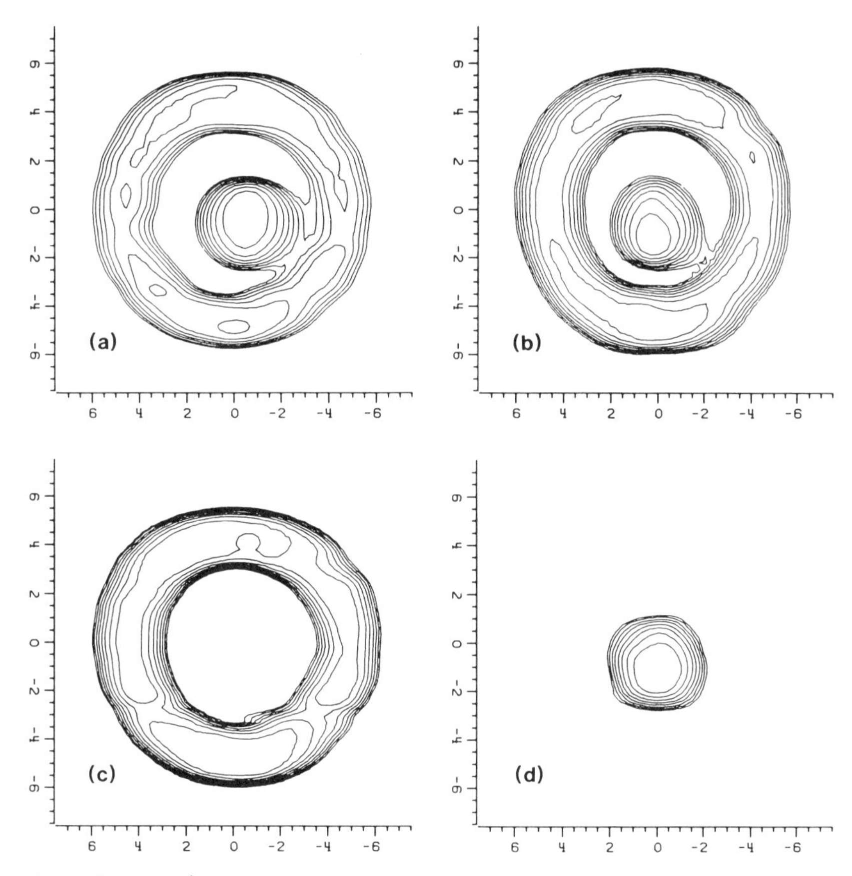

FIG. 3. Multiple-pulse ^1H NMR images of the sample shown in Fig. 2. The constant intensity contours forming the images differ by a factor of $2^{-1/4} = 0.84$. Figure 3a is obtained using 5 ms Zeeman storage and a 1 s repetition time, while Fig. 3b uses as a 5 ms spin lock and 2 s repetition time; these images are essentially equivalent, since both storage schemes have equal efficiency. In Fig. 3c the repetition is reduced to 250 ms, and the adamantane signal saturates, leaving only the faster recovery rubber image. In Fig. 3d, the repetition time is restored to 2 s, but the spin-lock time is extended to 50 ms; here the rubber signal decays because of its relatively short $T_{1\rho}$. The scales on the images are in millimeters.

same since they have nearly the same linewidths in these multiple-pulse experiments (approximately 200 Hz). This would not be the case under free precession conditions where the two materials' linewidths differ by more than an order of magnitude. The resolution in the ω_1 and ω_2 directions is not quite equal, but could be made so by increasing the value of t_1 in accord with Eq. [1]. This was not done because some signal intensity would have been lost.

The results of experiments employing storage along a spin-lock field are shown in Figs. 3b–d. In Fig. 3b the sample was spin-locked for 5 ms and the delay between successive acquisitions was 2 s. The result is comparable to that in Fig. 3a, demonstrating equivalence of the Zeeman and spin-lock storage results for this specimen.

Certain timing parameters of the experiment used to obtain Fig. 3b may be manipulated to discriminate between the two materials present in the sample. For example, in Fig. 3c the delay between successive acquisitions has been reduced to 250 ms. Because its T_1 is on the order of 1 s, adamantane is saturated and not observed in this experiment; this procedure could have been carried out using Zeeman storage with similar results. In contrast, Fig. 3d shows the image obtained where the spin-lock time is 50 ms while the delay between successive acquisitions is again 2 s. This time signal from the neoprene rubber vanishes because of $T_{1\rho}$ decay during the spin-lock storage.

We have demonstrated the feasibility of using coherent averaging techniques to obtain 2D NMR images of solids. The signal strength of the image can be weighted relative to molecular mobility. As it stands, the experiment is adaptable to most spectrometers having some line-narrowing capability and 2D FT NMR software. Only rudimentary gradient control facilities are required. Within reason, the experiment tolerates long settling times for gradient switching, so special magnet and gradient designs are unnecessary. The 2D FT NMR software used for image reconstruction is a feature supplied with most modern commercial spectrometers, so the only software development required is for simple drivers to set and increment gradients.

ACKNOWLEDGMENT

J.B.M. acknowledges an NRC/NRL cooperative postdoctoral associateship.

REFERENCES

1. (*a*) P. Mansfield and P. C. Morris, "Advances in Magnetic Resonance," Suppl. 2, "NMR Imaging in Biomedicine," Academic Press, New York, 1982; (*b*) E. R. Andrew, *Acc. Chem. Res.* **16,** 114 (1983).
2. B. H. Suits and B. White, *Solid State Commun.* **50,** 291 (1984).
3. (*a*) P. Mansfield, P. K. Grannell, A. N. Garroway, and D. C. Stalker, "Proceedings, 1st Spec. Colloque Ampere" (J. W. Hennel, Ed.), p. 16, Krakow, 1973; (*b*) P. Mansfield and P. K. Grannell, *Phys. Rev. B* **12,** 3618 (1975).
4. R. A. Wind and C. S. Yannoni, *J. Magn. Reson.* **36,** 269 (1979).
5. (*a*) M. Mehring, "NMR-Basic Principles and Progress," Vol. 11, "High Resolution NMR in Solids," 2nd ed., Springer-Verlag, New York, 1983; (*b*) U. Haeberlen, "Advances in Magnetic Resonance," Suppl. 1, "High Resolution NMR in Solids," Academic Press, New York, 1976.
6. N. M. Szeverenyi and G. Maciel, *J. Magn. Reson.* **60,** 460 (1984).
7. A. N. Garroway, J. Baum, M. G. Munowitz, and A. Pines, *J. Magn. Reson.* **60,** 337 (1984).
8. S. Emid and J. H. N. Creyghton, *Physica B* **128,** 81 (1985).
9. A. Kumar, D. Welti, and R. R. Ernst, *J. Magn. Reson.* **18,** 69 (1975).
10. (*a*) P. Mansfield, *J. Phys. C* **4,** 1444 (1971); (*b*) W.-K. Rhim, D. D. Elleman, and R. W. Vaughan, *J. Chem. Phys.* **58,** 1772 (1973); (*c*) W.-K. Rhim, D. D. Elleman, and R. W. Vaughan, *J. Chem. Phys.* **59,** 3740 (1973).
11. (*a*) A. Bax, A. F. Mehlkopf, and J. Smidt, *J. Magn. Reson.* **35,** 373 (1979); (*b*) A. Bax, R. Freeman, and G. A. Morris, *J. Magn. Reson.* **43,** 333 (1981).
12. P. Caravatti, G. Bodenhausen, and R. R. Ernst, *Chem. Phys. Lett.* **89,** 363 (1982).
13. W.-K. Rhim, D. P. Burum, and R. W. Vaughan, *Rev. Sci. Instrum.* **47,** 720 (1976).
14. G. C. Chingas, *J. Magn. Reson.* **54,** 153 (1983).

JOURNAL OF MAGNETIC RESONANCE 67, 258–266 (1986)

FLASH Imaging. Rapid NMR Imaging Using Low Flip-Angle Pulses

A. HAASE, J. FRAHM, D. MATTHAEI, W. HÄNICKE, AND K.-D. MERBOLDT

Max-Planck-Institut für biophysikalische Chemie, Postfach 2841, D-3400 Göttingen, Federal Republic of Germany

Received October 2, 1985; revised November 12, 1985

A new method for rapid NMR imaging dubbed FLASH (*f*ast *l*ow-*a*ngle *sh*ot) imaging is described which, for example, allows measuring times of the order of 1 s (64 × 128 pixel resolution) or 6 s (256 × 256 pixels). The technique takes advantage of excitation pulses with small flip angles eliminating the need of waiting periods in between successive experiments. It is based on the acquisition of the free induction decay in the form of a gradient echo generated by reversal of the read gradient. The entire imaging time is only given by the number of projections desired times the duration of slice selection and data acquisition. The method results in about a 100-fold reduction in measuring time without sacrificing spatial resolution. Further advantages are an optimized signal-to-noise ratio, the applicability of commercial gradient systems, and the deposition of extremely low rf power. FLASH imaging is demonstrated on phantoms, animals, and human extremities using a 2.3 T 40 cm bore magnet system. ^{1}H NMR images are obtained with variable relaxation time contrasts and without motional artifacts.

INTRODUCTION

Spatially resolved nuclear magnetic resonance techniques are now extensively used for noninvasive investigations of living matter in biology and medicine (*1*). However, measuring times of several minutes lead to motional artifacts within NMR images and give no access to the study of fast physiological processes. More rapid techniques are either limited in spatial resolution or signal-to-noise ratio (SNR) (*2*), or require the application of a large number of intense radiofrequency (rf) excitation pulses (*3*) exceeding safety guidelines (*4*) at magnetic field strengths of 1 to 2 T. The first rapid imaging technique applicable to high-field imaging is rapid stimulated-echo imaging (STEAM) (*5*). Although the method inherently has the advantages of the STEAM imaging method (*6*) such as giving access to rapid T_1 images or CHESS (chemical shift selective) images, it suffers from low SNR. This is mainly because subsequently excited stimulated echoes are attenuated by T_1 relaxation.

Here we present a new rapid imaging technique dubbed FLASH (*f*ast *l*ow-*a*ngle *sh*ot) imaging, which results in (i) optimized SNR, (ii) about a 100-fold reduction of the measuring time, (iii) no loss of spatial resolution, and (iv) low rf power deposition (*7, 8*). Since the FLASH sequence works continuously without internal waiting times, one may choose arbitrary compromises between spatial resolution, time resolution, and SNR. In addition, one may record sequential series of images ("movies") with image repetition times given by the individual measuring times.

0022-2364/86 $3.00

METHOD

NMR imaging experiments are based on the acquisition of free induction decays (*9*), spin echoes (*10*), or stimulated echoes (*6*). Since the FID requires only a single rf excitation pulse this type of signal turns out to be best-suited for rapid NMR imaging. An appropriate rf pulse and gradient sequence is shown in Fig. 1. The important difference between conventional imaging techniques and this rapid sequence is the use of rf pulses with low flip angles. Low-angle pulses have already been discussed for conventional NMR spectroscopy experiments in order to optimize the SNR per measuring time (*11, 12*). However, repetition times of the order of tens of milliseconds, i.e., flip angles of the order of 15°, have not been applied because spectroscopic FID signals normally have durations of several hundreds of milliseconds. This situation becomes different for NMR imaging where magnetic field gradients reduce the length of the FID to some milliseconds.

For example, using a flip angle of 15°, the intensity of the FID corresponds to 25% (sin 15°) of the maximum amplitude excited by a 90° pulse. In contrast to conventional imaging sequences, 96.5% (cos 15°) of the longitudinal magnetization remains unaffected and thus is available for immediate subsequent excitations (see Fig. 1). After termination of the rf pulse the slice-selective gradient (*G* slice) is inverted for proper refocusing of the transverse magnetization (*13*). The in-plane spatial discrimination may be achieved either by rotating a magnetic field gradient in a number of steps according to the projection reconstruction algorithm (*14*) or by applying a fixed "read" gradient (*G* read) and a perpendicular phase-encoding gradient (*G* phase) of variable strength according to the 2D Fourier imaging method (*15*). The read gradient is inverted prior to the data acquisition period leading to a so-called gradient or field echo. Immediately after acquisition of the data the experiment is repeated with a repetition time given by the time needed for slice selection and data acquisition. Thus the duration of the entire imaging experiment is reduced by the same factor as conventional repetition times of the order of 1 s are reduced to about 10–20 ms. After application of the first 20–40 excitation pulses, the spin system reaches a steady state where the loss

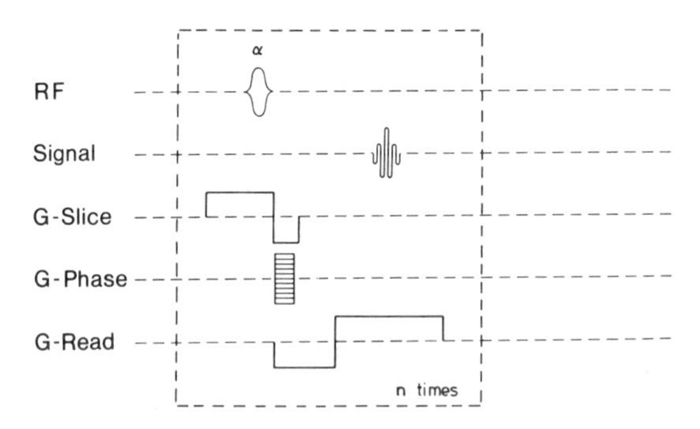

FIG. 1. Radiofrequency pulse and magnetic field gradient sequence for rapid FLASH NMR imaging. The method employs slice selective excitation pulses with flip angles of the order of 15°. The NMR free induction decay is detected in the form of a gradient echo after reversal of the read gradient. The sequence is repeated *n* times recording *n* projections with different phase-encoding gradients. No waiting times are required between subsequent excitations.

of longitudinal magnetization by excitation is compensated for by spin–lattice relaxation during the imaging sequence. The theoretical value of this steady state as well as its dependence on T_1, on the repetition time, and on the flip angle have already been discussed by Waugh (*12*). After reaching the steady state, experiments can be performed without time limits. For example, series of sequential images may be recorded describing the time course of physiological processes in tissues.

EXPERIMENTAL RESULTS

Biomedical applications of NMR imaging mainly rely on tissue contrasts due to differences in relaxation times. Obviously, new and in particular rapid NMR imaging sequences should retain this sensitivity to relaxation times. Tissue contrast in FLASH images is determined by the spin–lattice relaxation time T_1 and the effective spin–spin relaxation time T_{2eff}. Both effects will be experimentally demonstrated below. ^1H NMR images are recorded using a combined imaging/spectroscopy system supplied with a 40 cm bore 2.3 T magnet (Bruker Medizintechnik, Karlsruhe, West-Germany). The duration of the frequency-selective rf pulse (compare Fig. 1) is 2 ms. In the presence of a magnetic field gradient of strength 0.5 G/cm this pulse excites transverse magnetization of a plane of about 4 mm thickness perpendicular to the direction of the gradient. The time interval from the center of the rf pulse to the maximum of the gradient echo is about 9 ms. Prior to the recording of the first projection data we have employed dummy excitations for about 0.5 s to establish a steady-state magnetization. Dummy experiments can be avoided by adjusting the flip angles of the rf pulses such that equal amounts of transverse magnetizations are excited. Assuming an average value of T_1, this can be done by starting the experiment using low flip angles and increasing their values asymptotically to the desired steady-state value during the initial part of the imaging sequence. Since individual experiments have a duration of 18 ms, the imaging times are 1.15 and 2.3 s for a 64 × 128 image and a 128 × 128 image, respectively. 256 × 256 images may be recorded within 6 s. These imaging times may even be reduced using magnetic field gradients with higher field strengths (e.g., 2 G/cm) and/or better switching times (e.g., 1 ms or less).

Spin–Lattice Relaxation Time Contrast

FLASH imaging techniques take advantage of a high-level steady-state of the longitudinal magnetization. This level is determined by the repetition time of the rf pulses, their flip angles, and the spin–lattice relaxation time T_1. Assuming a constant measuring time, i.e., repetition time, the T_1 contrast is only dependent on the flip angle. Figure 2 demonstrates the effect of varying the flip angle from 10 to 40° (Figs. 2a–d) for a phantom with different T_1 values. For a selected T_1 value, increasing the flip angle

FIG. 2. Spin–lattice relaxation time contrast within 100 MHz ^1H NMR FLASH images (128 × 128 pixels corresponding to 1 mm resolution, 4 mm slice thickness, measuring time 2.3 s). The water phantom shown contains different T_1 values ranging from 2600 ms (upper row left) to 220 ms (upper row right), and 430, 860, and 1150 ms (lower row from left to right). (a–d) Images obtained with flip angles of 10, 20, 30, and 40°. (e) Same as (a) but with application of an 180° pulse immediately prior to the imaging sequence shown in Fig. 1. (f) Same as (e) but with a measuring time of 1.15 s according to a 64 × 128 pixel resolution.

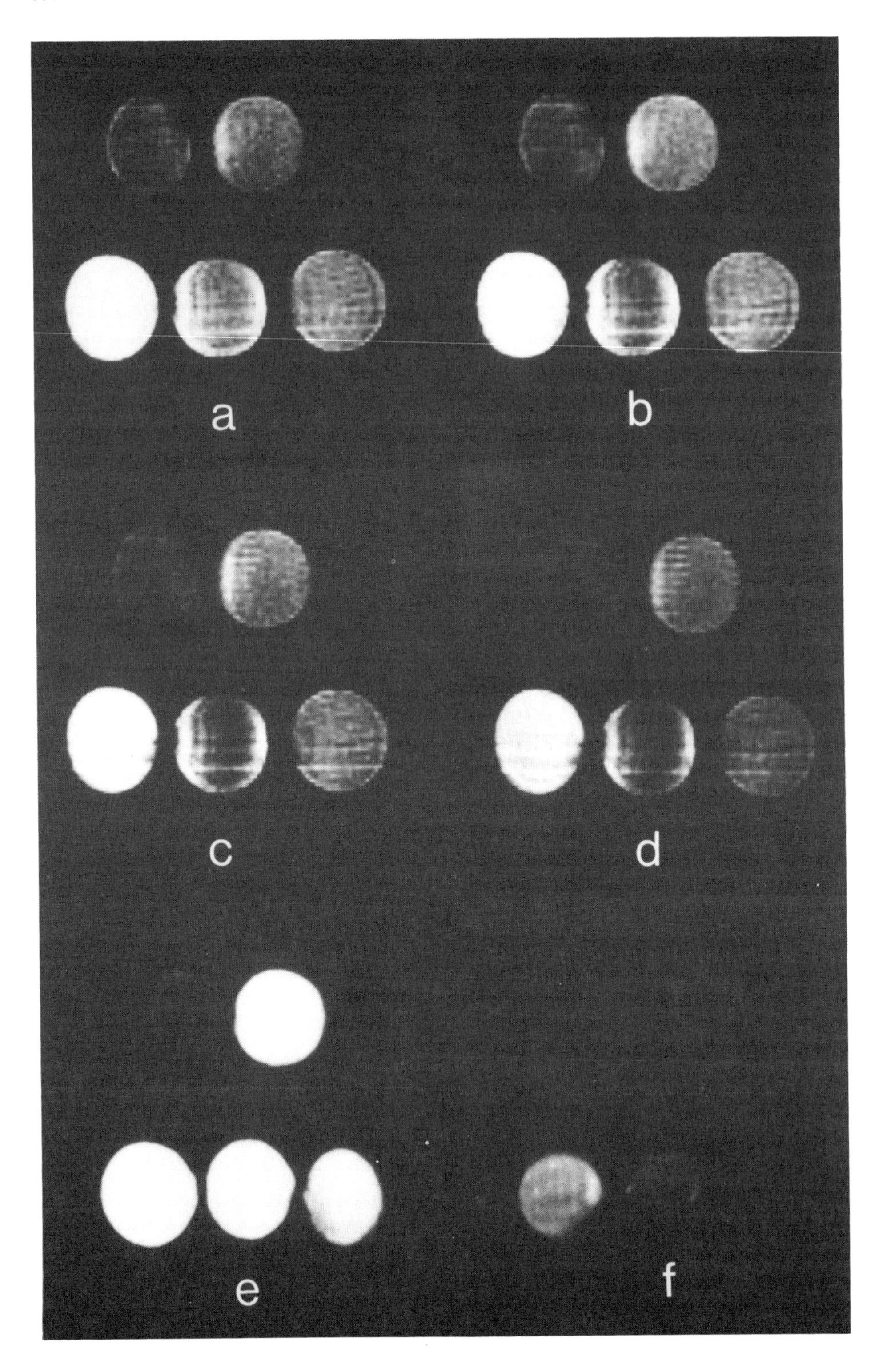
a
b
c
d
e
f

results in reduced signal intensities because of a more pronounced degree of saturation. Vice versa, for a given flip angle the intensity increases with decreasing T_1.

An additional way of enhancing T_1 contrast within FLASH images is due to the use of extra pulses, e.g., 90 or 180°, at selected positions during the imaging sequence. Since the measuring times of the rapid NMR images are of the order of the spin–lattice relaxation times, i.e., 1–2 s, even a single rf pulse will result in intensity changes. However, the point spread function along the phase-encoded direction might be changed due to imaging in the presence of a nonequilibrium state. As an example, Figs. 2e, f show the influence of a 180° pulse applied prior to the imaging sequence after termination of the dummy excitations. This pulse inverts the steady-state magnetization, so that the subsequent recovery period is probed by the FLASH imaging procedure. Depending on the imaging time the intensities of image contributions with different T_1 values are manipulated. For example, in the 2 s image shown in Fig. 2e the large water T_1 value of about 2 s is mainly affected. In addition, the 1 s image exhibits a further signal reduction of contributions with T_1 values of about 1 s leading to a strong contrast enhancement with respect to short T_1 values. In general, the accessible T_1 contrast within rapid images is about the same as within conventional NMR images, i.e., tissues with high T_1 values are represented by low intensities, while tissues with low T_1 values appear brighter.

Effective Spin–Spin Relaxation Time Contrast

Strong contrast within FLASH images is due to the attenuation of the gradient echo or FID by the effective spin–spin relaxation time $T_{2\text{eff}}$. In principle, $T_{2\text{eff}}$ reflects the occurrence of inhomogeneities of the static magnetic field either within the NMR magnet or the individual tissues. Problems of magnetic field inhomogeneities have been mainly overcome by the use of superconducting magnets, so that in the absence of gradients within the imaging object conventional T_2 contrasts are to be expected by varying the echo time. On the other hand, in the case of internal gradients in tissues *in vivo* $T_{2\text{eff}}$ contrasts give new access to structural information. This is demonstrated in Fig. 3 depicting horizontal FLASH images of the human hand at three different echo times. Using a short echo time of about 9 ms (Fig. 3a), the image appears more or less normal showing muscles, joints, marrow, vessels, and fat. However, as known from spin-echo or stimulated-echo images, the lipid signals from the bone marrow should be more intense due to their short T_1 values. Their intensity is considerably reduced because of the generation of internal gradients shortening $T_{2\text{eff}}$ mainly in the heterogeneously structured regions of the bones. In particular, the trabecular structure of the bones in the wrist and in the epiphysial parts of the metacarpalia leads to significant signal losses, whereas the lipid signals from the cavities exhibit strong intensities due to their short T_1 and relatively long T_2. These effects are even better seen in Fig. 3b using an echo time of 17 ms. In the 22 ms image shown in Fig. 3c one can already recognize the influence of B_0 inhomogeneities in the upper left part of the image.

Elimination of Motional Artifacts

NMR imaging *in vivo* often suffers from image artifacts in the direction of the phase-encoding gradient due to motions of the object under investigation. While periodic

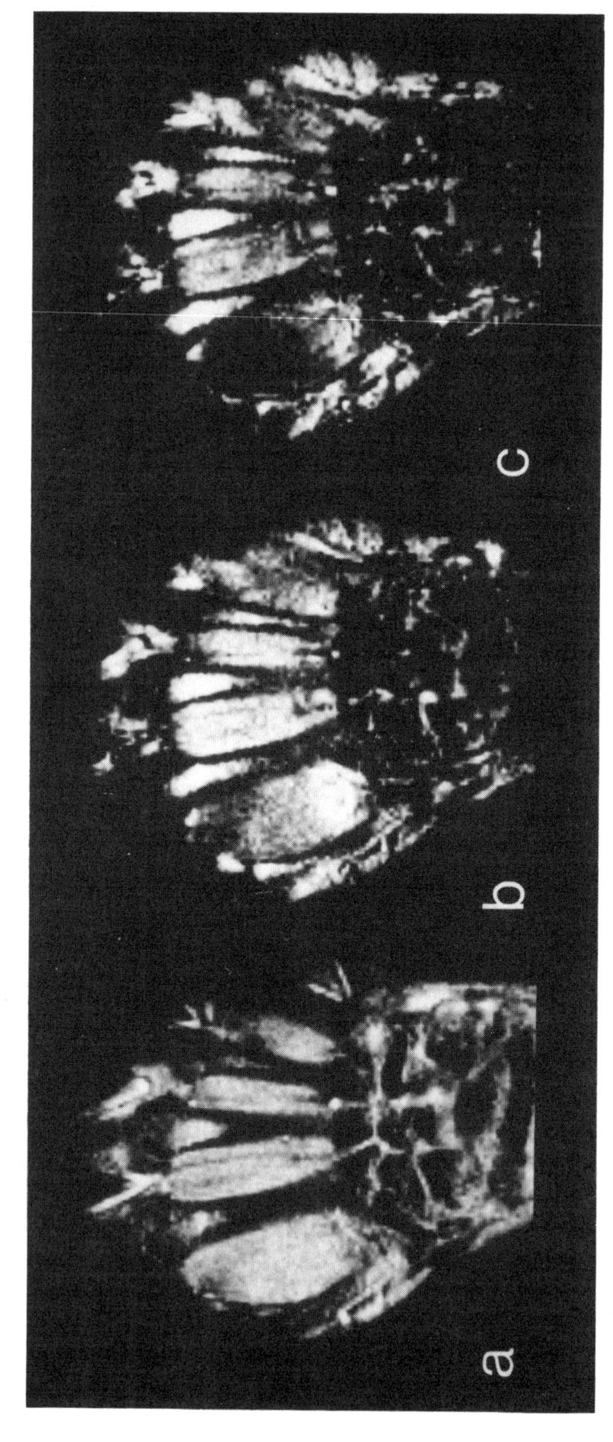

FIG. 3. Effective spin–spin relaxation time contrast within 100 MHz ^1H NMR FLASH images depicting a horizontal slice of the human hand (128×128 pixels corresponding to 1 mm resolution, 4 mm slice thickness, measuring time 2.3 s). The echo times of the images, i.e., the times between the center of the rf pulses and the maximum of the gradient echoes, are (a) 9 ms, (b) 17 ms, and (c) 22 ms.

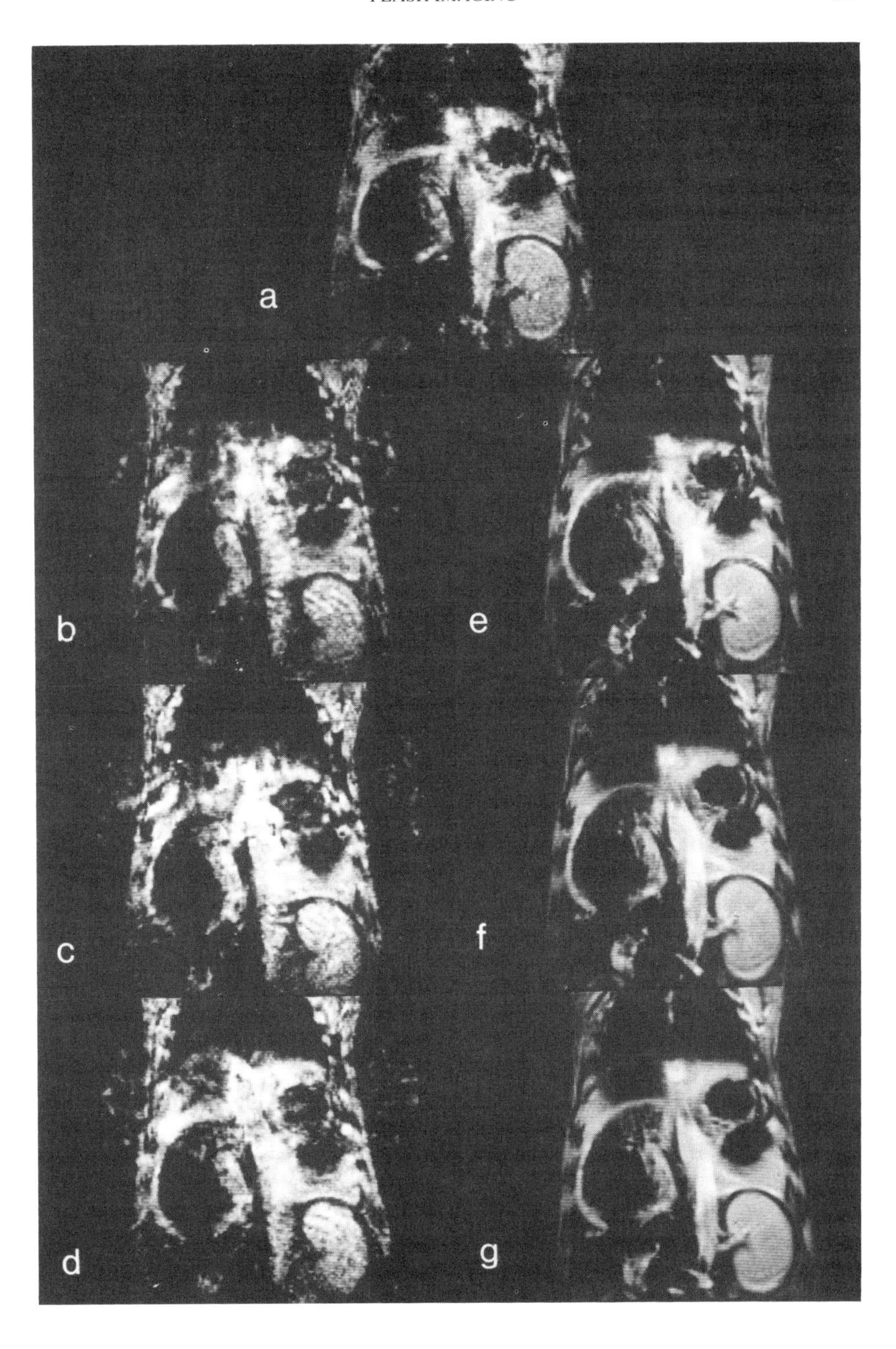
a
b
e
c
f
d
g

motions such as breathing and heart beat may be accounted for by gating or triggering techniques, artifacts due to nonperiodic motions such as peristaltic can hardly be avoided. Rapid FLASH imaging yields a simple solution to these problems. This is because fast movements with time constants of the order of the imaging time do not affect the imaging procedure. This is demonstrated in Fig. 4a for a horizontal slice through the abdomen of a live anesthesized rabbit depicting a kidney, the lungs, the diaphragm, the stomach, and parts of the liver and the dorsal mediastinum. The measuring time is 2.3 s. In Figs. 4b–d the measuring time has been increased to 9, 18, and 36 s by increasing the repetition time of the sequence. Obviously, the image quality is severely degraded.

On the other hand, averaging of accumulated FLASH images is possible to increase SNR at the expense of time resolution. This is demonstrated in Figs. 4e–g which show images with the same overall measuring time as in Figs. 4b–d. However, instead of increasing the repetition time, 4, 8, and 16 individual 2.3 s FLASH images have been summed up. Although the images exhibit some minor blurring effects, the gain in quality as compared to Figs. 4b–d is striking. It turns out that even periodic motions with time constants shorter than the measuring time may be imaged by averaging rapidly recorded FLASH images. This can be judged from the appearance of the dorsal mediastinum within Figs. 4e–g. This finding is currently explored for heart imaging without gating (*16*).

CONCLUSION

A new rapid NMR imaging method is presented (i) with optimized SNR, (ii) with the choice of a variable compromise between spatial, time resolution, and SNR, and (iii) with the possibility of recording NMR movies. Furthermore, the FLASH imaging sequence may easily be modified to allow three-dimensional NMR imaging by replacing the slice-selective rf pulse by a nonselective pulse and the slice selection gradient by an additional phase-encoding gradient perpendicular to the other gradients. 3D NMR images with a resolution of $128 \times 128 \times 128$ pixels may be obtained within a measuring time of about 4 min (*17*). However, a major advantage of the rapid 2D technique is the absence of motional artifacts within the images. Even dynamic imaging of rapid nonperiodic movements becomes possible (*18*) and fast physiological changes may be investigated by recording NMR "movies" comprising sequential images (*19*). When combined with rapid image reconstruction and display routines, sequential multislice FLASH images may be used to move the imaging plane "on line" through the body. In medical applications rapid NMR imaging will considerably improve the convenience

FIG. 4. Elimination of motion artifacts by FLASH NMR imaging. The images shown (128×128 pixels corresponding to 1 mm resolution, 4 mm slice thickness) refer to horizontal slices through the abdomen of a live anesthesized rabbit depicting a kidney, the lungs, the diaphragm, the stomach, and parts of the liver and the dorsal mediastinum. This cross section is affected by peristaltic, respiratory, and cardiac motions. (a) Conventional FLASH image with a measuring time of 2.3 s. From (b–d) as well as from (e–g) the measuring time increases from 9 s (b, e), to 18 s (c, f) and 36 s (d, g). The images (b–d) are obtained by increasing the repetition time, whereas the images (e–g) refer to averages of 4, 8, and 16 subsequent recordings of 2.3 s FLASH images, respectively.

of patients as well as lower the economic constraints to the use of NMR by today's public health care.

ACKNOWLEDGMENTS

Financial support by the Bundesminister für Forschung und Technologie (BMFT) of the Federal Republic of Germany (Grant 01 VF 242) is gratefully acknowledged.

REFERENCES

1. P. MANSFIELD AND P. G. MORRIS, *in* "Advances in Magnetic Resonance" (J. S. Waugh, Ed.), Suppl. 2, "NMR Imaging in Biomedicine," Academic Press, New York, 1982.
2. R. J. ORDIDGE, P. MANSFIELD, M. DOYLE, AND R. E. COUPLAND, *Br. J. Radiol.* **55,** 729 (1982).
3. J. HENNIG, A. NAUERTH, AND H. FRIEDBURG, *Magn. Reson. Med.,* **2,** in press (1985).
4. National Radiol. Protect. Board, *Br. J. Radiol.* **56,** 974 (1983).
5. J. FRAHM, A. HAASE, D. MATTHAEI, K. D. MERBOLDT, AND W. HÄNICKE, *J. Magn. Reson.* **65,** 130 (1985).
6. J. FRAHM, K. D. MERBOLDT, W. HÄNICKE, AND A. HAASE, *J. Magn. Reson.* **64,** 81 (1985).
7. German Patent Application P 3504734.8 (February 12, 1985).
8. A. HAASE, J. FRAHM, D. MATTHAEI, K.-D. MERBOLDT, W. HÄNICKE, Book of Abstracts, 4th Annual Meeting of the Society of Magnetic Resonance in Medicine, London, August 19–23, 1985, p. 980.
9. W. A. EDELSTEIN, J. M. S. HUTCHISON, G. JOHNSON, AND T. REDPATH, *Phys. Med. Biol.* **25,** 751 (1980).
10. P. A. BOTTOMLEY, W. A. EDELSTEIN, W. M. LEUE, H. R. HART, J. F. SCHENCK, AND R. W. REDINGTON, *Magn. Reson. Imaging* **1,** 69 (1982).
11. R. ERNST, *Adv. Magn. Reson.* **2,** 1 (1966).
12. J. S. WAUGH, *J. Mol. Spectrosc.* **35,** 298 (1970).
13. D. I. HOULT, *J. Magn. Reson.* **26,** 165 (1975); **35,** 69 (1979).
14. P. C. LAUTERBUR, *Nature* **242,** 190 (1973).
15. A. KUMAR, D. WELTI, AND R. R. ERNST, *J. Magn. Reson.* **18,** 69 (1975).
16. D. MATTHAEI, J. FRAHM, A. HAASE, W. HÄNICKE, AND K.-D. MERBOLDT, submitted.
17. J. FRAHM, A. HAASE, AND D. MATTHAEI, *J. Comput. Assist. Tomogr.,* in press.
18. J. FRAHM, A. HAASE, AND D. MATTHAEI, *Magn. Reson. Med.,* in press.
19. D. MATTHAEI, J. FRAHM, A. HAASE, AND W. HÄNICKE, *Lancet,* ii, 893 (1985).

JOURNAL OF MAGNETIC RESONANCE **67**, 501–518 (1986)

Spectral Rotation in Pulsed ESR Spectroscopy*

JOSEPH P. HORNAK

Department of Chemistry, Rochester Institute of Technology, Rochester, New York 14623

AND

JACK H. FREED

Baker Laboratory of Chemistry, Cornell University, Ithaca, New York 14853

Received October 29, 1985

The technique of pulsed Fourier transform NMR spectroscopy is well developed and understood. The equivalent technique for ESR has been slower to be developed, but recent developments in ESR instrumentation have now made FT ESR possible. In this paper we consider the requirements for rotating the magnetization associated with an ESR spectrum by a microwave pulse. In addition, techniques for reconstructing the frequency (field)-domain spectrum from the time-domain spectra are described. These techniques are applied to obtain a FT spectrum from the electron spin-echo decay of γ-irradiated quartz and a FT spectrum from the free induction decay of the 2,5-di-*t*-butyl-*p*-benzosemiquinone radical recorded in 50 and 25 ms, respectively, of real time.

I. INTRODUCTION

Modern Fourier transform and two-dimensional magnetic resonance techniques are based upon irradiating the spectrum with a resonant radiofrequency or microwave field. Ideally, one wishes to rotate the magnetization due to all portions of the spectrum by an equal angle, θ. In actual experiments it is not always possible to achieve this ideal, and different portions of a spectrum are rotated by different amounts depending on the extent to which they deviate from the resonance condition associated with the applied electromagnetic radiation (EMR). These matters are generally well understood in NMR (*1*), for which the modern FT and 2D techniques have been well-developed, but they appear to be inadequately understood in ESR. However, such matters are becoming increasingly relevant to ESR with the development of low Q, high B_1 resonators (*2–4*) in electron spin-echo (ESE) spectroscopy. In fact, it is becoming increasingly practical to rotate the magnetization from an entire ESR spectrum with a microwave pulse (*3*).

In this paper we wish to review the considerations regarding FT spectroscopy with a particular view to those that are relevant for pulsed ESR, given its current technical

* Supported by NSF Grant CHE8319826 and NIH Grant GM-25862.

0022-2364/86 $3.00

possibilities and limitations. We also demonstrate them with some experimental examples.

Let us now describe the appropriate approach by first considering simplified points of view. We use the standard convention that at equilibrium the magnetization $\mathbf{M} = M_z\hat{\mathbf{k}}$, i.e., is parallel to the applied static magnetic field $\mathbf{B} = B\hat{\mathbf{k}}$. The applied EMR with angular frequency $\omega_0 = 2\pi\nu_0$, which is applied along the x axis, is resolved into two counter-rotating components, only one of which is (near)-resonant. This component is static along the x' axis rotating in the xy plane at frequency ω_0. A pulse of EMR, applied for a period t will rotate the magnetization exactly at resonance by the angle (in radians):

$$\theta = \gamma B_1 t \qquad [1]$$

about the x' axis. Here γ is the gyromagnetic ratio of the spin-$\frac{1}{2}$ particle. With a spectrum of finite extent it is no longer true that "a $\pi/2$ pulse" (according to Eq. [1]) will rotate all portions of the spectrum by $\pi/2$. A very simple "rule of thumb" is that a $\pi/2$ pulse of microwave field strength B_1 will rotate "B_1 gauss" worth of spectrum by $\sim 90°$. More exactly the rotation angle is a function of ω and is about the effective $\mathbf{B}_1(\omega)$ vector with magnitude:

$$B_{\text{eff}}(\omega) = [B_1(\omega_0)^2 + ((\omega_0 - \omega)/2)^2]^{1/2}. \qquad [2]$$

The rotation angle about $B_{\text{eff}}(\omega)$ is $\theta(\omega) = \gamma B_{\text{eff}}(\omega)t$. The resultant magnetization vector $\mathbf{M}$ in the rotating frame coordinate system as a function of the initial magnetization $\mathbf{M}_0(\omega)$, is (*13*):

$$\mathbf{M}(\omega) = [\mathbf{M}_0(\omega)\cdot\hat{\mathbf{h}})\hat{\mathbf{h}} + [\mathbf{M}_0(\omega) - [\mathbf{M}_0(\omega)\cdot\hat{\mathbf{h}})\hat{\mathbf{h}}]\cos\theta - [\hat{h} \times M_0(\omega)]\sin\theta \qquad [3]$$

where $\hat{h}$ is the unit vector along $B_{\text{eff}}(\omega)$. Alternatively we may FT a square pulse of EMR extending from $t = 0$ to t to obtain the Fourier component of magnetic field:

$$B_1(B) = |B_1|\left[\frac{\sin(\gamma t(B_0 - B))}{(\gamma t(B_0 - B))} + i\,\frac{\sin^2(\gamma t(B_0 - B)/2)}{(\gamma t(B_0 - B)/2)}\right]. \qquad [4]$$

Here $|B_1|$ is the magnitude of the rotating field, and the real (imaginary) part of Eq. [4] is along the rotating x' (y') axis. In obtaining Eq. [4] we have equated $\omega_0 = \gamma B_0$ and $\omega = \gamma B$. Equation [4] shows that an "effective" B_1 actually varies continuously from its maximum at resonance (i.e., $B_0 = B$). If we choose the pulse width to satisfy Eq. [1], valid for resonance, then we may replace γt in Eq. [4] by θ/B_1 to obtain how the "effective" B_1 varies with the magnitude of B_1. In this form of Eq. [4] it is clear that by increasing B_1 (and correspondingly decreasing t), $B_1(B)$ will vary less over the spectrum.

The correct way to deal with this matter is, of course, to explicitly solve the quantum-mechanical equations of motion for a spin in the combined B and B_1 fields (*5*). More generally, we must include the effects of relaxation. In many cases the Bloch equations will apply, and one may integrate them directly (*6a*). In the usual notation we have

$$du/dt = -u/T_2 - (\omega - \omega_0)v \qquad [5]$$

$$dv/dt = -v/T_2 - \gamma B_1 M_z + (\omega - \omega_0)u \qquad [6]$$

$$dM_z/dt = -(M_z - M_{z0})/T_1 + \gamma B_1 v. \qquad [7]$$

Here u, v, and M_z are, respectively, the x', y', and z components of magnetization. T_1 is the spin–lattice relaxation time, T_2 the homogeneous spin–spin relaxation time, and ω the Larmor frequency of an electron at resonance in a magnetic field B. With the aid of these equations, the transverse magnetization present after the application of a square pulse of microwaves with frequency ω_0 and field strength B_1 can be determined. The u and v components of magnetization thus calculated represent the respective in- and out-of-phase amounts of magnetization rotated by the pulse and expressed as a function of ω. The results for the limit where T_1 and T_2 are long relative to t approximately match those described by Eq. [4]. With the appropriate form of B_1 as a function of time, Eqs. [5]–[7] describe the effects of a multi pulse sequence on the magnetization. Note that if there is significant inhomogeneous broadening present, then the Bloch equations [5]–[7] will have to be convoluted with the shape function for the inhomogeneous broadening. For the particular case of a Lorenztian shape function, of width $T_2^{\dagger -1}$, this will lead to an effective $T_2^{*-1} = T_2^{-1} + T_2^{\dagger -1}$ in the initial FID and in the spin-echo decay discussed in the next section, cf. Ref. (*6b*).

In reality, due to finite switching speeds (time constant $T' \simeq 10$ ns) of the microwave switches used to produce the microwave pulses and to the finite time required to form and remove power from a microwave cavity with a given Q, one should include a pulse shape function, $B_1(t)$, into the above equations to calculate the transverse magnetization. For microwaves with a magnetic field strength B_1 turned on between time $t = 0$ and t_1, $B_1(t)$ is given by

$$\begin{aligned} B_1(t) &= 0 && \text{for} \quad t \leqslant 0 \\ B_1(t) &= B_1(1 - \exp(-t/T')(1 - \exp(-t\omega_0/Q)) && \text{for} \quad 0 \leqslant t \leqslant t_1 \\ B_1(t) &= B_1(t_1)\exp(-t/T')\exp(-t\omega_0/Q) && \text{for} \quad t > t_1. \end{aligned} \qquad [8]$$

Of equally great importance in calculating the effects of a pulse on the magnetic moments is the distribution of B_1 in the sample. A distribution in the magnitude of B_1 will cause a corresponding distribution in rotation angle as described by Eq. [1]. Ideally B_1 should be constant along the sample axis in the sample cavity, in order that all spins of a given resonant frequency are rotated by the same angle.

In the spectral reconstruction section of this paper (Section II) we describe the rotation of spectral magnetization by different pulse shapes. Also addressed, is the procedure for transforming the resultant time domain spectrum into a field-domain spectrum by the application of separate phase and amplitude corrections. We have attempted to summarize some of the major instrumental requirements in Section III. In Section IV these instrumental techniques and transformations are applied to two simple ESR spectra. A summary appears in Section V.

II. SPECTRAL RECONSTRUCTION

In time-resolved magnetic resonance spectroscopy, two general categories of pulse experiments exist, viz., the free induction decay and the spin echo. The FID sequence employs a 90° pulse of EMR immediately after which the signal is recorded. The echo sequence moves the signal or echo away from the pulses, e.g., in the 90°–τ_1–180°–τ_2–echo sequence the maximum signal is approximately at τ_2 away from the last pulse.

In general, both pulse sequences contain the same basic information about the spin system (but see below). Information about the dynamics of the spin system is, however, more readily obtained from the echo sequence than in the FID. In general an FID will be seen when the inhomogeneous $T_2^\dagger$ is much greater than the spectrometer deadtime t_d, which is not always the case. Due to this deadtime after the application of a pulse of EMR (*7*), FIDs cannot be observed for all systems. These systems are frequently those in which spin echoes can be observed. Echoes will be seen in samples when the homogeneous $T_2 \gtrsim 2t_d$. In many samples in which there is substantial inhomogeneous broadening, it is often true that $T_2 \gg T_2^\dagger$, so it is often possible to see echoes. In fact, since the echo width $\sim T_2^*$ (i.e., the effective T_2, cf. Section I), it will be relatively sharp and intense when this inequality holds, so this is a favorable condition for detecting echoes. Electron spin-echo spectroscopy has evolved as the more popular form of pulsed ESR spectroscopy due to the nature of the systems investigated and instrumental constraints.

We can analyze the case for a one-pulse FID or a two-pulse ESE from the Bloch equations [5]–[7] by recognizing that the absorption signal is proportional to v, while the dispersion, detected along x', is proportional to u. That is, in-phase detection gives u and out-of-phase, v. The transverse magnetization present immediately after the application of a 29.75 ns, 3 G 90° square pulse, as a function of $\nu = \omega/2\pi$, was calculated from Eqs. [5]–[7]. This is a case of homogeneous broadening, i.e., $T_2^* = T_2$. Figure 1 represents the case with $T_1 = T_2 = 1$ μs. Much of the magnetization from spins away from those which resonate at ν_0, is located along the x' axis after the application of the 90° pulse (Fig. 1b). The similarities between Figs. 1b and c and, respectively, the real and imaginary parts of part of Eq. [2] should be noted. Quadrature detection (*8*) (simultaneous detection of both u and v components of magnetization) coupled with the proper form of signal reconstruction based on the shapes of the curves in Figs. 1b and c expands the width of a spectrum which might be recorded from a square pulse. A better representation of the amount of detectable signal is the resultant transverse magnetization,

$$M(\nu) = (v(\nu)^2 + u(\nu)^2)^{1/2} \qquad [9]$$

shown in Fig. 1d. The amount of spectrum rotated by greater than 65° (transverse magnetization greater than 0.9) is equivalent to B_1 ($B_1\gamma/2\pi$ s^{-1}) worth of static magnetic field.

The above example represents an idealized situation, since in reality $B_1(t)$ is not a square pulse but rather a function of the form described by Eq. [8]. As an example, Fig. 2 represents the transverse magnetization present in a spectrometer with a $Q = 100$ and an infinitely fast microwave switch. Parameters such as T_1, T_2, and the pulse width were kept at the values in the previous example. Since there is no sharp termination to such a pulse in terms of the B_1 inside the cavity, the three lines in each figure represent the magnetization at 0, 5, and 10 ns after t_1. The complexity of these curves would make the reconstruction of an ESR spectrum very difficult.

The use of a specially tailored pulse, which will produce a flat transverse magnetization curve over a width equivalent to a typical ESR spectrum, would simplify the reconstruction process. Such a function is the apodised sinc function ($\sin(x)/x$) (*9, 10*) commonly used in FT NMR spectroscopy. Figure 3 represents the transverse magnetization resulting from the application of a sinc(x) function pulse where $-3\pi \leqslant x$

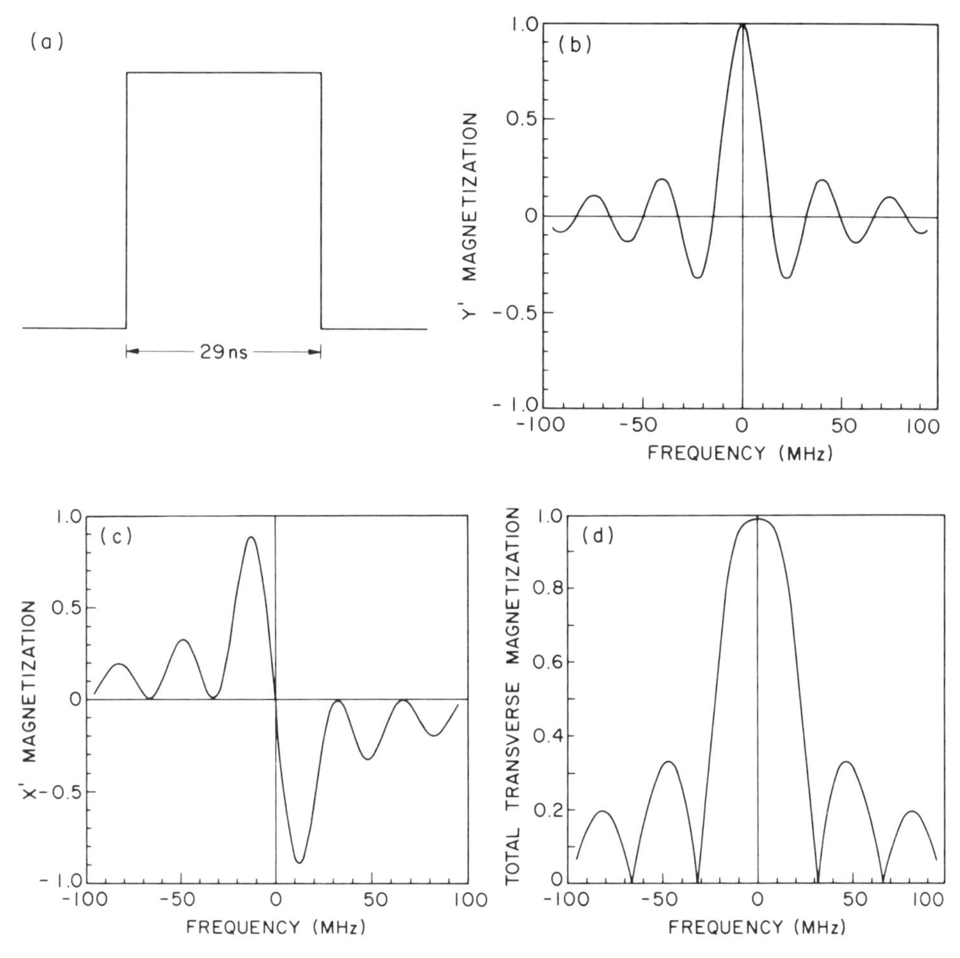

FIG. 1. Transverse magnetization resulting from the application of a square 29.75 ns $\pi/2$ pulse of B_1 at frequency ν_0 (a) to an electron spin system as a function of frequency offset from ν_0. The curves were calculated from a numerical integration of the Bloch equations with $T_1 = T_2 = 1\ \mu s$, $B_1 = 3$ G, and initially all magnetization along the z axis. The various components of magnetization in the rotating frame are v along the y' axis (b) u along the x' axis (c) and total transverse magnetization $(u^2 + v^2)^{1/2}$ (d).

$\leqslant 3\pi$. The maximum effective B_1 (i.e., $B_1(\nu_0)$) was 13 G, and 6π corresponded to 45 ns. As the microwave pulse becomes more like a sinc(x) pulse, the transverse magnetization becomes more like a square pulse (the FT of a sinc function). The width of the pulse is approximately $4\pi/t$, where t is the time between the $-\pi$ and π points of the sinc pulse.

As previously mentioned, quadrature detection should be employed to expand the range of signal detection and also to improve S/N. The time-domain signal is Fourier transformed to obtain the frequency-domain spectrum. Fast Fourier transform (FFT) routines require a real and an imaginary input and give a real and imaginary output array. If only v, the out-of-phase component of magnetization, is recorded, then it is typical to use it as the real entry into a FFT routine and an array of zeroes as the

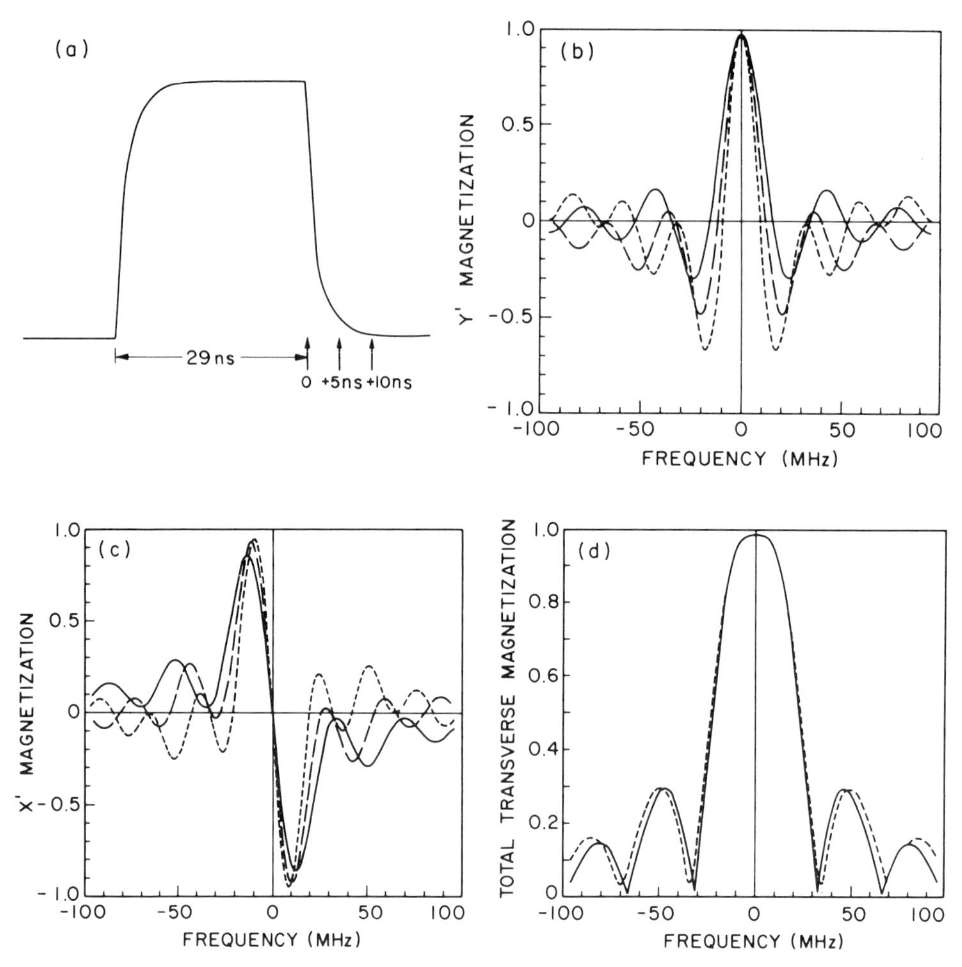

FIG. 2. Transverse magnetization resulting from the application of an exponentially rising and falling 29.75 ns $\pi/2$ pulse of B_1 at frequency ν_0 (a), typical of a square pulse sent into a cavity with a $Q = 100$, to an electron spin system as a function of frequency offset from ν_0. The magnetization at 0, 5, and 10 ns after the turn off of the pulse is represented by the (—), (---), and (· · ·) lines, respectively. The curves were calculated from a numerical integration of the Bloch equations with $T_1 = T_2 = 1$ μs, $H_1 = 3$ G, and initially all magnetization along the z axis. The v, u, and total transverse magnetization are represented in (b), (c), and (d), respectively.

imaginary entry. With these input arrays the FFT is not capable of distinguishing any difference between frequencies below and above ν_0. The resultant spectrum is even about ν_0 and therefore appears symmetric. When u is used as the imaginary input to the FFT routine, its resultant spectrum is odd or antisymmetric about ν_0. The output of the FFT routine in this case is the sum of these two curves and is a spectrum which distinguishes any asymmetries about ν_0. This summation results in an improvement in the S/N ratio of the resultant spectrum assuming the noise is random. If the time dependence of v and u are used as the real and imaginary parts respectively of the input to the FFT routine, then the absorption and dispersion spectra will be the re-

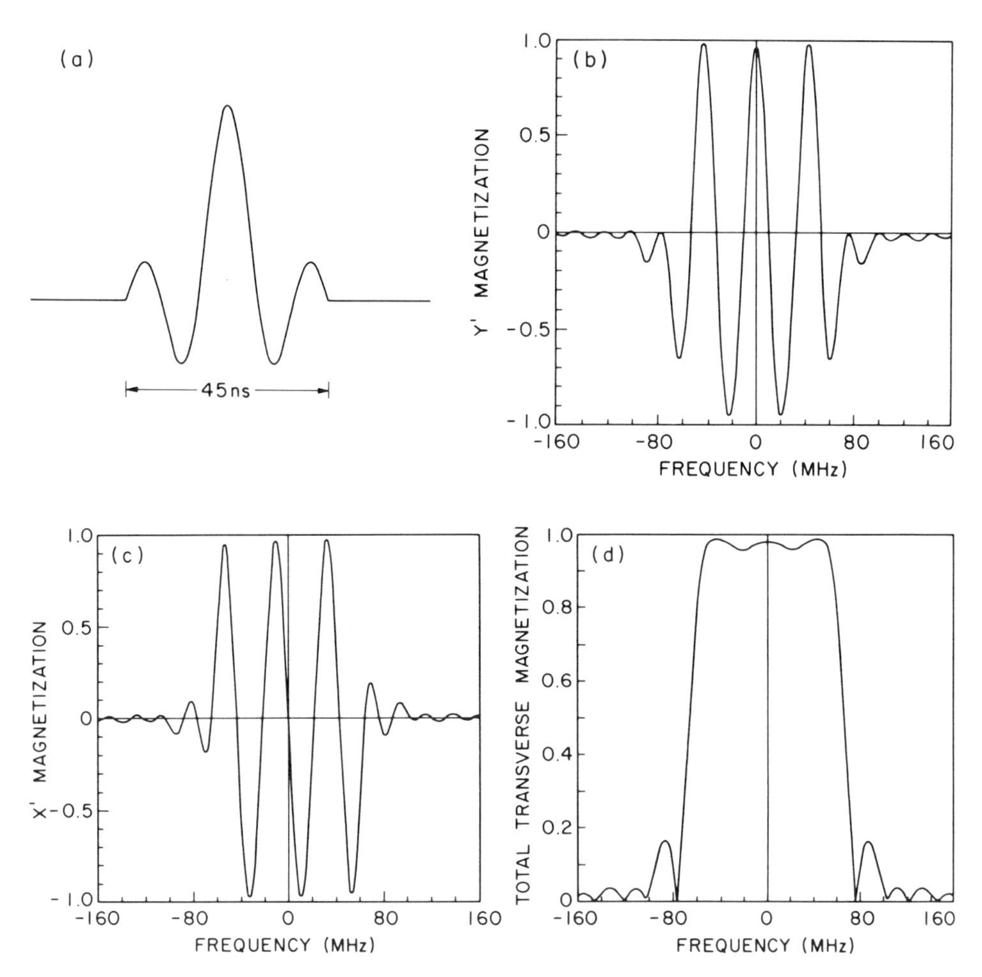

FIG. 3. Transverse magnetization resulting from the application of an apodised sinc function (see text) $\pi/2$ pulse of frequency ν_0 (a) to an electron spin system as a function of frequency offset from ν_0. The amounts of initial z magnetization $M_2(\nu)$ rotated into the y' axis, x' axis, and $x'y'$ plane are shown in (b), (c), and (d), respectively.

spective real and imaginary outputs. Collection of components of magnetization offset from the x' and y' axes by an angle α rather than the exact x' and y' components will result in spectra which are admixtures of absorption and dispersion. This problem is easily corrected by the use of a simple coordinate transformation on the real and imaginary outputs of the FFT.

$$\begin{vmatrix} \mathrm{Re}[S(\nu)] \\ \mathrm{Im}[S(\nu)] \end{vmatrix} = \begin{vmatrix} \cos(\alpha) & \sin(\alpha) \\ -\sin(\alpha) & \cos(\alpha) \end{vmatrix} \begin{vmatrix} \mathrm{Re}[S(\nu)]' \\ \mathrm{Im}[S(\nu)]' \end{vmatrix} . \quad [10]$$

The resultant spectrum from the FFT conversion has several experimental artifacts such as that from improper phasing of the x' and y' axes just described. These artifacts can be classed as those modifying the phase of the signals and those affecting the intensity. Both artifacts can be corrected for in the output arrays of the FFT, but the phase corrections must be applied before amplitude corrections.

The following discussion applies primarily to a 90° FID sequence. Phase adjustments can be classed as zeroeth, first, and second order and all are corrected for by the use of Eq. [8]. Zeroeth-order or constant phase corrections, where α equals a constant, usually result from improper adjustment of the detector to detect the true x' and y' magnetization. A first-order or linear phase correction, which is needed for FID but not echo techniques, is necessitated because of loss of early data points in the time-domain curve due to instrumental deadtime. Each point lost results in a linear phase deviation of π across the resultant frequency-domain spectrum (*1*), so $\alpha = \pi\nu/\Delta t$ where Δt is the time interval between digitized points. In an FID pulse sequence, second-order phase corrections are trigonometric in nature, and they arise from the distribution of the resultant transverse magnetization after a 90° pulse in the $x'y'$ plane as a function of frequency offset from ν_0. The phase correction angle α is equal to the angle between the resultant transverse magnetization and the y' axis immediately after the 90° pulse. The approximate form of the frequency dependence of this angle after the application of a square $\pi/2$ pulse is given by $\tan^{-1}$(Im(Eq. [4])/Re(Eq. [4])) and is shown in Fig. 4a. This implies that the initial u and v components have the same ratio as the Im and Re parts of Eq. [4].

The form of this function is best understood by looking at the path the transverse magnetization vector takes in the $x'y'$ plane as a function of offset from frequency ν_0 (Fig. 4b). The intensity correction adjusts the height of the lines in the resultant absorption spectrum for rotations of components of magnetization by less than 90°. It is accomplished by dividing the phase-corrected absorption spectrum by the resultant transverse magnetization function (Eq. [9]), which for a perfect square pulse is $\cong|\mathrm{sinc}(\pi t(\nu - \nu_0))|$. These corrections are important in deuterium NMR (*11*), where the spectral width is large relative to the frequency-domain spectrum of the 90° pulse, and in NMR, when the carrier frequency is substantially off resonance (*12, 13*). The trignometric and intensity corrections are generally not applied in conventional proton pulsed NMR due to the broad range of the spectrum which is rotated at least 90% of the expected rotation angle. The application of all of these corrections in ESR increase the extent over which a given spectrum may be recorded in a pulse experiment. The trignometric and intensity corrections are highly pulse-shape-dependent and, for imperfect pulses, may be difficult to implement.

The dynamics of the magnetization during a spin-echo sequence or multipulse sequence are more complex than for an FID. In a spin-echo sequence, a 90° pulse (at ω_0) is applied for a time t_1 followed by a 180° pulse (at ω_0) of length t_2 at a time τ_1 after the 90° pulse. At time $\tau_2 = (\tau_1 - 3t_1/2)$ after the 180° pulse, a maximum signal or echo is observed. Assuming the relaxation times are large relative to the time of the experiment, the 90° pulse rotates magnetization by an angle $(\gamma B_{\mathrm{eff}} t)$ about $B_{\mathrm{eff}}(\omega)$, which is directed off the z axis in the $x'z'$ plane, by an angle $\tan^{-1}(\gamma B_1/(\omega_0 - \omega))$ as given by Eq. [1]. The location of M after a time τ_1 is $(\omega_0 - \omega)\tau_1$ radians from its position after the 90° pulse. Application of a 180° pulse at time t_1 after the 90° pulse rotates $M(\omega)$ by $\gamma B_{\mathrm{eff}}(\omega)t_2$ about $B_{\mathrm{eff}}(\omega)$. The pulse does not perform a reflection of all $M(\omega)$ vectors through the $x'z$ plane as it does for $M(\omega_0)$. Therefore after a period of time τ_2 the magnetization is not identical to its value immediately after the application of the 90° pulse.

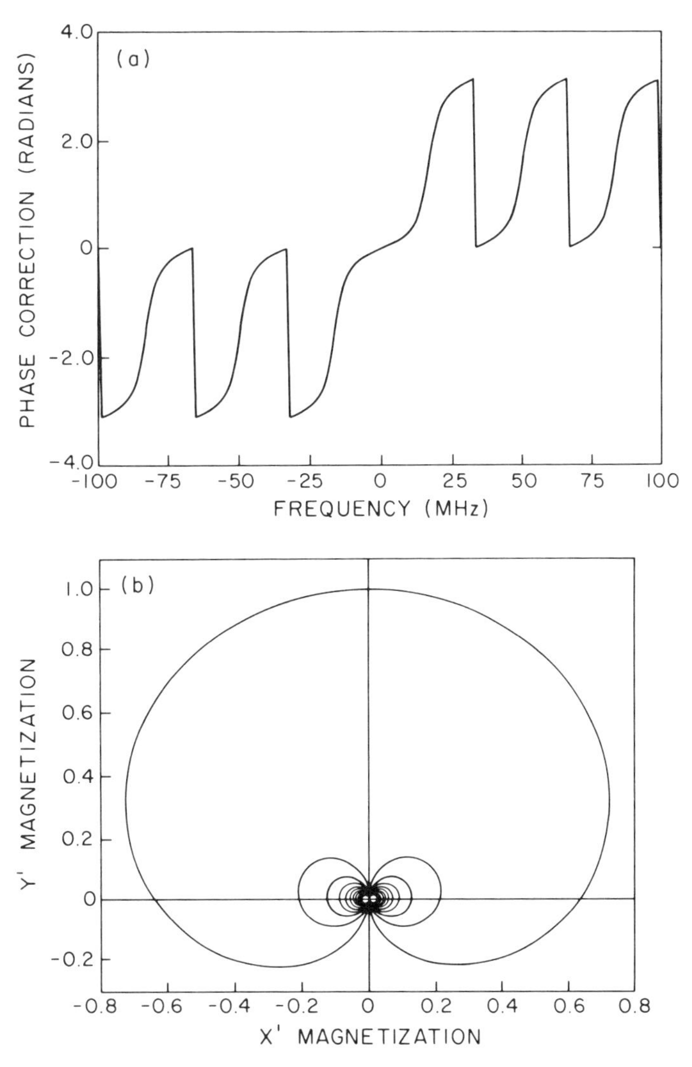

FIG. 4. (a) Frequency dependence of the trignometric phase correction angle, the angle between the y' axis and the transverse magnetization, after the application of 30 ns $\pi/2$ pulse. (b) The lengths of the transverse magnetization vectors in the $x'y'$ plane. Distance along the length of the curve from the y' axis is equivalent to frequency offset from ν_0. This figure explains the frequency dependence of (a).

The most appropriate way to reconstruct a frequency domain spectrum from an echo shape is to calculate the corresponding intensity and phase corrections from Eqs. [5]–[7]. For large values of τ_1 and τ_2 the resulting phase and intensity corrections for $t_1 \neq t_2$ are $(\omega - \omega_0)(t_1 - t_2/2)$ in radians and $|\text{sinc}(\pi t_1(\nu - \nu_0))\text{sinc}^2(\pi t_2(\nu - \nu_0))|$, respectively (*11*). For $t_1 = t_2$ the phase and intensity corrections become $(\omega - \omega_0) \times (t_1/2)$ and $|\text{sinc}^3(\pi t_1(\nu - \nu_0))|$ (Fig. 5), respectively. The interesting features of these corrections are that the phase correction is linear and that the amount of spectrum which may be conveniently reconstructed using the intensity correction factor is $\leqslant 2/t_1$ Hz.

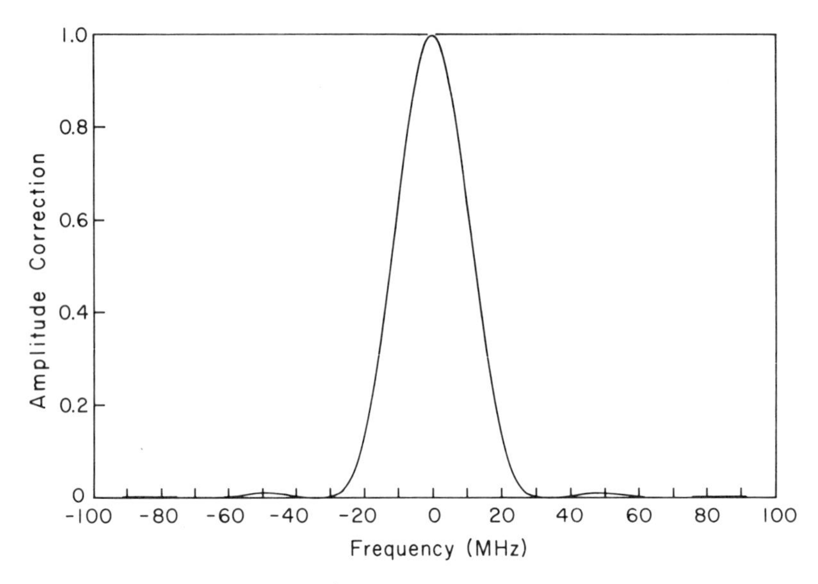

FIG. 5. The amplitude correction factor $|\mathrm{sinc}^3(\pi(\nu - \nu_0)t)|$ for $90°$–τ_1–$180°$ echo sequence where the width, t, of the 90 and 180° pulses is equal to 30 ns.

III. INSTRUMENTATION

Pulse generation for rotation of an entire ESR spectrum poses the biggest challenge in ESE spectroscopy. Perfectly square pulses of duration one to two nanoseconds or specially tailored pulses which can rotate ~50 G of spectrum by ~90° are desired. Minimum insertion loss and maximum isolation in the switched mode are also desired. Commercially available fast PIN diode microwave switches possess switching speeds of ~2–7 ns (*14*). Power-handling capability of the switches degrades with increasing switching speed. It is difficult to produce a perfectly square pulse of a few tens of nanoseconds with these switches. We have found that cascading two of these switches together could produce shorter more rounded pulses. Faster experimental microwave switches have been designed with 1 ns switching speed but lower power-handling capabilities (*15*). Emitter-coupled logic (ECL) switches can produce fairly square 3–4 ns pulses but have a 20 mW power-handling capability. Single-cycle (100 ps) X-band microwave pulses have been generated using experimental switches based on photoconductivity in silicon transmission-line structures (*16*). This switch is expected to have a 1 kW power rating with a pulsed input signal (*16*). Although the laser system required to pulse the semiconductor would add an extra degree of complexity to the ESE experiment, this switch represents the best prospect for short pulse generation.

Typical ESE spectrometers generate their π and $\pi/2$ pulses by varying the pulse length and keeping the pulse intensity constant. To assure rotation of equal amounts of spectrum by the two pulses and thus maximize the amount of reconstructable spectrum, the pulse width should be fixed and the pulse power of the π pulse adjusted to be four times that of the $\pi/2$ pulse. (We describe one method below). With the ratio of powers fixed, the absolute power was chosen as follows. By increasing the power from zero, the first maximum height echo was used to fix the power of the $\pi/2$ and π pulses. Higher powers would produce echoes corresponding to rotations of $(2n + 1) \times \pi/2$ and $(2n + 1)\pi$ with $n = 1, 2, 3 \cdots$. The pulse power in a 90° FID sequence was chosen such that a maximum FID intensity was produced with a minimum amount of power. This insured that the pulse was not a $(2n + 1)\pi/2$ pulse where $n = 1, 2, 3 \cdots$. As an added check of the $\pi/2$ power, an $n(\pi)$ pulse gave a minimum FID. At present, specially tailored pulse shapes such as sinc functions are not conveniently produced.

Much consideration should be given to the choice of a resonator in a pulsed ESR experiment. The resonator bandwidth (ν_0/Q) should be kept high so as to pass all the frequency components of the pulse. A pulse risetime of 200 ps in the resonator requires a $Q = 5$. Distortion of the pulse shape is most evident as resonator ringing after the application of a high-power pulse. Resonator ringing increases the instrument deadtime t_d, which determines the earliest time at which data can be collected after the pulse. Overcoupling the cavity or resonator decreases the ringdown time at the expense of S/N. The loop–gap resonator (LGR) has an advantage in this respect since its high efficiency (in converting incident microwave power to B_1 in the resonator) allows the use of a lower Q, thus diminishing the ringdown time and pulse distortion. Short pulse widths are often required in two-pulse relaxation studies especially when a minimum pulse separation is desirable. When this is not the case, the optimum Q condition (rotation of the entire spectrum with minimal distortion and maximum S/N) is when the spectral width in hertz is $\leqslant \nu_0/2Q$, i.e., half the resonator bandwidth.[1]

Ideally, the B_1 should be constant throughout the sample. B_1 along the sample axis (x axis) of a TE_{102} microwave cavity is approximately Gaussian with the maximum in the center of the cavity. Figure 6a depicts the normalized B_1 calculated by the method of perturbing spheres (*17a*). In ESE studies, samples are usually placed along the entire length of x axis. The $\pi/2$ and π pulse widths are adjusted to give maximum echo height, thus giving a spread in rotations from the different points along the sample. This is a major drawback to using this form of cavity in ESE experiments where complete spectral rotation is the goal. Admittedly one could use samples occupying a smaller length of the x axis, but serious attenuation of the signal-to-noise ratio would result (*3*) due to the lower filling factor. Another drawback with this resonator is the low efficiency of producing B_1 with incident EMR power. For an average B_1 of 3 G, approximately 200 W of power is necessary. This necessitates the use of a traveling wave tube amplifier (TWTA) to amplify the ~1 W from the microwave source. The slotted tube resonator (4b) has a high filling factor but a $\sin(x\pi/L)$ distribution of B_1 intensity along the sample of length L. A more suitable microwave sample cavity would be the loop–gap resonator (LGR) (*2, 3*). The B_1 distribution along the sample axis in this resonator is nearly uniform (Fig. 6b), thus assuring that all spins are rotated by nearly the same amount. Its superior signal-to-noise characteristics over a TE_{102} cavity for volume limited samples is another advantage in its favor (*3*). The high efficiency and low Q of the LGR make it ideal for use in ESE experiments where high power and short pulses are needed for complete spectral rotation.

Two important considerations concerning the data acquisition are the digitization period (i.e., time resolution) and the time over which the data are collected. The period of digitization in the time domain must be at least the reciprocal of the spread in the

[1] Although we discussed the effects of finite Q on the effective B_1 in the previous section, we did not consider its effect on attenuating the induced transverse magnetization components at ν for $|\nu - \nu_0| \gtrsim 2Q$. This is probably best determined in an empirical fashion by studying the variation in observed signal from a narrow-line signal as $\nu - \nu_0$ is varied.

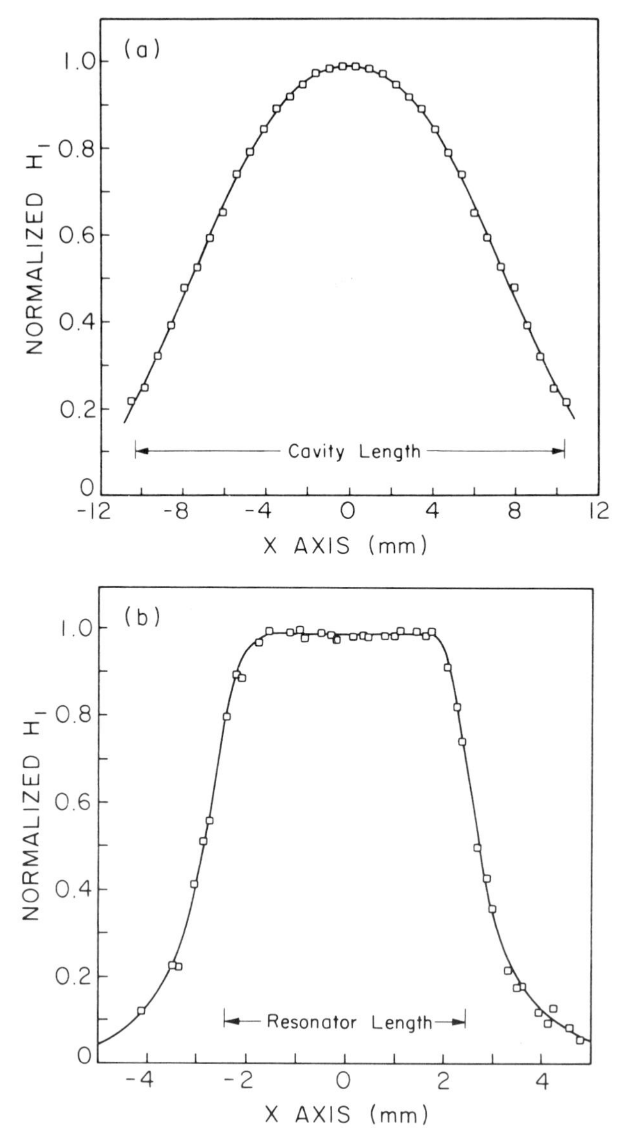

FIG. 6. Microwave magnetic field, B_1, as a function of position along the x axis of (a) a TE_{102} microwave cavity and (b) a loop–gap resonator. Data points are B_1 calculated by the method of perturbing spheres (see text). Solid lines are drawn to guide the eye.

frequencies to be recorded in the spectrum. The resolution in the frequency-domain spectrum will be just the reciprocal of the time over which the FID is recorded. Boxcar averagers and 1 ns digital delay generators commonly found on ESE spectrometers could be used to digitize a spectrum but would be time inefficient. Transient digitizers of 5–10 ns time resolution are usable with spectra <200–100 MHz. Such a digitizer (e.g., the Analogic Data 6000, with which we have experimented (*17b*)) would allow realization of the time efficiency of the FT experiment. Digitizing the prescribed amount of data is not necessary if the echo shape or the FID has diminished to zero well before all these data have been recorded. The typical practice in such cases is to record data until the signal has diminished to a point below the level recoverable by signal averaging and then to zero fill the remainder of the time-domain spectrum. The transition between actual data and zero-filled data should be made as smooth as possible to avoid artifacts in the frequency-domain spectrum close to ν_0. A dc offset or a sloping baseline, due to the detector recovering from a high-power pulse, should be eliminated before performing the FT. Alternatively, linear prediction methods (*17c*) may be applied, especially when the FID is well represented by a sum of decaying sinousoids.

The quadrature detection scheme for maximum data throughput would be a microwave bridge with a phase shifter which could shift by exactly 90 or 180° the phase of the reference signal into the phase-sensitive detector for alternate pulse sequences. The 90° shift is for quadrature detection, and the 180° shift is for noise reduction (*6, 7*). The use of phase shifters with a preset 90 or 180° shift is not preferable since the phase shift varies with microwave frequency. Two examples of alternative approaches are a continuously variable phase shifter consisting of a circulator-microwave switch-tunable short combination (cf. Fig. 7), or a pair of double balanced mixers, as noted below.

IV. EXPERIMENTAL

In this section, two examples of performing FT on time-domain data are given. In one, the latter half of a spin echo is collected, and in the other a FID is collected. In conventional ESE spectrometers, the π pulse is adjusted to be twice as long in time as the $\pi/2$ pulse. To assure rotation of equal amounts of spectrum by both pulses, our basic ESE spectrometer described by Stillman and Schwartz (*18*) was modified to produce pulses which would differ by a factor of 2 in intensity of B_1 or a factor of 4 in power. This was accomplished by inserting a three-port circulator between the usual pulse producing switch and the TWTA (Fig. 7). The circulator directed the usual power from the pulses produced by SW1 in Fig. 7 toward another microwave switch SW2. When SW2 was open, power was reflected from this switch and circulated towards the TWTA. In the closed position of SW2, a pulse is sent through a variable attenuator towards an adjustable short, which also serves as a phase shifter, and it is reflected back through the arm and toward the TWTA. In a 90–180° pulse sequence the pulse width was generated with SW1. The 180° pulse was produced with SW2 open, while the 90° pulse with SW2 closed, thus allowing it to be attenuated by the additional arm. By measuring the pulse power emanating from the TWTA from each pulse, the proper intensity could be set. The phase shift in the attenuated pulse, due to the increased length which it must travel, was corrected by using the adjustable short with a phase sensitive detection scheme to monitor the pulses.

The repetition rate of the pulse sequences used was 1 kHz, (but 10 kHz rates could readily be used). Time-domain data were collected using a boxcar averager stepped out at 1 or 2 ns intervals. Typically 25 averages were made by the boxcar for each point recorded in the time-domain spectrum. Once the 90° out-of-phase spectrum was recorded, the phase of the microwaves in the reference arm of the spectrometer was manually changed by 90° and the in-phase spectrum recorded. The use of a more automatic phase shifter is preferred in these types of experiments, but the present setup was suitable for the present purposes with stable chemical systems. (Another alternative is to split the signal, and send it into two double-balanced mixers whose

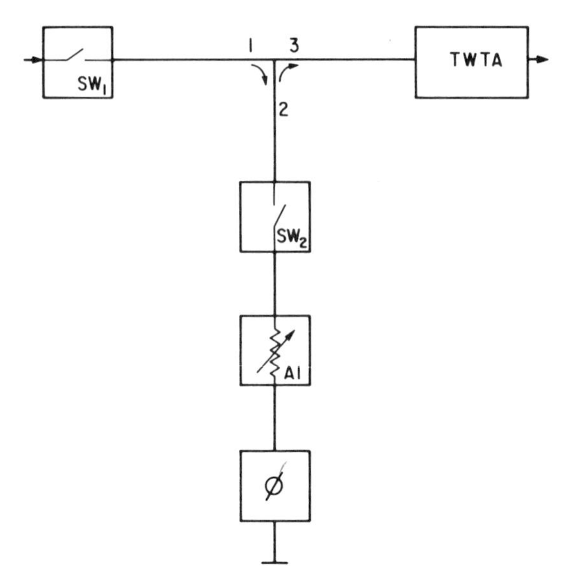

FIG. 7. A block diagram of the modifications made to the ESE spectrometer to allow it to produce variable power pulses. The modification consists of a three part circulator, a microwave switch (SW2), a variable attenuator (A1), and an adjustable short serving as a phase shifter (ϕ). The circulator was inserted between the normal low-power pulse-producing microwave switch and the input to the TWTA.

reference signal phases differ by 90°. We have previously used this approach in a 1 GHz pulse spectrometer (*17b*).) With a quadrature detection scheme capable of recording both the in- and out-of-phase signals simultaneously and with a transient digitizer, spectra of the intensity shown here could be recorded in ~25 ms of real time.

The basic loop–gap resonator described in (*3*) with a bandwidth of ~24 MHz was used in obtaining all the results. Typically, cavity resonators in ESE spectroscopy are overcoupled to minimize the cavity ringing after the application of the high-power pulses and to provide adequate signal. With the LGR, we found for these experiments that resonator ringing was not a problem and critical coupling was used. (However, overcoupling will increase the bandwidth). Samples were placed inside a 1.2 mm glass capillary such that only a length of capillary equal to the length of resonator with constant B_1 was filled.

As a demonstration of recovering an entire frequency-domain spectrum from the echo, a sample of crushed gamma irradiated quartz (E' centers with $g_{\parallel} = 2.0018$, $g_{\perp} = 2.004$) (*19*) was chosen. Figure 8a shows the cw absorption signal from γ-irradiated quartz as a function of frequency offset from the TE_{102} cavity resonant frequency.[2] Due to the short T_2^* of this sample and the large deadtime typical of most ESE spectrometers, the FID from this sample could not be FT to obtain a frequency-domain spectrum. The pulse widths were both ~30 ns and the total peak power needed to produce the 180° rotation was ~1.2 W. The pulse separation, τ_1, was ~150 ns. The magnetic field and the phase of the 90° out-of-phase spectrum were adjusted, so as to

[2] It should be noted that the frequency representation of this spectrum is the reverse of the magnetic field domain so the high-field end of the spectrum is at the left.

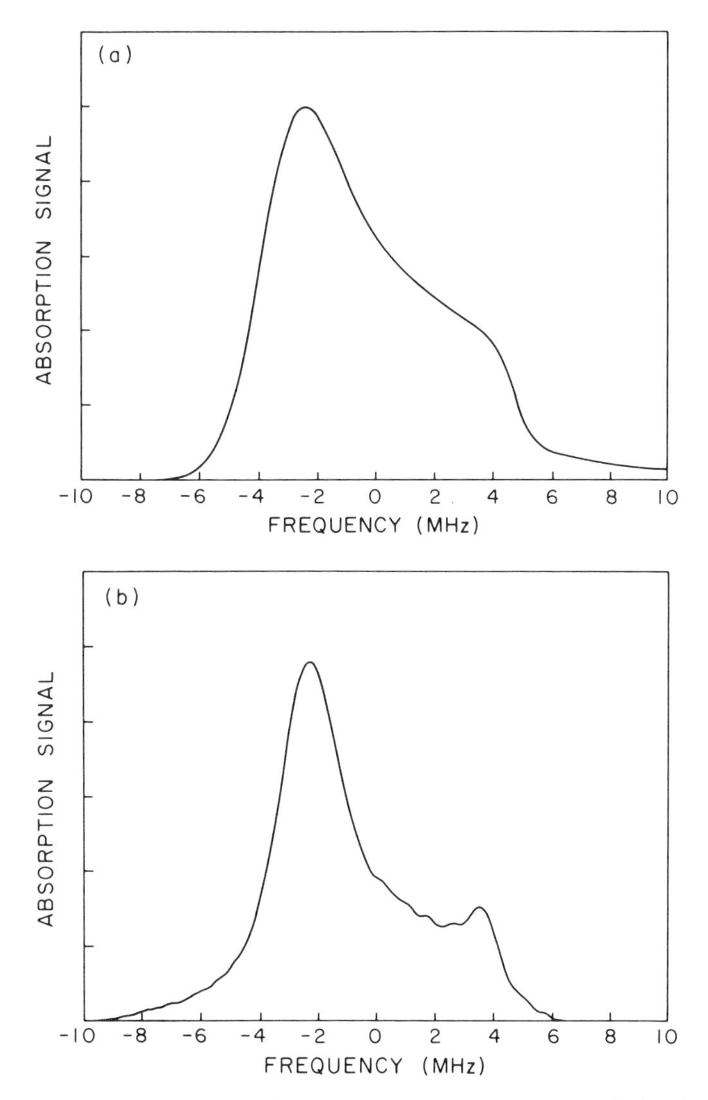

FIG. 8. The frequency-domain representation of the ESR absorption spectrum of γ-irradiated quartz ($g_{\parallel}$ = 2.0018, $g_{\perp}$ = 2.0004). Spectrum (a) was recorded using a TE_{102} cavity of a cw instrument with $\sim 2 \times 10^{-2}$ G, 100 kHz field modulation, and 32 mW of microwave power. Spectrum (b) was obtained using a loop-gap resonator by Fourier transforming the in- and out-of-phase latter half of the echo shape from a 90°–τ_1–180°–τ_2–echo sequence using pulse widths of 30 ns and B_1 = 3 and 6 G, respectively. The digitization period was 2 ns, and 2 μs of data were collected and zero filled to 8 μs. Amplitude and first-order phase corrections were applied to the transformed frequency-domain spectrum.

give the maximum symmetric echo height at a time τ_2 after the second pulse. The phase of the in-phase time-domain spectrum was adjusted so as to cause the echo height to be zero at the time when it was a maximum in the out-of-phase spectrum. The time-domain spectrum was recorded with a digitization period of 2 ns over a total time interval of 2 μs. After a baseline correction, each curve was zero filled to 4 μs and the FFT performed. Since data were collected starting at the middle (or maximum) of the echo, the true time zero data were included, and because the exact in-phase u and 90° out-of-phase v spectra were recorded, only the $(\omega - \omega_0)t_1/2$ phase correction was applied. An amplitude correction of $|\mathrm{sinc}^3(\pi(\nu - \nu_0)t_1)|$ was applied due to the relatively large frequency spectrum of the spin system (~10 MHz) compared with the width of the 30 ns pulse. The resultant absorption spectrum is shown in Fig. 8b. As expected, the Fourier-transformed spectrum did not look exactly like the cw spectrum may be due to differences in rate of relaxation of spectral components between the pulses and the echo.

Certain ESR spectra which are motionally narrowed and homogeneous do not show an echo.[3] These in turn often exhibit an intense FID which lasts much longer than the instrument deadtime and may be FT to obtain the frequency-domain spectrum. The 2,5-di-*t*-butyl-*para*-benzosemiquinone radical anion (2,5DTBSQ) (g = 2.0046, a(2H) = 2.14 G, a(18H) = 0.06 G.) (*20*) is a radical which shows no echo but a large FID. The phasing of the 90°-out-of-phase FID was chosen so as to give the maximum intensity FID. The in-phase was adjusted so as to cause what was a maximum excursion in the out-of-phase FID at a particular time to be zero. With this radical, where the ratio of the line intensities is 1:2:1, the above condition is easy to achieve when H_0 and ν_0 are set to the resonance condition for the central line of the spectrum. In this case the in-phase FID is equal to zero and the out-of-phase signal is an all-positive or all-negative FID depending on whether the phase is ±90°. The in-phase and 90° out-of-phase FIDs from the 2,5DTBSQ radical were digitized with a period of one point per ns for 1 μs. The out-of- and in-phase FIDs were input into the FFT routine as the real and imaginary parts, respectively. The output real and imaginary curves were phase corrected for a net linear phase error of $(\nu - \nu_0)\pi/14$ radians where ν is in megahertz. This was composed of a contribution from unrecorded data points at the beginning of the FID and from a trignometric correction which was nearly linear across the spectrum. Here again, the same amplitude correction factor as was used with the quartz spectrum had little influence over the width of this spectrum. The resultant spectrum is shown in Fig. 9. The shape of this spectrum is due to unresolved hyperfine lines and Heisenberg spin exchange. [To date we have obtained FT spectra from a nitroxide with a total spectral extent of 80 MHz with a 5 ns pulse width for which a $\pi/2$ pulse corresponds to B_1 = 18 G].

The corresponding time domain spectra of Figs. 7 and 8 were recorded in 50 and 25 ms worth of real time. We define real time as the total effective data collecting time. Each point in the time-domain spectra resulted from 25 averages. Based on similar experiments with these spin system in a TE_{102} microwave cavity, we calculate the S/N ratio to be a factor of 4 better using the loop–gap resonator. When the volume difference of the samples (*3*) is taken into account, this reflects a 64-fold improvement in S/N per unit volume of spins.

V. SUMMARY

In this paper we have discussed many of the aspects associated with performing pulsed FT ESR spectroscopy. Most of these are a direct carryover from FT NMR spectroscopy, where they have become so commonplace that they are now buried in automated computer software.

The application of a $\pi/2$ pulse of EMR at ν_0 to a spin system does not rotate all

[3] The usual technique of introducing field inhomogeneity by the use of a ferromagnetic object in the magnetic field would produce an echo but the resultant frequency-domain spectrum would be that of the added inhomogeneity rather than the true frequency-domain spectrum.

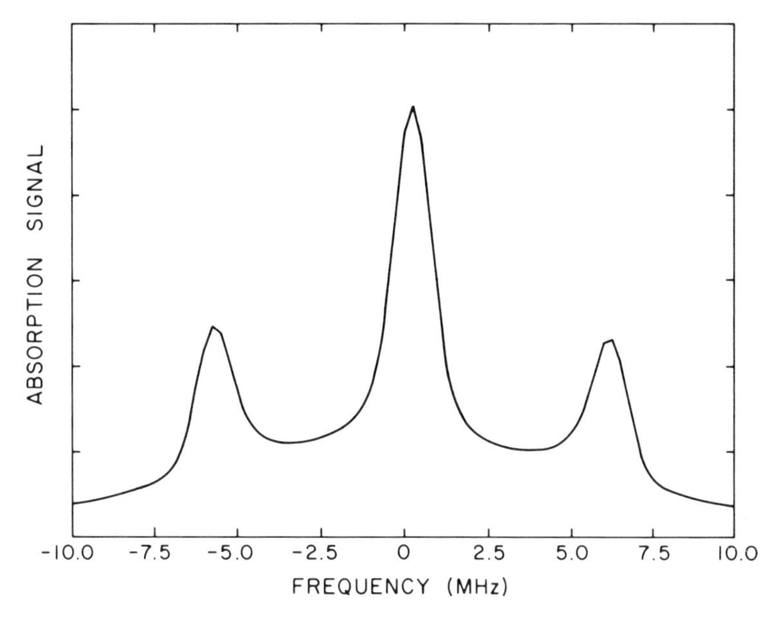

FIG. 9. The ESR absorption spectrum of the 2,5-di-*t*-butyl-*para*-benzosemiquinone radical anion [$g = 2.0046$, $a(2H) = 2.14$ G, $a(18H) = 0.06$ G]. The spectrum was obtained from the in- and out-of-phase components of the FID recorded after the application of a 30 ns pulse of $B_1 = 3$ G. The digitization period was 1 ns, and 1 μs worth of data were recorded and then zero filled to 4 μs. A linear phase correction of $(\nu - \nu_0)\pi/14$ radians and an amplitude correction factor of $|\text{sinc}^3((\nu - \nu_0)\pi/33.23)|$ with ν in MHz were used (see text). This spectrum is a good representation of the cw spectrum of this radical.

components of a spectrum equally. Those ν_0 are rotated by 90° about the x' axis. Other frequencies at $\pm\Delta\nu$ from ν_0 are rotated by a lesser extent, and the resultant transverse components are no longer along the y' axis, but their positions in the $x'y'$ plane may be estimated. The resultant FID or echo shape depends on these factors. Reconstruction of a frequency-domain spectrum requires both amplitude and phase corrections to the Fourier-transformed time-domain spectrum. Zeroeth- and any type of first-order phase corrections arise from instrumental features and are readily applied. In reconstructing a frequency-domain spectrum from an FID, a second-order phase correction arises from the angle between the rotated transverse magnetization (with frequencies displaced from resonance) and the y and the y' axis. In general the second-order phase corrections diverge when $t/\Delta\nu$ is an integer multiple of π. Reconstruction of a spectrum from an echo shape involves only a first-order phase correction. Amplitude corrections applied after phase corrections, account for magnetization at frequencies different than ν_0 being rotated by less than the ideal amount. The width of an ESR spectrum which may be reconstructed from an echo sequence, is less than that from an FID sequence for pulses of identical length.

The Fourier transform of echo and FID spectra need not be identical to the cw spectrum, even with the proper amplitude and phase corrections. In an echo experiment, spectral components with shorter T_2 may decay away during the period t of the pulse sequence, and thus not be present in the echo. The cw spectrum will not be affected by differences in relaxation times. A similar statement holds for the FID when some spectral components relax faster than others, and are lost during the spectrometer deadtime.

The advantages of FT ESR spectroscopy are, of course, the substantial time savings over cw techniques.

Such problems as are associated with deadspots in a spectrum reconstructed from an FID that are due to net rotations by 0° from the z axis may be overcome by performing two experiments with different pulse lengths. This would permit collection of a complete spectrum from all frequency components in that spectrum by moving the deadspot. The width limitation on the reconstructed echo spectrum may only be improved by decreasing the pulse widths (and/or by improving their shapes).

FT ESE spectroscopy would be expected to greatly increase the speed of two-dimensional ESE studies (*21*), since field scanning would not be necessary. It is conceivable that a 2-D spectrum recorded in 5 h by the old technique could take a few tens of minutes by the implementation of methods described here.

We have shown, in two separate examples, which are only intended to illustrate the phase and amplitude correction procedure, how an echo shape and a FID can be transformed to obtain frequency-domain spectra. These transformations, of course, require a knowledge of the true pulse shape at the sample.

REFERENCES

1. E. Fukushima and S. B. W. Roeder, "Experimental Pulse NMR, A Nuts and Bolts Approach," Addison-Wesley, Reading, Mass., 1981.

2. W. Froncisz and J. S. Hyde, *J. Magn. Reson.* **47,** 515 (1982).

3. J. P. Hornak and J. H. Freed, *J. Magn. Reson.* **62,** 311 (1985).

4. Papers presented at the 8th International EPR Symposium, Denver, Colorado, 1985: (*a*) W. Froncisz and J. S. Hyde, "Topologies of the Loop-Gap Resonator"; (*b*) A. Grupp, H. Seidel, P. Hofer, G. G. Maresch, and M. Mehring, "Design and Performance of Slotted Tube Resonators for Pulsed ESR and ENDOR"; (*c*) R. BoBrutto and J. S. Leigh, "Construction of a Compact X-Band Microwave Resonator for use with Liquid Helium Flow Systems"; (*d*) C. T. Lin, R. J. Massoth, J. R. Norris, and M. K. Bowman, "Microwave Resonator Performance In Photosynthetic Pulsed EPR Experiments."

5. A. Abragam, "The Principles of Nuclear Magnetism," Oxford Univ. Press, London, 1961.

6. (*a*) R. W. Fessenden, J. P. Hornak, and B. Venkataraman, *J. Chem. Phys.* **74,** 3694 (1981); (*b*) L. J. Schwartz, E. Meirovitch, J. A. Ripmeester, and J. H. Freed, *J. Phys. Chem.* **87,** 4453 (1983) (Appendix B).

7. L. Kevan and R. N. Schwartz, Eds., "Time Domain Electron Spin Resonance," Wiley, New York, 1979.

8. O. E. Stejskal and J. Schaefer, *J. Magn. Reson.* **13,** 249 (1974); **14,** 160 (1974).

9. M. S. Silver, R. I. Joseph, and D. I. Hoult, *J. Magn. Reson.* **59,** 347 (1984).

10. A. J. Temps, Jr., and C. F. Brewer, *J. Magn. Reson.* **56,** 355 (1984).

11. M. Bloom, J. H. Davis, and M. I. Valic, *Can. J. Phys.* **58,** 1510 (1980).

12. R. R. Ernst and W. A. Anderson, *Rev. Sci. Instrum.* **37,** 93 (1966).

13. P. Meakin and J. P. Jenson, *J. Magn. Reson.* **10,** 290 (1973).

14. "Fast Switching PIN Diodes," Hewlett Packard application note 929.

15. M. C. Young, H. W. Prinsen, R. J. Blum, and B. R. Tripp, *Proc. IEEE Lett.* **55,** 2185 (1967).

16. A. M. Johnson and D. H. Auston, *IEEE J. Quantum. Electron.* **QE-11,** 283 (1975).

17. (*a*) J. H. Freed, D. S. Leniart, and J. S. Hyde, *J. Chem. Phys.* **47,** 2762 (1967); (*b*) B. Johnson, Ph.D. Thesis, Cornell University, 1984; C. H. Barkhuijsen, R. de Beer, W. M. M. J. Bovee, and D. van Ormondt, *J. Magn. Reson.* **61,** 465 (1985).

18. A. E. Stillman and R. N. Schwartz, *J. Phys. Chem.* **85,** 3031 (1981).

19. R. H. Silsbee, *J. Appl. Phys.* **32,** 1459 (1961).

20. B. Venkataraman, Ph.D. Thesis, Columbia University, 1955.

21. (*a*) G. L. Millhauser and J. H. Freed, *J. Chem. Phys.* **81,** 37 (1984); (*b*) L. Kar, G. L. Millhauser, and J. H. Freed, *J. Phys. Chem.* **88,** 3951 (1984).

JOURNAL OF MAGNETIC RESONANCE **68**, 393–398 (1986)

High-Resolution Imaging. The NMR Microscope

C. D. ECCLES AND P. T. CALLAGHAN

Department of Physics and Biophysics, Massey University, Palmerston North, New Zealand

Received November 1, 1985; revised March 4, 1986

Nuclear magnetic resonance imaging has traditionally been applied to macroscopic objects with dimensions in excess of 1 cm. The submillimeter regime may be termed microscopic and to date has been largely unexplored. NMR microscopy is inherently difficult to perform due to the smaller signal which arises from each volume element as the resolution is enhanced. Indeed the problem is dramatically indicated by the dependence of the imaging time on the sixth power of the resolution (*1*).

Some of the early imaging experiments utilized the limited sample space of NMR spectrometers and were performed on small scale objects. Transverse resolutions of between 0.2 and 0.3 mm have been reported (*2, 3*). Since that time major developments in imaging techniques have centered on "scaling up" and the NMR imaging literature reports few, if any, examples in which microscopy is the principal objective. Hall *et al.* (*4*) have recently coined the term "chemical microscopy" to refer to the localization of chemical-shift information in small-scale samples. Using phantoms consisting of 1.6 mm i.d. capillaries they indicate a transverse resolution of 0.1 mm on long samples without slice selection. In this communication we report microscopic imaging studies carried out at 60 MHz on both phantom and plant stem samples in which the transverse resolution is 25 μm. A slice thickness of 1.5 mm takes advantage of the existing longitudinal symmetry. Plant stem microscopy is of course ideally suited to high resolution in two dimensions.

Signal-to-noise considerations are central to understanding limiting resolution. The time-domain signal to noise available in the free induction decay is simply expressed (*5*) in SI units as

$$\frac{S_t}{N_t} = \alpha^{-1/2}\left(\frac{V_s}{2V_c}\right)\left[\frac{\chi_0}{\gamma}\omega_0\right]\left[\frac{\mu_0 Q \omega_0 V_c}{4kT\Delta f}\right]^{1/2}\frac{1}{F_n} \qquad [1]$$

where α is a numerical factor of order unity which depends on the coil geometry, V_s and V_c are the sample and coil volumes, Q the quality factor, and Δf the bandwidth. F_n is the spectrometer noise figure. In an alternative expression for the signal to noise, due to Hoult and Richards, the inter-relationship of the parameters leads to a $\frac{7}{4}$ power dependence of the signal to noise on Larmor frequency, ω_0 (*6*). In small scale imaging rf penetration is not a problem and ω_0 may be maximized to advantage.

In NMR imaging the spatial dimension corresponds to the frequency domain in which signal to noise is related to the linewidth. An estimation of the resolution possible is straightforward once the limiting experimental constraints are determined. These

0022-2364/86 $3.00

are the residual line broadening, $(\pi T_2^*)^{-1}$, the receiver coil dimensions, the imaging time and the experimental repetition rate which is of order T_1. Mansfield and Morris (*7*) have shown that in the optimal case, the imaging time for a volume element $(\Delta x)^3$ is given by

$$t_{\text{vol}} = \left(\frac{S}{N}\right)^2 a^2 \left(\frac{T_1}{T_2}\right) \frac{2.8 \times 10^{-15}}{f^{7/2}} \left(\frac{1}{\Delta x}\right)^6 \qquad [2]$$

where S/N is the desired voxel signal to noise, a the coil radius, and f the spectrometer frequency in megahertz. The numerical factor arises from a presumption of ideal solenoidal geometry for the rf coil (*6*). Given a required S/N of 20, an optimistic linewidth of 3 Hz, repetition time 0.5 s, and the closest wound coil (radius $a = 128\Delta x$ for 256 pixels), it is apparent that Eq. [2] predicts a 12 μm voxel resolution for a 600 MHz spectrometer with an acquisition time of 1000 s. This resolution may be regarded as a lower limit. Of course Eq. [2] applies to the whole sample only if one uses an imaging technique (such as projection reconstruction (*8*) or Fourier zeugmatography (*9*), which simultaneously receives information from all volume elements.

In practice this resolution limit is difficult to achieve. It is, for example, based on an ideal solenoidal rf coil of radius 1.54 mm! At 600 MHz the signal may be degraded by the rf coil saddle geometry and by the difficulty in achieving low amplifier noise figures at such high frequencies. The use of high magnetic fields may also lead to line broadening or image distortion arising from chemical-shift variation or anisotropic magnetic susceptibility (*7*). While desirable, the gains from high-field NMR microscopy may not be as marked as Eq. [2] would suggest. Allowing for the factors indicated, a limiting voxel resolution of around 30 μm may be more reasonable. In contrast the presumptions inherent in Eq. [2] are more applicable at 60 MHz where an optimal voxel resolution of 45 μm is predicted. It is therefore quite reasonable to attempt microscopy at these lower frequencies.

It should be noted that microscopic imaging of biological samples such as plant stems differs from macroscopic studies in that transverse relaxation times may be considerably shorter, a consequence of proximity of water to solid surfaces. Plant stem water T_2 values are typically of order 10 ms in contrast with 50 ms values prevailing in animal tissue. The need to maximize signal to noise constrains the microscopic imaging technique in one major respect. That is, it necessitates a compression of the pulse sequence timescale in order to minimize depletion of the transverse magnetization before sampling begins. Furthermore, line broadening dictates a sampling bandwidth of order 10 kHz rather than the 1 kHz envisaged in Eq. [2]. Microscopy thus requires gradients to be larger by an order of magnitude than those available from shim coils.

A compact sequence implies the need for rectangular gradient pulses. Conventional imaging techniques employ echo refocusing to avoid problems associated with finite gradient rise times. Such methods require extra echo formation time during which the transverse magnetization is subject to relaxation. This can be avoided in microscopic applications where appropriate low inductance gradient coil design permits gradient pulse rise times shorter than one sampling interval. In our experiments we begin sampling at the point of gradient switching.

A time efficient sequence as used in this work is shown in Fig. 1. Projection reconstruction avoids the phase gradient period of Fourier zeugmatography and thus provides

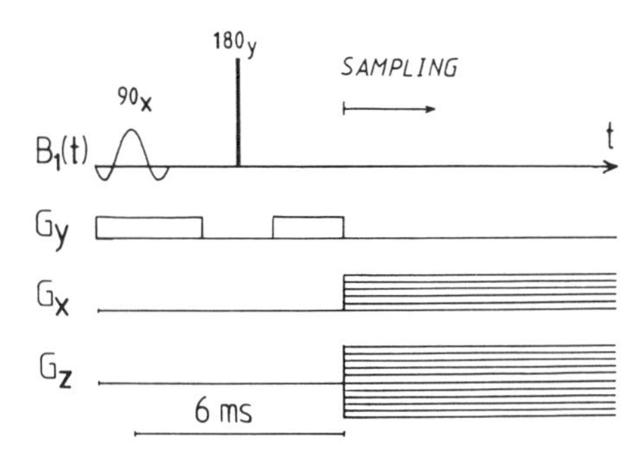

FIG. 1. Schematic of the compact pulse sequence used in the projection reconstruction imaging experiment. The start of data acquisition is simultaneous with the switching of the three gradients. No first-order phase correction is necessary and the zeroth-order phase is constant to within 5° throughout the gradient vector rotation.

the shortest time span. We have used since rf modulation in the 90° selection pulse (*10*) in order to stimulate a rectangular slice.

An alternative expression for the resolution may be obtained by considering the mutual dependence of transverse resolution, Δx, and voxel signal to noise (S/N) as the applied magnetic field gradient, G, is varied. The procedure consists in transforming the time-domain signal-to-noise expression to the frequency domain and eliminating G so as to formulate Δx in terms of desired (S/N).

We have calculated the resolution Δx resulting from the optimum signal to noise of Eq. [1] in which we take $Q = 100$ with $F_n = 1$ with the coil length equal to twice the radius a. We note that the signal to noise so calculated agrees well with the Hoult and Richards formula at an operating frequency of 60 MHz. We find

$$\Delta x = (3.1 \times 10^5 a^{1/2})\left(\frac{S}{N}\right)^{1/2} \Delta y^{-1/2} f^{-7/8} \Delta f^{1/4} N_{\text{acc}}^{-1/4} N_{\text{p}}^{-1/4} N^{-1/4}\, \mathcal{S}^{-1/2}\ \mu\text{m}. \qquad [3]$$

The equivalent voxel resolution is of course $(\Delta x^2 \Delta y)^{1/3}$ where Δy is the slice thickness in micrometers.

Equation [3] allows for the influence of coil radius, for variable filter and sampling bandwidth, Δf (in hertz), digitization points, N, number of projections, N_p, number of accumulations per projection, N_{acc}, and smoothing factor $\mathcal{S}$ arising from exponential filtering. For broadening corresponding to Lorentzian separation of 2 pixels, $\mathcal{S}$ is 2.99.

Our work employs a 60 MHz spectrometer (JEOL FX 60), modified for quadrature detection, in which the probe has been constructed specifically for microscopic imaging. A solenoidal rf coil of radius 2.8 mm is used and the probe incorporates two 20 mm diameter quadrupolar coils capable of providing up to 200 G cm^{-1}. These coils are used to provide the transverse (xz) imaging gradients. The y-axis shim provides slice selection gradients of up to 1.5 G cm^{-1}. The imaging system uses a specially constructed pulse programmer with 16 switching and 4 D/A outputs for rf, gradient and sampling control. Gradient currents are provided by three Kepco operational power supplies with current rise times of less than 50 μs. The data are processed as a 256 × 256 array using an 8088-based Hitachi MB 16 000 p.c. Reconstructions were typically performed

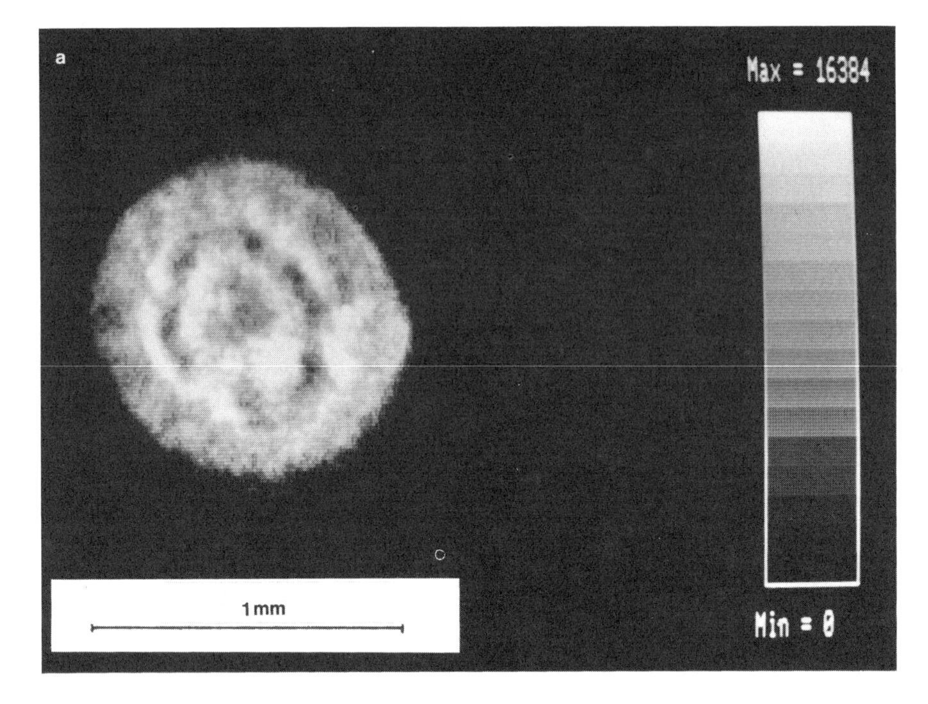

b

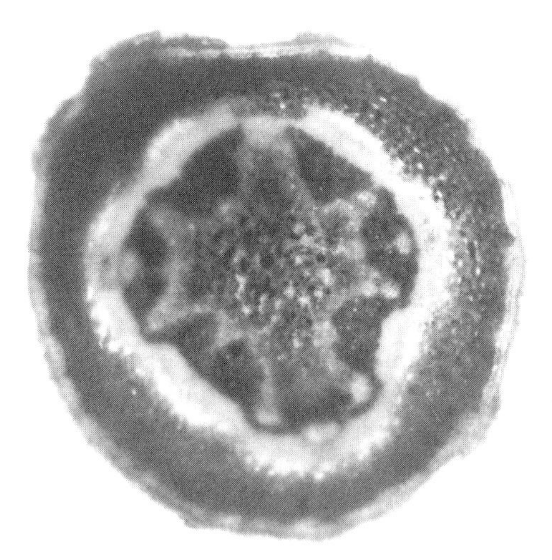

FIG. 2. (a) Proton density image from an intact stem of Alyssum Tenium. The slice thickness is 1.5 mm with a transverse resolution of 20 μm. Reconstruction was performed using 2° steps, the total imaging time being 36 min. (b) Optical microscope image obtained from a cut stem of Alyssum Tenium.

using 90 projections, the total acquisition time being 36 min for the images displayed here. The probe signal to noise is close to the optimum of Eq. [1] and corresponds to a spectrometer noise figure of order 3 dB.

Figure 2a shows the image obtained for a 1.5 mm slice from an intact stem of Alyssum Tenium. The imaging pulse sequence had a total extent of 7 ms, the water T_2 being 30 ms. The data were obtained at 20 kHz spectral width with transverse resolution set at 20 μm by the choice of exponential filtering. The vascular bundles are resolved and may be compared with the optical image (Fig. 2b) obtained with a cut slice from a similar stem.

Phantom construction is difficult on the microscopic scale. We have prepared fine capillaries by drawing glass tubes under a flame. Ten such capillaries were close packed and glued in two layers separated by paper of thickness 80 μm, filled with doped water (T_2 = 5.8 ms) and sealed. The capillary internal diameters ranged from 120 to 260 μm. Figure 3 shows the image obtained on a 1.5 mm slice using the same pulse sequence as in Fig. 2. The transverse resolution is about 25 μm which is of the order of the optimum predicted by Eq. [3]. In both images the signal to noise exceeded 20.

The voxel resolution apparent in our images is of order 90 μm but this could be improved by stimulating a narrower slice. We note that the tranverse resolution is limited here not by signal to noise but by artifacts associated with projection recon-

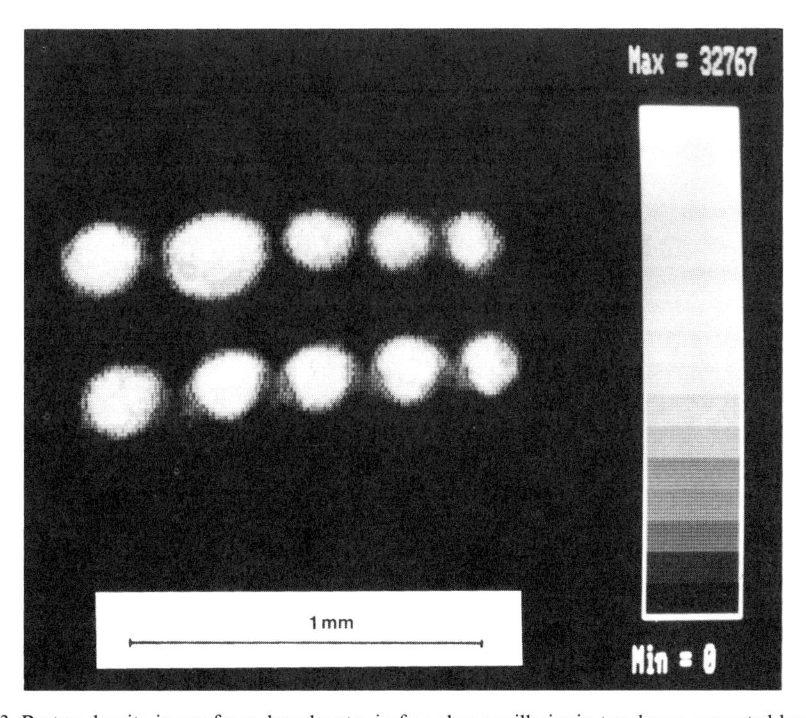

FIG. 3. Proton density image from doped water in fine glass capillaries in two layers separated by paper. The slice thickness is 1.5 mm and the transverse resolution 25 μm. Reconstruction was performed using 2° steps, the total imaging time being 36 min. The compact pulse sequence permits high signal strength despite the short T_2 value of 5.8 ms.

struction. While the gradient uniformity is better than 1% the coil orthogonality is sufficient to allow only 2% accuracy in reconstruction. Corrections can be applied but it may be preferable to apply Fourier zeugmatography in which such artifacts do not directly influence the image resolution. Despite these limitations the microscopic uses of the technique are apparent.

Finally we note that our imaging system has been designed to produce magnetic field gradients of sufficient magnitude for pulsed gradient spin echo experiments (*11, 12*). The capacity to perform localized time-resolved molecular translation measurements makes the NMR microscope ideally suited to the study of water and nutrient transport in intact plant tissue.

REFERENCES

1. P. Brunner and R. R. Ernst, *J. Magn. Reson.* **33,** 83 (1979).
2. W. S. Hinshaw, *J. Appl. Phys.* **47,** 3709 (1976).
3. P. C. Lauterbur, *Pure Appl. Chem.* **40,** 149 (1974).
4. L. D. Hall, V. Rajanayagam, and S. Sukumar, *J. Magn. Reson.* **60,** 199 (1984).
5. A. Abragam, The Principles of Nuclear Magnetism," Chap. 3. Oxford Univ. Press. London/New York, 1961.
6. D. I. Hoult and R. E. Richards, *J. Magn. Reson.* **34,** 71 (1976).
7. P. Mansfield and P. G. Morris, "Advances in Magnetic Resonance," Suppl. 2, "NMR Imaging in Biomedicine" (J. S. Waugh, Ed.), Academic Press, New York, 1982.
8. P. C. Lauterbur, *Nature (London)* **242,** 190 (1973).
9. A. Kumar, D. Welti, and R. R. Ernst, *J. Magn. Reson.* **18,** 69 (1975).
10. D. I. Hoult, *J. Magn. Reson.* **35,** 69 (1979).
11. E. O. Stejskal and J. E. Tanner, *J. Chem. Phys.* **42,** 288 (1965).
12. P. T. Callaghan, *Aust. J. Phys.* **37,** 359 (1984).

JOURNAL OF MAGNETIC RESONANCE 72, 502–508 (1987)

Localized Proton Spectroscopy Using Stimulated Echoes

JENS FRAHM, KLAUS-DIETMAR MERBOLDT, AND WOLFGANG HÄNICKE

Max-Planck-Institut für biophysikalische Chemie, Postfach 2841, D-3400 Göttingen, Federal Republic of Germany

Received October 9, 1986

This paper describes a new method for spatially resolved NMR spectroscopy that takes advantage of stimulated echo signals. STEAM (*st*imulated *e*cho *a*cquisition *m*ode) sequences, already used for a variety of imaging purposes, almost perfectly match the requirements of image-controlled localized ^{1}H NMR *in vivo.* Superior spatial discrimination as well as high flexibility with respect to location, size, and shape of the volume of interest is achieved by employing only three slice-selective 90° rf pulses in the presence of orthogonal gradients. The method is a single-step procedure minimizing rf power requirements and gradient switches. It further allows accurate determinations of localized T_1 and T_2 relaxation times simply by varying the length of corresponding intervals of the STEAM sequence. In fact, the inherent T_2 weighting may be used for water suppression and/or reduction of residual eddy current effects. Here we present first results on phantoms and human extremities demonstrating the ease of image selection, localized spectroscopy, and localized determinations of relaxation times. Future steps will deal with water/lipid-suppressed metabolic spectroscopy.

INTRODUCTION

Spatially resolved NMR, in particular localized high-resolution NMR spectroscopy *in vivo,* may become an important tool not only for noninvasive biochemical and biophysical research but also for medical diagnosis. In terms of technical approaches, one may generally distinguish between "global" and "local" methods of acquiring both spatial and spectral information in a combined imaging/spectroscopy experiment. Global methods are four- (or three-) dimensional chemical-shift imaging techniques yielding data sets with three (or two) spatial dimensions and one chemical-shift dimension. Obvious problems are the long measuring times, the handling and storage of very large data matrices, and the fact that there seldom is a real interest in truly global information. Moreover, spectroscopic *imaging* techniques often sacrifice spectral resolution and spectral quantification in order to keep imaging times short. Global methods also suffer from limited B_0 and B_1 homogeneities in cases where water suppression techniques have to be applied over large volumes.

Local methods comprise *spectroscopy* techniques that attempt to focus on a selected volume of interest (VOI) ideally defined by previous imaging investigations. The most successful technique up to now has been surface coil spectroscopy (*1*), where localization is achieved by radiofrequency (rf) gradients, i.e., the B_1 profile of the surface coil or arrangement of coils. Improvements with respect to a better spatial characterization of the surface coil spectra have been obtained using rotating frame imaging techniques

0022-2364/87 $3.00

(*2, 3*) including Fourier windowing (*4*). Surface coils have also been combined with magnetic field gradients for slice selection in the DRESS experiment (*5*). More recently, approaches using magnetic field gradients in combination with homogenous rf coils have regained interest in order to make the localization of a VOI more flexible with respect to location, size, and shape. The first promising techniques are ISIS (*6*) and SPARS (*7*), the latter of which being a derivative of the VSE experiment (*8*).

In most cases new sequences have been developed to be applicable for protons as well as for heteronuclei. However, the actual requirements for proton and, for example, phosphorus spectroscopy are very different and under certain circumstances even contradictory: extremely short T_2 weightings for phosphorus to keep ATP resonances visible versus strong T_2 weightings for water-suppressed proton spectroscopy of mobile metabolites. Consequently, it would be advisable to specifically design a localization technique for proton and phosphorus spectroscopy, respectively. In this paper we describe a new technique for localized ^{1}H NMR spectroscopy taking such considerations into account. It is able to provide strong T_1 and T_2 weightings and directly allows image selection and characterization of the selected volume.

METHODS

In order to spatially select a volume of interest within a three-dimensional object it is necessary to apply at least three selective rf pulses in the presence of orthogonal magnetic field gradients. This condition is almost perfectly met by stimulated echo sequences. In fact, STEAM (*st*imulated *e*cho *a*cquisition *m*ode) imaging sequences have already been used for a variety of purposes in improving the imaging specificity (*9–16*). For localization all three rf pulses of the STEAM sequence are taken as slice-selective pulses in the presence of orthogonal gradients as demonstrated in Fig. 1. Without the slice selection gradients in the read and phase-encoding direction, i.e., rf pulses No. 1 and No. 2, the sequence is a standard single-slice STEAM imaging sequence that may be used for selection of the VOI for the subsequent spectroscopic investigation. By inclusion of these gradients the sequence in Fig. 1 allows direct image control of the VOI if desired. Another interesting application might be zoom imaging

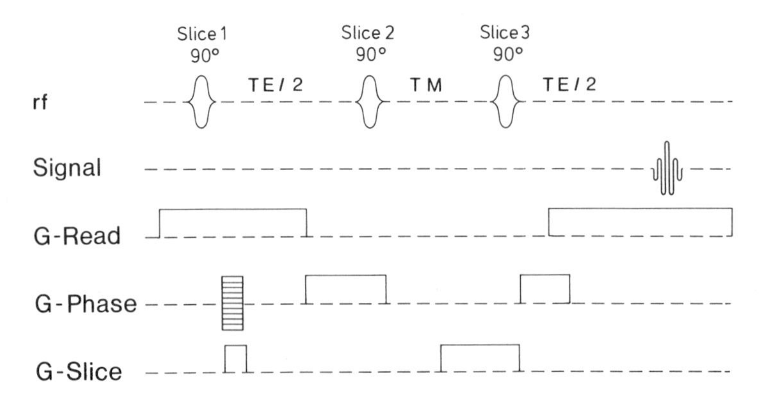

FIG. 1. Radiofrequency and magnetic field gradient sequence for volume-selective STEAM imaging.

or even localized microscopic imaging by restricting the field of view and using increased strengths for the read and phase-encoding gradients.

On the other hand, localized spectroscopy as shown in Fig. 2 is easily achieved by omitting both the phase-encoding gradient and the read gradient in Fig. 1 while retaining the slice-selective parts. In general, the procedure may include the following steps: conventional (multislice) STEAM imaging using a single slice-selective 90° pulse, selection of a VOI, imaging of the VOI using the sequence shown in Fig. 1, and localized spectroscopic examinations using the sequence shown in Fig. 2. Basically, localization on the basis of stimulated echoes is a direct "single-step" method that, for example, allows reshimming of the homogeneity over the VOI. The choice of a position or a certain size and shape of the VOI is under computer control and may be easily adjusted by changing the frequency and/or the shape of the selective rf pulses and/or the strength of the slice-selective gradients. Since only three selective 90° pulses are employed, localized STEAM spectroscopy will be applicable to any small-bore and whole-body NMR system with a very low rf load on animals or patients. Moreover, the spatial selectivity of the sequence is not affected by misadjustments of the pulse flip angles. Residual eddy current effects can be circumvented by long echo times.

At this stage the sequence shown in Fig. 2 offers elegant determinations of local T_1 and T_2 relaxation times. This can be accomplished simply by varying the length of the corresponding intervals TM or TE in a series of experiments without the need of additional pulses. Metabolic spectroscopy, however, requires a suppression of water and possibly also lipid proton signals. Also for this purpose, one may exploit the inherent suppression capabilities of the sequence, i.e., independent adjustments of strong T_1 (lipid resonances) and T_2 (water resonance) weightings. In addition, the basic STEAM sequence may be combined with any conventional suppression technique such as presaturation or frequency-selective spin-echo formation.

It should be noted that there are two extensions of the basic experiment that may improve the efficiency of the spectroscopic examinations in all cases where multiple accumulations are recommended. The first modification is a multislice or multipoint version of the sequence shown in Fig. 2 obtained by multiple applications of the third pulse while shifting its center frequency (at the expense of slightly increased TM values).

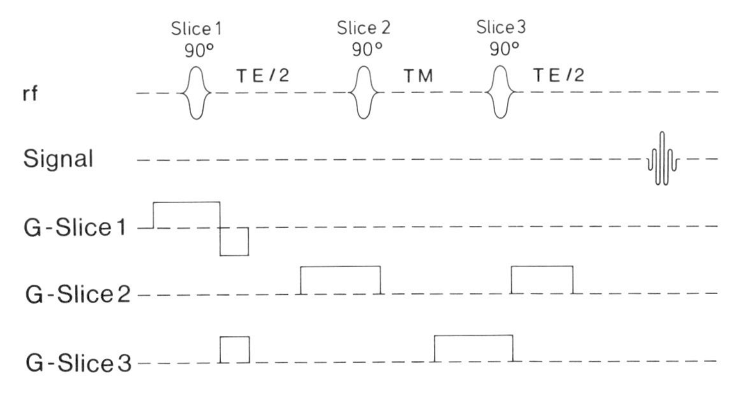

FIG. 2. Radiofrequency and magnetic field gradient sequence for volume-selective STEAM spectroscopy.

An alternative way of screening an extended VOI would be the 3D analog to multislicing: only two slice selection pulses (or a third with increased bandwidth/lowered gradient strength) are applied and combined with a phase-encoding gradient in the third direction, e.g., applied during the first interval of the STEAM sequence. This version simultaneously records spectra of an extended region ("line," compare Figs. 3b and 3c) in the form of a two-dimensional array with one (course) spatial dimension and one high-resolution chemical-shift dimension with the full signal-to-noise related to all excitations.

EXPERIMENTAL RESULTS

Proton NMR images and localized spectra (100 MHz) of phantoms and human extremities have been obtained using a 2.35 T 40 cm bore magnet system (Bruker Medspec). Both imaging and spectroscopy experiments were carried out using the standard imaging coil (slotted tube resonator) with a clear diameter of 22 cm. Figures 3a–d describe the first two steps of image-controlled STEAM spectroscopy, image selection, and characterization and gives an impression of the spatial selectivity. The images of a water bottle comprising five different test samples (2.5 cm outer diameter) show the entire object (Fig. 3a) and some selected volumes that have been recorded

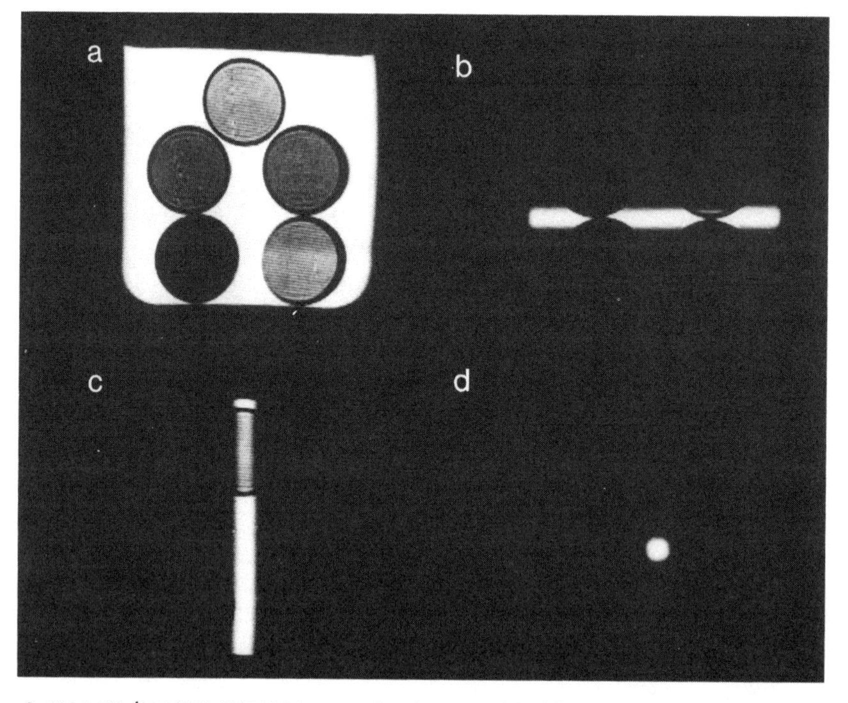

FIG. 3. 100 MHz ^{1}H NMR STEAM images of a phantom with different degrees of volume selection. (a) Conventional STEAM image with a slice thickness of 5 mm, (b and c) volume-selective STEAM images with a second slice-selective rf pulse either in the read or phase-encoding direction cutting out a line of image (a) with a thickness of 5 mm, and (d) full volume-selective STEAM image with three slice-selective rf pulses according to Fig. 1. The VOI is a 5 × 5 × 5 mm cube corresponding to a volume of 0.125 ml.

with two (Figs. 3b and c) and three (Fig. 3d) slice-selective rf pulses in accordance with the sequence shown in Fig. 1. The intervals correspond to TE = 30 ms and TM = 40 ms. For a measured resolution of 128 × 256 complex data points with a single excitation and a repetition time of TR = 2 s the imaging time was about 4.5 min. Of course, this time may be reduced using fast imaging techniques (*13, 17*). The slice thickness was adjusted to 5 mm using 2 ms Gaussian-shaped selective rf pulses and 7.5 mT/m slice selection gradients. The VOI in Fig. 3d is therefore limited to a 5 × 5 × 5 mm cube corresponding to a volume of 0.125 ml. The images clearly demonstrate the high degree of spatial discrimination achieved in a single experimental run without the need of compensation or add/subtract algorithms. In this example the spatial discrimination factor is more than 3000 with respect to the volume of the entire object. This finding also applies to the volume-selective spectra shown in Figs. 4a–e.

While the image shown in Fig. 4 represents an overlap of a conventional STEAM image with some volume-selective images, Figs. 4a–e contain the related localized proton spectra. The spectra are magnitude representations and have been recorded with two acquisitions using a 180° phase cycling of the final 90° pulse. Since our hardware has not yet been optimized to minimize eddy current effects in the magnet induced by switching magnetic field gradients, some of the spectra are slightly distorted due to the short echo time of TE/2 = 15 ms. In addition to the individual spectra, complete relaxation curves have been obtained by varying TM and TE in a series of

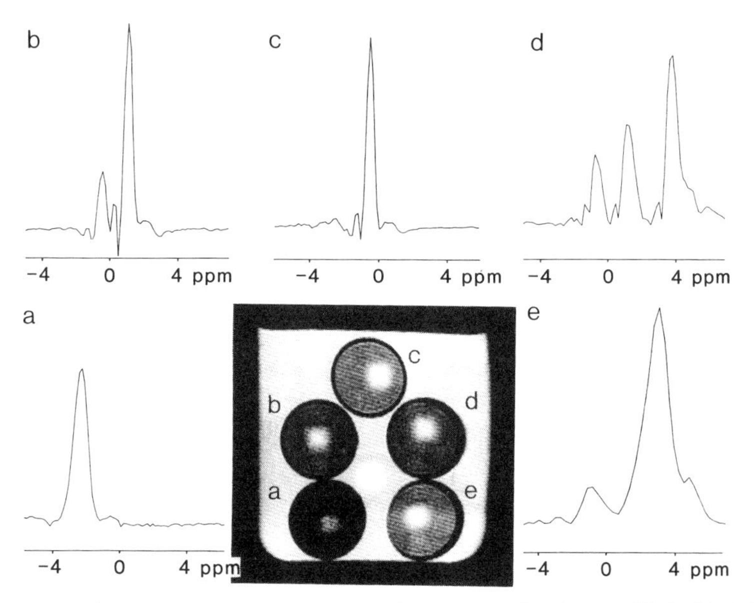

FIG. 4. 100 MHz ^{1}H NMR STEAM image and localized spectra of a phantom filled with $CuSO_4$-doped water and five different test samples containing (a) benzene, (b) methanol, (c) distilled water, (d) ethanol, and (e) vegetable oil. The conventional image is superimposed with volume-selective STEAM images to characterize the spatial selectivity of the spectroscopic examinations. The VOI is a 5 × 5 × 5 mm cube corresponding to a volume of 0.125 ml. (a)–(e) Localized 100 MHz ^{1}H NMR STEAM spectra (TE = 30 ms, TM = 40 ms, 2 acquisitions, 3 s repetition time, magnitude representations).

experiments with high accuracy. A first application *in vivo* is depicted in Fig. 5 where the VOI (0.125 ml) has been focused on the muscle tissue of a human forearm. The T_1 determination of muscle water protons yields 1.20 ± 0.05 s (2.35 T). The measuring time of the entire experiment was 30 s.

CONCLUDING REMARKS

A new method for image-controlled localized proton spectroscopy using stimulated echoes has been presented. Its initial steps include image selection and characterization, localized proton spectroscopy, and localized determinations of T_1 and T_2 relaxation times. Basically, the method is a single-step procedure with superior spatial selectivity insensitive to flip angle misadjustments and not dependent on add/subtract algorithms. The sequence allows reshimming of the homogeneity over a selected volume as well as computer-controlled adjustments of location, size, and form of the VOI. Since only three slice-selective 90° rf pulses are employed, the sequence is directly applicable to any small-bore and whole-body NMR system with a very low rf load on animals or patients.

Although the inherent T_1 and T_2 relaxation time weightings of the method become a severe drawback for the investigation of components with short relaxation times,

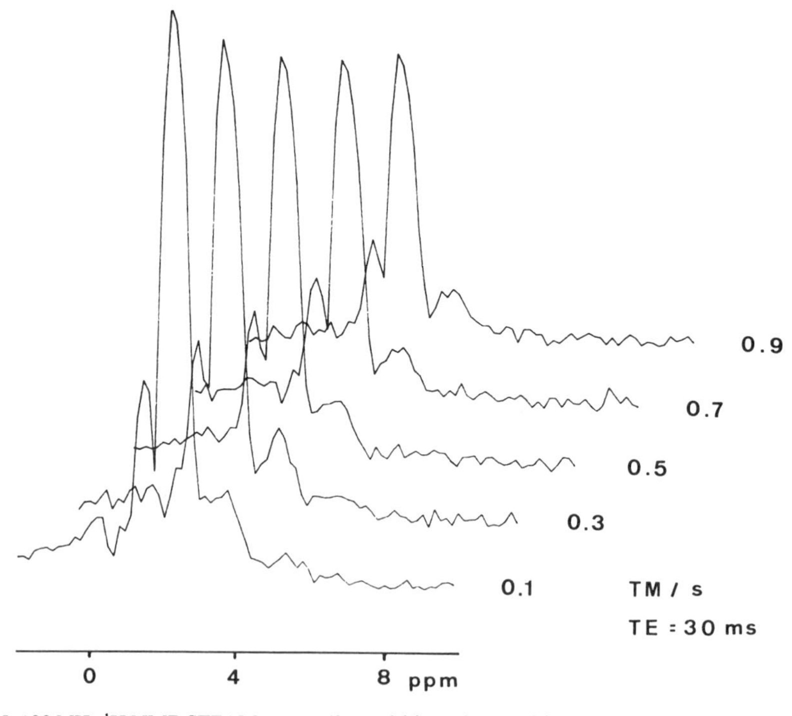

FIG. 5. 100 MHz ^{1}H NMR STEAM spectra (2 acquisitions, 3 s repetition time, magnitude representations) of a selected VOI (5 × 5 × 5 mm corresponding to 0.125 ml) within the muscle tissue of a human forearm. The series of spectra (TE = 30 ms, TM as indicated) describe the spin–lattice relaxation curve yielding T_1 = 1.20 ± 0.05 s.

e.g., ATP using phosphorus spectroscopy, they may be exploited for proton spectroscopy with respect to the suppression of water and/or lipid resonances as well as for eddy current stabilization. Our next steps in metabolic spectroscopy will include further combinations with presaturation and selective spin-echo formation as well as with the adaptation of more complex spectroscopic "editing" techniques.

A limitation in principle of stimulated echo sequences is the signal reduction by a factor of two when compared to a spin echo obtained at the same echo time. Although this does not seem to affect localized determinations of relaxation times for medical purposes, it may contribute to the sensitivity problem encountered for metabolic concentration levels. A possible alternative might be a 90°-180°-180° spin-echo sequence either using improved slice-selective refocusing pulses or compensating dual experiments (*18*).

ACKNOWLEDGMENT

Financial support by the Bundesminister für Forschung und Technologie (BMFT) of the Federal Republic of Germany (Grant 01 VF 242) is gratefully acknowledged.

REFERENCES

1. J. J. H. Ackerman, T. H. Grove, G. G. Wong, D. G. Gadian, and G. K. Radda, *Nature (London)* **283,** 167 (1980).
2. D. I. Hoult, *J. Magn. Reson.* **33,** 183 (1979).
3. P. Styles, C. A. Scott, and G. K. Radda, *Magn. Reson. Med.* **2,** 402 (1985).
4. M. J. Blackledge, P. Styles, and G. K. Radda, *J. Magn. Reson.,* in press.
5. P. A. Bottomley, T. H. Foster, and R. D. Darrow, *J. Magn. Reson.* **59,** 338 (1984).
6. R. J. Ordidge, A. Connelly, and J. A. B. Lohman, *J. Magn. Reson.* **66,** 283 (1986); R. Tycko and A. Pines, *J. Magn. Reson.* **60,** 156 (1984).
7. P. R. Luyten, A. J. H. Marien, B. Sijtsma, and J. A. Den Hollander, *J. Magn. Reson.* **67,** 148 (1986).
8. W. P. Aue, S. Müller, T. A. Cross, and J. Seelig, *J. Magn. Reson.* **56,** 350 (1984).
9. J. Frahm, K. D. Merboldt, W. Hänicke, and A. Haase, *J. Magn. Reson.* **64,** 81 (1985).
10. A. Haase and J. Frahm, *J. Magn. Reson.* **64,** 94 (1985).
11. K. D. Merboldt, W. Hänicke, and J. Frahm, *J. Magn. Reson.* **64,** 479 (1985).
12. A. Haase and J. Frahm, *J. Magn. Reson.* **65,** 481 (1985).
13. J. Frahm, A. Haase, K. D. Merboldt, W. Hänicke, and D. Matthaei, *J. Magn. Reson.* **65,** 130 (1985).
14. K. D. Merboldt, W. Hänicke, and J. Frahm, *J. Magn. Reson.* **67,** 336 (1986).
15. W. Sattin, T. H. Mareci, and K. N. Scott, *J. Magn. Reson.* **64,** 177 (1985).
16. W. Sattin, T. H. Mareci, and K. N. Scott, *J. Magn. Reson.* **65,** 298 (1985).
17. A. Haase, J. Frahm, D. Matthaei, W. Hänicke, and K. D. Merboldt, *J. Magn. Reson.* **67,** 258 (1986).
18. J. Frahm and W. Hänicke, *J. Magn. Reson.* **60,** 320 (1984).

JOURNAL OF MAGNETIC RESONANCE **73,** 574–579 (1987)

A Practical Approach to Three-Dimensional NMR Spectroscopy

C. GRIESINGER, O. W. SØRENSEN, AND R. R. ERNST

Laboratorium für Physikalische Chemie, Eidgenössische Technische Hochschule, 8092 Zürich, Switzerland

Received March 10, 1987

Ever since the first two-dimensional NMR experiments (*1, 2*) it has been clear that the principles of 2D spectroscopy (*3*) can be extended to three or even higher dimensions. Such experiments have indeed been performed in the context of NMR imaging, displaying all three spatial coordinates (*4*) or mapping spatially resolved chemical-shift information (*5*). On the other hand, it has been thought that analogous high-resolution 3D NMR experiments are impracticable because of huge data matrices and long measurement times. Also the potential usefulness of 3D NMR has been questioned as the information contained in a 3D spectrum can sometimes be extracted from two separate 2D spectra.

In this paper we describe a practical approach to 3D NMR spectroscopy, suggest a technique for reduction of data matrices, and mention applications where 3D spectra can provide information that would be difficult or impossible to obtain from a series of 2D experiments.

Three-dimensional pulse sequences can be derived by a combination of two 2D pulse sequences as illustrated in Fig. 1. The two experiments are linked together leaving out the detection period of the first experiment and the preparation period of the second. In analogy to the construction principle in Fig. 1, it seems natural to name 3D experiments according to the 2D experiments from which they arise. Interesting examples of 3D experiments are the combinations COSY–COSY, COSY–NOESY, and NOESY–COSY (see Fig. 2). This nomenclature is in accordance with the 3D experiment COSY-J proposed by Plant *et al.* (*6*).

In 3D experiments, the free induction decay is recorded as a function of t_3 for two additional time parameters t_1 and t_2 which are incremented independently from experiment to experiment. This leads to a 3D time-domain data matrix $s(t_1, t_2, t_3)$ that is subjected to triple Fourier transformation. The minimum number of individual experiments $N_1 \times N_2$, where N_1 and N_2 are the numbers of points in the t_1 and t_2 dimensions, respectively, is normally huge. A technique for reducing the number of experiments without sacrificing resolution is therefore needed. For this purpose we apply selective pulses for the selection of restricted volumes in 3D spectra in analogy to a recently proposed technique of selective 2D spectroscopy (*7*). This is demonstrated in the following for the COSY–COSY experiment.

While a 2D COSY spectrum contains two types of peaks, namely diagonal and cross peaks, there are three relevant types in 3D spectra. The 3D *diagonal peaks* arise from coherences which are not transferred by either of the two mixing processes. These peaks lie on the 3D diagonal $\omega_1 = \omega_2 = \omega_3$. The 3D *cross peaks* are peaks where all

0022-2364/87 $3.00

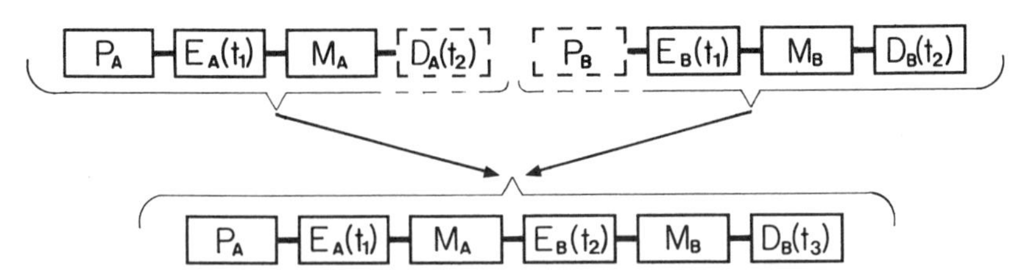

FIG. 1. General scheme for design of 3D experiment sequences by combination of two 2D experiment sequences.

three coordinates ω_1, ω_2, ω_3 are different; i.e., coherence of spin I_1 is transferred by the two mixing processes via a relay spin I_2 to a third spin I_3. For the third type of peaks two coordinates are identical, i.e., $\omega_1 = \omega_2$, $\omega_2 = \omega_3$, or $\omega_1 = \omega_3$. We call such peaks *cross-diagonal peaks.* The $\omega_1 = \omega_2$ and $\omega_2 = \omega_3$ cross-diagonal peaks arise when no coherence transfer takes place during the first and second mixing period, respectively. The $\omega_1 = \omega_3$ cross-diagonal peaks arise from back-transfer, i.e., represent coherence that starts and ends on the same spin, being on a different spin during the t_2 period.

For illustration we describe a soft COSY–COSY experiment applied to the three-spin system of 2,3-dibromopropionic acid. The full 3D spectrum contains $3 \times 3 \times 3 = 27$ multiplets each consisting of $4 \times 4 \times 4 = 64$ individual components. By multiplet-selective pulses it is possible to select one of the 27 3D multiplets. The pulse sequence is shown in Fig. 3. A selective excitation pulse on spin I_1 restricts the frequency range in the ω_1 dimension. The coherence transfer from I_1 to I_2 is effected by a pair of I_1- and I_2-selective pulses, and similarly the transfer I_2 to I_3 is effected by a pair of I_2- and I_3-selective pulses. Appropriate phase cycling selects the pathways indicated in the figure.

The two dotted pulses in Fig. 3 are optional. They influence the 3D multiplet pattern and the sensitivity. Inclusion of the two pulses results in a full 64-component multiplet

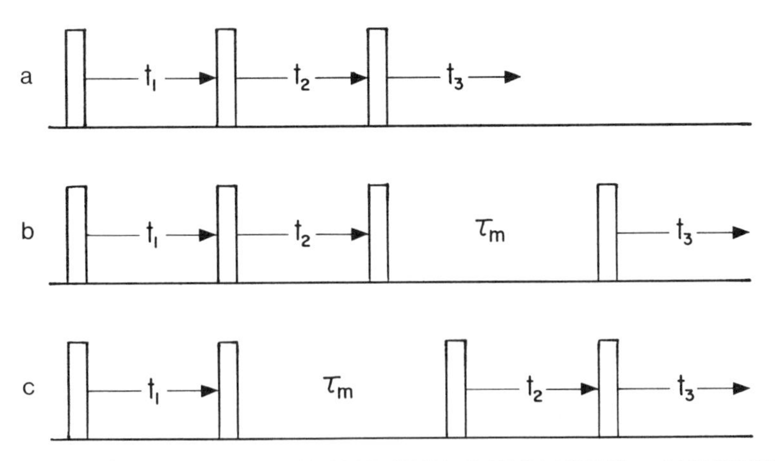

FIG. 2. Examples of 3D pulse sequences: (a) COSY–COSY, (b) COSY–NOESY, and (c) NOESY–COSY. Pure 3D absorption peakshapes can be obtained by extension of the procedures for obtaining pure 2D absorption in 2D spectra, e.g., by independent TPPI of the pulses preceding the first and the second mixing periods followed by real Fourier transformation in t_1 and t_2.

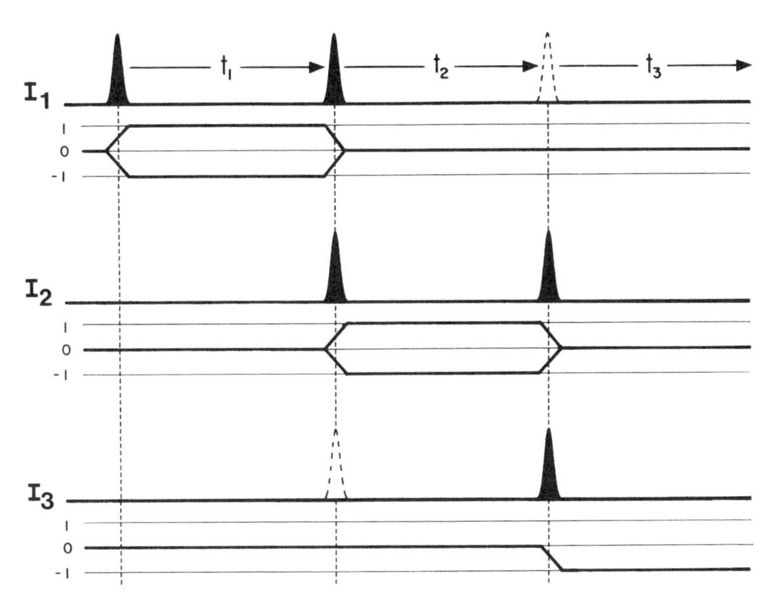

FIG. 3. Pulse sequences for soft COSY–COSY with coherence transfer pathways for the three individual spins. The basic five-pulse experiment (solid lines) results in a 16-component cross-peak multiplet for a three-spin system. All nominal pulse angles are 90° on resonance. Each of the two optional pulses (dotted lines) doubles the number of multiplet components. In the actual experiment, a four-step phase cycle consisting of independent 0, π phase alternation of the first I_1- and I_2-selective pulses is required.

identical to the result of a nonselective COSY–COSY experiment (Fig. 2a). By restricting coherence transfer with the selective five-pulse sequence, the number of multiplet components is reduced by a factor four in the case of a three-spin system in analogy to soft COSY (*7*) and E. COSY (*8, 9*). The sensitivity is thereby improved by a factor four in comparison with the nonselective experiment.

Figure 4b shows the 3D $I_1 \rightarrow I_2 \rightarrow I_3$ soft COSY–COSY cross peak in 2,3-dibromopropionic acid represented by four stacked 2D sections. It illustrates the characteristic arrangement of the 16 multiplet components. The construction principles become clearer by considering the three orthogonal sections indicated in Fig. 4b and shown in Fig. 5. Each section contains four peaks that arise from a particular coherence-transfer pathway tree as follows.

2D section with ω_1 fixed:

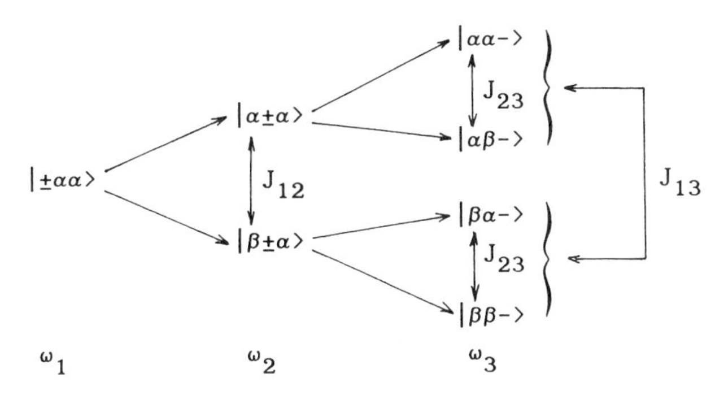

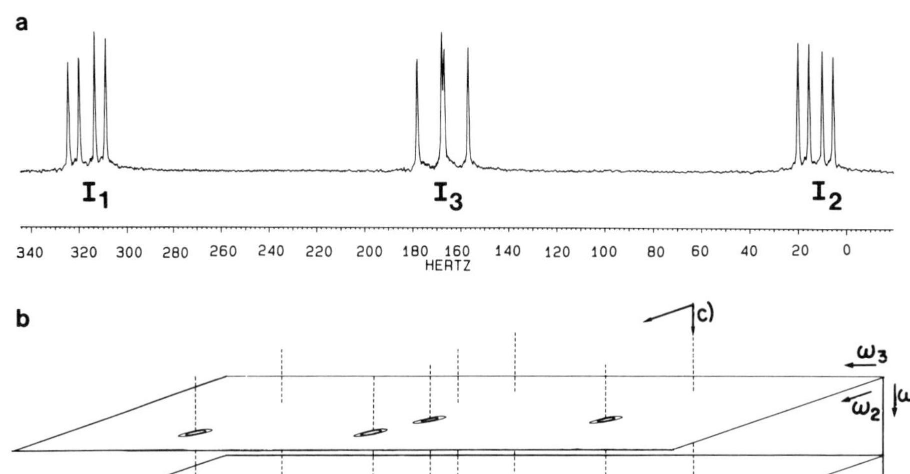

FIG. 4. (a) A 1D spectrum of 2,3-dibromopropionic acid. The period in which the spins are active in the 3D experiment of Fig. 3a is indicated in parentheses. (b) $I_1 \rightarrow I_2 \rightarrow I_3$ soft COSY–COSY cross-peak multiplet of 2,3-dibromopropionic acid consisting of 16 individual components. The 3D data matrix (48 × 48 × 128) was zero-filled to 128 × 512 × 1024 points prior to 3D Fourier transformation. Three 2D sections (a), (b), and (c) are plotted in Fig. 5.

2D section with ω_2 fixed:

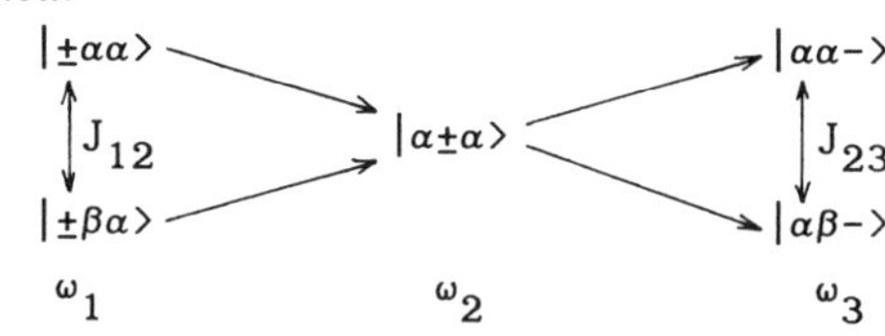

2D section with ω_3 fixed:

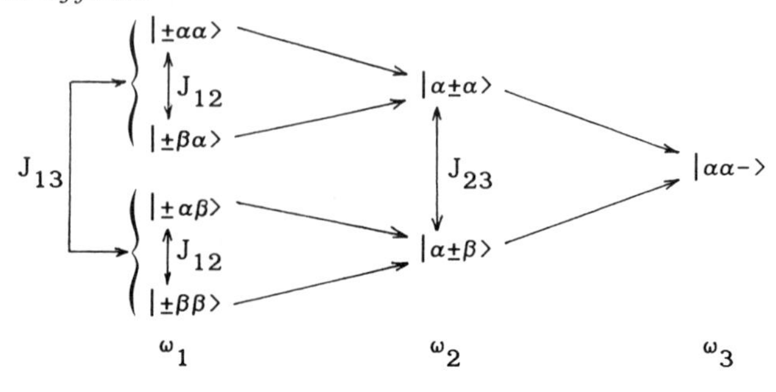

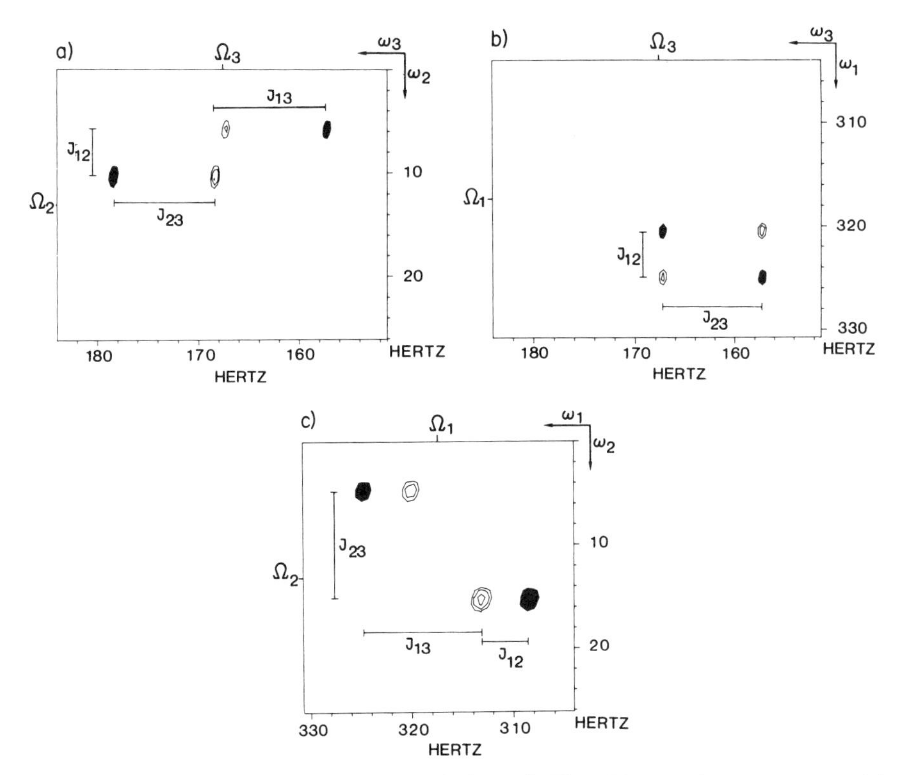

FIG. 5. Three 2D cross sections (a), (b), and (c) as indicated in Fig. 4. These sections were computed from the same data as in Fig. 4; however, zero-filling was employed only up to $128 \times 128 \times 256$ points. Filled contents indicate peaks of negative intensity.

When ω_1 is fixed, a single coherence in ω_1 branches into four coherences in ω_3. For ω_2 fixed, two coherences in ω_1 pass through the same coherence in ω_2 and branch into two different coherences in ω_3. In the case that ω_3 is fixed, finally, four coherences in ω_1 contract into a single coherence in ω_3.

The section of Fig. 5a with ω_1 fixed displays the second coherence transfer step with the active coupling J_{23}. It leads to an antiphase doublet splitting in ω_3. In addition, the spin I_1, passive in this transfer step, contributes a displacement vector (J_{12}, J_{13}) that produces a shifted and sign-inverted replica of the J_{23} doublet, leading to four peaks arranged in the form of a parallelogram. The slope is determined by the relative signs of the passive couplings J_{12} and J_{13} with the arrangement ▱ for equal and ⏢ for opposite signs. The J couplings relevant for the displacements are also indicated in the above coherence-transfer maps. For the section of Fig. 5b with ω_2 fixed both coherence-transfer steps contribute in full analogy to a relay experiment. Here the two active couplings J_{12} and J_{23} cause antiphase doublet splittings in ω_1 and ω_3, respectively, leading to a rectangular arrangement of peaks.

When ω_3 is fixed, as in Fig. 5c, the first coherence-transfer step is visualized with the active coupling J_{12} and the displacement vector (J_{13}, J_{23}). In analogy to Fig. 5a, again a parallelogram results that indicates relative signs of the two passive couplings with ▱ indicating equal and ⏢ opposite signs. It is noteworthy that in all three cases the area of the parallelogram is determined by the product of the two active couplings $J_{12} \times J_{23}$. Figures 5a and 5c show that $J_{12} \times J_{13} > 0$ and $J_{13} \times J_{23} < 0$ for 2,3-dibromopropionic acid in agreement with negative geminal and positive vicinal couplings.

In larger spin systems with further spins I_k that are passive in all three periods, additional splittings occur with 3D displacement vectors $J_k^{(123)}$ in analogy to the 2D displacement vectors described for E. COSY (*9*).

We have demonstrated that it is indeed possible to record 3D NMR spectra with reasonable measuring times and data matrices by application of pulse sequences that are volume-selective in the 3D frequency space. To conclude, we give hints in order to indicate the potential usefulness of 3D experiments. At first, 3D experiments can produce multiplet patterns that are even simpler than those in 2D experiments such as E. COSY and soft COSY. They can be obtained by selective excitation as demonstrated in this communication or by a 3D E. COSY–E. COSY experiment. This may ease the detailed analysis of the multiplet structure.

COSY–COSY cross peaks contain information about relayed coherence transfer whereby the relay spin is uniquely identified in contrast to 2D relay experiments (*10*). This direct access to relay spins may be useful in case of severe spectral overlap.

3D experiments are in general of practical utility when cross peaks in the corresponding 2D spectra are not fully resolved. Very often the spread in a third dimension then provides complete resolution. Especially biological macromolecules with numerous similar or identical building blocks seem to be a promising field of application for 3D NMR techniques, and further research along these lines is being pursued.

ACKNOWLEDGMENTS

This research has been supported by the Swiss National Science Foundation. The loan of an ASPECT 3000 computer by Spectrospin AG, Fällanden, is acknowledged. Help has been provided by R. Brüschweiler. The manuscript has been processed by Miss I. Müller.

REFERENCES

1. W. P. Aue, E. Bartholdi, and R. R. Ernst, *J. Chem. Phys.* **64,** 2229 (1976).
2. Anil Kumar, D. Welti, and R. R. Ernst, *J. Magn. Reson.* **18,** 69 (1975).
3. R. R. Ernst, G. Bodenhausen, and A. Wokaun, "Principles of Nuclear Magnetic Resonance in One and Two Dimensions," Clarendon, Oxford, 1987.
4. Ching-Ming Lai, *J. Appl. Phys.* **52,** 1141 (1981).
5. A. A. Maudsley, S. K. Hilal, W. H. Perman, and H. E. Simon, *J. Magn. Reson.* **51,** 147 (1983).
6. H. D. Plant, T. H. Mareci, M. D. Cockman, and W. S. Brey, 27th ENC, Baltimore, Maryland, 1986.
7. R. Brüschweiler, J. C. Madsen, C. Griesinger, O. W. Sørensen, and R. R. Ernst, *J. Magn. Reson.* **73,** 380.
8. C. Griesinger, O. W. Sørensen, and R. R. Ernst, *J. Amer. Chem. Soc.* **107,** 6394 (1985).
9. C. Griesinger, O. W. Sørensen, and R. R. Ernst, *J. Chem. Phys.* **85,** 6387 (1986).
10. G. W. Eich, G. Bodenhausen, and R. R. Ernst, *J. Amer. Chem. Soc.* **104,** 3731 (1982).

JOURNAL OF MAGNETIC RESONANCE 77, 274–293 (1988)

Iterative Schemes for Bilinear Operators; Application to Spin Decoupling

A. J. SHAKA, C. J. LEE, AND A. PINES

Department of Chemistry, University of California, Berkeley, California 94720, and Materials and Chemical Sciences Division, Lawrence Berkeley Laboratory, Berkeley, California 94720

Received June 23, 1987; revised August 13, 1987

We extend the idea of iterative schemes from the single-spin to the two-spin case. As an application we derive a new series of broadband heteronuclear decoupling sequences, called the DIPSI sequences. They give better *quality* decoupling of protons from carbon-13 than previous sequences like WALTZ-16 when there is scalar coupling among the protons. In the absence of such coupling, the DIPSI sequences offer the same high standard of performance as WALTZ-16, but over somewhat smaller bandwidths.

INTRODUCTION

Current schemes for broadband heteronuclear decoupling in NMR spectroscopy of isotropic liquids employ periodic sequences of radiofrequency pulses derived according to an iterative recipe (*1–5*). The idea behind these schemes rests on the observation (*6*) that, neglecting any homonuclear interactions among the irradiated (I) or observed (S) spins, decoupling can be analyzed by considering only the behavior of an isolated ensemble of I spins under the influence of a single period of the pulse sequence. To decouple I from S over a certain bandwidth, it is sufficient to design a propagator for the I spins which is nearly constant over the same bandwidth. In practice, and for a number of good reasons, iterative schemes all attempt to implement a propagator that is the identity operator over the desired bandwidth, corresponding to a Hamiltonian represented by the null operator. Almost all progress has centered on the simplest model of two (otherwise isolated) weakly coupled spins $I = S = \frac{1}{2}$; the resulting sequences work equally well for arbitrary spin nuclei in the absence of any quadrupolar interactions, however. We shall refer to this situation as the "single-spin case," as only a single I spin is involved.

The simple single-spin model can be expanded to a two-spin model by introducing a second I spin. The spins I_1 and I_2 are scalar coupled and I_1 is coupled to the heteronucleus S, but I_2 is not coupled to S. This last assumption ensures that any changes in the S-spin spectrum result only from the coupling between the I spins and not simply from an additional coupling between I_2 and S. In the practical case, the S spin will be carbon-13 and the I spins will be protons. The only role of S is, depending on its spin state, to augment or retard the local field experienced by I_1, giving it a slightly different chemical shift. Accordingly, most of our development will concentrate on the behavior of the two I spins under modulated irradiation.

In the presence of homonuclear scalar coupling among the I spins, it is not possible

0022-2364/88 $3.00

to produce the identity operator unless the pulse sequence pairwise decouples each I spin from every other. Such homonuclear decoupling in the case of arbitrary tightly coupled proton spin systems including large resonance offset effects appears to be an unrealistic goal at this time, so it becomes necessary to consider designing some other offset-independent propagator for the I spins. The natural choice for such a propagator corresponds to a Hamiltonian containing only the scalar interactions $\mathbf{I}_k \cdot \mathbf{I}_l$ between each pair of I spins.

Here we demonstrate that it is possible to design such a propagator for the case of two coupled spins $I = \frac{1}{2}$. Our sequences can be considered to be a generalization of the MLEV (*1*) and Waugh (*2*) schemes: in addition to averaging the terms involving linear spin operators to nearly zero, we also average all bilinear combinations to nearly zero except for the pure scalar $\mathbf{I}_1 \cdot \mathbf{I}_2$. As a result, the new sequences perform better than WALTZ-16 when there are inequivalent coupled I spins present. They may also have applications for the decoupling of spins $I = 1$ in anisotropic phase (*7*) and broadband homonuclear cross-polarization experiments (*8*). The theory needed to describe the removal of all linear and bilinear terms relates very closely to the theory of multiple-pulse experiments in zero field (*9*).

We begin by describing the design of composite pulses capable of removing the influence of I-spin resonance offsets from the effective Hamiltonian while leaving the scalar interaction largely unperturbed. In contrast to some previous discussions of this subject (*8, 10, 11*) and in agreement with a recent communication concerning broadband homonuclear cross polarization (*12*) and a paper describing the effect of homonuclear coupling on broadband heteronuclear decoupling (*13*), it is found that windowless sequences of RF pulses can produce an effective Hamiltonian containing a combination of bilinear operators $I_{1\alpha}I_{2\beta}$ ($\alpha, \beta = x, y, z$). In particular, the coefficients of $I_{1x}I_{2x}$, $I_{1y}I_{2y}$, and $I_{1z}I_{2z}$ can become unequal; that is, the scalar interaction becomes nonscalar under the influence of the modulated RF field. This behavior is markedly different than that discussed by Braunschweiler and Ernst (*8*) who assumed the δ-pulse limit in calculating the zeroth-order average Hamiltonian for their sequences: in that case no nonscalar behavior is predicted.

Given a composite pulse that largely removes resonance offset effects, we next show how to combine such pulses into sequences that, to first order, cancel all linear terms and bilinear cross terms. The initial sequences can be based on either composite 90° or 180° pulses combined into short cycles, as in the case of single I-spin sequences like MLEV or WALTZ (*1, 3*). We show by calculation that, over their somewhat smaller operational bandwidths, these new decoupling sequences can outperform previous sequences when there is scalar coupling among the I spins, while still maintaining the same high standard of performance in the absence of such coupling. The best candidates for proton decoupling, based on composite 180° pulses, are insensitive to spectrometer phase misadjustments or pulse miscalibration and perform well in practice, as we demonstrate by experiment. Finally, we close by discussing several closely related experiments and the advantages of bilinear averaging.

PHASE-ALTERNATING COMPOSITE PULSES IN THE PRESENCE OF SCALAR COUPLING

We consider a Hamiltonian for two coupled spins I_1 and I_2 in which I_1 is assumed to be on resonance and I_2 is off resonance by an amount $\Delta\omega$. We apply an RF decoupling field ω_2 to perturb the I spins. In the usual rotating frame, the Hamiltonian during the kth pulse, of duration τ_k, can be written

$$\mathcal{H}_k = \mathcal{H}_{k\mathrm{rf}} + V \quad [1]$$

where

$$\mathcal{H}_{k\mathrm{rf}} = (-1)^k \omega_2(I_{1x} + I_{2x}) \quad [2]$$

$$V = \Delta\omega I_{2z} + 2\pi J \mathbf{I}_1 \cdot \mathbf{I}_2. \quad [3]$$

We have assumed that each pulse is applied along the x or $-x$ axis of the rotating frame, and that its amplitude ω_2 and frequency are fixed. An earlier treatment of phase-alternating composite pulses has shown (*14*) that, providing $J = 0$, this class of composite pulses can produce a propagator at time $\tau = \sum_k \tau_k$ that approximates an ideal RF pulse; that is,

$$U(\tau) = \prod_k \exp(-i\tau_k \mathcal{H}_k) \approx \exp(i\alpha[I_{1x} + I_{2x}]). \quad [4]$$

The flip angle α can be selected at will; Eq. [4] is understood to be true for some range of $\Delta\omega$ values about exact resonance ($\Delta\omega = 0$) when the appropriate pulse widths τ_k are selected. We take the same approach here, but include the scalar coupling interaction in the calculation.

Using coherent averaging theory (*15*) we transform into an interaction picture dictated by $\mathcal{H}_{\mathrm{rf}}$ and treat V as a perturbation. The propagator separates into a product

$$U(\tau) = U_{\mathrm{rf}}(\tau) U_{\mathrm{v}}(\tau), \quad [5]$$

where $U_{\mathrm{rf}}(\tau)$ represents the ideal transformation and $U_{\mathrm{v}}(\tau)$ the imperfection,

$$U_{\mathrm{v}}(\tau) = T \exp\left\{-i \int \tilde{V}(t) dt\right\} \quad [6]$$

$$\tilde{V}(t) = U_{\mathrm{rf}}(t)^{-1} V U_{\mathrm{rf}}(t), \quad [7]$$

where T denotes time ordering. This decomposition is not an approximation. $U_v(\tau)$ is then approximated using the Magnus expansion (*16, 17*),

$$U_{\mathrm{v}}(\tau) = \exp(-i\tau[V^{(0)} + V^{(1)} + V^{(2)} + \cdots]), \quad [8]$$

in which the Hermitian operators $V^{(n)}$ rapidly become insignificant provided $\omega_2 \gg |\Delta\omega|$, $|J|$ as would be the case for proton decoupling in liquids.

The terms in the expansion are well-known. The first three terms in the series are given by

$$V^{(0)} = \frac{1}{\tau} \int_0^\tau dt_1 \tilde{V}_1 \quad [9]$$

$$V^{(1)} = \frac{-i}{2\tau} \int_0^\tau dt_1 \int_0^{t_1} dt_2 [\tilde{V}_1, \tilde{V}_2] \quad [10]$$

$$V^{(2)} = \frac{-1}{6\tau} \int_0^\tau dt_1 \int_0^{t_1} dt_2 \int_0^{t_2} dt_3 [\tilde{V}_1, [\tilde{V}_2, \tilde{V}_3]] + [\tilde{V}_3, [\tilde{V}_2, \tilde{V}_1]], \quad [11]$$

where we use the shorthand $\tilde{V}_n \equiv \tilde{V}(t_n)$. Higher-order terms in the Magnus expansion are available if necessary (*18–20*). Our goal is to arrange that all terms in $\Delta\omega$ vanish, so that $U_v(\tau)$ becomes the evolution due to pure scalar coupling,

$$U_v(\tau) = \exp(-i2\pi\tau J\mathbf{I}_1 \cdot \mathbf{I}_2), \qquad [12]$$

and the Magnus expansion is useful because it quickly allows us to predict the operators that the composite pulse may produce and their dependence on $\Delta\omega$ and J.

Consider an arbitrary phase-alternating sequence of m pulses $\alpha_1\bar{\alpha}_2\alpha_3 \ldots$, the overbars representing a phase shift of π and the flip angles α_i denoting those at exact resonance. Let Δ_n be defined by

$$\Delta_n = \sum_{k=0}^{n} (-1)^k \alpha_k \qquad [13]$$

with $\alpha_0 \equiv 0$. Carrying out the integration for $V^{(0)}$ we find

$$\tau V^{(0)} = 2\pi J\tau\mathbf{I}_1 \cdot \mathbf{I}_2 + \frac{\Delta\omega}{\omega_2} \sum_{j=1}^{m} (-1)^{(j+1)} \{I_{2z}(\sin \Delta_j - \sin \Delta_{j-1}) + I_{2y}(\cos \Delta_{j-1} - \cos \Delta_j)\}. \qquad [14]$$

To zeroth order in the Magnus expansion the coupling is unaffected by the composite pulse, while the terms under the summation correspond to those for the case of a single spin. A single 2π pulse causes these terms to vanish, so for $\Delta\omega/\omega_2$ sufficiently small a pure scalar operator is produced. This limit, corresponding to coherent irradiation with an extremely intense RF field, is not sufficiently discriminating to be of interest. Accordingly, the next term in the expansion must be examined.

After a perfectly straightforward but somewhat lengthy calculation, we find the following result for $\tau V^{(1)}$,

$$\tau V^{(1)} = \frac{\Delta\omega^2}{\omega_2^2} I_{2x}[\sum_j^m a_j + \sum_{k>j}^m a_{kj}] + \frac{\Delta\omega 2\pi J}{\omega_2^2} (I_{1y}I_{2x} - I_{1x}I_{2y})[\sum_j^m b_j + \sum_{k>j}^m b_{kj}] + \frac{\Delta\omega 2\pi J}{\omega_2^2} (I_{1x}I_{2z} - I_{1z}I_{2x})[\sum_j^m c_j + \sum_{k>j}^m c_{kj}] \qquad [15]$$

with coefficients defined by

$$a_j = (-1)^j\{\alpha_j - \sin \alpha_j\} \qquad [16]$$

$$a_{kj} = (-1)^{k+j}\{\sin(\Delta_k - \Delta_j) - \sin(\Delta_{k-1} - \Delta_j) - \sin(\Delta_k - \Delta_{j-1}) + \sin(\Delta_{k-1} - \Delta_{j-1})\} \qquad [17]$$

$$b_j = 2 \cos \Delta_{j-1} - 2 \cos \Delta_j + (-1)^j \alpha_j(\sin \Delta_j + \sin \Delta_{j-1}) \qquad [18]$$

$$b_{kj} = (-1)^{j+1} \alpha_k(\sin \Delta_j - \sin \Delta_{j-1}) + (-1)^k \alpha_j(\sin \Delta_k - \sin \Delta_{k-1}) \qquad [19]$$

$$c_j = 2 \sin \Delta_{j-1} - 2 \sin \Delta_j + (-1)^{j+1} \alpha_j(\cos \Delta_j - \cos \Delta_{j-1}) \qquad [20]$$

$$c_{kj} = (-1)^j \alpha_k(\cos \Delta_j - \cos \Delta_{j-1}) + (-1)^{k+1} \alpha_j(\cos \Delta_k - \cos \Delta_{k-1}). \qquad [21]$$

To first order in the Magnus expansion the action of the composite pulse creates additional bilinear operators not present in the unperturbed Hamiltonian. These terms, especially the combination $(I_{1y}I_{2x} - I_{1x}I_{2y})$, which is invariant to π phase shifts, can introduce an offset dependence into $U(\tau)$ that interferes with proper decoupling of I from S. The above equations provide a means to eliminate or at least minimize the contribution of these unwanted operators.

To explore the deviation of $\mathbf{I}_1 \cdot \mathbf{I}_2$ from its scalar form it is necessary to calculate the contribution of the operator $V^{(2)}$ to the propagator. $V^{(2)}$ has a very complicated form, so we restrict ourselves to the assessment of the operator combinations it can produce. These are as follows: the operators I_{2z} and I_{2y} with coefficients depending on $\Delta\omega^3/\omega_2^3$, the operators $I_{1x}I_{2x}$, $I_{1y}I_{2y}$, $I_{1z}I_{2z}$, $I_{1y}I_{2z}$, and $I_{1z}I_{2y}$ with coefficients depending on $\Delta\omega^2 J/\omega_2^3$, and the operators I_{1z}, I_{2z}, I_{1y} and I_{2y} with coefficients depending on $\Delta\omega J^2/\omega_2^3$. The interesting point here is that the coefficients of $I_{1\alpha}I_{2\alpha}$, $\alpha = x, y, z$, become unequal and that linear spin operators are produced for the first spin even though it is assumed to be exactly on resonance. These rather surprising linear terms arise due to the interaction of the RF field and the spin coupling producing an additional small field on the first spin.

ELIMINATING LINEAR AND BILINEAR SPIN OPERATORS

Assume we have found a composite pulse. We now explore methods to combine these into cycles that remove all linear and bilinear spin operators except for the scalar operator $\mathbf{I}_1 \cdot \mathbf{I}_2$. We write the propagator for the pulse in the form

$$U(\tau) = U_{\mathrm{rf}}(\tau)U_{\mathrm{v}}(\tau) \qquad [22]$$

and we assume nothing about $U_{\mathrm{v}}(\tau)$ except that it can be expressed as a complex exponential of the most general two-spin Hamiltonian, a sum of the 15 possible linear and bilinear operators. The only information we require is how U_{v} transforms under RF pulses and phase shifts. As the *original* perturbation V commutes with rotations about the z axis, it follows that U_{v} is tied to the phase of the RF irradiation. If U_0 denotes the propagator that results when the first pulse of the composite is applied with RF phase $\phi = 0$ then the propagator U_ϕ that results when all phases are shifted by ϕ is simply

$$U_\phi = R_z(\phi)U_0R_z(\phi)^{-1} = U_{\mathrm{rf}\phi}U_{\mathrm{v}\phi}. \qquad [23]$$

We wish to calculate the effect of a sequence of n such composite pulses, the kth pulse being applied with relative phase ϕ_k, with no detailed knowledge of the nature of U_v. In addition, we hope to find sequences that tend to cancel out the effect of the U_v operators. To do this we reorganize the product so that the RF terms are collected on the left,

$$\begin{aligned} U_{\phi_n}U_{\phi_{n-1}}\cdots U_{\phi_1} &= U_{\mathrm{rf}\phi_n}U_{\mathrm{v}\phi_n}U_{\mathrm{rf}\phi_{n-1}}U_{\mathrm{v}\phi_{n-1}}\cdots U_{\mathrm{rf}\phi_1}U_{\mathrm{v}\phi_1} \\ &= U_{\mathrm{rf}\phi_n}U_{\mathrm{rf}\phi_{n-1}}\cdots U_{\mathrm{rf}\phi_1}\{U_{\mathrm{rf}\phi_1}^{-1}\cdots U_{\mathrm{rf}\phi_{n-1}}^{-1}\}U_{\mathrm{v}\phi_n}\{U_{\mathrm{rf}\phi_{n-1}}\cdots U_{\mathrm{rf}\phi_1}\} \\ &\qquad \cdots\{U_{\mathrm{rf}\phi_1}^{-1}\}U_{\mathrm{v}\phi_2}\{U_{\mathrm{rf}\phi_1}\}U_{\mathrm{v}\phi_1} \\ &= U_{\mathrm{rf}\phi_n}U_{\mathrm{rf}\phi_{n-1}}\cdots U_{\mathrm{rf}\phi_1}\tilde{U}_{\mathrm{v}\phi_n}\tilde{U}_{\mathrm{v}\phi_{n-1}}\cdots\tilde{U}_{\mathrm{v}\phi_1}, \end{aligned} \qquad [24]$$

where the operators $\tilde{U}_{\rm v}$ are defined by

$$\tilde{U}_{{\rm v}\phi_k} = \{U^{-1}_{{\rm rf}\phi_1} \cdots U^{-1}_{{\rm rf}\phi_{k-1}}\} U_{{\rm v}\phi_k} \{U_{{\rm rf}\phi_{k-1}} \cdots U_{{\rm rf}\phi_1}\}. \qquad [25]$$

Since all the $\tilde{U}_{\rm v}$ are assumed to be small perturbations, they may be grouped into a single unitary operator by applying the Baker–Campbell–Hausdorff (BCH) formula:

$$\prod_k \exp(-iA_k) = \exp\left(-i\left\{\sum_k A_k + \frac{1}{2}\sum_{k>j}[A_k, A_j] + \cdots\right\}\right). \qquad [26]$$

Equation [26] is true for arbitrary linear operators A_k, the expansion consisting of an infinite series of nested commutators. Provided all A_k are of order ϵ, the commutator is of order ϵ^2 and can be neglected, along with all higher-order terms. We assume A_1 is arbitrary and that all other A_k are related by Eqs. [23] and [25]. Finding the appropriate pulse cycle then reduces to determining the sequence of phases ϕ_k such that

$$\frac{1}{n}\sum_{k=1}^{n} A_k = 0 \qquad [27]$$

for the single-spin case and

$$\frac{1}{n}\sum_{k=1}^{n} A_k = 2\pi J\tau \mathbf{I}_1 \cdot \mathbf{I}_2 \qquad [28]$$

for the two-spin case. Equations [27] and [28] form one possible criterion for primitive sequences in each of these cases.

The way this cancellation works can be illustrated by the single-spin case, where the imperfections A_k can be written in the form $\epsilon_k \cdot \mathbf{I}$. Consider a sequence of four composite 180° pulses, $U\ U\ \bar{U}\ \bar{U}$, the barred states denoting a 180° phase shift of all the constituent pulses, and assume U is equivalent to an x pulse on resonance. A table can be prepared listing each operator A_k underlying the $\tilde{U}_{{\rm v}\phi_k}$ for each value of k, as shown in Table 1. An even number of states is required to null the ϵ_z terms; the additional requirement that both the ϵ_x and ϵ_y terms vanish makes the minimum of four states, the origin of the MLEV-4 sequence proposed by Levitt and Freeman (*21*). Of course there are many sequences of composite 180° pulses that satisfy Eq. [27], but the MLEV-4 sequence is the shortest.

In the bilinear case the vectors ϵ_k have 15 components, reflecting all possible linear and bilinear terms. The MLEV-4 sequence cancels the linear terms ϵ_α and bilinear

TABLE 1

Cancellation of Linear Spin Operators by the Sequence $U\ U\ \bar{U}\ \bar{U}$

State	Operator		
	I_x	I_y	I_z
1	ϵ_x	ϵ_y	ϵ_z
2	ϵ_x	$-\epsilon_y$	$-\epsilon_z$
3	$-\epsilon_x$	$-\epsilon_y$	ϵ_z
4	$-\epsilon_x$	ϵ_y	$-\epsilon_z$

cross terms $\epsilon_{\alpha\beta}$ ($\alpha \neq \beta$), but fails to average the $\epsilon_{\alpha\alpha}$, as the operator $I_{1z}I_{2z}$ is invariant to 180° rotations. We considered two strategies: (1) find a composite 180° pulse that keeps the $\epsilon_{\alpha\alpha}$ nearly equal as a function of resonance offset and then use an MLEV-4 sequence to cancel all other terms; (2) devise another sequence that averages the $\epsilon_{\alpha\alpha}$ as well as cancels the other terms. The first approach has been more successful in the work described here, but we briefly describe the second to establish a clear connection between the single-spin and two-spin cases.

The averaging of the $\epsilon_{\alpha\alpha}$ requires that the number of states in the sequence be a multiple of three. With the independent requirement of four states to cancel the linear terms, a 12-state sequence seems to be required. Using only 90° (composite) pulses and 90° phase shifts three distinct sequences were found. With the shorthand X, Y, $\bar{X}$, and $\bar{Y}$ denoting (composite) 90° pulses equivalent to 90° pulses along these four axes, these can be written $(X\ Y\ \bar{X}\ Y)^3$, $(X\ Y\ X\ Y\ \bar{X}\ Y)^2$, and $(X\ Y\ X\ Y\ X\ \bar{Y})^2$. The behavior of the error terms can be examined in the same way as for the single-spin case. For example, Table 2 shows the systematic cancellation of the errors under the action of the 12-pulse sequence $(X\ Y\ \bar{X}\ Y)^3$. Since the operators $I_{1\alpha}$ and $I_{2\alpha}$ transform identically, only the former are listed. One advantage of this treatment is that we are free to substitute *any* composite 90° rotation into any of these sequences. This would not be the case if we had used the Magnus expansion alone and considered the operator V as the perturbation. In such a case a separate calculation over the entire sequence would be required each time a different composite pulse is considered. The price paid for this convenience is that the exact form of U_v must remain unspecified.

Table 2 shows both the strong and the weak points of the 12-pulse sequences. The strong point is that, to first order, a true scalar operator $\mathbf{I}_1 \cdot \mathbf{I}_2$ will be produced by such a sequence. The weak point is that the radius of convergence is poorer than MLEV-4, especially for the linear terms ϵ_α. These terms are canceled only at the end of the entire 12-pulse sequence, and all three components are mixed together along the way. The MLEV-4 sequence has a subcycle structure that cancels the ϵ_α much more effi-

TABLE 2

Cancellation of Nonscalar Spin Operators by the 12-Pulse Sequence $X\ Y\ \bar{X}\ Y\ X\ Y\ \bar{X}\ Y\ X\ Y\ \bar{X}\ Y$

State	Operator											
	I_{1x}	I_{1y}	I_{1z}	$I_{1x}I_{2y}$	$I_{1x}I_{2z}$	$I_{1y}I_{2z}$	$I_{1y}I_{2x}$	$I_{1z}I_{2x}$	$I_{1z}I_{2y}$	$I_{1x}I_{2x}$	$I_{1y}I_{2y}$	$I_{1z}I_{2z}$
1	ϵ_x	ϵ_y	ϵ_z	ϵ_{xy}	ϵ_{xz}	ϵ_{yz}	ϵ_{yx}	ϵ_{zx}	ϵ_{zy}	ϵ_{xx}	ϵ_{yy}	ϵ_{zz}
2	$-\epsilon_y$	$-\epsilon_z$	ϵ_x	ϵ_{yz}	$-\epsilon_{yx}$	$-\epsilon_{zx}$	ϵ_{zy}	$-\epsilon_{xy}$	$-\epsilon_{xz}$	ϵ_{yy}	ϵ_{zz}	ϵ_{xx}
3	ϵ_z	$-\epsilon_x$	$-\epsilon_y$	$-\epsilon_{zx}$	$-\epsilon_{zy}$	ϵ_{xy}	$-\epsilon_{xz}$	$-\epsilon_{yz}$	ϵ_{yx}	ϵ_{zz}	ϵ_{xx}	ϵ_{yy}
4	$-\epsilon_x$	$-\epsilon_y$	ϵ_z	ϵ_{xy}	$-\epsilon_{xz}$	$-\epsilon_{yz}$	ϵ_{yx}	$-\epsilon_{zx}$	$-\epsilon_{zy}$	ϵ_{xx}	ϵ_{yy}	ϵ_{zz}
5	$-\epsilon_y$	ϵ_z	$-\epsilon_x$	$-\epsilon_{yz}$	ϵ_{yx}	$-\epsilon_{zx}$	$-\epsilon_{zy}$	ϵ_{xy}	$-\epsilon_{xz}$	ϵ_{yy}	ϵ_{zz}	ϵ_{xx}
6	ϵ_z	ϵ_x	ϵ_y	ϵ_{zx}	ϵ_{zy}	ϵ_{xy}	ϵ_{xz}	ϵ_{yz}	ϵ_{yx}	ϵ_{zz}	ϵ_{xx}	ϵ_{yy}
7	ϵ_x	$-\epsilon_y$	$-\epsilon_z$	$-\epsilon_{xy}$	$-\epsilon_{xz}$	ϵ_{yz}	$-\epsilon_{yx}$	$-\epsilon_{zx}$	ϵ_{zy}	ϵ_{xx}	ϵ_{yy}	ϵ_{zz}
8	ϵ_y	ϵ_z	ϵ_x	ϵ_{yz}	ϵ_{yx}	ϵ_{zx}	ϵ_{zy}	ϵ_{xy}	ϵ_{xz}	ϵ_{yy}	ϵ_{zz}	ϵ_{xx}
9	$-\epsilon_z$	$-\epsilon_x$	ϵ_y	ϵ_{zx}	$-\epsilon_{zy}$	$-\epsilon_{xy}$	ϵ_{xz}	$-\epsilon_{yz}$	$-\epsilon_{yx}$	ϵ_{zz}	ϵ_{xx}	ϵ_{yy}
10	$-\epsilon_x$	ϵ_y	$-\epsilon_z$	$-\epsilon_{xy}$	ϵ_{xz}	$-\epsilon_{yz}$	$-\epsilon_{yx}$	ϵ_{zx}	$-\epsilon_{zy}$	ϵ_{xx}	ϵ_{yy}	ϵ_{zz}
11	ϵ_y	$-\epsilon_z$	$-\epsilon_x$	$-\epsilon_{yz}$	$-\epsilon_{yx}$	ϵ_{zx}	$-\epsilon_{zy}$	$-\epsilon_{xy}$	ϵ_{xz}	ϵ_{yy}	ϵ_{zz}	ϵ_{xx}
12	$-\epsilon_z$	ϵ_x	$-\epsilon_y$	$-\epsilon_{zx}$	ϵ_{zy}	$-\epsilon_{xy}$	$-\epsilon_{xz}$	ϵ_{yz}	$-\epsilon_{yx}$	ϵ_{zz}	ϵ_{xx}	ϵ_{yy}

ciently. We note that the opposite situation can prevail for a single spin $I = 1$ in a liquid crystal environment, due to the bilinear terms generated by the large quadrupolar interaction (*7*). In such a case, rapid cancellation of the bilinear terms could be more important. For proton decoupling in liquids, however, the resonance offset and RF inhomogeneity terms are so much larger than the proton scalar coupling that increased compensation for the latter at the expense of the former is unacceptable. As a result, we have found that sequences based on composite 180° pulses are still the method of choice.

THE S-SPIN SPECTRUM UNDER A PURE SCALAR OPERATOR

The expected form of the S-spin spectrum under periodic irradiation to the I spins has already been described in detail for the case of two coupled I spins (*13*) and also treated generally for any multilevel spin system (*22*), so we restrict ourselves to a brief summary of the conclusions relevant to our discussion. We assume stroboscopic observation of the S-spin free induction decay synchronized with the decoupler cycling rate.

The evolution operator for the joint spin system at the time t_s the first point of the S-spin free induction decay is sampled separates into a sum of commuting operators depending on the spin state of S,

$$U(t_s) = P_+ U_+(t_s) + P_- U_-(t_s), \tag{29}$$

where

$$P_\pm = \tfrac{1}{2}(\mathbf{1} \pm S_z). \tag{30}$$

In a rotating frame on resonance for S and assuming quadrature detection after an initial 90° pulse applied to the S spins, the S-spin signal can be calculated at any multiple of t_s using the formula

$$\begin{aligned} \langle S^-(nt_s)\rangle &= \mathrm{Tr}\{(S_x - iS_y)U^n(t_s)S_x U^{-n}(t_s)\} \\ &= \mathrm{Tr}\{U_+^n(t_s)U_-^{-n}(t_s)\}. \end{aligned} \tag{31}$$

The unitary operators $U_\pm$ can be considered to arise from fictitious time-independent Hamiltonians $\mathscr{H}_\pm$ that would have caused the same change in the quantum states over the time t_s,

$$U_\pm(t_s) = \exp(-i\mathscr{H}_\pm t_s). \tag{32}$$

The only difference in $\mathscr{H}_\pm$ arises from the different resonance frequency of I_1, by an amount J_{IS}. In the regime of interest, and for the sequences we consider, it is safe to assume that the eigenvectors of $\mathscr{H}_+$ and $\mathscr{H}_-$ are nearly coincident, and Eq. [31] can thus be evaluated in a basis in which both operators are diagonal. The result is

$$\langle S^-(nt_s)\rangle = \sum_{j=1}^{4} \exp(it_s(\Omega_{j+} - \Omega_{j-})), \tag{33}$$

where $\Omega_{j\pm}$ are the jth eigenvalues of $\mathscr{H}_\pm$. In this limit, Fourier transformation of the time-domain signal of Eq. [31] gives an S-spin spectrum containing four lines of unit intensity with frequencies given by the appropriate energy difference. The eigenvalues are functions of the chemical shift of I_1, δ_1: $\Omega_j = \Omega_j(\delta_1)$. The only role of the coupling is to give I_1 a slightly different chemical shift in each subspectrum. Since the perturbation is small, it is valid to approximate the difference by the derivative. Putting $E_{j\pm} = 2\pi\Omega_{j\pm}$ and expressing all offsets and energies in hertz, we have

$$E_{j+} - E_{j-} = E_j(\delta_1 + J_{IS}/2) - E_j(\delta_1 - J_{IS}/2) \approx J_{IS}\frac{\partial E_j}{\partial \delta_1}. \qquad [34]$$

Accordingly, we can define a set of j scaling factors by

$$\lambda_j = \frac{\partial E_j}{\partial \delta_1}. \qquad [35]$$

The frequency of the jth S-spin transition is then given by $J_{IS}\lambda_j$. Note that this definition of the scaling factor differs from that given by Waugh (*6*) in terms of the net rotation angle induced by the pulse sequence. In the current nomenclature there would be two scaling factors for the case of a single IS pair. The difference arises because the present scaling factors give the actual line *positions* in the S-spin spectrum, whereas the single scaling factor as defined by Waugh gives the frequency *difference* between the two parent transitions that occur in the case of a single I spin coupled to S. In multilevel systems there is no convenient generalization of the Waugh scaling factor, and it becomes necessary to plot derivatives of the energy levels of the fictitious Hamiltonian instead (*22*). In the single-spin case we plot the Waugh scaling factor to facilitate comparison with previous work.

A complete description of the decoupling performance would seem to require the construction of the complete two-dimensional energy level surface $\Omega_j(\delta_1, \delta_2)$ as a function of the resonance offsets of I_1 and I_2 for each value of j, followed by partial differentiation with respect to δ_1. This time-consuming calculation is circumvented by taking a single cross section at $\delta_2 = 0$. Sequences performing well over this slice also perform well for any values of δ_1 and δ_2 within their effective bandwidth.

Suppose the decoupling sequence manages to produce a pure scalar operator between the I spins. The S-spin spectrum is particularly easy to predict in this case. The correct eigenstates are the familiar triplet and singlet states with energies $(1/4)J(\delta_1)$ and $(-3/4)J(\delta_1)$, respectively. In the case of a perfect sequence $J(\delta_1)$ is independent of the offset δ_1 and is equal to the unperturbed coupling J. All four S-spin transitions become degenerate and a sharp singlet is observed at the chemical shift of S, corresponding to perfect decoupling.

Although it may seem somewhat surprising to the uninitiated, it is quite possible for a sequence to produce a propagator whose underlying Hamiltonian is a pure scalar operator with a coupling constant that depends on δ_1 (or more generally, on the difference between the chemical shifts of I_1 and I_2). In fact, this is invariably what happens. In such a case the S-spin spectrum consists of two lines with the intensity ratio 3:1 and with positions

$$\frac{1}{4}J_{IS}\frac{\partial J(\delta_1)}{\partial \delta_1}; \qquad \frac{-3}{4}J_{IS}\frac{\partial J(\delta_1)}{\partial \delta_1} \qquad [36]$$

corresponding to the S spin evolving in the local field of the triplet or singlet state, respectively. Figure 1 shows this behavior schematically.

We can guess the minimum achievable splitting between these two lines by estimating

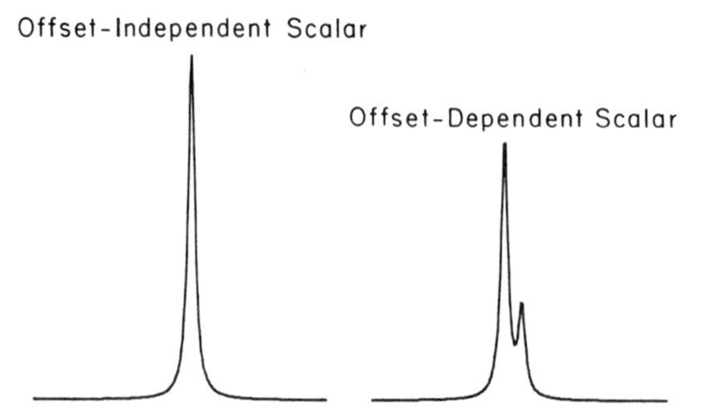

FIG. 1. The expected form of the S-spin spectrum under a decoupling sequence that produces an underlying Hamiltonian that is a pure scalar operator $2\pi J \mathbf{I}_1 \cdot \mathbf{I}_2$. If the effective coupling constant is offset-independent then a sharp singlet is observed. When the coupling constant is offset-dependent a 3:1 pattern emerges, in which the S spin experiences the local field of the triplet or singlet states, respectively.

the minimum rate of change of J with δ_1 for a certain decoupling field amplitude ω_2. During each single pulse of the complicated sequence, the spins I_1 and I_2 precess about their effective fields in the rotating frame. The angle θ between the effective fields remains constant throughout and, assuming $\delta_2 = 0$, is given by

$$\theta = \tan^{-1}(2\pi\delta_1/\omega_2). \quad [37]$$

Since ω_2 is fixed for the sequences we are considering, the angle θ is constant throughout the sequence: changing the phase of the RF field changes both effective fields in exactly the same way. If the decoupling sequence functions perfectly, all linear and bilinear cross terms are removed, leaving only the projection of I_1 on I_2 intact. As a result the effective Hamiltonian becomes

$$\mathscr{H} = 2\pi J \cos\theta \mathbf{I}_1 \cdot \mathbf{I}_2 \quad [38]$$

leading directly to a splitting ΔS between the singlet and triplet lines of

$$\Delta S = J_{\mathrm{IS}}(2\pi J/\omega_{\mathrm{e}})\cos\theta \sin\theta, \quad [39]$$

where ω_{e} is the effective field felt by I_1:

$$\omega_{\mathrm{e}} = (\omega_2^2 + (2\pi\delta_1)^2)^{1/2}. \quad [40]$$

When δ_1 is small compared with the RF field amplitude ω_2, Eq. [39] reduces to

$$\Delta S = J_{\mathrm{IS}}(2\pi J/\omega_2)(2\pi\delta_1/\omega_2). \quad [41]$$

For the representative values $J_{\mathrm{IS}} = 200$ Hz, $J = 10$ Hz, $\omega_2/2\pi = 2$ kHz, and $\delta_1 = 200$ Hz we find a residual splitting $\Delta S = 0.1$ Hz. The maximum splitting ($\Delta S = 0.385$ Hz) occurs at the offset $\delta_1 = 1414$ Hz, and at $(2^{1/2}/2)\omega_2/2\pi$ for arbitrary ω_2. These rough estimates should be regarded as very optimistic. Most actual decoupling sequences tend to decouple the *two I spins* at a somewhat faster rate as a function of δ_1. That is, the effective coupling constant between the two I spins is smaller than $J \cos\theta$. For some very simple sequences, a detailed evaluation of Eq. [11] predicts a reduction of about $(8/3)^{1/2}$ larger, quite close to the value observed for many of our actual pulse sequences. This larger reduction leads to a larger gradient and larger splitting ΔS for small δ_1, resulting in poorer performance. As δ_1 approaches the edge of the decoupling bandwidth, the sequence is unable to average away the other operators properly, and the S-spin spectrum breaks up into four lines.

The signature of the aforementioned hypothetical sequence producing a pure scalar operator is shown in Fig. 2 as a function of δ_1 for the case $\delta_2 = 0$. Both the energy eigenvalues and the corresponding λ scaling factors are shown. The triplet state gives rise to a line three times more intense than that from the singlet state. When $\delta_1 = \delta_2 = 0$ there is no residual splitting, but there is a finite residual splitting at all other values of δ_1. This irremovable residual splitting is the direct result of the inability of the sequence to produce a completely offset-independent propagator. We have found no broadband sequence capable of producing a smaller residual splitting near $\delta_1 = 0$, regardless of the degree of iteration, and so regard this estimate as a theoretical lower limit. In contrast to the case of a single I spin, where iterative schemes can produce arbitrarily small residual splittings in the neighborhood $\delta_1 = 0$, two coupled I spins with different chemical shifts result in a small but irremovable residual splitting in the S-spin spectrum.

From Eq. [41] we see that the residual splitting depends on ω_2, and this property carries over to all the sequences we have investigated, whether or not a pure scalar is produced. By contrast, changing ω_2 does *not* change the residual splitting appreciably in the single-spin case as long as the I spin remains within the bandwidth of the decoupling sequence and the flip angles of the pulses are correctly set for each value of ω_2 (*5*). A simple way to check for the effect of homonuclear coupling is to increase ω_2, adjusting the pulse widths appropriately, while observing the S-spin resonance. If homonuclear coupling is playing a role, the linewidth should decrease. By the same token, RF inhomogeneity gives rise to a superposition of closely spaced multiplets, making the structure difficult to discern.

Cross sections along δ_2 at the point $\delta_1 = 0$ are somewhat more difficult to characterize for practical sequences. Provided δ_2 is within the decoupling bandwidth and a scalar

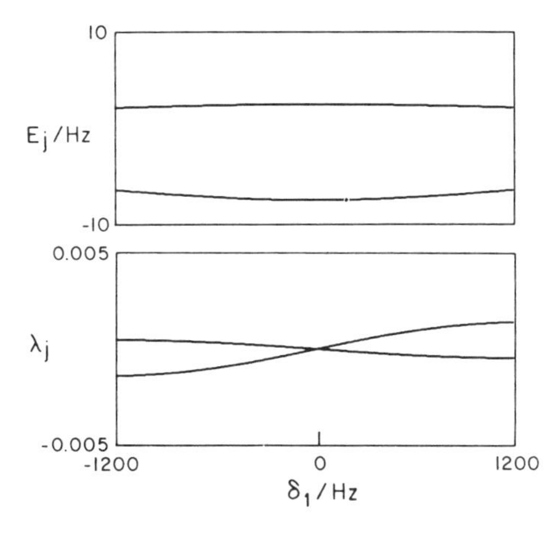

FIG. 2. The energy eigenvalues, E_j, of the Hamiltonian of Eq. [62], assuming an unperturbed homonuclear coupling of 10 Hz and a 2 kHz RF field. The derivatives of the energies, λ_j, shown below, are the scaling factors giving the line positions in the S-spin spectrum.

operator is produced, there will be a residual splitting bounded from below by Eq. [39], in which the angle θ now refers to δ_2 rather than to δ_1. As δ_2 nears the edge of the effective decoupling bandwidth the compensatory properties of the sequence fail and four lines are generally produced in the S-spin spectrum. When δ_2 is much larger than the bandwidth of the sequence, I_1 and I_2 become decoupled and the S-spin spectrum reverts to a sharp singlet. It follows that any additional *heteronuclear* coupling to I_1 has no important role in the proper decoupling of I_1 from S, since I_1 will simultaneously be decoupled from all heteronuclei.

Any real decoupling sequence will produce a mix of other operators in addition to $\mathbf{I}_1 \cdot \mathbf{I}_2$. It is important to assess the influence of these unwanted operators on the S-spin spectrum. As the scalar operator $\mathbf{I}_1 \cdot \mathbf{I}_2$ is approached more closely, the Cartesian product operators become less convenient, and we employ instead the symmetric and antisymmetric linear combinations

$$I_{z+} = I_{1z} + I_{2z} \quad [42]$$

$$I_{z-} = I_{1z} - I_{2z} \quad [43]$$

$$I_{xy+} = I_{1x}I_{2y} + I_{2x}I_{1y} \quad [44]$$

$$I_{xy-} = I_{1x}I_{2y} - I_{2x}I_{1y}, \quad [45]$$

etc. All the symmetric operators $I_{\alpha+}$, $I_{\alpha\beta+}$ commute with $\mathbf{I}_1 \cdot \mathbf{I}_2$ and so can be eliminated by cyclic permutations and phase shifts, in the same way as in the single-spin case. The antisymmetric operators $I_{\alpha-}$, $I_{\alpha\beta-}$ *evolve* under $\mathbf{I}_1 \cdot \mathbf{I}_2$ and can therefore pose problems as the cycle time increases. Fortunately, none of the antisymmetric operators have any nonzero diagonal elements in the eigenbasis of $\mathbf{I}_1 \cdot \mathbf{I}_2$. They also have no nonzero matrix elements connecting the degenerate states within the triplet manifold. If the decoupling sequence produces coefficients for these operators that are small compared to the proton–proton coupling constant the small terms are quenched by J and have very little effect on the decoupling performance. This stabilization of offset-dependent small terms by a larger residual term in the effective I-spin Hamiltonian has been discussed previously for the quadrupolar case (*22*).

PRACTICAL SEQUENCES

It is useless to start with an inferior composite pulse and then attempt to refine away all the errors with the iterative scheme alone. Such an approach misses the point that the *numerical* value of the scalar must be held constant: eliminating all the other operators does not guarantee perfect performance. Ironically, a certain gradient of the scalar part is "built in" during the initial stages, when there are still other terms present in the effective I-spin Hamiltonian. After most of the error terms have been removed it is not possible to alter the numerical value of the scalar much anymore: the effective Hamiltonian, precisely because it is very nearly scalar, becomes invariant to cyclic permutations and phase shifts, and commutes with itself over the various segments of the combined sequence. For this reason, special care must be taken in the design of the composite pulse.

We were able to find suitable composite 180° pulses by using a hybrid approach. Using Eqs. [14]–[21], composite 180° pulses could be discovered which offered compensation for resonance offset and which minimized the production of bilinear cross terms. The composite pulse could then be improved by an exact calculation of its performance over a limited variation of the constituent pulse flip angles, followed by extraction of the underlying Hamiltonian. An improvement was registered whenever the deviation of the scalar part, as a function of offset, was decreased. Using the well-known methods to iteratively improve composite 180° pulses for decoupling sequences by cyclic permutation of 90° pulses (*2–5*) then allowed the bandwidth to be extended without losing the desirable property that the scalar remain as unperturbed as possible. The final stage is then to assemble the decoupling sequence in the form $R\ \bar{R}\ \bar{R}\ R$, which again attenuates the linear and bilinear cross terms without affecting the scalar part much. A selection of the composite 180° pulses we found is set out in Table 3. We distinguish the corresponding decoupling sequences with the label DIPSI-*n*, the index *n* referring to the composite pulse R_n in Table 3. DIPSI stands for "*d*ecoupling *i*n the *p*resence of *s*calar *i*nteractions."

The scaling factors for DIPSI-1, DIPSI-2, and DIPSI-3 are shown in Fig. 3, assuming a 10 Hz proton–proton coupling and a 2 kHz RF field. The scaling factors for WALTZ-16 are included for comparison. Larger bandwidths are offered by the more complex 180°'s. Except for WALTZ-16, the scaling factors show that the underlying Hamiltonian is nearly a pure scalar operator as a function of offset, for the three "triplet" states are nearly degenerate. WALTZ-16 shows the largest deviation from pure scalar behavior, giving a spectrum of four lines over much of the calculated range.

Figure 4 shows the single-spin Waugh scaling factor for WALTZ-16 and each of the DIPSI sequences. The predicted quality of decoupling is extremely good for all these sequences, the scaling factor remaining well below $\lambda = 0.001$ over their respective bandwidths. DIPSI-1 has a cycling rate of 130.4 Hz using a 2 kHz field, and so is comparable to WALTZ-8 in length and complexity. DIPSI-2 and -3 have cycling rates and complexities comparable to WALTZ-16 and WALTZ-32, respectively.

EXPERIMENTAL

Spectra were obtained both on a Bruker AM-400 spectrometer using a standard 10 mm broadband probe and on an AM-500 spectrometer using a 5 mm broadband probe. The 10 mm probe, with its larger sample volume and substantially worse RF homogeneity, provided a searching test of the single-spin bandwidths of the DIPSI

TABLE 3

Phase-Alternating Composite 180° Pulses For Two Coupled Spins $I = \frac{1}{2}$

Label	Sequence	Bandwidth[a]	Length[b]
R_1	$365\ \overline{295}\ 65\ \overline{305}\ 350$	±0.4	1380
R_2	$320\ \overline{410}\ 290\ \overline{285}\ 30\ \overline{245}\ 375\ \overline{265}\ 370$	±0.6	2590
R_3	$\overline{245}\ 395\ \overline{250}\ 275\ \overline{30}\ 230\ \overline{360}\ 245\ \overline{370}\ 340\ \overline{350}$		
	$260\ \overline{270}\ 30\ \overline{225}\ 365\ \overline{255}\ 395$	±0.8	4890

[a] Approximate, in terms of the dimensionless offset parameter $\Delta\omega/\omega_2$.
[b] Total rotation of the composite pulse in degrees.

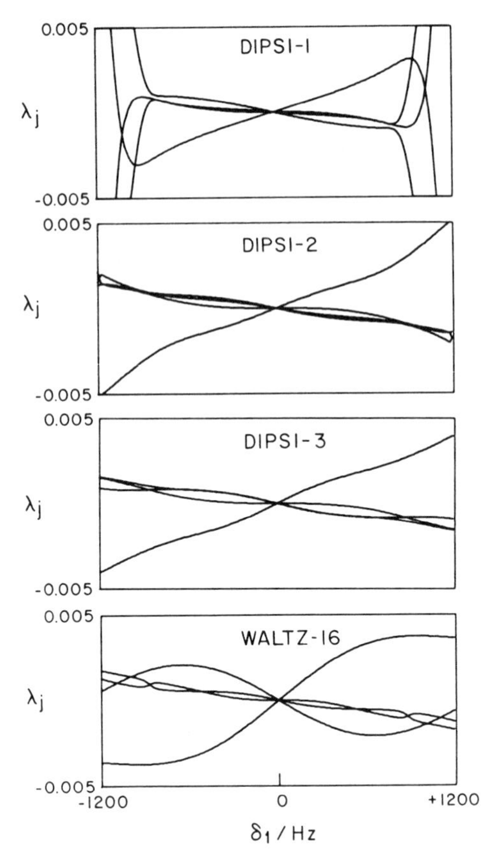

FIG. 3. Scaling factors for DIPSI-1, -2, -3, and WALTZ-16. The scaling factors are shown as a function of δ_1, the offset of the directly coupled I spin, for the case $\delta_2 = 0$, and assuming a homonuclear coupling of 10 Hz and a 2 kHz RF field. Even though the DIPSI sequences use only 180° phase shifts, a pure scalar propagator is approached quite closely. By contrast, WALTZ-16 gives a different signature, showing nonscalar behavior and resulting in four different transitions.

sequences under routine operating conditions. The combination of the 5 mm probe and higher B_0 field available on the AM-500 allowed us to explore the expected fine structure of the carbon-13 resonances due to homonuclear scalar coupling among the protons. An undoped mixture of methyl iodide and ethyl iodide in acetone-d_6 provided one convenient test sample. The width of the carbon-13 resonance of methyl iodide provided an internal standard, and that of the methyl resonance of ethyl iodide revealed the effect of proton–proton coupling. WALTZ-16 and DIPSI-2, since they are of similar length and complexity, should be directly comparable without worrying too much about any differential effects of sample spinning or relaxation (*5*) between the two sequences.

Figure 5 shows the observed 100 MHz carbon-13 resonance (10 mm probe) of methyl iodide (J_{CH} = 151 Hz) as a function of the proton decoupler offset for WALTZ-16, DIPSI-2, and DIPSI-3. As the methyl protons have identical chemical shifts, there

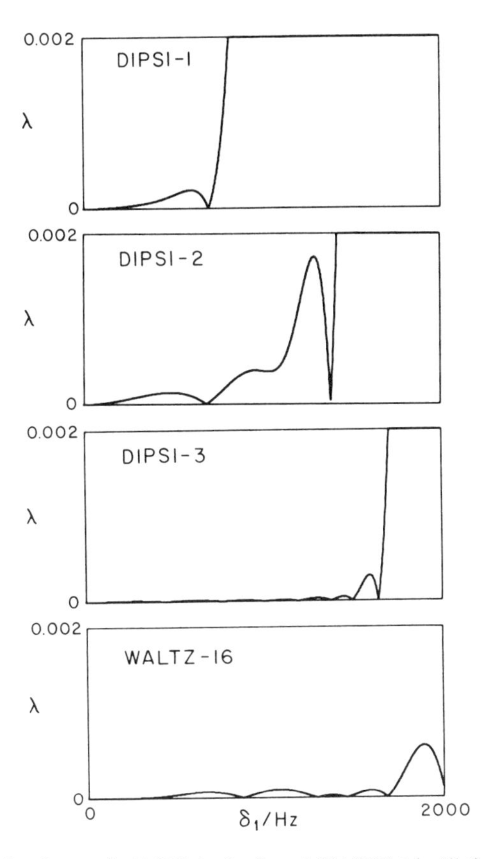

FIG. 4. Single-spin scaling factors for DIPSI-1, -2, -3, and WALTZ-16. All the sequences offer excellent single-spin performance over their bandwidths, but WALTZ-16 gives the largest bandwidth.

is no influence of proton–proton coupling on the heteronuclear decoupling performance. With the sample spinning at 6 Hz, the B_0 field was shimmed until a full width at half-height of 0.20 Hz was obtained using coherent on-resonance decoupling. A sensitivity enhancement function added 0.05 Hz of line broadening, to bring the linewidth to 0.25 Hz. The same settings were retained to investigate the broadband sequences. The carbon-13 signal was not sampled synchronously with the decoupler cycling.

The decoupler field was calibrated and set to the value $\omega_2/2\pi = 1480$ Hz. This relatively low value provides a stringent test of each sequence. The decoupler offset was incremented in 200 Hz steps over a range ±1400 Hz about exact resonance; a single transient was acquired at each offset. Under these conditions all three sequences give narrow linewidths over their respective bandwidths, and in particular for on-resonance irradiation of the protons. However, WALTZ-16, designed expressly for this single-spin case, gives the largest bandwidth. The observed bandwidths for WALTZ-16 and DIPSI-2 agree well with the theoretical predictions. DIPSI-3 gives an enhanced tolerance to RF inhomogeneity, evident as a slight increase in peak height, but does

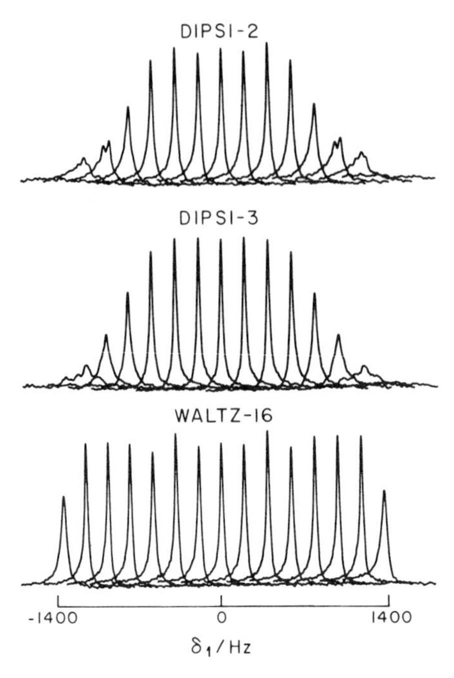

FIG. 5. Carbon-13 resonance of methyl iodide showing the offset dependence of DIPSI-2, DIPSI-3, and WALTZ-16. The decoupler offset has been stepped in 200 Hz increments over a ±1400 Hz range about exact resonance. All three sequences give narrow resonances over their bandwidths, but WALTZ-16 decouples over the largest range. The variations in peak height are attributable to poor ω_2 homogeneity over the sample volume.

not improve much on DIPSI-2 in terms of bandwidth. The cycling rate for DIPSI-3 is only 27.2 Hz under these conditions and is one factor causing the discrepancy between theory and experiment (*5*).

The situation is completely different when there are coupled protons, as there are in most molecules of interest. Figure 6 shows the 125 MHz carbon-13 methyl resonance (5 mm probe) of ethyl iodide in a mixture of methyl and ethyl iodide under conditions of broadband proton decoupling, using decoupling fields of 1100, 1460, and 1930 Hz and a spinning rate of 15 Hz. The decoupler offset was adjusted to the resonance frequency of the methyl protons of ethyl iodide; the methylene protons are then off resonance by 690 Hz. The value of J_{HH} is 7 Hz. Each spectrum is the result of 64 transients; no line broadening has been applied. Coherent on-resonance decoupling gave a linewidth of 0.20 Hz for methyl iodide.

Using WALTZ-16, a very broad multiplet is obtained at $\omega_2/2\pi = 1100$ Hz. Some line narrowing is achieved with a 1460 Hz field, and a 1930 Hz field narrows the resonance further, but a lineshape approaching a singlet is never obtained. The bizarre "wings" in the lineshape, most apparent at the intermediate field strength, are not artifacts due to poor homogeneity of the static magnetic field B_0. They result from the outer lines of the methyl quartet, which experience an effective heteronuclear

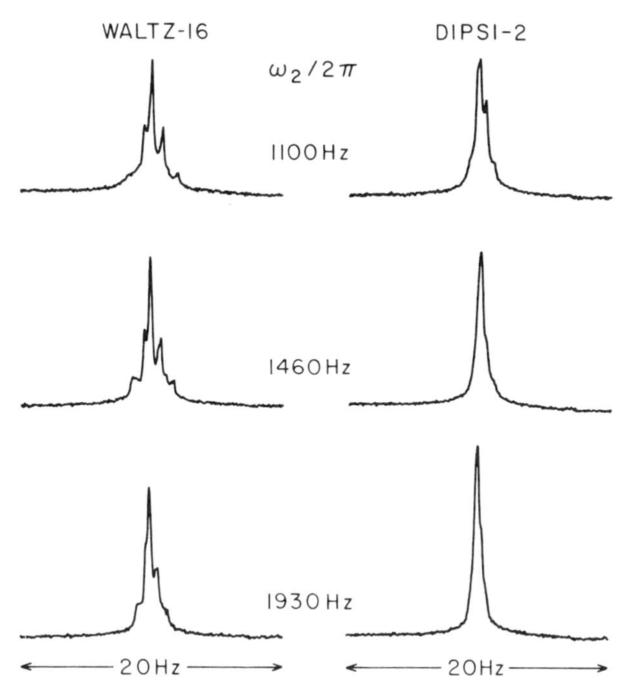

FIG. 6. Carbon-13 methyl resonance of ethyl iodide using three different values of ω_2. Due to the effect of scalar coupling between the protons, distorted lineshapes are obtained with WALTZ-16 (left-hand spectra). DIPSI-2 gives better results, as shown on the right.

coupling constant three times as large as the inner lines (*23*) and which are three times more sensitive to RF field inhomogeneity (*24*).

By using the sequence DIPSI-2 instead, a substantial improvement is obtained, as shown by the right-hand series of spectra. At the lowest field strength of 1100 Hz there is a resolved residual splitting, and some evidence of the pedestal from the outer lines, but the performance is already as good as that obtained with WALTZ-16 at the highest field strength tested. The multiplet pattern shows that a propagator close to a pure scalar operator is being produced by DIPSI-2. At the intermediate value of 1460 Hz the splitting becomes unresolved, showing only as a slight shoulder on the lineshape. With the highest level of 1930 Hz the line narrows still more to give a slightly distorted singlet. Aside from the more pleasing lineshape, DIPSI-2 gives an increase in peak height, and hence carbon-13 sensitivity, of about 25% in this example.

Figure 7 shows the low-field carbon-13 ethylenic resonance of *trans*-cinnamic acid, a molecule previously used to illustrate the effect of homonuclear coupling on broadband decoupling (*13*). The ethylenic protons form an almost isolated two-spin system with J_{HH} = 16 Hz, providing an ideal test for the effects of proton–proton coupling. The decoupler offset was once again adjusted to the resonance frequency of the directly attached proton. The other proton is then off resonance by −584 Hz. At $\omega_2/2\pi$ = 1100 Hz, WALTZ-16 gives a broad multiplet with the predicted pattern of four lines. DIPSI-

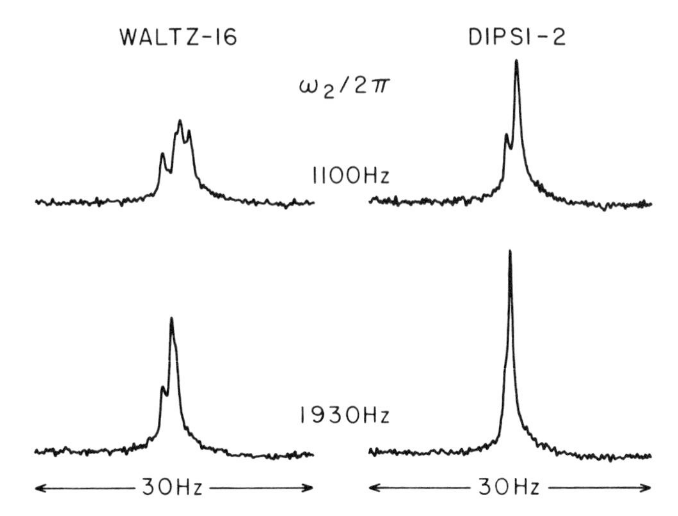

FIG. 7. Low-field ethylene resonance of *trans*-cinnamic acid under conditions of broadband decoupling. WALTZ-16 gives broad multiplets, and at the lowest decoupler level all four lines are resolved. DIPSI-2 narrows the resonance considerably, resulting in better sensitivity and resolution.

2 gives a 3:1-type pattern, almost twice as narrow and twice as intense, under the same conditions. Once again, increasing the field strength to 1930 Hz narrows the resonance line in each instance, but there is still a 50% increase in peak height using DIPSI-2. We have obtained similar results with DIPSI-3. In Fig. 8 we show that the main features of the multiplet patterns, using both WALTZ-16 and DIPSI-2, can be reproduced by simulation even though the latter assumes a perfectly homogeneous RF field.

CONCLUSIONS

Past improvements in broadband decoupling aimed to improve the bandwidth for a given value of ω_2, and much progress has been made. For the single-spin case, sequences like WALTZ-16 deliver very high quality decoupling over bandwidths $\pm\Delta\omega/\omega_2$, and larger bandwidths can be attained if the quality is allowed to deteriorate to some extent (*5*). While very wide bandwidths may look impressive on paper, they usually offer no particular advantage for proton decoupling and so are not much used. The lowest value of the decoupling field that can be used is dictated by the requirement that the modulation sidebands in the carbon-13 spectrum be acceptably small, so that values much less than 2 kHz are rarely employed. Well-designed probes can achieve fields $\omega_2/2\pi = 3$ kHz with acceptable sample heating. Consequently, for all except the very highest B_0 fields, the entire proton chemical-shift range is decoupled with sequences like WALTZ-16. Furthermore, many molecules do not have proton resonances that span the entire 10 ppm range. In these cases, further improvement in bandwidth is beside the point.

We have shown here that scalar coupling among the protons can also set a lower limit to the decoupling field that can be used and, using WALTZ-16, values substan-

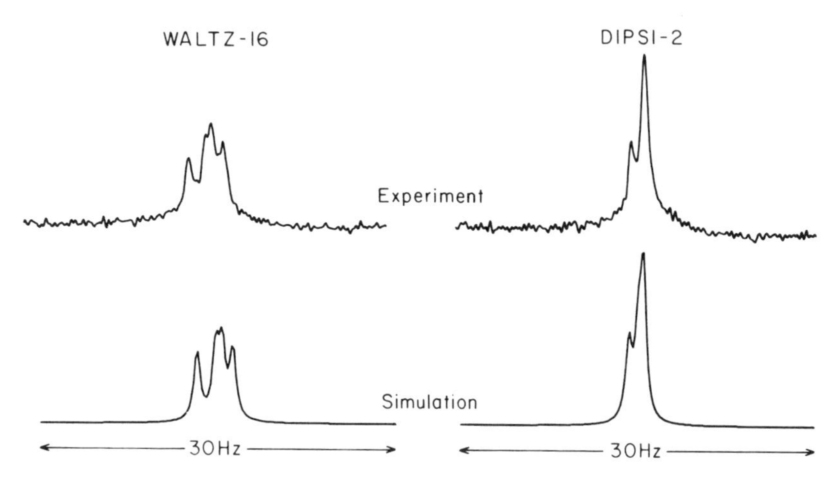

FIG. 8. Comparison between simulation and experiment for *trans*-cinnamic acid using WALTZ-16 (left) and DIPSI-2 (right). The parameters used in the simulation are $\omega_2/2\pi = 1100$ Hz, $\delta_1 = 0$ Hz, $\delta_2 = -584$ Hz, $^1J_{CH} = 150$ Hz, $^2J_{CH} = 0$ Hz, and $J_{HH} = 16$ Hz. The simulations assume a completely homogeneous ω_2 field and have been artificially line broadened to match the linewidths of the experimental spectra. No attempt has been made to fit the experimental spectra using an ω_2 distribution.

tially larger than 2 kHz may have to be employed if the true instrumental linewidth is the desired resolution. In some cases the multiplet patterns could prove confusing, especially if they happen to overlap or if the carbon-13 nucleus is coupled to other heteronuclei. Isotopic substitution of deuterium for protons changes the proton coupling network; some change in the carbon-13 lineshape may be observed in addition to any splittings and isotope shifts. It goes without saying that experiments like INADEQUATE (*25, 26*) can demand the narrowest carbon-13 resonances to achieve the best sensitivity. The advent of less costly computer memory and array processors increasingly allows very fine digitization of even routine survey spectra, while improvements in temperature stability and B_0 homogeneity (*27–29*) have produced impressively narrow carbon-13 linewidths in favorable cases.

In the aforementioned cases it may be possible to improve resolution and sensitivity by using the DIPSI sequences instead of existing single-spin sequences. Using a 2 kHz field, a bandwidth of 2.4 kHz is obtained with DIPSI-2, large enough for all normal applications at B_0 fields up to 200 MHz for protons and for many applications up to 300 MHz. When scalar coupling among the protons is limiting the carbon-13 resolution, DIPSI-2 can offer a factor of two in line narrowing over WALTZ-16 or, equivalently, a fourfold reduction in RF power.

This work has implications for homonuclear Hartmann–Hahn experiments (*8, 11*). In such experiments a broadband decoupling sequence is often used during the mixing period, causing a coherent exchange of magnetization among the protons. Elegant calculations (*30*) suppose that a pure scalar operator determines the evolution during the mixing sequence. While sequences like MLEV-16, WALTZ-16, or MLEV-17 (*31*) suppress the effects of resonance offset to a large extent and undoubtedly result in magnetization transfer, none of them produces a pure scalar operator. In addition, calculations show that *both* the MLEV sequences are sensitive to small errors in the RF phase shifts, in disagreement with claims in the literature (*31*). We expect the DIPSI sequences to be of potential use in these experiments.

ACKNOWLEDGMENTS

The authors thank Dieter Suter for stimulating discussions concerning fictitious Hamiltonians. This work has been supported by the Director, Office of Energy Research, Materials Science Division of the U.S. Department of Energy under Contract DE-AC03-76SF00098.

REFERENCES

1. M. H. LEVITT, R. FREEMAN, AND T. FRENKIEL, *in* "Advances in Magnetic Resonance" (J. S. Waugh, Ed.), Vol. 12, p. 47, Academic Press, New York, 1982.
2. J. S. WAUGH, *J. Magn. Reson.* **49,** 517 (1982).
3. A. J. SHAKA, J. KEELER, T. FRENKIEL, AND R. FREEMAN, *J. Magn. Reson.* **52,** 335 (1983).
4. A. J. SHAKA, J. KEELER, AND R. FREEMAN, *J. Magn. Reson.* **53,** 313 (1983).
5. A. J. SHAKA AND J. KEELER, *Prog. NMR Spectrosc.* **19,** 47 (1986).
6. J. S. WAUGH, *J. Magn. Reson.* **50,** 30 (1982).
7. K. V. SCHENKER, D. SUTER, AND A. PINES, *J. Magn. Reson.* **73,** 99 (1987).
8. L. BRAUNSCHWEILER AND R. R. ERNST, *J. Magn. Reson.* **53,** 521 (1983).
9. C. J. LEE, D. SUTER, AND A. PINES, *J. Magn. Reson.* **75,** 110 (1987).
10. P. B. BARKER, A. J. SHAKA, AND R. FREEMAN, *J. Magn. Reson.* **65,** 361 (1986).
11. D. G. DAVIS AND A. BAX, *J. Magn. Reson.* **65,** 355 (1985).
12. J. S. WAUGH, *J. Magn. Reson.* **68,** 189 (1986).
13. A. J. SHAKA, P. B. BARKER, AND R. FREEMAN, *J. Magn. Reson.* **71,** 520 (1987).
14. A. J. SHAKA AND A. PINES, *J. Magn. Reson.* **71,** 495 (1987).
15. U. HAEBERLEN AND J. S. WAUGH, *Phys. Rev.* **175,** 453 (1968).
16. W. MAGNUS, *Commun. Pure Appl. Math.* **7,** 649 (1954).
17. P. PECHUKAS AND J. C. LIGHT, *J. Chem. Phys.* **44,** 3897 (1966).
18. R. M. WILCOX, *J. Math. Phys.* **8,** 962 (1967).
19. I. BIALYNICKI-BIRULA, B. MIELNIK, AND J. PLEBANSKI, *Ann. Phys.* **51,** 187 (1969).
20. W. R. SALZMAN, *J. Chem. Phys.* **82,** 822 (1985).
21. M. H. LEVITT AND R. FREEMAN, *J. Magn. Reson.* **43,** 502 (1981).
22. D. SUTER, K. V. SCHENKER, AND A. PINES, *J. Magn. Reson.* **73,** 90 (1987).
23. W. A. ANDERSON AND R. FREEMAN, *J. Chem. Phys.* **37,** 85 (1962).
24. R. FREEMAN, J. B. GRUTZNER, G. A. MORRIS, AND D. L. TURNER, *J. Am. Chem. Soc.* **100,** 5637 (1978).
25. A. BAX, R. FREEMAN, AND S. P. KEMPSELL, *J. Am. Chem. Soc.* **102,** 4849 (1980).
26. A. BAX, R. FREEMAN, AND T. FRENKIEL, *J. Am. Chem. Soc.* **103,** 2102 (1981).
27. A. ALLERHAND, R. E. ADDLEMAN, AND D. OSMAN, *J. Am. Chem. Soc.* **107,** 5809 (1985).
28. A. ALLERHAND AND M. DOHRENWEND, *J. Am. Chem. Soc.* **107,** 6684 (1985).
29. A. ALLERHAND, R. E. ADDLEMAN, D. OSMAN, AND M. DOHRENWEND, *J. Magn. Reson.* **65,** 361 (1985).
30. N. CHANDRAKUMAR, *J. Magn. Reson.* **71,** 322 (1987).
31. S. SUBRUMANIAN AND A. BAX, *J. Magn. Reson.* **71,** 325 (1987).

JOURNAL OF MAGNETIC RESONANCE **78**, 292–301 (1988)

Linear Prediction and Projection of Pure Absorption Lineshapes in Two-Dimensional FTESR Correlation Spectroscopy*

JEFF GORCESTER AND JACK H. FREED

Baker Laboratory of Chemistry, Cornell University, Ithaca, New York 14853-1301

Received October 26, 1987; revised November 18, 1987

New ESR experiments based on techniques of two-dimensional correlation spectroscopy have been shown to be useful in the determination of magnetization transfer rates in motionally narrowed nitroxides. Spectral enhancement based upon linear prediction with singular value decomposition (LPSVD) is applied in the present work to project 2D absorption lineshapes and to dramatically improve the signal/noise ratio. Heisenberg spin exchange rates obtained from volume integrals of LPSVD-projected 2D absorption lineshapes compare well with those obtained from absolute value peak amplitudes in the case of 2,2,6,6-tetramethyl-4-piperidone-*N*-oxyl-d_{16} (pd-tempone) dissolved in toluene-d_8 (where theory predicts that the two should agree).

INTRODUCTION

In a recent report we have demonstrated the application of two-dimensional correlation spectroscopy to ESR (*1*). Such techniques are based upon the irradiation of the entire ESR spectrum with a broadband microwave (MW) pulse (*2, 3*) and the recording of the free precession signal (i.e., free induction decay or FID) that follows. More recently we have demonstrated the quantitative capability of ESR correlation spectroscopy with a three-pulse experiment (2D ELDOR) for the observation of magnetization transfer between hyperfine (hf) lines (*4*). In (*4*) we determined the rate of Heisenberg exchange by comparison of peak heights in the absolute value representation of the 2D ELDOR spectrum. The theory indicates, however, that it is the volume integrals of the 2D absorption lines, not the absolute value peak heights, which correctly reflect the electron spin population differences in the general case. In (*4*) we depended on the fact that ratios of absolute value peak heights give the same information as ratios of 2D absorption volume integrals when T_2 is the same for each hf line. In general there is significant variation of T_2 across the spectrum, in which case volume integrals of the 2D absorption lines are essential for accurate measurement of magnetization transfer rates.

The ESR spectrum obtained upon Fourier transformation of the FID is, in general, an admixture of absorption and dispersion. In order to obtain the pure absorption spectrum, numerical phase corrections are required (*3*). The phase of each resonance line depends on the resonance offset of that line as well as on the dead time of the spectrometer (*4*). We observe an almost linear variation of the phase of the recorded signal on frequency (*5*). During the spectrometer dead time, a component of the free

* Supported by NIH Grant GM-25862 and NSF Grant CHE 8703014.

0022-2364/88 $3.00

precession signal may undergo several periods of oscillation, leading to a phase shift of as much as 6π radians for a resonance offset of 50 MHz. For a spectrum of bandwidth 100 MHz, this implies a total phase variation of 12π radians across the spectrum. Numerical correction of this phase variation is accomplished by a simple coordinate transformation in the frequency domain of the real and imaginary outputs of the FFT. Given a complex spectral function $S(\nu)$, the expression for this coordinate transformation is

$$\begin{bmatrix} \mathrm{Re}[S(\nu)] \\ \mathrm{Im}[S(\nu)] \end{bmatrix} = \begin{bmatrix} \cos\alpha & \sin\alpha \\ -\sin\alpha & \cos\alpha \end{bmatrix} \begin{bmatrix} \mathrm{Re}[S(\nu)]' \\ \mathrm{Im}[S(\nu)]' \end{bmatrix}, \qquad [1]$$

where α consists of a frequency-independent correction (the so-called zeroth-order correction) and a frequency-dependent correction (a first-order or linear correction), i.e., $\alpha = \alpha_0 + \alpha_1\nu$. Such corrections are feasible in one-dimensional FTESR (*3*), but are very cumbersome when applied in both dimensions of a two-dimensional spectrum because of the large required linear corrections.

Two-dimensional correlation spectroscopy ("COSY") in ESR is performed in much the same way as for NMR. The pulse sequence $\pi/2$–t_1–$\pi/2$–t_2 constitutes the simplest of the COSY experiments. The initial $\pi/2$ pulse generates the transverse magnetization components which precess during the evolution period t_1 becoming amplitude encoded according to their precessional frequencies in the rotating frame. The FID is recorded during the detection period of duration t_2, which begins with the final $\pi/2$ pulse. The closely related experiment, 2D ELDOR, uses three MW pulses in the sequence $\pi/2$–t_1–$\pi/2$–T–$\pi/2$–t_2. Magnetization transfer is revealed by cross correlations which give rise to 2D ELDOR cross peaks. Peaks which appear along the diagonal in 2D COSY type experiments are associated with autocorrelations and are thus called auto peaks. In each type of 2D experiment an FID is collected for each t_1; then the phase of the first pulse is advanced by 90°, and a second FID is collected [call them $s'(t_1, t_2)$ and $s''(t_1, t_2)$]. These two signals depend on terms oscillatory in t_1 that are in phase quadrature (*6*). Fourier transformation with respect to t_2 of each data set yields the spectral functions $\hat{s}'(t_1, \omega_2)$ and $\hat{s}''(t_1, \omega_2)$. Baseplane corrections facilitate suppression of axial peaks (*4*) after which we form

$$\hat{s}(t_1, \omega_2) = \mathrm{Re}[\hat{s}'] + i\,\mathrm{Re}[\hat{s}''] \qquad [2]$$

which yields the 2D spectrum $S(\omega_1, \omega_2)$ upon final FT.

This two-step 2D quadrature phase alternation sequence provides the phase information in the t_1 domain necessary for the pure absorption representation of the 2D spectrum; i.e., we obtain four quadrant phase information (*7*). $S(\omega_1, \omega_2)$ obtained by the above procedure (known as hypercomplex FT) is not, in general, a 2D absorption spectrum, but an admixture of absorption and dispersion in both ω_1 and ω_2 axes. In order to project out the 2D absorption representation, one would apply phase corrections as described by Eq. [1] to each 1D spectrum obtained upon FT with respect to t_2 at a given t_1. Analogous phase corrections would then be applied to each spectrum obtained upon FT with respect to t_1 at a given ω_2. As already noted, this procedure is difficult in ESR because of the large phase variation across the spectrum in both frequency domains.

To facilitate accurate projection of 2D absorption lineshapes and to suppress certain artifacts that appear in 2D ELDOR spectra, we have applied a linear prediction method developed by Kumaresan and Tufts (*8*) that was introduced to magnetic resonance by Barkhuijsen *et al.* (*9*). In that initial application to magnetic resonance, the emphasis was on obtaining good frequency information in one-dimensional data. Millhauser and Freed (*10*) demonstrated the application of these methods to two-dimensional electron spin-echo (2D ESE) spectroscopy. The 2D ESE method is distinct from COSY and related techniques in that COSY requires irradiation of the entire spectrum (i.e., nonselective pulses), whereas in 2D ESE one irradiates only a narrow region of the spectrum (i.e., selective pulses), recording the maximum ESE signal voltage as the dc field is swept through the spectrum. Real valued linear prediction is applied to the ESE envelope at each field position in the 2D ESE spectrum, facilitating extrapolation of the time series to zero dead time. In this report we demonstrate the use of complex linear prediction in both axes of a 2D ELDOR spectrum. We show how this new application of the general technique, which we call hypercomplex linear prediction, facilitates the projection of 2D absorption lineshapes as well as the rejection of residual axial peaks and much of the noise.

LINEAR PREDICTION FOR 2D CORRELATION SPECTROSCOPY

The basis for (autoregressive) linear prediction is that a discrete time series

$$\{x_1, x_2, \ldots, x_N\} \qquad [3]$$

can be modeled by the expression

$$x_n = \sum_{i=1}^{M} b_i x_{n-i}, \qquad [4]$$

where the set $\{b_i\}$ are called the forward linear prediction (lp) coefficients, and the order M is less than N. The implication of Eq. [4] is that each sampling of the time series can be expressed as a linear combination of the M previous ones. With the application of backward lp, described by writing Eq. [4] in the backward sense, i.e.,

$$x_n = \sum_{i=1}^{M} a_i x_{n+i}, \qquad [5]$$

one can model an FID in terms of exponentially damped sinusoids and determine all of the relevant parameters: frequency, time constant, amplitude, and phase. In such a procedure one first generates a set of coupled equations obtained upon writing Eq. [5] for the $N - M$ possible values of n. The least-squares solution for the set of backward lp coefficients, $\{a_i\}$, in terms of the N complex data points, is written as

$$\begin{bmatrix} x_2 & x_3 & \cdots & x_{M+1} \\ x_3 & x_4 & \cdots & x_{M+2} \\ \vdots & \vdots & \ddots & \vdots \\ x_{N-M+1} & x_{N-M+2} & \cdots & x_N \end{bmatrix} \begin{bmatrix} a_1 \\ a_2 \\ \vdots \\ a_M \end{bmatrix} = \begin{bmatrix} x_1 \\ x_2 \\ \vdots \\ x_{N-M} \end{bmatrix}. \qquad [6]$$

The least-squares problem is solved by determination of the singular values of the data matrix in Eq. [6]. Noise rejection is achieved by subtraction of the root mean square of the singular values attributed to noise from singular values associated with the FID. There are two parameters which must be selected in order to model the FID: (1) the number of lp coefficients (i.e., the order M) and (2) the number of singular values attributable to signal, which we refer to in the standard fashion as the reduced order, K. Linear prediction yields the frequencies and T_2's, after which a linear least-squares (LS) procedure is used to obtain amplitudes and phases. Once the frequency, T_2, amplitude, and phase of each FID component are known, reconstruction of the FID is facilitated by the expression

$$x'_n \equiv x'(n\Delta t) = \sum_{j=1}^{K} c_j \exp(-n\Delta t/T_{2j})\cos(\omega_j n\Delta t + \phi_j), \qquad [7]$$

representing one quadrature component and an analogous expression for the other quadrature component. As noted above, the phases, ϕ_j, may vary substantially across the spectrum, so that the spectrum obtained upon FT of the reconstructed FID described by Eq. [7] is an admixture of absorption and dispersion. To remedy this problem we zero the K phases, $\{\phi_j, j = 1, K\}$, prior to reconstruction of the FID, so that upon FT of the LPSVD result we obtain the pure absorption lineshapes. This procedure is easily generalized to two frequency dimensions by applying complex LPSVD to each FID prior to FT with respect to t_2, collecting the results of the FT as described by Eq. [2], then applying complex LPSVD once again in the t_1 domain. If the phases, ϕ_j, are zeroed at each call to LPSVD, the pure absorption representation of the 2D spectrum is retrieved.

In addition to projection of pure 2D absorption lineshapes, LPSVD facilitates removal of distortions found in all of our 2D FTESR spectra near $\omega_1 = 0$. The source of these distortions is a combination of extra amplitude modulation in t_1 due to variations in the MW pulse flip angle as a function of t_1 (arising from distortions in the second MW pulse) and incomplete cancellation of axial peaks (*4*). To remove these distortions we reject components whose frequencies fall within a band centered at $\omega_1 = 0$ by excluding them from the reconstructed FID; i.e., we apply a narrow reject filter of desired width (usually ± 3 MHz) in ω_1. Careful experimental adjustment of the MW carrier frequency ensures that all three hf lines are off resonance by at least 3 MHz, so that no spectral information is destroyed upon application of the narrow reject filter. Elimination of these distortions improves the accuracy of our measurements of 2D absorption volume integrals, especially for those peaks close in frequency, ω_1, to $\omega_1 = 0$.

The application of LPSVD to 2D ELDOR data is based upon the following theoretical expressions which describe the experiment (*4, 11*). Let $s'(T, t_1, t_2)$ represent the time-domain 2D ELDOR spectral function obtained with three MW pulses of equal phase. Then

$$s'(T, t_1, t_2) = B' \sum_{nml} c'_{nml} \exp(-\Lambda_n t_2) \exp(-T/\tau_m) \mathrm{Re}\left(\sum_j b_{lj} \exp(-\Lambda_j t_1)\right), \qquad [8]$$

where the τ_m represent the decay times associated with magnetization transfer and the c'_{nml} and b_{lj} are the relative amplitudes; the imaginary parts of the Λ_j are the precessional frequencies whereas the real parts represent the homogeneous widths;

i.e., $\Lambda_j = (T_{2,j}^{-1} - i\omega_j)$. Equation [8] shows that the 2D ELDOR spectra are, in the t_1 and t_2 time domains, sums of decaying sinusoids as assumed by LPSVD. In the case of motionally narrowed nitroxides with magnetization transfer driven by Heisenberg spin exchange we can derive from Eq. [8] the simple expression

$$\omega_{\mathrm{HE}} = \frac{1}{T} \ln\left(\frac{2a_{mj}V_j + V_m}{V_m - a_{mj}V_j}\right), \qquad [9]$$

where the V_j are determined from volumes of auto peaks (which go as V_j^3) and the a_{mj} are the appropriate ratios of observed cross-peak to auto-peak volumes, and ω_{HE} is the Heisenberg exchange rate. Equation [9] emphasizes the importance of obtaining accurate relative peak volumes from the 2D absorption spectrum.

RESULTS

In our 1D FTESR experiment we record the in-phase and quadrature components of the FID following a $\pi/2$ MW pulse, taking care to cancel image peaks with the four-step CYCLOPS image cancellation sequence (*12*). The spectrum of 2,2,6,6-tetramethyl-4-piperidone-*N*-oxyl-d_{16} (pd-tempone) in toluene-d_8 obtained upon FT of the FID is shown in Fig. 1a (only one of the quadrature components is shown). Superimposed on these data is the spectrum obtained upon FT of the LPSVD-treated FID. The absorption spectrum projected from the LPSVD-treated data is given in Fig. 1b. LPSVD enabled a clear distinction in this example between singular values associated with noise and those associated with the three ^{14}N hf components. The "signal" singular values were at least an order of magnitude greater than the rms of the "noisy" singular values. Weak ^{13}C sidebands present in the data were not recovered by LPSVD because of their smaller amplitudes relative to the noise; some of these components could be recovered by increasing the order M.

In our 2D ELDOR experiments we record the in-phase and quadrature components of the FID during the detection period for each of 16 different phase alternation sequences at 90 different values of t_1. The 16-step phase alternation scheme incorporates 2D quadrature detection as described above, as well as sequences for the cancellation of axial peaks and transverse interference (*4*). The 2D ELDOR spectrum of pd-tempone in toluene-d_8 at 21°C is shown in Fig. 2a; note that this is an absolute value plot (i.e., the square root of the power spectrum). The application of Eq. [9] to the spectrum of Fig. 2a gives the result $\omega_{\mathrm{HE}} = 4.89 \times 10^6\ \mathrm{s}^{-1}$, where we have compared relative amplitudes, rather than volumes, of cross peaks and auto peaks. Since all three hf lines in the spectrum of Fig. 2a have about the same T_2, we can expect reasonable agreement between ω_{HE} determined from volume integrals and from peak amplitudes as described.

We applied linear prediction to this same data set by using 24 complex lp coefficients in the t_2 domain, and up to 60 coefficients in the t_1 domain. In the t_2 domain six of the singular values were attributed to signal, whereas in the t_1 domain up to 12 were attributed to signal. The required CPU time was approximately 4 h on a Prime 9955 computer, including the time required for FFTs. LPSVD consistently recovered three components in the t_2 domain, corresponding to the three hf lines, and no greater than six components in the t_1 domain. Projection of pure absorption lineshapes was per-

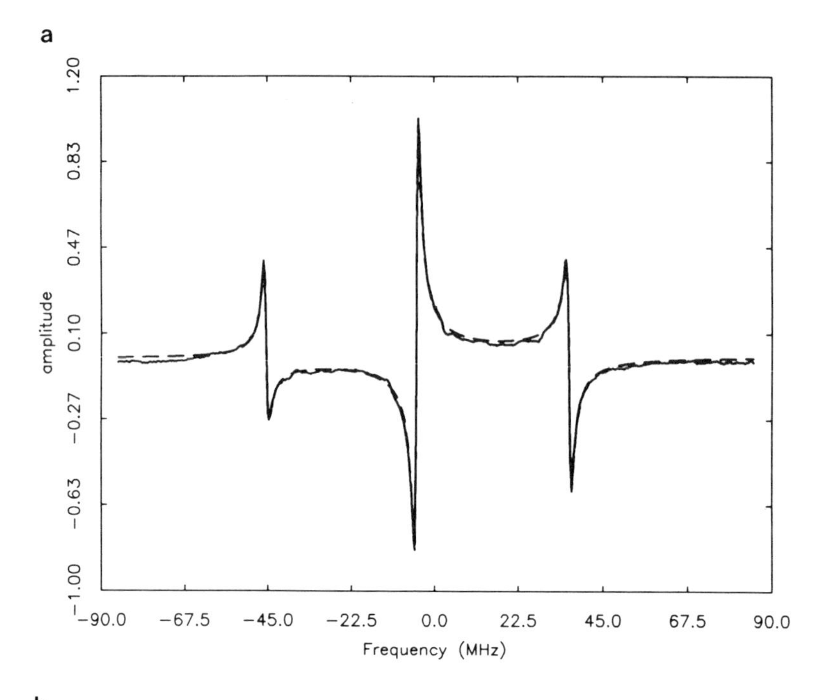

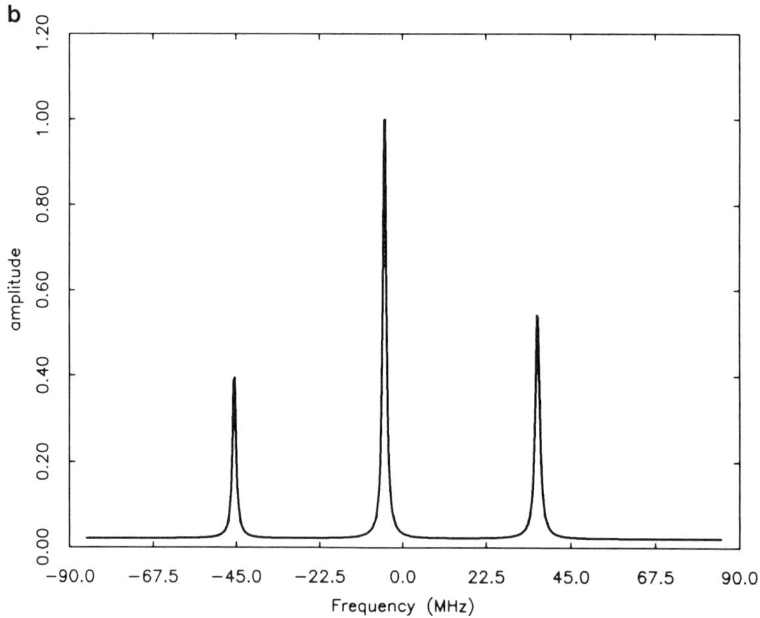

FIG. 1. (a) FT spectrum of 5.1×10^{-4} M pd-tempone in toluene-d_8 at 21°C (solid line) obtained from the average of 40 FIDs, each consisting of 256 complex data points extending to 1.2 μs; FT of the same FID after LPSVD (dashed line) with $M = 20$ and $K = 6$; (b) FT spectrum after LPSVD with projection of pure absorption lineshapes.

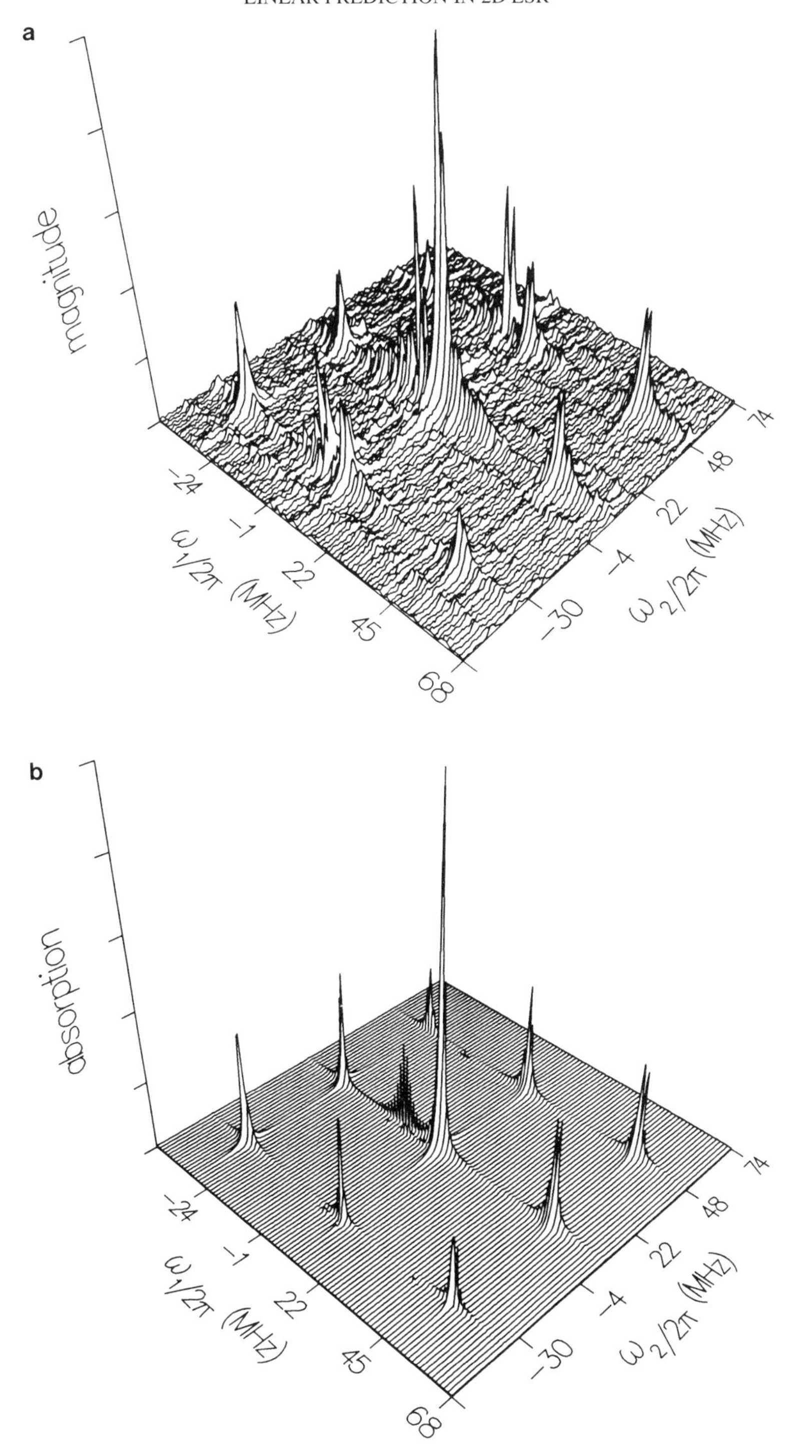

FIG. 2. (a) Absolute value 2D ELDOR spectrum of 1.17×10^{-3} M pd-tempone in toluene-d_8 at 21°C; mixing time $T = 3.10 \times 10^{-7}$ s; 90 t_1 steps; 256 complex data points per FID extending to 1 μs, data from Ref. (*4*). (b) LPSVD-projected pure 2D absorption representation of the same spectrum; $M = 24$, $K = 6$ in the t_1 domain, $M = 60$, $K = 12$ in the t_2 domain; the broad peak near the center of the spectrum (at $\omega_1/2\pi = -8$ MHz) has zero width in ω_2 (hence zero volume) and is apparently an artifact of the computation.

formed in both time domains. Extrapolation of the time series in t_1 to 128 points eliminated artifacts caused by t_1 truncation and enabled a more accurate determination of baseplane offset. In Fig. 2b we illustrate the LPSVD result obtained after eliminating components for which $|\omega_1/2\pi| < 3$ MHz (i.e., narrow reject filtering); note the considerable improvement in signal/noise ratio. The volume integral of each 2D absorption line was measured by numerically integrating in the ω_2 domain and summing the results over the discrete values of ω_1. Estimation of the Heisenberg exchange rate from ratios of volume integrals with the use of Eq. [9] gave the result $\omega_{HE} = 4.59 \times 10^6\ s^{-1}$, in good agreement with the result obtained from Fig. 2a and with the ESE result, $\omega_{HE} = 4.25 \pm 0.60 \times 10^6\ s^{-1}$ (*4*).

DISCUSSION

Many possibilities exist for improvement of the 2D linear prediction method presented in this report. First, we consider potential improvements in the computational method for solving Eq. [6]. The SVD algorithm is generally the most stable method of determining the singular values of the rectangular lp data matrix [call it $\mathbf{A}$], but it is also the most computationally inefficient. An alternative approach is to solve for the eigenvalues of the matrix $\mathbf{A}^\dagger\mathbf{A}$, which are simply related to the singular values of $\mathbf{A}$ (*13*). This method is computationally more efficient than SVD but may become unstable in the case of large order M. Barkhuijsen *et al.* have nevertheless found this technique to be generally successful in LPSVD applications (*9, 14*). Related computational methods for the solution of Eq. [6] and variations on the linear prediction theme have been demonstrated by Tang *et al.* (*15*). We are exploring the use of iterative Lanczos approximation methods for singular value decomposition which are known to predict the large singular values very accurately, and which could be modified to exploit the Hankel structure of $\mathbf{A}$ (i.e., the property $\mathbf{A}_{ij} = \mathbf{A}_{i+j}$).

We have illustrated that linear prediction may be applied to 2D data simply by applying LPSVD to the constituent 1D spectra (in both t_1 and t_2 time domains). We now propose a substantially different approach, which can truly be considered 2D linear prediction, because it exploits all of the symmetries of 2D COSY type spectra and yields all of the 2D spectral information while performing the analysis entirely in the time domain, i.e., without any Fourier transformation. The procedure, which we will refer to as 2D LPSVD, is carried out as follows: (1) calculate the M lp coefficients in Eq. [6] with the FID obtained at the initial t_1 (for convenience) and determine the frequencies and T_2's; (2) recognizing that frequencies and T_2's are the same for all of the FIDs (i.e. along the t_2 axis) irrespective of the value of t_1, perform the linear LS procedure on each FID to determine amplitudes and phases at each t_1 and for each of the quadrature components in t_1 (i.e., s' and s'', cf. Eq. [2]), utilizing the frequencies and T_2's obtained in step 1; (we now define a time series with parametric dependence on t_1 by the expression

$$x_{nm} \equiv x(n\Delta t_1, m\Delta t_2) = \sum_{j=1}^{K} c_j(n\Delta t_1)\exp(-m\Delta t_2/T_{2j})\cos(\omega_{2j}m\Delta t_2 + \phi_{2j}), \quad [10]$$

which reflects the dependence on t_1 of the complex valued amplitudes, c_j, determined in step 2); (3) recognizing that the $c_j(n\Delta t_1)$ in Eq. [10] are themselves time series of the same form as $x(n\Delta t)$ in Eq. [7] (cf. Eq. [8]), perform LPSVD on the $c_j(n\Delta t_1)$

for every component j (corresponding to the different frequency components in the ω_2 frequency domain); and (4) reconstruct the 2D time-domain data with the expression

$$x_{nm} \equiv x(n\Delta t_1, m\Delta t_2) = \sum_{j=1}^{K_2} \sum_{i=1}^{K_1(j)} c_{ij} \exp(-n\Delta t_1/T_{2,1i}) \cos(\omega_{1i} n\Delta t_1 + \phi_{1i}) \times \exp(-m\Delta t_2/T_{2,2j}) \cos(\omega_{2j} m\Delta t_2 + \phi_{2j}), \quad [11]$$

which is the 2D LPSVD analog of Eq. [7], where c_{ij}, $T_{2,1i}$, $T_{2,2j}$, ω_{1i}, ω_{2j}, ϕ_{1i}, and ϕ_{2j} are the parameters returned by 2D LPSVD, and K_2 is the reduced order in t_2 with $K_1(j)$ the reduced order in t_1 associated with $c_j(n\Delta t_1)$ (cf. Eq. [10]). Equation [11] represents one quadrant of the hypercomplex data, and analogous expressions may be written for the other three quadrants.

For the spectrum of Fig. 2a, consisting of three resonance lines in the ω_2 domain, only four singular value decompositions (as well as some additional computation, primarily LS) would be necessary for the 2D LPSVD calculation. Since SVD is the primary computational burden in LPSVD, this technique represents a dramatic reduction in CPU time (we estimate 90%) relative to the technique demonstrated in Figs. 1b and 2b. In the general case of a 2D spectrum consisting of n resonance lines in the ω_2 domain, 2D LPSVD would require $n + 1$ SVD calculations.[1] Finally, we note that the techniques presented in this report are equally well suited to spectral analysis of COSY type 2D NMR data.

CONCLUSION

We have demonstrated the application of LPSVD to two-dimensional ESR correlation spectroscopy for the projection of the pure 2D absorption lineshapes as well as for spectral enhancement and narrow reject filtering. The recovery of phase-sensitive 2D information from the raw data and the removal of distortions near $\omega_1 = 0$ improve the accuracy of our 2D techniques and enable better comparison between theory and experiment. We obtained good agreement between the Heisenberg exchange rate obtained from absolute value peak amplitudes and that obtained from volume integrals of LPSVD-projected 2D absorption lineshapes in a 2D ELDOR spectrum of pd-tempone in toluene-d_8. We have outlined a two-dimensional formulation of linear prediction which exploits the special symmetries of 2D correlation spectroscopy, and which dramatically reduces the computational burden of LPSVD in these applications.

ACKNOWLEDGMENT

We thank Mr. Nick Bigelow for contributions to the software.

[1] Note added in proof. We have implemented this algorithm on a Convex C1 super minicomputer, and we find that it successfully yields all of the 2D spectral information outlined above in 63 s of CPU time for 256 × 128 point hypercomplex data with 50 lp coefficients in t_2 and 90 lp coefficients in t_1 while using a vectorized SVD algorithm.

REFERENCES

1. J. Gorcester and J. H. Freed, *J. Chem. Phys.* **85,** 5375 (1986).
2. J. Gorcester, G. L. Millhauser, and J. H. Freed, *in* "Proceedings, XXIII Congress Ampere on Magnetic Resonance, Rome, 1986" p. 562.
3. J. P. Hornak and J. H. Freed, *J. Magn. Reson.* **67,** 501 (1986).
4. J. Gorcester and J. H. Freed, *J. Chem. Phys.* **88,** 4678 (1988).
5. See Ref. (*4*), Fig. 4.
6. J. Keeler and D. Neuhaus, *J. Magn. Reson.* **63,** 454 (1985).
7. D. J. States, R. A. Haberkorn, and D. J. Ruben, *J. Magn. Reson.* **48,** 286 (1982).
8. R. Kumaresan and D. W. Tufts, *IEEE Trans.* **ASSP-30,** 833 (1982).
9. H. Barkhuijsen, R. de Beer, W. M. M. J. Bovee, and D. van Ormondt, *J. Magn. Reson.* **61,** 465 (1985).
10. G. L. Millhauser and J. H. Freed, *J. Chem. Phys.* **85,** 63 (1986).
11. G. L. Millhauser, J. Gorcester, and J. H. Freed, *in* "Electron Magnetic Resonance of the Solid State" (J. A. Weil, Ed.), p. 571, Can. Chem. Soc. Pub., 1987.
12. D. I. Hoult and R. E. Richards, *Proc. R. Soc. London, A* **344,** 311 (1975).
13. G. H. Golub and C. F. van Loan, "Matrix Computations," Johns Hopkins Univ. Press, Baltimore, 1983.
14. H. Barkhuijsen, R. de Beer, and D. van Ormondt, *J. Magn. Reson.* **64,** 343 (1985).
15. (*a*) J. Tang, C. P. Lin, M. K. Bowman, and J. R. Norris, *J. Magn. Reson.* **62,** 167 (1985); (*b*) J. Tang and J. R. Norris, *J. Chem. Phys.* **84,** 5210 (1986).

JOURNAL OF MAGNETIC RESONANCE **84**, 14–63 (1989)

Three-Dimensional Fourier Spectroscopy. Application to High-Resolution NMR

C. GRIESINGER, O. W. SØRENSEN, AND R. R. ERNST

Laboratorium für Physikalische Chemie, Eidgenössische Technische Hochschule, 8092 Zurich, Switzerland

Received October 13, 1988

The principles of three-dimensional Fourier spectroscopy, applied to high-resolution NMR, are developed. The main emphasis is placed on the design of experimental techniques, the appearance of 3D spectra in terms of multiplet structures and peak shapes, information content, and assignment procedures. The sensitivity of 3D experiments is analyzed. The potential of 3D experiments is discussed in view of applications to biomolecules. © 1989 Academic Press, Inc.

I. INTRODUCTION

It was recognized in the early days of nuclear magnetic resonance that the information content of nuclear spin systems for the analysis of molecular structure and dynamics is by no means exhausted by conventional one-dimensional spectroscopic measurements. Additional information has been obtained by various types of double-resonance experiments (*1–3*) where explicitly nonlinear properties of the spin system are tested. This provided access to connectivity within spin coupling networks through tickling effects and polarization transfer, and to internuclear distances via cross relaxation. In addition, double resonance also allowed the simplification of spectra by homonuclear and, in particular, heteronuclear spin decoupling (*4*).

More recently, double-resonance experiments have to a large extent been replaced by two-dimensional spectroscopy (*5–7*) which leads to a particularly enlightening representation of internuclear relations that cannot be deduced from one-dimensional spectra. In the course of the last 10 years, 2D spectroscopy has developed into one of the most important and most powerful tools for the elucidation of molecular structure and dynamics, in particular for the investigation of biomolecules (*8*).

The next step, the introduction of a third frequency dimension, has been already anticipated on several occasions. Triple-resonance experiments, where a system strongly perturbed by two frequencies is tested by the response to a third frequency, have been applied occasionally in detailed studies of spin systems (*9*). Two-dimensional experiments with an additional continuous wave radiofrequency perturbation (*10*) also lead to three frequency variables. Several two-dimensional experiments already contain a third time parameter that could be converted into a time variable for 3D spectroscopy. This applies in particular to experiments that involve two coherence transfer steps, separated by a time delay, such as relay experiments (*11*), or an extended cross-relaxation or exchange period, such as NOESY and EXSY. Finally,

0022-2364/89 $3.00

indirectly detected 2D spectra, such as 2D correlation in zero field with high-field detection (*12, 13*), require a 3D experimental technique, although the third dimension necessary for the high-field detection is seldom exploited for spectral features.

In addition to these experiments whose inherent 3D character is normally not utilized, several genuine 3D experiments have recently been performed. They include 3D *J*-resolved COSY (*14*), 3D correlation experiments (*15, 16*) and 3D combinations of correlation and cross-relaxation experiments (*16, 17*). These few prototype experiments have shown that the potential of 3D spectroscopy is significant in view of the analysis of even larger biomolecules than have been investigated by 2D NMR.

In this paper we present a unified treatment of 3D NMR experiments in liquid phase in order to provide a basis for the development of further 3D techniques and to pave the way toward practical applications in molecular biology. The following section provides a classification scheme of 3D spectra and 3D experiments. The different classes of peaks as they appear in 3D spectra are discussed in Section III. A theoretical foundation of 3D experiments is provided in Section IV. Section V is devoted to the resulting peak shapes. In 3D spectroscopy it turns out that the appearance of combination and multiple-quantum peaks, not observed in the common 2D experiments, can often not be inhibited. These features are discussed in Section VI. Similar to the situation in two dimensions, 3D spectra also exhibit symmetry properties as shown in Section VII. Section VIII is devoted to a discussion of the inherent sensitivity of 3D spectroscopy. The practical aspects of 3D experiments are the subject of Section IX and the prospects for practical applications are summarized in the concluding Section X. The connections of 3D spectroscopy to stochastic resonance are discussed in the Appendix.

II. CLASSIFICATION OF 3D SPECTRA AND EXPERIMENTS

Multidimensional spectroscopy pursues two major goals, (i) the simplification and unraveling of complex spectra, and (ii) the elucidation of spin systems and molecules in terms of connectivity (via *J* couplings) and proximity of spins (via dipolar couplings and cross relaxation) and the characterization of molecular dynamics in terms of large-amplitude motion and chemical exchange (*6*). In *2D spectroscopy,* this leads to two classes of spectra:

(i) *2D separation spectra* achieve a 2D spread by the evolution of coherence under two different Hamiltonians that determine the frequency spread in the ω_1 and ω_2 dimensions. Normally no coherence transfer or mixing process is involved. Examples are 2D spectra in liquid phase with separation of chemical shifts and *J* couplings (*18, 19*) and separated-local-field spectra in solids (*20*).

(ii) *2D transfer spectra* exploit a transfer of spin order for elucidating structure or dynamics. (a) The use of coherent transfer processes through *J* coupling or dipolar coupling leads to *2D correlation spectroscopy.* Examples are COSY (*5*), heteronuclear shift correlation (*21*), relay COSY (*11*), total correlation spectroscopy (TOCSY) (*22*), and 2D multiple-quantum spectroscopy. (*23*). (b) Incoherent transfer processes are involved in *2D exchange spectroscopy* where the frequencies before and after the exchange process are displayed. Representative examples are 2D chemical-exchange spectroscopy (EXSY) (*24, 25*), 2D cross-relaxation spectroscopy (NOESY, ROESY) (*26–28*), and 2D spin-diffusion spectroscopy (*29*).

In *3D spectroscopy,* the variety of possible experiments is larger, as two processes are involved that pairwise relate three frequency coordinates. A homonuclear or heteronuclear 3D experiment combines the features of two 2D experiments:

(i) *3D separation spectra* result from two distinct separation processes. A conceivable example is the separation of chemical shifts and homonuclear and heteronuclear spin couplings in three separate dimensions. Recently, a 3D separation experiment in solid-state NMR has been described where dipolar couplings and anisotropic and isotropic chemical shifts are displayed in three separate dimensions (*30*).

(ii) *3D transfer spectra* correlate three resonance frequencies by means of two transfer processes that may either involve coherent or incoherent transfer of spin order. They can both be of the same or of different type:

(a) For 3D correlation spectra, two coherence-transfer processes through J coupling are used to correlate origin, relay, and destination nuclei in partial analogy to relay COSY experiments. The two correlation steps may be homonuclear or heteronuclear. The correlation can be achieved by transfer of antiphase coherence (COSY) or by transfer of in-phase coherence (TOCSY). This leads to experiments such as COSY–COSY, hetero-COSY–TOCSY, etc. A more exotic application could involve multiple-quantum coherence, relating, for example, double-quantum coherence in ω_1 with triple-quantum coherence in ω_2 and single-quantum coherence in ω_3.

(b) In *3D exchange spectra,* two successive exchange processes are mapped. These processes may be chemical exchange (EXSY) and cross relaxation in the laboratory (NOESY) or in the rotating frame (ROESY). Examples are EXSY–EXSY and NOESY–ROESY.

(c) Many of the most useful 3D experiments combine coherent and incoherent transfer steps, such as NOESY–COSY or NOESY–TOCSY (*16, 17, 31*). The practical importance of these combinations stems from the fact that the two types of techniques required for the assignment of resonances in biological macromolecules are combined here in a single experiment. In particular NOESY–TOCSY holds the potential to extend the analysis of biomolecules to even larger ones than has been possible so far.

(iii) Finally, it is also possible to combine a separation and a transfer step in one 3D experiment. We have already mentioned 3D J-resolved COSY spectra where scalar couplings are utilized to spread a 2D COSY spectrum in a third dimension (*14*).

The most general approach to 3D spectroscopy is *3D time-domain pulse Fourier spectroscopy.* Here two 2D pulse experiments are combined into a 3D experiment as indicated in Fig. 1, where the 3D experiment is made up of preparation, evolution, and mixing periods of 2D experiment 1 and evolution, mixing, and detection periods of 2D experiment 2. In a few cases, as, for example, NOESY as experiment 1 and multiple-quantum spectroscopy as experiment 2, parts of the preparation sequence of the second experiment must be retained. This is because a conventional NOESY sequence does not excite multiple-quantum coherences. The 3D signal is then recorded as a function of the two evolution times t_1 and t_2 and the detection time t_3.

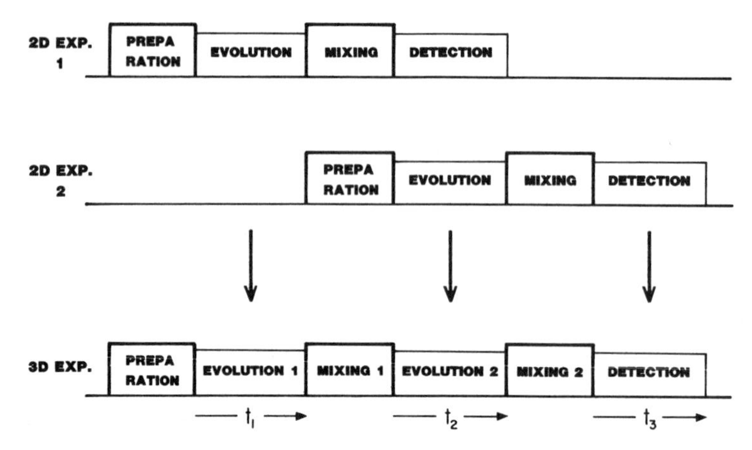

FIG. 1. Design procedure of a 3D pulse sequence by merging together two 2D sequences. The detection period of the first sequence and the preparation period of the second sequence are omitted.

3D time-domain experiments are time-consuming as the two time variables t_1 and t_2 must be incremented stepwise in a two-dimensional set of experiments. In order to minimize performance time, frequency-band-selective pulses are often used to focus the experiment on a specific volume in the 3D frequency space. It is also possible to replace one evolution time interval, for example, t_1, by a selective pulse that excites only a single resonance. A sequence of experiments is then necessary, stepping this pulse through the frequency region of interest. It has also been proposed (*32*) to replace both evolution time intervals by single-resonance selective pulses. Then, the two selective pulses must be stepped independently through the spectrum, leading to a very simple experiment but to much reduced sensitivity per unit time. Finally, a 3D spectrum can also be computed from the stochastic response of a spin system (*33*), as discussed in the Appendix.

III. CLASSIFICATION OF PEAKS IN 3D SPECTRA

The five different classes of peaks that occur in 3D spectra can be defined, based on a three-spin system A–B–C with the Larmor frequencies Ω_A, Ω_B, and Ω_C, and on the transfer pathways induced by the two mixing processes M_1 and M_2:

$$\text{Cross peaks} \quad \Omega_A \overset{M_1}{\rightarrow} \Omega_B \overset{M_2}{\rightarrow} \Omega_C \quad [1a]$$

$$(\omega_1 = \omega_2)\ \text{cross-diagonal peaks} \quad \Omega_A \rightarrow \Omega_A \rightarrow \Omega_B \quad [1b]$$

$$(\omega_2 = \omega_3)\ \text{cross-diagonal peaks} \quad \Omega_A \rightarrow \Omega_B \rightarrow \Omega_B \quad [1c]$$

$$\text{Back-transfer peaks} \quad \Omega_A \rightarrow \Omega_B \rightarrow \Omega_A \quad [1d]$$

$$\text{Diagonal peaks} \quad \Omega_A \rightarrow \Omega_A \rightarrow \Omega_A. \quad [1e]$$

For cross-diagonal peaks only one mixing process (which is sometimes also used to name these peaks) leads to transfer, for diagonal peaks there is no transfer at all, while for cross peaks and back-transfer peaks both mixing processes cause transfer. The latter two classes are therefore the most informative ones.

Cross peaks relate three distinct Larmor frequencies and contain information on connectivity or proximity of triples of nuclei. These peaks are relevant for two identical mixing processes as well as for two nonidentical processes. In a COSY–COSY spectrum, for example, they indicate a three-spin fragment within a possibly extended coupling network. In a NOESY–COSY spectrum, they can be used to assign the backbone protons of proteins, as described in more detail in Sections IX and X.

Back-transfer peaks, on the other hand, are of interest mainly in spectra originating from two unequal mixing processes. These peaks in a NOESY–TOCSY spectrum, for example, identify protons close in space that belong to the same spin system (i.e., the same amino acid residue).

The general features of 3D spectra are sketched in Figs. 2–5. A three-spin subsystem with mutually interacting spins leads to 27 peaks, as is shown in Fig. 2a, with 6 cross peaks, 6 ($\omega_1 = \omega_2$) and 6 ($\omega_2 = \omega_3$) cross-diagonal peaks, 6 back-transfer peaks, and 3 diagonal peaks. The diagonal peaks lie on the space diagonal indicated in Figs. 2b–

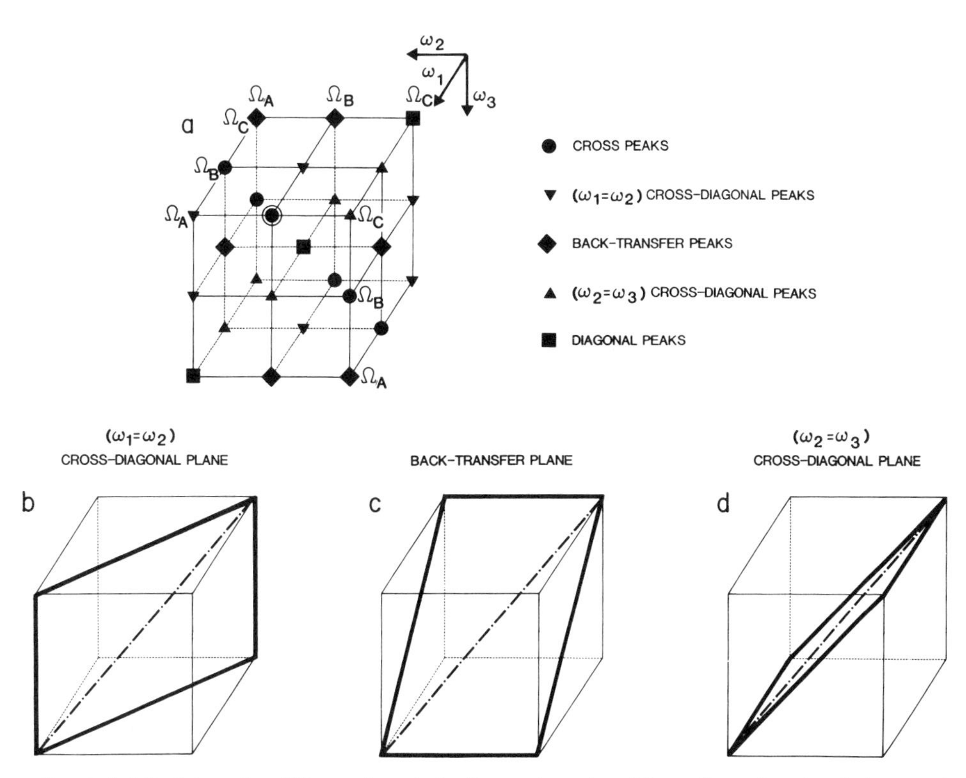

FIG. 2. (a) Peak types in a 3D experiment of a three-spin system. Three diagonal peaks, six ($\omega_1 = \omega_2$) and six ($\omega_2 = \omega_3$) cross-diagonal peaks, six back-transfer peaks, and six cross peaks are present. The cross peak A → B → C has been marked in this figure as well as in Figs. 3 and 4. (b–d) Three planes ($\omega_1 = \omega_2$), ($\omega_1 = \omega_3$), ($\omega_2 = \omega_3$) in 3D frequency space containing the corresponding types of peak (compare with (a)). The 3D diagonal $\omega_1 = \omega_2 = \omega_3$ is included in all figures.

2d while the ($\omega_1 = \omega_2$) and ($\omega_2 = \omega_3$) cross-diagonal and back-transfer peaks are situated each in a diagonal plane that is indicated in Figs. 2b–2d.

For a linear interaction network, a reduced set of peaks occurs, whereby the occurrence may differ for equal and unequal mixing processes. For two equal mixing processes, each of which would produce a symmetric 2D spectrum, such as in COSY–COSY or NOESY–NOESY, two-way transfers of the type

$$\Omega_A \overset{M_1}{\rightleftarrows} \Omega_B \overset{M_2}{\rightleftarrows} \Omega_C \qquad [2]$$
$$\Omega_C \overset{M_1}{\rightleftarrows} \Omega_B \overset{M_2}{\rightleftarrows} \Omega_A$$

lead to a 3D spectrum symmetric with respect to the ($\omega_1 = \omega_3$) plane as shown in Fig. 3a with two cross peaks, four ($\omega_1 = \omega_2$) and four ($\omega_2 = \omega_3$) cross-diagonal peaks, four back-transfer peaks, and three diagonal peaks. Obviously, peaks on the four lines ($\omega_1 = \Omega_A, \omega_2 = \Omega_C$), ($\omega_1 = \Omega_C, \omega_2 = \Omega_A$), ($\omega_2 = \Omega_A, \omega_3 = \Omega_C$), and ($\omega_2 = \Omega_C, \omega_3 = \Omega_A$) are missing. Symmetry with respect to the ($\omega_1 = \omega_3$) plane holds more generally for mixing sequences M_1 and M_2, each of which produces an asymmetric 2D spectrum, if they are time- and phase-reversal (TPR) counterparts (*34*) (cf. Section VII). For non-TPR-symmetric mixing-sequence pairs, on the other hand, it can happen that the only allowed transfers are

$$\Omega_A \overset{M_1}{\rightleftarrows} \Omega_B \overset{M_2}{\rightleftarrows} \Omega_C. \qquad [3]$$

This situation is typical for NOESY–COSY where spins A and B can be close in space but belong to different coupling networks, while spins B and C are within the same network and are connected by a nonvanishing J coupling but are distant in space. This leads to a 3D spectrum as sketched in Fig. 3b without back-transfer peaks, with

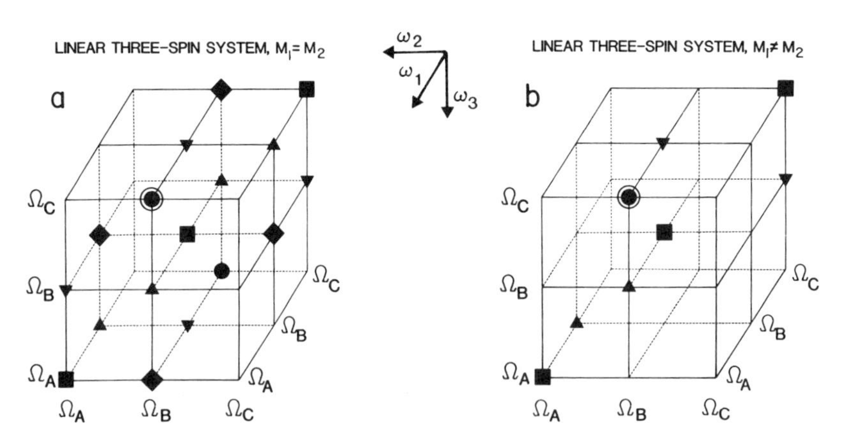

FIG. 3. (a) Schematic 3D spectrum of a linear spin system A–B–C with identical mixing processes M_1 and M_2. "Linear" means that transfer of magnetization between A and C is forbidden for both M_1 and M_2. (b) Schematic 3D spectrum of a linear spin system A–B–C where transfer with M_1 is possible exclusively between A and B and with M_2 exclusively between B and C.

one cross peak, two ($\omega_1 = \omega_2$) and two ($\omega_2 = \omega_3$) cross-diagonal peaks, and three diagonal peaks.

Most frequently, it is necessary to represent 3D spectra by 2D sections, normally perpendicular to one of the three axes. The features of 2D sections are illustrated in Fig. 4 by sections through the cross peak $\Omega_A \rightarrow \Omega_B \rightarrow \Omega_C$ of a three-spin system. Here, cross-diagonal and back-transfer planes appear as line sections. The (ω_2, ω_3) section for $\omega_1 = \Omega_A$ contains all those peaks that stem from magnetization originating from spin A and are distributed to further spins in the course of the two transfers. This 2D section could also be obtained by a 2D experiment with a selective excitation pulse as preparation sequence followed by the evolution period and nonselective mixing. The (ω_1, ω_3) section at $\omega_2 = \Omega_B$ can be visualized as a "log book" of an observer at the relay station $\omega_2 = \Omega_B$ monitoring the inflow and outflow of coherence through spin B. This section contains in addition to the peak (Ω_A, Ω_B, Ω_C) also its symmetry-equivalent peak (Ω_C, Ω_B, Ω_A) (open circle), provided this pathway is allowed. Finally, the (ω_1, ω_2) section at $\omega_3 = \Omega_C$ displays the history of all magnetization components that arrive at frequency $\omega_3 = \Omega_C$. Each of these sections has its own merits.

As an illustration, the full 3D COSY–COSY spectrum of the three-spin system of 2,3-dibromopropionic acid is given in Fig. 5. The experiment employed the pulse sequence outlined in Fig. 6a with constant delays $\Delta = \Delta'' = 25$ ms and $\Delta' = 35$ ms inserted in order to obtain partially in-phase multiplet structure (no refocusing π pulses were applied in the delay intervals and a magnitude representation was used to avoid the appearance of inherent phase distortions). Digital line broadening by 6 Hz and the magnitude representation produce almost structureless 3D multiplets. All 27 peaks expected are visible.

IV. THEORETICAL FORMULATION OF 3D EXPERIMENTS

The theoretical treatment of 3D experiments can proceed in analogy to that of 2D experiments (*5*, *6*). In order to simplify the description we assume in the following

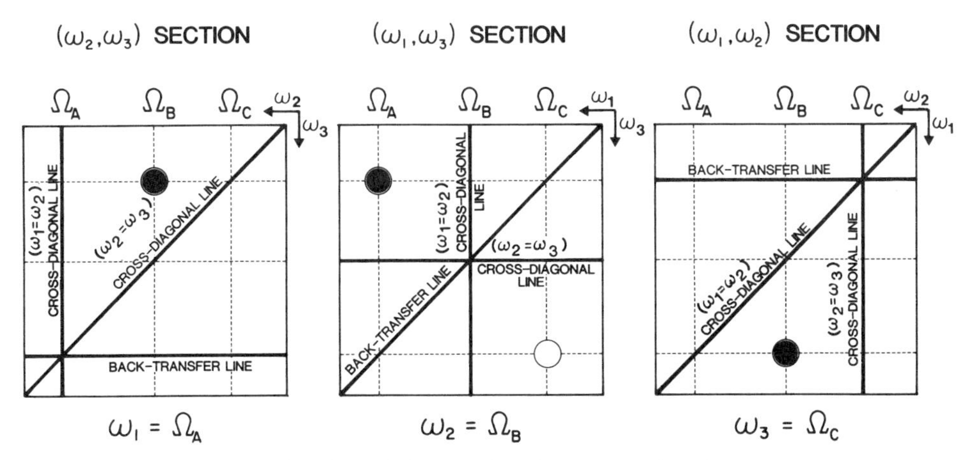

FIG. 4. Three 2D cross sections through the A → B → C cross peak of a 3D spectrum taken at constant ω_1, ω_2, and ω_3 frequencies, respectively. The intersecting lines with the three planes in Figs. 2b–2d are indicated. A symmetry-related peak with respect to the back-transfer diagonal (vide infra) is indicated by the open circle.

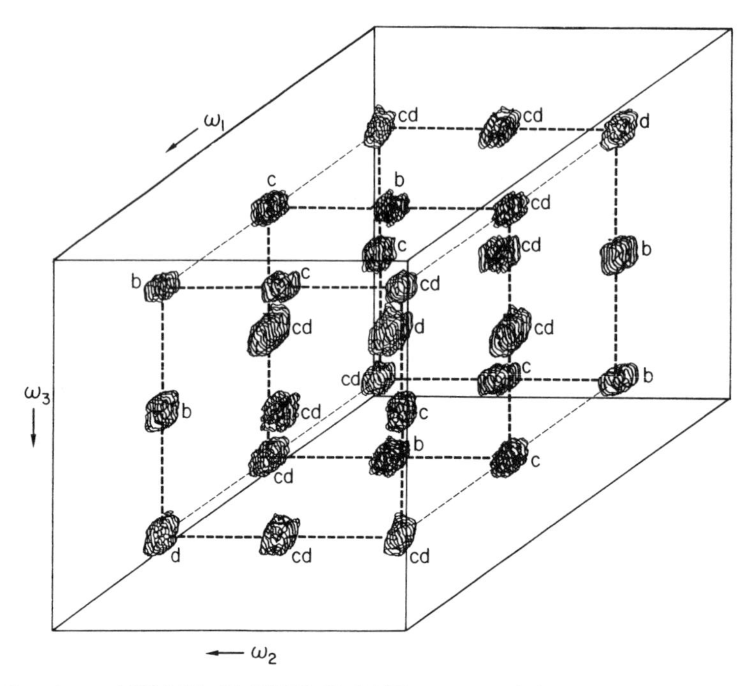

FIG. 5. Experimental 300 MHz 3D COSY–COSY ^{1}H spectrum of dibromopropionic acid showing the expected 27 peaks. In order to keep the spectrum simple, the spectral width in ω_2 was adjusted to let combination and multiple-quantum peaks (vide infra) fold on top of the 27 single-spin coherence peaks. The data matrix of 96 × 96 × 512 was zero-filled to 256 × 256 × 1K prior to 3D Fourier transformation. An absolute value representation is shown. The peaks are labeled with d, diagonal peak; c, cross peak; cd, cross-diagonal peak; b, back-transfer peak.

that the same Hamiltonian is acting during the three evolution periods t_1, t_2, and t_3, and we disregard 3D separation experiments where the effective Hamiltonian is different in the three time intervals. The hypothetical experiment to be treated is sketched in Fig. 7. It starts with a preparation period, represented by the superoperator $\hat{\hat{P}}(\Phi_P)$ that generates the initial density operator $\sigma(0)$

$$\sigma(0) = \hat{\hat{P}}(\Phi_P)\sigma_0 \qquad [4]$$

from the equilibrium density operator σ_0. The preparation period is followed by three free-precession periods where the eigenmodes of the Liouville superoperator (Eq. [8]) are indicated by $\{tu\}$, $\{rs\}$, and $\{pq\}$. They are separated by two mixing processes, represented by the superoperators $\hat{\hat{R}}^{(1)}(\Phi_M^{(1)})$ and $\hat{\hat{R}}^{(2)}(\Phi_M^{(2)})$. The observable operator is denoted by $D(\Phi_D)$. The various operators and superoperators, respectively, depend on four phase variables Φ_P, $\Phi_M^{(1)}$, $\Phi_M^{(2)}$, and Φ_D, of which three are linearly independent.

The observed signal $s(t_1, t_2, t_3)$ can, as usual, be decomposed into contributions of individual coherence-transfer pathways (*35*). The sequence of coherence orders p_{tu},

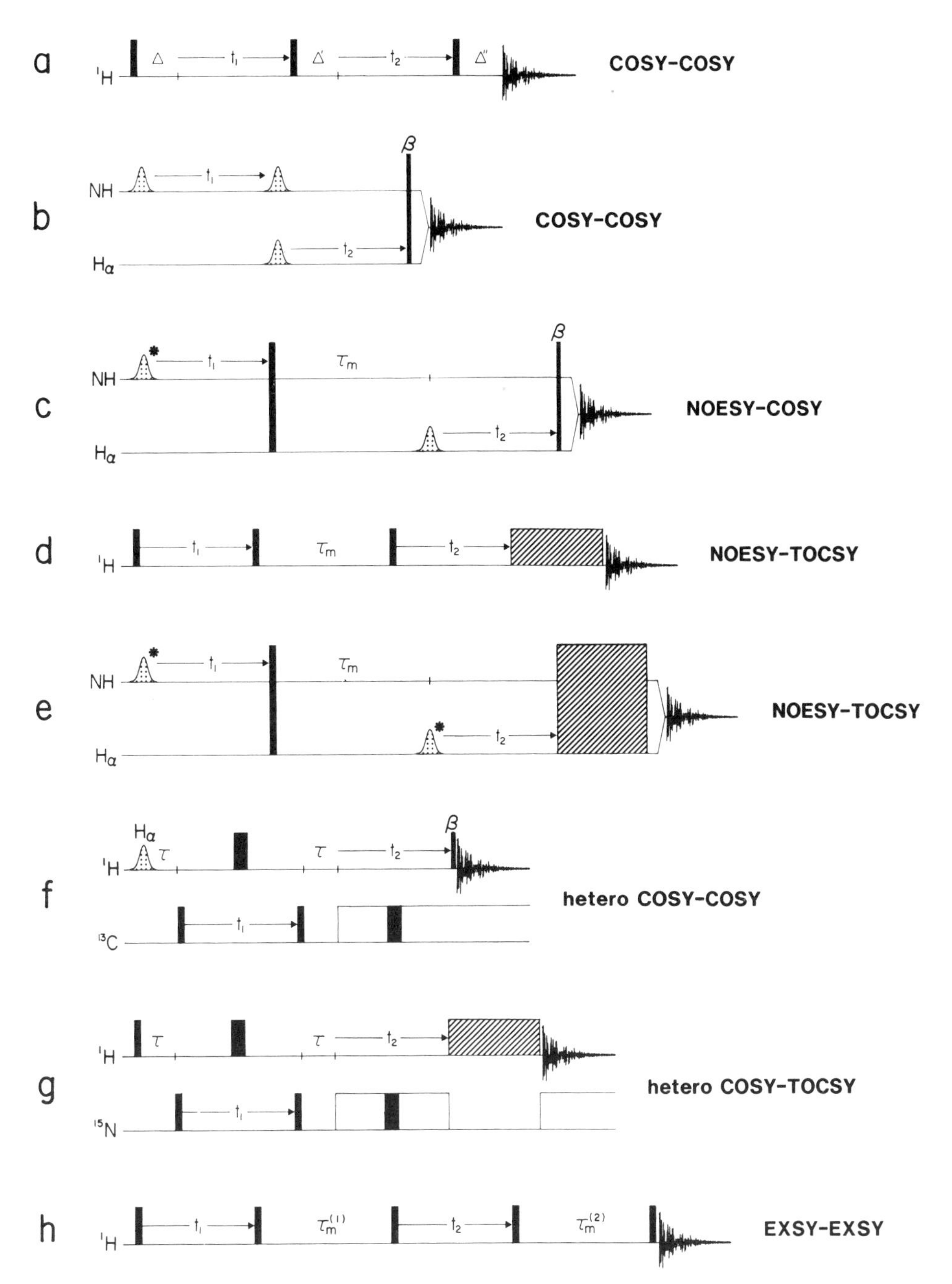

FIG. 6. Pulse sequences for 3D time-domain NMR spectroscopy. Nonselective pulses are indicated by filled bars with $\pi/2$ and π pulses being distinguished by their widths. Nonselective pulses of variable flip angle are indicated by the flip angle β. Frequency-selective pulses, all nominally $\pi/2$, are drawn with dotted Gaussian shapes. (a) COSY–COSY sequence with constant delays inserted in the three time periods for refocusing of antiphase magnetization to obtain in-phase multiplet structure. (b) COSY–COSY sequence with frequency restriction in ω_1 and ω_2. Depending on the choice of flip angle β, it is possible to emphasize

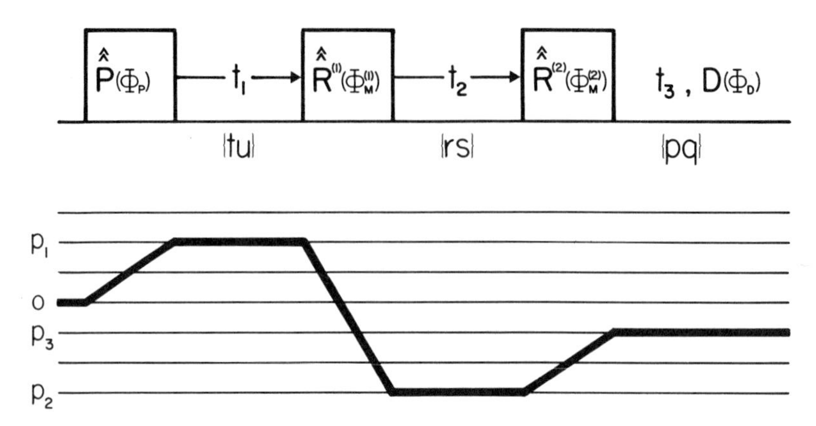

FIG. 7. Coherence-transfer pathways in a schematic 3D experiment correlating eigenmodes $\{tu\}$ with $\{rs\}$ and finally $\{pq\}$ in t_1, t_2, and t_3, respectively. The phases Φ_P, $\Phi_M^{(1)}$, $\Phi_M^{(2)}$, and Φ_D are cycled to select the desired coherence-transfer pathways as explained in the text.

p_{rs}, and p_{pq} of a particular pathway is represented in Fig. 7. This leads to the following general expression for the observed signal

$$s(t_1, t_2, t_3) = \sum_{\{pq\}} \sum_{\{rs\}} \sum_{\{tu\}} s_{\{pq\}\{rs\}\{tu\}}(t_1, t_2, t_3) \qquad [5]$$

with

$$s_{\{pq\}\{rs\}\{tu\}}(t_1, t_2, t_3) = Z_{\{pq\}\{rs\}\{tu\}} \exp\{-(i\omega_{\{pq\}} + \lambda_{\{pq\}})t_3\} \times \exp\{-(i\omega_{\{rs\}} + \lambda_{\{rs\}})t_2\} \exp\{-(i\omega_{\{tu\}} + \lambda_{\{tu\}})t_1\} \qquad [6]$$

---

connected or anticonnected transitions in the spectrum. (c) NOESY–COSY pulse sequence with selective excitation in ω_1 and ω_2. Because the nonselective pulse after t_1 excites all resonances, phase cycling is required to select zero-quantum coherence during the mixing time τ_m. The asterisk indicates a refocusing sequence consisting of a nonselective π pulse followed by a delay Δ (see text). (d, e) Nonselective and selective NOESY–TOCSY pulse sequences, respectively. The nonselective TOCSY mixing sequences are represented by hatched areas. (f) Heteronuclear COSY–COSY sequence exemplified by ^{1}H and ^{13}C correlation. In applications to peptides and proteins, the selective $\pi/2$ pulse may, for example, be adjusted to excite the H_α resonance region. No further selective pulses are required for frequency restriction in ω_2. The delay τ must be tuned to $(2\ ^1J_{CH})^{-1}$ for optimum sensitivity. The first τ delay includes about half the duration of the selective pulse. If possible, ^{13}C decoupling in the C_α resonance region is applied during t_2 and during acquisition. Alternatively, a central π pulse can ensure decoupling at least during t_2. (g) Heteronuclear COSY–TOCSY sequence arranged for ^{1}H and ^{15}N correlation. In peptides and proteins, the frequency range in t_2 can be restricted to amide protons without any selective pulses. The selectivity is achieved by tuning the τ delays to the large one-bond couplings ($\tau = (2\ ^1J_{NH})^{-1}$). Note that, during the homonuclear TOCSY mixing, heteronuclear couplings are ineffective. (h) EXSY–EXSY pulse sequence for studies of multisite chemical-exchange processes. The same pulse sequence is employed for NOESY–NOESY, EXSY–NOESY, and NOESY–EXSY.

and

$$Z_{\{pq\}\{rs\}\{tu\}} = D_{\{qp\}} R^{(2)}_{\{pq\}\{rs\}} R^{(1)}_{\{rs\}\{tu\}} \sigma(0)_{\{tu\}}$$
$$\times \exp\{-i[p_{\{tu\}}(\Phi_P - \Phi_M^{(1)}) + p_{\{rs\}}(\Phi_M^{(1)} - \Phi_M^{(2)}) + p_{\{pq\}}(\Phi_M^{(2)} - \Phi_D)]\}. \quad [7]$$

The frequencies $\omega_{\{pq\}}$ and the damping constants $\lambda_{\{pq\}}$ are imaginary and real parts, respectively, of the eigenvalues of the Liouville superoperator

$$\hat{\hat{L}} = -i\hat{\hat{\mathscr{H}}} - \hat{\hat{\Gamma}} + \hat{\hat{\Xi}}, \quad [8]$$

where $\hat{\hat{\mathscr{H}}}$, $\hat{\hat{\Gamma}}$, and $\hat{\hat{\Xi}}$ are as usual Hamiltonian, relaxation, and chemical-exchange superoperators. $Z_{\{pq\}\{rs\}\{tu\}}$ is the complex intensity of a peak in the 3D frequency space.

Equation [7] involves the three independent phase variables, associated with the three frequency dimensions, that must be taken into account and adjusted in order to control the peak shape in all three dimensions. Among numerous possible choices of three independent phases, there are two systematic ones, indicated in Fig. 8. The phases can be introduced as "preparation phases" ϕ_1, ϕ_2, and ϕ_3 that affect all operations *preceding* a particular evolution period, with the identifications (see Fig. 8a)

$$\begin{aligned} \phi_1 &= \Phi_P - \Phi_M^{(1)}, & \Phi_P &= \phi_3 + \phi_2 + \phi_1, \\ \phi_2 &= \Phi_M^{(1)} - \Phi_M^{(2)}, & \Phi_M^{(1)} &= \phi_3 + \phi_2, \\ \phi_3 &= \Phi_M^{(2)}, & \Phi_M^{(2)} &= \phi_3. \end{aligned} \quad [9]$$

Three independent phase cycles of ϕ_k, $k = 1, 2, 3$, are applied to select the coherence orders, $p_{\{tu\}} = p_1$, $p_{\{rs\}} = p_2$, and $p_{\{pq\}} = p_3$, in the respective evolution periods; i.e.,

$$\phi_{k,j_k} = j_k \frac{2\pi}{N_k}, \qquad j_k = 0, 1, \ldots, N_k - 1, \quad [10]$$

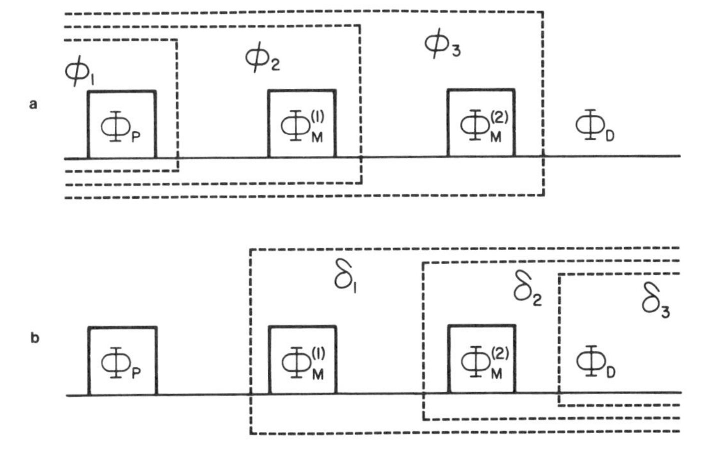

FIG. 8. Two different definitions of the pulse and detection phases in 3D experiments. In (a) the three phases ϕ_i are cycled independently against the receiver phase, and in (b) the three phases δ_i are cycled independently against the preparation phase.

where N_k is the number of steps in the phase cycle of ϕ_k. Selection of the desired coherence-transfer pathway is obtained by coadding the $N_1 \times N_2 \times N_3$ individual experiments with detection phase $\Phi_D(j_1, j_2, j_3) = (1/p_3) \sum_{k=1}^{3} \phi_{k,j_k} p_k$. Alternatively, the opposite phase $-\Phi_D(j_1, j_2, j_3)$, can be added to all pulses of the sequence with the receiver phase kept constant: $\phi'_{k,j_k} = \phi_{k,j_k} - (1/p_3) \sum_{l=1}^{3} \phi_{l,j_l} p_l$.

A different identification of the independent phases is shown in Fig. 8b where the phases δ_1, δ_2, and δ_3 are introduced as "detection phases" and affect all operations *following* the particular evolution period. Here we have

$$\begin{aligned} \delta_1 &= \Phi_M^{(1)}, & \Phi_M^{(1)} &= \delta_1, \\ \delta_2 &= \Phi_M^{(2)} - \Phi_M^{(1)}, & \Phi_M^{(2)} &= \delta_1 + \delta_2, \\ \delta_3 &= \Phi_D - \Phi_M^{(2)}, & \Phi_D &= \delta_1 + \delta_2 + \delta_3. \end{aligned} \quad [11]$$

The preparation phase Φ_P is used as reference phase for the entire experiment. The phase cycles for δ_k, $k = 1, 2, 3$, are equivalent to those for ϕ_k, $k = 1, 2, 3$, in Eq. [10]:

$$\delta_{k,j_k} = j_k \frac{2\pi}{N_k}, \qquad k = 0, 1, \ldots, N_k - 1. \quad [12]$$

The desired coherence-transfer pathway selection is now obtained by cycling the preparation phase according to $\Phi_P = (1/p_1) \sum_{k=1}^{3} \delta_{k,j_k} p_k$. On the other hand if it is desirable to keep the preparation phase constant, $-\Phi_P$ must be added to δ_k: $\delta'_{k,j_k} = \delta_{k,j_k} - (1/p_1) \sum_{l=1}^{3} \delta_{l,j_l} p_l$.

The three independent phase cycles can be used in full analogy to 2D spectroscopy for the selection of coherence-transfer pathways (*35*). This allows separation of positive and negative frequencies in all three dimensions, possibly combined with multiple-quantum filtering (*36*) (using additional pulses for order selection) and recording of 3D spectra with pure-phase cross peaks (see Section V).

The distinction of positive and negative frequencies, when placing the carrier frequency in the center of the spectrum, can also be achieved by time-proportional phase incrementation (TPPI) (*6, 37, 38*) independently for the two phase variables ϕ_1 and ϕ_2 or δ_1 and δ_2 according to

$$\phi_1, \delta_1 = (\pi/2p_1)(t_1/\Delta t_1), \qquad \phi_2, \delta_2 = (\pi/2p_2)(t_2/\Delta t_2). \quad [13]$$

The effective frequency zero is then shifted to the edge of the (ω_1, ω_2) frequency plane.

In 2D spectroscopy, it is also possible to distinguish positive and negative ω_1 frequencies based on the processing of two 2D spectra recorded with two phase-shifted pulse sequences (*39, 40*). This procedure can be extended to 3D spectroscopy. Here separate data processing of 3D spectra recorded with four phase-shifted 3D pulse sequences is required. They correspond to the four combinations of ϕ_1 or $\delta_1 = \{0, \pi/2p_1\}$ and ϕ_2 or $\delta_2 = \{0, \pi/2p_2\}$. The general scheme for quadrature phase detection in ω_1 and ω_2 domains also holds for the case that pulses during evolution periods induce changes of the coherence orders, such as in constant time experiments (*41, 42*) or for more complicated pulse schemes. When pulses during an evolution period induce a change of coherence order, p_k in Eq. [13] refers to the desired coherence order at the beginning of the associated evolution period when the ϕ_k scheme is used. On the other hand p_k refers to the coherence order at the end of the associated evolution period for the δ_k scheme.

V. PEAK SHAPES IN 3D NMR

The resolution achievable in 3D spectra is often severely limited by the number of available sampling points in the three dimensions. It is therefore essential to minimize the linewidth by plotting phase-sensitive absorption-mode spectra whenever possible. It is known from 2D spectroscopy that the feasibility of pure absorption mode depends on the experimental technique (*6*). While 2D separation experiments usually produce mixed-phase peak shapes (sometimes described as phase-twisted peaks), many 2D correlation and 2D exchange techniques generate pure-phase cross peaks that can be phased to pure absorption by adjusting the three phase variables introduced in the previous section. It has also been recognized (*6, 35*) that the attainment of pure-phase peaks in 2D spectra is related to the symmetry of coherence transfer pathways that leads to a sinusoidal amplitude modulation of the acquired time-domain signals.

The condition, in terms of coherence-transfer pathway symmetry, that must be fulfilled for pure-phase spectra can be expressed by the complex intensities $Z_{\{pq\}\{rs\}\{tu\}}$ of Eq. [7]:

$$Z_{\{pq\}\{rs\}\{tu\}} = Z_{\{pq\}\{rs\}\{ut\}}e^{i\phi} = Z_{\{pq\}\{sr\}\{tu\}}e^{i\psi} = Z_{\{pq\}\{sr\}\{ut\}}e^{i(\phi+\psi)}. \quad [14]$$

The values $\phi = 0$ or $\phi = \pi$ correspond to a cosine or sine modulation, respectively, in the t_1 domain, and ψ is the analogous phase associated with t_2. Comparison with Eq. [7] identifies the phases $\phi = 2p_{\{ut\}}(\Phi_P - \Phi_M^{(1)})$ and $\psi = 2p_{\{sr\}}(\Phi_M^{(1)} - \Phi_M^{(2)})$. Equation [14] implies that four coherence-transfer pathways must have equal amplitudes in order to obtain pure phase spectra. They can then be combined to a time-domain signal

$$s_{\{pq\}\{rs\}\{tu\}}(t_1, t_2, t_3) = 4Z_{\{pq\}\{rs\}\{tu\}}\exp\{-i\omega_{\{pq\}}t_3\}\cos\left[\left((\omega_{\{rs\}} + \frac{\psi}{2}\right)t_2\right] \times \cos\left[\left(\omega_{\{tu\}} + \frac{\phi}{2}\right)t_1\right]\exp\{-\lambda_{\{pq\}}t_3 - \lambda_{\{rs\}}t_2 - \lambda_{\{tu\}}t_1\}. \quad [15]$$

Its 3D Fourier transform can be phase-adjusted to pure 3D absorption:

$$S(\omega_1, \omega_2, \omega_3) = \frac{\lambda_{\{pq\}}}{\lambda_{\{pq\}}^2 + (\omega_3 - \omega_{\{pq\}})^2} \cdot \frac{\lambda_{\{rs\}}}{\lambda_{\{rs\}}^2 + (\omega_2 - \omega_{\{rs\}})^2} \cdot \frac{\lambda_{\{tu\}}}{\lambda_{\{tu\}}^2 + (\omega_1 - \omega_{\{tu\}})^2}. \quad [16]$$

When expressing Eq. [14] in terms of the individual factors given in Eq. [7] with all phases set equal to zero, we find the conditions

$$\sigma(0)_{\{tu\}} = \sigma(0)_{\{ut\}} \quad [17a]$$

$$R^{(2)}_{\{pq\}\{rs\}} = R^{(2)}_{\{pq\}\{sr\}} \quad [17b]$$

$$R^{(1)}_{\{rs\}\{tu\}} = R^{(1)}_{\{rs\}\{ut\}}. \quad [17c]$$

These relations are identical to the conditions imposed upon a 2D experiment with a single coherence-transfer process. In other words, a peak in a 3D spectrum can be phased to pure absorption when the corresponding peaks in the two 2D experiments from which the 3D experiment is constructed can be phased to pure absorption.

For an exhaustive discussion of peak phases in a 3D spectrum it is necessary to compute the complex factors $Z_{\{pq\}\{rs\}\{tu\}}$ in Eq. [7]. However, for many common 2D experiments in the weak coupling limit, as, for example, COSY with a 90° mixing pulse, or NOESY, simple rules allow a quick determination of peak phases in 3D spectra based on the phases in the spectra of the constituent 2D experiments. Rules can be stated when the corresponding peaks in the two 2D spectra exhibit pure phases, i.e., a_1a_2, a_1d_2, d_1a_2, or d_1d_2, where a_i and d_i mean absorption and dispersion in the ω_i dimension, respectively. In the case of mixed peak shapes in the two 2D spectra, mixed peak shapes normally also occur in the 3D spectrum.

The phases χ_1, χ_2, and χ_3 of a 3D peak, in the three domains ω_1, ω_2, and ω_3 can be obtained from the phases $\chi_1^{(1)}$ and $\chi_2^{(1)}$, and $\chi_1^{(2)}$ and $\chi_2^{(2)}$ of the corresponding peaks in the spectra of the constituent two 2D experiments 1 and 2, provided evolution of one-quantum coherence is selected during ω_2. One finds

$$\chi_1 = \chi_1^{(1)} \quad [18a]$$

$$\chi_2 = \chi_2^{(1)} + \chi_1^{(2)} \quad [18b]$$

$$\chi_3 = \chi_2^{(2)}. \quad [18c]$$

This implies than when using the same phase setting in the 3D spectrum as that in the 2D spectra the phase in the ω_1 domain and the phase in the ω_3 domain is equal to the phase in the ω_1 domain of the first 2D spectrum and the phase in the ω_2 domain of the second 2D spectrum, respectively. The phase in the ω_2 domain is the sum of the phases in the ω_2 domain of spectrum 1 and the ω_1 domain of spectrum 2; i.e., absorption in these two frequency domains of the two 2D spectra leads to absorption in ω_2 of the 3D spectrum while dispersion in the two spectra leads to negative absorption in ω_2 of the 3D spectrum. Table 1 lists explicitly the peak shapes for the 3D experiments COSY–COSY, COSY–NOESY, and NOESY–COSY in the weak coupling limit, using 90° mixing pulses in the COSY parts.

In practice, the computer phase correction of spectra obtained with nonselective experiments can easily be determined in advance. The phase correction in ω_3 is identical to that required to phase correct a normal 1D spectrum. Sometimes, a 90° phase shift is necessary depending on the phase of the mixing pulses. In ω_1 and ω_2, sometimes linear frequency-dependent phase corrections are needed that are due to un-

TABLE 1

Peak Shapes in 3D Spectra Using 90° Mixing Pulses in the COSY Parts[a]

Peak type	3D experiment		
	COSY–COSY	COSY–NOESY	NOESY–COSY
Cross and back-transfer	$a_1a_2a_3$	$a_1a_2a_3$	$a_1a_2a_3$
Cross-diagonal ($\omega_1 = \omega_2$)	$d_1d_2a_3$	$d_1d_2a_3$	$a_1a_2a_3$
Cross-diagonal ($\omega_2 = \omega_3$)	$a_1d_2d_3$	$a_1a_2a_3$	$a_1d_2d_3$
Diagonal	$d_1a_2d_3$	$d_1d_2a_3$	$a_1d_2d_3$

[a] The table assumes that cross and back-transfer peaks are phased to pure 3D absorption $a_1a_2a_3$.

avoidable pulse delays in the evolution periods for t_1 or $t_2 = 0$. They can be determined from 1D spectra with identical delay of the start of acquisition.

If selective pulses are employed, linear frequency-dependent phase corrections are necessary which can be determined from a 1D spectrum with selective excitation provided that only single selective pulses are applied at the beginning of the evolution periods.

In the case of more than one selective pulse per evolution period (Fig. 6b) or mixing sequences that by their nature introduce frequency-dependent phase shifts (*44*), a procedure based on the separately recorded constituent 2D spectra may be employed. The phase correction in ω_1 can be set equal to that in $\omega_1^{(1)}$ of 2D experiment 1, and that in ω_3 is set equal to the phase correction in $\omega_2^{(2)}$ of 2D experiment 2. The required phase correction in ω_2 is identical to that in $\omega_1^{(2)}$, apart from a frequency-independent phase shift that depends on the relative phase of the mixing of the first and the preparation of the second 2D experiment. This frequency-independent additive phase shift can be determined by a product operator calculation (*45*) or experimentally as the phase correction in $\omega_2^{(1)}$ of 2D experiment 1 minus the phase correction of a nonselective 1D experiment in which the pulse phase must be the same as the phase of the preparation pulse of 2D experiment 2.

The peak shapes of individual multiplet components in 3D spectra of weakly coupled spin systems can be calculated using the product operator formalism (*45*). The results of such a calculation for the COSY–COSY experiment with arbitrary mixing pulse angles β_1 and β_2 are summarized in Table 2. The notation is adopted from Ref. (*43*) with the following explicit expressions:

TABLE 2

Peak-Shape Contributions in a COSY–COSY Spectrum with Mixing Pulses β_1 and β_2 Applied to an Arbitrary Weakly Coupled Spin System[a]

	Peak-shape contribution			
Peak type	$a_1a_2a_3$	$a_1d_2d_3$	$d_1a_2d_3$	$d_1d_2a_3$
Cross $\Omega_A \rightarrow \Omega_B \rightarrow \Omega_C$	$F_{r_1;g_1}^{K_c(AB)}(\beta_1)F_{r_2;g_2}^{K_c(BC)}(\beta_2)$			
Cross-diagonal $(\omega_1 = \omega_2)$ $\Omega_A \rightarrow \Omega_A \rightarrow \Omega_B$	$A_{r_1}^{K_d(A)}(\beta_1)F_{r_2;g_2}^{K_c(AB)}(\beta_2)$			$D_{r_1}^{K_d(A)}(\beta_1)F_{r_2;g_2}^{K_c(AB)}(\beta_2)$
Cross-diagonal $(\omega_2 = \omega_3)$ $\Omega_A \rightarrow \Omega_B \rightarrow \Omega_B$	$F_{r_1;g_1}^{K_c(AB)}(\beta_1)A_{r_2}^{K_d(B)}(\beta_2)$	$F_{r_1;g_1}^{K_c(AB)}(\beta_1)D_{r_2}^{K_d(B)}$		
Back-transfer $\Omega_A \rightarrow \Omega_B \rightarrow \Omega_A$	$F_{r_1;g_1}^{K_c(AB)}(\beta_1)F_{r_2;g_2}^{K_c(AB)}(\beta_2)$			
Diagonal $\Omega_A \rightarrow \Omega_A \rightarrow \Omega_A$	$A_{r_1}^{K_d(A)}(\beta_1)A_{r_2}^{K_d(A)}(\beta_2)$	$A_{r_1}^{K_d(A)}(\beta_1)D_{r_2}^{K_d(A)}(\beta_2)$	$D_{r_1}^{K_d(A)}(\beta_1)D_{r_2}^{K_d(A)}(\beta_2)$	$D_{r_1}^{K_d(A)}(\beta_1)A_{r_2}^{K_d(A)}(\beta_2)$

[a] Indices 1 and 2 refer to the first (β_1) and second (β_2) mixing pulse, respectively. Superscript K_c(AB) refers to the K_c value of the A–B spin pairs, and K_d(A) refers to the K_d value of spin A (*43*).

$$F_{r,g}^{K_c(AB)}(\beta) = g \cdot 2^{-K_c} \sin^2\beta \left\{ \sin^2 \frac{\beta}{2} \right\}^r \left\{ \cos^2 \frac{\beta}{2} \right\}^{K_c - 2 - r} \qquad [19]$$

$$A_r^{K_d(A)}(\beta) = 2^{1-K_d} \left\{ \sin^2 \frac{\beta}{2} \right\}^r \left\{ \cos^2 \frac{\beta}{2} \right\}^{K_d - r - 1} \cos\beta \qquad [20]$$

$$D_r^{K_d(A)}(\beta) = 2^{1-K_d} \left\{ \sin^2 \frac{\beta}{2} \right\}^r \left\{ \cos^2 \frac{\beta}{2} \right\}^{K_d - r - 1}, \qquad [21]$$

given as Eqs. [7], [29a], and [29b] in Ref. (*43*).

$F_{r,g}^{K_c(AB)}(\beta)$ is the intensity of a 2D COSY cross-peak multiplet component where r of a total of $K_c - 2$ passive spins, all coupled to both the two active spins A and B, have had their spin state changed by the mixing pulse of flip angle β. The index g distinguishes progressively ($g = 1$) and regressively ($g = -1$) connected transitions (*5, 43*). The phases of all these multiplet components in 2D COSY spectra can be adjusted to pure 2D double absorption a_1a_2. $A_r^{K_d(A)}(\beta)$ and $D_r^{K_d(A)}(\beta)$ are absorptive and dispersive intensities, respectively, of a diagonal-peak multiplet component of spin A where r of a total of $K_d - 1$ passive spins have had their spin state changed during mixing.

Time-domain signals that are purely amplitude-modulated as functions of t_1 and t_2 are a necessary but not sufficient condition to obtain pure 3D absorption peak shapes. It may be that frequency-dependent phase shifts of nonsystematic nature prevent pure absorption profiles. This arises, for example, when constant delays with central π refocusing pulses are introduced into evolution periods. They cause a phase gradient in the multiplets that is different from the phase gradient between resonances with different chemical shifts. A similar type of phase distortion is encountered in the purely phase-modulated t_3 dimension if a refocusing interval is introduced before data acquisition.

A number of guidelines as to the type of Fourier transformation and spectrum display can be stated in order to obtain maximum benefits from 3D time-domain spectroscopy: (i) A real Fourier transformation should be employed in the dimensions with pure amplitude modulation irrespective of possible frequency-dependent phase distortion of the type mentioned above. (ii) A complex Fourier transformation is employed in dimensions with pure phase modulation. (iii) The imaginary (dispersive) part should be discarded in the dimensions where pure absorption profiles can be obtained. (iv) If pure absorption profiles can be obtained in all three dimensions, a phase-sensitive spectral display is the display of choice. (v) If one or more dimensions suffer from phase distortions or mixed peak shape, absolute value spectra should be calculated after discarding the imaginary parts in the dimensions exhibiting pure absorption profiles.

As examples we indicate the attainable peak shapes in four variants of the COSY–COSY experiment in Fig. 6a with different selections of the refocusing delays Δ, Δ', and Δ'':

$$\Delta = \Delta' = \Delta'' = 0: \quad S(\omega_1, \omega_2, \omega_3) = a_1a_2a_3 \qquad [22]$$

$$\Delta \neq 0, \Delta' = \Delta'' = 0: \quad S(\omega_1, \omega_2, \omega_3) = \sqrt{a_1^2 + d_1^2}\,|a_2a_3| \qquad [23]$$

$$\Delta \neq 0, \Delta' \neq 0, \Delta'' = 0: \qquad S(\omega_1, \omega_2, \omega_3) = \sqrt{a_1^2 + d_1^2}\sqrt{a_2^2 + d_2^2}|a_3| \qquad [24]$$

$$\Delta \neq 0, \Delta' \neq 0, \Delta'' \neq 0: \qquad S(\omega_1, \omega_2, \omega_3) = \sqrt{a_1^2 + d_1^2}\sqrt{a_2^2 + d_2^2}\sqrt{a_3^2 + d_3^2}. \qquad [25]$$

Permutations in the identification of the delays do not change the qualitative forms of Eqs. [22]–[25].

The COSY–COSY experiment with $\pi/2$ mixing pulses is an example that leads to pure amplitude modulation. The same holds for arbitrary flip angles as far as cross and back-transfer peaks are concerned whereas the other types of peaks may exhibit both amplitude and phase modulation. An example of pure-phase modulation in t_1 and pure amplitude modulations in t_2 is *J*-resolved COSY (*14*) employing the pulse sequence

$$\left(\frac{\pi}{2}\right) - \frac{t_1}{2} - (\pi) - \frac{t_1}{2} - t_2 - \left(\frac{\pi}{2}\right) - t_3.$$

In this case the preferred spectral display is

$$S(\omega_1, \omega_2, \omega_3) = \sqrt{a_1^2 + d_1^2}\sqrt{a_2^2 + d_2^2}|a_3|. \qquad [26]$$

As a final example we consider the case of mixed amplitude and phase modulation of the type $\cos(\Omega t + \varphi)\cdot f(t)$, where $f(t)$ represents the phase modulation. If the frequencies in $f(t)$ are small compared to $(t^{max})^{-1}$, it is advantageous to ignore the phase-modulation part and apply a real Fourier transformation. A prominent example is indirect detection of ^{13}C or ^{15}N resonance via multiple-quantum coherence (*46, 47*) as in the first parts of the pulse sequences in Figs. 6f and 6g, where the phase modulation from proton–proton couplings is negligible at the short t_1^{max} value usually employed (*48*).

VI. COMBINATION AND MULTIPLE-QUANTUM PEAKS

Two-dimensional spectra that result from experiments with a single nonselective preparation pulse are free from multiple-quantum peaks and, when applied to weakly coupled spins, also free from one-quantum combination peaks because these "forbidden" transitions are not excited by a nonselective pulse. In 3D experiments, however, all these transitions may become excited by the combined action of preparation and first mixing pulse sequences and persist through the t_2 period. The second mixing process can convert them to observable single-quantum coherences which are detected in the t_3 period. Thus 3D spectra will generally exhibit cross peaks involving all possible coherences during the t_2 period. The situation is reminiscent of 2D multiple-quantum spectroscopy where a $(\pi/2)$–$\tau/2$–(π)–$\tau/2$–$(\pi/2)$ pulse sequence also excites combination coherences. For example, 2D double-quantum spectra exhibit, in addition to peaks with $\omega_1 = \Omega_A + \Omega_B$, also four-spin double-quantum coherences at $\omega_1 = \Omega_A + \Omega_B - \Omega_C + \Omega_D$.

The undesired multiple-quantum coherences can in principle be eliminated by a suitable phase cycle, selecting, for example, single-quantum coherence during the t_2 period. On the other hand, multiple-spin single-quantum coherences are more difficult and, in some cases, impossible to suppress. As an example, we consider the conversion of multiple-spin single-quantum coherence evolving during t_2 by a single mixing pulse of flip angle β into observable one-spin single-quantum coherence. We focus on n-spin operator terms and obtain the following transformations under a β_y pulse (disregarding all nonessential terms):

One-spin coherences

$$(xzz\cdots z) \xrightarrow{\beta_y} (\cos\beta)^n(xzz\cdots z) - \sin^2\beta(\cos\beta)^{n-2}(zxz\cdots z)$$

$$(yzz\cdots z) \xrightarrow{\beta_y} (\cos\beta)^{n-1}(yzz\cdots z)$$

Three-spin coherences

$$(xxxzz\cdots z) \xrightarrow{\beta_y} \sin^2\beta(\cos\beta)^{n-2}(xzzzz\cdots z) - \sin^4\beta(\cos\beta)^{n-4}(zzzxz\cdots z)$$

$$(xxyzz\cdots z) \xrightarrow{\beta_y} \sin^2\beta(\cos\beta)^{n-3}(zzyzz\cdots z)$$

$$(xyyzz\cdots z) \xrightarrow{\beta_y} 0$$

$$(yyyzz\cdots z) \xrightarrow{\beta_y} 0$$

Five-spin coherences

$$(xxxxxzz\cdots z) \xrightarrow{\beta_y} \sin^4\beta(\cos\beta)^{n-4}(xzzzzzz\cdots z) - \sin^6\beta(\cos\beta)^{n-6}(zzzzzxz\cdots z)$$

$$(xxxxyzz\cdots z) \xrightarrow{\beta_y} \sin^4\beta(\cos\beta)^{n-5}(zzzzyzz\cdots z).$$

We use the notation $(xxyz) = I_{1x}I_{2x}I_{3y}I_{4z}$ and remember that permutations of the spin labels lead to operator products with identical transformation properties.

Operator terms with unequal transformation properties under a β_y rotation, leading to different trigonometric factors in the above equations, can be distinguished by performing experiments for several values β and forming appropriate linear combinations in analogy to heteronuclear spin-editing procedures such as DEPT and SEMUT (*49–51*). It is apparent that selection of the desired one-spin cross peaks originating from the operators $(xzz\cdots z)$ with the flip-angle dependence $\sin^2\beta(\cos\beta)^{n-2}$ also retains cross peaks that stem from the three-spin coherences $(xxxzz\cdots z)$ for n-operator terms and $(xxyzz\cdots z)$ for $(n+1)$-operator terms. They have the same flip-angle dependence. The higher spin coherences can, however, be fully suppressed (see the Appendix of Ref. (*43*)).

Each triple of spins A, B, and C can give rise to three combination frequencies $\Omega_{\text{comb}} = \Omega_A + \Omega_B - \Omega_C$, $\Omega_A - \Omega_B + \Omega_C$, and $-\Omega_A + \Omega_B + \Omega_C$. Two of these frequencies fall outside the range of the three Larmor frequencies while one of them lies inside and could lead to misinterpretations. In practice, the ω_2 spectral width is rarely chosen large enough to encompass the outer combination peaks because of the limited number of t_2 experiments. Hence the outer combination peaks will fold in and might further obscure the 3D spectrum.

Figure 9 presents a general 3D spectrum of a three-spin system with 27 peaks relating single-quantum one-spin transitions, as shown in Fig. 2a, and a total of 27 additional peaks that involve single-quantum combination transitions (so-called combination peaks). Apart from different multiplet structures, amplitudes, and phases, this leads to planes of similar appearance at the six possible ω_2 frequencies. The number of combination peaks actually observed depends on the pulse sequence and the spin-coupling network. For COSY–COSY applied to a linear A–B–C spin system ($J_{AC} = 0$) only one combination peak (that with $\omega_1 = \omega_3 = \Omega_B$) occurs in each of the three combination-peak planes, while the $\omega_2 = \Omega_B$ plane is complete (9 peaks), and the $\omega_2 = \Omega_A$ and $\omega_2 = \Omega_C$ planes contain 4 peaks each (compare Fig. 3a).

Combination peaks that arise from interactions between three spins can be distinguished from one-spin peaks by their different phase. When the latter are adjusted to pure 3D absorption, the former show $d_1a_2d_3$ peak shapes as can be inferred from the following two sets of transformations:

$$(xz1) \xrightarrow{(\pi/2)_y} -(zx1) \xrightarrow{t_2} -(1xz) \xrightarrow{(\pi/2)_y} (1zx)$$

$$(yzz) \xrightarrow{(\pi/2)_y} (yxx) \xrightarrow{t_2} (xxy) \xrightarrow{(\pi/2)_y} (zzy).$$

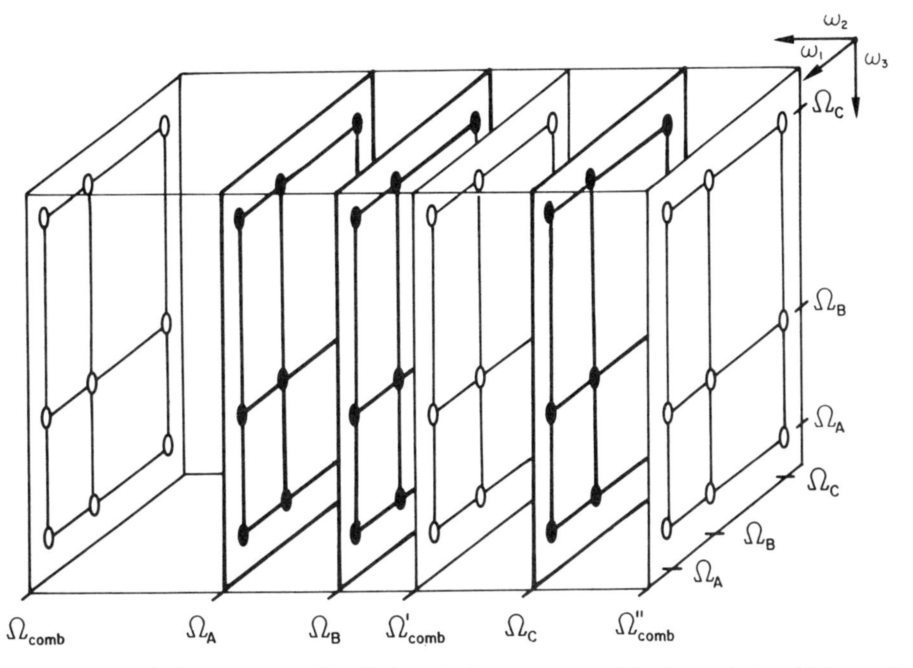

FIG. 9. Schematic 3D spectrum which displays single-quantum combination peaks in addition to the one-spin peaks shown in Fig. 2a. The three combination-peak planes occur at $\omega_2 = \Omega_{comb} = \Omega_A + \Omega_B - \Omega_C$, $\omega_2 = \Omega'_{comb} = \Omega_A - \Omega_B + \Omega_C$, and $\omega_2 = \Omega''_{comb} = -\Omega_A + \Omega_B + \Omega_C$.

It should, however, be noted that the shapes of these two peaks transform according to the same irreducible representation of the symmetry group D_{2h} and can therefore be distinguished only with difficulty in case of broad peaks. The phase relations of higher-order combination coherences are more complicated. Combination peaks of the type $(\Omega_A, \Omega_{comb}, \Omega_B)$ can also be recognized by the absence of corresponding cross-diagonal peaks $(\Omega_{comb}, \Omega_{comb}, \Omega_B)$ and $(\Omega_A, \Omega_{comb}, \Omega_{comb})$, and of diagonal peaks $(\Omega_{comb}, \Omega_{comb}, \Omega_{comb})$.

In some types of 3D experiments, it is easy to suppress undesired combination frequency peaks. During an extended cross-relaxation or chemical- exchange mixing period, coherences die out due to transverse relaxation or they can be suppressed by phase cycling and by a systematic variation of the position of a refocusing π pulse (*26, 52*). Then only operator terms such as (z), (zz), $(zz\cdots)$ remain. Assuming accurate $\pi/2$ pulses at the beginning and end of the mixing period, only pure z magnetization is created or converted into observable coherence. No combination peaks occur in these circumstances.

It should be mentioned that extensive phase cycling and shifting of π pulses is out of the question in practice because of excessive performance time. When applying nonselective mixing pulses, it is therefore unavoidable to live with the undesired combination and multiple-quantum peaks, and appropriate care is necessary when interpreting 3D spectra. The application of selective mixing pulses often prevents the formation of combination peaks because combination coherences are excited only when at least two spins, passive in the preceding period, are transformed by the mixing process.

VII. SYMMETRY OF 3D SPECTRA

In analogy to 2D spectra, it is possible to discuss global as well as local symmetry properties of 3D spectra. Local symmetry refers to the 3D multiplet structure of cross and diagonal peaks for weakly coupled spins. It shall not be treated here.

Global symmetry in 2D spectra refers to reflection symmetry about the diagonal $\omega_1 = \omega_2$. The conditions that a pulse sequence must fulfill to cause symmetric 2D spectra have been stated earlier (*34*). They can be summarized in the requirement that the mixing sequence must be time- and phase-reversal symmetric. In 3D spectra, there are three mirror planes that lead to symmetry-related peak positions: $\omega_1 = \omega_2$, $\omega_2 = \omega_3$, and $\omega_1 = \omega_3$ (Fig. 10). However, the requirement for reflection symmetry of peak amplitudes can only be fulfilled with respect to the $\omega_1 = \omega_3$ (back-transfer) plane.

Investigation of symmetry with respect to the plane $\omega_1 = \omega_3$ requires comparison of transfers of the type

$$\Omega_A \xrightarrow{R^{(1)}} \Omega_B \xrightarrow{R^{(2)}} \Omega_C$$

$$\Omega_C \xrightarrow{R^{(1)}} \Omega_B \xrightarrow{R^{(2)}} \Omega_A.$$

The symmetry condition amounts to equality of two Z factors (see Eq. [7]). This condition may be formulated

$$Z_{\{pq\}\{rs\}\{tu\}} = Z_{\{tu\}\{rs\}\{pq\}} \qquad [27]$$

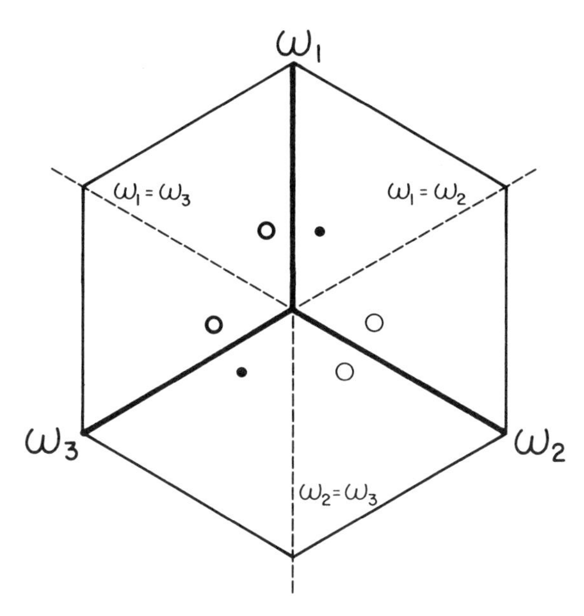

FIG. 10. View along the 3D diagonal ($\omega_1 = \omega_2 = \omega_3$) in a schematic 3D spectrum. The six peak positions with permuted coordinates Ω_A, Ω_B, and Ω_C are mutually transformed by reflection at the ($\omega_1 = \omega_2$) cross-diagonal plane, back-transfer plane ($\omega_1 = \omega_3$), and ($\omega_2 = \omega_3$) cross-diagonal plane. However, pairwise symmetry in the peak amplitudes occurs only with respect to the ($\omega_1 = \omega_3$) back-transfer plane for certain 3D pulse sequences.

or explicitly

$$D_{\{qp\}} R^{(2)}_{\{pq\}\{rs\}} R^{(1)}_{\{rs\}\{tu\}} \sigma(0)_{\{tu\}} = D_{\{ut\}} R^{(2)}_{\{tu\}\{rs\}} R^{(1)}_{\{rs\}\{pq\}} \sigma(0)_{\{pq\}}$$
$$\times \exp\{ i(p_{\{tu\}} - p_{\{pq\}})(\Phi_P - \Phi_M^{(1)} - \Phi_M^{(2)} + \Phi_D)\}. \quad [28]$$

In principle Eq. [28] can be fulfilled in numerous ways but in practice pairwise equality of related terms is generally required (*34*), leading to the conditions

$$\sigma(0)_{\{ab\}} = D_{\{ba\}} \quad [29]$$

$$R^{(1)}_{\{ab\}\{cd\}} = R^{(2)}_{\{cd\}\{ab\}}. \quad [30]$$

The phase factor in Eq. [28] induces a separate condition that can be fulfilled without difficulty. Equation [29] imposes the equivalence of the density operator after the preparation period and the detection operator, which is fulfilled for preparation by a nonselective pulse and detection of single-quantum coherence (*34*). On the other hand this condition can normally not be fulfilled for more complicated preparation sequences. The second condition, Eq. [30], requires the two mixing sequences to be related by TPR symmetry (*34*). This condition is obviously fulfilled when the two mixing sequences are identical and by themselves TPR symmetric. It is, however, important to note that 3D sequences obtained by merging together two 2D sequences which produce asymmetric 2D spectra lead to 3D spectra symmetric with respect to reflection about the $\omega_1 = \omega_3$ plane if the two mixing sequences are TPR counterparts. In terms of appearance of the spectrum this means merging together two 2D sequences which yield spectra that are mutual mirror images with respect to $\omega_1 = \omega_2$. An example of this is a relayed-NOESY transfer step combined with a NOESY-relayed transfer step (*53, 54*).

In summary we have the following consequences for the symmetry of specific 3D spectra: (i) 3D correlation spectra employing two identical TPR-symmetric mixing sequences such as COSY–COSY or ROESY–ROESY are symmetric with respect to the plane $\omega_1 = \omega_3$. (ii) 3D experiments with unequal mixing processes such as COSY–NOESY or NOESY–TOCSY lead to 3D spectra without global symmetry. (iii) 3D experiments with non-TPR-symmetric mixing sequences but together forming a TPR pair produce 3D spectra symmetric with respect to the $\omega_1 = \omega_3$ plane; however, no such experiments with great practical potential are currently in sight.

As an example of an $\omega_1 = \omega_3$ symmetric 3D spectrum, a 3D EXSY–EXSY spectrum of heptamethylbenzenonium sulfate in sulfuric acid is shown in Fig. 11. The peaks which lie in the same (ω_1, ω_3) planes are connected by lines. They are symmetric with respect to reflection at the respective ($\omega_1 = \omega_3$) line in the planes. The possible exchange processes are due to the 1,2 methyl group shifts: A $\rightleftarrows$ B, B $\rightleftarrows$ C, and C $\rightleftarrows$ D (*25*).

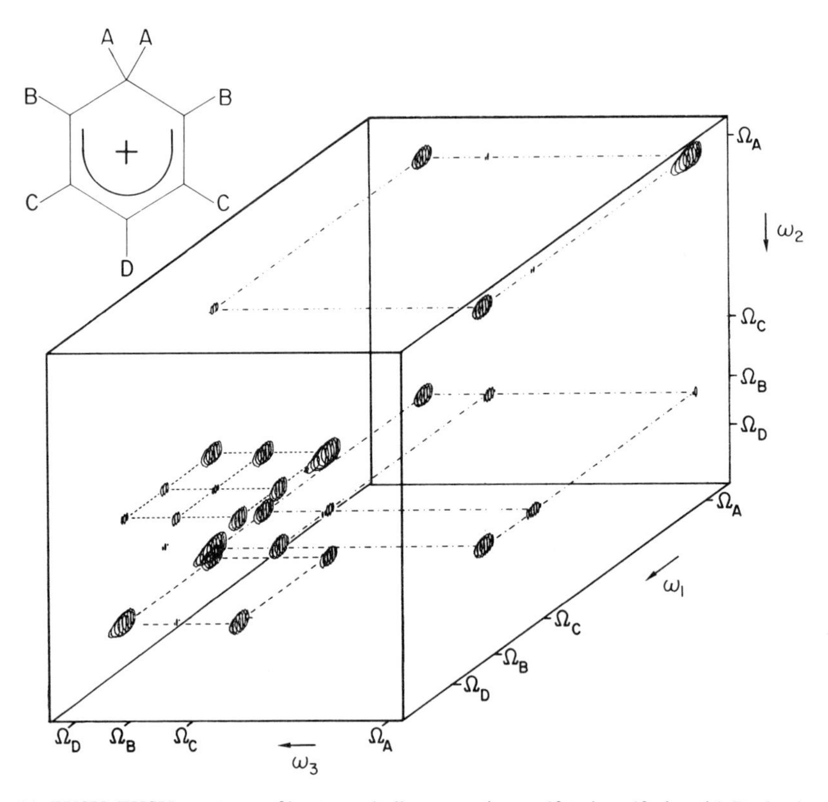

FIG. 11. EXSY–EXSY spectrum of heptamethylbenzenonium sulfate in sulfuric acid. Peaks that lie on identical (ω_1,ω_3) planes are connected by lines. The $\omega_1 = \omega_3$ reflection symmetry is apparent from the spectrum.

VIII. SENSITIVITY OF 3D SPECTROSCOPY

It has been shown earlier that the loss in sensitivity per unit time when going from 1D to 2D spectroscopy is not a serious limitation of 2D spectroscopy (*55*). The equivalent is found when extending 2D spectroscopy to three dimensions as will be shown in the following.

It is known that the maximum signal-to-noise ratio (S/N) achievable with a matched filter is given by the square root of the ratio of signal energy and noise power per unit bandwidth (*6, 55*). As long as the relaxation decay during evolution in the 2D and 3D extensions of a 1D experiment is negligible, the signal energies in all three experiments are comparable. Factors that can significantly reduce sensitivity are the distribution of signal intensity across numerous multiplet lines or over several cross peaks either in two or three dimensions and losses in intensity due to inefficient coherence transfer in a particular mixing process.

The sensitivity, i.e., the S/N achievable in a unit time interval, has been given for a 2D experiment in Refs. (*6, 55*),

$$\frac{S}{\sigma_N T_{\mathrm{tot}}^{1/2}} = \frac{\overline{sh}}{[\overline{h^2}]^{1/2} \rho_n} \left(\frac{t_2^{\max}}{T} \right)^{1/2} \tag{31}$$

with

$$\overline{sh} = \frac{1}{t_1^{\max}} \frac{1}{t_2^{\max}} \int_0^{t_1^{\max}} \int_0^{t_2^{\max}} s^e(t_1, t_2) h(t_1, t_2) dt_1 dt_2 \tag{32}$$

$$[\overline{h^2}] = \frac{1}{t_1^{\max}} \frac{1}{t_2^{\max}} \int_0^{t_1^{\max}} \int_0^{t_2^{\max}} h(t_1, t_2)^2 dt_1 dt_2. \tag{33}$$

The corresponding expression for a 3D experiment is

$$\frac{S}{\sigma_N T_{\mathrm{tot}}^{1/2}} = \frac{\overline{sh}}{[\overline{h^2}]^{1/2} \rho_n} \left(\frac{t_3^{\max}}{T} \right)^{1/2} \tag{34}$$

with

$$\overline{sh} = \frac{1}{t_1^{\max}} \frac{1}{t_2^{\max}} \frac{1}{t_3^{\max}} \int_0^{t_3^{\max}} \int_0^{t_2^{\max}} \int_0^{t_1^{\max}} s^e(t_1, t_2, t_3) h(t_1, t_2, t_3) dt_1 dt_2 dt_3 \tag{35}$$

$$[\overline{h^2}] = \frac{1}{t_1^{\max}} \frac{1}{t_2^{\max}} \frac{1}{t_3^{\max}} \int_0^{t_3^{\max}} \int_0^{t_2^{\max}} \int_0^{t_1^{\max}} h^2(t_1, t_2, t_3) dt_1 dt_2 dt_3. \tag{36}$$

The noise power per unit bandwidth (measured in hertz) is σ_N^2, $h(t_1, t_2, t_3)$ is the 3D filter function applied, and $s^e(t_1, t_2, t_3)$ is the signal envelope function.

For a comparison of the sensitivities of 2D and 3D experiments we discuss a 3D experiment that is derived from a 2D pulse sequence by insertion of a (t_1, mixing) moiety between preparation and evolution periods:

$$\text{2D:}\quad \mathrm{P} \qquad\qquad —t_2—\mathrm{M}_2—t_3$$

$$\text{3D:}\quad \mathrm{P}—t_1—\mathrm{M}_1—t_2—\mathrm{M}_2—t_3.$$

This comparison has the advantage that the sensitivity of 3D experiments can be compared directly with current 2D experiments, like COSY, NOESY, or TOCSY, the sensitivity of which is usually known.

We assume that the signal envelope function s^{e}_{ijk} for a cross peak connecting spins I_i, I_j, and I_k in ω_1, ω_2, and ω_3, respectively, and the filter function h applied in the 3D experiment are separable into products of one-dimensional functions,

$$s^{e}_{ijk}(t_1, t_2, t_3) = s^{e}_{1i}(t_1)s^{e}_{2j}(t_2)s^{e}_{3k}(t_3) \quad [37]$$

$$h(t_1, t_2, t_3) = h_1(t_1)h_2(t_2)h_3(t_3) \quad [38]$$

and similarly for the cross peak between I_j and I_k in ω_2 and ω_3 in the 2D experiment:

$$s^{e}_{jk}(t_2, t_3) = s^{e}_{2j}(t_2)s^{e}_{3k}(t_3) \quad [39]$$

$$h(t_2, t_3) = h_2(t_2)h_3(t_3). \quad [40]$$

The identity of the signal envelope function in the evolution time t_2 of the 2D and 3D experiment restricts the following discussion to the simple and, at the same time, relevant 3D experiments which employ in-phase coherence transfer in the first mixing element. We further assume that the inserted mixing process M_1 has the transfer efficiency T_{ij} for the transfer from spin I_i to I_j. Then we obtain for the sensitivity ratio of 2D and 3D experiments with equal overall performance time:

$$\frac{(S/N)^{3D}_{ijk}}{(S/N)^{2D}_{jk}} = T_{ij}\frac{(1/t_1^{\max})\int_0^{t_1^{\max}}(1/N_i)s^{e}_{1i}(t_1)h_1(t_1)dt_1}{\left[(1/t_1^{\max})\int_0^{t_1^{\max}}h_1(t_1)^2dt_1\right]^{1/2}}, \quad [41]$$

where N_i is the number of resolved multiplet components of spin I_i caused by the additional evolution period t_1 in the 3D experiment. The division by N_i takes into account that the multiplet splitting reduces the peak intensity by this factor.

Equation [41] can be further simplified by the assumption that $s^{e}_{1i}(t_1)$ is not affected by relaxation in the range of $0 < t_1 < t_1^{\max}$. Thus, when $h_1(t_1)$ is selected to be constant as well one arrives at

$$\frac{(S/N)^{3D}_{ijk}}{(S/N)^{2D}_{jk}} = T_{ij}/N_i. \quad [42]$$

T_{ij} depends on the first mixing process employed. For COSY, we have $T_{ij} = 1$ between coupled spins whereas for NOESY and TOCSY T_{ij} is smaller than unity, depending on cross-relaxation rates and the J-coupling network, respectively, and on the mixing time. As an example we consider the combination of NOESY with COSY under the above conditions of well-resolved multiplet lines. For the (NH,H_α,NH) peaks, there results a loss in S/N of only a factor of 2 compared to conventional NOESY, because NH resonances whose multiplet splittings are observed in the third dimension are normally doublets. The loss in S/N for (H_α,NH,H_α) peaks may be larger because of more extensive multiplet splittings of H_α resonances. In general, multiplet splittings will reduce peak intensities in 3D spectra in comparison to 2D spectra by a factor 2^{N-1}, where N is the number of mutually coupled spins. The antiphase structure of overlapping multiplet components due to the active coupling causes a further loss in the S/N of a 3D experiment. In the case of vanishing or ill-resolved passive couplings, the loss in S/N is, on the contrary, smaller.

It is possible to enhance sensitivity of 3D experiments by selective mixing pulses that reduce the distribution of intensity (*56*). Another more generally applicable possibility is selective spin decoupling by refocusing during t_1 and t_2 by the insertion of selective inversion pulses. Depending on the pulse sequence, the selective π pulses may either be applied to the spins active during a certain evolution period or to a spectral region of spins which one wants to decouple (*57*). In favorable cases, the collapse of the entire multiplet structure in t_1 can be achieved and $N_i = 1$, e.g., in the combination of a soft-NOESY experiment (*57*) with another transfer experiment. A further example without additional multiplet structure in ω_1 is the combination of hetero-COSY (*46, 47*) with a second transfer sequence. The fully refocused experiments of Figs. 6f and 6g lead to tilted multiplets much like in a 2D *J*-resolved spectrum without additional multiplet lines. Given the usual assumption in this section that in-phase magnetization is transferred during the first mixing, Eq. [41] simplifies to

$$\frac{(S/N)^{3D}_{ijk}}{(S/N)^{2D}_{jk}} = T_{ij} \frac{(1/t_1^{max}) \int_0^{t_1^{max}} e^{-t_1/T_{2i}} h_1(t_1) dt_1}{\left[(1/t_1^{max}) \int_0^{t_1^{max}} h_1(t_1)^2 dt_1 \right]^{1/2}}. \qquad [43]$$

This then yields for the ratio of sensitivities achievable in the same total time, using matched filters $h(t)$ in situations without multiplet splittings in the additional dimension, i.e., for $N_i = 1$,

t_1^{max}/T_{2i}	0.5	1	2	3	4
$((S/N)^{3D}_{ijk}/(S/N)^{2D}_{jk}) \cdot (1/T_{ij})$	0.8	0.66	0.5	0.4	0.35

[44]

The sensitivity loss due to the third dimension is negligible in this special case, especially if one takes into account the circumstance that the digitization density in frequency space in 2D experiments normally must be chosen higher than in 3D in order to resolve overlap, as the latter profits from the dispersion in a third dimension. This requires longer t_1^{max} values in 2D than in 3D experiments such that the two techniques become virtually equivalent in terms of sensitivity.

Finally we analyze the sensitivity ratio for macromolecules, where the lines are so broad that the homonuclear couplings are not resolved and the entire multiplet is represented by a single broad peak of width $(\pi T^{\Delta}_{2i})^{-1}$. Then for in-phase transfer, as in the practically important combinations of NOESY, TOCSY, ROESY, or heteronuclear shift correlation, it is permissible to neglect factors distributing intensity within the multiplets and to set $s_1^e(t_1) = e^{-t_1/T^{\Delta}_{2i}}$ in Eq. [41]:

$$\frac{(S/N)^{3D}_{ijk}}{(S/N)^{2D}_{jk}} = T_{ij} \frac{(1/t_1^{max}) \int_0^{t_1^{max}} e^{-t_1/T^{\Delta}_{2i}} h_1(t_1) dt_1}{\left[(1/t_1^{max}) \int_0^{t_1^{max}} h_1(t_1)^2 dt_1 \right]^{1/2}}. \qquad [45]$$

This expression is equivalent to Eq. [43] with T_{2i} replaced by T^{Δ}_{2i}, the width of the whole multiplet. The data in Eq. [44] are then again applicable, showing that the sensitivity losses due to the introduction of a third frequency domain are often negligible.

In 3D experiments with two-step in-phase magnetization transfer, it is also possible to relate the sensitivity of the 3D experiment to the sensitivity of the two constituent 2D experiments (henceforth called $2D^{(1)}$ and $2D^{(2)}$). On the basis of Eqs. [31]–[36] and using the expression of the S/N of a 1D spectrum (*6*), we obtain for experiments with matched filters the following sensitivity ratio for the cross peak between spins I_i, I_j, and I_k in the 3D experiment compared to the I_i, I_j cross peak in the $2D^{(1)}$ and the I_j, I_k cross peak in the $2D^{(2)}$ experiment:

$$\frac{(S/N)^{3D}_{ijk}(S/N)^{1D}_{j}}{(S/N)^{2D^{(2)}}_{jk}(S/N)^{2D^{(1)}}_{ij}} = \left[\frac{(1/t_1^{\max}t_2^{\max}t_3^{\max})\int_0^{t_3^{\max}}\int_0^{t_2^{\max}}\int_0^{t_1^{\max}}[s^{e3D}_{ijk}(t_1,t_2,t_3)]^2\,dt_1\,dt_2\,dt_3}{(1/t_1^{(1)\max}t_2^{(1)\max})\int_0^{t_1^{(1)\max}}\int_0^{t_2^{(1)\max}}[s^{e2D^{(1)}}_{ij}(t_1,t_2)]^2\,dt_1\,dt_2} \times \frac{(1/t^{\max})\int_0^{t^{\max}}[s^{e1D}_{j}(t)]^2\,dt}{(1/t_1^{(2)\max}t_2^{(2)\max})\int_0^{t_1^{(2)\max}}\int_0^{t_2^{(2)\max}}[s^{e2D^{(2)}}_{jk}(t_1,t_2)]^2\,dt_1\,dt_2}\right]^{1/2}. \quad [46]$$

We again assume that the signal function can be separated into products of one-dimensional functions (Eq. [37]). For equal detection times for the 1D and the $2D^{(1)}$ as well as for the $2D^{(2)}$ and 3D experiment,

$$t^{\max} = t_2^{(1)\max}, \qquad t_2^{(2)\max} = t_3^{\max}, \quad [47]$$

and equal signal envelope in the detection periods,

$$s^{e3D}_{ijk}(t_3) = s^{e2D^{(2)}}_{jk}(t_2), \qquad s^{e2D^{(1)}}_{ij}(t_2) = s^{e1D}_{j}(t), \quad [48]$$

we obtain

$$(S/N)^{3D}_{ijk} = \frac{(S/N)^{2D^{(1)}}_{ij}(S/N)^{2D^{(2)}}_{jk}}{(S/N)^{1D}_{j}} \times \left[\frac{(1/t_1^{\max}t_2^{\max})\int_0^{t_1^{\max}}\int_0^{t_2^{\max}}[s^{e3D}_{ijk}(t_1,t_2)]^2\,dt_1\,dt_2}{(1/t_1^{(1)\max}t_1^{(2)\max})\int_0^{t_1^{(1)\max}}[s^{e2D^{(1)}}_{ij}(t_1)]^2\,dt_1\int_0^{t_1^{(2)\max}}[s^{e2D^{(2)}}_{jk}(t_1)]^2\,dt_1}\right]^{1/2}. \quad [49]$$

If we again assume unresolved peaks for the spins with a width of e.g. $(\pi T^{\Delta}_{2i})^{-1}$ for I_i we can evaluate the integrals and obtain

$$(S/N)^{3D}_{ijk} = \frac{(S/N)^{2D^{(1)}}_{ij}(S/N)^{2D^{(2)}}_{jk}}{(S/N)^{1D}_{j}} \times \left[\frac{(1-e^{-2t_1^{\max}/T^{\Delta}_{2i}})(1-e^{-2t_2^{\max}/T^{\Delta}_{2j}})t_1^{(1)\max}t_1^{(2)\max}}{(1-e^{-2t_1^{(1)\max}/T^{\Delta}_{2i}})(1-e^{-2t_1^{(2)\max}/T^{\Delta}_{2j}})t_1^{\max}t_2^{\max}}\right]^{1/2}.$$

This expression shows that the loss in sensitivity when going from 2D to 3D, keeping the overall performance time constant, is similar to the loss in going from 1D to 2D. In many cases this loss is small. In addition, it is favourably weighted by the square root factor in the above equation. This factor is usually >1 as the maximum evolution times in a 3D experiment must be kept short in order to keep performance time within reasonable bounds.

IX. PRACTICAL ASPECTS OF 3D SPECTROSCOPY AND ASSIGNMENT PROCEDURES IN PROTEINS

Three-dimensional experiments are invariably time-consuming, both in terms of data acquisition and to a lesser extent in terms of computation time. Optimization of data acquisition is therefore of primary practical interest. A further aspect of great practical importance is the visual representation of the enormous amount of data in a form that is comprehensible for an eye accustomed to the interpretation of two-dimensional displays.

(a) Nonselective Approach to 3D Spectroscopy

In order to exploit the full 3D multiplex advantage, it is desirable to record 3D spectra that cover the entire spectral range in all three dimensions. The procedure is in principle straightforward and does not pose conceptual problems. However, the amount of data can become enormous when a resolution comparable to that usual in 2D spectra shall be achieved.

In order to cover a frequency range of 5 kHz (e.g., proton resonance at 500 MHz) with a resolution of 5 Hz, a data matrix of $1000 \times 1000 \times 1000 = 1$ G words must be acquired. Assuming a repetition rate of two FIDs per second, leading to 7200 experiments/h, 172,800 experiments/day, and 1,209,600 experiments/week, a full week is necessary for data acquisition. Each additional step in a phase cycle would add a further week of acquisition time. Although this time consumption might be tolerable in certain situations, the amount of data will be difficult to handle with present day laboratory computer systems. It can, however, be expected that data processing will cease to be a problem in the future based on improved computing equipment.

At present, the number of points in t_1 and t_2 must be limited to about 256, leading to $256 \times 256 = 65{,}536$ experiments multiplied by the number of steps in the phase cycle. The number of points in the t_3 dimension can be chosen freely without any time penalty. Data matrices of $256 \times 256 \times 512$ or $128 \times 128 \times 2048$ can be handled with modest computing equipment.

For a nonselective 3D experiment a minimum phase cycle of four steps is needed to suppress axial peaks in ω_1 and ω_2. This applies to combinations of COSY, TOCSY, ROESY, and hetero-COSY. More extensive phase cycles are required for NOESY or multiple-quantum-filtered COSY where coherence order selection is necessary during mixing.

As an example, we show a ROESY–TOCSY spectrum of buserilin (Fig. 12), a linear nonapeptide with the sequence pyro-Glu–His–Trp–Ser–Tyr–(*O-tert.*-butyl–)D-Ser–Leu–Arg–Pro–$NHCH_2CH_3$(*p*–E–H–W–S–Y–S̲–L–R–P–NH–Et), acquired with nonselective pulses and 256×256 experiments in t_1 and t_2. A stereographic view of the spectrum is given together with assignment of groups of cross peaks in the boxes A, B_1, and B_2. Box A encloses the (H_α, NH, aliphatic-H) region, box B_1 the (NH, aliphatic-H, NH) region, and box B_2 the (NH, aliphatic-H, side chain-H) region. These regions are useful for sequence analysis in proteins as is discussed in Section IX (d). The diagonal and the TOCSY cross-diagonal peaks ($\omega_1 = \omega_2$) appear in blue, whereas the ROESY cross-diagonal ($\omega_2 = \omega_3$), back-transfer, and cross peaks are red.

(b) Selective Approach to 3D Spectroscopy

In order to optimize resolution, it is often necessary to record partial 3D spectra, whereby advance information is required to localize the relevant volumes of a 3D spectrum. Such a volume is displayed in Fig. 13.

The selected subvolume in a 3D spectrum can be chosen by the application of frequency-selective pulses (*58, 59*). The preparation pulse selects the frequency range for the t_1 evolution. The first mixing process is, in the case of a COSY-type experiment, replaced by a pair of selective pulses, one for the ω_1 and one for the ω_2 frequency range (Fig. 6b). In the case of a NOESY-type transfer, the first pulse of the mixing sequence can be nonselective while the second pulse selects the desired ω_2 frequency range (Figs. 6c and 6e). In most cases, the second mixing process can be chosen to be nonselective because frequency selectivity in ω_3 can easily be attained after Fourier transformation of each FID by software.

The shape of the frequency-selective pulses is not very critical, and Gaussian (*58*), Hermite (*60*), or other shapes can be employed. The pulses of extended duration, however, introduce strong frequency-dependent phase shifts that cannot be properly corrected by a linear phase shift of the resulting 3D spectrum and lead to severe line-shape and baseline distortions. This can be avoided by applying a nonselective π pulse followed by a refocusing interval Δ after each selective pulse. The delay Δ is about half the duration of the selective pulse. When the purpose of the selective pulse is to convert coherence into polarization, a Δ–π sequence should be inserted *before* the selective pulse.

The effect of a refocusing pulse inserted after a Hermite excitation pulse of duration 11.9 ms in a 1D experiment applied to the H_α region of buserilin is demonstrated in Fig. 14. Figure 14a shows the reference spectrum without refocusing while a refocusing sequence was employed for the spectrum in Fig. 14b leading to a flat baseline. Practical work has shown that the refocusing sequence is essential when the multiplets are in-phase in the associated frequency dimension. In the case of antiphase multiplets, as, for example, in COSY-type experiments, the signal distortions tend to cancel each other. The selective pulses that should be followed by a refocusing sequence are marked with an asterisk in Figs. 6c and 6e. The H_α-selective pulse in Fig. 6c does not require refocusing because the multiplets are antiphase in ω_2. In the sequence of Fig. 6f, the refocusing is achieved by the π pulse in the center of the t_1 period. An alternative possibility, circumventing the need for refocusing sequences, is the use of self-refocused selective pulses (*61*).

It is in general advantageous to keep the number of selective pulses as small as possible because relaxation during these long pulses leads to intensity losses. The use of nonselective mixing pulses at the beginning of the NOESY mixing sequence is exemplified in Figs. 6b, 6c, and 6e. Although such a pulse excites the entire proton

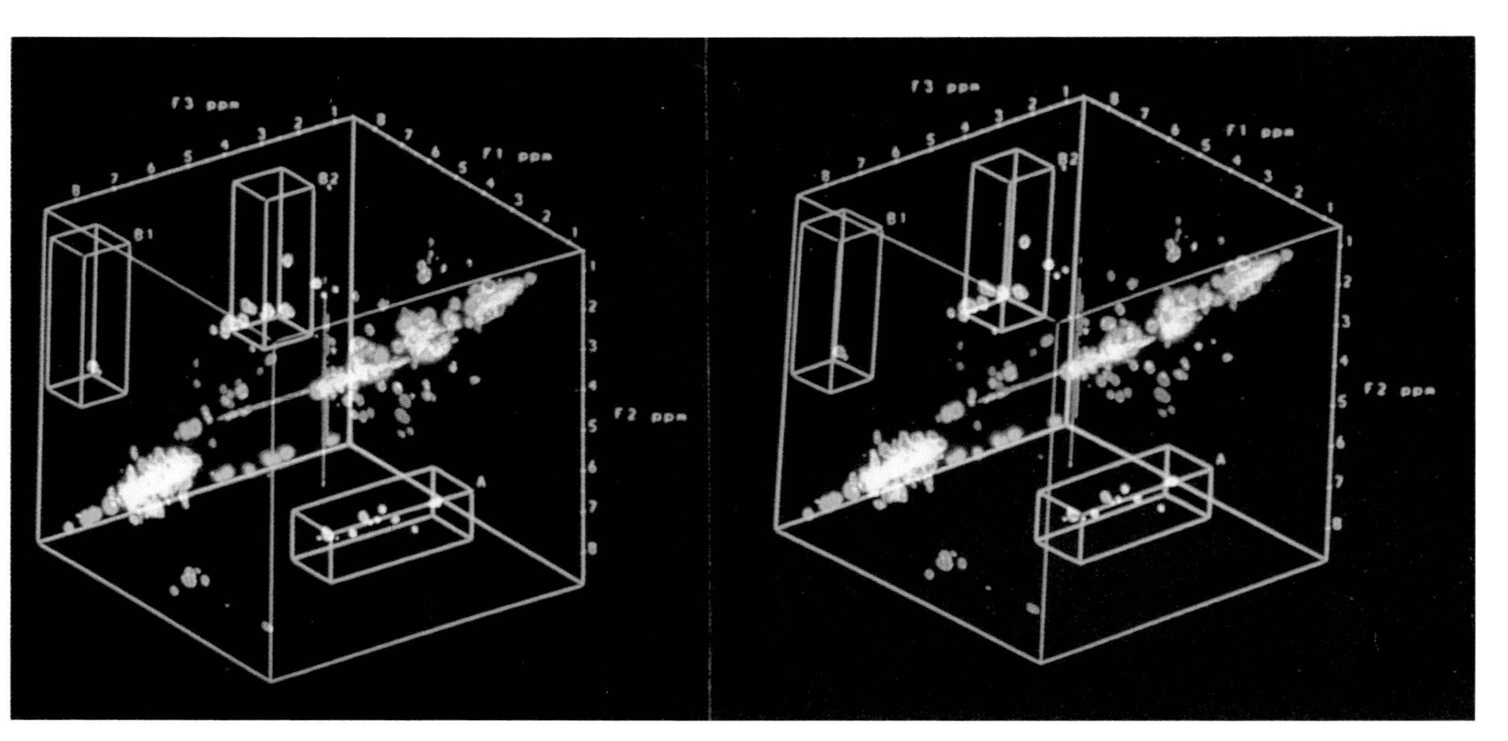

FIG. 12. Stereographic representation of a nonselective 300 MHz ROESY–TOCSY spectrum of buserilin in DMSO-d_6 photographed from an Evans and Sutherland picture system. Assignment of peak regions: (H_α, NH, aliphatic-H) in box A, (NH, aliphatic-H, NH) in box B_1, and (NH, aliphatic-H, side chain-H) in box B_2. The data matrix of $256 \times 256 \times 512$ was Fourier-transformed to $128 \times 128 \times 256$ real points to accommodate the data in the picture system. Four scans were performed for each combination of t_1, t_2 values. The ROESY mixing time was 200 ms, and the TOCSY mixing achieved by MLEV-17 (*65, 66*) with $\gamma B_1/2\pi = 4.2$ kHz lasting for 95.04 ms. The duration of one scan was 2.3 s.

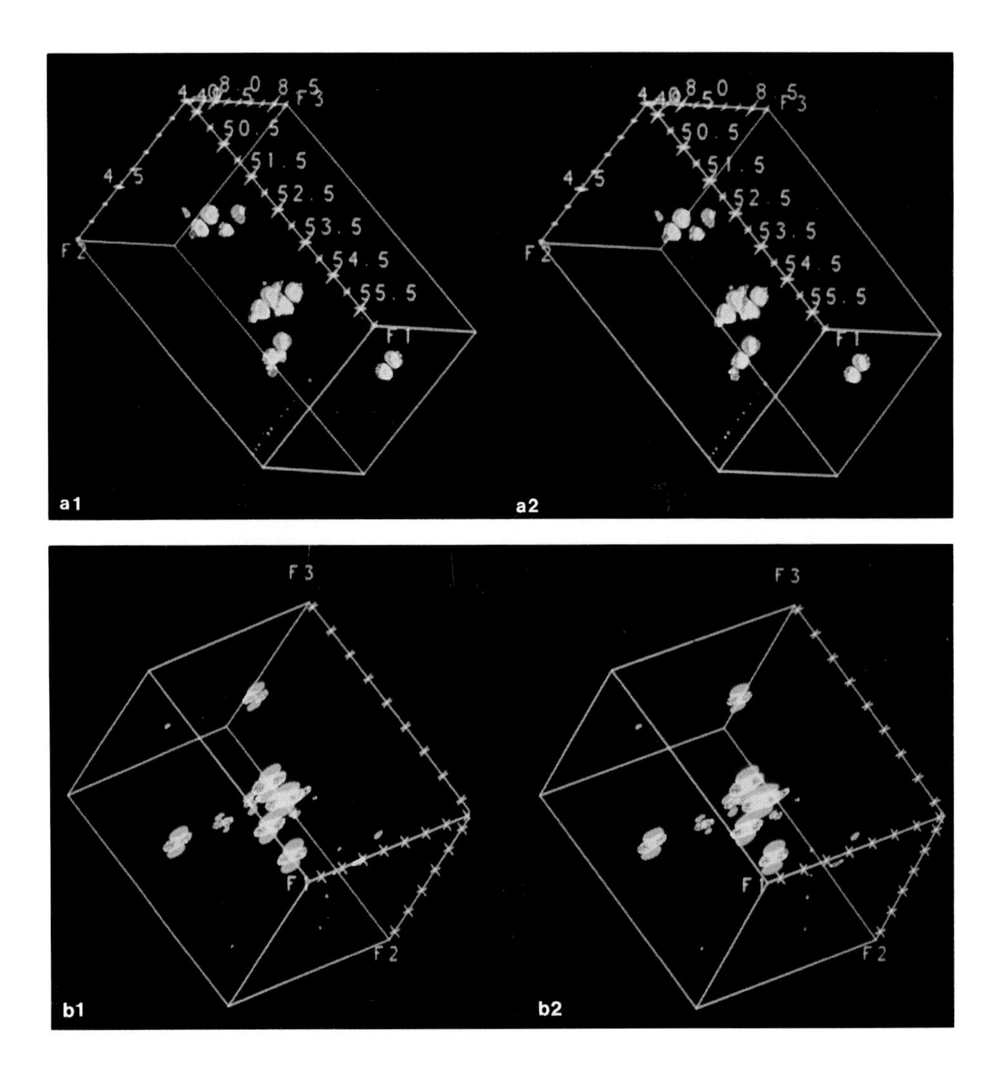

FIG. 22. (a1, a2) Stereographical view of the C_α,H_α,NH region in the C,H-COSY–COSY spectrum of buserilin in dimethyl sulfoxide-d_6. Photographs were taken from an Evans and Sutherland display. (a3 on next page) A sketch of the picture contains the peak assignment. Figure 17 gives another representation of the same spectrum. (b1, b2) Stereographical view of the NH,H_α,NH region of a NOESY–TOCSY spectrum of buserilin. (b3 on next page) Assignments are given in a sketch of the picture. The notation Y/S indicates a cross peak with tyrosine (Y) NH chemical shift in ω_1 and serine (S) H_α and NH chemical shifts in ω_2 and ω_3, respectively.

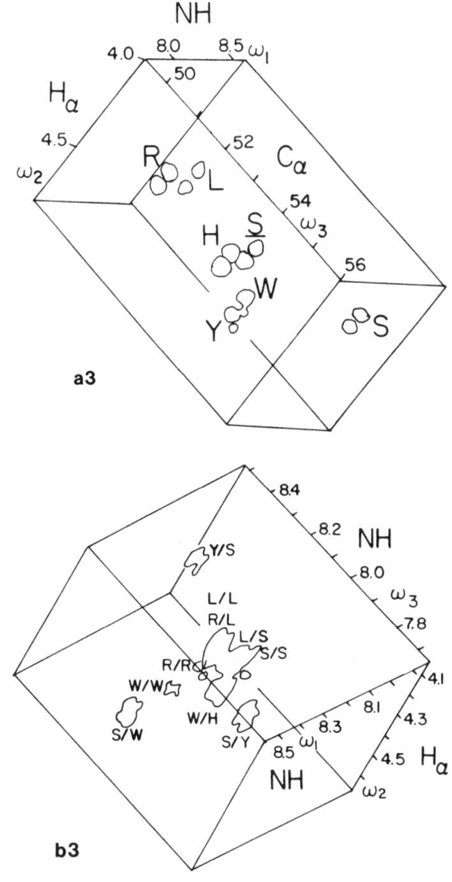

FIG. 22—*Continued*

spectrum, phase cycling makes it possible to reject all magnetization components except for those that are transverse during the preceding evolution period. In addition, it is recommended that axial peaks in ω_1 and ω_2 be suppressed by a phase cycle with 0, π phase shifts of all pulses preceding the respective evolution period.

A NOESY–TOCSY spectrum of buserilin shall illustrate selective 3D experiments. To correlate the NH protons through H_α with all protons the 3D volume shown in Fig. 15 was chosen. An overview of the spectrum is given in the form of stacked 2D contour plots in Fig. 16. The sequential information of the backbone is contained in the NH–H_α–NH region of the NOESY–TOCSY spectrum.

(c) Heteronuclear 3D Experiments

Shift correlation of proton 2D spectra with heteronuclei, like ^{13}C and ^{15}N, which have a large shift dispersion, is very promising to increase resolution of crowded 2D proton spectra. In addition, correlation via large one-bond couplings is efficient, and almost no signal loss occurs for isotope-enriched samples. Natural abundance hetero-

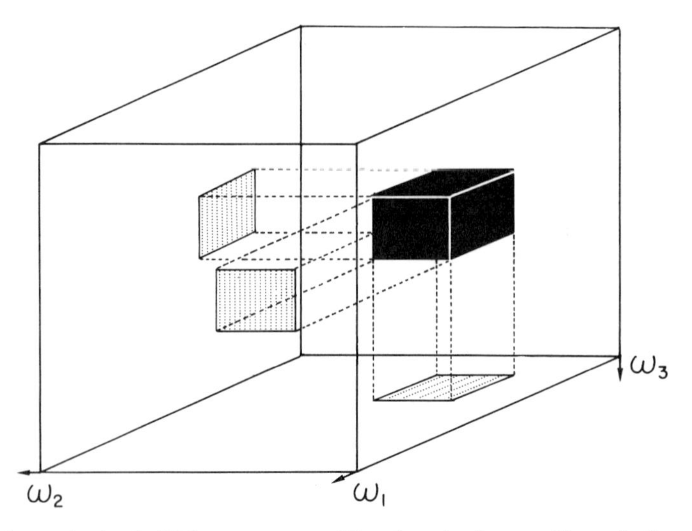

FIG. 13. Volume selection in 3D frequency space. The selected volume and its projections are shown.

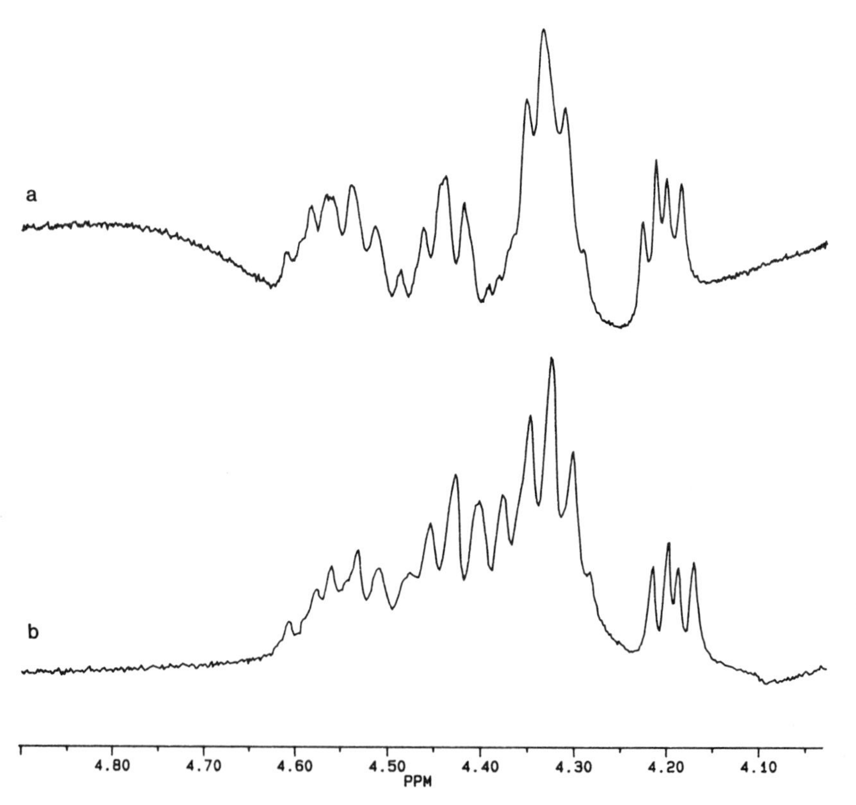

FIG. 14. (a) 1D spectrum of the H_α region of buserilin acquired with a Hermitian pulse of 11.9 ms duration. The strong peak-shape distortions are clearly visible. (b) Same as (a) but with a nonselective π–Δ (Δ = 5.5 ms) sequence between the selective pulse and acquisition.

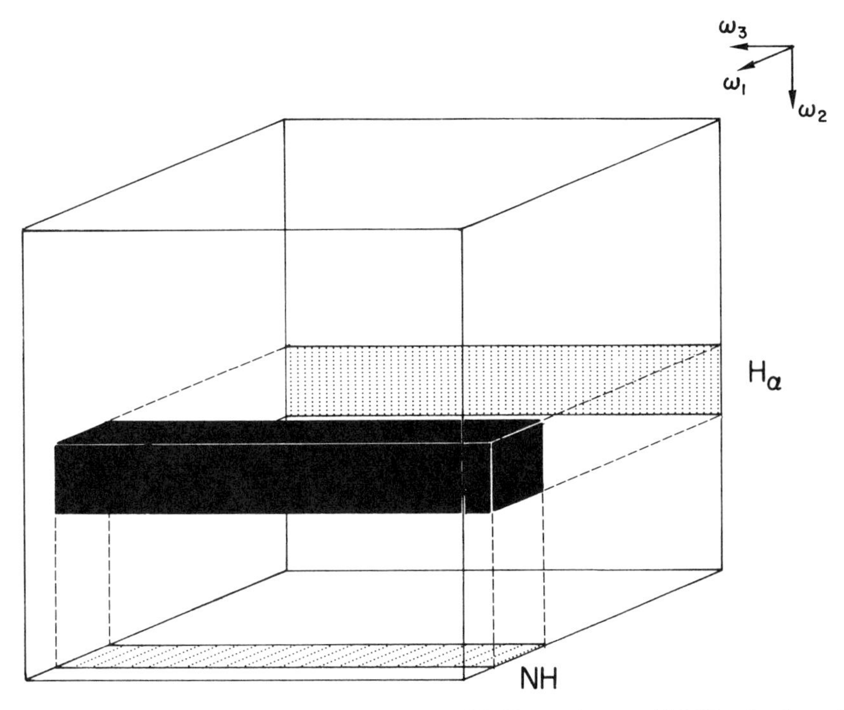

FIG. 15. Volume selection for selective 3D NOESY–TOCSY experiments with NH region in ω_1, H_α in ω_2, and all protons in ω_3.

nuclear 3D spectra of large molecules, however, are often low in sensitivity, making isotopic enrichment almost indispensable.

Heteronuclear correlation experiments applied to proteins and involving nuclei such as ^{15}N are facilitated by the fact that no selective pulses are required for frequency selectivity. Protons bound to heteronuclei often lie in a restricted chemical-shift range and the "selective excitation" of this shift range is achieved via the J couplings to the heteronucleus. Pulse sequences for hetero-COSY–COSY and hetero-COSY–TOCSY experiments are shown in Figs. 6d and 6e. The first part of the sequence is the proton-detected shift correlation sequence proposed by Müller (*46, 47*) with a delay $\tau = (2J_{HX})^{-1}$ in order to allow for refocusing of the heteronuclear coupling. During t_2, heteronuclear decoupling is achieved by a central π pulse or by broadband irradiation applied to the heteronucleus. A homonuclear coherence transfer concludes the 3D pulse sequence. The combination of heteronuclear shift correlation with homonuclear transfer experiments does not impede the quantification of the latter as is explained in the next section.

As an illustration of a heteronuclear 3D NMR experiment, we show a hetero-COSY–COSY spectrum of buserilin which correlates C_α, H_α, and NH chemical shifts in ω_1, ω_2, and ω_3, respectively (Fig. 17). Slices through the 3D spectrum taken at the positions indicated in the 2D hetero-COSY spectrum are shown. The separation of the H_α,NH cross peaks due to the correlation with the C_α resonance is obvious upon

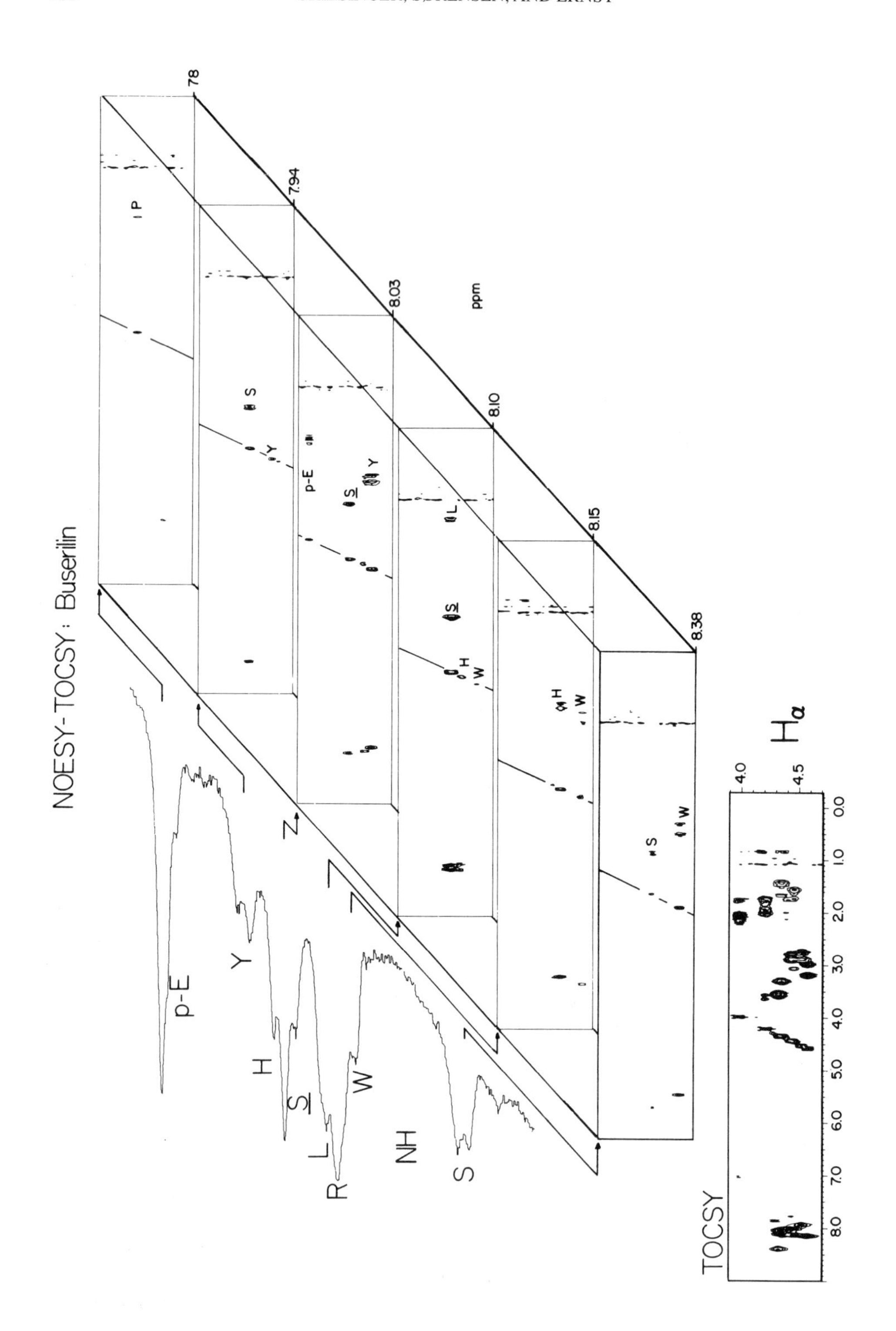
NOESY-TOCSY: Buserilin
TOCSY
ppm
7.78
7.94
8.03
8.10
8.15
8.38
p-E
Y
H
S
L
R
W
NH
S
Z
H_α
4.0
4.5
0.0
1.0
2.0
3.0
4.0
5.0
6.0
7.0
8.0

comparison with the NH,H_α section of the 2D COSY spectrum. This can also be appreciated from the 3D representation of the spectrum in Fig. 21.

It should be noted that, for a complete assignment of proton resonances, normally two hetero 3D experiments are required: a hetero-COSY–COSY or a hetero-COSY–TOCSY and a hetero-COSY–NOESY experiment. When heteronuclear three-bond couplings can be evaluated, it is sometimes sufficient to work with a single hetero-COSY–NOESY spectrum.

(d) Procedure for Sequential Assignments in Proteins

Sequential assignment procedures in proteins based on 3D spectra that combine an incoherent and a coherent transfer step (NOESY–COSY, NOESY–TOCSY, ROESY–TOCSY, or hetero-COSY–NOESY) can take advantage of the fact that three nuclei in the peptide chain are related by a single cross peak, two of which belong to the same amino acid residue and are related by the coherent transfer while the third spin may belong to the following or preceding residue in the chain. Such a cross peak constitutes the sequential connectivity of two amino acids. Assuming that one cross peak has been assigned, the next related peak is preferably found from the first one by keeping fixed two of its frequency coordinates and searching along a line in the third dimension of the 3D frequency space. This procedure has the advantage that assignment is unambiguous as long as there are no two pairs of resonances with both chemical shifts being equal. This condition is less likely to be violated than the requirement in 2D spectroscopy that there be no two resonances with degenerate chemical shifts.

The assignment strategies differ as usual for β sheets and α helices due to the fact that in β sheets the NH_i,$H_{\alpha,i-1}$ NOE cross peaks are dominant while for α helices NH_i,NH_{i+1} NOE cross peaks are strongest (*8*). We treat in the following both cases based on 3D spectra recorded with nonselective or with selective pulses.

(1) Assignments based on nonselective 3D experiments. Possible assignment strategies using homonuclear or heteronuclear NOESY–COSY-type experiments are indicated in Table 3. We concentrate at first on the assignment within β sheets based on NH and H_α resonances. Two regions of the 3D spectrum, schematically shown in Fig. 18, are relevant, the (NH,H_α,NH) and the (H_α,NH,H_α) regions that represent restricted subvolumes of the entire 3D spectrum.

We assume that the first cross peak ($H_\alpha^{(1)}$,$NH^{(2)}$,$H_\alpha^{(2)}$) has been found. In order to proceed along the sequence of the peptide, we first reflect this cross peak at the ω_2

FIG. 16. The 300 MHz selective 3D NOESY–TOCSY of buserilin in DMSO-d_6 according to the frequency selection of Fig. 15. The data matrix of $96 \times 96 \times 2K$ was zero-filled to $256 \times 256 \times 4K$ prior to Fourier transformation. Thirty-two scans were performed per t_1, t_2 value. The NOESY mixing time was 250 ms, the TOCSY mixing, achieved with MLEV-17 (*65, 66*), lasted 42.66 ms. The duration of one scan was 2.66 s. Selective excitation at the beginning of t_1 and t_2 was achieved with a 10.0 ms Gaussian pulse followed by a nonselective π pulse and a 4.0 ms delay. (ω_2,ω_3) slices are taken at interesting positions in ω_1 which are marked by arrows in the 1D spectrum of the NH resonances. The enhanced resolution in the TOCSY sections of the 3D spectrum is obvious by comparison with the 2D TOCSY spectrum of buserilin at the bottom of the figure. The assignment of the peaks is given with the one-letter codes for the amino acids.

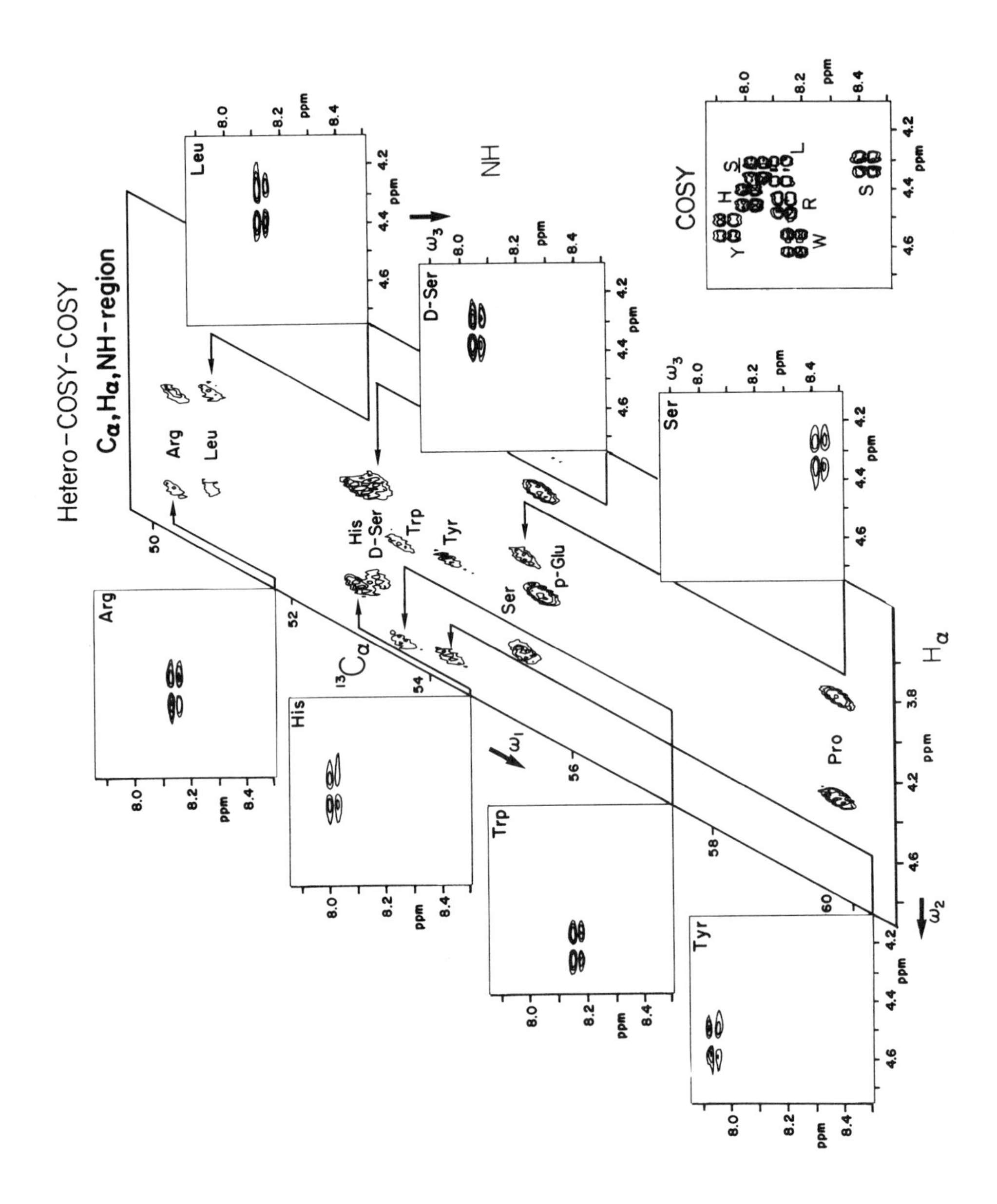
Hetero-COSY-COSY
Cα,Hα,NH-region
COSY
NH
Hα
^{13}Cα
ω1
ω2
ω3
ppm
Arg
Leu
His
D-Ser
Trp
Tyr
Ser
p-Glu
Pro
Y
H
S
L
R
W

$= \omega_3$ diagonal within the (ω_2, ω_3) section that contains this cross peak (line 1 in Fig. 18). This gives a peak location determined by $\omega_3 = \Omega_{NH^{(2)}}$, $\omega_2 = \Omega_{\alpha^{(2)}}$ within the above-mentioned (ω_2, ω_3) section (usually with a very small or absent peak due to the weak $H_\alpha^{(1)}H_\alpha^{(2)}$ NOE). Now following a line (2) through this peak location parallel to the ω_1 axis, we encounter the ($NH^{(3)}$,$H_\alpha^{(2)}$,$NH^{(2)}$) cross peak. Reflection at the $\omega_1 = \omega_2$ line in the ω_1, ω_2 section (line 3) containing the latter peak leads then to a location ($H_\alpha^{(2)}$,$NH^{(3)}$,$NH^{(2)}$) which cannot occur in the spectrum because of the absence of a $NH^{(2)}$,$NH^{(3)}$ J coupling. However, following a ω_3 line parallel to the ω_3 axis through the ($H_\alpha^{(2)}$,$NH^{(3)}$,$NH^{(2)}$) peak location (line 4), one finds finally the ($H_\alpha^{(2)}$,$NH^{(3)}$,$H_\alpha^{(3)}$) cross peak. Thus one step in the sequential assignment is completed and one can continue analogously.

This type of analysis is not restricted to NH and H_α resonances but β and γ resonances can also be taken into account. In a NOESY–TOCSY experiment where often additional resonances within intraresidual spin systems are correlated in the TOCSY part, the assignment strategy of Table 4 may be employed. It has been employed in the sequence analysis of buserilin to establish the sequence of the three residues Ser–Tyr–D-Ser. This procedure is represented in Fig. 19 by four sections through the 3D spectrum alternating between (ω_1,ω_2) slices and (ω_2,ω_3) slices. The two regions of the 3D spectrum that are covered by the four sections are shown schematically in (e). Arrows indicate the logical path along which the sequence analysis proceeds, in which the transition from one section to the next occurs at points indicated with equal symbols. We suppose that the NH, H_α, and H_β resonances of Ser have been assigned. These resonances fix the vertical lines through the open circles in Fig. 19a at coordinates $(\omega_2,\omega_3) = (\Omega_{S_\alpha}, \Omega_{S_\beta})$ and $(\omega_2,\omega_3) = (\Omega_{S_\alpha}, \Omega_{S_{NH}})$. Along these lines, we find the two cross peaks (Y_{NH},S_α,S_{NH}) and (Y_{NH},S_α,S_β). They define a horizontal line parallel to ω_3: $(\omega_1,\omega_2) = (\Omega_{Y_{NH}}, \Omega_{S_\alpha})$ indicated by a diamond sign on the front of the 3D subvolume. There is a related line starting at the diamond sign in Fig. 19b, also parallel to ω_3, but with reversed coordinates $(\omega_1,\omega_2) = (\Omega_{S_\alpha}, \Omega_{Y_{NH}})$. Following this line, the two cross peaks (S_α,Y_{NH},Y_α) and (S_α,Y_{NH},Y_β) are encountered, which define two new lines parallel to ω_1: $(\omega_2,\omega_3) = (\Omega_{Y_{NH}}, \Omega_{Y_\alpha})$ and $(\omega_2,\omega_3) = (\Omega_{Y_{NH}}, \Omega_{Y_\beta})$ (indicated by squares on the front of Fig. 19b). From those lines the related lines parallel to ω_1 at $(\omega_2,\omega_3) = (\Omega_{Y_\alpha}, \Omega_{Y_{NH}})$ and $(\omega_2,\omega_3) = (\Omega_{Y_\alpha}, \Omega_{Y_\beta})$ are derived (squares on top of the 3D subvolume in Fig. 19c) and one step of the sequence analysis establishing the S–

FIG. 17. The 300 MHz 2D soft-COSY (*56*) spectrum and 3D hetero-COSY–COSY of buserilin acquired with the pulse sequence of Fig. 6f correlating $^{13}C_\alpha$, H_α, and NH in ω_1, ω_2, and ω_3, respectively. Prior to the actual pulse sequence, selective inversion (BIRD sequence (*67*)) of protons not attached to ^{13}C, followed by a 320 ms delay, served to zero the magnetization of the NH resonances; 96×57 t_1, t_2 experiments, each with 32 scans and 2K data acquisition in t_3, were performed. The delays τ were tuned to $^1J_{CH} = 143$ Hz. The duration of one scan was 2.4 s. Selective excitation of the H_α resonances was achieved with a 6.5 ms Gaussian pulse. The data matrix was zero-filled to $256 \times 128 \times 4K$ prior to Fourier transformation. The interesting NH region in ω_3 was cut out in order to accommodate the data on disk. Slices through the spectrum taken at positions in ω_1 indicated by arrows in the 2D hetero-COSY spectrum are shown. The effective spectral separation of peaks in the 3D spectrum is reflected in the (H_α,NH)-COSY sections, each showing only one cross peak. Assignments are given with the three-letter codes for amino acids. The improved separation of the peaks is obvious upon comparison with the inserted soft-COSY spectrum which corresponds to superposition of the seven 2D sections from the 3D spectrum.

TABLE 3

Assignment Procedures Based on Nonselective 3D Experiments[a]

	$NH^{(i)}$–$H_\alpha^{(i)}$ ...		$NH^{(i+1)}$–$H_\alpha^{(i+1)}$ ...		$NH^{(i+2)}$–$H_\alpha^{(i+2)}$	
β sheet		1	2	3		
			3	2	1	
				1	2	3
Helix	1		2	3		
			2	3	1	
			1		2	3
	$NH^{(i)}$–$^{15}N^{(i)}$ ...		$NH^{(i+1)}$–$^{15}N^{(i+1)}$ ...		$NH^{(i+2)}$–$^{15}N^{(i+2)}$	
Helix	1		2	3		
			2	3	1	
			1		2	3

[a] The numbers 1, 2, 3 indicate the frequency coordinates ω_1, ω_2, ω_3 of a 3D cross peak.

Y connectivity is completed. Following the same procedure, the next amino acid D-Ser is linked to the S–Y dipeptide fragment in Figs. 19c and 19d. It is likely that this approach with some built-in redundancy can be extended to quite large molecules.

For helical parts of a protein where NH,NH distances of neighboring amino acids are short (*8*), sequence analysis can be based on NH,NH-NOESY cross peaks as indicated in Table 3. Coupling partners for the coherent-transfer step may be H_α or side-chain protons. Particularly helpful, when available, are ^{15}N nuclei in isotope-labeled compounds where the spectral range for protons may be restricted because only NH protons are employed as mentioned in the previous subsection. The corresponding sequence of cross peaks used in the assignment procedure is shown in Table 3.

(2) Assignments based on selective experiments. In principle, it is possible to develop assignment strategies for β sheets (as well as for α helices) that require only one of the two informative 3D spectral regions (NH,H_α,NH) and (H_α,NH,H_α) which can be recorded by a selective 3D experiment. Obviously, based on two selective experiments, the strategy described above could be used. The modified strategies using only one selective experiment are summarized in Table 5.

The procedure based on proton resonances alone can be explained by the schematic 3D spectrum of a tripeptide fragment with the six resonances $NH^{(1)}$, $H_\alpha^{(1)}$, $NH^{(2)}$, $H_\alpha^{(2)}$, $NH^{(3)}$, and $H_\alpha^{(3)}$, shown in Fig. 20. The sequence analysis starts from the $(NH^{(1)},H_\alpha^{(1)},NH^{(1)})$ back-transfer peak. Following the ω_1 line at $(\omega_2,\omega_3) = (\Omega_{\alpha^{(1)}},\Omega_{NH^{(1)}})$, the $(NH^{(2)},H_\alpha^{(1)},NH^{(1)})$ cross peak establishing the residue (1)–residue (2) connectivity is found. The next cross peak $(NH^{(3)},H_\alpha^{(2)},NH^{(2)})$, establishing the residue (2)–residue (3) connectivity, has the $NH^{(2)}$ chemical shift in common with the $(NH^{(2)},H_\alpha^{(1)},NH^{(1)})$ cross peak. This cross peak can be found by proceeding parallel to ω_3 to the back-transfer plane intersected at $(\Omega_{NH^{(2)}},\Omega_{\alpha^{(1)}},\Omega_{NH^{(2)}})$. Advanc-

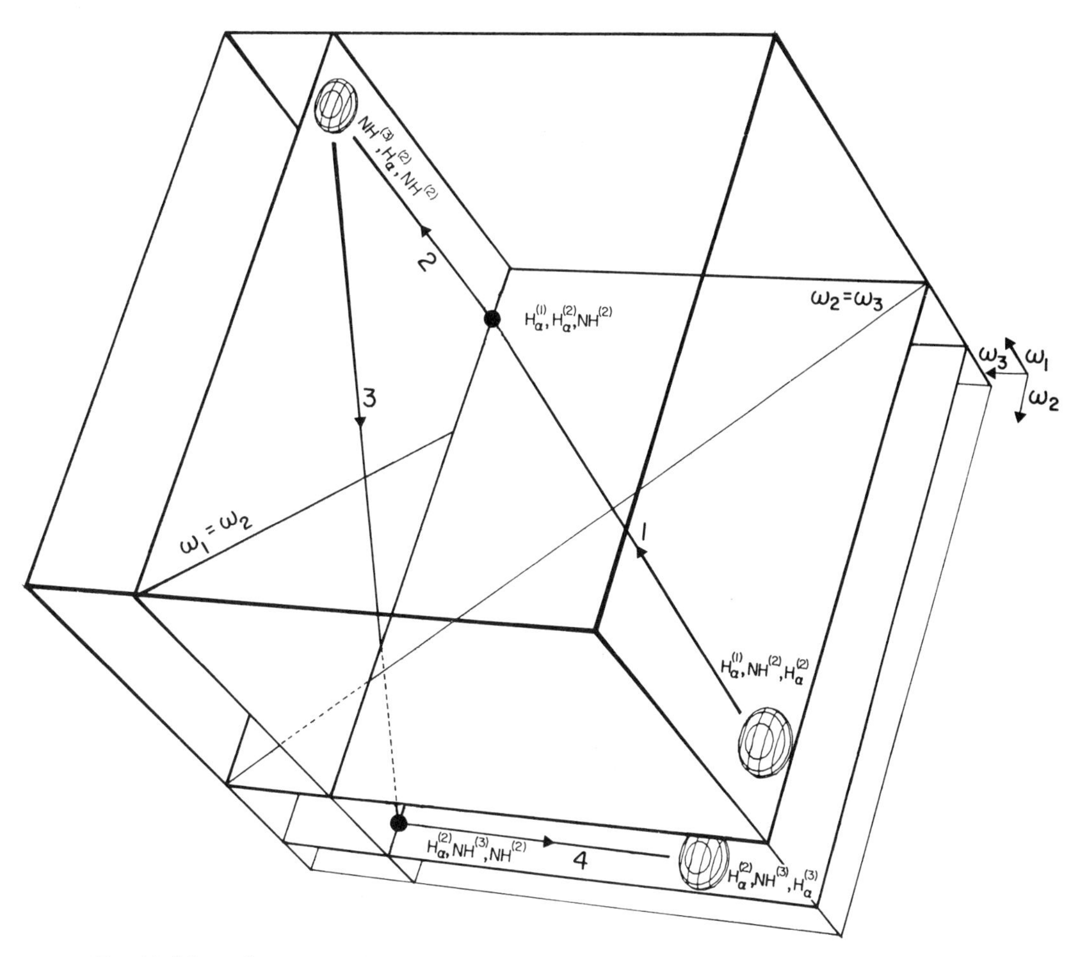

FIG. 18. Schematic representation of the sequence analysis in a NOESY–TOCSY spectrum. Two (ω_2,ω_3) planes intersected by a (ω_1,ω_2) plane are shown. Starting from the $(H_\alpha^{(1)},NH^{(2)},H_\alpha^{(2)})$ cross peak, the point $(H_\alpha^{(1)},H_\alpha^{(2)},NH^{(2)})$ is reached by reflection at the $(\omega_2 = \omega_3)$ diagonal (1). Following a line along ω_1 (2) the $(NH^{(3)},H_\alpha^{(2)},NH^{(2)})$ cross peak is found. By reflection at the $(\omega_1 = \omega_2)$ line in the (ω_1,ω_2) section, the latter peak position is transformed into the point $(H_\alpha^{(2)},NH^{(3)},NH^{(2)})$ from which the cross peak $(H_\alpha^{(2)},NH^{(3)},H_\alpha^{(3)})$ is found by a search parallel to ω_3.

ing then parallel to ω_2, the back-transfer peak $(NH^{(2)},H_\alpha^{(2)},NH^{(2)})$ is encountered. From there the $(NH^{(3)},H_\alpha^{(2)},NH^{(3)})$ cross peak is found by moving again parallel to ω_1.

A stepwise procedure of this kind allows a complete sequence analysis based on the (NH,H_α,NH) region alone. The procedure, however, hinges on nondegenerate NH resonances necessary to distinguish different back-transfer peaks. Back-transfer peaks are determined by two frequencies and inherently have lower resolution as they are all located within a single plane while cross peaks are dispersed in three dimensions. In full analogy, the (H_α,NH,H_α) region could also be used, where a corresponding requirement demands nondegenerate H_α resonances. Clearly, the combination

TABLE 4

Assignment Strategy Including H_β Protons Exemplified by a Ser–Tyr–D-Ser Fragment Illustrated in Fig. 19

	S			Y			$\underline{S}$			
	$NH^{(i)}$–	$H_\alpha^{(i)}$–	$H_\beta^{(i)}\ldots$	$NH^{(i+1)}$–	$H_\alpha^{(i+1)}$–	$H_\beta^{(i+1)}\ldots$	$NH^{(i+2)}$–	$H_\alpha^{(i+2)}$–	$H_\beta^{(i+2)}\ldots$	
Step (1)	3	2		1						(Fig. 19a)
		2	3	1						
Step (2)		1		2	3					(Fig. 19b)
		1		2		3				
Step (3)				3	2		1			(Fig. 19c)
					2	3	1			
Step (4)					1		2	3		(Fig. 19d)
					1		2		3	

of both subdomains eases the accidental degeneracy problem as described in subsection (1).

Two steps of the procedure of Table 5 are illustrated for the NH–H_α–NH subvolume of the NOESY–TOCSY spectrum of buserilin in Fig. 21 establishing the segments of the sequence W–S–Y and S–Y–$\underline{S}$. A pair of (ω_1,ω_3) slices connected by an (ω_2,ω_3) slice according to the schematic Fig. 20 is shown in Figs. 21a and 21b. The peaks are assigned in such a way that the first letter refers to the NH active in ω_1 and the second to the H_α and NH active in ω_2 and ω_3, respectively. In Fig. 21a the (W_{NH},W_α,W_{NH}) back-transfer peak and the (S_{NH},W_α,W_{NH}) cross peak in the lower (ω_1,ω_3) plane have the same (ω_2,ω_3) coordinates and are easily identified to belong together. A (ω_2,ω_3) section through the (S_{NH},W_α,W_{NH}) cross peak shows the back-

FIG. 19. Sequence analysis of the Ser–Tyr–D-Ser tripeptide fragment of buserilin in a 3D NOESY–TOCSY spectrum similar to the 3D ROESY–TOCSY shown in Fig. 12. The data matrix of $128 \times 128 \times 2K$ was Fourier-transformed to give $128 \times 128 \times 1024$ real points. Sixteen scans were performed for each combination of t_1, t_2 values. The NOESY mixing time was 250 ms, and the TOCSY mixing achieved with MELV-17 (*65, 66*) and $\gamma B_1/2\pi = 4.2$ kHz lasting for 93.7 ms. The duration of one scan was 2.75 s. The four 3D subvolumes represent consecutive steps in the sequence analysis described in detail in the text. They are taken from two regions in the 3D spectrum shown in the schematic (e). (a) The two open circles define frequency pairs $(\omega_2 = \Omega_{S_\alpha},\omega_3 = \Omega_{S_\beta})$ and $(\omega_2 = \Omega_{S_\alpha},\omega_3 = \Omega_{S_{NH}})$ that are known from the previous assignment step. Following the lines parallel to ω_1 through the open circles, the Y_{NH},S_α,S_{NH} and Y_{NH},S_α,S_β cross peaks are found which define the coordinate pair $(\omega_1 = \Omega_{Y_{NH}},\omega_2 = \Omega_{S_\alpha})$ represented by a diamond symbol. (b) The two frequencies determined above define a new search line $(\omega_1 = \Omega_{S_\alpha},\omega_2 = \Omega_{Y_{NH}})$ marked with a diamond symbol for the next sequencing step. The two cross peaks $(S_\alpha,Y_{NH},Y_\alpha)$ and $(S_\alpha,Y_{NH},Y_\beta)$ are found along this line. They define new frequencies $\Omega_{Y_{NH}}$, Ω_{Y_α}, and Ω_{Y_β} leading to the two lines indicated by square symbols. They serve as new starting points in (c) in analogy to the open circle in (a). The two sequencing steps in (a) and (b) establish the S–Y connectivity. (c, d) The two diagrams establish the Y–$\underline{S}$ connectivity in analogy to the previous two steps. The robustness of the procedure against degeneracy of reasonances can be appreciated by comparing (a) and (d) where S_α and $\underline{S}_\alpha$ chemical shifts are almost identical. Nevertheless the relevant cross peaks can easily be distinguished on the basis of their different NH chemical shifts.

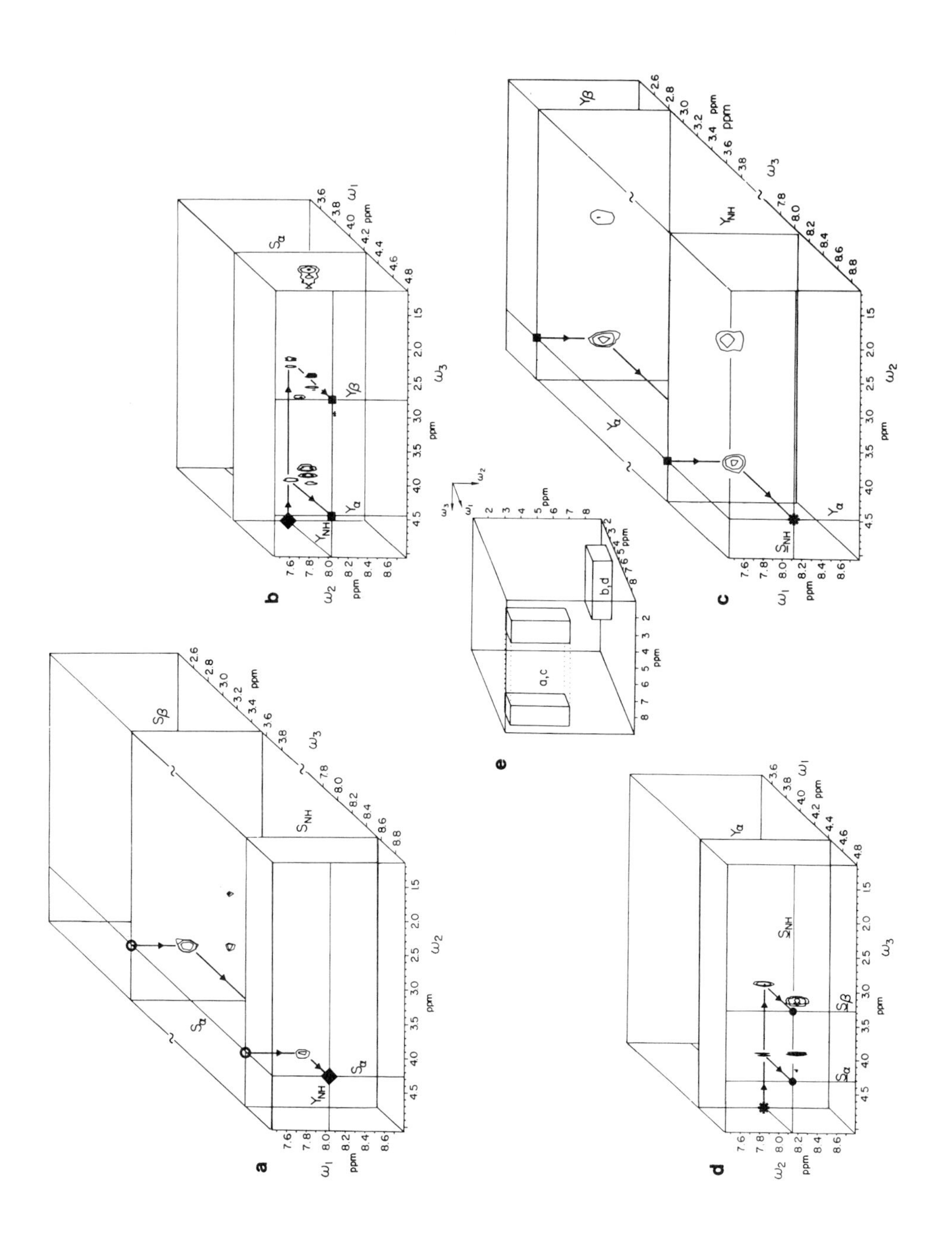
a
b
c
d
e
a,c
b,d
ppm

TABLE 5

Assignment Strategies Based on Selective 3D Experiments

	$^{15}N^{(i-1)}$–	$NH^{(i-1)}$–	$H_\alpha^{(i-1)}$...	$^{15}N^{(i)}$–	$NH^{(i)}$–	$H_\alpha^{(i)}$...	$^{15}N^{(i+1)}$–	$NH^{(i+1)}$–	$H_\alpha^{(i+1)}$...
NH,H_α,NH region of NOESY–TOCSY		3	2		1				
					1, 3	2			
					3	2		1	
H_α,NH,H_α region of NOESY–TOCSY			1		2	3			
					2	1, 3			
						1		2	3
a	1	2	3						
b			3	1	2				
a				1	2	3			
b						3	1	2	

[a] Intraresidual cross peak in ^{15}N,H–COSY–TOCSY or ^{15}N,H–COSY–COSY.
[b] Interresidual cross peak in ^{15}N,H–COSY–NOESY.

transfer peak (S_{NH},S_α,S_{NH}) which is found as indicated by a search along a line parallel to ω_2 within the back-transfer plane which is indicated only by dotted lines in the (ω_1,ω_3) planes. Taking a (ω_1,ω_3) section through the serine back-transfer peak now allows one to find the (Y_{NH},S_α,S_{NH}) peak by a search along ω_1. Note that the other cross peaks in this plane do not disturb the identification of the (Y_{NH},S_α,S_{NH}) cross peak. The sequence segment W–S–Y has then been established. In Fig. 21b the sequence assignment is continued for one further amino acid residue. Starting from the (Y_{NH},S_α,S_{NH}) cross peak we find the (Y_{NH},Y_α,Y_{NH}) back-transfer peak by a linear search in the back-transfer plane. From there the ($\underline{S}_{NH}$,Y_α,Y_{NH}) cross peak is found which establishes the S–Y–$\underline{S}$ connectivity.

A similar type of assignment could be based on a ^{15}N,H–COSY–NOESY spectrum combined with a ^{15}N,H–COSY–TOCSY (or COSY–COSY) spectrum as indicated in Table 5. This procedure hinges again on nondegenerate H_α resonances.

(e) Quantification of 3D Spectra

While for assignment purposes a qualitative interpretation of 3D spectra is sufficient, quantification of peak intensities and of multiplet splittings is needed for the determination of cross-relaxation rates and coupling constants, respectively.

(i) Cross-relaxation rates. Cross-peak intensities in 2D NOESY or ROESY spectra, measured as peak integrals, are a direct measure for cross-relaxation efficiency. In a 3D spectrum containing in addition a cross-relaxation transfer step, however, the peak intensities are also affected by the second (usually coherent) transfer process. As long as the cross-relaxation rate is determined from a full build-up curve, deduced from several 3D experiments with different mixing times, the results remain unaffected by the second transfer. However, it will often be necessary to obtain an estimate from a single 3D spectrum for reasons of performance time. Then special precautions must be taken. In favorable situations, the second transfer process intro-

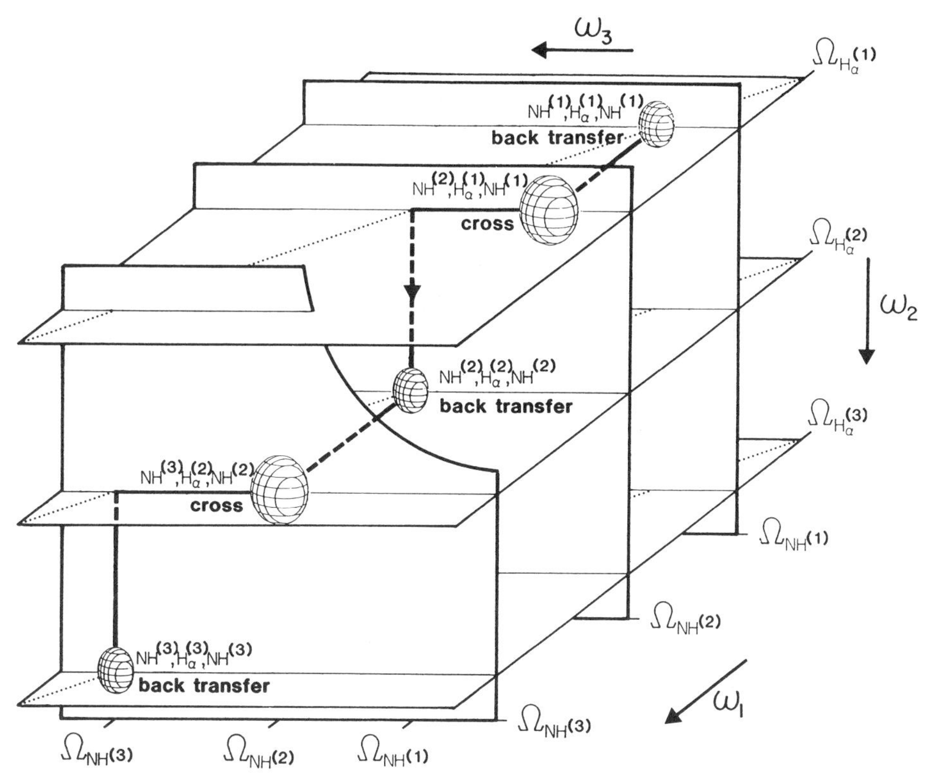

FIG. 20. Schematic representation of the sequential assignment procedure in a selective 3D NOESY–TOCSY experiment as described in the text. The procedure starts with back-transfer peak ($NH^{(1)},H_\alpha,NH^{(1)}$). A linear search parallel to ω_1 leads to the cross peak $NH^{(2)}H_\alpha^{(1)}NH^{(1)}$. In a second step, a linear search at coordinate $\omega_1 = \Omega_{NH^{(2)}}$ within the back-transfer plane which is indicated by dotted lines in the (ω_1,ω_3) planes allows one to reach the next back-transfer peak ($NH^{(2)},H_\alpha^{(2)},NH^{(2)}$). This procedure can be applied repetitively as explained in the text.

duces uniform scaling factors for all peaks in the spectrum. An example of this case is provided by a coherent heteronuclear transfer step through one-bond couplings as in hetero-COSY–NOESY. Their magnitude normally varies little and the transfer efficiency will be nearly constant. Furthermore, because of the large couplings, the two transfer periods $\tau = (2J_{HX})^{-1}$ will be short and relaxation losses small. For example, with an assumed heteronuclear coupling constant $J = 90$ Hz and a linewidth varying between 1 and 16 Hz, the relaxation loss will result in a maximum variation of the apparent internuclear distance of ±5% when the delay τ is set to $(2J_{XH})^{-1}$.

(ii) Spin–spin coupling constants. The determination of accurate values of J from 3D spectra is feasible only in exceptional cases as high resolution is needed at least in two dimensions while the third one with reduced resolution can be used for separating overlapping multiplet patterns. For the two high-resolution domains ω_2 and ω_3, it is convenient to employ an E.COSY-type experiment (*43, 62, 63*) or soft COSY experiment (*56, 57*), while the first transfer step should produce in-phase magnetization transfer as in TOCSY or in refocused hetero-COSY.

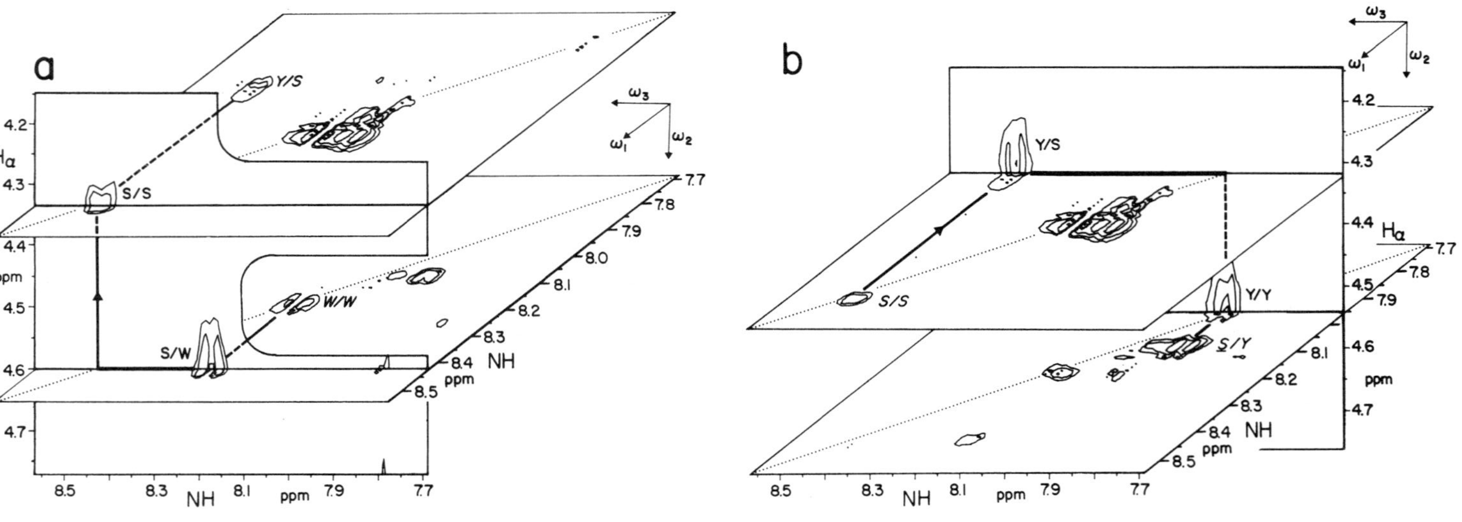

FIG. 21. Sequence analysis of buserilin based on the NH–H_α–NH region of a NOESY–TOCSY experiment; (a) establishes connectivity of the segment Trp–Ser–Tyr (W–S–Y) and (b) establishes connectivity of the segment Ser–Tyr–D–Ser (S–Y–S). In the assignment of the peaks, the first letter refers to NH active in ω_1 and the second letter to H_α and NH active in ω_2 and ω_3, respectively. Details of the sequencing procedure are given in the text.

(f) Visual Representation of 3D Spectra

Even based on the present limited practical experience with 3D spectroscopy, it has become evident that the visual representation of 3D spectra is crucial for its practical success. Obviously 2D sections cut at various angles through 3D spectra and represented by contour plots are indispensable tools for working with 3D spectra. However, they are insufficient for visualizing more complex three-dimensional relationships between peaks, for example, for identifying symmetry-related peak pairs.

The use of sophisticated display equipment that allows 3D representation and manipulation of data, such as Evans–Sutherland or Silicon Graphics displays, is indispensable for speedy work as well as for gaining new insights and for stimulating intuition. An ideal display system should have the entire 3D data matrix at instantaneous disposal, allow the display of the entire data matrix with high resolution, and permit the rotation of the 3D spectrum as well as the blowup of details. Of great utility is a cursor that can be guided in three dimensions through the spectrum, whose coordinates are constantly displayed, and that can be used to measure distances in 3D frequency space and to define subvolumes for data manipulation.

In the longer run, it is also desirable to have computer procedures available that allow pattern recognition in 3D data space and relieve the spectroscopist from some of the time-consuming pedestrian work in three dimensions.

To give an impression of what is now becoming routine in 3D spectroscopy, Fig. 22a shows the stereographic view of the C_α,H_α,NH region of the C,H–COSY–COSY spectrum of buserilin discussed in Section IX (c). An assignment of the cross peaks is given below in a schematic figure. Figure 22b shows the NH,H_α,NH subvolume of a NOESY–COSY spectrum of buserilin which is similar to the NOESY–TOCSY spectrum of Fig. 16. Again, peak assignments are given in a schematic representation of the left picture.

X. THE POTENTIAL OF 3D SPECTROSCOPY

Three-dimensional spectroscopy is just at its emergence and its success in practice cannot yet be appraised with certainty although several advantages (and limitations) are already evident.

The initial motivation to develop 3D spectroscopy was its use to unravel complex 2D spectra and to separate overlapping cross peaks by introducing a third frequency dimension. The exploitation of the large chemical-shift range of "heteronuclei," such as ^{13}C and ^{15}N, for spreading crowded proton spectra springs immediately to one's mind. The increasing availability of isotope-enriched macromolecules by biological methods removes the severe sensitivity problems inherent in experiments on natural abundance samples. Also proton shifts can be useful for the spread in a third dimension. In most cases a coherent-transfer process is used for this purpose, for small molecules usually of COSY-type, while for large molecules a TOCSY transfer is preferentially utilized. It can be applied to all conceivable types of 2D experiments, in particular to COSY, NOESY, and ROESY. These 3D experiments not only disentangle peaks overlapping in 2D NMR spectra but also allow improved assignment strategies that are less susceptible to degeneracies of chemical shifts. It may well turn out that the correlation with a heteronuclear chemical shift will turn out to be the most useful feature of 3D spectroscopy.

An advantage of 3D spectra that should not be underestimated is the possibility to combine two different 2D experiments into one single experiment. One single cross peak contains now the information of two peaks taken from two 2D spectra. The most interesting experiment in this respect is the combination of a coherent- and an incoherent-transfer step. In particular NOESY–COSY, NOESY–TOCSY, and ROESY–TOCSY combinations contain all the information necessary for a complete assignment of protein and nucleic acid proton resonance spectra.

It should, however, not be concealed that the combination of two experiments where the peak intensities reflect the product of two transfer coefficients does not facilitate their separate measurement. This is particularly troublesome for the accurate determination of cross-relaxation rates from 3D spectra with NOESY or ROESY as one constituent. Combination with an experiment like COSY or TOCSY, where the transfer efficiency of the coherence-transfer step depends on the coupling constants and, for COSY, also on spectral resolution, will be difficult to analyze quantitatively. The quantitative analysis of combinations of hetero-COSY with an exchange process is easier, as has been described in Section IX (d). In many situations, however, qualitative information on cross-relaxation rates is sufficient to establish constraints for internuclear distances that allow a quantitative determination of molecular geometry (*8*).

The choice between selective and nonselective 3D experiments depends greatly on the available previous knowledge. When the expected cross peaks can be confined to a reduced volume, it is of advantage to record frequency-selective 3D spectra in order to maximize resolution. Using present day equipment, the resolution of nonselective 3D experiments is necessarily low. This disadvantage is compensated by the fact that analysis of nonselective 3D spectra is less prone to erratic assignments as numerous cross-checks, comparing different spectral regions, are feasible.

It is obvious that so far only the surface of 3D spectroscopy has been scratched. Many further possibilities for the design of interesting 3D experiments exist, not only in liquid-state NMR but also in application to liquid crystals and solids. Applications to other fields of spectroscopy are also conceivable.

APPENDIX. STOCHASTIC MULTIDIMENSIONAL NMR

Stochastic NMR spectroscopy (*33*) provides alternatives to pulsed multidimensional NMR spectroscopy (*5–8*). It is the purpose of this appendix to work out the equivalence of and differences between pulsed and stochastic multidimensional NMR. The main result will be that the information present in 3D transfer spectra as described in this paper is available only from 3D sections cut through 5D or even 7D stochastic spectra while it is absent in straight 3D stochastic spectra.

Following references (*33*) and adapting the notation to that used in this paper, the n-dimensional stochastic NMR spectrum has the form

$$S(\omega_1', \omega_2', \ldots, \omega_n') = \sum_{\{rs\}_1 \cdots \{rs\}_n} D_{\{rs\}_n} \left[\prod_{j=2}^{n} \frac{1}{i(\omega_j' - \omega_{\{rs\}_j}) - \lambda_{\{rs\}_j}} (\hat{\hat{F}}_x)_{\{rs\}_j \{rs\}_{j-1}} \right] \times \frac{1}{i(\omega_1' - \omega_{\{rs\}_1}) - \lambda_{\{rs\}_1}} P_{\{rs\}_1},$$

where $\{rs\}_j$ is an eigenmode leading to a peak in frequency domain ω'_j with the resonance frequency $\omega_{\{rs\}_j}$ and the half-width $\lambda_{\{rs\}_j}$. D is the observable operator as in Section IV, and the initial state operator P is defined by

$$P_{\{rs\}_1} = \sum_q (\hat{\hat{F}}_x)_{\{rs\}_1\{qq\}}(\sigma_0)_{qq}.$$

$\hat{\hat{F}}_x$ is the commutator superoperator associated with the operator F_x. The primed frequencies refer to the fact that $S(\omega'_1, \omega'_2, \ldots, \omega'_n)$ is the result of shearing of the original spectrum $S(\omega_1, \omega_2, \ldots, \omega_n)$ according to $\omega'_j = \sum_{i=1}^{j} \omega_i$. Using $\hat{\hat{F}}_x = (e^{i\beta\hat{\hat{F}}_x} - 1)/(i\beta)$ for $\beta \rightarrow 0$, this yields the equivalence of an n-dimensional stochastic NMR spectrum and an n-dimensional pulsed NMR spectrum which employs for every mixing sequence $R^{(i)}$ a β_x pulse with $\beta \rightarrow 0$, imposing in addition the selection rule $\Delta p = \pm 1$ (*33*). Obviously, the equivalent number of pulses in the pulsed analog of an n-dimensional stochastic NMR spectrum is exactly n. However, since the equivalent mixing sequence is restricted to small-flip-angle pulses $\beta \rightarrow 0$ with $\Delta p = \pm 1$ (which we henceforth call "stochastic mixing"), possibly several of these stochastic mixing processes are needed to achieve the same result as that with a single $\pi/2$ pulse. We deduce in the following the number of stochastic mixing processes necessary to achieve a certain transformation. In order to make the argument clear, we use a slightly different description of the stochastic mixing process than above. Assuming N weakly coupled spins $I = \frac{1}{2}$, and using the identity $\hat{\hat{F}}_x = \sum_{k=1}^{N} \hat{\hat{I}}_{kx} = \frac{1}{2} \sum_{k=1}^{N} (e^{i\pi/2\hat{\hat{I}}_{kx}} - e^{-i(\pi/2)\hat{\hat{I}}_{kx}})$ "stochastic mixing" is equivalent to a set of N independent experiments in each of which a selective pulse $(\pi/2)_k$ is applied to a single spin I_k using for the change in coherence order the constraints that only transfers with $\Delta p = \pm 1$ are selected. An n-dimensional stochastic experiment, in consequence, corresponds to N^{n-1} independent selective pulse experiments, each with a different sequence of selective pulses.

We then obtain the results presented in Table 6. It shows which type of 2D spectra can be extracted from 3D stochastic NMR. For example, E.COSY cross peaks arise when zero- or double-quantum coherence is present between the virtual $(\pi/2)_k$ and $(\pi/2)_l$ pulses (first line in Table 6); their chemical shifts in ω_2 lie at the sum or difference frequencies of the two resonance lines that appear in ω_1 and ω_3, respectively. Thus a section taken at $\omega_2 = \omega_1 + \omega_3$ or $\omega_2 = \omega_1 - \omega_3$ yields E.COSY patterns (taking into account the restriction $\Delta p = \pm 1$).

NOE and exchange peaks at Ω_k in ω_1 and Ω_l in ω_3 connecting two spins I_k and I_l arise when longitudinal magnetization is present between the two $\pi/2$ pulses. Thus the NOESY spectrum is found in the (ω_1, ω_3) plane at $\omega_2 = 0$, and the shape of the cross peaks in ω_2 is the Fourier transform of their build-up curve. It should however be noted that each multiplet peak in the (ω_1, ω_3) plane has a different build-up curve much like in a small-flip-angle NOESY experiment (*64*). Zero-quantum contributions to the I_k, I_l peak at $\omega_1 = \Omega_k$, $\omega_3 = \Omega_l$, that are familiar artifacts in pulsed NOESY, are modulated with $\Omega_k - \Omega_l$ in ω_2 and are therefore identical with the E.COSY peaks at $\omega_2 = \omega_1 - \omega_3$ mentioned above.

Transfer of coherence from one spin to another one and to a third spin, such as in COSY–COSY, requires at least four stochastic mixing processes of the form

TABLE 6

Required Dimension of a Stochastic Spectrum in Order to Be Capable of Representing Various 2D Features

Transformation	Expressed in product operators	Equivalent pulse sequence employing "stochastic mixing"	Number of pulses = dimensions of equivalent stochastic spectrum
Coherence transfer (CT)	$I_{kx}I_{lz} \rightarrow -I_{kz}I_{lx}$	$(\pi/2)_k(\pi/2)_l$	2
CT with 2QF	$I_{kx}I_{lz} \rightarrow -I_{kz}I_{lx}$	$(\pi/2)_k(\pi/2)_l$	2
CT with 3QF	$I_{kx}I_{lz}I_{mz} \xrightarrow{\text{TQF}} I_{kz}I_{lx}I_{mz}$	$(\pi/2)_l(\pi/2)_m(\pi/2)_k(\pi/2)_m$	4
CT with nQF	$I_{kx}I_{lz}\prod_{m=1}^{n-2} I_{mz} \xrightarrow{\text{nQF}} I_{kz}I_{lx}\prod_{m=1}^{n-2} I_{mz}$	$(\pi/2)_l(\pi/2)_k \prod_{m=1}^{n-2} [(\pi/2)_m(\pi/2)_m]$	$2(n-1)$
Relayed CT	$I_{kx}I_{lz} \rightarrow I_{lz}I_{mx}$	$(\pi/2)_k(\pi/2)_l(\pi/2)_m(\pi/2)_l$	4
NOESY	$I_{kx} \rightarrow I_{lx}$	$(\pi/2)_k(\pi/2)_l$	2
TOCSY			∞
ROESY			∞
π inversion	$I_k^+ \rightarrow I_k^-$	$(\pi/2)_k(\pi/2)_k$	2

$(\pi/2)_i(\pi/2)_j(\pi/2)_j(\pi/2)_k$ to correlate spins i, j, and k. Such a cross peak can be found in suitable 3D sections through a 5D stochastic spectrum. For multidimensional correlation spectroscopy in its simplest form, the general rule is that pulsed n-dimensional spectroscopy is equivalent to stochastic $(2n - 1)$-dimensional spectroscopy. Inclusion of further pulses in 3D pulse sequences, such as π pulses for decoupling, further increases the dimensionality of the equivalent stochastic experiment. In general, the number of dimensions in a stochastic NMR experiment equivalent to a pulsed NMR experiment is equal to the number of selective $\pi/2$ pulses that would be needed to build up the pulsed experiment.

ACKNOWLEDGMENTS

This project has been supported in part by the Swiss National Science Foundation (Project 2.075-0.86). The loan of an ASPECT 3000 and an ASPECT 1000 computer by Spectrospin AG, Fällanden, is gratefully acknowledged. Discussions with Dr. A. M. Gronenborn, Dr. G. M. Clore, and Dr. H. Oschkinat, MPI Martinsried, have been helpful in an early phase of this project. Support from C. Cieslar and Dr. H. Oschkinat, providing a 3D FT package for a Convex computer and conversion software for the representation of 3D spectra with the program FRODO on an Evans and Sutherland picture system, is gratefully acknowledged. The authors are grateful to Dr. W. Bermel (Bruker Analytische Messtechnik, Karlsruhe), R. Brüschweiler, and Z. L. Mádi for support. The sample of buserilin, a gift from Hoechst AG, Frankfurt, was kindly provided by Prof. Dr. H. Kessler, Frankfurt. The manuscript has been processed by Mrs. I. Müller.

REFERENCES

1. A. C. Bloom and J. N. Shoolery, *Phys. Rev.* **97,** 1261 (1955).
2. R. Freeman and W. A. Anderson, *J. Chem. Phys.* **37,** 2053 (1962).

3. R. A. HOFFMAN AND S. FORSÉN, *Prog. NMR Spectrosc.* **1,** 15 (1966).
4. R. R. ERNST, *J. Chem. Phys.* **45,** 3845 (1966).
5. W. P. AUE, E. BARTHOLDI, AND R. R. ERNST, *J. Chem. Phys.* **64,** 2229 (1976).
6. R. R. ERNST, G. BODENHAUSEN, AND A. WOKAUN, "Principles of NMR in One and Two Dimensions," Clarendon, Oxford, 1987.
7. N. CHANDRAKUMAR AND S. SUBRAMANIAN, "Modern Techniques in High Resolution FT-NMR," Springer, New York, 1987.
8. K. WÜTHRICH, "NMR of Proteins and Nucleic Acids," Wiley–Interscience, New York, 1986.
9. A. D. COHEN, R. FREEMAN, K. A. MCLAUCHLAN, AND D. H. WHIFFEN, *Mol. Phys.* **7,** 45 (1963).
10. K. NAGAYAMA, *J. Chem. Phys.* **71,** 4404 (1979).
11. G. W. EICH, G. BODENHAUSEN, AND R. R. ERNST, *J. Am. Chem. Soc.* **104,** 3731 (1982).
12. A. M. THAYER, J. M. MILLER, AND A. PINES, *Chem. Phys. Lett.* **129,** 55 (1986).
13. R. KREIS, A. THOMAS, W. STUDER, AND R. R. ERNST, *J. Chem. Phys.* **89,** 6623 (1988).
14. H. D. PLANT, T. H. MARECI, M. D. COCKMAN, AND W. S. BREY, "27th Experimental NMR Conference, Baltimore, Maryland, April 13–17, 1986"; G. W. VUISTER AND R. BOELENS, *J. Magn. Reson.* **73,** 328 (1987).
15. C. GRIESINGER, O. W. SØRENSEN, AND R. R. ERNST, *J. Magn. Reson.* **73,** 574 (1987).
16. C. GRIESINGER, O. W. SØRENSEN, AND R. R. ERNST, *J. Am. Chem. Soc.* **109,** 7227 (1987).
17. H. OSCHKINAT, C. GRIESINGER, P. J. KRAULIS, O. W. SØRENSEN, R. R. ERNST, A. M. GRONENBORN, AND G. M. CLORE, *Nature (London)* **332,** 374 (1988).
18. L. MÜLLER, ANIL KUMAR, AND R. R. ERNST, *J. Chem. Phys.* **63,** 5490 (1975).
19. W. P. AUE, J. KARHAN, AND R. R. ERNST, *J. Chem. Phys.* **64,** 4226 (1976).
20. R. K. HESTER, J. L. ACKERMAN, B. L. NEFF, AND J. S. WAUGH, *Phys. Rev. Lett.* **36,** 1081 (1976).
21. A. A. MAUDSLEY AND R. R. ERNST, *Chem. Phys. Lett.* **50,** 368 (1977).
22. L. BRAUNSCHWEILER AND R. R. ERNST, *J. Magn. Reson.* **53,** 521 (1983).
23. L. BRAUNSCHWEILER, G. BODENHAUSEN, AND R. R. ERNST, *Mol. Phys.* **48,** 535 (1983).
24. J. JEENER, B. H. MEIER, P. BACHMANN, AND R. R. ERNST, *J. Chem. Phys.* **71,** 4546 (1979).
25. B. H. MEIER AND R. R. ERNST, *J. Am. Chem. Soc.* **101,** 6441 (1979).
26. S. MACURA AND R. R. ERNST, *Mol. Phys.* **41,** 95 (1980).
27. ANIL KUMAR, R. R. ERNST, AND K. WÜTHRICH, *Biochem. Biophys. Res. Commun.* **95,** 1 (1980).
28. A. A. BOTHNER-BY, R. L. STEPHENS, J. LEE, C. D. WARREN, AND R. W. JEANLOZ, *J. Am. Chem. Soc.* **106,** 811 (1984).
29. P. CARAVATTI, P. NEUENSCHWANDER, AND R. R. ERNST, *Macromolecules* **18,** 119 (1985).
30. T. NAKAI, J. ASHIDA, AND T. TERAO, *J. Chem. Phys.* **88,** 6049 (1988).
31. G. W. VUISTER, R. BOELENS, AND R. KAPTEIN, *J. Magn. Reson.* **80,** 176 (1988).
32. S. DAVIES, J. FRIEDRICH, AND R. FREEMAN, *J. Magn. Reson.* **76,** 555 (1988).
33. B. BLÜMICH AND D. ZIESSOW, *J. Chem. Phys.* **78,** 1059 (1983); B. BLÜMICH, *Prog. NMR Spectrosc.* **19,** 331 (1987).
34. C. GRIESINGER, C. GEMPERLE, O. W. SØRENSEN, AND R. R. ERNST, *Mol. Phys.* **62,** 295 (1987).
35. G. BODENHAUSEN, H. KOGLER, AND R. R. ERNST, *J. Magn. Reson.* **58,** 370 (1984).
36. U. PIANTINI, O. W. SØRENSEN, AND R. R. ERNST, *J. Am. Chem. Soc.* **104,** 6800 (1982).
37. G. DROBNY, A. PINES, S. SINTON, D. WEITEKAMP, AND D. WEMMER, *Faraday Disc. Chem. Soc. Symp.* **13,** 49 (1979).
38. G. BODENHAUSEN, R. L. VOLD, AND R. R. VOLD, *J. Magn. Reson.* **37,** 93 (1980).
39. P. BACHMANN, W. P. AUE, L. MÜLLER, AND R. R. ERNST, *J. Magn. Reson.* **28,** 29 (1977).
40. D. J. STATES, R. A. HABERKORN, AND D. J. RUBEN, *J. Magn. Reson.* **48,** 286 (1982).
41. A. BAX AND R. FREEMAN, *J. Magn. Reson.* **44,** 542 (1981).
42. M. RANCE, G. WAGNER, O. W. SØRENSEN, K. WÜTHRICH, AND R. R. ERNST, *J. Magn. Reson.* **59,** 250 (1984).
43. C. GRIESINGER, O. W. SØRENSEN, AND R. R. ERNST, *J. Chem. Phys.* **85,** 6837 (1986).
44. C. GRIESINGER AND R. R. ERNST, *J. Magn. Reson.* **75,** 261 (1987).
45. O. W. SØRENSEN, G. W. EICH, M. H. LEVITT, G. BODENHAUSEN, AND R. R. ERNST, *Prog. NMR Spectrosc.* **16,** 163 (1983).
46. L. MÜLLER, *J. Am. Chem. Soc.* **101,** 4481 (1979).
47. M. R. BENDALL, D. T. PEGG, AND D. M. DODDRELL, *J. Magn. Reson.* **52,** 81 (1983).

48. A. BAX AND D. MARION, *J. Magn. Reson.* **78,** 186 (1988).
49. D. T. PEGG, D. M. DODDRELL, AND M. R. BENDALL, *J. Chem. Phys.* 77, 2745 (1982).
50. H. BILDSØE, S. DØNSTRUP, H. J. JAKOBSEN, AND O. W. SØRENSEN, *J. Magn. Reson.* **53,** 154 (1983).
51. O. W. SØRENSEN, S. DØNSTRUP, H. BILDSØE, AND H. J. JAKOBSEN, *J. Magn. Reson.* **55,** 347 (1983).
52. M. RANCE, G. BODENHAUSEN, G. WAGNER, K. WÜTHRICH, AND R. R. ERNST, *J. Magn. Reson.* **62,** 497 (1985).
53. G. WAGNER, *J. Magn. Reson.* **57,** 497 (1984).
54. O. W. SØRENSEN, C. GRIESINGER, AND R. R. ERNST, *Chem. Phys. Lett.* **135,** 313 (1987).
55. M. H. LEVITT, G. BODENHAUSEN, AND R. R. ERNST, *J. Magn. Reson.* **58,** 462 (1984).
56. R. BRÜSCHWEILER, J. C. MADSEN, C. GRIESINGER, O. W. SØRENSEN, AND R. R. ERNST, *J. Magn. Reson.* **73,** 380 (1987).
57. R. BRÜSCHWEILER, C. GRIESINGER, O. W. SØRENSEN, AND R. R. ERNST, *J. Magn. Reson.* **78,** 178 (1988).
58. C. J. BAUER, R. FREEMAN, T. FRENKIEL, J. KEELER, AND A. J. SHAKA, *J. Magn. Reson.* **58,** 442 (1985).
59. H. KESSLER, H. OSCHKINAT, C. GRIESINGER, AND W. BERMEL, *J. Magn. Reson.* **70,** 106 (1986).
60. W. S. WARREN, *J. Chem. Phys.* **81,** 5437 (1984).
61. F. LOAIZA, M. MCCOY, S. L. HAMMER, AND W. S. WARREN, *J. Magn. Reson.* **77,** 175 (1988); L. EMSLEY AND G. BODENHAUSEN, *J. Magn. Reson.* **82,** 211 (1989).
62. C. GRIESINGER, O. W. SØRENSEN, AND R. R. ERNST, *J. Am. Chem. Soc.* **107,** 6394 (1985).
63. C. GRIESINGER, O. W. SØRENSEN, AND R. R. ERNST, *J. Magn. Reson.* **75,** 474 (1987).
64. H. OSCHKINAT, A. PASTORE, AND G. BODENHAUSEN, *J. Am. Chem. Soc.* **109,** 4110 (1987).
65. A. BAX AND D. G. DAVIS, *J. Magn. Reson.* **65,** 355 (1985).
66. C. GRIESINGER, G. OTTING, K. WÜTHRICH, AND R. R. ERNST, *J. Am. Chem. Soc.* **110,** 7870 (1988).
67. J. R. GARBOW, D. P. WEITEKAMP, AND A. PINES, *Chem. Phys. Lett.* **93,** 505 (1982).

JOURNAL OF MAGNETIC RESONANCE **84**, 72–84 (1989)

Practical Aspects of 3D Heteronuclear NMR of Proteins

LEWIS E. KAY, DOMINIQUE MARION,* AND AD BAX

Laboratory of Chemical Physics, NIDDK, National Institutes of Health, Bethesda, Maryland 20892

Received November 22, 1988

Practical aspects regarding the acquisition and processing of 3D heteronuclear data sets are discussed, with particular emphasis on the 3D NOESY-HMQC experiment which combines the 2D NOE and the heteronuclear multiple-quantum coherence (HMQC) experiments. Slices through the 3D spectrum are equivalent to ^{15}N-filtered 2D NOESY spectra and exhibit sensitivity similar to that obtained in regular 2D NOE experiments. We discuss experimental procedures for obtaining maximum resolution with a relatively small number of t_1 and t_2 increments. In addition, a simple and efficient procedure for performing the third Fourier transformation that permits use of standard 2D software for processing in the other dimensions is described. Other important aspects of the data processing concern optimal digital filtering, noninteractive phasing, and minimization of the space needed for the processed data. © 1989 Academic Press, Inc.

Recently a number of groups have reported the use of three-dimensional NMR techniques for the simplification of complex spectra (*1–8*). For example, homonuclear (^{1}H) 3D techniques, combining *J* connectivity and NOE information, have been applied to a number of protein systems, clearly demonstrating the applicability of such techniques (*5, 7*). However, there is a fundamental limitation associated with the application of homonuclear techniques to the study of increasingly larger proteins: Proton magnetization decays rapidly in the time required for the scalar transfer of magnetization between spins, necessary for the labeling of the magnetization. In this regard, heteronuclear 3D techniques can be much more sensitive (provided that isotopic enrichment is used) since the transfer of magnetization between the heteronucleus and a directly coupled proton proceeds via a scalar coupling which is much larger than typical ^{1}H linewidths (*6, 8*). Also, the number of resonances in the heteronuclear 3D experiment is similar to that for two-dimensional NMR spectra; in contrast, the number of resonances in homonuclear 3D spectra is much larger. The relatively high sensitivity and the reduction in spectral overlap make the heteronuclear 3D experiment ideally suited for the study of proteins that are too large for detailed studies with the conventional 2D approach. Heteronuclear 3D experiments that separate conventional 2D spectra such as NOESY (*9, 10*), TOCSY/HOHAHA (*11, 12*), or COSY (*13–15*) according to the ^{15}N chemical shift of the amide nitrogen are particularly useful because of the relatively large ^{15}N chemical-shift dispersion (*16*).

Heteronuclear 3D NMR techniques require isotopic labeling of the system to be studied. For proteins this may be easily achieved by addition of suitably labeled nutri-

* On leave from Centre de Biophysique Moleculaire, Centre National de la Recherche Scientifique, 45071 Orleans Cedex 2, France.

0022-2364/89 $3.00

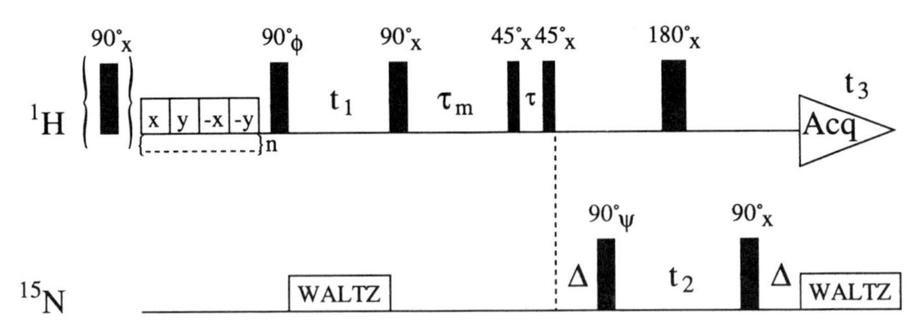

FIG. 1. Pulse scheme of the 3D NOESY-HMQC experiment. The ^{1}H carrier is positioned in the center of the amide region, and a DANTE type presaturation of water is used (*31*). The phase cycling used is as follows: $\phi = x, y, -x, -y$; $\psi = 4(x), 4(-x)$; Acq. $= 2(x), 4(-x), 2(x)$, with data for odd and even numbered scans stored separately. The entire sequence is repeated with ψ incremented by 90° to obtain complex data in the F_2 dimension. The first ^{1}H pulse is applied only for the first scan of each new t_1 increment and serves to establish a steady state of magnetization. Δ is set to a value slightly less than $1/2J_{NH}$ (4.5 ms) to minimize relaxation losses. The delay τ is carefully adjusted to maximize water suppression ($\tau \approx 1/(2\delta_{H_2O})$), where δ_{H_2O} is the offset of water. Depending on the spectrometer convention, the phases of the off-resonance presaturation pulses may have to be $(x, -y, -x, y)$ or $(x, y, -x, -y)$.

ents to the growth medium of microorganisms that contain the expression system for the molecule of interest (*16, 17*).

In a recent communication (*8*), we reported on the application of a NOESY-heteronuclear multiple-quantum coherence (NOESY-HMQC) 3D experiment to the protein staphylococcal nuclease (S. Nase). The final 3D spectrum obtained from this experiment displays NOE information between amide protons (F_3) and other protons in the protein (F_1) that are in close spatial proximity to the amide protons. The 3D spectrum is a collection of 2D (F_1–F_3) NOE maps separated in the F_2 dimension according to the ^{15}N chemical shift of the corresponding amide nitrogen. In addition to providing information necessary for sequential assignment of the backbone resonances (NH, CαH), the spectrum provides many of the distance constraints necessary for a structure determination.

As will be discussed below, high-quality heteronuclear 3D spectra can be recorded on samples of relatively low concentration (1–2 m*M*). In this paper the experimental aspects of the NOESY-HMQC pulse scheme will be discussed first, followed by an outline of several important practical considerations for the minimization of both data storage requirements and measuring time. Finally, a very effective and simple new strategy for the processing of 3D data will be presented, largely using existing 2D processing software.

THE NOESY-HMQC PULSE SCHEME

Figure 1 shows the NOESY-HMQC pulse sequence used in our laboratory. In the discussion that follows the convention of Griesinger et al. (*2, 3*) will be used, whereby the detection period is called t_3 and the two preceding evolution periods are referred to as t_1 and t_2. The variables t_1 and t_3 represent the "regular" time variables in the NOESY experiment and t_2 and t_3 are the "regular" time variables in the HMQC experiment. As in the NOESY experiment, ^{1}H magnetization excited by application of the ^{1}H 90°_ϕ pulse is chemical-shift labeled during t_1 and is transferred during the mixing time (τ_m) to all spins in close spatial proximity. ^{15}N decoupling is employed during t_1 to remove heteronuclear scalar couplings. A homospoil pulse is applied in the middle of the mixing time to minimize coherent transfer of magnetization between scalar-coupled spins (*18*). For reasons to be discussed later, the carrier is positioned in the center of the amide proton region, and to obtain coherent presaturation of the H_2O resonance (*19*) a novel DANTE-type sequence is used. An analysis of this scheme is presented in the Appendix. Only a very weak presaturating RF field (10–15 Hz) is used, providing attenuation of the intense H_2O resonance by about a factor of 20, and saturating only a very narrow band of CαH resonances (<0.1 ppm) that are very close to the H_2O frequency. After the NOE mixing time, τ_m, the application of an off-resonance jump-and-return read pulse, 45°_x–τ–$45^\circ_{\bar{x}}$, generates transverse ^{1}H magnetization. The delay τ is adjusted to minimize excitation of water magnetization that has recovered during the mixing time or escaped saturation ($\tau \approx 1/(2\delta_{H_2O})$). The small amount of spurious transverse H_2O magnetization generated by the carefully calibrated 180° pulse at the midpoint of the t_2 period does not present any dynamic range problems and is removed from the 2D spectrum by the ^{15}N filtering effect of the HMQC sequence. The sequence that follows the NOE mixing period is similar to the regular ^{1}H–^{15}N HMQC shift correlation scheme (*20, 21*) (sometimes named forbidden echo (*16*)), and with phase cycling of the first 90° ^{15}N pulse, only signals from protons directly attached to ^{15}N remain during the detection period, t_3. For every t_2 value, a 2D data set is obtained with only amide protons in the detected (F_3) dimension and all protons in close spatial proximity to the amide protons in the F_1 dimension. The intensity of an H–NH cross peak in such a 2D spectrum is modulated as a function of t_2 by the ^{15}N chemical shift, and Fourier transformation with respect to t_2 of the (F_1, t_2, F_3) spectra results in the final 3D spectrum. For reasons to be discussed later, the Fourier transformations are executed in a different order.

At first sight, it may appear undesirable to record only the NH protons in F_3 and all protons in the F_1 dimension because this requires a larger minimum number of experiments to obtain sufficient digitization. Indeed, Fesik and Zuiderweg have recorded a 3D spectrum of a tripeptide dissolved in a deuterated solvent using an HMQC-NOESY 3D scheme (*6*), instead of a NOESY-HMQC scheme (*8*). In the former sequence, the NH proton shifts are recorded in the F_2 dimension. Note however, that the sensitivity of the experiment (given identical total measuring times) is essentially independent of this choice (*22*). Our choice for the scheme of Fig. 1 is based on the necessity to observe the crucial NH–Hα connectivities that otherwise would be buried in the wings of the intense H_2O resonance. At the low concentrations used (1–2 m*M*) typically a width of at least 0.7 ppm centered about the water is unobservable in the (F_1, F_2) = (NH, CαH) region of 2D NOESY spectra.

Since only NH resonances are detected during t_3, it is desirable to position the carrier at the center of the NH region. This minimizes the size of the data matrix that needs to be recorded.

Because of the large number of (t_1, t_2) experiments needed for a 3D spectrum with sufficient digital resolution, only a limited number of scans can be acquired for every (t_1, t_2) value. Therefore, only a relatively short phase cycle can be utilized. In practice, we find it sufficient to use a 16-step cycle, incrementing the phase of ϕ in 90° steps to

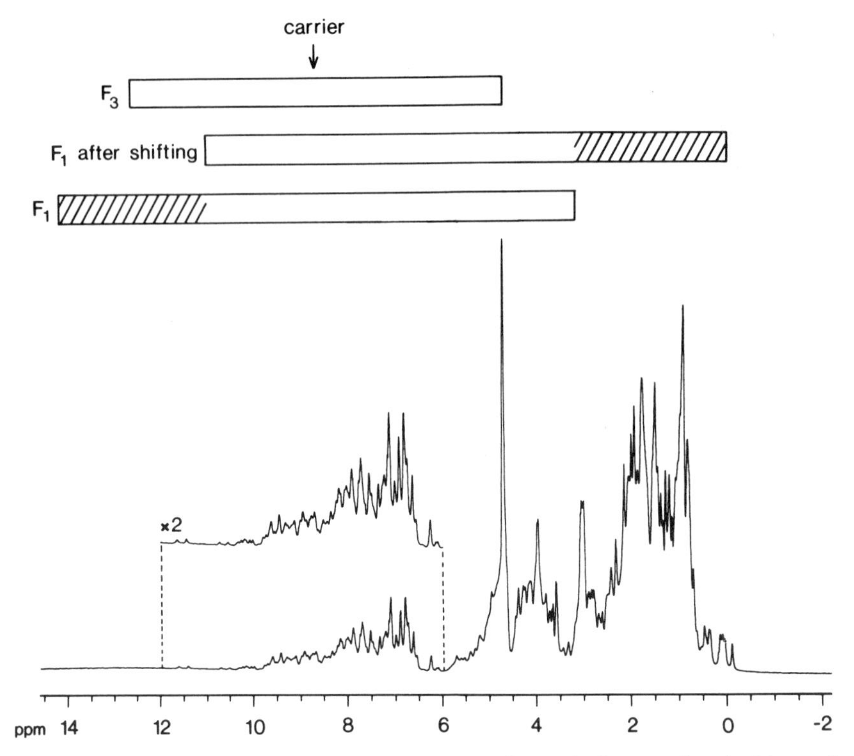

FIG. 2. Spectrum of S. Nase indicating the positioning of the ^{1}H carrier for recording the 3D experiment. The spectral windows employed in F_1 and F_3 are indicated in the top of the figure. Using either an acquisition (real or complex data in t_1) or a processing procedure (complex data in t_1) the carrier position is shifted in the F_1 dimension and an unfolded spectrum can be obtained. This is illustrated by the middle spectral window where the hatched region indicates the portion of the window that has been shifted.

obtain complex data in the F_1 dimension and to suppress axial peaks at $F_1 = 0$; the phase ψ is incremented in 90° steps to obtain quadrature information in the F_2 dimension and to eliminate signals during t_3 from protons not attached to ^{15}N. Because of the small number of scans per (t_1, t_2) value, the use of dummy scans to establish a steady state of magnetization prior to each new (t_1, t_2) value is undesirable.[1] Instead, we use a single 90° pulse (Fig. 1) prior to the start of each new (t_1, t_2) experiment.

PRACTICAL CONSIDERATIONS FOR MAXIMIZING F_1 AND F_2 RESOLUTION

Figure 2 illustrates the regular 1D spectrum of S. Nase with the carrier positioned in the center of the NH region. At first glance, it might appear that with the carrier in this position, a very large spectral window would be required in the F_1 dimension since resonances showing NOE connectivities to the NH protons lie in all regions of

[1] The number of dummy scans required and hence the actual savings in acquisition time is highly dependent on the dead time between each experiment (i.e., the time spent in disk I/O transfers and reloading of the pulse programmer) and varies for different types of spectrometers.

the 1D spectrum. A large spectral window would, of course, result in the need to acquire a large number of t_1 increments for sufficient resolution. In fact the F_1 spectral window that we employ is much smaller than might be anticipated and is indicated by the window marked F_1 in the figure. Use of this smaller spectral window is possible without introducing confusing folding artifacts by employing one of two different procedures depending on whether real data, acquired by TPPI (*23*), or complex data, acquired by the method of States *et al.* (*24*), are obtained in t_1.

For the TPPI case, the phase of the proton pulse immediately preceding the t_1 period can be incremented by $\pi/4$ for successive t_1 increments, instead of the usual $\pi/2$ increment.[2] Incrementing the phase by $\pi/2$ decreases the measured precession frequencies of all spins by SW/2 Hz (SW, spectral width) relative to the carrier, such that after a real Fourier transformation no folding is obtained for resonances that are within $\pm$SW/2 of the carrier. By incrementing the phase by $\pi/4$ the precessional frequencies are decreased by SW/4 Hz (instead of SW/2 for regular TPPI), and resonances between SW/4 and $-$3SW/4 Hz of the carrier are not folded. In other words, by incrementing the phase of the first ^{1}H pulse by $\pi/4$ the carrier is effectively moved downfield by SW/4 and folding of the data does not occur. In this way spectra can be acquired in which the position of the carrier need not be fixed from one ^{1}H dimension to the next.

If complex data are acquired in t_1 (using the method of States *et al.*) the precessional frequencies of all signals must be increased by SW/4 so that the carrier is positioned in the center of the spectrum prior to Fourier transformation of the data. This can be achieved using a procedure analogous to that discussed for TPPI data acquisition with the exception that the phase of the acquired signal must now be shifted by $-\pi/2$ for each t_1 increment.[2] Note that the sign of this phase shift is opposite to that required for real data since the carrier must be shifted in the opposite direction. In addition, the magnitude of the phase shift is doubled since for complex data the phase shift per data point is twice as large as for real data. Alternatively, shifting the carrier upfield can be performed during the data processing stage by multiplication of the time domain data $S(t_1, t_2, t_3)$ by $\exp\{i\pi t_1/(2\Delta t_1)\}$, where Δt_1 is the t_1 increment value. Such multiplication can conveniently be accomplished by using the routine that normally is used for the linearly frequency-dependent phase correction of frequency-domain data (*20, 25, 26*).

In the ^{15}N dimension (F_2), the carrier is positioned near the center of the amide region. Folding of a few resonances near the outer edges of the ^{15}N chemical-shift region can be used in the 3D experiment provided that a 2D ^{1}H–^{15}N correlation map with a sufficiently large ^{15}N spectral width is initially recorded. If the first t_2 increment (including the 180° ^{1}H pulse plus $4/\pi$ times the 90° ^{15}N pulse, as will be discussed in the processing section) is set to $\Delta t_2/2$, the linear phase correction will be 180°, and the folded resonances will appear with opposite phase relative to the unfolded resonances (provided that complex data are acquired in the t_2 dimension) making them easily identified. Use of extensive folding in the F_2 dimension may lead to signal cancellation of resonances with opposite sign and therefore should be avoided. As mentioned

[2] Depending on the convention of the spectrometer, the sign of the phase increment may have to be reversed.

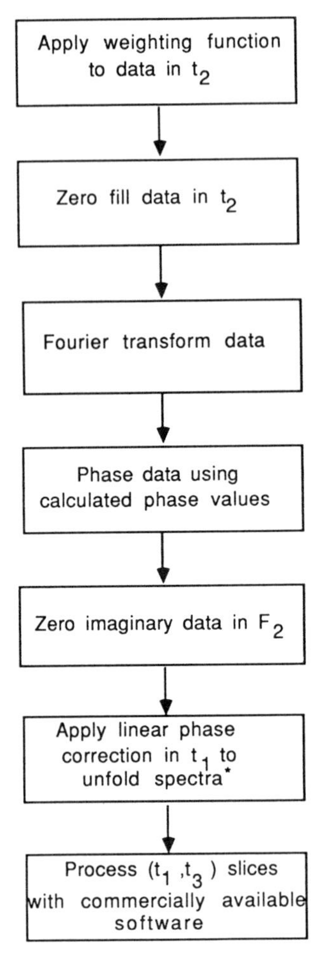

FIG. 3. Flowchart indicating the steps taken in processing the 3D data matrix. *If complex t_1 data were recorded.

before, the amide ^{15}N resonances show a substantial chemical-shift dispersion ($\approx$35 ppm). However, because of the low magnetogyric ratio of ^{15}N, the spectral width required in the ^{15}N dimension is relatively small, and only a modest number of t_2 increments is needed for a t_2 acquisition time on the order of T_2. Longer t_2 acquisition times would result in poorer sensitivity per unit of measuring time.

We have recorded 3D experiments on our Bruker-AM and Nicolet-NT spectrometers as sets of 2D experiments, using standard 2D pulse programming procedures. The t_2 delay is incremented after acquisition of each 2D data set is completed.

PROCESSING STRATEGY

Figure 3 shows a block diagram of the steps involved in the processing of the 3D data set. As mentioned above, the data for each t_2 value are acquired and stored as separate 2D data sets. For each t_2 value, two 2D data sets are acquired, corresponding to the real ($\psi = x, -x$) and the imaginary ($\psi = y, -y$) data in the t_2 dimension. The processing strategy that we use considers the data as a set of $2N$ 2D (t_1, t_3) NOESY matrices, where N is the number of complex t_2 points. To this end, the data are first processed in the t_2 (^{15}N) dimension, using simple software routines to be discussed later. The t_2 Fourier transformation is performed efficiently, without changing the data structure by transposition routines. This facilitates the use of any commercially available software for the subsequent transformations in t_1 and t_3. The software developed for processing in the t_2 dimension is written in modular form, with all manipulations that would normally be applied to each point in a one-dimensional process now applied to each (t_1, t_3) 2D matrix. The software is written for complex t_2 data, although, with a straightforward modification, real data can be processed as well.

The first steps in processing of the data in the t_2 dimension are the application of a weighting function and zero-filling. In the t_2 dimension, we typically use a doubly phase shifted sine-bell window function, shifted by about 60° at the beginning of the window and by about 10° at the end. The phase shift at the end of the sine bell is used to avoid too low a weighting factor for the last t_2 increments, improving sensitivity and resolution slightly, at the expense of a small (in practice not noticeable) increase in truncation artifacts in the F_2 dimension. Zero-filling in the F_2 dimension is essential; doubling the length of the time-domain data by adding zeroes to the end of the time domain is sufficient for providing adequate digitization in the frequency domain.

Fourier transformation in the t_2 dimension is accomplished using a modified version of a one-dimensional fast Fourier transform routine (FFT) which calculates the discrete complex Fourier transform of a set of N points, where N is a power of 2 (*27*). Figure 4 illustrates the manipulations that must be performed on the (t_1, t_3) planes in carrying out a t_2 transform. The FFT routine that we employ consists of two distinct parts: the bit reversal routine and the Danielson-Lanczos (butterfly) algorithm. During the first step, each (t_1, t_3) plane is reshuffled according to the prescription given for each point in the 1D transform. Because of the convenient storage of each (t_1, t_3) plane as an individual file this only amounts to a renaming of the planes (without the lengthy process of physically moving the data). This is indicated schematically in Fig. 4 where the labels of the two planes highlighted are interchanged with their relative positions in the data matrix remaining fixed. When points N_k and N_m are combined in the subsequent section of the 1D transform (butterfly algorithm), the equivalent combination of all points in planes corresponding to the points N_k and N_m in t_2 must be performed. A detailed description of the mechanics of the discrete 1D Fourier transform is provided in reference (*27*). Upon completion of the Fourier transformation, a (t_1, F_2, t_3) matrix is obtained with the zero frequency at one end of the spectrum. The data are reshuffled so that the carrier frequency is placed in the middle of the spectrum with positive and negative frequencies to the left and right of the carrier, respectively, in accord with NMR convention.

After Fourier transformation, the data are phased using a noninteractive phasing procedure (*28*). To a good approximation, the first 90° ^{15}N pulse can be replaced by a δ pulse followed by a delay of $2\tau_{90°}(^{15}N)/\pi$, where $\tau_{90°}(^{15}N)$ is the duration of the 90° ^{15}N pulse. Similarly, the final 90° ^{15}N pulse may be replaced by a δ pulse preceded

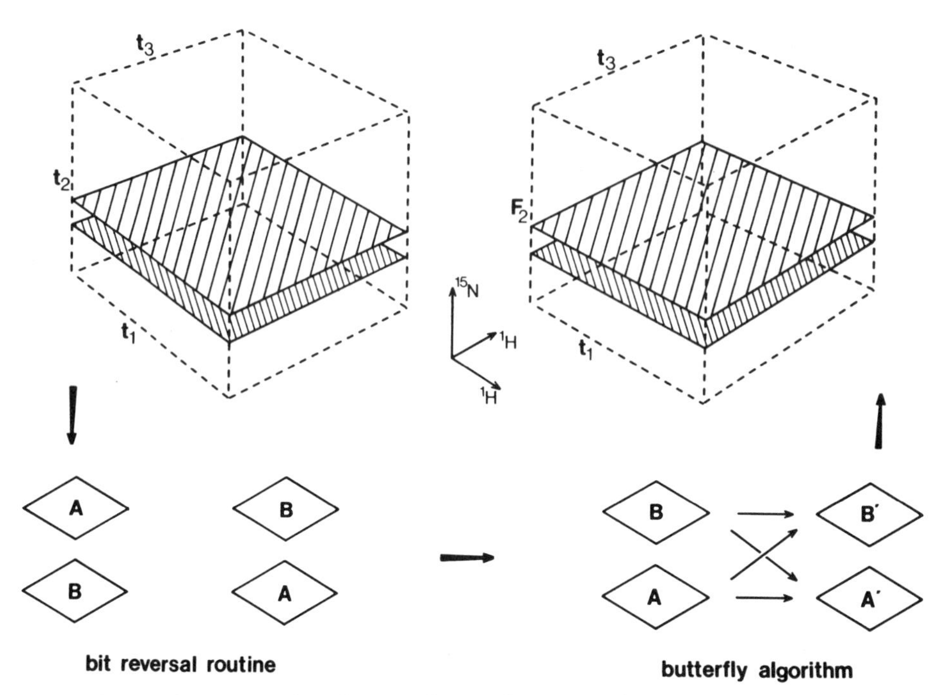

FIG. 4. Schematic representation of the Fourier transform process in t_2. Each (t_1, t_3) NOESY plane is manipulated according to the prescription for each point in a 1D transform. The transform is divided into two sections. The first consists of a bit reversal routine whereby the planes are renamed appropriately. The planes are then recombined according to the "butterfly algorithm" (*27*) to generate a (t_1, F_2, t_3) data set.

by the same delay. In the sequence depicted in Fig. 1, the first value of t_2 is effectively equal to $T = \tau_{180°}(^1\text{H}) + (4/\pi)\tau_{90°}(^{15}\text{N}) + t_2(0)$, where $t_2(0)$ is twice the value of the initial delay between the end of the 90° ^{15}N pulse and the beginning of the 180° ^{1}H pulse. Therefore, a linear phase shift correction across the spectrum of $(T/\Delta t_2) \times 360°$ is required in F_2 (*28*), where Δt_2 is the dwell time in the t_2 dimension. A zero-order phase correction must be applied in order to ensure that the phase at the center of the spectrum is zero. In a similar fashion, a linear phase correction of $(T'/\Delta t_1) \times 360°$, with $T' = (4/\pi)\tau_{90°}(^1\text{H}) + t_1(0)$, is required in F_1, where Δt_1 is the dwell time in t_1 and $t_1(0)$ is the initial value of the delay between the first two 90° ^{1}H pulses. In our experience, the calculated phase parameters are at least as accurate as phase parameters obtained manually. At this stage of the process, the imaginary data in F_2 can be discarded, reducing the size of the data matrix to the size it had before zero-filling in t_2. In the case of data acquired in the complex mode in t_1, an additional linear phase correction of the t_1 time-domain data is applied to shift the carrier position to unfold the spectrum, as discussed above. Since the parameters for both the phasing and the unfolding processes can be calculated without visual inspection of any of the Fourier-transformed spectra, all of the processes described in Fig. 3 can be executed automatically. On the SUN-4-110 system used for data processing, a $(2 \times 128) \times (2 \times 32) \times (2 \times 256)$ (t_1, t_2, t_3) matrix consisting of single precision floating point data requires approximately 1.5 hours to reach this stage of processing. As the diagram in Fig. 3 indicates, existing commercial software is used to process the 2D NOESY planes which are then stored as separate 2D files on disk.

Several points pertaining to the software that we have developed deserve attention. First, it is not necessary to do the initial processing of the data in the dimension in which it is acquired (F_3 in this case, F_2 in the case of 2D experiments). Having the flexibility of processing the data in another dimension first may have certain advantages for both 3D and 2D experiments. For example, for the 3D experiment of Fig. 1, processing of the data in t_2 generates a set of standard NOESY spectra. If particular NOE information is required it is not necessary to process all the data; the pertinent information can simply be extracted from a single (t_1, t_3) slice, taken at the appropriate ^{15}N chemical shift. Extensive zero-filling may then be used for this relatively small 2D matrix. A second advantage lies in the fact that for a 2D spectrum the F_1 phasing parameters and for a 3D spectrum both the F_1 and the F_2 phasing parameters can be calculated exactly. Obtaining the exact phasing is sometimes more difficult in F_3, in particular if one must use a t_3 slice through the (t_1, t_2, t_3) time-domain data, where a single FID typically yields a spectrum with a poor signal-to-noise ratio. In the case of 2D NMR data similar problems with phasing in F_2 exist. Most existing 2D software packages either save real and imaginary data in both dimensions, which requires substantial storage capabilities, or require that processing be first carried out in F_2 followed by discarding of the imaginaries. In the latter case, it is not possible to rephase the data in F_2 once the imaginaries are discarded. By processing the data first in the F_1 dimension using the calculated phasing parameters and discarding the imaginaries, followed by processing of the data in F_2, it is possible to rephase in this dimension several times if necessary. Our approach for processing of 3D data does not require the use of a time-consuming data transposition, and moreover, commercial software can be used to process the two "standard" dimensions of the 3D data set after the third dimension is processed. All modules are written in the C programming language and have been implemented on a SUN 4 computer. They are available upon request.

Figure 5 presents two NOESY slices taken from a crowded region of the 3D spectrum at ^{15}N chemical shifts of 116.8 and 118.2 ppm. The 3D spectrum was recorded in 2.5 days on a NT-500 spectrometer, on a 1.8 m*M* solution of S. Nase complexed with pdTp and Ca^{2+}, in 90% H_2O/10% D_2O at pH 7.5, 35°C. The sensitivity of the 3D spectrum is similar to that of the conventional NOESY spectrum (data not shown), but the severe overlap of cross peaks that is present in the 2D spectrum is almost completely removed.

The heteronuclear 3D experiment described above generates absorptive 3D spectra that have in-phase multiplets. This is particularly important for applications to large molecules where short T_2 values and signal overlap often cause extensive cancellations of antiphase signals (*29*). During the HMQC portion of the 3D experiment, favorable multiple-quantum T_2 relaxation rates are operative, due to the fact that to first order the ^1H–^{15}N dipolar interaction does not contribute to the relaxation process (*30*). This may be particularly important for proteins significantly larger than S. Nase, where T_2's rather than digitization will determine spectral resolution. Moreover, heteronuclear 3D experiments have the advantage that the flow of magnetization can be restricted without the need for selective pulses (*2*). For proteins significantly larger than about 10 kDa, techniques relying on the ^1H–^1H J coupling such as

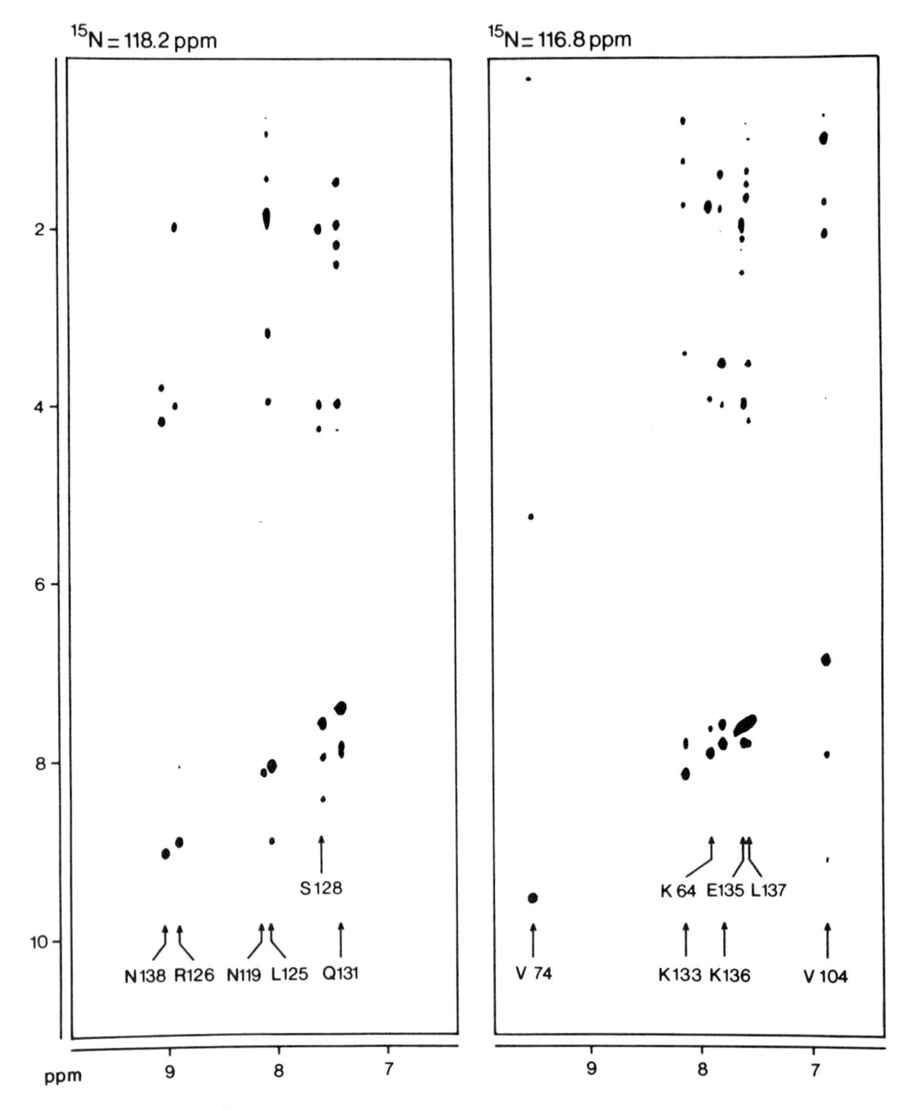

FIG. 5. F_1/F_3 slices at ^{15}N chemical shifts of 116.8 and 118.2 ppm recorded with the scheme of Fig. 1, for an NOE mixing time of 125 ms. The spectrum results from a 128 × 32 × 256 complex data matrix, with acquisition times of 23, 21, and 64 ms in the t_1, t_2, and t_3 dimensions, respectively. The total measuring time was 2.5 days. Zero-filling was used in all three dimensions to yield a 256 × 64 × 512 matrix for the absorptive part of the 3D spectrum. Note that in the NH–NH region of the spectra the cross peaks are asymmetrically distributed about the diagonal. Only in cases where the ^{15}N chemical shifts of both amide nitrogens overlap in the F_2 dimension (e.g., L125 and R126 in the left panel) will cross peaks be symmetrically positioned about the diagonal.

COSY and HOHAHA/TOCSY rapidly become less efficient. However, preliminary results obtained with the HOHAHA-HMQC 3D experiment for two different proteins in the range 16–18 kDa indicate that this experiment also is capable of generating high-quality spectra for proteins in this molecular weight range. Based on our experience with 2D HOHAHA experiments, it is expected that for proteins larger than about 20 kDa the sensitivity of the HOHAHA-HMQC 3D experiment will become too low for practical use. In contrast, the NOE experiment still works very well for many of these proteins and we therefore believe that the NOESY-HMQC experiment will be applicable to proteins significantly larger than 20 kDa.

APPENDIX

This appendix provides a theoretical description of the effects of the off-resonance water presaturation sequence used in the pulse scheme of Fig. 1. First will be considered the excitation versus frequency profile of a rectangular pulse. Its frequency profile can be calculated explicitly from the transient solutions of the Bloch equations (*31*). For a pulse of length, t_p, and of magnitude B_1, an absorption-mode component is obtained which has a dependence on the resonance offset, $\gamma\Delta B$, given by

$$A(\nu) = (B_1/B_{eff})\sin(\gamma B_{eff} t_p) \quad [A1]$$

and a dispersion-mode signal,

$$D(\nu) = (B_1 \Delta B / B_{eff}^2)[1 - \cos(\gamma B_{eff} t_p)], \quad [A2]$$

where $B_{eff}^2 = B_1^2 + (\Delta B)^2$ and $\nu = \gamma(\Delta B)/2\pi$.

An *approximate* excitation spectrum of a square pulse can be obtained by Fourier analysis, yielding

$$A'(\nu) = (B_1/\Delta B)\sin(\gamma \Delta B t_p) \quad [A3]$$

and

$$D'(\nu) = (B_1/\Delta B)[1 - \cos(\gamma \Delta B t_p)]. \quad [A4]$$

Comparing Eqs. [A1] and [A2] with Eqs. [A3] and [A4] indicates that in the limit $\Delta B \gg B_1$, Fourier analysis of a square pulse gives the correct frequency response. Moreover, it is also easily verified that in this limit the magnitude of the frequency distribution, $M(\nu) = [A(\nu)^2 + D(\nu)^2]^{1/2}$, obtained by the solution of the Bloch equations is the same as that obtained by Fourier analysis. It follows that the magnitude of the frequency distribution of any pulse train can be evaluated correctly by Fourier analysis as long as $\Delta B \gg B_1$.

The off-resonance water suppression sequence considered in this appendix is of the form $(\theta_0\theta_\phi\theta_{2\phi}\theta_{3\phi}\cdots\theta_{(k-1)\phi})_n$, where $e^{ik\phi} = 1$ and where the duration of each θ pulse, D, is set to $1/(k\delta_{H_2O})$. In the sequence used for presaturation in Fig. 1, $\phi = \pi/2$ and $k = 4$.

In order to evaluate the magnitude of the frequency distribution, $M_{WS}(\nu)$, of the sequence $(\theta_0\theta_\phi\theta_{2\phi}\theta_{3\phi}\cdots\theta_{(k-1)\phi})_n$, $n \to \infty$, this sequence is rewritten as

$$\sum_{m=-\infty}^{\infty} (\theta_0\theta_\phi\theta_{2\phi}\theta_{3\phi}\cdots\theta_{(k-1)\phi})_n \circledast \delta(t - mkD)$$

$$= F(\theta) \circledast \sum_{m=-\infty}^{\infty} \delta(t - mkD) = A(t), \quad [A5]$$

where δ is a dirac delta function defined as

$$\delta(t - mkD) = 0 \quad \text{if} \quad t \neq mkD$$
$$= 1 \quad \text{if} \quad t = mkD, \qquad [A6]$$

$F(\theta) = (\theta_0\theta_\phi\theta_{2\phi}\theta_{3\phi}\cdots\theta_{(k-1)\phi})_n$, and $\circledast$ is the convolution operator (*32*). The Fourier transform of $A(t)$ gives

$$\mathrm{FT}[A(t)] = \mathrm{FT}[F(\theta)]/(kD) \sum_{m=-\infty}^{\infty} \delta[\nu - m/(kD)], \qquad [A7]$$

where $\mathrm{FT}(F(\theta))$ is the Fourier transform of $F(\theta)$.

In the derivation of Eq. [A7] we have made use of the fact that the Fourier transform of the convolution of two functions is the product of their Fourier transforms (*32*) and that

$$\mathrm{FT}[\sum_{m=-\infty}^{\infty} \delta(t - mkD)] = 1/(kD) \sum_{m=-\infty}^{\infty} \delta[\nu - m/(kD)]. \qquad [A8]$$

The expression for $M_{\mathrm{WS}}(\nu)$ is given by

$$M_{\mathrm{WS}}(\nu) = \{\mathrm{FT}[F(\theta)]\mathrm{FT}[F(\theta)]^*\}/(kD) \sum_{m=-\infty}^{\infty} \delta[\nu - m/(kD)], \qquad [A9]$$

where $\mathrm{FT}[F(\theta)]^*$ is the complex conjugate of $\mathrm{FT}[F(\theta)]$. Equation [A9] can be readily evaluated to give

$$M_{\mathrm{WS}}(\nu) = \sum_{m=-\infty}^{\infty} |(B_1/\pi m)\sin(\omega_{\mathrm{m}} D/2)|$$
$$\times \{k + 2\sum_{p=1}^{k-1} (k-p)\cos[p(\omega D + \phi)]\}^{1/2}\delta[\nu - \omega_{\mathrm{m}}/(2\pi)], \qquad [A10]$$

where $\omega_{\mathrm{m}} = 2\pi m/(kD)$. For the values of ϕ and k used, $\phi = \pi/2$ and $k = 4$, Eq. [A10] reduces to a series of sidebands, centered $1/(4D)$ Hz upfield from the carrier and spaced $1/D$ Hz apart with intensities that decrease asymmetrically with respect to offset from the central band. In particular, the intensity of the B_1 field associated with the mth sideband upfield from the carrier is given by

$$B_{1_{\mathrm{eff}}}(m) = 2\sqrt{2}B_1/[\pi(4m - 3)], \qquad [A11]$$

while the intensity of the B_1 field associated with the mth sideband downfield of the carrier is given by

$$B_{1_{\mathrm{eff}}}(m) = 2\sqrt{2}B_1/[\pi(4m - 1)]. \qquad [A12]$$

The carrier frequency is chosen such that only the central band lies within the spectrum. In the applications of this sequence, typically $\gamma B_1/2\pi \approx$ 10–15 Hz and the carrier is positioned 2000 Hz from the water resonance so that $\gamma\Delta B/2\pi = 2000$ Hz. Therefore the condition that $\Delta B \gg B_1$, which is required for Eq. [A10] to be valid, is satisfied. Equation [A10] has been derived in the limit as $n \to \infty$. For finite values of n, Eq. [A10] must be convoluted with a sinc function which has a width of $2/t_{\mathrm{d}}$

between the first zero-crossing points, where t_d is the duration of the pulse train (*31*). For typical values of t_d (1–2 s), this additional factor has no significant effect and can be neglected.

ACKNOWLEDGMENTS

We thank Rolf Tschudin for technical support and Dennis A. Torchia and Steven W. Sparks for providing us with the sample of S. Nase and for making available to us the assignments indicated in Fig. 5. We thank New Methods Research, Inc. (Syracuse, NY) for providing a copy of NMR2 used to process the individual 2D planes of the 3D data set and Julie Forman-Kay for a critical reading of the manuscript. This work was supported by the Intramural AIDS Antiviral Program of the Office of the Director of the National Institutes of Health. We acknowledge financial support from the Medical Research Council of Canada (L.E.K.), the Alberta Heritage Trust Foundation (L.E.K.), and from a CNRS/NIH exchange agreement (D.M.).

REFERENCES

1. H. D. Plant, T. H. Mareci, M. D. Cockman, and W. S. Brey, "27th ENC Conference," Poster A23, Baltimore, Maryland, 1986.
2. C. Griesinger, O. W. Sørensen, and R. R. Ernst, *J. Magn. Reson.* **73,** *S*74 (1987).
3. C. Griesinger, O. W. Sørensen, and R. R. Ernst, *J. Am. Chem. Soc.* **109,** 7227 (1987).
4. G. W. Vuister and R. Boelens, *J. Magn. Reson.* **73,** 328 (1987).
5. H. Oschkinat, C. Griesinger, P. J. Kraulis, O. W. Sørensen, R. R. Ernst, A. M. Gronenborn, and G. M. Clore, *Nature (London)* **332,** 374 (1988).
6. S. W. Fesik and E. P. R. Zuiderweg, *J. Magn. Reson.* **78,** 588 (1988).
7. G. W. Vuister, R. Boelens, and R. Kaptein, *J. Magn. Reson.* **80,** 176 (1988).
8. D. Marion, L. E. Kay, S. W. Sparks, D. A. Torchia, and A. Bax, *J. Am. Chem. Soc.* **111,** 1515 (1989).
9. J. Jeener, B. H. Meier, P. Bachmann, and R. R. Ernst, *J. Chem. Phys.* **71,** 4546 (1979).
10. S. Macura and R. R. Ernst, *Mol. Phys.* **41,** 95 (1980).
11. L. Braunschweiler and R. R. Ernst, *J. Magn. Reson.* **53,** 521 (1983).
12. A. Bax and D. G. Davis, *J. Magn. Reson.* **65,** 355 (1985).
13. W. P. Aue, E. Bartholdi, and R. R. Ernst, *J. Chem. Phys.* **64,** 2229 (1976).
14. A. Bax and R. Freeman, *J. Magn. Reson.* **44,** 542 (1981).
15. K. Nagayama, A. Kumar, K. Wüthrich, and R. R. Ernst, *J. Magn. Reson.* **40,** 321 (1980).
16. L. P. McIntosh, R. H. Griffey, D. C. Muchmore, C. P. Nielson, A. G. Redfield, and F. W. Dahlquist, *Proc. Natl. Acad. Sci. USA* **84,** 1244 (1987).
17. M. A. Weiss, A. Jeitler-Nillson, N. J. Fischbein, M. Karplus, and R. T. Sauer, *in* "NMR in the Life Sciences" (E. M. Bradbury and C. Nicolini, Eds.), pp. 37–48, NATO ASI Series, Vol. 107, Plenum, London, 1986.
18. J. Jeener, B. H. Meier, P. Bachmann, and R. R. Ernst, *J. Chem. Phys.* **71,** 4546 (1979).
19. E. R. P. Zuiderweg, K. Hallenga, and E. T. Olejniczak, *J. Magn. Reson.* **70,** 336 (1986).
20. A. Bax, R. H. Griffey, and B. L. Hawkins, *J. Magn. Reson.* **55,** 301 (1983).
21. M. R. Bendell, D. T. Pegg, and D. M. Doddrell, *J. Magn. Reson.* **52,** 81 (1983).
22. M. H. Levitt, G. Bodenhausen, and R. R. Ernst, *J. Magn. Reson.* **58,** 462 (1984).
23. D. Marion and K. Wüthrich, *Biochem. Biophys. Res. Commun.* **113,** 967 (1983).
24. D. J. States, R. A. Haberkorn, and D. J. Ruben, *J. Magn. Reson.* **48,** 286 (1982).
25. A. A. Bothner-By and J. Dadok, *J. Magn. Reson.* **72,** 540 (1987).
26. R. R. Ernst, G. Bodenhausen, and A. Wokaun, "Principles of Nuclear Magnetic Resonance in One and Two Dimensions," Clarendon, Oxford, 1987.
27. W. H. Press, B. P. Flannery, S. A. Teukolsky, and W. T. Vetterling, "Numerical Recipes, The Art of Scientific Computing," Cambridge Univ. Press, London/New York, 1986.
28. A. Bax and D. Marion, *J. Magn. Reson.* **78,** 186 (1988).
29. K. Wüthrich, "NMR of Proteins and Nucleic Acids," Wiley, New York, 1986.
30. A. Bax, L. E. Kay, S. W. Sparks, and D. A. Torchia, *J. Am. Chem. Soc.,* in press.
31. A. G. Morris and R. Freeman, *J. Magn. Reson.* **29,** 433 (1978).
32. R. M. Bracewell, "The Fourier Transform and Its Applications," McGraw–Hill, New York, 1965.

JOURNAL OF MAGNETIC RESONANCE **87**, 422–428 (1990)

Gradient-Enhanced Spectroscopy

RALPH E. HURD

GE NMR Instruments, 255 Fourier Avenue, Fremont, California 94539

Received December 19, 1989

Although the use of gradients to select specific coherences was introduced several years ago (*1, 2*), widespread application in high-resolution NMR appears to have been impeded by limits in gradient technology. Improvements afforded by the recent introduction of actively shielded gradients (*3*), however, have made it practical to take advantage of gradient-based coherence selection. Our initial efforts were focused on *in vivo* applications in which gradient selection of double-quantum coherence was used to obtain spectra and images of metabolites such as lactate (*4, 5*). It is also possible to devise gradient-enhanced versions of the many useful high-resolution 1D and multidimensional homonuclear and heteronuclear NMR experiments. A major advantage of these methods is that because only the desired coherences are selectively rephased, phase cycling and, in some cases, traditional water suppression methods are not required. The two experiments discussed in this Communication are gradient-enhanced double-quantum COSY (ge-2qcosy) and gradient-enhanced TOCSY (ge-tocsy).

A double-quantum COSY spectrum of 8 m*M* angiotensin II in H_2O (Fig. 1) was obtained using the gradient-enhanced experiment (ge-2qcosy) shown in Fig. 2. Data were collected on a GE NMR Instruments Omega 400 WB equipped with Microstar gradients (maximum strength ±137 G/cm). The probe used consisted of a 5 mm diameter Helmholtz radiofrequency coil tuned to 400.06 MHz. The potential for eddy current generation within the probe was minimized by use of nonconductive materials where possible. In this experiment, only a single acquisition per evolution time (t_1) increment is required to select the +1/+2 coherence pathway. The large population of water protons remains dephased, thus avoiding loss of signal which can result from misuse of presaturation, selective excitation, and/or attenuation. The receiver never detects this unwanted signal, while desired coherences, even from resonances at the solvent chemical shift, are observed. Water suppression by this method is independent of lineshape or shimming (B_0 homogeneity). Another benefit of this approach is that, because the unwanted signals remain dephased, they contribute very little to coherent t_1 noise as illustrated by the noise floor contour plot shown in Fig. 3. These data were collected nonspin, as a 1K by 1K matrix with a spectral width of 4000 Hz in both ω_1 and ω_2. The average recycle time was 900 ms. Gradients (x, y, and z) were 1 ms half-sinusoid with maximum amplitudes of 10, 10, and 30 G/cm. These gradient strengths were determined empirically to be the minimum required to completely eliminate the water signal in this sample, which had a diameter of 5 mm

0022-2364/90 $3.00

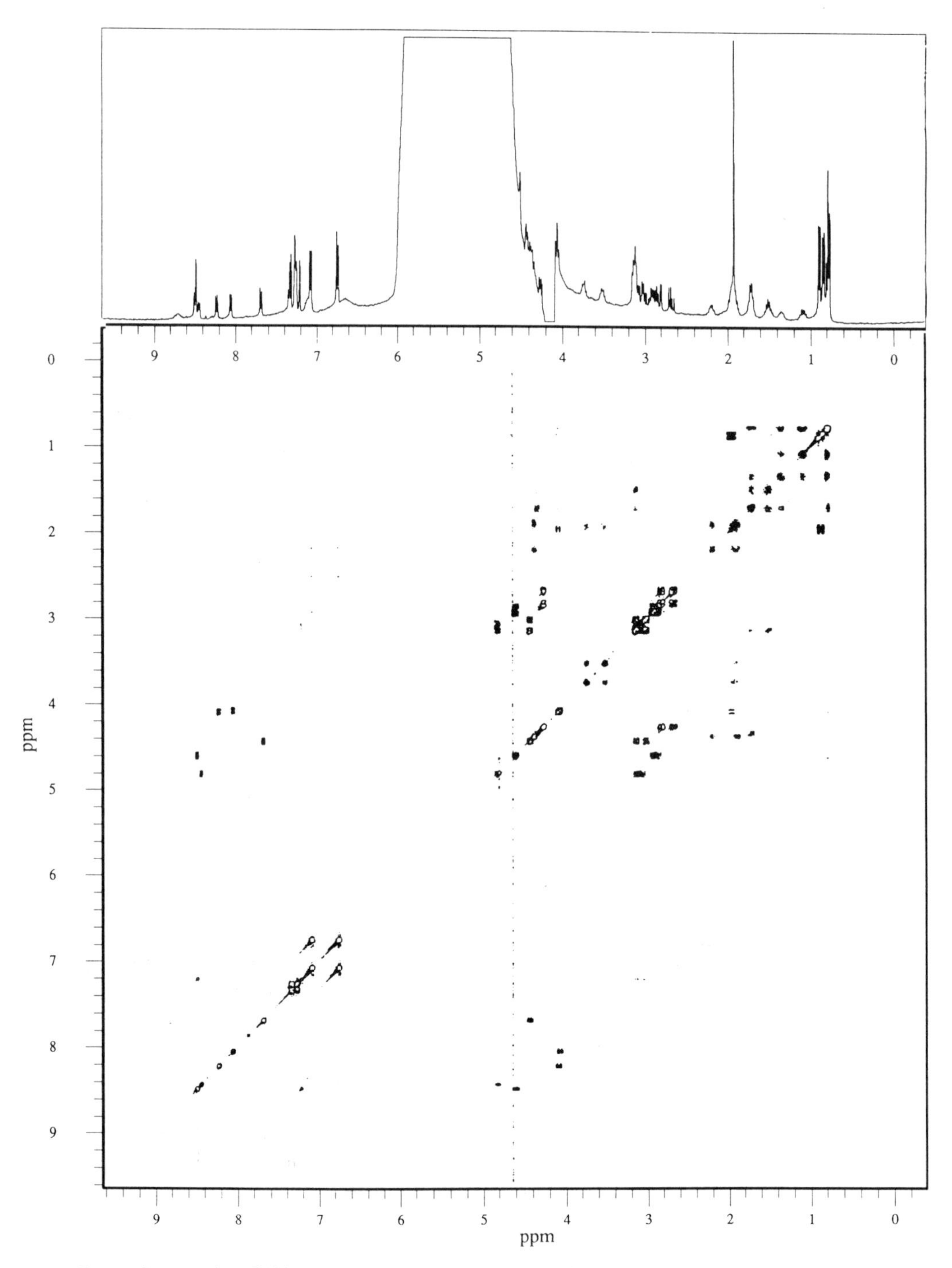

FIG. 1. Contour plot of 400 MHz proton ge-2qcosy spectrum of 8mM angiotensin II in H_2O. There were 1024 t_1 increments collected without phase cycling. Acquisition times of 256 ms provided 4 Hz resolution in both ω_1 and ω_2. One millisecond half-sinusoid gradients of 10, 10, and 30 G/cm were used. An 800-fold vertical expansion of the reference one-pulse spectrum is plotted across the top of the contour plot.

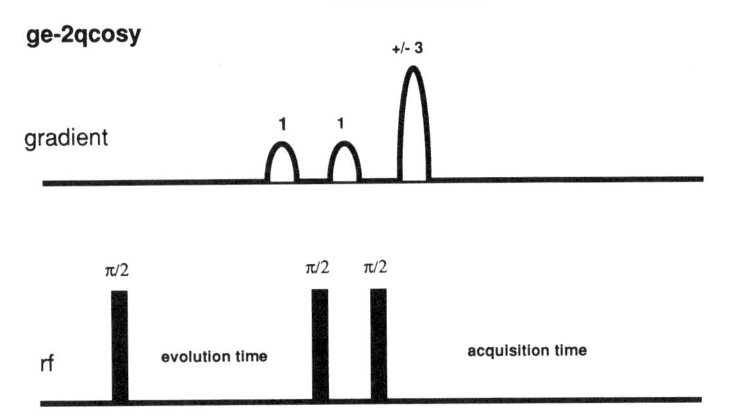

FIG. 2. Gradient-enhanced double-quantum COSY (ge-2qcosy) pulse sequence. Evolution time includes the first gradient pulse and acquisition time starts after the final gradient pulse.

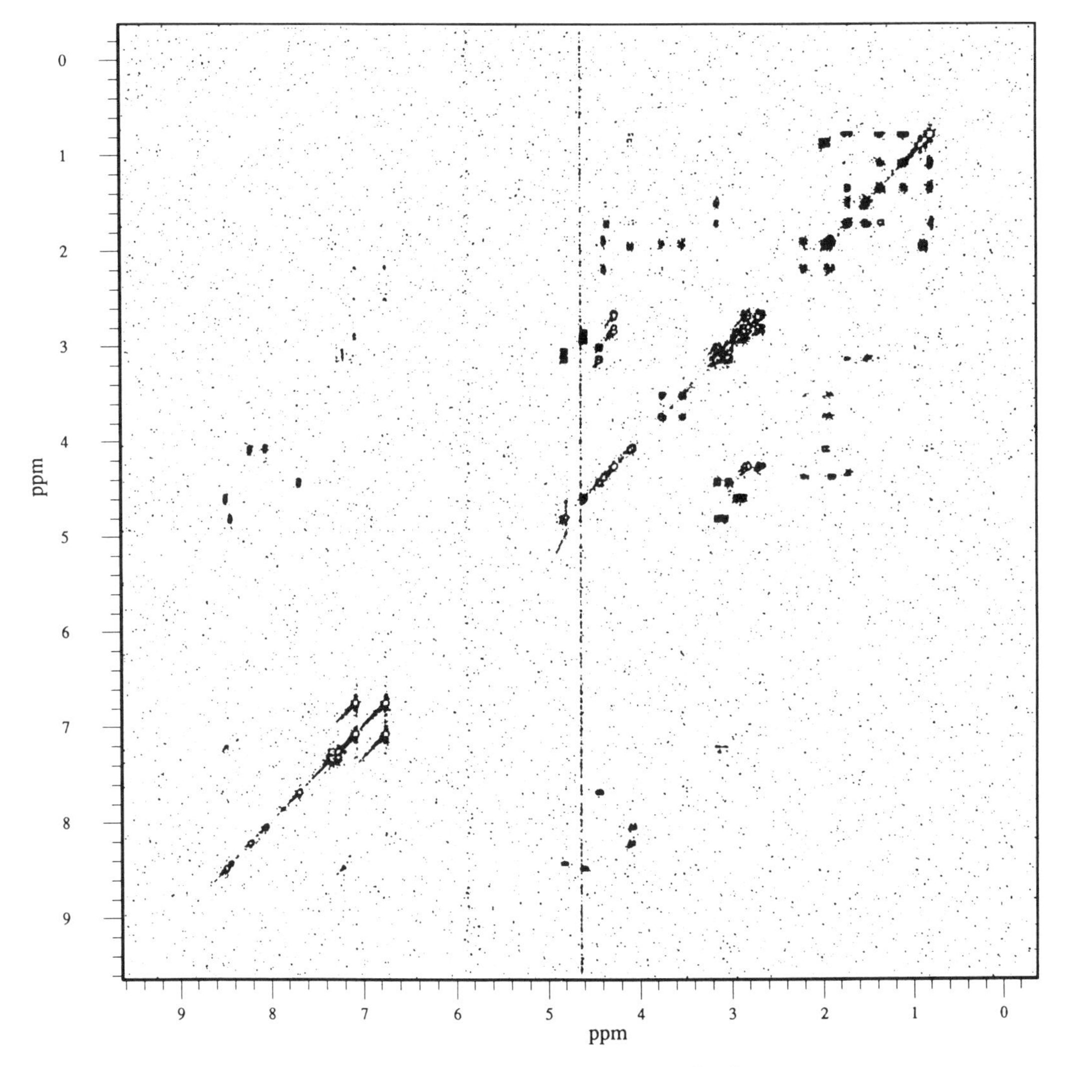

FIG. 3. Noise floor of the contour plot shown in Fig. 1.

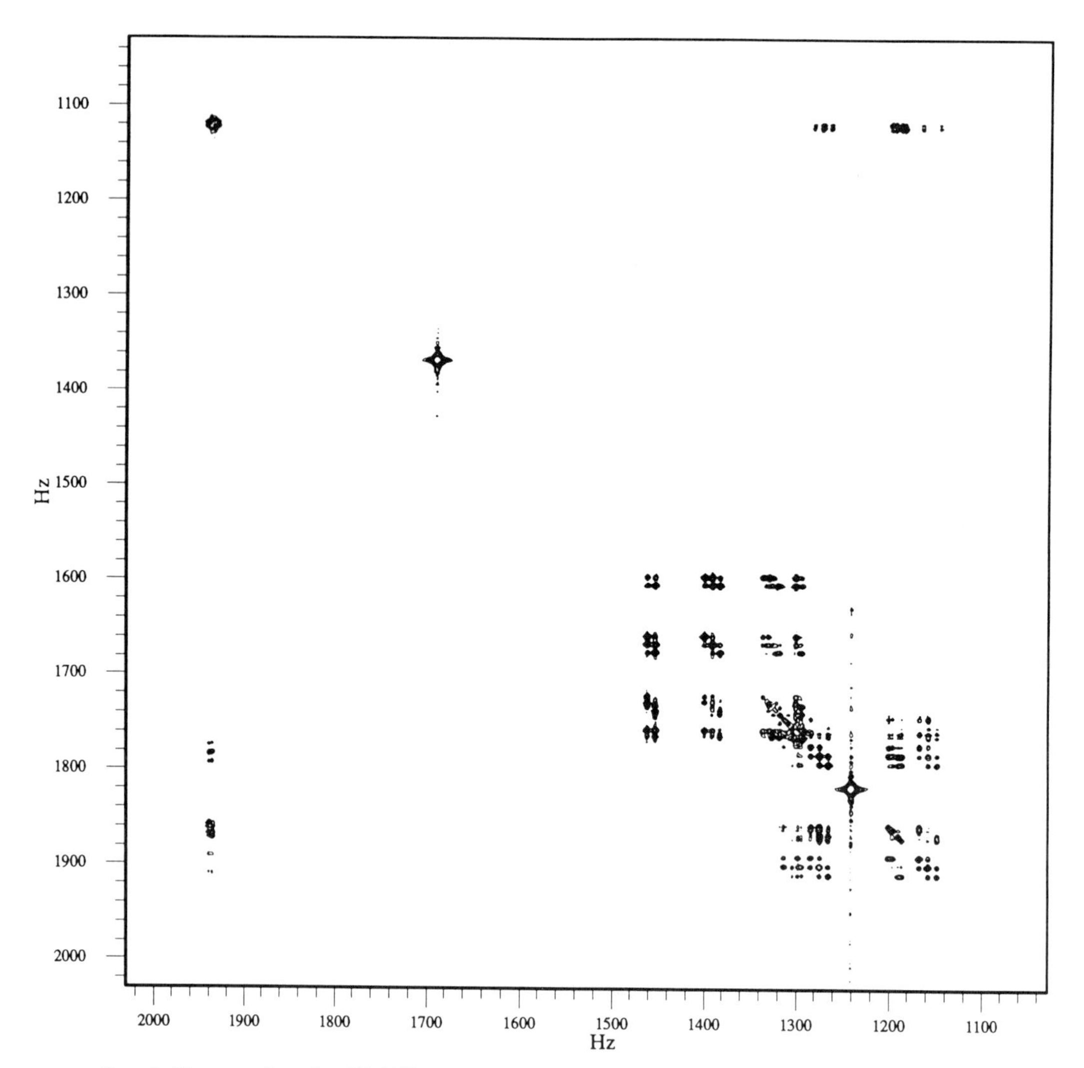

FIG. 4. Contour plot of a 400 MHz proton ge-tocsy spectrum of 10 mM sucrose in D_2O. A single acquisition was collected for each of 512 t_1 increments. Acquisition times of 512 ms provided 2 Hz resolution in both dimensions. A 20 ms MLEV-16 spin-lock pulse with a $\gamma B_1/2\pi$ strength of 4 kHz was bracketed by a 1 ms half-sinusoid gradient pair with maximum amplitudes of +2 and −2 G/cm.

and a length of 35 mm. Signal loss due to self-diffusion (*6*) was calculated to be less than 1% for the observed spins. The elimination of phase cycling resulted in a small zero-frequency artifact at the transmitter setting (4.65 ppm) in ω_2. Minor, yet unexplained, artifacts were observed for the resonances from the tyrosine ring protons at 7.05 and 6.76 ppm. Compromise of quadrature in ω_1 and some peak distortion along the diagonal were noted.

Total scalar correlation spectra (*7*) of 10 mM sucrose in D_2O (Figs. 4 and 5) were obtained to illustrate the gradient-enhanced version of HOHAHA (*8*) using both

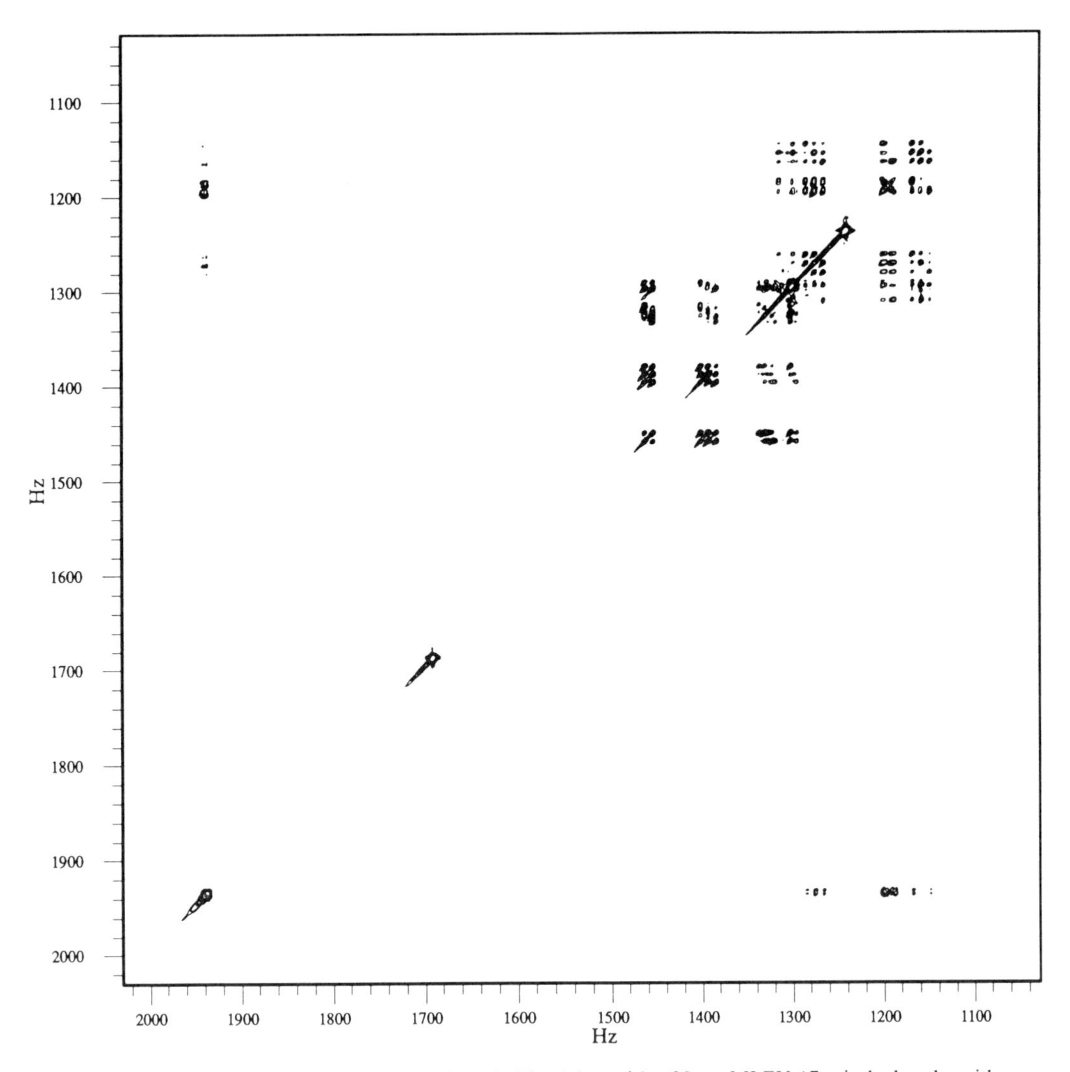

FIG. 5. Same sample and conditions as those in Fig. 4, but with a 20 ms MLEV-17 spin-lock pulse with a $\gamma B_1/2\pi$ strength of 4 kHz and bracketed by a 1 ms half-sinusoid same-sign gradient pair with maximum amplitudes of +2 G/cm.

MLEV-16 (*9*) and MLEV-17 (*10*) as spin-lock pulses. Gradient-enhanced versions of these pulse sequences are shown in Fig. 6. Unlike COSY, the gradient pair can only be of opposite sign for the MLEV-16-based ge-tocsy sequence. The additional pulse (60°) in the MLEV-17 cycle reestablishes both pathways and makes it possible for either a same-sign or an opposite-sign gradient pair to rephase coherence. Data were collected as 512 × 512 matrices with a single acquisition per evolution time (t_1) increment. Spectral width was 1000 Hz in both ω_1 and ω_2 and a 20 ms spin-lock pulse with a B_1 field strength of 4 kHz was used. A 1 ms half-sinusoid gradient pair with maximum amplitudes of 2 and −2 G/cm gave nearly identical results for both

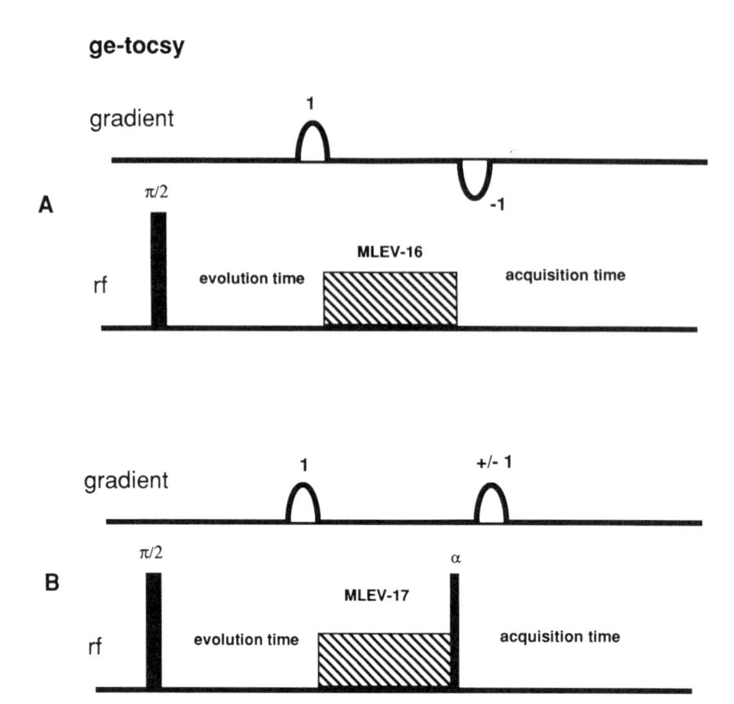

FIG. 6. Gradient-enhanced total scalar correlation experiments (ge-tocsy) with (A) MLEV-16 spin lock and (B) MLEV-17 spin lock. Evolution time includes the first gradient pulse and acquisition time begins after the final gradient pulse.

MLEV-16- and MLEV-17-based experiments (the MLEV-16 result is shown in Fig. 4). A same-sign gradient pair was used to obtain the ge-tocsy spectrum shown in Fig. 5. The most notable difference in the spectra is that peak distortion is observed along ω_1 for the opposite-sign gradient pair and along the diagonal in the same-sign case.

The examples presented here illustrate the potential of gradient-enhanced spectroscopy. The list of advantages includes a significant reduction in measuring time, reduced t_1 artifacts, the elimination of phase cycling and difference methods (which makes these methods less susceptible to vibration), the potential for pure three- and four-quantum editing, and the ability to detect resonances at the same chemical shift as a strong solvent resonance.

Limitations include a requirement for field-frequency-lock blanking during long runs and the possible elimination of useful signals along with those that are unwanted. Also, in the absence of phase cycling, it is important to avoid quadrature artifacts in ω_2 by keeping the two receiver channels balanced.

The benefit of gradient-enhanced spectroscopy should be of particular importance in high-resolution 3D experiments via the reduction in total measurement time associated with the elimination of phase cycling and in proton-detected heteronuclear correlation experiments via the elimination of subtraction-based selection of coherence.

REFERENCES

1. A. A. MAUDSLEY, A. WOKAUN, AND R. R. ERNST, *Chem. Phys. Lett.* **55,** 9 (1978).
2. A. BAX, P. G. DE JONG, A. F. MEHLKOPF, AND J. SMIDT, *Chem. Phys. Lett.* **69,** 567 (1980).
3. P. B. ROEMER, W. A. EDELSTEIN, AND J. S. HICKEY, "Book of Abstracts, 5th Annual Meeting of the Society of Magnetic Resonance in Medicine, Montreal, August 19–22, 1986," p. 1067.
4. C. H. SOTAK, D. M. FREEMAN, AND R. E. HURD, *J. Magn. Reson.* **78,** 355 (1988).
5. R. E. HURD AND D. M. FREEMAN, *Proc. Natl. Acad. Sci. USA* **86,** 4402 (1989).
6. E. O. STEJSKAL AND J. E. TANNER, *J. Chem. Phys.* **42,** 288 (1965).
7. L. BRAUNSCHWEILER AND R. R. ERNST, *J. Magn. Reson.* **53,** 521 (1983).
8. D. G. DAVIS AND A. BAX, *J. Am. Chem. Soc.* **107,** 2821 (1985).
9. M. H. LEVITT, R. FREEMAN, AND T. FRENKIEL, *J. Magn. Reson.* **47,** 328 (1982).
10. A. BAX AND D. G. DAVIS, *J. Magn. Reson.* **65,** 355 (1985).

Index